atomic number	7
atomic weight	14.0067
Allred-Rochow electronegativity	3.07
covalent radius	0.75
Radii in Å { Shannon crystal radius (common chg. and coord. no.)	1.32(3−)
name	nitrogen

1 H
1.0079
2.20
0.37
1.22(1−)
hydrogen

3 Li
6.941
0.97
1.34
0.73(1+)
lithium

4 Be
9.01218
1.47
0.90
0.41(2+)
beryllium

11 Na
22.98977
1.01
1.54
1.16(1+)
sodium

12 Mg
24.305
1.23
1.30
0.86(2+)
magnesium

19 K
39.0983
0.91
1.96
1.52(1+)
potassium

20 Ca
40.08
1.04
1.74
1.14(2+)
calcium

21 Sc
44.9559
1.20
1.44
0.885(3+)
scandium

22 Ti
47.90
1.32
1.32
1.00(2+)
titanium

23 V
50.9415
1.45
1.25
0.78(3+)
vanadium

24 Cr
51.996
1.56
1.27
0.755(3+)
chromium

25 Mn
54.9380
1.60
1.46
0.97(2+)
manganese

26 Fe
55.847
1.64
1.20
0.92(2+)
iron

27 Co
58.93
1.70
1.26
0.885(
cobalt

37 Rb
85.4678
0.89
2.11
1.75(1+)
rubidium

38 Sr
87.62
0.99
1.92
1.32(2+)
strontium

39 Y
88.9059
1.11
1.62
1.04(3+)
yttrium

40 Zr
91.22
1.22
1.48
0.86(4+)
zirconium

41 Nb
92.9064
1.23
1.37
0.78(5+)
niobium

42 Mo
95.94
1.30
1.45
0.79(4+)
molybdenum

43 Tc
(97)
1.36
1.56
0.785(4+)
technetium

44 Ru
101.07
1.42
1.26
0.82(3+)
ruthenium

45 Rh
102.9
1.45
1.35
0.805(
rhodi

55 Cs
132.9054
0.86
2.25
1.88(1+)
cesium

56 Ba
137.33
0.97
1.98
1.49(2+)
barium

57 La
138.9055
1.08
1.69
1.17(3+)
lanthanum

72 Hf
178.49
1.23
1.49
0.85(4+)
hafnium

73 Ta
180.9479
1.33
1.38
0.78(5+)
tantalum

74 W
183.85
1.40
1.46
0.74(6+)
tungsten

75 Re
186.207
1.46
1.59
0.77(4+)
rhenium

76 Os
190.2
1.52
1.28
0.77(4+)
osmium

77 Ir
192.2
1.55
1.37
0.82(
iridium

87 Fr
(223)
0.86
——
1.94(1+)
francium

88 Ra
226.0254
0.97
——
1.62(2+)
radium

89 Ac
227.0278
1.00
1.70
1.26(3+)
actinium

58 Ce
140.12
1.08
1.65
1.15(3+)
cerium

59 Pr
140.9077
1.07
1.65
1.13(3+)
praseodymium

60 Nd
144.24
1.07
1.64
1.123(3+)
neodymium

61 Pm
(145)
1.07
1.63
——
promethium

62 Sm
150.4
1.07
1.62
1.098(3+)
samarium

90 Th
232.0381
1.11
1.65
1.19(4+)
thorium

91 Pa
231.0369
1.14
——
1.15(4+)
protactinium

92 U
238.029
1.22
1.42
1.14(4+)
uranium

93 Np
237.0482
1.22
——
1.12(4+)
neptunium

94 Pu
(244)
1.22
——
1.10(4+)
plutonium

Periodic Table (fragment)

			(13) IIIa	(14) IVa	(15) Va	(16) VIa	(17) VIIa	(18) 0

(18) 0

2 He — 4.00260 — — — helium

(13) IIIa | (14) IVa | (15) Va | (16) VIa | (17) VIIa

5 B	6 C	7 N	8 O	9 F	10 Ne
10.81	12.011	14.0067	15.9994	18.998403	20.179
2.01	2.50	3.07	3.50	4.10	—
0.82	0.77	0.75	0.73	0.72	—
0.25(3+)	—	1.32(3−)	1.24(2−)	1.19(1−)	—
boron	carbon	nitrogen	oxygen	fluorine	neon

(10) | (11) Ib | (12) IIb

13 Al	14 Si	15 P	16 S	17 Cl	18 Ar
26.98154	28.0855	30.97376	32.06	35.453	39.948
1.47	1.74	2.06	2.44	2.83	—
1.18	1.17	1.06	1.02	0.99	—
0.675(3+)	—	—	1.70(2−)	1.67(1−)	—
aluminum	silicon	phosphorus	sulfur	chlorine	argon

28 Ni	29 Cu	30 Zn	31 Ga	32 Ge	33 As	34 Se	35 Br	36 Kr
58.70	63.546	65.38	69.72	72.59	74.9216	78.96	79.904	83.80
1.75	1.75	1.66	1.82	2.02	2.20	2.48	2.74	—
1.20	1.38	1.31	1.26	1.22	1.20	1.16	1.14	—
0.83(2+)	0.87(2+)	0.74(2+)	0.76(3+)	0.53(4+)	0.72(3+)	1.84(2−)	1.82(1−)	—
nickel	copper	zinc	gallium	germanium	arsenic	selenium	bromine	krypton

46 Pd	47 Ag	48 Cd	49 In	50 Sn	51 Sb	52 Te	53 I	54 Xe
106.4	107.868	112.41	114.82	118.69	121.75	127.60	126.9045	131.30
1.35	1.42	1.46	1.49	1.72	1.82	2.01	2.21	2.4?
1.31	1.53	1.48	1.44	1.41	1.40	1.36	1.33	1.25
0.78(2+)	1.14(1+)	1.09(2+)	0.94(3+)	1.36(2+)	0.90(3+)	2.07(2−)	2.06(1−)	—
palladium	silver	cadmium	indium	tin	antimony	tellurium	iodine	xenon

78 Pt	79 Au	80 Hg	81 Tl	82 Pb	83 Bi	84 Po	85 At	86 Rn
195.09	196.9665	200.59	204.37	207.2	208.9804	(209)	(210)	(222)
1.44	1.42	1.44	1.44	1.55	1.67	1.76	1.90	—
1.28	1.43	1.51	1.52	1.47	1.46	1.46	1.45	—
0.74(2+)	1.51(1+)	1.10(2+)	1.64(1+)	1.33(2+)	1.17(3+)	—	—	—
platinum	gold	mercury	thallium	lead	bismuth	polonium	astatine	radon

Eu	64 Gd	65 Tb	66 Dy	67 Ho	68 Er	69 Tm	70 Yb	71 Lu
.96	157.25	158.9254	162.50	164.9304	167.26	168.9342	173.04	174.967
	1.11	1.10	1.10	1.10	1.11	1.11	1.06	1.14
5	1.61	1.59	1.59	1.58	1.57	1.56	1.72	1.56
87(3+)	1.078(3+)	1.063(3+)	1.052(3+)	1.041(3+)	1.030(3+)	1.020(3+)	1.008(3+)	1.00(3+)
opium	gadolinium	terbium	dysprosium	holmium	erbium	thulium	ytterbium	lutetium

Am	96 Cm	97 Bk	98 Cf	99 Es	100 Fm	101 Md	102 No	103 Lr
3)	(247)	(247)	(251)	(252)	(257)	(258)	(259)	(260)
	—	—	—	—	—	—	—	—
?	—	—	—	—	—	—	—	—
9(4+)	—	—	—	—	—	—	—	—
ericium	curium	berkelium	californium	einsteinium	fermium	mendeleevium	nobelium	lawrencium

Inorganic Chemistry

Inorganic Chemistry
A Unified Approach

Second Edition

William W. Porterfield
Hampden-Sydney College
Hampden-Sydney, Virginia

Academic Press, Inc.
A Division of Harcourt Brace & Company
San Diego   New York   Boston   London   Sydney   Tokyo   Toronto

This book is printed on acid-free paper. ∞

Academic Press, Inc.
1250 Sixth Avenue, San Diego, California 92101-4311

United Kingdom Edition published by
Academic Press Limited
24–28 Oval Road, London NW1 7DX

Library of Congress Cataloging-in-Publication Data

Porterfield, William W.
    Inorganic chemistry : a unified approach / William W. Porterfield.
  -- 2nd ed.
        p.   cm.
    Includes index.
    ISBN 0-12-562981-8   (Papercover)
    1. Chemistry, Inorganic.   I. Title.
QD151.2.P67   1993
546--dc20   92-26704
                                                        CIP

PRINTED IN THE UNITED STATES OF AMERICA
93 94 95 96 97 98   MM   9 8 7 6 5 4 3 2 1

Contents ─────────────────────────

v

Tables

xi

Preface

When the first edition of this text appeared in 1984, I suggested in the preface that its novel plan of organization was a response to the growing integration of conceptual structures within inorganic chemistry. This integration makes it more helpful, in my judgment, to treat cluster compounds together than to treat boron cluster compounds together (but separately from, for instance, platinum cluster compounds); many such overlaps make the periodic table less useful than it has been in the past as an organizational principle for a text.

In the ensuing decade, the intellectual integration of inorganic chemistry has proceeded apace, and at the same time direct applications of inorganic chemistry have spread farther than before into areas such as materials science and biomedical chemistry. As a result, I have attempted to expand and improve the rather skimpy coverage of these areas in the second edition. The overall plan of the book has been preserved, but Chapter 3 now presents a good deal more background on inorganic chemistry as it applies to solid-state chemistry and novel materials such as ferrites, ceramics, and optical fibers. Similarly, Chapter 10 now includes material on magnetic alloys and cuprate high-temperature superconductors. The almost explosive growth of understanding in bioinorganic chemistry has made it desirable to add to Chapters 16 and 17 a great deal of that material, which I hope is fairly presented. Of course, a large number of changes and amplifications have been made throughout the book, including the use of density-of-states diagrams and Tanabe–Sugano diagrams as well as much new descriptive material.

The breadth and pace of change in inorganic chemistry make it impossible to keep up with new discoveries and, indeed, with entire areas in which important work is being done. I can only apologize for the inevitable sins of omission that will have occurred. I take consolation from the fact that this is not a new problem. In 1789 or 1790 the French chemist J. A. Chaptal wrote what is probably the first comprehensive inorganic chemistry book in what might be considered the modern form; near the end of his preface, two centuries ago, he says (in Nicholson's contemporary translation), "I was well aware that the pretension of knowing, discussing, and methodically distributing the whole of our present science of chemistry was an enterprise beyond my ability. This science has made so great a progress, and its applications are so multiplied, that it is impossible to attend to the whole with the same care." The task has not become simpler since; in fact, the journal *Inorganic Chemistry* alone has published over 43,000 pages of research results since the first edition of this book appeared. So I am even more conscious than Chaptal that I cannot "attend to the whole with the same care." Colleagues have read sections of this revision and have offered helpful suggestions. I am particularly grateful to Bill Anderson and Geoff Sykes for help with the bioinorganic material, and also to numerous speakers at Gordon Research and Royal Society of Chemistry conferences over many years. But errors of fact, interpretation, and emphasis will still be found, and of course they are my own.

Watching our discipline grow and interpenetrate an increasing number of facets of our culture is a fascinating process. I hope this text will be able to convey some of that excitement and perhaps lead new inorganic chemists into some of the areas still unexplored.

Part I

Elements and Atoms

1

Elements, Atoms, and Periodicity _____

Inorganic chemists work with an incredible variety of compounds, techniques, and purposes. They study the matter in interstellar space, strengthen concrete structures, convert solar energy to usable power, catalyze the reactions of the polymer industry, and model the functional molecules of life itself. The technology that characterizes the earliest modern human cultures—pottery, copper smelting—is inorganic chemical processing, sometimes quite sophisticated. So is the technology that characterizes our most advanced glimpses of the future—platinum anticancer drugs binding into nucleic acids to attack cancers, heat-resistant ceramic tiles on the space shuttle. The studies of modern inorganic chemists over two centuries have given us a coherent picture of many of these substances and technologies: Why are their properties as they are? What chemical choices give us the greatest power over nature? At the same time, these studies have opened exciting, even amazing, possibilities before us. No textbook can forecast research, but we can look at the principles underlying many of the areas in which some of the most surprising advances are being made.

More than anything else, inorganic chemistry is distinguished from other branches of the science by the fact that it studies all the elements. We need, then, to begin by appreciating the elements present in our universe and on our planet. We can look in turn at the way in which the elements were formed through nuclear processes, the sequence of discovery of the elements and the ways in which this influenced and was influenced by technology, and finally at the atomic structure of the elements as we now understand both its basic theory and the influences it has on the chemical properties of the elements, which we shall begin to study in Chapter 2.

1.1 THE ORIGIN OF THE ELEMENTS

Current cosmological theories hold that the universe evolved from a "big bang" in which an incredibly small aggregation of mass–energy (perhaps having a radius no larger than 0.01 mm and an energy density so large as to correspond to a temperature of about 10^{32} K) suddenly began to expand adiabatically. According to this model, very nearly all of the initial mass–energy was present in the form of radiation, whose effective mass could be inferred from the relativistic equation $E = mc^2$. As the adiabatic expansion occurred, the temperature dropped rapidly. For about the first microsecond, both protons and antiprotons existed in abundance because the thermal-energy density was greater than that needed to create such a pair. After the first microsecond, nearly all of the proton–antiproton pairs annihilated each other, leaving a very small surplus of protons and perhaps neutrons flooded by electrons and positrons (another matter–antimatter pair whose smaller mass allowed stable pairs until cooling had progressed further). Under these circumstances, protons and neutrons interconverted in a sort of equilibrium determined by a Boltzmann energy distribution $e^{-\Delta E/kT}$, where ΔE is the energy equivalent of their mass difference. This interconversion occurred for perhaps the first two seconds, by which time cooling reduced the temperature to about 10^{10} K. Below that temperature, electrons and positrons annihilate each other, "freezing" the proton–neutron equilibrium at about 25% neutrons, 75% protons. Until this time, no complex nuclei existed because the energy of the radiation flux was high enough to have dissociated any that formed. However, we can write an equation for the formation of a deuterium (^{2}H) nucleus from a proton and a neutron:

$$^1_1\text{H} \quad + \quad ^1_0n \quad \rightarrow \quad ^2_1\text{H} \quad + \quad 2.2\,\text{MeV}$$

$$\text{Proton} \qquad \text{Neutron} \qquad \text{Deuterium} \qquad \text{Binding energy}$$

Here the subscript indicates the atomic number (units of positive charge) and the superscript indicates the weight in atomic mass units, ignoring differences caused by the release of the nuclear binding energy. Once the temperature had dropped to a point at which its energy equivalent was less than 2.2 Me V, a deuterium nucleus could exist. That temperature is also about 10^{10} K, so deuterium formation began at about two seconds.

With deuterium nuclei to work with, the formation of the second element, helium, began. For the next three or four minutes, protons, neutrons, and deuterium nuclei reacted by schemes such as

$$^2_1\text{H} + ^1_0n \rightarrow ^3_1\text{H} \qquad\qquad\qquad ^2_1\text{H} + ^1_1\text{H} \rightarrow ^3_2\text{He}$$
$$\qquad\qquad\qquad\qquad\qquad\qquad\text{or}$$
$$^3_1\text{H} + ^1_1\text{H} \rightarrow ^4_2\text{He} + \text{Energy} \qquad ^3_2\text{He} + ^1_0n \rightarrow ^4_2\text{He} + \text{Energy}$$

These reactions stopped producing helium when essentially all of the neutrons had been used. up. Since equal numbers of protons and neutrons are present in a $^4_2$He nucleus, the presence of about 25% neutrons in the total matter mass of the nucleus led to about 25% helium in the cosmic abundance of the elements, which is near the observed abundance (see Table 1.1). Since nuclei with masses of 5 and 8 are unstable, the ^4_2He could not react further with either a proton, a neutron, or another helium nucleus; only traces of

Table 1.1   Abundance of Elements in the Earth's Crust, Including Oceans and Atmosphere, and in the Universe (ppm by weight)

Atomic number	Symbol	Earth's crust	Universe	Atomic number	Symbol	Earth's crust	Universe
1	H	1,400	739,000	37	Rb	70	0.01
2	He	0.01	240,000	38	Sr	450	0.04
3	Li	20	0.006	39	Y	35	0.007
4	Be	2	0.001	40	Zr	140	0.05
5	B	7	0.001	41	Nb	20	0.002
6	C	200	4,600	42	Mo	1	0.005
7	N	20	970	44	Ru	0.001	0.004
8	O	464,000	10,700	45	Rh	0.001	0.0006
9	F	460	0.4	46	Pd	0.003	0.002
10	Ne	0.005	1,340	47	Ag	0.08	0.0006
11	Na	23,200	22	48	Cd	0.2	0.002
12	Mg	27,700	580	49	In	0.2	0.0003
13	Al	80,000	55	50	Sn	2	0.004
14	Si	272,000	650	51	Sb	0.2	0.0004
15	P	1,010	7	52	Te	0.002	0.009
16	S	300	440	53	I	0.5	0.001
17	Cl	190	1	54	Xe	0.00003	0.01
18	Ar	3	220	55	Cs	2	0.0008
19	K	16,800	3	56	Ba	380	0.01
20	Ca	50,600	67				
21	Sc	22	0.03	57–71	La–Lu	225	0.014
22	Ti	8,600	3			total	total
23	V	170	0.7	72	Hf	4	0.0007
24	Cr	96	14	73	Ta	2.4	0.00008
25	Mn	1,000	8	74	W	1	0.0005
26	Fe	58,000	1,090	75	Re	0.004	0.0002
27	Co	28	3	76	Os	0.0002	0.003
28	Ni	72	60	77	Ir	0.0002	0.002
29	Cu	58	0.06	78	Pt	0.01	0.005
30	Zn	82	0.3	79	Au	0.002	0.0006
31	Ga	17	0.01	80	Hg	0.02	0.001
32	Ge	1	0.2	81	Tl	0.5	0.0005
33	As	2	0.008	82	Pb	10	0.01
34	Se	0.05	0.03	83	Bi	0.004	0.0007
35	Br	4	0.007	90	Th	6	0.0004
36	Kr	0.0002	0.04	92	U	2	0.0002

$^7_3$Li were formed, and only traces of deuterium remained. After about the first million years, the adiabatic expansion and cooling had progressed far enough to allow the electrons present to combine with helium and hydrogen nuclei, forming the first atoms of those elements and eliminating the electrostatic repulsion between nuclear ions that had previously prevented atoms from aggregating.

At this point, the density of the expanding universe was about 10^7 atoms per liter, which is at least 1000 times the density of an average galaxy. In this relatively dense (by astronomical standards) gas, chance regions of higher density began to condense because of the gravitational attraction of the atoms for each other. The gradual compression of enormous gas clouds into proto-stars or protogalaxies reheated them. When the process of condensation and heating had reached the level of star formation, nuclear synthesis began again.

Within a star, there is a great deal more matter mass–energy than there is radiation mass–energy, so radiation pressure does not force expansion and cooling. Gravitational contraction can produce temperatures on the order of 10^8 K and matter densities on the order of 10^5 g/cm^3, under which conditions nuclear collisions occur so frequently and at such high energy that even unstable product nuclei can react again before they decay.

$$^4_2\text{He} + {}^4_2\text{He} \rightarrow {}^8_4\text{Be} \qquad {}^8_4\text{Be} + {}^4_2\text{He} \rightarrow {}^{12}_6\text{C}$$

$$\text{Unstable} \qquad\qquad\qquad \text{Stable}$$

As long as the star retains some helium, reactions such as this will continue, forming nuclei that are multiples of helium:

$$^{12}_6\text{C}, \ {}^{16}_8\text{O}, \ {}^{20}_{10}\text{N}, \ \ldots \ , {}^{40}_{20}\text{Ca}$$

This process is called *helium burning*. With carbon present, the carbon–nitrogen cycle can operate, as shown in Fig. 1.1, forming nitrogen. At still higher temperatures, processes such as carbon burning and oxygen burning can occur, producing elements that are not helium multiples:

$$^{12}_6\text{C} + {}^{12}_6\text{C} \rightarrow {}^{24}_{12}\text{Mg} + \text{Energy} \qquad {}^{16}_8\text{O} + {}^{16}_8\text{O} \rightarrow {}^{32}_{16}\text{S} + \text{Energy}$$

$$^{12}_6\text{C} + {}^{12}_6\text{C} \rightarrow {}^{23}_{11}\text{Na} + {}^1_1\text{H} \qquad {}^{16}_8\text{O} + {}^{16}_8\text{O} \rightarrow {}^{31}_{15}\text{P} + {}^1_1\text{H}$$

$$^{12}_6\text{C} + {}^{12}_6\text{C} \rightarrow {}^{20}_{10}\text{Ne} + {}^4_2\text{He} \qquad {}^{16}_8\text{O} + {}^{16}_8\text{O} \rightarrow {}^{31}_{16}\text{S} + {}^1_0 n$$

All of these processes release energy, because the binding energy per nucleon (proton or neutron) increases with atomic number Z up to $Z = 26$ (iron), as shown in Fig. 1.2. Thus there are "burning" processes for most elements lighter than calcium analogous to those shown above for carbon and oxygen, but occurring at still higher temperatures and thus later in the star's evolution.

Past $Z = 20$ (calcium), the increasing concentration of positive charge in the nucleus must be stabilized by extra neutrons, the atomic weight increasing faster than the atomic number. Usually a helium-multiple nucleus with

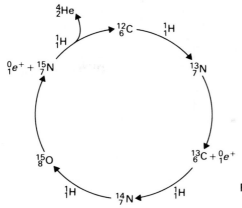

Figure 1.1   Nuclear carbon–nitrogen cycle.

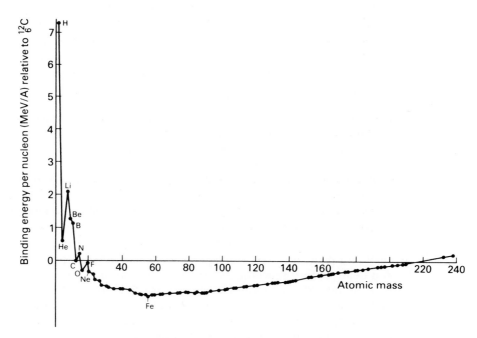

Figure 1.2   Stability of naturally occurring nuclei (expressed as binding energy per nucleon) as a function of atomic mass.

$Z > 20$ will decay either by electron capture or positron emission, forming a new element:

$$\,^{52}_{26}\text{Fe} \rightarrow \,^{52}_{25}\text{Mn} + \,^{0}_{1}e^+ \qquad\qquad \,^{44}_{22}\text{Ti} + \,^{0}_{-1}e \rightarrow \,^{44}_{21}\text{Sc}$$

Beyond iron in the periodic table, there is no longer any energy yield from nuclear fusion reactions. It is believed that under the conditions of very high temperature and pressure late in the development of red-giant stars, a slow process of neutron capture occurs, interrupted by electron emission (the *s*-process):

$$\,^{68}_{30}\text{Zn} + \,^{1}_{0}n \rightarrow \,^{69}_{30}\text{Zn} \rightarrow \,^{69}_{31}\text{Ga} + \,^{0}_{-1}e$$

$$\,^{68}_{30}\text{Zn} + \,^{1}_{0}n \rightarrow \,^{70}_{31}\text{Ga} \rightarrow \,^{70}_{32}\text{Ge} + \,^{0}_{-1}e$$

and so on.

Still later, the temperature can become so high that stable nuclei can photodissociate to yield many particles:

$$\,^{56}_{26}\text{Fe} + \text{Radiation} \rightarrow \,^{28}_{14}\text{Si} + 6\,^{4}_{2}\text{He} + 4\,^{1}_{0}n$$

This produces a very fast chain reaction of energy release by lighter elements within the star, creating a supernova. In the fantastically intense neutron flux within the supernova, multiple neutron captures can occur within a single nucleus before an electron emission (β^- decay) can occur. This is the rapid *r*-process, which has the same effect as the *s*-process in increasing the atomic number of the nucleus as neutrons are added progressively to the nucleus and partially converted to protons. The *s*-process can make nuclei up to $Z = 83$ (bismuth), whereas the *r*-process can go as far as the heaviest known elements. Beyond bismuth, many isotopes are unstable to fission, which breaks the nucleus into two roughly equal parts (plus leftover neutrons). This somewhat increases the abundance of elements between about $Z = 30$ and $Z = 60$ in such stars.

Let us return at this point to reflect on the earth's composition. Table 1.1 shows that, at least in the earth's crust, heavy elements are more abundant than they are in the universe as a whole. This suggests that the matter of the earth has probably been through several cycles of star formation, evolution, and death before reaching its present composition. It is at least conceivable that the periodic table established by investigators on a planet made of less mature matter might go no farther than the first-row transition elements! Our solar system may be quite rare in having achieved a full range of elements; we are the fortunate inheritors of a fantastic past.

1.2 THE ABUNDANCE OF THE ELEMENTS

Earth's elements are available to us in the widely varying quantities presented in Table 1.1. It is perhaps helpful to see those data in a graphic presentation, as in Fig. 1.3. The erratic shape of that graph is the result of many factors, not

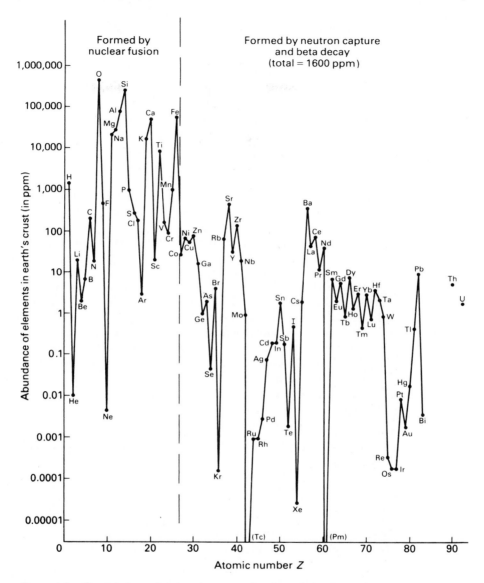

Figure 1.3   Crustal elemental abundance as a function of atomic number.

all of which are known. Several generalizations are possible, however. The noble gases He, Ne, Ar, Kr, and Xe are far less abundant than neighboring elements because their chemical unreactivity means that they had to be acquired on an atom-by-atom basis during the gravitational condensation of the earth, instead of by chemically binding to crystal lattices in space dust.

Argon is much more abundant than the other noble gases because much of it was produced by the radioactive decay of potassium-40 within the earth's crust after the planet was formed.

Another observation is that the elements formed by nuclear fusion— helium burning, carbon burning, and the like—are much more abundant than those with an atomic number higher than iron, which must be formed by the relatively rare neutron-capture processes. All of these heavy elements combined (three-quarters of the periodic table) have a total abundance of only about 1600 ppm, little more than manganese alone. Nine of the abundant light elements have abundances near or above 10,000 ppm (or 1%) and can be thought of as "framework elements" for the earth's crust: O, Na, Mg, Al, Si, K, Ca, Ti, and Fe. Since only one of these (iron) was known to the ancients, it is not surprising that they had a quite different idea of what constituted an element than we do.

Part of the abundance pattern is a reflection of nuclear stabilities. The very low abundance of beryllium, for example, presumably results from the instability of the double-helium nucleus ^8_4Be; the only existing beryllium isotopes are those with extra neutrons. Similarly, in the vicinity of the lanthanide rare earths ($Z = 57$ to 71) we see an alternation in abundances, in which an element with an even atomic number is more abundant than its two neighbors with odd atomic numbers. Since these nuclei are all produced by comparable processes and their atoms have similar chemical properties, the pattern of abundances suggests that, in the absence of other factors, nuclei with even atomic numbers are more stable (or at least are more likely to be produced) than those with odd atomic numbers. In what is apparently an extension of this pattern of symmetry, nuclei having both an odd number of protons and an odd number of neutrons are usually unstable and always rare; the only exception to this is $^{14}_7N$. All other elements with odd atomic numbers have a predominant isotope with an even number of neutrons, and indeed no odd–odd nucleus is stable past $^{14}_7N$.

It is important to distinguish between overall abundance in the earth's crust and economic availability. Cobalt is a good example. It is present in the earth's crust to the extent of about 28 ppm, which does not seem high but is greater than that of nitrogen. Nitrogen is in effect available without limit in the atmosphere, but cobalt is exceedingly rare in economic concentrations. It is found principally in Zaire, Zambia, and Canada; and even where it is produced, it is usually a recovered by-product of copper or nickel ore treatment. Although cobalt is found in minor concentrations in Idaho and Missouri, the United States produced no cobalt from domestic sources at all in 1988, although it used some 8900 tons in magnetic and machine-tool alloys. Another striking example is aluminum, which with an overall abundance of some 80,000 ppm (8%) is one of the framework elements of the earth's crust. It is as abundant in the United States as elsewhere, in aluminosilicate rock. Unfortunately, the economic production of aluminum metal requires that the

Table 1.2 U.S. Production, Consumption, and Reserves of Key Metals (tons)

Element (or ore)	Production (1988)	Consumption (incl. recycling)	Reserves 1975	Reserves Potential	Chief import sources
Iron and steel associated metals					
Iron	99,900,000	120,000,000	5.4 billion	60 billion	Canada, Brazil
Chromium (chromite)	0	551,000	100,000	5,400,000	South Africa, Turkey
Manganese (ferromanganese)	0	468,000	2,000,000	65,000,000	S. Africa, France
Tungsten	0	7,800	120,000	300,000	China
Cobalt	0	8,900	90,000		Zaire, Canada
Nickel	0	196,000	700,000		Canada, Norway
Molybdenum	47,500	39,300	3,100,000	14,400,000	—
Vanadium	3,200	5,300	115,000	3,000,000	S. Africa, China
Nonferrous metals					
Aluminum	3,944,000	5,350,000	9,000,000	70,000,000	Canada
(bauxite)	590,000	10,070,000			Australia
Copper	1,850,000	2,210,000	84,000,000	60,000,000	Canada, Chile
Lead	370,000	1,230,000	38,000,000		Canada, Mexico
Tin	0	45,000	0		Brazil, China
Silver	1,830	4,570	25,000		Mexico, Canada
Mercury	700	1,750	15,000		Spain, China
Magnesium	156,000	111,000	(large)		—
Titanium	24,500	23,100			(ore) Australia, Canada, S. Afr.

ore be bauxite, a mixture of hydrated aluminum oxides containing very little silica. The United States' supply of bauxite is equal to roughly four years' consumption, even though the natural abundance of aluminum makes it essentially inexhaustible. We are, under these circumstances, dependent on Australia and other countries for an element that is literally all around us.

As suggested in the preceding paragraph, the United States is dependent on foreign sources for most of the commercially important elements. Table 1.2 indicates the extent of this dependence by comparing U.S. production, consumption, and reserves as quoted for the year 1988. If we assume that a reserve equal to a 20-year supply at present consumption rates is adequate, only iron, tungsten, molybdenum, copper, magnesium, and lead are available on that basis. Chromium, manganese, cobalt, nickel, vanadium, aluminum, uranium, tin, silver, and mercury all must be imported, either entirely or in large measure.

1.3 THE DISCOVERY OF THE ELEMENTS

A number of chemical elements existed as such in human technology long before their elemental nature was recognized. Still more elements were discovered by experimenting alchemists or early chemists and were recognized as new substances even though the chemical theory of the time did not classify them as elements. If we take Lavoisier's publication of *Traité élémentaire de chimie* in 1789 as the beginning of modern chemistry, it found an audience among chemists who already knew of 27 elements by the modern definition. Table 1.3 gives a list of the elements in the approximate order of

Table 1.3   Elements in Their Approximate Order of Discovery

Ancient times (before 1000 B.C.)
  Au Ag Cu Sn Pb Fe S Hg C Sb

Alchemical period (1000 B.C. to A.D. 1700)
  Zn As Bi Pt P

Chemical extraction (1700–1900)
  Co Ni H N O Cl Mn Mo W Zr U Ti Y Cr Te Nb Ta Pd Os
  Rh Ir K Na Ca Sr Ba B I Cd Se Si Al Ce Br Be Mg Th V
  La Er Tb Ru Li Sc Yb Sm Tm Gd Dy Ge F Ar Eu Lu(1907)

Spectroscopic or radioactive identification (1860–1925)
  Cs Rb Tl In He Ga Ho Pr Nd Kr Ne Xe Ra Po Ac Rn Pa
  Hf Re

Synthetic elements (1937–1961)
  Tc Fr At Np Pu Pm Am Cm Bk Cf Es Fm Md No Lr

their discovery, arranged by eras that coincide fairly well with experimental techniques.

We have no documentary evidence about the discovery of the elements known in ancient times. Archaeological inferences allow us to place some of them very far back. We can, rather arbitrarily, define an *alchemical period* from about 1000 B.C. to A.D. 1700, during which more elements were discovered and given at least rudimentary documentation. For both of these two earliest periods, it should be noted that several of the elements occur as the native elements, requiring no processing. Under these circumstances it seems likely that a kind of discovery may have occurred far back in the Stone Age, but we will in most cases be concerned with the first production of the element from its compounds or ores.

About the beginning of the eighteenth century, chemical techniques had reached a level at which we recognize the laboratory procedures to be similar to our own. During the two centuries from 1700 to 1900, the traditional "wet chemical" techniques of decomposition by acid or base, oxidation or reduction, recrystallization, and so on produced most of the elements in the periodic table from the mineralogical specimens that were becoming available. In the mid-nineteenth century, spectroscopy became the first instrumental technique for examining unknown specimens, followed about 1900 by the detection of characteristic radioactive decay. These two techniques were responsible for identifying a large number of additional elements that were hard to separate or identify by wet methods, up to 1925 when the last naturally occurring element (rhenium) was identified. Beginning in 1937, the production of synthetic elements by nuclear reactions filled the two gaps (technetium and promethium) in the periodic table up to bismuth, the heaviest element with a stable isotope, and also produced astatine, francium, and the transuranic elements. This process can presumably continue, though with increasing atomic number the synthetic elements are apparently less and less stable.

It seems likely that gold was the first element human beings recognized. It occurs as the native element (only rarely as a telluride ore) in almost every region of the earth's surface and has an unusually high density and a striking appearance. The earliest documents of every culture mention it, and dazzling examples of the goldsmith's art exist at least as far back as 4500 B.C. Of course, no chemical processing was involved in either its production or its use, so there was no true chemical discovery, but it is probably the oldest element. The other elements of the ancient period also occur as the native element (except tin), but most are quite rare geographically. Perhaps the rarest is iron, which occurs as the native metal only in meteorites. There is, however, evidence that meteorites were used as talismans in early societies. Primitive peoples certainly found native silver and native copper, but they seem to have begun quite early to produce the metal chemically by heating the sulfide ores with charcoal in a limited supply of air. Archaeological evidence has dated copper mines worked as long ago as 6000 B.C. Bronze, which is a copper–tin alloy containing 10–15% tin, is much harder than copper and is thus a better material for tools. Bronze articles, which imply the production of tin, date to about 4000 B.C., and tin–lead alloys were used as solder in Greek and Roman times.

Zinc ornaments in which the zinc is 80–90% pure are known back to perhaps 500 B.C. However, the metal disappeared from European culture until about the seventeenth century, when it began to appear in trade from China and India. In these areas, it had been extracted from oxide ores since around A.D. 1000. Arsenic and bismuth gradually appeared in commerce between the thirteenth and fifteenth centuries, although their compounds had been known in Roman times. The first element for which a specific date of discovery can be quoted is phosphorus. In 1669, a German merchant and amateur alchemist named Brand prepared white phosphorus by distilling it from urine, a preparation method that has since been improved upon. Native platinum was discovered earlier (in Central American gold mines after the Spanish conquest), but was regarded as a troublesome impurity and ignored chemically until the middle of the eighteenth century. The first chemical fortune was made by William Wollaston, who developed a secret process for making malleable platinum and manufactured platinum laboratory ware from about 1800 to 1820.

At this stage, with chemical laboratory operations well established and widely known, the pace of discovery accelerated. Most of the metallic elements were prepared by purifying their oxides and then reducing them either by electrolysis or by treating them with sodium or potassium metals. These experiments were usually deliberate attempts to discover new elements, but accidental discoveries still occurred, as in the case of iodine. In 1811, Courtois was trying to economize on gunpowder production for Napoleon's armies by drying and burning seaweed to form soda ash (sodium carbonate). He found he was producing a violet vapor that corroded his copper equipment. He experimented briefly with it, but failed to recognize it as a new element. He subsequently manufactured it for sale, but left Gay-Lussac to report its elemental nature.

By 1860, approximately 60 elements were recognized, many of which had physical and chemical similarities. A number of chemists began to evolve groupings of similar elements or overall classifications that might suggest the number and character of missing elements. Several classifications had some merit, but the one generally adopted was, of course, the periodic table published by Mendeleev in 1871, which listed 63 elements and confidently predicted three others. Within five years, one of these (gallium) had been found, with properties corresponding closely to those Mendeleev had predicted. With the guidance of the periodic table, 22 elements were discovered in the 30 years after Mendeleev's publication; only four naturally occurring elements remained unknown.

This period coincided with the introduction of spectroscopic techniques to chemical analysis. Although it had been realized for about a century that certain metallic elements produced characteristic flame colors, it was not until 1860, when Bunsen and Kirchhoff developed the spectroscope, that qualitative analysis of elements from their emission lines became an established practice. Bunsen designed his gas burner for the specific purpose of providing heat without light, to reduce interference by the flame in the observed spectra. The two men discovered cesium by observing its previously unknown spectral lines in 1860. This analytical tool was extensively applied by themselves and others through the next 40 years. Another spectroscopic technique used in discovering the last few naturally occurring elements was the measurement of the wavelength of x-rays emitted by a target metal in an x-ray tube. Since these involve inner-core electron transitions, their wavelength or frequency is directly related to the nuclear charge or atomic number. The number of rare earths was finally fixed using this technique, and hafnium and rhenium were conclusively identified to complete the list of naturally occurring elements.

Rhenium ($Z=75$) lies under manganese in the periodic table, but directly under manganese lies the missing element $Z = 43$. The Noddacks, discoverers of rhenium, claimed also to have found it on the strength of very faint lines in the x-ray spectrum, but their claim could not be substantiated. In 1937, Segre and Perrier isolated an extremely small amount of it from a molybdenum target that had been bombarded with deuterium nuclei in a cyclotron for several months. Its chemical behavior resembled that of manganese and rhenium rather than molybdenum, as indicated by the distribution of its radioactivity after passing through chemical separations. The new element was eventually named technetium, since it was the first synthetic element. Other radioactive elements were prepared during the 1940s and 1950s, primarily by bombarding heavy-element targets with neutrons or accelerated light-element nuclei:

$$^{238}_{92}\text{U} + ^{1}_{0}n \rightarrow ^{239}_{92}\text{U} \rightarrow ^{231}_{93}\text{Np} + ^{0}_{-1}e$$

$$^{238}_{92}\text{U} + ^{2}_{1}\text{H} \rightarrow 2^{1}_{0}n + ^{239}_{92}\text{Np} \rightarrow ^{238}_{94}\text{Pu} + ^{0}_{-1}e$$

$$^{239}_{94}\text{Pu} + ^{4}_{2}\text{He} \rightarrow ^{242}_{96}\text{Cm} + ^{1}_{0}n$$

The extreme instability of the heaviest elements has slowed the pace of discovery. The final actinide, lawrencium ($Z = 103$), was synthesized in 1961, and a system has been established for naming future elements simply by their atomic number. Element 104 is named *un*(one)*nil*(zero)*quad*(four)*ium*, for example, though discoverers may still propose names in the time-honored way after their priority of discovery is established.

_____ 1.4 ATOMIC STRUCTURE FOR ONE-ELECTRON ATOMS

All of the elements in the previous discussion are, of course, made of nuclear atoms. It is appropriate at this point to discuss our theoretical understanding of atomic structure. In any such discussion, it is important to distinguish between exact mathematical consequences of the theory and approximations introduced to avoid mathematically intractable situations. In this section we shall deal solely with the mathematically exact results of quantum mechanical derivations, leaving approximate methods to the next section.

The nucleus of an atom has a diameter of about 10^{-12} cm, compared to a diameter of 10^{-8} cm for the atom with its inner-core and valence electrons. Almost all the volume of the atom is thus occupied by electrons. The nucleus contains protons and neutrons, each with a mass very near 1 atomic mass unit (amu) or 1 g/mol. The number of protons present—the total number of units of positive electrical charge—is the atomic number, which is the basis of periodic behavior. All of the atoms of a given element have the same number of protons, but they can differ in the number of neutrons and thus in total mass. Atoms differing on this basis are called *isotopes* of the element. The experimental atomic weight of an element is the number average of the weights of the naturally occurring isotopes of that element. The atomic weight of any one isotope is very close to an integer equal to the total number of protons plus neutrons. No naturally occurring isotope differs in mass from that integer by more than 0.10 amu ($^{127}_{53}$I mass = 126.9004 amu).

The electrons in these atoms are believed to be distributed in a three-dimensional wave pattern; this belief results from the fact that each electron has an individually observable wave character. This amounts to a restatement of de Broglie's hypothesis that there is an equivalence between the momentum of a particle and the wave number (reciprocal wavelength) of a wave that may be said to be associated with the particle: $\lambda = h/mv$, where λ is the wavelength. The electron-diffraction experiments of Davisson and Germer confirmed de Broglie's hypothesis for the electron, so that the electron must be thought to possess an intrinsic wave character in addition to its particle characteristics.

The wave character of the electron is described in one dimension by the Schrödinger equation,

$$\frac{d^2\psi}{dx^2} + \frac{8\pi^2 m}{h^2}(E - V)\psi = 0$$

The amplitude quantity ψ is proportional to the perceived intensity of the wave when squared, by analogy with other wave phenomena:

$$\text{Intensity} = k \cdot \psi^2$$

The "intensity" of an electron in a bound system is usually interpreted by saying that ψ^2 constitutes a probability distribution function for the electron. Where ψ^2 is zero, the electron will never be found; but where ψ^2 is a small finite number, there is a small probability of finding the electron.

For the electron in a hydrogen atom, the Schrödinger equation above must be expanded to a three-dimensional form. The second derivative

$d^2\psi/dx^2$ must be replaced by $\nabla^2\psi$, defined for Cartesian coordinates as

$$\nabla^2\psi \equiv \frac{\partial^2\psi}{\partial x^2} + \frac{\partial^2\psi}{\partial y^2} + \frac{\partial^2\psi}{\partial z^2}$$

and for spherical polar coordinates as

$$\nabla^2\psi \equiv \frac{1}{r^2}\left[\frac{\partial}{\partial r}\left(r^2\frac{\partial\psi}{\partial r}\right)\right] + \frac{1}{r^2\sin^2\theta}\frac{\partial^2\psi}{\partial\phi^2} + \frac{1}{r^2\sin\theta}\frac{\partial}{\partial\theta}\left(\sin\theta\frac{\partial\psi}{\partial\theta}\right)$$

In spite of the awesome appearance of the $\nabla^2\psi$ quantity, polar coordinates are the most convenient choice for describing the hydrogen atom because of the radial dependence of the potential energy: $V = -e^2/r$, where e is the charge on the electron and proton and r is the distance of the electron at any moment from the nucleus (the proton). The Schrödinger equation becomes

$$\nabla^2\psi + \frac{8\pi^2 m}{h^2}\left(E + \frac{e^2}{r}\right)\psi = 0$$

By assuming that ψ is the product of a radial wave function $R(r)$ and an angular wave function $\Theta(\theta) \cdot \Phi(\phi)$, the Schrödinger equation can be separated into three one-dimensional equations, each of which can be solved for the form of the one-dimensional wave function described by that equation. The ϕ equation can be solved by inspection, but the θ and r equations require series solutions that must be truncated at a finite number of terms to remain physically reasonable. This truncation introduces quantization in a natural way to the hydrogen atom. Three quantum numbers result: the *principal quantum number n* (originating in the r series), the *azimuthal quantum number l* (originating in the cos θ series), and the *magnetic quantum number m* (originating in the ϕ solution). These are linked by the three-dimensional nature of the system, so that $n > l \geq |m|$. Any set of quantum-number integers for n, l, and m meeting the above condition defines a satisfactory solution to the Schrödinger wave equation for the hydrogen atom. The set of these solutions constitutes the hydrogen-atom wave functions.

The wave functions for $n = 1, 2, 3$ are given in Table 1.4. Conventional pictures of the angular wave functions are shown in Fig. 1.4. These pictures are usually described as atomic orbital shapes, but it must be remembered that they do not include the radial dependence of the electron distribution. That radial dependence is shown in Fig. 1.5 for the wave functions or orbitals of Table 1.3. It is not possible to draw adequate pictures of three-dimensional hydrogen-atom wave functions, but the angular-wave function diagrams of Fig. 1.4 are usually adequate for describing atomic-orbital overlap in molecules.

It is worth noting in the context of the angular-dependence sketches of Fig. 1.4 that all of the atomic orbitals except the s orbitals have nodes, or two-dimensional surfaces in space around the nucleus where the electron will

Table 1.4   Normalized Wave Functions for a One-Electron Atom

Quantum numbers			
n	l	m	Wave function
1	0	0	$\psi_{1s} = \dfrac{1}{\sqrt{\pi}}\left(\dfrac{4\pi^2 mZe^2}{h^2}\right)^{3/2} e^{-\rho}$
2	0	0	$\psi_{2s} = \dfrac{1}{4\sqrt{2\pi}}\left(\dfrac{4\pi^2 mZe^2}{h^2}\right)^{3/2}(2-\rho)e^{-\rho/2}$
2	1	0	$\psi_{2p_z} = \dfrac{1}{4\sqrt{2\pi}}\left(\dfrac{4\pi^2 mZe^2}{h^2}\right)^{3/2}\rho e^{-\rho/2}\cos\theta$
2	1	±1	$\psi_{2p_x} = \dfrac{1}{4\sqrt{2\pi}}\left(\dfrac{4\pi^2 mZe^2}{h^2}\right)^{3/2}\rho e^{-\rho/2}\sin\theta\cos\phi$
			$\psi_{2p_y} = \dfrac{1}{4\sqrt{2\pi}}\left(\dfrac{4\pi^2 mZe^2}{h^2}\right)^{3/2}\rho e^{-\rho/2}\sin\theta\sin\phi$
3	0	0	$\psi_{3s} = \dfrac{2}{81\sqrt{3\pi}}\left(\dfrac{4\pi^2 mZe^2}{h^2}\right)^{3/2}(27-18\rho+2\rho^2)e^{-\rho/3}$
3	1	0	$\psi_{3p_z} = \dfrac{2}{81\sqrt{\pi}}\left(\dfrac{4\pi^2 mZe^2}{h^2}\right)^{3/2}(6\rho-\rho^2)e^{-\rho/3}\cos\theta$
3	1	±1	$\psi_{3p_x} = \dfrac{2}{81\sqrt{\pi}}\left(\dfrac{4\pi^2 mZe^2}{h^2}\right)^{3/2}(6\rho-\rho^2)e^{-\rho/3}\sin\theta\cos\phi$
			$\psi_{3p_y} = \dfrac{2}{81\sqrt{\pi}}\left(\dfrac{4\pi^2 mZe^2}{h^2}\right)^{3/2}(6\rho-\rho^2)e^{-\rho/3}\sin\theta\sin\phi$
3	2	0	$\psi_{3d_z^2} = \dfrac{1}{81\sqrt{6\pi}}\left(\dfrac{4\pi^2 mZe^2}{h^2}\right)^{3/2}\rho^2 e^{-\rho/3}(3\cos^2\theta-1)$
3	2	±1	$\psi_{3d_{xz}} = \dfrac{\sqrt{2}}{81\sqrt{\pi}}\left(\dfrac{4\pi^2 mZe^2}{h^2}\right)^{3/2}\rho^2 e^{-\rho/3}\sin\theta\cos\theta\cos\phi$
			$\psi_{3d_{yz}} = \dfrac{\sqrt{2}}{81\sqrt{\pi}}\left(\dfrac{4\pi^2 mZe^2}{h^2}\right)^{3/2}\rho^2 e^{-\rho/3}\sin\theta\cos\theta\sin\phi$
3	2	±2	$\psi_{3d_{x^2-y^2}} = \dfrac{1}{81\sqrt{2\pi}}\left(\dfrac{4\pi^2 mZe^2}{h^2}\right)^{3/2}\rho^2 e^{-\rho/3}\sin^2\theta\cos 2\phi$
			$\psi_{3d_{xy}} = \dfrac{1}{81\sqrt{2\pi}}\left(\dfrac{4\pi^2 mZe^2}{h^2}\right)^{3/2}\rho^2 e^{-\rho/3}\sin^2\theta\sin 2\phi$

Note: For simplicity, $(4\pi^2 mZe^2/h^2)r$ has been indicated by ρ.

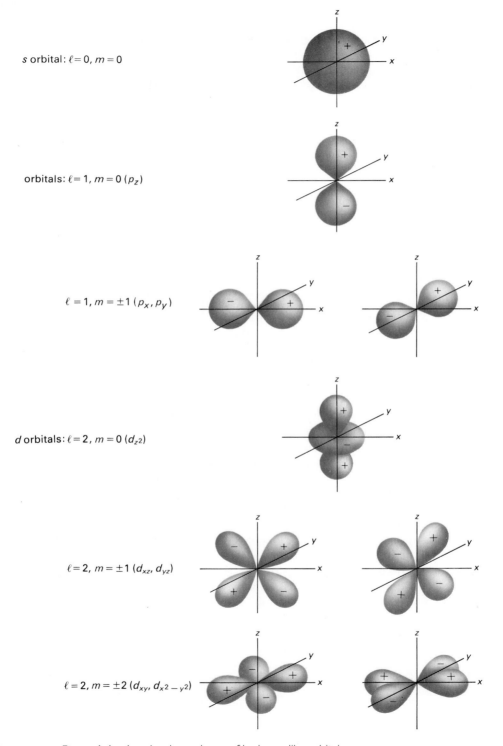

Figure I.4 Angular dependence of hydrogenlike orbitals.

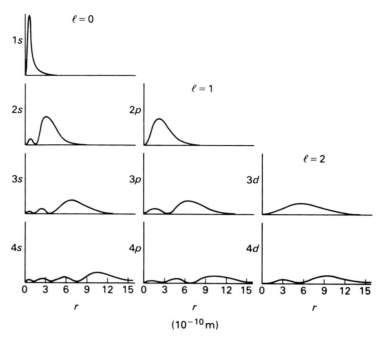

Figure I.5 Radial dependence of hydrogenlike orbitals.

never be found. Thus, the xy plane is a node for the $2p_z$ orbital: The electron will be found half of the time above the xy plane and half of the time below it, but never in the plane itself. A *nodal surface* is the set of points in space at which the wave function changes sign, thereby passing through zero and yielding zero probability of finding the electron. So the wave function has opposite signs on the two sides of a nodal surface, which has important consequences when we bring atomic orbitals from two nuclei together to let them overlap and form a bond. The number of angular nodes is given by the l quantum number: An s orbital has zero angular nodes, a p orbital has one, a d orbital has two, and so on.

The total energy E of the electron in a hydrogen atom is built into the Schrödinger equation in such a way that it ultimately appears in the $R(r)$ equation. When that equation is solved for the radial wave function, the quantum number n that arises from the truncation of the power series in r is equal to a group of constants that contains E. If the expression defining n in this way is rearranged to solve for E, we get the quantum-mechanical expression for the energy of the electron in a hydrogen atom or in any other one-electron atom:

$$E = -\frac{2\pi^2 m Z^2 e^4}{n^2 h^2}$$

In this expression, m is the mass of the electron, Z is the atomic number, e is the charge on the electron, and h is Planck's constant. Since n must be a positive integer, the energy is quantized; only certain values are possible. The fact that n can range from 1 to infinity means that there is an infinite number of energy levels. In experimental atomic spectroscopy, we can in principle populate any of these levels. In practice, however, levels higher than about $n = 8$ or 9 are almost never seen. The fact that the total energy is negative for any value of n simply defines a bound state for the electron, since zero energy is defined as if the electron and nucleus were separated by an infinite distance (so that they have zero potential energy with respect to each other) and are motionless with respect to each other (so that they have zero kinetic energy). Positive values for the energy are possible; they represent ionized states in which the electron is no longer bound to the nucleus. For positive values of E, the quantum-mechanical limitations no longer apply, and a continuum of positive energies is possible.

I.5 ATOMIC STRUCTURE FOR POLYELECTRONIC ATOMS

When one goes beyond one-electron atoms, the Schrödinger equation is no longer soluble in closed form, and approximate methods become necessary. One formal condition on polyelectronic systems that is rigorously met regardless of the specific approach, however, is the Pauli exclusion principle. In its simplest form, the exclusion principle says that no two electrons can have identical sets of quantum numbers. This includes a new number, called the *spin quantum number*. An electron has an intrinsic angular momentum or spin (independent of its spatial distribution) whose axis can adopt either of two orientations with respect to an external reference direction. The quantum number m_s is used for spin, by analogy with the magnetic quantum number m, which describes the orbital distribution of the electron around the z axis. The number m_s can have either of two values, corresponding to the two possible spin orientations: $+\frac{1}{2}$ or $-\frac{1}{2}$. Electrons in a polyelectronic atom, then, are described by four quantum numbers: $n, l, m,$ and m_s. The exclusion principle requires that each electron in a single atom differ in at least one of these quantum numbers from every other electron in that atom.

A more sophisticated form of the exclusion principle, applicable to atoms or molecules where specific quantum numbers may not be readily definable, requires that the complete polyelectronic wave function for the atom or molecule be antisymmetric with respect to electron exchange. This means that the wave function, in which each electron is numbered and dealt with separately, must change its sign if any two electron numbers or labels are interchanged. Changing the sign of the wave function does not change the electron distribution; the square of the wave function, which eliminates any sign effect, determines that distribution. If a large number of electrons is involved in the atom or molecule, the number of possible permutations of electron labels becomes very large, which complicates theoretical calculations.

When we attempt to deal with polyelectronic atoms in calculations—even the helium atom with its two electrons—we find that the potential energy of repulsion between the two moving electrons makes it impossible to separate the variables in the Schrödinger equation to yield one-dimensional differential equations. Therefore, the problem cannot be solved exactly and approximate solutions are necessary. As a first step, we normally make the orbital approximation—that is, we assume that each electron occupies one of the orbitals that would exist for a one-electron atom. It is by no means obvious that this should be true, since multiple electrons will correlate their angular arrangement about the nucleus in a way that has no analogy in a one-electron atom. In practice, however, it proves to be a sound approximation.

The next step in handling polyelectronic atoms is to devise a method of allowing for the energy effect of electron–electron repulsion. The radial dependence of hydrogenlike orbitals (Fig. 1.5) suggests an approach, since the exclusion principle requires that electrons occupy different orbitals (except for pairing). For example, a 1s electron in a hydrogenlike atom is almost always much closer to the nucleus of the atom (which is assumed to be at the origin of the coordinate system) than a 3d electron. This means that for the 3d electron, to a good approximation, the repulsion of the 1s electron can be described by treating it as simply a reduction of the nuclear positive charge. Of course, even using the approximation of hydrogenlike orbitals there will be situations in which two electrons repel each other in orbitals at about the same average distance from the nucleus. In the latter case, we can still assume that the repulsion can be described as a reduction of the net attraction of each electron for the nucleus, but only as a fractional reduction. The result is to create for each electron in a polyelectronic atom an *effective nuclear charge,* Z_{eff}, equal to the true nuclear charge (the atomic number) minus a screening constant that is the sum of all the reductions of the nuclear charge due to electrons closer to the nucleus than the average for the one being considered. The 1s electrons will thus experience a greater effective nuclear charge than the 2s electrons, because the 2s electrons are shielded by the 1s electrons, but not vice versa. In general, energy increases with n for this reason. However, it can also be seen from Fig. 1.5 that s and p electrons penetrate closer to the nucleus than d electrons; f electrons, with $l = 3$, penetrate still less. Therefore, for a given polyelectronic atoms and a given n value, s electrons will have a lower energy than p electrons, which in turn are lower than d and f electrons.

The most common approximate atomic orbitals used in descriptions of inorganic systems are Slater-type orbitals (STOs). These are constructed according to the pattern just described.

1. There is an STO corresponding to each hydrogenlike orbital, with its angular dependence identical to that of the hydrogenlike orbital.
2. The radial dependence of the STO is given by

$$R(r) = N \cdot r^{n^{*}-1} \cdot e^{-Z_{eff} r/n^{*}}$$

where N is a normalization constant scaled to make the probability distribution function R^2 integrate to a total probability of 1 over all space; n^* is an effective quantum number for the orbital derived from the true $n(n^* = n$ for $n = 1, 2, 3$, but $n = 4$ requires $n^* = 3.7$, $n = 5$ requires $n^* = 4.0$, and $n = 6$ requires $n^* = 4.2$); and Z_{eff} is equal to the true atomic number Z minus a screening constant σ.

3. The screening constant σ is calculated for a given electron by grouping all electrons in the atom and considering contributions from each group. The groups, from the nucleus out, are $(1s)$, $(2s, 2p)$, $(3s, 3p)$, $(3d)$, $(4s, 4p)$, $(4d)$, $(4f)$, $(5s, 5p)$, $(5d)$, ... Each group contributes to the screening constant as follows:

   (a) Electrons outside the one being considered contribute 0.00 screening.
   (b) Each electron in the same group as the one being considered contributes 0.35 screening, except that a $1s$ electron screens 0.30.
   (c) If the electron being considered is a d or f electron, each electron in an inner group contributes 1.00 screening—that is, screens completely.
   (d) If the electron being considered is an n s or n p electron each inner-group electron with quantum number $n - 1$ contributes 0.85 screening, and each electron farther in (that is, $n - 2$, $n - 3$, and so on) contributes 1.00 screening.

Slater orbitals have no radial nodes (where the wave function vanishes) such as those displayed by the hydrogenlike orbitals of Fig. 1.5. This is generally considered a conceptual defect, but it does make computations much more convenient.

Most medium-to-high-level calculations at the present time assume STO wave functions for the individual atomic orbitals, but each STO is simulated as the sum of several Gaussian functions. That is, the STO function with its $\exp(-Z_{eff}r/n^*)$ is replaced by a series of terms: $c_1\exp(\zeta_1 r^2) + c_2\exp(\zeta_2 r^2) + \cdots$. The reason for this seemingly awkward substitution is that it greatly simplifies calculations for molecules in which two atomic orbitals overlap each other but are centered on different nuclei. Such calculations yield integrals of the product of two orbitals, and, conveniently, the product of two Gaussian functions on different centers is a single Gaussian—much easier to evaluate. If each STO is being simulated by six Gaussians, the calculation will be described as STO-6G, and so on.

Clementi and Raimondi have produced a set of Z_{eff} values for the first 36 elements by fitting approximate orbitals having the general STO form to the electron densities predicted by self-consistent-field calculations. The Clementi–Raimondi values are given in Table 1.5. It is interesting to compare one of their values with that predicted by Slater's rules. For example, chlorine has 17 electrons arranged $(1s)^2(2s, 2p)^8(3s, 3p)^7$, so the Slater Z_{eff} value for a $3p$ electron would be calculated as follows:

$$Z_{eff} = Z - \sigma = 17 - [2(1.00) + 8(0.85) + 6(0.35)] = 17 - 10.90 = 6.10$$

Table 1.5   Clementi–Raimondi Effective Nuclear Charges for Light Elements

Element	1s	2s	2p	3s	3p	4s	3d	4p
H	1.000							
He	1.688							
Li	2.691	1.279						
Be	3.685	1.912						
B	4.680	2.576	2.421					
C	5.673	3.217	3.136					
N	6.665	3.847	3.834					
O	7.658	4.492	4.453					
F	8.650	5.128	5.100					
Ne	9.642	5.758	5.758					
Na	10.626	6.571	6.802	2.507				
Mg	11.619	7.392	7.826	3.308				
Al	12.591	8.214	8.963	4.117	4.066			
Si	13.575	9.020	9.945	4.903	4.285			
P	14.558	9.825	10.961	5.642	4.886			
S	15.541	10.629	11.977	6.367	5.482			
Cl	16.524	11.430	12.993	7.068	6.116			
Ar	17.508	12.230	14.008	7.757	6.764			
K	18.490	13.006	15.027	8.680	7.726	3.495		
Ca	19.473	13.776	16.041	9.602	8.658	4.398		
Sc	20.457	14.574	17.055	10.340	9.406	4.632	7.120	
Ti	21.441	15.377	18.065	11.033	10.104	4.817	8.141	
V	22.426	16.181	19.073	11.709	10.785	4.981	8.983	
Cr	23.414	16.984	20.075	12.368	11.466	5.133	9.757	
Mn	24.396	17.794	21.084	13.018	12.109	5.283	10.528	
Fe	25.381	18.599	22.089	13.676	12.778	5.434	11.180	
Co	26.367	19.405	23.092	14.322	13.435	5.576	11.855	
Ni	27.353	20.213	24.095	14.961	14.085	5.711	12.530	
Cu	28.339	21.020	25.097	15.594	14.731	5.858	13.201	
Zn	29.325	21.828	26.098	16.219	15.369	5.965	13.878	
Ga	30.309	22.599	27.091	16.966	16.204	7.067	15.093	6.222
Ge	31.294	23.365	28.082	17.760	17.014	8.044	16.251	6.780
As	32.278	24.127	29.074	18.596	17.850	8.944	17.378	7.449
Se	33.262	24.888	30.065	19.403	18.705	9.758	18.477	8.287
Br	34.247	25.643	31.056	20.218	19.571	10.553	19.559	9.028
Kr	35.232	26.398	32.047	21.033	20.434	11.316	20.626	9.769

Since the Clementi–Raimondi Z_{eff} value is 6.116 for a Cl 3p electron from Table 1.5, the two systems are quite close in this case. Note, however, that Slater's rules would give the same value of Z_{eff} for a Cl 3s electron, while the Clementi–Raimondi value is 7.068, significantly different.

In the *self-consistent-field* (SCF) calculations just referred to, electron–electron repulsion is dealt with directly, although approximate orbitals such as STOs are used as a starting point. Such a calculation begins by assigning electrons to approximate orbitals as was done for chlorine in the example above. It is then assumed for one electron that its potential energy consists of the attraction of the bare nucleus and the repulsion of the other electrons, each distributed according to its approximate orbital. A new set of improved approximate orbitals is calculated, making this assumption in each individual case. In a second pass, these new improved approximate orbitals are used as the electron-distribution function for the potential energy of repulsion, and a third set of approximate orbitals results. The calculation is carried out in a repetitive or iterative fashion until the calculated electron distribution is essentially identical to the distribution assumed in that iteration; the potential field is thus self-consistent (if not, perhaps, correct). Angular correlation of electron positions is not dealt with in this approach. The energy difference between an SCF calculation and experiment—which is usually fairly small— is thus called the *correlation energy*. The specific electron-by-electron consideration of repulsion that is included in SCF calculations makes them a significantly better fit to experimental electron distributions than STOs, which is the basis for Clementi and Raimondi's SCF-adjusted orbitals.

Even when working with approximate atomic orbitals, it is assumed that the hydrogenlike orbital quantum numbers n, l, m, and m_s apply for an electron distributed according to that orbital, and that the exclusion principle can be applied in the simple sense in which quantum numbers are not duplicated. Making the orbital approximation, we build up the overall electron distribution of a polyelectronic atom by inserting electrons one at a time into the approximate atomic orbitals, placing each electron in the lowest-energy orbital available but following the exclusion principle. The final arrangement of electrons in orbitals is called the electron configuration of the atom in its ground state. The orbitals fill in the order $1s$ $2s$ $2p$ $3s$ $3p$ $4s$ $3d$ $4p$ $5s$ $4d$ $5p$ $6s$ $4f$ $5d$ $6p$ $7s$ $5f$ $6d$. Figure 1.6 shows the sequence in a mnemonic diagram in which the orbitals fill in the order shown by the diagonal arrows, starting at the bottom. Since there is only one s orbital for any given n, it can contain only two electrons ($m_s = +\frac{1}{2}$, $-\frac{1}{2}$). However, the three p orbitals for any m can each contain two electrons for a total of six, the five d orbitals can hold a total of 10 electrons, and the seven f orbitals can hold 14 electrons. Using these limits, we build up the electron configuration for any element simply from its atomic number. This process is sometimes given Pauli's German title, the "Aufbau Prinzip." For the element tin, with $Z = 50$, the electron configuration (following Fig. 1.6) would be $1s^2$ $2s^2$ $2p^6$ $3s^2$ $3p^6$ $4s^2$ $3d^{10}$ $4p^{10}$ $5s^2$ $4d^{10}$ $4p^2$, in which the number of electrons in any given set of orbitals is indicated by a superscript to the right of the symbol for that set. Table 1.6 gives the configurations, built up in this way, for all the elements.

In building up electron configurations, it frequently occurs that the ground state has an unfilled set of orbitals. One must then decide how to place the partial set of electrons in the orbitals. The fairly obvious principle is that the electron will adopt distributions that minimize their mutual repulsion.

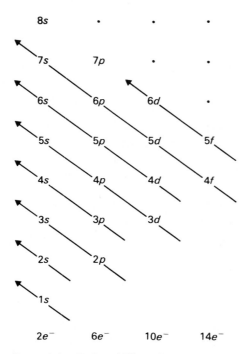

Figure 1.6   Order of filling of atomic orbitals in polyelectronic atoms.

Table 1.6   Ground-State Electron Configurations of the Elements

Atomic number	Element	Electronic configuration	Atomic number	Element	Electronic configuration
1	H	$1s$	13	Al	$-3s3p$
2	He	$1s^2$	14	Si	$-3s3p^2$
3	Li	[He] $2s$	15	P	$-3s3p^3$
4	Be	$-2s^2$	16	S	$-3s3p^4$
5	B	$-2s^22p$	17	Cl	$-3s3p^5$
6	C	$-2s^22p^2$	18	Ar	$-3s3p^6$
7	N	$-2s^22p^3$	19	K	[Ar] $4s$
8	O	$-2s^22p^4$	20	Ca	$-4s^2$
9	F	$-2s^22p^5$	21	Sc	$-3d4s^2$
10	Ne	$-2s^22p^6$	22	Ti	$-3d^24s^2$
11	Na	[Ne] $3s$	23	V	$-3d^34s^2$
12	Mg	$-3s^2$	24	Cr	$-3d^54s$

Table 1.6 (*Continued*)

Atomic number	Element	Electronic configuration	Atomic number	Element	Electronic configuration
25	Mn	$-3d^5 4s^2$	65	Tb	$-4f^9 6s^2$
26	Fe	$-3d^6 4s^2$	66	Dy	$-4f^{10} 6s^2$
27	Co	$-3d^7 4s^2$	67	Ho	$-4f^{11} 6s^2$
28	Ni	$-3d^8 4s^2$	68	Er	$-4f^{12} 6s^2$
29	Cu	$-3d^{10} 4s$	69	Tm	$-4f^{13} 6s^2$
30	Zn	$-3d^{10} 4s^2$	70	Yb	$-4f^{14} 6s^2$
31	Ga	$-3d^{10} 4s^2 4p$	71	Lu	$-4f^{14} 5d 6s^2$
32	Ge	$-3d^{10} 4s^2 4p^2$	72	Hf	$-4f^{14} 5d^2 6s^2$
33	As	$-3d^{10} 4s^2 4p^3$	73	Ta	$-4f^{14} 5d^3 6s^2$
34	Se	$-3d^{10} 4s^2 4p^4$	74	W	$-4f^{14} 5d^4 6s^2$
35	Br	$-3d^{10} 4s^2 4p^5$	75	Re	$-4f^{14} 5d^5 6s^2$
36	Kr	$-3d^{10} 4s^2 4p^6$	76	Os	$-4f^{14} 5d^6 6s^2$
37	Rb	[Kr] $5s$	77	Ir	$-4f^{14} 5d^7 6s^2$
38	Sr	$-5s^2$	78	Pt	$-4f^{14} 5d^9 6s$
39	Y	$-4d 5s^2$	79	Au	[Xe $4f^{14} 5d^{10}$]$6s$
40	Zr	$-4d^2 5s^2$	80	Hg	$-6s^2$
41	Nb	$-4d^4 5s$	81	Tl	$-6s^2 6p$
42	Mo	$-4d^5 5s$	82	Pb	$-6s^2 6p^2$
43	Tc	$-4d^5 5s^2$	83	Bi	$-6s^2 6p^3$
44	Ru	$-4d^7 5s$	84	Po	$-6s^2 6p^4$
45	Rh	$-4d^8 5s$	85	At	$-6s^2 6p^5$
46	Pd	$-4d^{10}$	86	Rn	$-6s^2 6p^6$
47	Ag	$-4d^{10} 5s$	87	Fr	[Rn] $7s$
48	Cd	$-4d^{10} 5s^2$	88	Ra	$-7s^2$
49	In	$-4d^{10} 5s^2 5p$	89	Ac	$-6d 7s^2$
50	Sn	$-4d^{10} 5s^2 5p^2$	90	Th	$-6d^2 7s^2$
51	Sb	$-4d^{10} 5s^2 5p^3$	91	Pa	$-5f^2 6d 7s^2$
52	Te	$-4d^{10} 5s^2 5p^4$	92	U	$-5f^3 6d 7s^2$
53	I	$-4d^{10} 5s^2 5p^5$	93	Np	$-5f^4 6d 7s^2$
54	Xe	$-4d^{10} 5s^2 5p^6$	94	Pu	$-5f^6 7s^2$
55	Cs	[Xe] $6s$	95	Am	$-5f^7 7s^2$
56	Ba	$-6s^2$	96	Cm	$-5f^7 6d 7s^2$
57	La	$-5d 6s^2$	97	Bk	$-5f^9 7s^2$
58	Ce	$-4f 5d 6s^2$	98	Cf	$-5f^{10} 7s^2$
59	Pr	$-4f^3 6s^2$	99	Es	$-5f^{11} 7s^2$
60	Nd	$-4f^4 6s^2$	100	Fm	$-5f^{12} 7s^2$
61	Pm	$-4f^5 6s^2$	101	Md	$-5f^{13} 7s^2$
62	Sm	$-4f^6 6s^2$	102	No	$-5f^{14} 7s^2$
63	Eu	$-4f^7 6s^2$	103	Lr	$-5f^{14} 6d 7s^2$
64	Gd	$-4f^7 5d 6s^2$			

This means that an electron will not pair with one already present—which means adopting the same spatial distribution—if a vacant orbital of the same energy is present. Thus, a $2p_x$ electron repels a $2p_y$ electron less than two $2p_x$ electrons repel each other. Beyond this simple spatial arrangement, it is also true that repulsion will be minimized if the two electrons adopt the same spin orientation. This is an experimental fact, but also comes directly out of the antisymmetric-wave-function form of the exclusion principle when it is applied to SCF wave functions. Figure 1.7 gives the relative energies of three possible arrangements of two electrons in a set of three p orbitals. It can be seen that the total electronic energy of an atom depends not only on its electron configuration but on the arrangement of electrons in unfilled shells. Two p electrons yield three overall electronic states for the atom: the ground state (with parallel spins) and two excited states (with opposite spins). We shall return to the question of establishing atomic states for atoms when we discuss spectroscopy in Chapter 11.

The electronic configuration of the atoms of an element obviously has a major effect on the chemical properties of that element. However, chemical properties are ultimately reflections of the availability of electrons for chemical bonding. For any polyelectronic atom, then, we need to know how tightly its electrons are bound to the nucleus. If they are weakly bound they will be freely available, whereas if they are very tightly bound they cannot take part at all in chemical bonding. The binding energy of an electron in a polyelectronic atom is influenced by two factors: the effective nuclear charge it experiences, and its average distance from the nucleus. Figures 1.8 and 1.9 show these two properties of the electron distribution as a function of atomic number for the first 36 elements.

Figure 1.8 is plotted with effective nuclear charge increasing toward the bottom of the graph to suggest the effect of nuclear charge on the corresponding energy levels, which become more stable—lie at lower energy—as Z_{eff}

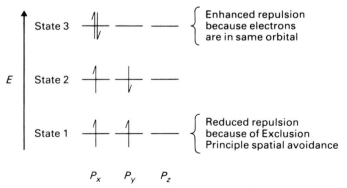

Figure 1.7 Relative energies of states from the p^2 configuration.

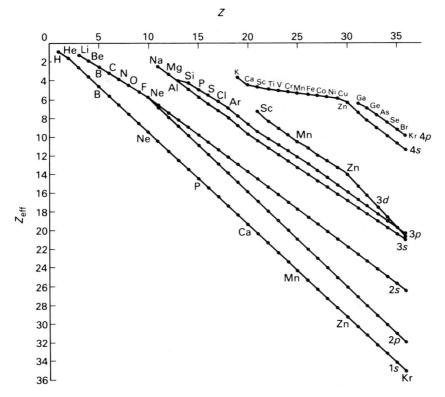

Figure 1.8 Clementi–Raimondi effective nuclear charges as a function of atomic number through Kr.

increases. The plotted data are simply those from Table 1.5. Several trends are evident. After the 1s orbital has been filled (at He, $Z = 2$), adding outer electrons has virtually no shielding effect on the 1s electrons, and the effective nuclear charge increases almost exactly one unit for each proton added to the nucleus. Interestingly, the same appears to be true for the 2p orbitals after they are completely filled (at Ne, $Z = 10$), but not for the 2s orbitals, whose shallower slope indicates that the addition of outer electrons is having at least a modest shielding effect on them. It is also true for the 3d orbitals after they are filled (at Zn, $Z = 30$). The pattern seems to be that the orbitals corresponding to the highest l value for a given n are not shielded by outer electrons after they are filled in building up the electron shells of the periodic table. However, within any set of orbitals there is a substantial shielding effect as the set fills. Therefore, although Z_{eff} increases slowly when the number of protons in the nucleus increases by one and the number of electrons at the same average distance from the nucleus also increases by one, Z_{eff} increases much

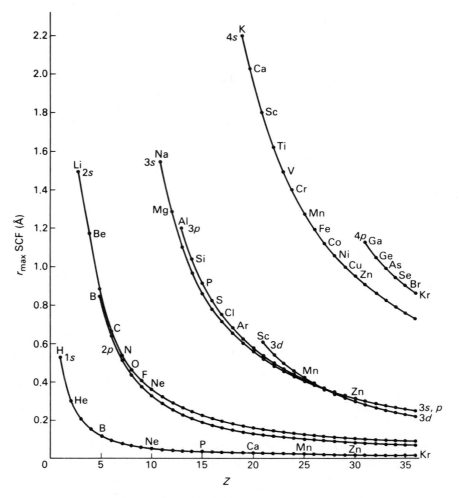

Figure 1.9   Radii of maximum electron density for each atomic orbital (from SCF calculation) as a function of atomic number through Kr.

faster when the electron being added is, on the average, farther from the nucleus than the one being considered. It should also be noted that the Z_{eff} of the 4s electrons is almost constant from Sc to Zn—that is, while the 3d orbitals are being filled. The 3d electrons almost perfectly shield the 4s electrons because, on the average, they lie only about a third as far from the nucleus as the 4s electrons, as Fig. 1.9 indicates.

   Figure 1.9 shows the radii corresponding to maximum electron density for the same group of elements as Fig. 1.8. These are theoretical maxima, taken from SCF calculations on the atoms. Figure 1.9 should be compared with Fig. 1.5, which suggests orbital sizes for hydrogenlike atoms. In general, for hydrogenlike atoms, the average radius (not the maximum-density radius) increases as the square of the quantum number n. Thus an orbital with $n = 2$ should be four times as large as one with $n = 1$, one with $n = 3$ nine times as large, and so on. For polyelectronic atoms, however, the increasing effective nuclear charge tends to shrink the orbital, so that the lithium $2s$ orbital is slightly less than three times as large as the hydrogen $1s$, not four times as large. Similarly, the sodium $3s$ is three times as large, not nine times, and the potassium $4s$ is about four times as large, not 16. The effect is to keep all atoms roughly the same size.

The electrons in a polyelectronic atom have widely differing energies and stabilities. Remembering that a greater radius of maximum density will in general mean lower attraction to the nucleus, we can see that only the electrons highest in the Fig. 1.9 diagram for any given element are loose enough to be further stabilized by forming a chemical bond of some sort ($3d$ electrons, however, are much more readily involved in bonding than the diagram would suggest). It can be seen that for any given orbital distribution, the energy of attraction increases rapidly after the orbital is first populated. The high-energy electrons are obviously available for chemical combination, but it is equally true that low-energy vacant orbitals can accept electrons.

We customarily think of all the electrons in an unfilled set of orbitals as *valence electrons* and of electrons below that curve as *inner-core electrons*. In addition, for the transition metals having unfilled d orbitals, the higher-energy s orbitals from the next higher value of n are considered valence electrons, and the same is true for the rare-earth metals having unfilled f orbitals. Of course, not all valence electrons are removable. Whether or not a given oxidation state can be reached depends on many factors, but primarily on whether the electrons being considered can be further stabilized by bond formation. Thus fluorine has seven valence electrons ($2s$ and $2p$), but it is never found as a positive ion because there is no element that can stabilize those electrons more by removing them, and it is rarely found covalently bonded to more than two atoms. There is, therefore, no fluorine-containing species in which all seven valence electrons are involved in bonding. Similarly, no oxidation state greater than 4+ is found in the lanthanide rare earths (elements $Z = 57–71$), even though there may be as many as 14 $4f$ electrons.

Once we can identify the valence electrons for any element, we can correlate the chemical properties of all the elements to a considerable extent by listing them in order of atomic number, starting a new row of the list when a set of orbitals having a new value of n is occupied for the first time. Such a listing is, of course, the periodic table. A table having this form is given in Fig. 1.10, with the set of orbitals being filled indicated in each region of the table. This form of the periodic table indicates well the continuity of chemical properties across the bottom two rows of the table, and also coincides well

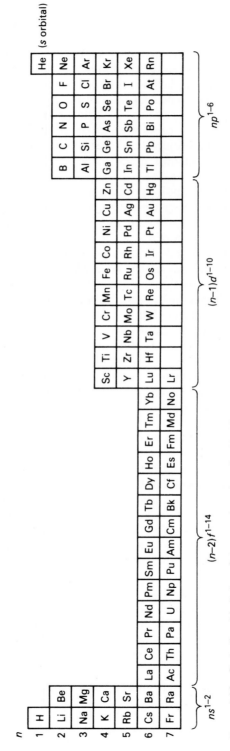

Figure 1.10 Periodic-table listing of elements in order of orbital filling by valence electrons.

with the electron configurations that form the theoretical basis of any periodic table. Unfortunately, the large number of elements in the bottom two rows gives it an awkward shape, and also separates the top five rows by such large distance that it is very misleading about the continuity of physical and chemical properties that exists from Be to B, Mg to Al, and so on. However, it is valuable as the logical extension of our ideas about atomic structure, and it gives us a starting point for discussing periodicity in the properties of the elements.

In the discussion that follows, it will be important to bear in mind that the elements in Fig. 1.10 can be naturally divided into what we can call the s-block, the p-block, the d-block, and the f-block. The s-block elements (including He) have either one or two s electrons, and normally lose them in chemical combination to elements with more stable orbitals (except for He, which is inert). The s-block elements lie at the very top of Fig. 1.9, with their valence electrons farthest from the nucleus and least tightly held.

The p-block elements have the n s electrons as valence electrons in addition to the appropriate number of n p electrons. All of them except oxygen, fluorine, and the ligher noble gases can show oxidation states corresponding to the loss of all valence electrons, but the ones with one or two vacancies in the set of p orbitals more often serve as electron acceptors in chemical combination and adopt negative oxidation states. When they do have positive oxidation states, it is not unusual that they—especially the heavier p-block elements—retain the s valence electrons, which are progressively more stable relative to the p electrons as atomic number increases.

The d-block elements are metals and normally adopt positive oxidation states, which usually correspond to the loss of the $(n + 1)$ s electrons and all, some, or none of the n d electrons. Figure 1.9 suggests (correctly) that the d electrons become much more stable as the atomic number increases across a row of the periodic table, so that the later transition metals (d-block elements) rarely lose many d electrons in compounds.

The f-block elements are also metals that adopt positive oxidation states. They all can enter a 3+ oxidation state in which both s electrons are lost and either a d or f electron as well, but some of the lanthanide rare earths (f-block elements; $Z = 57$–71) also show 2+ or 4+ oxidation states, while the actinide rare earths ($Z = 89$–103) have a wider variety of positive oxidation states, insofar as they are known.

1.6 PERIODIC PROPERTIES: IONIZATION POTENTIAL AND ELECTRON AFFINITY

Although in the previous discussion we, in effect, derived the periodic table from the principles and approximations of quantum mechanics, it is important to remember that the periodic table was originally derived from empirical experimental results, not theoretical formulations. The concept of atomic number was unknown to Mendeleev, so that he had to order the elements by their atomic weight (although he inverted cobalt and nickel, and tellurium and iodine, to achieve a better fit of chemical properties). His earliest predecessors, however, did not even use that quantitative guide. For instance, in 1829 Döbereiner had grouped some of the elements in triads, such as Li/Na/K or S/Se/Te, that showed strong chemical similarities. Within a triad, the middle member's atomic weight was approximately the mean of the other two, but there was no relationship between the atomic weights of different triads.

Mendeleev apparently did not set out to develop an ordering based on atomic weights, either. He prepared a set of cards each containing tabulated physical and chemical properties of an element, then attempted to arrange the cards so as to produce groupings of uniform properties (like Döbereiner's triads) as well as smooth trends from one grouping to another. His first attempt (1869) produced a vertical-format table in which atomic weights increased going down the table and families such as F/Cl/Br/I lay in a horizontal row. In the few cases in which chemical resemblances argued against a strict atomic-weight order (such as Te/I), he had no qualms about adopting the chemical order and arguing that the atomic weight was wrong. His 1871 table had approximately the modern format, with eight columns that he called "group I," "group II," and so on. Within each group there was an A and a B subgroup, as shown in Fig. 1.11, listed at the left and right sides of their group column, respectively. This led to some minor difficulties, such as the implication that sodium resembled copper more than it did potassium, but it did offer a complete structure for the overall pattern of chemical elements. The table had predictive power, as seen in the four blanks provided in the top

Reihen	Gruppe I. — R^2O	Gruppe II. — RO	Gruppe III. — R^2O^3	Gruppe IV. RH^4 RO^2	Gruppe V. RH^3 R^2O^5	Gruppe VI. RH^2 RO^1	Gruppe VII. RH R^2O^7	Gruppe VIII. — RO^4
1	H=1							
2	Li=7	Be=9,4	B=11	C=12	N=14	O=16	F=19	
3	Na=23	Mg=24	Al=27,3	Si=28	P=31	S=32	Cl=35,5	
4	K=39	Ca=40	—=44	Ti=48	V=51	Cr=52	Mn=55	Fe=56, Co=59, Ni=59, Cu=63.
5	(Cu=63)	Zn=65	—=68	—=72	As=75	Se=78	Br=80	
6	Rb=85	Sr=87	?Yt=88	Zr=90	Nb=94	Mo=96	—=100	Ru=104, Rh=104, Pd=106, Ag=108.
7	(Ag=108)	Cd=112	In=113	Sn=118	Sb=122	Te=125	J=127	
8	Cs=133	Ba=137	?Di=138	?Ce=140	—	—	—	— — —
9	(—)	—	—	—	—	—	—	
10	—	—	?Er=178	?La=180	Ta=182	W=184	—	Os=195, Ir=197, Pt=198, Au=199.
11	(Au=199)	Hg=200	Tl=204	Pb=207	Bi=208	—	—	
12	—	—	—	Th=231	—	U=240	—	— — —

Figure 1.11   Mendeleev's horizontal table of 1871, from Annalen der Chemie, supplemental vol. 8 (1972). [Courtesy of the Trustees of the Boston Public Library.]

six rows for unknown elements whose atomic weights Mendeleev predicted in the table. The three lightest (Sc, Ga, and Ge) were discovered in 1879, 1875, and 1886 respectively, and it is interesting to compare their current atomic weights against Mendeleev's predictions. The fourth element, Tc, also has very nearly the properties he predicted, but he had no way of foreseeing the problems that its synthesis would present.

There will be numerous occasions in later discussions to examine periodic trends in chemical properties, but it might be appropriate here to look at one of the most strikingly periodic physical properties of the elements. In 1870—even before Mendeleev's horizontal-format table had been published—Meyer described the *atomic-volume* quantity and its dependence on atomic weight. If one takes the atomic weight (in g/mol) and divides it by the density of the element in a condensed phase (in g/cm^3), the result is the volume of one mole of the element (in cm^3/mol). Figure 1.12 shows the result, using modern

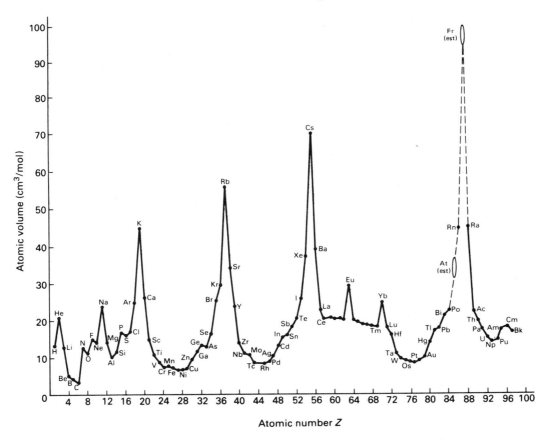

Figure 1.12   Periodic dependence of atomic volume on atomic number.

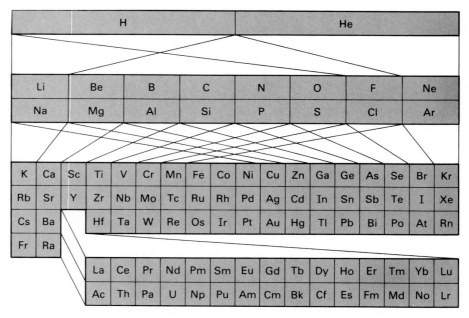

Figure 1.13   Von Antropov periodic table.

atomic weights and densities. Meyer only had about two-thirds as many elements to work with and was plotting against atomic weight rather than atomic number, but it is not surprising that Meyer's graph, sketchy as it was, provided a strong impetus to the analysis of periodic relationships.

The strong emphasis that the atomic volume and chemical properties place on continuity of chemical properties across any given row of the periodic table is perhaps best represented by the von Antropov periodic table, shown in Fig. 1.13. The diagonal bands connecting families of elements indicate resemblances, particularly in oxidation states, and thus preserve Mendeleev's A and B subgroups. In general, the wider the band, the closer the resemblance. This form of the periodic table also allows a clearer view of what is sometimes called the "diagonal relationship" between the light metals. We expect the chemical properties of lithium to resemble those of sodium, and so they do; but they also strongly resemble those of magnesium. Thus, for example, the two common organic organometallic reagents are organolithium compounds and Grignard (organomagnesium halide) reagents. In the same sense, the two most energetic additives to metallized solid rocket propellants are beryllium and aluminum, and boron and silicon have obvious physical similarities as metalloids. The relationship may be summarized:

$$\begin{array}{cccc} \text{Li} & \text{Be} & \text{B} & \text{C} \\ \searrow & \searrow & \searrow & \\ \text{Na} & \text{Mg} & \text{Al} & \text{Si} \end{array}$$

The von Antropov table is less clearly related to electron configurations than the more conventional form inside the front cover of this book, however, and we will in general use the latter for periodic correlations.

If, at this point, we choose to concentrate on the periodicity of electronic properties, perhaps the most obvious is the *ionization potential* of the elements, or the voltage at which one or more electrons can be stripped from the atom, leaving a charged particle. Ionization potentials are measured in the gas phase at very low pressure using a mass spectrometer and are isolated as much as possible from influences from the surroundings. Table 1.7 gives values for ionization potential in volts; these are numerically equal to the ionization energy in electronvolts, since one electron is being removed in each case. Values in parentheses refer to the removal of electrons beyond the highest oxidation state normally shown by that element. A few of these potentials are plotted in Fig. 1.14, from which the periodic nature of the ionization potential can be clearly seen. The correlation with formal oxidation state is apparent—not only for Li, Na, and K, which have very low first-ionization potentials, but for Be, Mg, and Ca with low second-ionization potentials and B, Al, and Sc with low third-ionization potentials. The enormous difference between the first and second IPs of Li or Na, or between the second and third IPs of Mg and Ca, comes about because an inner-core electron experiencing a much greater effective nuclear charge is being removed. For most of the transition metals ($Z = 21–30$) no such wild zigzags appear for any element in going from one curve to the next, and this is reflected in the fact that all of the transition metals after scandium have a reasonably stable $2+$ oxidation state and all except zinc show a $3+$ oxidation state. Zinc, of course, has a relatively high third-ionization potential that limits its further ionization, but it is worth noting that Cu^{3+} and Ni^{3+} are very strong oxidizing agents (electron acceptors), Co^{3+} and Mn^{3+} are moderately strong oxidizing agents, and Cr^{3+} and Fe^{3+} are approximately neutral in their redox behavior. This coincides nicely with the relative position of their third-ionization potentials. Other similar arguments can be constructed, and we will encounter them at several points in later discussions.

Another quantity, *valence-orbital ionization potential* (VOIP), is analogous to ionization potential but is more directly related to the atomic orbitals that result from approximate quantum-mechanical calculations. Unlike the true ionization potential of an atom, the VOIP is not measured by removing an electron from the atom. Instead, the energy differences between different states of the atom (as revealed in the atom's spectrum) are correlated with the electron configurations responsible for the states, up to an energy that corresponds to photoionization. The results can be interpreted as an approximation

Table I.7 Ionization Potentials (IP) of the Elements (in volts)

Ionization potentials in volts are numerically equal to ionization energies in electronvolts.

Value in kJ/mol = tabulated value × 96.487.

Period 1

	H	He
IP₁ =	13.598	(24.59)
IP₂ =		(54.42)

Period 2

	Li	Be	B	C	N	O	F	Ne
IP₁ =	5.39	9.32	8.30	11.26	14.53	13.62	17.42	21.56
IP₂ =	(75.64)	18.21	25.15	24.38	29.60	35.12	(34.97)	(40.96)
IP₃ =	(122.45)	(153.89)	37.93	47.89	47.45	(54.93)	(62.71)	(63.45)
IP₄ =		(217.71)	(259.37)	64.49	77.47	(77.41)	(87.14)	(97.11)

Period 3

	Na	Mg	Al	Si	P	S	Cl	Ar
IP₁ =	5.14	7.65	5.99	8.15	10.49	10.36	12.97	15.76
IP₂ =	(47.29)	15.04	18.83	16.35	19.73	23.33	23.81	27.63
IP₃ =	(71.64)	(80.14)	28.45	33.49	30.18	34.83	39.61	40.74
IP₄ =	(98.91)	(109.24)	(119.99)	45.14	51.37	47.30	53.46	59.81

Period 4

	K	Ca	Sc	Ti	V	Cr	Mn	Fe	Co	Ni	Cu	Zn	Ga	Ge	As	Se	Br	Kr
IP₁ =	4.34	6.11	6.54	6.82	6.74	6.77	7.44	7.87	7.86	7.64	7.73	9.39	6.00	7.90	9.81	9.75	11.81	14.00
IP₂ =	(31.63)	11.87	12.80	13.58	14.65	15.50	15.64	16.18	17.06	18.17	20.29	17.96	20.51	15.93	18.63	21.19	21.8	24.36
IP₃ =	(45.72)	(50.91)	24.76	27.49	29.31	30.96	33.67	30.65	33.50	35.17	36.83	(39.72)	30.71	34.22	28.35	30.82	36.	(36.95)
IP₄ =	(60.91)	(67.10)	(73.43)	43.27	46.71	49.1	51.2	(54.8)	(51.3)	(54.9)	(55.2)	(59.4)	(64.)	45.71	50.13	42.94	47.3	(52.5)

Period 5

	Rb	Sr	Y	Zr	Nb	Mo	Tc	Ru	Rh	Pd	Ag	Cd	In	Sn	Sb	Te	I	Xe
IP₁ =	4.18	5.70	6.38	6.84	6.88	7.10	7.28	7.37	7.46	8.34	7.58	8.99	5.79	7.34	8.64	9.01	10.45	12.13
IP₂ =	(27.28)	11.03	12.24	13.13	14.32	16.15	15.26	16.76	18.08	19.43	21.49	16.91	18.87	14.63	16.53	18.6	19.13	21.21
IP₃ =	(40.)	(43.6)	20.52	22.99	25.04	27.16	29.54	28.47	31.06	32.93	(34.83)	(37.48)	28.03	30.50	25.3	27.96	33.	32.1
IP₄ =	(52.6)	(57.)	(61.8)	34.34	38.3	46.4							(54.)	40.73	44.2	37.41		46.

Period 6

	Cs	Ba	La	Hf	Ta	W	Re	Os	Ir	Pt	Au	Hg	Tl	Pb	Bi	Po	At	Rn
IP₁ =	3.89	5.21	5.58	7.0	7.98	7.98	7.88	8.7	9.1	9.0	9.23	10.44	6.11	7.42	7.29	8.48	9.4	10.75
IP₂ =	(23.1)	10.00	11.06	14.9	16.2	17.7	16.6	16.9		18.56	20.5	18.76	20.43	15.03	16.69			
IP₃ =	(35.)		19.18	23.3								(34.2)	29.83	31.94	25.56			
IP₄ =	(51.)			33.3								(72.)	(50.8)	42.32	45.3			

Period 7

	Fr	Ra	Ac
IP₁ =	4.0	5.28	6.9
IP₂ =		10.15	12.1

Lanthanides

	Ce	Pr	Nd	Pm	Sm	Eu	Gd	Tb	Dy	Ho	Er	Tm	Yb	Lu
IP₁ =	5.47	5.42	5.49	5.55	5.63	5.67	5.85	5.85	5.93	6.02	6.10	6.18	6.25	5.43
IP₂ =	10.85	10.55	10.72	10.90	11.07	11.25	11.52	11.52	11.67	11.80	11.93	12.05	12.17	13.9
IP₃ =	20.20	21.62												
IP₄ =	36.72	(38.95)												

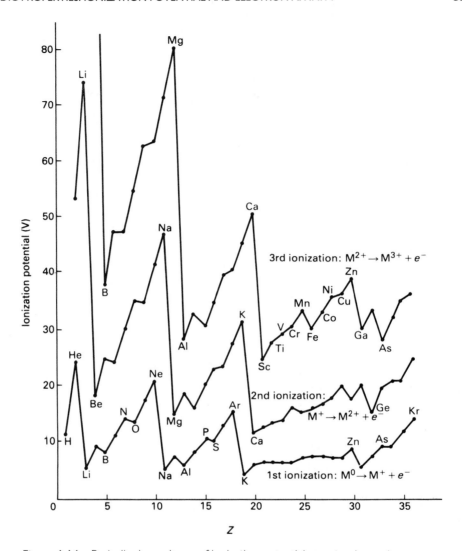

Figure 1.14 Periodic dependence of ionization potentials on atomic number.

of a unique energy level for each set of orbitals in the atom (each n and l combination). That energy, expressed in electronvolts, is equated to that valence orbital's ionization potential. Table 1.8 gives values of these for the valence orbitals of the elements through krypton. Trends in the VOIP values parallel those of the potential energy (Z_{eff}/r) that might be inferred from the effective nuclear charge and orbital radius values in Figs. 1.8 and 1.9. In particular, as the $3d$ orbitals fill from Sc to Zn, their energy drops fairly

Table 1.8 Valence Orbital Ionization Potentials (VOIP) for Z = 1–36 (in volts)

	H												**He**
	1s 13.6												24.5

Value in kJ/mol = tabulated value × 96.5

	Li	**Be**											**B**	**C**	**N**	**O**	**F**	**Ne**
2s =	5.5	9.3											14.0	19.5	25.5	32.4	46.4	48.5
2p =													8.3	10.7	13.1	15.9	18.7	21.6

	Na	**Mg**											**Al**	**Si**	**P**	**S**	**Cl**	**Ar**
3s =	5.2	7.7											11.3	15.0	18.7	20.7	25.3	29.3
3p =													6.0	7.8	10.2	11.7	13.8	15.8

	K	**Ca**	**Sc**	**Ti**	**V**	**Cr**	**Mn**	**Fe**	**Co**	**Ni**	**Cu**	**Zn**	**Ga**	**Ge**	**As**	**Se**	**Br**	**Kr**
4s =	4.3	6.1	5.7	6.1	6.3	6.6	6.8	7.1	7.3	7.6	7.7	9.4	12.7	15.6	17.6	20.8	24.0	27.5
3d =			4.7	5.6	6.3	7.2	7.9	8.7	9.4	10.0	10.7							
4p =				3.3	3.5	3.5	3.6	3.7	3.8	3.9	4.0	5.0	6.0	7.6	9.1	10.8	12.5	14.0

Table I.9 Electron Affinities of the Elements (in eV)[a]

Charge transitions for the multi-valued (parenthetical) entries:
0 → −1
−1 → −2
−2 → −3

1	2	3	4	5	6	7	8	9	10	11	12	13	14	15	16	17	18
H 0.754209																	**He** (−0.22)
Li 0.6180	**Be** (−2.5)											**B** 0.277	**C** 1.2629	**N** <0.0 (−8.3) (−13.4)	**O** 1.46112 −8.08	**F** 3.399	**Ne** (−0.3)
Na 0.54793	**Mg** (−2.4)											**Al** 0.441	**Si** 1.385	**P** 0.7465 (−4.8) (−9.2)	**S** 2.07712 −6.11	**Cl** 3.617	**Ar** (−0.36)
K 0.50147	**Ca** (−1.62)	**Sc** 0.188	**Ti** 0.079	**V** 0.525	**Cr** 0.666	**Mn** <0.0	**Fe** 0.163	**Co** 0.661	**Ni** 1.156	**Cu** 1.228	**Zn** <0.0	**Ga** 0.30	**Ge** 1.2	**As** 0.81 (−4.5) (−8.3)	**Se** 2.0207 −4.35	**Br** 3.365	**Kr** (−0.40)
Rb 0.4859	**Sr** (−1.74)	**Y** 0.307	**Zr** 0.426	**Nb** 0.893	**Mo** 0.746	**Tc** 0.55	**Ru** 1.05	**Rh** 1.137	**Pd** 0.557	**Ag** 1.302	**Cd** <0.0	**In** 0.3	**Sn** 1.2	**Sb** 1.07	**Te** 1.9708	**I** 3.0591	**Xe** (−0.42)
Cs 0.4716	**Ba** <0.0	**La** 0.5	**Hf** ≈0.0	**Ta** 0.322	**W** 0.815	**Re** 0.15	**Os** 1.1	**Ir** 1.565	**Pt** 2.128	**Au** 2.3086	**Hg** <0.0	**Tl** 0.2	**Pb** 0.364	**Bi** 0.946	**Po** 1.9	**At** (2.8)	**Rn** (−0.42)
Fr (0.456)	**Ra**	**Ac**															

Ce	**Pr**	**Nd**	**Pm**	**Sm**	**Eu**	**Gd**	**Tb**	**Dy**	**Ho**	**Er**	**Tm**	**Yb**	**Lu**

[a]Value in kJ/mol = tabulated value × 96.487. Parenthetical values are theoretical or extrapolated.

sharply because the nuclear charge is increasing and the $3d$ electrons do not shield each other very well. However, the energy of the $4s$ electrons in these same elements changes only modestly because the $4s$ distribution only penetrates the $3d$ distribution to a small extent and, therefore, is fairly thoroughly shielded from the increasing nuclear charge. These VOIP values will be useful to us in constructing qualitative molecular-orbital energy-level diagrams, because they provide a starting point for the atomic orbitals that combine into molecular orbitals.

Another quantity analogous to the ionization potential of an atom is its *electron affinity*, which is the energy released when an electron is added to a neutral atom to form a negative ion:

$$M^0 + e^- \rightarrow M^-$$

Electron affinities tend to be much smaller than ionization potentials, because they describe the relationship between an electron and a neutral atom rather than that between an electron and a positive ion. Since the reverse of the electron-affinity reaction would be the ionization of an atom's anion, it is possible to think of the negative of the electron affinity as the zeroth ionization potential, which places it in a more familiar perspective. Electron affinity is difficult to measure because the product of the reaction is electrically neutral, but a number of values have been fairly well established (see Table 1.9) by providing a source of gaseous negative ions and photoionizing them. Note that a positive electron affinity represents energy released *from* the atom when an electron is added, whereas all ionization potentials are positive and represent energy that must be supplied *to* the atom when an electron is removed. This distinction will be important in the thermodynamic applications of these quantities.

An important point to consider from Table 1.9 is that no atom has a positive second electron affinity and that, in fact, no atom releases energy on going from oxidation state 0 to oxidation state -2 or -3. This means that, for instance, the oxide ion is unstable in the gas phase with respect to electron emission; it can only exist in the stabilizing environment of a crystal lattice or a strongly solvating liquid such as a molten salt.

1.7 ATOMIC BONDING: DIFFERENTIAL IONIZATION ENERGIES AND ELECTRONEGATIVITY

There is an interesting property of single atoms that has a direct bearing on the extent to which atoms transfer electrons in forming chemical bonds. This property comes directly from the combined properties of ionization energy and electron affinity. Suppose we prepare a plot of the energy of an atom as a function of its net charge, arbitrarily defining the energy of the neutral atom as zero. Figure 1.15a shows such a plot for the atom Na. At -1 charge, the

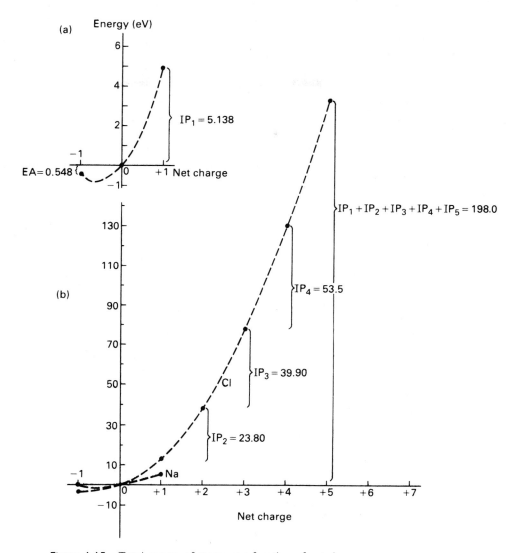

Figure 1.15 Total energy of atoms as a function of net charge.

atom's energy corresponds to its electron affinity, which is negative because it represents additional stability for the atom. At $+1$ charge, however, the atom's energy is just equal to the ionization energy for removing the first electron. The dashed line is the best-fit parabola or quadratic function connecting these points. The function may seem a bit presumptuous for two reasons: first, because there are only three points to be connected, and second, because it is not physically clear how a fractional electron could be removed from the atom. However, Fig. 1.15b shows that successive ionization energies yield a smooth quadratic curve for at least seven points for the atom Cl; the energy of the Cl atom at any net charge is the *sum* of all ionization energies to that charge. As to the second concern, it is precisely at this point that these properties of isolated atoms become relevant to chemical bonding. If two unlike atoms form a diatomic molecule, their respective ownership of the bonding electrons will be unequal, and a fractional net charge (representing a statistical partial ownership of the electron) is possible. The quadratic functions of Fig. 1.15 give us a way to describe the dependence of the energy of each bonded atom on that atom's charge. Each function is quadratic to a very good approximation as long as the successive ionizations are removing electrons with the same n and l values, but there is a discontinuity when a different type of electron is removed. This is the reason the Na function only extends to a charge of $+1$, while Cl can be extended to $+5$.

In Fig. 1.15b, the Na energy function is reproduced on the same scale as the Cl function. It can be seen that the two functions do not behave the same near zero charge. Figure 1.16 enlarges this section of the plot to allow us to

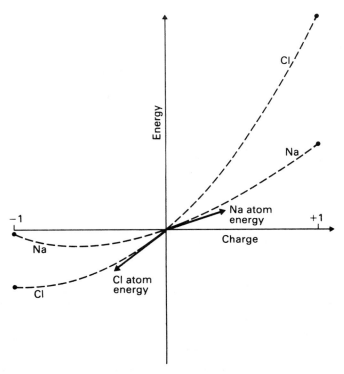

Figure 1.16   Energy relationships for electron transfer between Na and Cl.

consider what happens to the valence electrons when a neutral gaseous Na atom meets a neutral gaseous Cl atom. Near zero charge, both atoms would lower their energy by acquiring a slight negative charge. However, this is impossible—if one atom is to become negative, the other must release electron ownership and become positive. The total energy of the atom pair is lowered if the Cl becomes negative and the Na positive, as the arrows indicate on the figure. When both atoms are near zero charge, the Cl is stabilized more by a small charge increment than the Na is destabilized by losing it. The energy minimum for the atom pair—that is, the point at which the two atoms have the greatest stability—occurs at a partial charge Na^{+q}, Cl^{-q}, where the slopes of the functions are equal. The slope is represented by a derivative of energy with respect to charge, known as the *differential ionization energy* $I(q)$:

$$E = a_0 q + \frac{a_1}{2} q^2 \qquad \text{(Quadratic energy function)}$$

$$\frac{dE}{dq} = a_0 + a_1 q \equiv I(q) \qquad \text{(Differential ionization energy)}$$

We can calculate, in a simple approximation, the equilibrium charges on two atoms in a gaseous diatomic molecule by equating their $I(q)$ functions at the appropriate charges ($+q$ for the more positive, $-q$ for the less positive). For example, consider CO. We expect that C will be the more positive atom, so we assign it $+q$ and O $-q$. Table 1.10 gives the a_0 and a_1 constants in the $I(q)$ functions for a number of atoms. Using the values for C and O, we have

$$6.27 + 9.99(+q) = 7.54 + 12.15(-q)$$
$$22.14q = 1.27$$
$$q = 0.0574$$

or

$$C^{+0.057}O^{-0.057}$$

It must be emphasized that this is only a crude estimate of the net charges, because the electrostatic attraction of the charged atoms for each other has not been considered, nor has the change in repulsion for electrons occupying a two-atom volume rather than a one-atom volume. These effects, however, tend to oppose each other, and the $I(q)$ charge prediction is an interesting, if imprecise, estimate of the extent of charge transfer in gaseous compounds.

For polyatomic molecules, a similar approach can be used, making allowance for the fact that if several outer atoms are withdrawing electrons from a single central one, the central atom's charge will increase proportionately faster than that of each outer atom. Consider the two molecules SO_2 and

Table 1.10 Differential Ionization Energy—$I(q)$—Constants and Electronegativities (χ)

H
Pauling χ: 2.20
Allred–Rochow χ: 2.20

Legend of rows per element: $I(q)$ a_0; $I(q)$ a_1; Pauling χ; Allred–Rochow χ.

Element	$I(q)\ a_0$	$I(q)\ a_1$	Pauling χ	Allred–Rochow χ
H			2.20	2.20
He				
Li	3.00	4.77	0.98	0.97
Be	3.41	11.82	1.57	1.47
B	4.27	8.06	2.04	2.01
C	6.27	9.99	2.55	2.50
N	7.27	14.54	3.04	3.07
O	7.54	12.15	3.44	3.50
F	10.38	14.08	3.98	4.10
Ne				
Na	2.84	4.59	0.93	1.01
Mg	2.62	10.04	1.31	1.23
Al	3.22	5.52	1.61	1.47
Si	4.69	6.91	1.90	1.74
P	5.89	10.23	2.19	2.06
S	5.56	9.59	2.58	2.44
Cl	8.31	9.40	3.16	2.83
Ar				
K	2.42	3.84	0.82	0.91
Ca	2.25	7.73	1.00	1.04
Sc			1.36	1.20
Ti	3.61	6.44	1.54	1.32
V	3.84	5.80	1.63	1.45
Cr	3.71	6.10	1.66	1.56
Mn			1.55	1.60
Fe	4.24	7.32	1.83	1.64
Co	4.40	6.92	1.88	1.70
Ni	4.39	6.48	1.91	1.75
Cu	4.50	6.44	1.90	1.75
Zn	4.24	10.29	1.65	1.66
Ga	3.19	5.63	1.81	1.82
Ge	4.54	6.68	2.01	2.02
As	5.31	9.01	2.18	2.20
Se	5.89	7.73	2.55	2.48
Br	7.60	8.48	2.96	2.74
Kr				
Rb	2.33	3.69	0.82	0.89
Sr	1.98	7.43	0.95	0.99
Y			1.22	1.11
Zr			1.33	1.22
Nb			1.60	1.23
Mo	4.19	6.38	2.20	1.30
Tc			1.90	1.36
Ru			2.20	1.42
Rh			2.28	1.45
Pd			2.20	1.35
Ag	4.44	6.27	1.93	1.42
Cd	3.84	10.30	1.69	1.46
In	3.09	5.39	1.78	1.49
Sn	4.32	6.03	1.88	1.72
Sb	4.85	7.59	2.05	1.82
Te	5.49	7.04	2.10	2.01
I	6.75	7.38	2.66	2.21
Xe				
Cs	2.18	3.42	0.79	0.86
Ba	2.33	5.75	0.89	0.97
La			1.10	1.08
Hf			1.30	1.23
Ta	4.20	6.80	1.50	1.33
W	4.24	7.48	2.36	1.40
Re	4.03	7.65	1.90	1.46
Os			2.20	1.52
Ir			2.20	1.55
Pt	5.56	6.87	2.28	1.44
Au	5.77	6.91	2.54	1.42
Hg			2.00	1.44
Tl	3.30	5.61	1.8	1.44
Pb	4.23	6.37	2.1	1.55
Bi	4.17	6.24	2.02	1.67
Po				
At				
Rn				
Fr			0.7	0.86
Ra			0.9	0.97
Ac			1.1	1.00
Ce			1.12	1.08
Pr			1.13	1.07
Nd			1.14	1.07
Pm				1.07
Sm			1.17	1.07
Eu				1.01
Gd			1.20	1.11
Tb				1.10
Dy			1.22	1.10
Ho			1.23	1.10
Er			1.24	1.11
Tm			1.25	1.11
Yb				1.06
Lu			1.27	1.14

SO_3:

SO_2	SO_3
S $I(2q)$ = O $I(-q)$	S $I(3q)$ = O $I(-q)$
$5.56 + 9.59(2q) = 7.54 + 12.15(-q)$	$5.56 + 9.59(3q) = 7.54 + 12.15(-q)$
$31.33q = 1.98$	$40.92q = 1.98$
$q = 0.0632$	$q = 0.0484$
$S^{+2q}(O^{-q})_2 = S^{+0.126}(O^{-0.063})_2$	$S^{+3q}(O^{-q})_3 = S^{+0.145}(O^{-0.048})_3$

The charge on the sulfur atom is higher in SO_3 than in SO_2, even though the charge on each oxygen atom is lower.

Looking at the a_0 and a_1 quantities in Table 1.10, we see a clear periodic trend in which a_0 increases fairly uniformly across any given row of the periodic table to the right. It also increases fairly uniformly toward the top of any given column or group. This is the pattern shown by the *electronegativity* quantity, also given in that table. Electronegativity is a more familiar quantity than $I(q)$; it was defined by its originator, Linus Pauling, as the power of an atom to attact electrons to itself in a molecule. A wide variety of numerical measures of electronegativity has been developed on both empirical and theoretical bases, although the empirical scales are not free from theoretical assumptions and the theoretical scales rely on experimental data.

Pauling's original definition of electronegativity was based on patterns discernible in the bond energies of single bonds in molecules, which are derived from thermochemical measurements and from assumptions about bond order in the molecules. He noted that a bond energy E_{AB} between unlike atoms was uniformly greater than the geometric mean of the homonuclear bond energies E_{AA} and E_{BB}. He proposed that a hypothetical ideal covalent bond between A and B should have the geometric-mean energy, and that the excess bond energy was caused by the electrostatic attraction between the partially charged atoms. This extra ionic contribution to the bond energy could be reproduced fairly well by adding a term proportional to the square of the electronegativity difference between the bonded atoms:

$$E_{AB} + \sqrt{E_{AA} \cdot E_{BB}} + (\Delta\chi_{AB})^2 \qquad \text{in eV units}$$

where χ (chi) has been used to designate electronegativity and $\Delta\chi$ the electronegativity difference.

The largest electronegativity difference on this scale was that between Cs and F, 3.3 electron volts. Pauling therefore set the arbitrary absolute electronegativity of F at 4.0 and scaled other elements down from that figure. Single-bond energies are predictable to within about 10 kJ/mol using this approach, and there are other interesting correlations as well. One comes directly from the extra-bond-energy basis of Pauling's scale and from the fact that most reactions are enthalpy controlled and exothermic at room temperature: Reactions between molecular species will tend to go in the direction that forms bonds between the most electronegative and least electronegative atoms. This principle can be seen at work in the following spontaneous reactions:

$$SiCl_4 + LiAlH_4 \rightarrow SiH_4 + LiCl + AlCl_3$$
$$CS_2 + 2Ni(CO)_4 \rightarrow 2NiS + C + 8CO$$
$$2HF + Fe \rightarrow FeF_2 + H_2$$

Another useful correlation allows bond lengths in small molecules to be predicted from covalent radii. We will have more to say about these in Chapter 4, but it is commonly accepted that a rough estimate of molecular-bond length can be had by taking the sum of the covalent radii for the atoms involved: $r_{AB} = r_A + r_B$. The Schomaker–Stevenson relationship is considerably more accurate, however: $r_{AB} = r_A + r_B - 0.09(\Delta\chi_{AB})$ if radii are in angstrom units (Å). A still better fit uses the square of the electronegativity difference: $r_{AB} = r_A + r_B - 0.07(\Delta\chi_{AB})^2$. Bond lengths can usually be predicted within about 0.03 Å using one of these relationships.

An interesting feature of the electronegativity quantity is that although widely varying bases for calculating the individual electronegativities have been proposed, all the resulting scales can be brought to a common set of numbers very close to the Pauling electronegativities. An early alternate definition was provided by Mulliken, who suggested that the electronegativity of an element should be the average of its ionization potential and its electron affinity. If the energy-versus-charge curve for an atom were perfectly quadratic, this definition would make the Mulliken electronegativity equal to the a_0 differential-ionization-energy parameter. An important difference, however, is that Mulliken electronegativities are usually quoted on the basis of individual orbitals. This means assuming a valence-state electron configuration and/or hybridization for the atom and calculating from spectroscopic data what the valence-state ionization potential and electron affinity would be. Since $n\,p$ orbitals lie at higher energies than $n\,s$ orbitals, sp hybrid orbitals for an atom are more electronegative than sp^2 or sp^3 orbitals for that same atom. Single values of the Mulliken electronegativity for an atom are usually quoted for the most common molecular geometry and hybridization for that atom.

The last set of electronegativity values we will consider is that of Allred and Rochow, who developed a relatively theoretical set in which the electronegativity of an atom is equated to the force of attraction it develops for an electron near it. Specifically, the coulomb law of electrostatic force is q_1q_2/r^2. For q_1, Allred and Rochow used the Slater Z_{eff} calculated with all electrons present; for r, they used the accepted covalent radius of the atom. This has a more theoretical appearance than Pauling's thermochemical electronegativities, even though the covalent radius is derived from diffraction or spectroscopic experiments. Because of the ease of computation, a full set of Allred–Rochow electronegativities has been calculated and converted to lie in the same 0–4 range as Pauling's; these are also provided in Table 1.9. The Allred–Rochow values have been widely adopted by inorganic chemists and are probably the most frequently used set of electronegativities, but for most of the semiquantitative purposes to which electronegativity is applied, any single set of values is satisfactory. (It's important, though, not to mix values from two different sets.) Figure 1.17 is a diagrammatic representation of the periodic table in which the horizontal scale is Allred–Rochow electronegativity rather than atomic number; it suggests the nonlinearity of nearly all the electronegativity trends within the periodic table (except that the top row is spaced about 0.5 apart from Li at 1.0 to F at 4.1). Although electronegativity primarily describes electron transfer between unlike atoms in compounds, there is a fairly good correlation between electronegativity and the nature of the bonding in the bulk element, as the figure suggests.

The idea that electronegativity differences between bonded atoms lead to partial charges on those atoms makes it attractive to calculate dipole moments for simple molecules. An electric dipole is a combination of a positive charge and an equal negative charge, $+q$ and $-q$, separated by a distance R. The *dipole moment* μ is defined as $\mu = qR$. In SI units μ would be C m, but it is usually expressed in debyes (D). The debye was originally defined as 1 D = 10^{-18} esu cm, which upon conversion to SI yields 1 D = 3.336×10^{-30} C m.

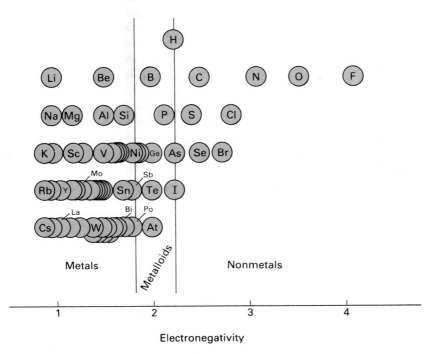

Figure 1.17 Periodic table presented in electronegativity format.

Apart from the differential-ionization-energy approach, simple empirical relationships have been developed for predicting dipole moments from electronegativity differences by treating each bond as two point charges separated by the observed average internuclear distance. Of course, comparison with experimental dipole moments is only possible for molecules whose symmetry is low enough to prevent cancellation of individual bond dipoles. Thus the bent H_2O molecule has a dipole moment of 1.85 debye, resulting from the vector addition of the two O—H bond dipoles, but the linear $HgBr_2$ molecule has a zero dipole moment even though the Hg—Br bond dipoles are presumably nonzero.

Unfortunately, the point-charge approximation for each atom is too simple to be successful in most cases. When the overlapping atomic orbitals that form a bond are not of equal size, the centroid of electron negative charge in the covalent bond does not coincide with the centroid of positive charge from the nuclei, and the result is an *asymmetry dipole* within the bond in addition to the point-charge dipole. Furthermore, the point-charge approximation has no way of accounting for lone pairs of electrons, and a lone-pair dipole can be fully as large as a bond dipole. For example, we calculated atomic charges for the CO molecule as $C^{+0.057}$ $O^{-0.057}$ using differential ionization energies. Using the experimental bond distance of 1.131 Å, we would calculate a dipole moment of

$$\mu = qR = 0.057 \times 1.602 \times 10^{-19} \text{ C} \times 1.131 \times 10^{-10} \text{ m} \times 1 \text{ debye}/3.336 \times 10^{-30} \text{ C m}$$

$$\mu = 0.312 \text{ debye}$$

The experimental dipole moment for CO is 0.112 debye, which does not seem too far off—but it is in the opposite sense. That is, the dipole moment of CO is oriented so as to make the carbon end of the molecule negative. To rationalize this observation, it is necessary to credit the carbon lone-pair with more influence than the oxygen lone-pair in the overall charge distribution. This is not unreasonable in view of the more diffuse carbon orbital, but it reveals the inadequacies of the point-charge model calculated from electronegativities.

1.8 ELECTRICAL INTERACTIONS BETWEEN ATOMS AND MOLECULES

Although our simple model does not calculate electric-dipole moments very well, their effects are extremely important in chemistry. All interactions between atoms are electrical in their nature, and whether we are dealing with the strong chemical bonding interactions present in ionic crystals or covalent molecules or with the weak interactions that make gases nonideal, the mathematical model uses electric monopoles (point charges) and dipoles to provide a reasonably accurate picture of them. Ionic bonding in crystals at its simplest is the interaction between point-charge atoms; we shall look at that model in some detail in Chapter 2. Covalent bonding in molecules at its simplest is the quantum mechanically governed interaction between a point-charge electron and two or more point-charge nuclei; we shall look at that model in Chapter 4. Here, however, we shall examine several kinds of interactions involving electric dipoles that are important in chemical systems. There is a whole hierarchy of these, which Table 1.11 presents in order of their dependence on interatomic distance. Let us take them one at a time. In a monopole–dipole interaction, there is a net potential energy of attraction under the right geometric circumstances, even though there is no net charge on the dipole. Figure 1.18 indicates this favorable situation: Even though the coulomb law is nondirectional, the monopole–dipole interaction is directional in the sense that it orients the dipole. The net attraction exists only because the opposite-charged end of the dipole is closer to the monopole than the repelling like-charged end. When an ion is hydrated by water molecules, for example, the attraction is strong enough that thermal agitation (kT) can only cause the water molecules in the first layer around the ion to oscillate slightly around the preferred direction. In the second layer, the greater distance in the denominator of the potential-energy expression reduces the net attraction, and a good deal of thermal scrambling occurs. For water molecules farther away, the attraction is small enough relative to kT at ordinary temperature that very little orientation occurs. The strength of the interaction indicated in Table 1.11 is calculated assuming a +1 charge on the monopole ion, a dipole moment of 1 D (debye), and a distance of 3 Å between the center of the ion and the center of the dipole molecule:

$$E = \frac{-q\mu}{r^2(4\pi\varepsilon_0)}$$

$$= -\frac{(1 \times 1.602 \times 10^{-19}\ \text{C e}^{-1})(1 \times 3.336 \times 10^{-30}\ \text{C m})}{(3 \times 10^{-10}\ \text{m})^2(4)(3.142)(8.854 \times 10^{-9}\ \text{C}^2\ \text{kJ}^{-1}\ \text{m}^{-1})} \times 6.022 \times 10^{23}\ \text{mol}^{-1}$$

$$E = -32\ \text{kJ/mol}$$

By comparison, the attraction between a +1 ion and a −1 ion 3 Å apart is 460 kJ/mol, and that attraction drops off only as $1/r$ rather than as $1/r^2$. At a separation of 10 Å, the ion–dipole attraction drops off to about 3 kJ/mol, which is comparable to kT at room temperature (2.48 kJ/mol at 298 K).

The interaction between two dipoles is shown in Fig. 1.19 for the special case of parallel dipoles. There is also another stable orientation of two dipoles in which they are collinear with opposite-charged ends touching. In general, the orientation of Fig. 1.19 will be more stable for slender dipole molecules, and the collinear orientation will be more stable for ones nearly spherical. Note that the introduction of an extra distance quantity (as part of the dipole moment) in the numerator of the potential energy expression requires that

Table 1.11 Electrical Interactions Present in Chemical Systems

Interaction	Potential energy	Directional character	Strength kJ/mol	Occurrence
Monopole–monopole	$E = \dfrac{q_1 q_2}{r}$	Nondirectional	~400	Ionic and covalent bonding
Monopole–dipole	$E = \dfrac{-q_1 \mu_2}{r^2}$	Orients dipole	~25	Solvation energies
Dipole–dipole	$E = \dfrac{-\mu_1 \mu_2}{r^3}\left(1 - 3\cos^2\theta\right)$ (for parallel dipoles)	Orients dipole	~3	Polar liquid structures
Monopole–induced dipole	$E = \dfrac{q_1^2 \alpha_2}{2r^4}$	May orient if α is anisotropic	~10	Solvation, clathrates
Dipole–induced dipole	$E = \dfrac{\mu_1^2 \alpha_2}{r^6}$	May orient if α is anisotropic	~0.2	Nonideal solutions
Induced dipole–induced dipole	$E = \dfrac{-3}{2}\left(\dfrac{I_1 I_2}{I_1 + I_2}\right)\dfrac{\alpha_1 \alpha_2}{r^6}$	Nondirectional	~1	Nonideal gases

Note: q_1 is charged on monopole 1; μ_2 is dipole moment of dipole 2; α_2 is polarizability of species 2; I_1 is ionization energy of species 1.

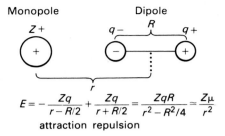

$$E = -\frac{Zq}{r - R/2} + \frac{Zq}{r + R/2} = \frac{ZqR}{r^2 - R^2/4} \simeq \frac{Z\mu}{r^2}$$

attraction   repulsion

Figure 1.18 Monopole–dipole attraction.

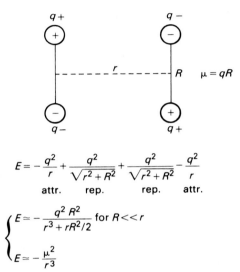

$$E = -\frac{q^2}{r} + \frac{q^2}{\sqrt{r^2 + R^2}} + \frac{q^2}{\sqrt{r^2 + R^2}} - \frac{q^2}{r}$$

attr.   rep.   rep.   attr.

$$\begin{cases} E \simeq -\dfrac{q^2 R^2}{r^3 + rR^2/2} \text{ for } R \ll r \\[2mm] E \simeq -\dfrac{\mu^2}{r^3} \end{cases}$$

Figure 1.19 Dipole–dipole attraction.

the denominator contain a higher power of r to maintain dimensional consistency. The result is a sharper dependence on intermolecular distance and a smaller energy of interaction at ordinary distances:

$$E = \frac{-\mu^2}{r^3(4\pi\varepsilon_0)} = -\frac{(1 \times 3.336 \times 10^{-30} \text{ C m})^2}{(3 \times 10^{-10} \text{ m})^3 (4)(3.142)(8.854 \times 10^{-9} \text{ C}^2 \text{ kJ}^{-1} \text{ m}^{-1})}$$
$$\times 6.022 \times 10^{23} \text{ mol}^{-1}$$
$$E = -2.2 \text{ kJ/mol}$$

This value is comparable to kT at room temperature, however, and a realistic assessment of the interaction should allow for a Boltzmann distribution of orientations as the dipole molecules tumble. Using the approximation that E is less than kT, the average net energy of interaction is given by

$$E = -\frac{2}{3}\frac{\mu^4}{r^6(kT)}$$

The radial dependence of this potential energy has now become $1/r^6$, which reduces the range of the interaction sharply. At the distance of 3 Å used in our earlier calculations, the net energy for tumbling dipoles is only -1.3 kJ/mol. Energies of this order must be overcome, however, when a nonpolar liquid mixes with a polar one. A large difference in polarity will make the liquids immiscible, and even a fairly small one will lead to nonideal solution behavior.

If an atom with its mobile negatively charged electrons and positive nucleus is placed near a positive point charge, its electrons will be attracted toward the point charge and its nucleus will be repelled away, causing the atom's centers of positive and negative charge to no longer coincide. The presence of the point charge has induced a dipole where none existed before, and the atom is said to be *polarized*. The induced dipole moment is proportional to the strength of the electric field $\mathscr{E}$ produced by the point charge, and the proportionality constant is called the *polarizability*, α:

$$\mu_{\text{induced}} = \alpha \cdot \mathscr{E}$$

The interaction of an induced dipole with another charge can only be an attraction, since it is automatically created with the correct geometry. If the species being polarized is an atom, there can be no orienting effect; however, if the species is a molecule, the numerical value of the polarizability may be different in one direction relative to other directions, and the molecule will tend to orient itself to create the largest induced dipole moment. Most energies involving induced dipoles are small, however, and the orienting effect may be lost in thermal tumbling. However, thermal tumbling does not remove the net attraction because the dipole is immediately and always induced in the correct direction no matter what the overall orientation of the molecule is. Dimensional analysis of the defining equation above indicates that polarizability must have units of "volume," which can be thought of as the volume swept out by the electrons as they deform away from their original distribution. Typical values of α for atoms and small molecules range from 0.2×10^{-30} m^3 to 10×10^{-30} m^3, or very roughly one cubic angstrom unit (Å^3). The polarizability increases with the volume of the atom or molecule itself, but there is also a fairly obvious dependence on the electronegativity of the atom whose electrons are being deformed. A very electronegative atom will have a small polarizability, since it does not allow its electrons to deform much for a given electric-field intensity. In general, an atom with a large a_1 differential-ionization-energy constant will have a small polarizability, cations will have small polarizabilities, and anions will have large polarizabilities. The a_1 quantity can be considered the "hardness" of the atom, a concept that will be useful to us in later chapters.

When a monopole (point-charge ion) induces a dipole in a nearby atom or molecule, the potential energy of their interaction varies as $1/r^4$ because the attraction of the two varies as $1/r^2$ and the magnitude of the induced dipole-moment also varies as $1/r^2$. The range of this attraction is therefore quite short, and the attraction energy is not great. For a $+1$ ion, an induced dipole 3 Å away, and a polarizability of 1 Å^3, we calculate

$$E = -\frac{q^2\alpha}{2r^4(4\pi\varepsilon_0)} = -\frac{(1 \times 1.602 \times 10^{-19} \text{ C})^2(1 \times 10^{-30} \text{ m}^3)}{2(3 \times 10^{-10} \text{ m})^4(4)(3.142)(8.854 \times 10^{-9} \text{ C}^2 \text{ kJ}^{-1} \text{ m}^{-1})}$$
$$\times 6.022 \times 10^{23} \text{ mol}^{-1}$$
$$E = -8.6 \text{ kJ/mol}$$

This number is independent of whether or not the molecule has a permanent dipole moment. If there is a permanent moment, the attraction will be increased by this amount. Since this number is roughly one-fourth the attraction of an ion for a typical permanent dipole moment, it can be seen that the induced moment makes an important contribution to such phenomena as solvation energies.

A permanent dipole will also induce a dipole moment in nearby atoms or molecules. This effect, like that between thermally averaged permanent dipoles, varies as $1/r^6$, so its range is extremely short. Although by the nature of the induced moment it is always a stabilizing influence, the energy of the interaction is quite small:

$$E = -\frac{2\mu^2\alpha}{r^6(4\pi\varepsilon_0)} = -\frac{2(1 \times 3.336 \times 10^{-30} \text{ C m})^2(1 \times 10^{-30} \text{ m}^3)}{(3 \times 10^{-10} \text{ m})^6(4)(3.142)(8.854 \times 10^{-9} \text{ C}^2 \text{ kJ}^{-1} \text{ m}^{-1})}$$
$$\times 6.022 \times 10^{23} \text{ mol}^{-1}$$
$$E = -0.17 \text{ kJ/mol}$$

One might expect that this effect would stabilize solutions of polar and nonpolar liquids by creating energy to replace the lost dipole–dipole attractions, but the effect is only about one-tenth large enough. It must be included in a careful calculation, but is not large enough to change any fundamental chemical behavior.

The last possibility to be considered is that the dipole formed at any moment by an electron and its nucleus (whatever the electron's instantaneous location) will induce a dipole in a nearby unbonded atom or molecule. This is actually a cooperative correlation of electron positions between adjacent unbonded atoms in which both atoms are polarizing and both are being polarized. The effect is that of an induced-dipole/induced-dipole attraction. As with the other induced-dipole phenomena, it can only be an attraction. Because the initial electron's position is quantum mechanically governed, it is not possible to write a classical expression for the potential energy of this *London dispersion force*. The quantum mechanical result, however, from Table 1.11, yields a significant energy. If two identical atoms, each having an ionization energy of 10 eV or 965 kJ/mol and a polarizability of 1 Å^3, are placed 3 Å apart,

$$E = -\frac{3I\alpha^2}{4r^6} = -\frac{3(965 \text{ kJ mol}^{-1})(1 \times 10^{-30} \text{ m}^3)^2}{4(3 \times 10^{-10} \text{ m})^6}$$

$$E = -1.0 \text{ kJ/mol}$$

This is the only one of the forces considered that can produce a net attraction between two electrically neutral atoms or nonpolar molecules. It constitutes the entire binding energy in genuinely covalent molecular crystals such as those of the noble gases or elemental halogens. Accordingly, such crystals should sublime or boil at temperatures at which kT is comparable to this energy:

$$kT \simeq 1.0 \text{ kJ/mol} = 1.67 \times 10^{-21} \text{ J/molecule}$$

$$T \simeq \frac{1.67 \times 10^{-21} \text{ J molecule}^{-1}}{1.38 \times 10^{-23} \text{ J K}^{-1}}$$

$$\simeq 120 \text{ K}$$

In fact, argon boils at 87 K, krypton at 121 K, xenon at 165 K, fluorine at 85 K, and chlorine at 239 K, so the theoretical result is consistent with the behavior of these gases.

Because the binding energies involving monopole ions are so large, we do not see gaseous monopoles under any ordinary conditions. The intermolecular forces that are responsible for nonideality in gases are thus limited to those from Table 1.11 that involve only dipoles and induced dipoles. There are three of these, all relatively weak: The dipole–dipole orientation energy, the dipole/induced-dipole energy, and the London dispersion energy. All of these have $1/r^6$ dependence, so that it is particularly easy to lump them together algebraically as van der Waals forces in creating potential-energy function for the approach of two molecules. The best known approximate function of this type is the Lennard–Jones potential shown in Fig. 1.20, in which the attractive forces are represented by a $1/r^6$ potential and the repulsive forces due to the interpenetration of electron clouds at very short range are represented by a $1/r^{12}$ potential (both because it is a reasonable empirical fit and because mathematical operations with the potential are easier if one exponent is twice the other). Since the van der Waals attractions have a slightly longer range than the repulsion term, a shallow potential well results. The bottom of this well lies at an intermolecular distance that corresponds to the distance between atoms in a crystal held together only by van der Waals forces, such as a crystal of a noble-gas element. The depth of the well corresponds to the enthalpy of sublimation of the crystal, or roughly to the enthalpy of vaporization. Since dipole forces are included, the Lennard–Jones potential can be extended to cover any system that forms identifiable molecules in both the crystal and vapor phase. The size of the atom or

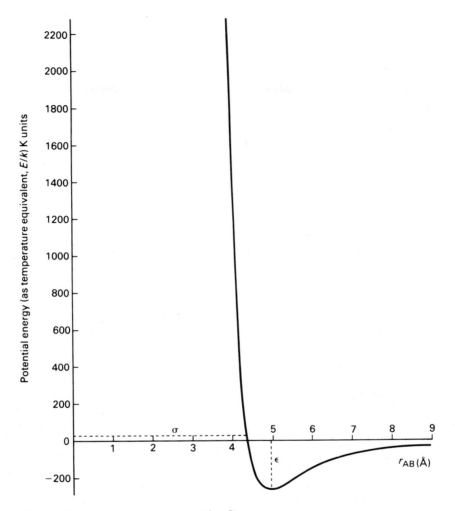

Figure I.20   Lennard–Jones potential for Cl_2:

$$E = 4\epsilon\left[\left(\frac{\alpha}{r_{AB}}\right)^{12} - \left(\frac{\alpha}{r_{AB}}\right)^{6}\right]$$

$$\frac{\epsilon}{k} = 257\ K \quad \sigma = 4.40\ Å$$

molecule as it is revealed by the Lennard–Jones potential or van der Waals radius is substantially larger than that revealed by other traditional measures of atomic size, such as covalent radius or ionic radius. This reflects the fact that the exclusion principle prevents electron clouds from penetrating each other significantly in nonbonding interactions between two atoms (such as the van der Waals interactions), whereas in bonding interactions involving electron sharing or transfer the outer layer of each atom consists of valence electrons whose distributions can overlap within the requirements of the exclusion principle. The size and energy requirements for such bonding interactions will be a major topic of the next three chapters.

PROBLEMS

A. DESCRIPTIVE

A1. Since direct nuclear fusion could not produce stable isotopes of Li, Be, and B, how could those elements have been formed?

A2. How many examples can you find in Fig. 1.3 of an element with even Z that is less abundant in the earth's crust than both its neighbors with odd Z?

A3. Why are most of the elements discovered in ancient times quite low in absolute abundance?

A4. Write the equation for the nuclear reaction that produces technetium from molybdenum.

A5. Why is the atomic-volume curve in Fig. 1.12 nearly flat from Ce to Lu? Why are Eu and Yb markedly higher?

A6. Suggest an electronic reason for the diagonal relationship between Li and Mg, Be and Al, and so on.

A7. In Fig. 1.14, why does the third-ionization curve increase more sharply from Al to Ca than the first-ionization curve does from Na to Ar, even though the electron configurations are comparable?

A8. On the basis of the atomic electronegativities involved, in which direction should the following reactions proceed spontaneously?

(a) $AlCl_3 + POCl_3 \rightleftharpoons POCl_2^+ + AlCl_4^-$

(b) $Si(OCH_3)_4 + 2H_2S \rightleftharpoons SiS_2 + 4CH_3OH$

(c) $2CH_3COCl + Cd(CH_3)_2 \rightleftharpoons 2CH_3COCH_3 + CdCl_2$

A9. The heavier atoms near the bottom of any given column of the periodic table have larger inner cores, which lower the ionization energy of their valence electrons but generally increase their polarizability. What influence does this pattern have on the boiling points of the elments?

A10. Why don't the electron-affinity values in Table 1.9 show a clear periodic trend going across a row of the periodic table, as ionization potentials do?

A11. Account for the following sequence of first-ionization potentials by describing the atomic-structure effects at work:

F: 17.42 V   Ne: 21.56 V   Na: 5.14 V   Mg: 7.65 V   Al: 5.99 V

A12. The first-ionization energies and electron affinities of the atoms C, Ca, Cl, Cr, and Cs are

IE: 3.89 eV,   6.11 eV,   6.76 eV,   11.26 eV,   13.01 eV

EA: −1.62 eV,   0.47 eV,   0.66 eV,   1.27 eV,   3.61 eV

though not necessarily in that order. Assign an ionization energy and an electron affinity to each element, using a periodic table but no other data. Explain your reasoning.

B. NUMERICAL

B1. Calculate the temperature equivalent of the mass of a proton.

B2. Calculate the ionization energy of H using tabular data for fundamental constants and the expression for the energy of the electron in a hydrogen atom. Express your answer in electronvolts.

B3. Plot the radial dependence of the Slater $3d$ orbital for Co. (The unit for r is the Bohr radius for the hydrogen $1s$ orbit, 0.529 Å.) Discuss the relationship between this graph and the covalent radius for Co given in Table 4.4.

B4. Compare the Slater and Clementi–Raimondi Z_{eff} values for a $4s$ electron in K, Cr, Zn, and Br. Comment on the level of agreement you observe.

B5. Convert the Z_{eff} and r_{max} values in Figs. 1.8 and 1.9 to electrostatic potential energies for the $3s$ electron in Na, Si, and Cl and compare your results with the ionization energies in Table 1.7 and the VOIP values in Table 1.8.

B6. Use the IP and EA values from Tables 1.7 and 1.9, respectively, to calculate the energy liberated (or absorbed) by the reaction

$$Ca(g) + O(g) \rightarrow Ca^{2+}(g) + O^{2-}(g)$$

Comment on the stability of calcium oxide.

B7. Use $I(q)$ functions to estimate atomic charges in $SnCl_2$ and $SnCl_4$. On this basis alone, which molecule is more likely to be soluble in a nonpolar solvent such as CCl_4?

B8. From the experimental bond energies below, calculate the electronegativity difference between carbon and chlorine. What other data would be necessary to establish a single value for the Pauling electronegativity of chlorine? $E_{Cl-Cl} = 239.7$ kJ/mol; $E_{CH3-CH3} = 345.6$ kJ/mol; $E_{CH3-Cl} = 327.2$ kJ/mol.

B9. The normalized $1s$ wavefunction for a hydrogen atom (see Table 1.4) can be written more simply as

$$\psi = \frac{1}{\sqrt{\pi a_0^3}} e^{-r/a_0}$$

where a_0 is the Bohr radius, 0.529 Å. The volume element dr in spherical polar coordinates is $r^2 \sin \theta \, dr \, d\theta \, d\phi$. Calculate the most probable radius $\langle r \rangle$ for the $1s$

electron in a hydrogen atom by evaluating the integral

$$\langle r \rangle = \int\limits_{0}^{\infty} \int\limits_{0}^{\pi} \int\limits_{0}^{2\pi} \psi \cdot r \cdot \psi \, d\tau$$

C. EXTENDED REFERENCE

C1. Referring to a quantum-mechanics text or *Quanta* [P. W. Atkins (Clarendon Press, Oxford, 1974)], explain the relationship between the number of nodes in the atomic-orbital diagrams of Figs. 1.4 and 1.5 and the energy of those orbitals.

C2. Consider Mendeleev's original vertical-format periodic table [*Z. Chemie* (1869), *12*, 405], or, in English, *A Source Book in Chemistry 1400–1900* [H. M. Leicester and H. S. Klickstein (Harvard University Press, Cambridge, Mass., 1952)]. Between Ca and Ti he inserts not only the then-known scandium with an atomic weight of 45 g/mol, but also two rare earths and indium, although these fall far out of the atomic-weight order. He also places uranium between cadmium and tin. What has gone wrong? Discuss his suppositions and his data.

C3. R. T. Sanderson has proposed an electronegativity scale based on atomic-electron densities [*J. Chem. Educ.* (1952), *29*, 539]. Compare his scale to the Allred–Rochow scale. What similarities persist through the algebraic manipulations necessary to make the comparison? Can the spontaneous formation of XeF_6 from Xe and F_2 be reconciled with Sanderson's approach?

Part II

Main-Group Compounds

2

Ions and Their Environments ━━━━━

The wide electronegativity differences that can occur between neighboring bonded atoms in inorganic systems lead to substantial transfer of valence electrons from one atom to another under the appropriate circumstances. In this chapter we will examine what those circumstances are, what the resulting ionic bonding is like, and some physical and chemical features of ionic crystals.

━━━━━ ## 2.1 IONS IN GASES

If we bring together two atoms isolated as in the gas phase and having quite different electronegativities, we might expect the ions to form:

$$Na(g) + Cl(g) \rightarrow Na^+(g) + Cl^-(g)$$

However, this reaction does not occur. Since there is little entropy difference between reactants and products, $\Delta G°$ for the reaction is essentially equal to $\Delta H°$. The enthalpy change is the thermodynamic sum of the ionization energy for sodium and the electron affinity for chlorine; from Tables 1.7 and 1.9, we calculate $\Delta H° = +5.138$ eV $- 3.614$ eV $= +1.524$ eV $= +147.0$ kJ/mol. This large positive enthalpy change guarantees a positive free-energy change and thus no significant number of free ions—and this result holds for any pair of atoms in the periodic table. No ionic compound, when vaporized, yields free gaseous ions.

When NaCl is vaporized, it forms what we can call diatomic NaCl "molecules" without specifying anything about the bonding. From microwave rotational spectra, the internuclear distance in gaseous NaCl is 2.3606 Å. At that distance (and at any other between about 1.7 and 10.0 Å) the Na^+Cl^- ion pair is stable, in spite of the unfavorable enthalpy change for the reaction above. We must consider two additional energy terms for the ion pair as opposed to individual free ions: the coulomb monopole–monopole attraction of the ions for each other at finite distances, and the strong repulsion of interpenetrating electron shells under the exclusion principle at very close approach. The attraction (U_2) is simply

$$U_2 = -\frac{q_{Na}q_{Cl}}{r}$$

where the q values are net charges on the ions and r is the internuclear separation for the ions, which are assumed to be spherical. The repulsion (U_3) is a bit more complicated. An early approach assumed the $1/r^{12}$ repulsion of the Lennard–Jones potential, or a variable $1/r^n$ ($n = 4$–12). A more successful algebraic representation recognizes the exponential nature of radial wave functions by using two parameters, b and ρ, in an exponential form of the repulsion.

$$U_3 = +be^{-r/\rho}$$

A common value of ρ, 0.345 Å, can be used for all alkali halides with good accuracy for the crystalline materials. The potential energy U of the ion pair thus becomes

$$U = 35.15 + U_2 + U_3 = 35.15 - \frac{q_{Na}q_{Cl}}{r} + be^{-r/\rho}$$

The value of b can be established by recognizing that U is at a minimum when r is equal to r_0 (2.3606 Å). At that r, $dU/dr = 0$.

$$\frac{dU}{dr} = 0 = +\frac{q_{Na}q_{Cl}}{r_0^2} - \frac{b}{\rho}e^{-r_0/\rho}$$

$$b = \frac{\rho}{r_0}\left(\frac{q_{Na}q_{Cl}}{r_0}\right)e^{r_0/\rho} = 80609 \text{ kJ/mol for NaCl}$$

Figure 2.1 shows these three components of the potential energy plotted as a function of r (Fig. 2.1a) and the sum of these as the total energy U (Fig. 2.1b). The obvious result is that there is a potential well near 2.4 Å that stabilizes the Na^+Cl^- ion-pair, not only relative to free ions but also relative to the neutral atoms.

Another approach to the stability of ionic species in the gas phase uses differential-ionization energies to calculate the approximate charges on each atom. Using a_0 and a_1 from Table 1.10, this approach yields a predicted charge of 0.39 on each atom: $Na^{+0.39}Cl^{-0.39}$. However, that technique has been based on minimizing the electronic energies of the two individual atoms

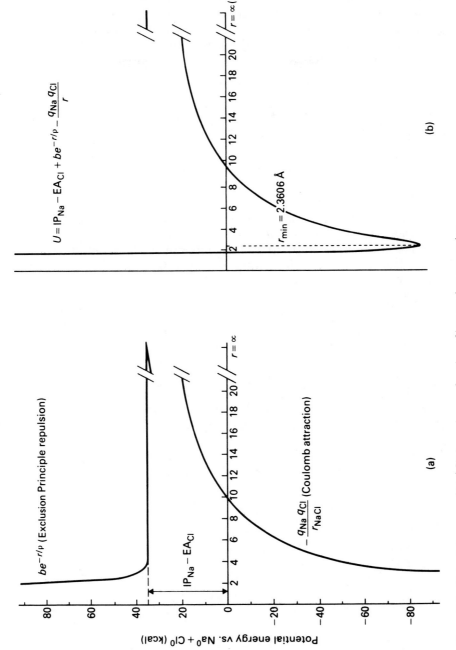

Figure 2.1 Potential energy of Na^+Cl^- ion pairs as a function of internuclear separation.

Na and Cl with respect to charge transfer, without considering the stabilization of the pair by their increasing coulomb attraction for each other as the charges increase. When that energy term is included, quite a different prediction results. For the total energy of the pair of atoms, we can write

$$U = E_{Na} + E_{Cl} - \frac{q^2}{r_0} + be^{-r_0/\rho}$$

where q is the magnitude of the charge on each atom. At a charge-transfer equilibrium, this energy will be at a minimum with respect to charge: $dU/dq = 0$.

$$\frac{dU}{dq} = 0 = I_{Na}(q) - I_{Cl}(-q) - \frac{2}{r_0}q$$

The $ClI(q)$ is negative because we are moving to the left on its energy-versus-charge curve as q increases. The $I(q)$ values from Table 1.10 are in volts and volts/electron, so the coulomb-attraction term must be expressed in the same units. If r_0 is taken in angstrom units (Å), the conversion factor to volts is 14.399 V Å/electron:

$$\frac{dU}{dq} = 0 = (2.84 + 4.59q) - (8.31 - 9.40q) - \frac{2}{2.3606}q \times 14.399 \frac{\text{volt}}{\text{e}^-/\text{Å}}$$

$$0 = -5.47 + 1.79q$$

$$q = 3.06$$

This number is not realistic, of course, because there are discontinuities in the Na and Cl energy-versus-charge curves at $+1$ and -1 respectively. What this result predicts is complete valence-electron transfer, to Na^+Cl^-, which is compatible with Fig. 2.1b. It might be noted that we have omitted any energy term for the atom pair that would express covalent binding energy as a function of net charge. Because this energy is necessarily zero when electron transfer is complete, such a term presumably acts to minimize electron transfer, counteracting the ionic-binding energy term. Chapter 4 will suggest some ways to calculate the covalent binding energy.

There is, apparently, some covalent bonding in gaseous NaCl. If we use Hess's law with experimental enthalpies to obtain an experimental equivalent of the energy calculated above as U, we write

$$Na(g) \xrightarrow{-\Delta H_{\text{atom}}} Na(s)$$

$$Cl(g) \xrightarrow{-\frac{1}{2}BE} \tfrac{1}{2}Cl_2(g)$$

$$\underline{Na(s) + \tfrac{1}{2}Cl_2(g) \xrightarrow{\Delta H_f} NaCl(g)}$$

$$Na(g) + Cl(g) \xrightarrow{U} NaCl(g)$$

From this summation, we have

$$U = -\Delta H_{atom} - \tfrac{1}{2}BE + \Delta H_f$$
$$= -100.4 - 121.3 - 182.0 = -403.7 \text{ kJ/mol}$$

From Fig. 2.1b, the theoretical U for an ion pair with no covalent bonding is -355.6 kJ/mol, which is not a bad match for the experimental number but clearly implies that there is some additional bonding at work in the gaseous NaCl molecule. Quite generally we find, for even the most predominantly ionic systems, that some electron sharing occurs; there is no purely ionic compound. Frequently, however, we can reproduce experimental binding energies under the assumption of ionic bonding to an accuracy even better than that shown here, and under these circumstances it is appropriate to describe such compounds as ionic.

To the chemist interested in the molecular and crystal structures of inorganic compounds, the absolute and relative sizes of atoms are important. They are, however, difficult to define uniquely, both because the measured size of an atom will depend on the forces compressing it (atoms are squishy) and because the size is strongly influenced by the net charge on the atom. To continue our example of gaseous NaCl, we can define a theoretical radius of an atom of Na or Cl as the distance from the nucleus to the maximum valence-electron probability density as determined by an SCF calculation on the atom. For the Na $3s$ electron, that distance is 1.713 Å, and for the Cl $3p$ electrons it is 0.723 Å. On this basis the Na atom is $2\tfrac{1}{2}$ times as large as Cl, although it is very fluffy because most of its volume is occupied by the single $3s$ electron. However, as the two atoms approach each other and electron transfer begins, the Na atom shrinks drastically and the Cl atom expands by about the same amount. A reasonable estimate for the ionic radius of Na$^+$ in a gaseous ion pair (that is, with coordination number 1) is 0.82 Å. For Cl$^-$ with coordination number 1, the corresponding radius is 1.54 Å. The chloride ion is now roughly twice as large as the sodium ion, and the size relationship has changed completely.

While this electron transfer is taking place and the sizes of the atoms are changing so drastically, the polarizabilities of the atoms are changing as well. A neutral sodium atom is quite large and does not hold its valence electron at all tightly; thus, it is very polarizable. Conversely, the chlorine atom is small and quite electronegative; thus, it is only slightly polarizable. After electron transfer has occurred, however, the Na$^+$ is small and holds its remaining inner-core electrons very tightly (it has the same electron configuration as F$^-$ but has two *additional* protons to bind them), so it is only very slightly polarizable. By a similar argument, the Cl$^-$ ion is more polarizable than the Cl atom, mostly because it is larger and the electrons can deform through a greater volume. A sodium atom, then, is soft, meaning that it deforms readily in the presence of an external electric charge, but a sodium ion is quite hard. Correspondingly, a chlorine atom is fairly hard, but a chloride ion is softer.

2.2 IONS IN CRYSTALS

The ideas of the previous section apply equally well to ions in crystals, but the energy computations are complicated by the presence of a large three-dimensional array of ions, all attracting and repelling each other. Fortunately, the problem is only a geometric one and we can consider ways to solve it.

Suppose we wish first to calculate the net energy of attraction of a one-dimensional array of sodium and chloride ions arranged in alternating fashion, as in Fig. 2.2. For our initial purposes we will not consider the contact repulsion between filled shells of electrons, but we must consider the long-range repulsion between two Cl^- ions or two Na^+ ions. We begin by considering the total potential energy of the Na^+ in the center of the infinitely long one-dimensional array shown in Fig. 2.2. It is attracted by the two Cl^- on either side of it at a distance r_0, but it is also repelled by the two Na^+ on either side of it at a distance $2r_0$:

$$U = -2\frac{q_{Na}q_{Cl}}{r_0} + 2\frac{q_{Na}q_{Cl}}{2r_0} = -2\frac{q^2}{r_0}\left(1 - \frac{1}{2}\right)$$

Here the q values refer only to the magnitude of the charges; the signs of the terms emphasize the alternating attraction and repulsion. If we carry the sum of attractions and repulsions out a few more terms we can see a series forming that converges on 0.693147 (ln 2):

$$U = -2\frac{q^2}{r_0} + 2\frac{q^2}{2}r_0 - 2\frac{q^2}{3}r_0 + 2\frac{q^2}{4}r_0 - \cdots = -2\frac{q^2}{r_0}\left(1 - \frac{1}{2} + \frac{1}{3} - \frac{1}{4} + \cdots\right)$$

$$= -2(0.693147)\frac{q^2}{r_0} = -1.38629\frac{q^2}{r_0}$$

The number 1.38629 is the geometrical weighting of the sum of attractions and repulsions in this one-dimensional array of ions. It is known as the

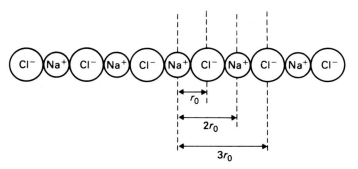

Figure 2.2   Geometry of a one-dimensional NaCl crystal.

Madelung constant for the array. If we changed to a one-dimensional array of K^+Br^- or $Mg^{2+}O^{2-}$, the value of the Madelung constant M would not change, but q and r_0 would. If, on the other hand, we changed to a square packing of Na^+ and Cl^-, q and r_0 would not change, but M would be different.

The series that converges to give the Madelung constant does so very slowly. A much quicker convergence can be had by slicing the surface ions at each stage of the summation so that the array being built up has no overall net charge, as in Fig. 2.3. If the ordinary series is carried out to 10 terms, the resulting Madelung constant is 1.2913, whereas if the series derived from the neutral array is so carried out, the resulting constant is 1.3913, a much better approximation.

For a three-dimensional crystal the approach is exactly the same. A series generated for a particular crystal sums to a Madelung constant characteristic of the geometric symmetry of that crystal type. As in the one-dimensional case, we sum beginning with the nearest-neighbor ions, then the next-nearest, and so on. In the NaCl lattice shown in Fig. 2.4, the central Na^+ (shown with darker shading) has six Cl^- nearest neighbors (in the centers of the faces of the cube shown). The next-nearest neighbor ions are the twelve Na^+ along the edges of the cube, and the third-nearest are the eight Cl^- at the cube corners. The potential energy on this basis would be

$$U = -6\frac{q_{Na}q_{Cl}}{r_0} + 12\frac{q_{Na}q_{Cl}}{\sqrt{2}r_0} - 8\frac{q_{Na}q_{Cl}}{\sqrt{3}r_0} = -2.134\frac{q^2}{r_0}$$

Again the series does not converge rapidly, and we resort to summing a series of larger and larger electrically neutral cubes with fractional ions on the surface. In the first of these, the nearest-neighbor Cl^- ions are one-half inside

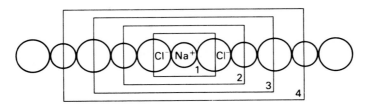

Figure 2.3 Summing attractions and repulsions to a value for the Madelung constant within electrically neutral segments of a one-dimensional crystal:

Stage 1: $M = 2 \times 1/2 = 1.0000$

Stage 2: $M = (2 \times 1) - (2 \times 1/2 \times 1/2) = 1.5000$

Stage 3: $M = (2 \times 1) - (2 \times 1/2) + (2 \times 1/3 \times 1/2) = 1.3333$

Stage 4: $M = (2 \times 1) - (2 \times 1/2) + (2 \times 1/3) - (2 \times 1/4 \times 1/2) = 1.4167$

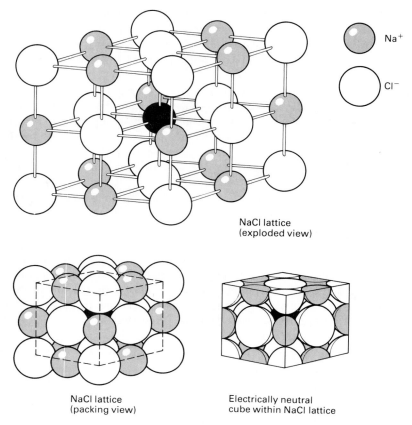

NaCl lattice
(exploded view)

Na$^+$

Cl$^-$

NaCl lattice
(packing view)

Electrically neutral
cube within NaCl lattice

Figure 2.4   Three-dimensional ion symmetry in the NaCl lattice.

the cube (in a face), so half their charge is counted. One-fourth of the charge
of the next-nearest Na$^+$ is counted, since they are on an edge of the cube, and
one-eight of the charge of the third-nearest Cl$^-$ (at corners of the cube) is
counted:

$$U = \frac{1}{2}\left(-6\frac{q^2}{r_0}\right) + \frac{1}{4}\left(+12\frac{q^2}{\sqrt{2}r_0}\right) + \frac{1}{8}\left(-8\frac{q^2}{\sqrt{3}r_0}\right)$$

$$= -1.4561\frac{q^2}{r_0}$$

The next approximation involves a cube containing $5 \times 5 \times 5$, or 125, ions.
Of these, the 27 we have already considered are inside the cube and the other

98 are on the faces, edges, or corners of the cube:

$$U = 1\left(-6\frac{q^2}{r_0} + 12\frac{q^2}{\sqrt{2}r_0} - 8\frac{q^2}{\sqrt{3}r_0}\right) + \frac{1}{2}\left(+6\frac{q^2}{2r_0} - 24\frac{q^2}{\sqrt{5}r_0} + 24\frac{q^2}{\sqrt{6}r_0}\right)$$
$$+ \frac{1}{4}\left(+12\frac{q^2}{\sqrt{8}r_0} - 24\frac{q^2}{3r_0}\right) + \frac{1}{8}\left(+8\frac{q^2}{\sqrt{12}r_0}\right)$$

$$U = -1.7518\frac{q^2}{r_0}$$

The final value of the Madelung constant for NaCl and other compounds having the same lattice symmetry is $M = 1.74756$, so we are already quite close.

Table 2.1 gives values of the Madelung constant for a number of different commonly observed crystal lattices, some of which are shown later in this chapter. Unfortunately, several conventions exist for literature values of Madelung constants, because different authors have included different factors from the $q_1 q_2 / r_0$ fraction within the Madelung constant. For example, many sources make $q_1 q_2$ the square of the greatest common divisor of the two ion charges. For TiO_2, this would be $(2e)^2$ because the charges are 4+ and 2−; but for Al_2O_3 it would be $(e)^2$ since the charges are 3+ and 2−. Such a Madelung constant for TiO_2 (rutile) would be twice that given in Table 2.1, or 4.8160, but that for Al_2O_3 would be six times the value in Table 2.1, or 25.0312. For less symmetric crystals, other authors quote r in the denominator in terms of the dimension a_0 of the crystal's unit cell. In this case, M must

Table 2.1 Madelung Constants for Common Crystal Lattices

Lattice	Madelung constant (M)
NaCl	1.74756
CsCl	1.76267
ZnS (α, wurtzite)	1.64132
ZnS (β, sphalerite)	1.63805
PtS	1.58021
CaF$_2$ (fluorite)	2.51939
CaCl$_2$	2.365
CdI$_2$	2.1915
TiO$_2$ (rutile)	2.4080
TiO$_2$ (anatase)	2.400
SiO$_2$ (β-quartz)	2.2197
Al$_2$O$_3$ (corundum)	4.17187

include the conversion factor from that length to the true internuclear distance. For both of these reasons, great care must be exercised in using M values from different sources. In this book, values will all fit the following relationship:

$$U = -M\frac{(\text{actual cation charge})(\text{actual anion charge})}{\text{cation–anion internuclear distance}}$$

A calculation done in the above manner will yield potential energy in joules per cation if charges are in coulombs (C), distances are in meters (m), and the permittivity of free space is included in the denominator as $4\pi\epsilon^0$. A more useful number to think about is the *lattice energy*, which is the potential energy (calculated as above) per mole of compound. The conversion in-involves Avogadro's number and a factor converting C^2/m directly to kilo-joules (kJ). A convenient conversion factor is 5.4124×10^{30} kJ/mol per C^2/m-ion.

We do not yet have a complete expression for the lattice energy of an array of ions, however. No provision has been made for the short-range repulsion that prevents the lattice from collapsing under its electrostatic attraction. We can treat it as for gaseous ions, using what is called the *Born–Mayer repulsion potential*:

$$U_{\text{rep}} = +be^{-r/\rho} \qquad (\rho = 0.345 \text{ Å})$$

As for the case of the gaseous ions, we can avoid having to determine a value for the parameter b by using the equilibrium internuclear distance as the value of r for which U is a minimum, so that $dU/dr = 0$:

$$U = -M\frac{q_1 q_2}{r} + be^{-r/\rho}$$

$$\frac{\partial U}{\partial r} = 0 = M\frac{q_1 q_2}{r_0^2} - \frac{b}{\rho}e^{-r_0/\rho}$$

$$M\frac{q_1 q_2}{r_0^2} = \frac{b}{\rho}e^{-r_0/\rho}$$

$$b = \left(M\frac{q_1 q_2}{r_0}\right)\frac{\rho}{r_0}e^{+r_0/\rho}$$

$$U = -M\frac{q_1 q_2}{r_0}\left(1 - \frac{\rho}{r_0}\right)$$

Two other energy influences on the array of ions in a crystal should be considered in a careful calculation, but they are both small and are of opposite sign, so to a good approximation they cancel each other and need not be considered. One is the small attraction due to the London dispersion force between an instantaneous dipole and its induced dipole in a neighbor atom, which we discussed in Chapter 1. The other is the slight destabilization of the atoms in the crystal due to their quantum-mechanical zero-point energy, a small kinetic energy that they cannot lose by thermal transfer. The difference between these two energies is rarely larger than 3 or 4 kJ/mol, so neglecting them does not seriously damage calculated lattice energies. For most predominantly ionic compounds, the equation calculates lattice energies accurate to within about 10 kJ/mol.

Another approximation, that of Kapustinskii, is particularly useful because we can use it without knowing the crystal structure of the ionic solid. Kapustinskii noted that when the values of Madelung constants were tabulated as in Table 2.1, they were nearly proportional to the number of ions in the "molecular" formula of the compound. That is, where NaCl has two ions in its formula, Al_2O_3 has five, and M for Al_2O_3 is very nearly five-halves as large as M for NaCl. Because deviations from this rule closely parallel changes in r_0 for two given ions when they change their lattice symmetry, we can use a standard r_0 (the sum of the ionic radii for the cation and anion) and the reduced Madelung constant due to Kapustinskii ($M \div$ number of ions in formula) to calculate the lattice energy of an ionic crystal to a good approximation, no matter what its geometry may be. Kapustinskii collected the reduced Madelung constant, the length conversion factor, and the energy conversion factor into a particularly simple form of the lattice-energy equation:

$$U = \frac{-1202 v z_C z_A}{r_C + r_A} \left(1 - \frac{0.345}{r_C + r_A}\right) \quad \text{kJ/mol}$$

In this expression, v is the total number of ions in the formula of the ionic compound, z_C is the charge on the cation in multiples of the electronic charge, z_A is the charge on the anion in the same units, and r_C and r_A are the ionic radii off the cation and anion in angstrom units. Note that v counts ions, not atoms: Ammonium perchlorate, NH_4ClO_4, has 10 atoms but only two ions, so $v = 2$ for NH_4ClO_4.

As both Table 1.11 and Fig. 2.1 suggest, lattice energies are quite large, whether calculated directly from the Madelung constant of a crystal or through the Kapustinskii equation. We can calculate the lattice energy of NaCl both ways to compare the results and to see the general magnitude of ionic lattice energies:

Madelung

$$U = -1.74756 \frac{(1.60219 \times 10^{-19} \text{ C})^2}{2.814 \times 10^{-10} \text{ m}} \left(1 - \frac{0.345}{2.814}\right) \times 5.4124 \times 10^{30} \frac{\text{kJ} - \text{m}}{\text{C}^2 - \text{mol}}$$

$$= -757.0 \text{ kJ/mol}$$

Kapustinskii

$$U = -\frac{1202 \times 2 \times 1 \times 1}{1.16 + 1.67}\left(1 - \frac{0.345}{1.16 + 1.67}\right)$$

$$= -745.9 \text{ kcal/mol}$$

Note that for both calculations, the numerical accuracy is limited by the error in the internuclear distance. In the Madelung calculation, r_0 is the experimental distance as derived from x-ray studies, whereas in the Kapustinskii calculation, r_{Na+} and r_{Cl-} are taken from Shannon's compilation in Table 2.3 (see p. 82).

In general, lattice energies for $+1/-1$ ionic crystals such as NaCl, NaH, KCN, KNO_3, and the like tend to lie in the 600–750 kJ/mol range. The lattice energy is influenced far more by the charges on the ions than by their sizes, since the sizes do not differ too widely. For CaF_2 with its $2+$ ion, the calculated lattice energy is 2470 kJ/mol; for MgO with a charge product of 4, the calculated lattice energy is 3835 kJ/mol. These are fairly typical values—1:2 compounds, whether $+1/-2$ or $+2/-1$, usually have lattice energies of roughly 2300–2700 kJ/mol; and $+2/-2$ compounds usually have lattice energies of 2900–3500 kJ/mol. These are very large energies, representing a massive stabilization of the ionic system. The NaCl lattice energy of 750 kJ/mol may be compared with the maximum possible ion-pair stabilization in the gas phase, 355 kJ/mol.

When we calculated the ion-pair stabilization energy, we used Hess's law to provide an experimental thermochemical value of the energy for the same process. We can do the same thing to compare theoretical lattice-energy values with experimental results. It is important to realize that no direct measurement of lattice energies is possible, since it refers to the hypothetical process

$$M^{n+}(g) + X^{n-}(g) \rightarrow MX(s)$$

Even vaporization is not the reverse of the lattice-energy process, because as we have seen, the vapor is not free ions but ion pairs. However, we can create a series of thermodynamic processes that are equivalent to the lattice-energy process, as shown in the following diagram:

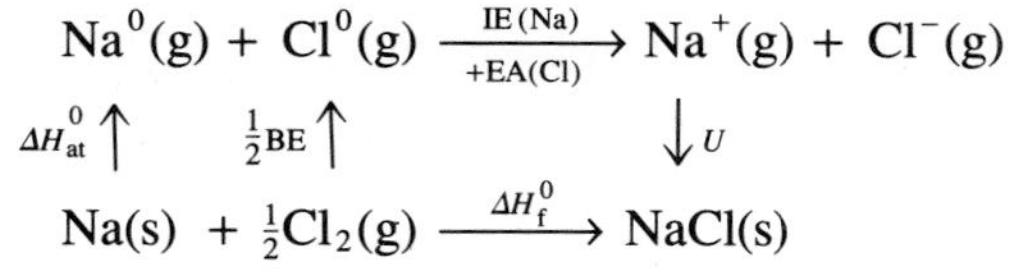

$$Na^0(g) + Cl^0(g) \xrightarrow[+EA(Cl)]{IE(Na)} Na^+(g) + Cl^-(g)$$

$$\Delta H_{at}^0 \uparrow \qquad \tfrac{1}{2}BE \uparrow \qquad\qquad\qquad \downarrow U$$

$$Na(s) + \tfrac{1}{2}Cl_2(g) \xrightarrow{\Delta H_f^0} NaCl(s)$$

Here ΔH_{at}^0 is the enthalpy of atomization of Na metal, BE is the bond energy of Cl_2 gas, IE is the ionization energy of Na, EA is the electron affinity of Cl, and ΔH_f^0 is the enthalpy of formation of solid NaCl. This diagram is called a *Born–Haber cycle*. It is equivalent to the following Hess's law summation:

$$Na(g) \xrightarrow{-\Delta H^0_{at}} Na(s)$$

$$Na^+(g) \xrightarrow{-IE} Na(g)$$

$$Cl(g) \xrightarrow{-\frac{1}{2}BE} \tfrac{1}{2}Cl_2(g)$$

$$Cl^-(g) \xrightarrow{-EA} Cl(g)$$

$$\underline{Na(s) + \tfrac{1}{2}Cl_2(g) \xrightarrow{\Delta H^0_f} NaCl(s)}$$

$$Na^+(g) + Cl^-(g) \xrightarrow{U} NaCl(s)$$

From this summation, we have $U = \Delta H^0_f - \Delta H^0_{at}(Na) - \tfrac{1}{2}BE(Cl_2) -$ IE(Na) $-$ EA(Cl). Each of the values on the right can be measured, so that there is a direct experimental equivalence for U, the lattice energy: $U = -411.0 - 108.7 - \tfrac{1}{2}(239.7) - 495.8 + 348.5 = -786.9$ kJ/mol. This is not a bad match for the theoretical value we calculated previously of -757.0 kJ/mol, especially considering that we omitted induced-dipole effects and the zero-point energy. The difference suggests for the solid (as it did for the gas) that there is a small contribution to the total energy from covalent bonding of the atoms in the lattice. The difference to be made up is only about half as big for the crystal, however, which suggests that the larger number of neighbors in the crystal tends to favor electron transfer and ionic bonding.

Any Hess's law summation written to use energies for processes involving individual atoms is usually called a Born–Haber cycle. Usually a Born–Haber cycle can be set up easily by writing the desired laboratory reaction as the bottom line, writing an equivalent gas-phase reaction featuring individual ions or atoms above the laboratory reaction, and using vertical reaction arrows to indicate the transitions to and from the gas-phase, isolated-atom reaction. An important component of many Born–Haber cycles is the atomization energy for various elements, which is equivalent to the heat of sublimation for many solids and to half the bond energy for diatomic gases. Table 2.2 gives values for the enthalpy of formation of gaseous atoms from the elements in their standard states.

It is important to realize that numerical agreement between calculated lattice energies and Born–Haber cycle experimental energies does not constitute proof that the substance is ionic. There is no way, conceptually, to distinguish the previous cycle from the following one, in which only partial electron transfer occurs from the metal M to the halogen X and both electrostatic attraction and covalent bonding occur between the partially charged atoms.

$$Na^0(g) + Cl^0(g) \xrightarrow{\substack{\int_0^{+q} I(q)dq(Na) \\ +\int_0^{-q} I(q)dq(Cl)}} Na^{q+}(g) + Cl^{q-}(g)$$

$$\Delta H^0_{at}\uparrow \qquad \tfrac{1}{2}BE\uparrow \qquad\qquad\qquad\qquad \downarrow U_{ionic} + BE_{covalent}$$

$$Na(s) + \tfrac{1}{2}Cl_2(g) \xrightarrow{\Delta H^0_f} NaCl(s)$$

Table 2.2 Enthalpies of Atomization (ΔH_f^0, Gaseous Atoms) (in kJ/mol atoms at 298 K)

1	2	3	4	5	6	7	8	9	10	11	12	13	14	15	16	17	18
H 218																	**He** 0
Li 162	**Be** 324											**B** 563	**C** 717	**N** 473	**O** 249	**F** 79	**Ne** 0
Na 107	**Mg** 148											**Al** 326	**Si** 456	**P** 315	**S** 279	**Cl** 122	**Ar** 0
K 90	**Ca** 178	**Sc** 389	**Ti** 471	**V** 514	**Cr** 397	**Mn** 281	**Fe** 416	**Co** 425	**Ni** 430	**Cu** 338	**Zn** 131	**Ga** 277	**Ge** 377	**As** 303	**Se** 227	**Br** 112	**Kr** 0
Rb 81	**Sr** 164	**Y** 431	**Zr** 611	**Nb** 772	**Mo** 658	**Tc** 678	**Ru** 643	**Rh** 557	**Pd** 378	**Ag** 285	**Cd** 112	**In** 244	**Sn** 302	**Sb** 262	**Te** 197	**I** 107	**Xe** 0
Cs 79	**Ba** 180	**La** 368	**Hf** 607	**Ta** 782	**W** 849	**Re** 770	**Os** 791	**Ir** 665	**Pt** 565	**Au** 366	**Hg** 62	**Tl** 182	**Pb** 195	**Bi** 207	**Po** 142	**At**	**Rn** 0
Fr	**Ra**	**Ac**															

Ce	**Pr**	**Nd**	**Pm**	**Sm**	**Eu**	**Gd**	**Tb**	**Dy**	**Ho**	**Er**	**Tm**	**Yb**	**Lu**
													U 523

There is no satisfactory way to calculate what the covalent contribution to the overall bonding should be as a function of charge, so we cannot make a numerical comparison between the two cycles. However, the assumption of perfect ionicity in the usual Born–Haber cycle should be recognized as an approximation. Even so, lattice energies and Born–Haber cycles provide us with useful approximations for the thermodynamic energies of many processes involving compounds that can reasonably be thought of as ionic.

Lattice energies are particularly helpful in understanding the stability of oxidation states in ionic compounds. For example: Why, thermodynamically, does the following reaction not occur?

$$NaCl(s) + \tfrac{1}{2}Cl_2(g) \rightarrow NaCl_2(s)$$

This is equivalent to asking what limits the oxidation state of Na to +1. ΔS for the given reaction will be negative, because the reacting system is becoming more orderly. Only if ΔH is negative can ΔG be negative and the process spontaneous. We can calculate an approximate value of ΔH using a Born–Haber cycle in which the two lattices are disassembled into their component ions:

$$Cl^0(g) + Na^+(g) + Cl^-(g) \xrightarrow[+EA(Cl)]{IE_2(Na)} Na^{2+}(g) + 2Cl^-(g)$$

$$\tfrac{1}{2}BE \uparrow \qquad\quad U_1 \uparrow \qquad\qquad\qquad\quad \downarrow U_2$$

$$\tfrac{1}{2}Cl_2(g) \quad + \quad NaCl(s) \xrightarrow{\Delta H^0} NaCl_2(s)$$

Using this cycle, we have $\Delta H = \tfrac{1}{2}BE(Cl_2) + EA(Cl) + IE_2(Na) + U_2(NaCl_2) - U_1(NaCl)$. All of these numbers except U_2 are readily available for the unknown lattice of the hypothetical $NaCl_2$. Kapustinskii's lattice-energy expression allows us to calculate U_2, however, if we estimate that the ionic radius of Na^{2+} is the same as that of Mg^{2+}, or 0.86 Å for a coordination number of six.

$$U_2 = -\frac{1202 v z_C z_A}{r_C + r_A}\left(1 - \frac{0.345}{r_C + r_A}\right) = -\frac{1202 \times 3 \times 2 \times 1}{0.86 + 1.67}\left(1 - \frac{0.345}{0.86 + 1.67}\right)$$

$$= -2461 \text{ kJ/mol}$$

Using the Born–Haber cycle value for U_1, we have $\Delta H = \tfrac{1}{2}(240) - 349 + 787 + 4563 - 2461 = +2660$ kJ/mol. We do not have to look far to see the reason for this very unfavorable enthalpy change: The second-ionization energy of sodium is too high a price for the chemical surroundings of the sodium ion to pay.

We can ask the same kind of question in reverse. Alkaline earth metals (Group IIa) normally show only the +2 oxidation state. If we could make, for instance, the +1 salt CaCl, should it disproportionate to Ca^0 and $CaCl_2$?

$$2CaCl(s) \rightarrow Ca^0(s) + CaCl_2(s)?$$

ΔS for this reaction will be very small (2 moles solid = 2 moles solid), and it would have to be driven by a negative ΔH. We can, again, set up to a Born–Haber cycle for the reaction:

$$2Ca^{+}(g) + 2Cl^{-}(g) \xrightarrow[-IE_1]{IE_2} Ca^{0}(g) + Ca^{2+}(g) + 2Cl^{-}(g)$$

$$\Big\uparrow{\scriptstyle -2U_1} \qquad\qquad \Big\downarrow{\scriptstyle \Delta H_{at}^{0}} \qquad\qquad \Big\downarrow{\scriptstyle U_2}$$

$$2CaCl(s) \xrightarrow{\Delta H^{0}} Ca(s) + \qquad CaCl_2(s)$$

Now $\Delta H = U_2(CaCl_2) + IE_2(Ca) - IE_1(Ca) - \Delta H_{at}^{0}(Ca) - 2U_1(CaCl)$, and we can estimate both lattice energies using Kapustinskii's expression. Using the true ionic radius for Ca^{2+}, and assuming it to be equal to that of K^{+} for the chemically hypothetical Ca^{+}, we have

$$CaCl$$

$$U_1 = -\frac{1202 \times 2 \times 1 \times 1}{1.52 + 1.67}\left(1 - \frac{0.345}{1.52 + 1.67}\right)$$

$$= -672 \text{ kJ/mol}$$

$$CaCl_2$$

$$U_2 = -\frac{1202 \times 3 \times 2 \times 1}{1.14 + 1.67}\left(1 - \frac{0.345}{1.14 + 1.67}\right)$$

$$= -2251 \text{ kJ/mol}$$

ΔH for the disproportionation reaction is thus $-2251 + 1145 - 590 - 178 - 2(-672) = -530$ kJ/mol. This large negative value indicates that we should never see the intermediate oxidation state Ca(I) unless there is some very large kinetic barrier to its disproportionation.

These two cases reveal that, for ionic compounds, the stability of a given oxidation state is determined by the energy balance between the lattice energy (which becomes more favorable as net charge on the cation increases) and the ionization energy (which represents an increasingly great energy cost as the net charge on the cation increases). Figure 2.5 indicates this relationship for some elements in the second row of the periodic table and for a fairly typical transition metal, Mn. For each element, the curves shown are similar to those of Figure 1.15, but a discontinuity is shown for each element when the electrons being removed change from valence electrons to inner-core electrons. The lattice-energy curve follows the points calculated from Kapustinskii's expression using average radii for each charge. The U value for zero charge is the negative of the enthalpy of atomization for the cation elements. Although the figure is an oversimplification of the Born–Haber cycle, it can be seen that the stability of an oxidation state is determined by whether the lattice-energy of its crystalline environment is great enough to exceed the ionization-energy cost of forming the ion. The slope (which for an

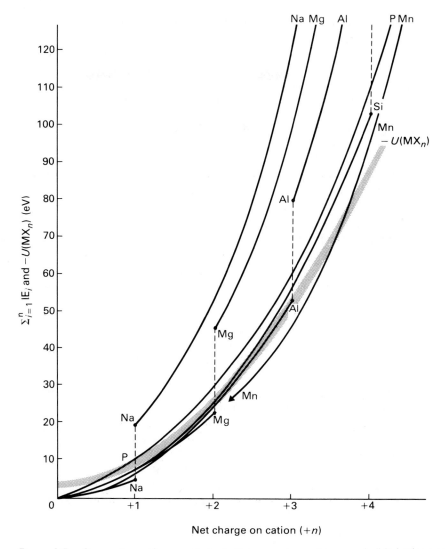

Figure 2.5 Comparison of aggregate ionization energies and average halide lattice energies, both as a function of cation charge.

element is its differential ionization energy) is also important. If at a given cation charge the slope of the element's curve is less than that of the lattice-energy curve, that oxidation state of the cation will be unstable toward disproportionation in its crystalline environment. Thus although Mg^+ lies below the lattice-energy curve and is therefore presumably stable, the system can become even more stable by disproportionating, just as CaX does.

The Si(IV) halides cannot be ionic, because the lattice energy cannot meet the enthalpy cost of ionizing Si to Si^{4+}. In fact, all SiX_4 compounds are liquids or gases at room temperature except SiI_4, which is a low-melting solid. Apparently no phosphorus compound at any positive oxidation state of P can be ionic, which also agrees with our experience. Furthermore, although the most stable oxidation state of Mn appears to be Mn^{3+}, which is somewhat unfamiliar, MnF_3 is synthesized by fluorinating $MnCl_2$ but MnF_4 spontaneously decomposes to MnF_3 and F_2.

2.3 ION SIZES IN CRYSTALS

Within a very few years after the first accurate determination of internuclear distances in ionic crystals, W. L. Bragg, who was one of the first x-ray crystallographers, proposed a set of atomic radii for ionic and metallic crystals. When added together, these radii reproduced experimental distances to about ± 0.06 Å on the assumption that all atoms could be regarded as hard spheres with a constant radius in any environment. This proved to be an oversimplification, but in 1926 V. M. Goldschmidt produced a widely used set of ionic radii based on an assumed radius for the oxide ion of 1.32 Å. A year later, Linus Pauling proposed another set based on a scaling of the sizes of isoelectronic ions (such as Na^+ and F^-, which have the same number of electrons) according to their effective nuclear charges. Some assumption of this sort is necessary if the only available experimental information is a set of internuclear distances; otherwise there is no unique way to apportion the internuclear distance between the cation and the anion. For example, from the information below on internuclear separations an assumed radius for Li^+ of 1.000 Å leads to $r_{F^-} = 1.01$, $r_{Cl^-} = 1.57$, $r_{Br^-} = 1.75$, and $r_{I^-} = 2.03$ Å. These values in turn lead to calculated values for r_{Na^+} of 1.30, 1.24, 1.23, and 1.20 Å for the four sodium halides. These values seem reasonably consistent, but equal consistency can be had by assuming $r_{Li^+} = 2.000$ Å. The halide radii are then 0.01, 0.57, 0.75, and 1.03 Å, and the calculated r_{Na^+} would then be 2.30, 2.24, 2.23, and 2.20 Å. Neither of these sets is accurate, but it can be seen that they are equally consistent.

LiF	LiCl	LiBr	LiI
2.009	2.566	2.747	3.025 Å
NaF	NaCl	NaBr	NaI
2.307	2.814	2.981	3.231 Å

Goldschmidt's values of ionic radii rested on the assumption that the molar refractivities of the ions in alkali halides MX and alkaline-earth oxides MO could be apportioned on the basis of ionic volume, which led to $r_{F^-} = 1.33$ Å and $r_{O^{2-}} = 1.32$ Å. On the other hand, Pauling's values apportioned the distance for isoelectronic ions as follows:

$$\frac{r_{Na^+}}{r_{F^-}} = \frac{Z_{eff}(F^-)}{Z_{eff}(Na^+)}$$

Using Slater effective nuclear charges and taking the experimental internuclear distance to be the sum of the two radii, this yields

$$\frac{2.307 - r_{F^-}}{r_{F^-}} = \frac{4.55}{6.55}$$

$$r_{F^-} = 1.36 \text{ Å}$$

By a similar calculation, $r_{O^{2-}} = 1.40$ Å. Remembering (from Fig. 1.17) the extreme electronegativities of O and F, it seems likely that oxide and fluoride crystals should be most nearly ionic and should therefore have the most reliable ionic radii as anions. The two sets of radii based on these r_{F^-} and $r_{O^{2-}}$ values found extensive use, but some adjustments were necessary.

The most important adjustment to tabulated ionic radii allowed the radius of a cation to vary with coordination number. Cations are in general smaller than anions, and the more anions there are near a given cation, the more their repulsion for each other will prevent them from approaching the cation as closely as in a gaseous ion pair. On this basis, we expect to see the quoted radius for a cation increase with increasing coordination number. From the form of the repulsion potential in the lattice-energy expression, we can approximate that when a cation is four-coordinate its radius is related to that of the same cation in a six-coordinate lattice by

$$\frac{r_4}{r_6} = \left(\frac{4}{6}\right)^{\rho/r_0 - \rho}$$

For other coordination numbers a similar expression holds, where r_0 is the predicted distance between ions for six-coordination. Tables for ionic radii normally refer to six-coordination for all ions, and a simple rule of thumb says that a four-coordinate cation is 0.95 as large as the tabulated value, while an eight-coordinate ion has a radius 1.04 times as large as the tabulated value.

The crystal radii given in Table 2.3 are a recent compilation by Shannon based on an analysis of high-resolution x-ray studies of alkali halide crystals that shows that the minimum of electron density occurs at a spot along the internuclear axis corresponding to smaller fluoride and oxide ions (and correspondingly larger cation). These radii, based on $r_{F^-} = 1.19$ Å and $r_{O^{2-}} = 1.26$ Å, are about 0.14 Å larger for cations and the same amount smaller for anions than the traditional Pauling values. However, they are probably a truer picture of sphere packing in crystals. They predict internuclear distances equally well. Table 2.3 is based on an exhaustive analysis of x-ray data for different structures, so the radii given for different coordination numbers are

Table 2.3   Ionic Radii (Å)

Ion	Coord. number	Crystal radius	SPI radius	Ion	Coord. number	Crystal radius	SPI radius
Ag^+	4	1.14		Co^{2+}	4HS	0.72	
	4SQ	1.16			5	0.81	
	5	1.23			6LS	0.79	0.98
	6	1.29	1.24		6HS	0.885	
Al^{3+}	4	0.53			8	1.04	
	5	0.62		Co^{3+}	6LS	0.685	0.97
	6	0.675	0.92		6HS	0.75	
As^{3+}	6	0.72		Cr^{2+}	6LS	0.87	1.08
As^{5+}	4	0.475			6HS	0.94	
	6	0.60	0.88	Cr^{3+}	6	0.755	0.975
Au^+	6	1.51	1.26	Cr^{4+}	4	0.55	
Au^{3+}	4SQ	0.82			6	0.69	0.91
	6	0.99		Cr^{6+}	4	0.40	
B^{3+}	3	0.15			6	0.58	0.80
	4	0.25		Cs^+	6	1.81	1.68
	6	0.41			8	1.88	
Ba^{2+}	6	1.49	1.40	Cu^+	2	0.60	
	7	1.52			4	0.74	
	8	1.56			6	0.91	1.10
Be^{2+}	3	0.30		Cu^{2+}	4	0.71	
	4	0.41			4SQ	0.71	
	6	0.59	0.69		5	0.79	
Bi^{3+}	6	1.17			6	0.87	1.00
	8	1.31		Cu^{3+}	6LS	0.68	
Br^-	6	1.82	1.79	F^-	2	1.145	
Ca^{2+}	6	1.14	1.26		3	1.16	
	8	1.26			4	1.17	
Cd^{2+}	4	0.92			6	1.19	1.12
	5	1.01		Fe^{2+}	4HS	0.77	
	6	1.09	1.20		4SQ HS	0.78	
	7	1.17			6LS	0.75	1.04
	8	1.24			6HS	0.920	
Ce^{3+}	6	1.15	1.17		8HS	1.06	
	8	1.283		Fe^{3+}	4HS	0.63	
Ce^{4+}	6	1.01	1.06		6LS	0.69	1.00
	8	1.11		Fe^{4+}	6	0.725	0.91
Cl^-	6	1.67	1.64	Fe^{6+}	4	0.39	

Notes: The notation "4SQ" in the coordination-number column refers to the square-planar coordination; "3PY" refers to pyramidal coordination; "3" alone is trigonal planar; "4" alone is tetrahedral; and other numbers do not differentiate between different geometries for each coordination number.

Table 2.3   (*Continued*)

Ion	Coord. number	Crystal radius	SPI radius	Ion	Coord. number	Crystal radius	SPI radius
Fr^+	6	1.94		Mn^{2+}	4HS	0.80	
Ga^{3+}	4	0.61			5HS	0.89	
	6	0.760	1.00		6LS	0.81	1.09
Ge^{2+}	6	0.87			6HS	0.970	
Ge^{4+}	4	0.530			7HS	1.04	
	6	0.670	0.90		8	1.10	
H^+	1	−0.24		Mn^{3+}	5	0.72	
	2	−0.04			6LS	0.72	0.99
H^-	4	1.22			6HS	0.785	
	6	1.40	1.18	Mn^{4+}	4	0.53	
Hf^{4+}	4	0.72			6	0.670	0.90
	6	0.85	1.04	Mo^{3+}	6	0.83	1.12
Hg^+	3	1.11		Mo^{4+}	6	0.790	1.01
	6	1.33		Mo^{5+}	4	0.60	
Hg^{2+}	2	0.83			6	0.75	
	4	1.10		Mo^{6+}	4	0.55	
	6	1.16	1.34		6	0.73	0.87
	8	1.28		N^{3-}	4	1.32	
I^-	6	2.06	2.02	Na^+	4	1.13	
In^{3+}	4	0.76			5	1.14	
	6	0.940	1.16		6	1.16	1.18
Ir^{3+}	6	0.82	1.10		7	1.26	
Ir^{4+}	6	0.765	1.05		8	1.32	
Ir^{5+}	6	0.71		Nb^{3+}	6	0.86	1.125
K^+	4	1.51		Nb^{4+}	6	0.82	1.015
	6	1.52	1.45	Nb^{5+}	4	0.62	
	7	1.60			6	0.78	0.92
	8	1.65			8	0.88	
La^{3+}	6	1.172	1.20	Ni^{2+}	4	0.69	
	8	1.300			4SQ	0.63	
Li^+	4	0.730			5	0.77	
	6	0.90	0.92		6	0.830	0.94
	8	1.06		Ni^{3+}	6LS	0.70	0.93
Lu^{3+}	6	1.001	1.12		6HS	0.74	
	8	1.117		Ni^{4+}	6LS	0.62	0.91
Mg^{2+}	4	0.71		O^{2-}	2	1.21	
	5	0.80			3	1.22	
	6	0.860	1.02		4	1.24	
	8	1.03			6	1.26	1.16
					8	1.28	

Notes: The notation "4SQ" in the coordination-number column refers to the square-planar coordination; "3PY" refers to pyramidal coordination; "3" alone is trigonal planar; "4" alone is tetrahedral; and other numbers do not differentiate between different geometries for each coordination number.

Table 2.3   (*Continued*)

Ion	Coord. number	Crystal radius	SPI radius	Ion	Coord. number	Crystal radius	SPI radius
Os^{4+}	6	0.770	1.045	Tc^{4+}	6	0.785	1.01
Pb^{2+}	6	1.33	1.55	Te^{2-}	6	2.07	1.92
PB^{4+}	6	0.915	1.12	Th^{4+}	6	1.08	1.14
Pd^{2+}	4SQ	0.78			8	1.19	
	6	1.00	1.175		9	1.23	
Pd^{3+}	6	0.90	1.11	Ti^{2+}	6	1.00	1.14
Pd^{4+}	6	0.755	1.00	Ti^{3+}	6	0.810	1.03
Pt^{2+}	4SQ	0.74		Ti^{4+}	4	0.56	
	6	0.94	1.175		6	0.745	0.94
Pt^{4+}	6	0.765	1.05	Tl^{+}	6	1.64	1.60
Ra^{2+}	8	1.62		Tl^{3+}	4	0.89	
Rb^{+}	6	1.66	1.56		6	1.025	1.2
	7	1.70		U^{3}	6	1.165	1.20
	8	1.75		U^{4+}	6	1.03	1.1
Re^{4+}	6	0.77	1.045		8	1.14	
Rh^{3+}	6	0.805	1.10		9	1.05	
Rh^{4+}	6	0.74	1.00	V^{2+}	6	0.93	1.07
Rh^{5+}	6	0.69		V^{3+}	6	0.780	1.00
Ru^{3+}	6	0.82	1.105	V^{4+}	6	0.72	0.925
Ru^{4+}	6	0.760	1.005	V^{5+}	4	0.495	
Ru^{5+}	6	0.705			6	0.68	0.84
S^{2-}	6	1.70	1.58	W^{4+}	6	0.80	1.045
Sb^{3+}	4PY	0.90		W^{5+}	6	0.76	
	5	0.94		W^{6+}	4	0.56	
	6	0.90			5	0.65	
Sc^{3+}	6	0.885	1.06		6	0.74	0.88
	8	1.010		Xe^{8+}	4	0.54	
Se^{2-}	6	1.84	1.74		6	0.62	
Si^{4+}	4	0.40		Y^{3+}	6	1.040	1.14
	6	0.540			8	1.159	
Sn^{2+}	8	1.36		Zn^{2+}	4	0.74	
Sn^{4+}	4	0.69	1.03		5	0.82	
	6	0.830			6	0.88	1.07
Sr^{2+}	6	1.32	1.38	Zr^{4+}	4	0.73	
	7	1.35			6	0.86	1.02
	8	1.40			7	0.92	
Ta^{3+}	6	0.86	1.10		8	0.98	
Ta^{4+}	6	0.82	1.04				
Ta^{5+}	6	0.78	0.95				
	8	0.88					

Notes: The notation "4SQ" in the coordination-number column refers to the square-planar coordination; "3PY" refers to pyramidal coordination; "3" alone is trigonal planar; "4" alone is tetrahedral; and other numbers do not differentiate between different geometries for each coordination number.

Table 2.3   *(Continued)*

Ion	Coord. number	Crystal radius	SPI radius	Ion	Coord. number	Crystal radius	SPI radius
Polynuclear ions				**Polynuclear ions**			
BO_3^{3-}		1.77		IO_4^-		2.35	
BF_4^-		2.14	2.12	NH_2^-		1.16	
BrO_3^-		1.77		NO_2^-		1.41	
CH_3COO^-		1.45		NO_3^-		1.75	2.0
CN^-		1.68	1.72	O_2^{2-}		1.66	
CNO^-		1.45		OH^-	2	1.21	
CNS^-		1.81			3	1.20	
CO_3^{2-}		1.71			4	1.21	
ClO_3^-		1.86			6	1.23	
ClO_4^-		2.22	2.26	PO_4^{3-}		2.24	
$HCOO^-$		1.44		SH^-		1.81	1.84
HCO_3^-		1.49		SO_4^{2-}		2.16	
IO_3^-		1.68					

Notes: The notation "4SQ" in the coordination-number column refers to the square-planar coordination; "3PY" refers to pyramidal coordination; "3" alone is trigonal planar; "4" alone is tetrahedral; and other numbers do not differentiate between different geometries for each coordination number.

experimental, not calculated. Shannon has also included for O^{2-} and F^- the effect of coordination number on anion radii, and has further included the effect of differing spin states for transition-metal cations, to which we shall return in Chapter 11. Some values for oxoanions and fluoroanions are included, even though those ions are nonspherical and thus do not really have a uniquely defined radius. These are adjusted from a table by Yatsimirskii; they are useful in the sense that they permit lattice-energy calculations with these ions. For this reason these radii are sometimes called *thermochemical radii*.

The SPI radii given in Table 2.3 represent an interesting alternative approach to the question of apportioning internuclear distance into ionic radii, proposed by Johnson. If the minimum electron density along the internuclear axis is taken as the boundary between cation and anion, the cation radii that result for ions with noble-gas electron configurations (such as Na^+, Mg^{2+}) show an almost constant relation to the metallic radius of the same element, which is half the internuclear distance in the metal:

$$r_{M^{n+}} = 0.64 r_{M^0}$$

Such radii can thus be thought of as the radii of ion cores in a spherically symmetric potential field of the electrons in the metal, or as spherical potential ion (SPI) radii. Since cations are much less polarizable than anions, these metal ion SPI radii are taken to be constant regardless of the anion surroundings. The anions, on the other hand, are fairly polarizable and are squeezed closer to the cations in the crystals whose cations have a high polarizing power (high positive charge, small radius), which is roughly equivalent to a high Allred–Rochow electronegativity. The anion SPI radii in Table 2.3 are those quoted for the Na^+ salt, but the Cl^- ion, for example, shrinks from a radius of 1.87 Å in CsCl to a radius of 1.25 Å in CuCl. The contraction is less severe if only cations with noble-gas configurations are considered. Although this approach is not as convenient as Shannon's extensive table for predicting internuclear distances in crystals, it offers more insight into the electronic processes that accompany crystal formation.

2.4 PACKING SYMMETRIES FOR SPHERICAL IONS

The question of determining or predicting crystal symmetries involves not only establishing the sizes of the spherical ions involved, but also establishing the symmetry with which they pack together. To begin with, let us limit our consideration to arrays of spheres that are all the same size. If we did not have ionic attractions to consider, the spheres would still attract each other through induced-dipole forces and, in the absence of thermal agitation, would tend to move together to the bottom of the Lennard–Jones potential well shown in Fig. 1.20. In a one-dimensional array, this would simply lead to a row of touching spheres, where "touching" refers to establishing an internuclear separation that makes attraction and repulsion equal. However, if the spheres are charged, the very strong electrostatic force requires that positive and negative ions alternate along the row. They would still touch, but the stronger attraction would mean that the internuclear separation would be smaller than for uncharged spheres.

In two dimensions, uncharged spheres will tend to pack in such a way as to maximize the number of touching neighbor spheres (the coordination number), since that produces the greatest stabilization of the array through van der Waals attractions. Of the two possible arrays shown in Fig. 2.6, that with square packing is less favorable than that with triangular (hexagonal) packing, because the latter has more near-neighbor spheres. It should also be noted that triangular packing (also called *close packing*) leads to a higher density for the bulk solid, since the same number of spheres cover a smaller area. Accordingly, square packing is extremely rare in nature (α-polonium is reported to have a three-dimensional form of it), but hexagonal packing is quite common in planes of atoms in most metals and noble-gas crystals.

For layers of ions, however, the packing situation is more complicated, because stability requires that ions with like charge not touch. If the stoichiometry of the ionic compound is 1:1, as in NaCl or MgO, square packing allows alternating charges, both within the layer and from one layer

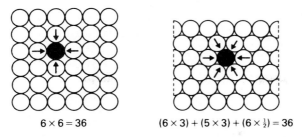

$6 \times 6 = 36$ $(6 \times 3) + (5 \times 3) + (6 \times \frac{1}{2}) = 36$

Figure 2.6   Two-dimensional packing symmetries for uncharged spheres.

to the next, and is thus commonly observed. The hexagonal packing, however, forces neighboring ions to have like charges and is correspondingly less stable. If the stoichiometry of the ionic compound is 2:1, as in CaF_2 and Na_2O, some neighbor repulsions of the 1+ or 1− ions can be accepted as long as there are none between the 2+ or 2− ions. A layer of this structure is shown in Fig. 2.7. This is a hexagonally packed or close-packed layer with half of the positions that might be occupied by 2+ or 2− ions vacant, which eliminates the 2+/2+ repulsions and preserves the stoichiometry. There is also a close-packed layer structure with no vacancies that preserves the MX_2 stoichiometry, as shown in Fig. 2.8, but it is not displayed by any strongly ionic compound. $TiSi_2$ and $CrSi_2$ have this structure; presumably the small charges on the atoms in these substances prevent the repulsions from being

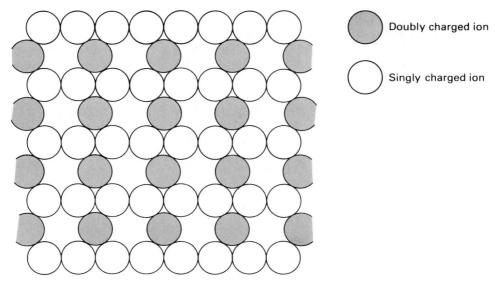

Doubly charged ion

Singly charged ion

Figure 2.7   Sphere packing in a layer of the CaF_2 or Na_2O lattices.

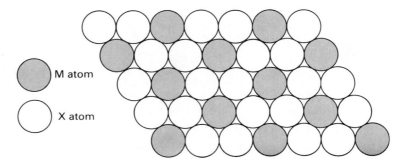

Figure 2.8   MX$_2$ layer structure in partly ionic crystals with low charges.

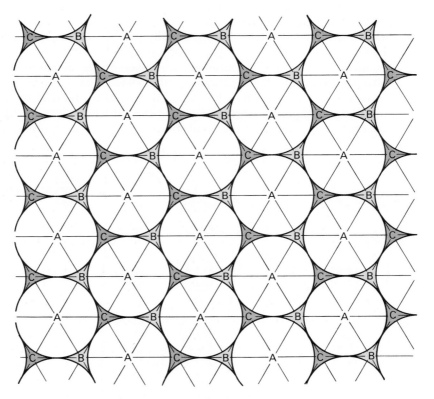

Figure 2.9   Sphere-packing positions in a triangular network.

too severe. The next chapter will consider in greater detail favorable structures for crystals that are only partly ionic.

Close packing is still advantageous in a three-dimensional structure. Even
if all atoms are identical and uncharged, there are still two ways to construct a
close-packed lattice, and many more ways when those constraints are removed. Figure 2.9 shows a triangular network on which a close-packed layer
can be constructed—we will create a three-dimensional close-packed structure by stacking layers that are themselves close-packed. In that figure, there
are three distinct symmetry sets of intersections of the triangular network,
labeled A, B, and C. If we place the centers of spheres at the A positions, a
close-packed layer results—spheres cannot be squeezed in at the B and C
positions. An all-B layer or an all-C layer is equally close-packed, however.
When we place a second layer on top of the first, each atom in the second
layer will establish a maximum coordination number by centering the atoms
of the layer at either the B positions (lightly shaded) or the C positions
(heavily shaded), so either of those symmetries will be favored over, for
example, direct superposition of an A layer over an A layer. If the first two
layers are, say, AB, then the third layer can be either A or C and maintain
maximum coordination number in either case. Random orders are not usually
found: Most crystals are either ABABABABAB . . . or ABCABCABC . . . .
Figure 2.10 shows several layers of each of these. ABABAB is called *hexagonal close-packing*, while ABCABC is called *cubic close-packing*.

As for the two-dimensional case, if the atoms are charged ions of equal
size the close-packed lattices are not stable because the high coordination
number (12) forces too many repulsions between touching ions with like
charge. We shall return shortly to examine some of the commonly found
lattices, but we need first to consider the effects of allowing cations and anions
to have different sizes in the lattice. We have already seen that in general
anions tend to be larger than cations, the more so the higher the charge on the
cation. If the difference is great enough, the cation can fit in the holes, or
interstices, in a lattice composed of close-packed anions. Such a cation is said
to be *interstitial*. Many ionic or partly ionic compounds form an approximation
to this type of lattice.

In such a case, the close-packed ions all have the same charge and have a
large repulsion potential as a result—but its effect is overcome by the attraction exerted by cations regularly disposed in the open space within the anion
lattice. If the lattice is an ideal close-packed array of spheres, that open space
takes the form of one octahedral and two tetrahedral lattice sites for each ion
in the array. The terms "octahedral" and "tetrahedral" refer to the geometric
arrangement of the nuclei of the ions surrounding the site, not to the shape of
the hole itself. Each anion has a tetrahedral site above and below it in the
lattice, because it touches three other ions in the layer below it and in the
layer above it. If the ion is in, say, an A layer, there will be tetrahedral sites
between the A layer and the B or C layer above or below it. In addition, there

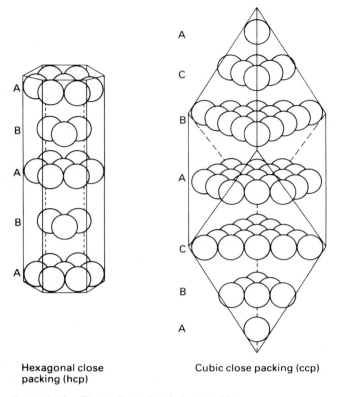

Hexagonal close   Cubic close packing (ccp)
packing (hcp)

Figure 2.10 Three-dimensional close packing.

will be an octahedral site at the C position between an A and B layer or at the B position between an A and C layer.

The geometry of these sites is indicated in Fig. 2.11, from which we can deduce the limits on the size of cations that occupy such sites. It is clear that a cation has to be smaller to occupy a tetrahedral site than to occupy an octahedral one. The limiting radius ratio r/R (where r is the radius of the small interstitial cation and R the radius of the large close-packed anion) will be larger the greater the coordination number of the cation; for cubic eight-coordinate cations, the greatest possible radius ratio is 0.7321. Such coordination is not found in close-packed lattices, but it is seen in the CsCl and CaF_2 lattices.

Pauling, using his original set of ionic radii, proposed radius-ratio rules under which the ratios 0.225, 0.414, and 0.732 represented *lower* limits (so that the anions would be forced apart by the cations, reducing repulsion) for four-, six-, and eight-coordination, respectively. His crystal-structure rules thus predicted that a cation would be four-coordinate if its radius ratio r/R

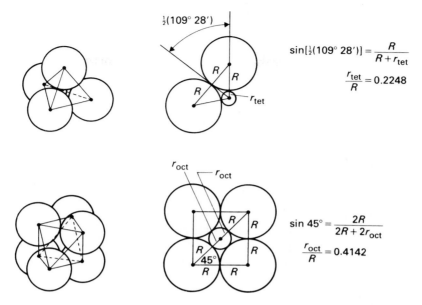

$$\sin[\tfrac{1}{2}(109° 28')] = \frac{R}{R + r_{tet}}$$

$$\frac{r_{tet}}{R} = 0.2248$$

$$\sin 45° = \frac{2R}{2R + 2r_{oct}}$$

$$\frac{r_{oct}}{R} = 0.4142$$

Figure 2.11   Geometry of tetrahedral and octahedral sites in a close-packed lattice.

were between about 0.2 and 0.4, six-coordinate if r/R were between 0.3 and 0.7, and eight-coordinate if r/R were above about 0.75. However, this is essentially a hard-sphere model in which covalent bonding plays no role. On examining the radii in Table 2.3, it will be seen that for any system in which the cation is small enough and the anion large enough to yield $r/R < 0.414$, the electronegativity difference between the two atoms is small enough that covalent bonding must be a significant part of the lattice energy. Because covalent bonding, unlike ionic bonding, is directional, it is likely that the arrangement of "anions" around the "cation" is determined by the orbital-overlap requirements of covalent bonding rather than by ion-size fitting to available sites. Either set of radii in Table 2.3 fails in a number of cases to meet Pauling's radius-ratio rules, usually by predicting a larger coordination number for the cation than is actually observed. For example, consider the four compounds LiBr, NaBr, KBr, and CsBr, using SPI radii:

LiBr	NaBr	KBr	CsBr
$\dfrac{r}{R} = \dfrac{0.92}{1.79}$	$\dfrac{r}{R} = \dfrac{1.18}{1.79}$	$\dfrac{r}{R} = \dfrac{1.45}{1.79}$	$\dfrac{r}{R} = \dfrac{1.68}{1.79}$
$= 0.514$	$= 0.659$	$= 0.810$	$= 0.939$

Pauling's rules predict that LiBr and NaBr should be six-coordinate, while KBr and CsBr should be eight-coordinate. These predictions are correct except for KBr, which is six-coordinate in the NaCl lattice. On the other hand, Pauling's original radii predicted that LiBr should have been four-coordinate, but it too has the six-coordinate NaCl lattice. We must conclude that the factors influencing the choice of crystal-lattice symmetry by an ionic compound are too numerous and subtle to be covered by a simple rule.

Several lattice symmetries are sufficiently common to deserve individual discussion. The most familiar is that of NaCl (see Fig. 2.4). The NaCl symmetry is adopted by all the alkali halides except CsCl, CsBr, and CsI, by all of the silver halides except AgI, by most of the $M^{2+}O^{2-}$ monoxides, by the alkaline earth monosulfides except BeS, by many transition-metal monosulfides, and by a wide variety of interstitial carbides, nitrides, phosphides, and other similar compounds. The NaCl lattice can be thought of either as simple cubic (square) packing with alternating charges, or as cubic close-packing (ccp) of the larger anions, pushed apart somewhat by cations that are too large to fit in the octahedral holes. The NaCl lattice shows a very high degree of symmetry. In addition to the compounds that adopt it in full symmetry, there are many that show essentially the same structure, but with reduced symmetry. For example, the FeS_2 (pyrites) structure is that of NaCl with the nonspherical S_2^{2-} ion disposed at an angle to the Fe–Fe axis, but centered on that axis as a Cl^- would be. The CaC_2 structure, which is also adopted by several peroxides and superoxides, also has the C_2^{2-} ion at the Cl^- positions, but symmetrically arranged along the Ca–Ca axis. There are also several defect structures in metal oxides and sulfides that resemble NaCl with fractions of the cation positions vacant, either on a random basis in non-stoichiometric crystals, or on an ordered basis in compounds such as Sc_2S_3 or NbO, which also has an equal number of anion vacancies. The next chapter will consider defect structures on a more comprehensive basis.

The CsCl structure, shown in Fig. 2.12, is the only other structure

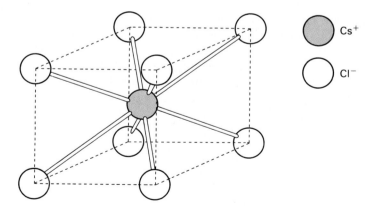

Figure 2.12 Cesium chloride lattice.

adopted by alkali halides, and then only by CsCl, CsBr, and CsI. It is also seen in TlCl, TlBr, and TlI, in a few intermetallic compounds, and in a few complexes such as $Be(H_2O)_4^{2+}SO_4^{2+}$ and $K^+SbF_6^-$. It is perhaps surprising that there are so few compounds with the CsCl structure, since radius-ratio rules predict many and are fairly successful in predicting other structures. The CsCl structure is not close packed either in terms of all ions or in terms of anions only; it should be apparent from Fig. 2.12 that the anions are simple cubic with cations in the eight-coordinate sites, although when the ions are of comparable size the anions do not touch. Wells has suggested that perhaps the only reason it occurs at all is that the large coordination number (8 for both ions) is favored by the London dispersion forces in a compound both of whose ions have high polarizabilities, since that force is proportional to the product of the polarizabilities of the nearest-neighbor ions.

We have so far examined a six-coordinate and an eight-coordinate structure for ionic compounds whose stoichiometry is 1:1. If the system adopts tetrahedral four-coordination, it is likely to show one of two structures of ZnS: *zincblende*, which is a ccp array of sulfide ions with zinc ions in half the tetrahedral holes, or *wurtzite*, which is an hcp (hexagonal close-packed) array of sulfide ions with zinc ions in half the tetrahedral holes. These are shown in Fig. 2.13, along with the diamond structure, the symmetrical equivalent of the zincblende lattice. (Note that there is a third ZnS structure, *sphalerite*, in which the boat-configuration six-membered Zn_3S_3 rings of wurtzite are in the chair configuration.) BeS, ZnS, CdS, and HgS adopt the zincblende structure, and the last three also have stable forms with the wurtzite structure. AgI, CuCl, CuBr, and CuI have the zincblende structure, and so do some compounds that cannot possibly be considered ionic, such as BN and BP. Together with the obvious geometric identity between diamond and zincblende, these latter suggest strongly that covalent bonding is important in the lattice energy of any crystalline compound having one of these structures.

A lattice closely related to that of wurtzite is the NiAs lattice shown in Fig. 2.14, which to a good approximation consists of hcp arsenic atoms with nickel atoms in the octahedral holes. Like the wurtzite structure, the NiAs lattice is found only in compounds in which covalent bonding is important in overall lattice stability. It is adopted by many transition-metal sulfides, arsenides, selenides, and other such compounds having 1:1 stoichiometry. The coordination octahedra around the transition-metal atoms are arranged in stacks in which the octahedra share a face, which places the central metal atoms so close together that they are partially bonded and thus produce metallic properties in many of these compounds. In the compounds with the most pronounced metallic properties, the octahedra are distorted so that each metal atom has eight equidistant neighbors: six nonmetal atoms and two metal atoms.

There are many examples of defect NiAs structures in which there are vacant cation positions, sometimes randomly and sometimes in an ordered

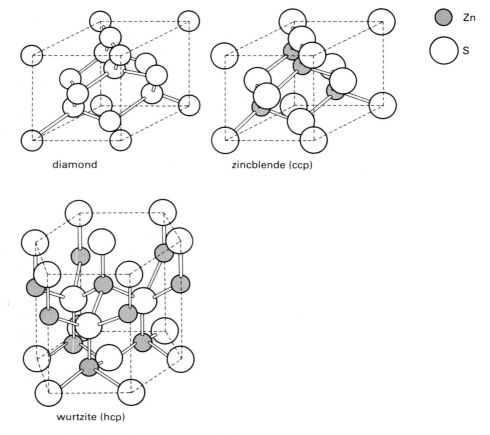

diamond

zincblende (ccp)

Zn

S

wurtzite (hcp)

Figure 2.13   ZnS four-coordinate structures.

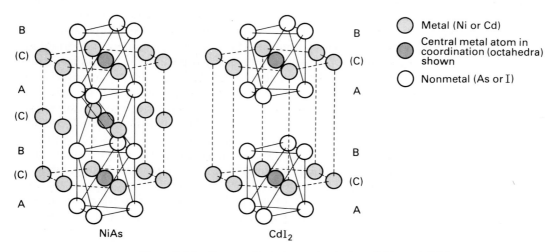

B
(C)
A
(C)
B
(C)
A

NiAs

B
(C)
A
B
(C)
A

CdI$_2$

Metal (Ni or Cd)

Central metal atom in coordination (octahedra) shown

Nonmetal (As or I)

Figure 2.14   Six-coordinate hcp-anion lattices (NiAs and CdI$_2$).

way. FeS is usually slightly deficient in iron atoms, but when exactly one-eighth of them are missing, the resulting vacancies are ordered in alternate layers and the result can be called a distinct compound Fe_7S_8. If this process continues to the extreme in which half the metal atoms are lost, the resulting compound has 2:1 stoichiometry and forms the CdI_2 lattice also shown in Fig. 2.14. The CdI_2 lattice has an obvious layered nature, but it is essentially hcp in nonmetal atoms with metals filling half the octahedral holes—that half being chosen to yield the layered structure. The cobalt–tellurium system is continuously variable betweeen CoTe (NiAs lattice) and $CoTe_2$ (CdI_2 lattice), and thus is an interesting example both of nonstoichiometry and of the symmetry relationships between lattice structures. In Fig. 2.14, the nonmetal atoms are stacked ABABAB... and the metal atoms are in all (or half) of the C positions, which we can distinguish from the anion positions by calling them the c positions. If the packing of all the atoms is considered, NiAs shows a kind of *superstructure* in which layers of close-packed atoms are stacked AcBcAcBcAcBc... and the CdI_2 structure is essentially ccp, although the cubes are distorted by the differing radii of the two kinds of atoms.

As one might by this time expect, there is a ccp-anion structure analogous to the hcp-anion CdI_2 structure, in which alternate planes of metal ions are missing to give the 2:1 stoichiometry. This is the $CdCl_2$ lattice, in which metal ions are missing from the NaCl lattice in the same way that vacancies in the NiAs lattice yield the CdI_2 lattice. Both of the layer structures seem to require considerable covalent character in the lattice bonding.

For systems having 2:1 stoichiometry that are clearly ionic, such as oxides and fluorides, two lattices are frequently found: the *rutile* (TiO_2) lattice and the *fluorite* (CaF_2) lattice. These are shown in Fig. 2.15. In the rutile structure, each titanium ion is octahedrally coordinated with six oxide ions. The lattice symmetry is such that the coordination octahedra form chains sharing opposite edges, the chains form a sheet in which alternate chains use their apical oxygens to link these shared edges, and these sheets stack so that every octahedron shares two opposite edges and the two remaining corners. A view of one layer of octahedra is shown in Fig. 2.15. The octahedra in the first and third chains are seen from the top and those in the second and fourth chains are seen from the side. An important point to note here is that the cation coordination number and the overall stoichiometry determine what the anion coordination number is to be. If, as in the rutile lattice, the cation is six-coordinate and the stoichiometry allows only half as many cations as anions, then each anion can touch only half as many cations and its coordination number will be three. Ideal three-coordination would be trigonal planar with bond angles of 120°, but this is not geometrically possible if the octahedron is ideal. Most compounds form the rutile lattice with one angle at the oxygen position near 90° and the other two near 135°. Many metal difluorides and dioxides adopt the rutile lattice, but sulfides and other dihalides do not, which seems to emphasize its applicability to strongly ionic systems.

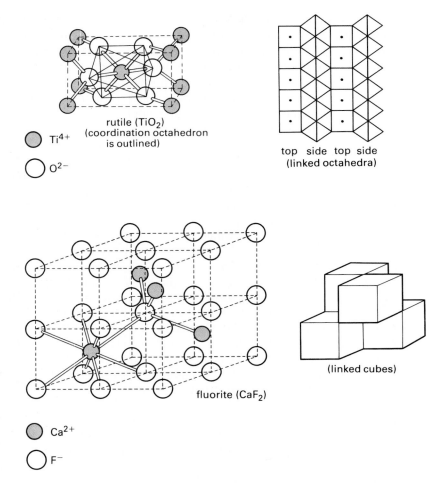

rutile (TiO$_2$)
(coordination octahedron
is outlined)

Ti^{4+}

O^{2-}

top side top side
(linked octahedra)

fluorite (CaF$_2$)

(linked cubes)

Ca^{2+}

F^{-}

Figure 2.15   Cube and octahedron stacking in fluorite and rutile lattices.

The fluorite lattice is essentially the CsCl lattice with alternate cations removed. Each cation still has cubic coordination (8-coordinate), but the stoichiometry now requires that each anion be four-coordinate, which in the fluorite lattice is tetrahedral. Figure 2.15 indicates the cubic and tetrahedral coordination for one Ca^{2+} and one F^{-}, and also shows the stacking of coordination cubes. Many dioxides and difluorides of heavier metals adopt the fluorite lattice, but so do some hydrides (TiH$_2$) and intermetallic compounds such as SnMg$_2$ and PtAl$_2$. Most of these contain fairly polarizable ions, and the remarks about the stability of the CsCl lattice for large polarizable ions may apply here as well. If the positions of the cations and anions are reversed, the lattice is called the *antifluorite* structure. Since the coordination numbers are unchanged, the stoichiometry must be reversed so that there are two cations for every anion. Most alkali metal oxides and sulfides (Na_2O, Rb_2S, and so on) adopt the antifluorite lattice, along with more covalent compounds such as Mg_2Si and other group II–group IV compounds.

2.5 IONIC OXIDES

Shannon and others have derived ionic radii based on the assumption that the only compounds likely to show strongly ionic bonding in a lattice are oxides and fluorides. Not only are these two elements the most electronegative and thus most likely to yield electron transfer when bonded to a neighbor atom [as indicated by their high $I(q)$ a_0 value], they also have two of the highest a_1 values, indicating that they are extremely hard (unpolarizable) and unlikely to share electrons in that sense. Of these two anions, oxides are by far more widespread and more thoroughly studied. Because we live in an oxidizing atmosphere containing the dioxygen molecule, many elements are found in the earth's crust as oxides. Those oxides are important both as ores and as pigments and construction materials, and their formation is the central problem of corrosion study.

A convenient way to examine ionic and partly ionic oxides is by the charge (or formal oxidation state) of the metal involved. The M_2O oxides, with a $+1$ cation, are the most completely ionic in their lattice bonding. Of the alkali metals, only lithium forms the normal oxide, Li_2O, on reacting with O_2; sodium, on burning in oxygen or air, forms the peroxide Na_2O_2, and the other alkali metals form the superoxides KO_2, RbO_2, and CsO_2 respectively. Even lithium oxide is frequently prepared by heating lithium peroxide. The other alkali oxides can be prepared by heating the metal azide or the metal itself with the metal nitrate:

$$5NaN_3 + NaNO_3 \rightarrow 3Na_2O + 8N_2$$
$$10Rb + 2RbNO_3 \rightarrow 6Rb_2O + N_2$$

All the alkali-metal oxides adopt the antifluorite lattice (that of CaF_2 with cation and anion locations reversed) except Cs_2O, which crystallizes in a layered lattice like CdI_2 but with cations and anions reversed and with the packing ccp rather than hcp. It has been speculated that this structure permits the greatest induced-dipole/induced-dipole stabilization by the adjacent layers of Cs^+ ions. All these oxides are extremely strong bases, forming hydroxide ions in water and reacting at high temperatures with other oxides, thereby serving as a metal flux in welding:

$$K_2O + H_2O \rightarrow 2K^+(aq) + 2OH^-(aq)$$
$$Li_2O + Fe_2O_3 \rightarrow 2LiFeO_2$$

In group Ib, copper and silver both form the M_2O monoxide on exposure to air, but only as a thin, closely adhering film. Although Au_2O has been reported, its existence is doubtful. The red Cu_2O is the visible product in the Fehling's solution test for reducing sugars, appearing because alkaline solutions of Cu^{2+} are quite easily reduced to Cu^+. Cu_2O and Ag_2O have an interesting and unusual crystal structure analogous to the diamond structure of Fig. 2.13: O atoms occupy the C locations, a metal atom occupies the center of each O–O axis in this diamond-like symmetry, and the O—O distance is thereby expanded so much that a second identical lattice can interpenetrate the first. These two compounds are much less basic than the alkali-metal oxides and are only sparingly soluble in water.

At least 28 metals form an MO monoxide in which the metal is presumably M^{2+}. The obvious candidates are the alkaline-earth metals of group IIa and the metals of group IIb, but all the first-row transition metals also form a monoxide, and so do a few heavier transition metals and the rare earths Eu, Th, Pa, U, Np, Pu, and Am. The MO oxide is formed when most of these metals oxidize in air, although in some transition metals internal redox equilibria lead to higher oxidation states or to mixed oxides like $Fe_3O_4(Fe^{2+}Fe_2^{3+}O_4)$. Nearly all of these oxides have the NaCl crystal lattice. Be, Zn, and Cr adopt the wurtzite structure, and a few others have lattices of lower symmetry. HgO forms a one-dimensional —O—Hg—O—Hg—O— polymer with angles of 107° at O and 180° at Hg, and PbO has an interesting two-dimensional buckled layer structure in which the lead atoms form the outside of the buckled layer so that there is extensive Pb—Pb contact in the crystal.

The alkaline-earth oxides are all strong bases except for BeO, which is quite unreactive but shows some slight amphoterism by dissolving in hot KOH:

$$BeO + 2OH^- + H_2O \rightarrow Be(OH)_4^{2-}$$

The other oxides not only hydrolyze to the hydroxide ion in water, but also absorb CO_2 to form the carbonates. Most of them are made commercially by *calcination* (strong heating) of the carbonate, which is the reverse of the CO_2 absorption reaction driven by entropy at high temperatures:

$$CaCO_3(s) \rightarrow CaO(s) + CO_2$$

The M_2O_3 oxides, such as Al_2O_3 and the rare-earth oxides, present an interesting structural problem. Stoichiometry requires that the coordination numbers of the metal and oxide be in a ratio $3:2$, which immediately suggests that the metal should be in an MO_6 octahedron and the oxygen in an OM_4 tetrahedron. Unfortunately, this is not geometrically possible. One or the other has to yield, and the two most common structures both involve some compromise. In the very complex Al_2O_3 (*corundum*) structure, the oxygens are more or less hcp with the metal in two-thirds of the octahedral holes; but the symmetry is distorted to bring the bond angles around the oxygen closer to the tetrahedral 109°. As a result, some of the AlO_6 octahedra share corners, others edges, and still others faces. In the other structure, adopted by most rare-earth oxides, the lattice is similar to that of fluorite (Fig. 2.15) with one-fourth of the anions missing. The metal is thus six-coordinate, but the six oxides are at six of the eight corners of a cube, not at the corners of an octahedron.

Bauxite, the principal aluminum ore, is a partially hydrated aluminum oxide. There are numerous stages of hydration between Al_2O_3 and $Al(OH)_3$, and in general, the less hydroxide is present, the less chemically reactive the oxide appears. Pure corundum is extremely unreactive; in synthetic single-crystal form it is called white sapphire. Activated aluminas, on the other hand, are in fact incompletely dehydrated aluminum hydroxide. Corundum is extremely hard, which results from the very high lattice energy arising from the $+3/-2$ charge product and the directional covalent bonding also present in the crystal lattice. A trace of Cr^{3+} substituted for Al^{3+} in the corundum lattice yields a deep red color, and a trace of Fe^{3+} (together with Fe^{2+} and Ti^{4+} for two Al^{3+}) in corundum yields a smoky blue color. These are the two gems ruby and sapphire, respectively. Besides the naturally occurring stones, large quantities of synthetic ruby and sapphire are produced worldwide.

Although there are many MO_2 oxides, the $+4$ oxidation state requires such a large energy input to ionize the cation that few such oxides can be thought of as strongly ionic in their bonding. The obvious candidates are the group-IVb metals Ti, Zr, and Hf, along with the heavier, less electropositive metals from group IVa, Sn and Pb (although most transition metals also form a dioxide). The 1:2 stoichiometry suggests either 8:4 coordination or 6:3 coordination, and the major structures that are found for dioxides are rutile (6:3) and fluorite (8:4) (see Fig. 2.15). In general, lighter elements tend to form rutile lattices and heavier elements fluorite lattices. ZrO_2, however, represents an interesting intermediate case. Each Zr is seven-coordinate with a geometry shown in somewhat idealized form in Fig. 2.16. The stoichiometry

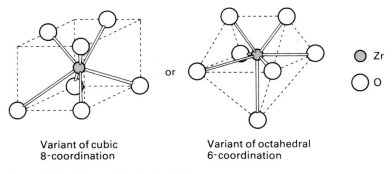

Variant of cubic
8-coordination

or

Variant of octahedral
6-coordination

Zr
O

Figure 2.16 Zr coordination in ZrO_2.

requires an average coordination number of oxygen of 3.5; accordingly, half of the oxygens have tetrahedral coordination while the other half have trigonal planar coordination (though with irregular bond angles). When ZrO_2 is heated, it changes to the rutile lattice at about 1100°C and to the fluorite lattice at about 2300°C, thereby displaying the full range of possibilities.

Among the MO_2 oxides, several are commercially important even without considering SiO_2, which will be deferred to Chapter 3. TiO_2 as synthetic rutile is a major industrial chemical; it is almost the only pure white pigment used in commercial paints. MnO_2, found as the mineral *pyrolusite* with the rutile structure, is widely used in dry cells and as a heterogeneous catalyst. ThO_2, with the fluorite structure, is used in the burner mantles of gasoline lanterns because of its very high melting point (3390°C) and thermal stability.

The MO_2 oxides nearly all show acidic properties, in that they react with strong-base oxides (or even carbonates) to yield mixed-oxide lattices, many of which are stoichiometric and are known as titanates, stannates, zirconates, and so on:

$$Na_2CO_3 + TiO_2 \rightarrow Na_2TiO_3 + CO_2$$
$$2MgO + TiO_2 \rightarrow Mg_2TiO_4$$
$$BaO + ZrO_2 \rightarrow BaZrO_3$$
$$2K_2O + SnO_2 \rightarrow K_4SnO_4$$
$$La_2O_3 + 2ZrO_2 \rightarrow La_2Zr_2O_7$$

It should be clear that the bonding in these "metallates" and even in the MO_2 oxides themselves is at most partly ionic, with covalent bonding being significant in determining lattice geometry and energy relationships. This combination of energy factors can lead to very complex structures and superstructures; Chapter 3 will consider these oxides in more detail.

2.6 OXOANIONS—FIRST-ROW ELEMENTS

The mixed oxides in the preceding examples usually still have a substantial degree of ionic bonding between the O^{2-} and both cations. When such an oxide is dissolved in a strong acid, the coordination polyhedra usually lose their identity and there is no trace of the original crystal structure. However, if there is substantial covalent bonding between the atoms in an oxide coordination polyhedron, the polyhedron is likely to persist in solution or in other crystal forms as an *oxoanion*. This will occur when there is only a modest difference in electronegativity between the central "cation" and oxygen, either because the "cation" is a metalloid or nonmetal or because it has a high formal oxidation state and thus would be strongly polarizing if it were a true cation.

Identifiable oxoanions formed by elements in the first row of the periodic table include the ions BO_3^{3-} (and polyborates), CO_3^{2-} (and organic anions such as oxalate), NO_3^-, NO_2^- and other lower-oxidation-state nitrites, O_2^-, O_2^{2-}, and O_3^-. We can begin our study of these with boron, which occurs in nature only as boron-oxygen compounds of various sorts. In most borates, the boron shows trigonal planar BO_3 coordination mixed with tetrahedral BO_4 coordination; the proportion of tetrahedral coordination increases with the formal charge on the borate anion. Boron is in group III, and one expects the simplest oxo- or hydroxo-compound to be boron hydroxide, $B(OH)_3$, or *orthoboric acid*. Of course, this is more commonly written as H_3BO_3, the precursor of the BO_3^{3-} ion. This planar ion is found in $Mg_3(BO_3)_2$, $Co_3(BO_3)_2$, and a few other compounds. Boric acid functions as a very weak acid in water ($K_a = 7 \times 10^{-10}$), not by donating protons but by accepting hydroxides:

$$B(OH)_3 + 2H_2O \rightarrow B(OH)_4^- + H_3O^+$$

The tetrahedral tetrahydroxoborate anion is found in a few solid borates, notably $LiB(OH)_4$. An obviously related species, BO_4^{5-}, is found in the partly ionic $TaBO_4$.

When it is heated, orthoboric acid loses water in steps:

$$2H_3BO_3 \rightarrow 2HBO_2 + 2H_2O \rightarrow B_2O_3 + H_2O$$

The HBO_2 species, *metaboric acid*, yields the metaborate ion BO_2^-; but since this formula leaves at least one coordination position open around the boron, the metaborate ion is always found as a polymer $(BO_2)_n^{n-}$. Most frequently, metaborate forms the planar cyclic trimer $B_3O_6^{3-}$, as in the sodium compound $Na_3B_3O_6$ (usually written $NaBO_2$), although long-chain structures are also found in $Ca(BO_2)_2$ and a few other metaborates. Although the reaction is only hypothetical it is possible to imagine a dehydration in which two ortho-boric acid units lose only one water:

$$[2H_3BO_3 \rightarrow (OH)_2BOB(OH)_2 + H_2O]$$

While the acid $H_4B_2O_5$, *pyroboric acid*, does not exist, the pyroborate ion $B_2O_5^{4-}$ is found in some crystals: $Mg_2B_2O_5$, $Co_2B_2O_5$, and a few others. Once the polymerization of boric acid reaches the stage of three borons, the basic structural unit for most polyborates is the six-membered ring found in $Na_3B_3O_6$:

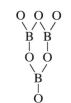

Boric acid itself consists of 20-membered rings of alternating B and O atoms, folded like a two-bladed propeller and cross-linked into a three-dimensional polymer at each B atom (see Fig. 2.17).

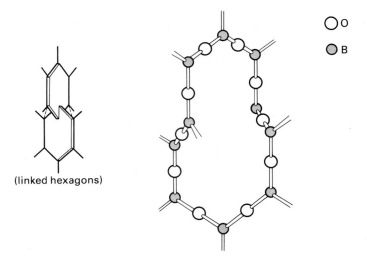

Figure 2.17   The B_2O_3 structure.

After the rather complex nature of borates, it is somewhat surprising to discover that the carbonate ion has no polymers. It has been speculated that this reflects the strong C=O pi bonding possible for these two atoms—better for carbon/oxygen than for boron/oxygen, because the electronegativity difference is less for C and O. Organic chemists, however, taking advantage of the catenating (chain-forming) ability of carbon, have prepared the squarate, croconate, and the rhodizonate ions (shown in Fig. 2.18 along with the structures of some metal compounds of these ions). Because of the formation and deposition of carbonate minerals in the oceans over geologic time, several elements are found in nature as carbonates: primarily alkaline-earth metals (limestone, $CaCO_3$; dolomite, $CaCO_3 \cdot MgCO_3$) but also Ce, Mn, Fe, Cu, and Pb in various minerals. When these are mined, they are usually roasted to the oxide,

$$MCO_3 \rightarrow MO + CO_2$$

a reaction (calcination) that will be considered more fully in Chapter 8. In a somewhat similar reaction, insoluble oxides or minerals are sometimes converted to soluble alkali-metal salts by fusion in an alkali carbonate:

$$SiO_2 + Na_2CO_3 \rightarrow Na_2SiO_3 + CO_2$$

Nitrates are much less commonly found in nature than carbonates, essentially because they are so water soluble. In extremely arid northern Chile, however, massive deposits of $NaNO_3$ and KNO_3 have been mined for nearly two centuries for use in fertilizers and explosives. The nitrate ion is trigonal

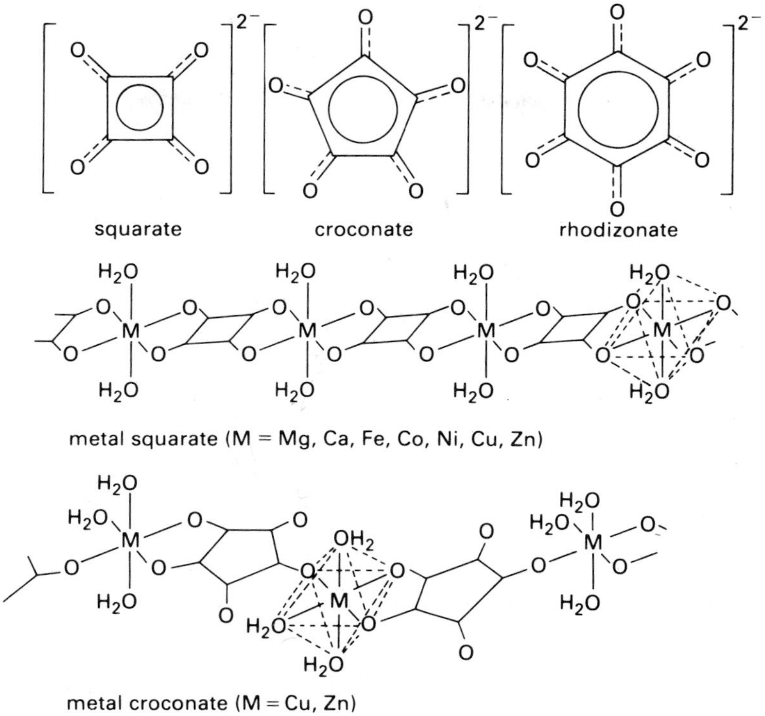

squarate   croconate   rhodizonate

metal squarate (M = Mg, Ca, Fe, Co, Ni, Cu, Zn)

metal croconate (M = Cu, Zn)

Figure 2.18 Carbon oxyanions and their compounds.

planar, like BO_3^{3-} and CO_3^{2-}, with which it is isoelectronic. There are no polynitrates, even though several nitrogen oxides with more than one N are known. Commercially, the nitrate ion is prepared by the catalytic oxidation of ammonia, followed by the disproportionation of NO_2 in water solution:

$$4NH_3 + 5O_2 \rightarrow 4NO + 6H_2O$$

$$2NO + O_2 \rightarrow 2NO_2$$

$$3NO_2 + H_2O \rightarrow 2HNO_3 + NO$$

Ammonium nitrate is used in very large quantities both as fertilizer and (with fuel oil) as a mining explosive.

The BO_2^- ion occurs only as a polymer in crystals, and there is no CO_2^- ion, but the NO_2^- ion is relatively stable. It has a bent conformation (ONO angle 115°) that is presumably caused by the presence of a nonbonding pair of electrons in the third trigonal-planar coordination position. This distinguishes it from the metaborate ion, which polymerizes because the B accepts electrons in that position from the O of a neighboring BO_2^-. (CO_2^-, as a free radical, would dimerize to the oxalate ion.) Nitrites are prepared commercially by a reverse disproportionation, sometimes known as *comproportionation*,

$$NO + NO_2 + 2NaOH \rightarrow 2NaNO_2 + H_2O$$

even though in acidic solution the disproportionation occurs:

$$3HNO_3 \rightarrow H_3O^+ + 2NO = NO_3^-$$

The pyrolysis of alkali nitrates also yields nitrites fairly conveniently:

$$NaNO_3 \rightarrow NaNO_2 + \tfrac{1}{2}O_2$$

Nitrites are readily susceptible to redox reactions, both oxidation and reduction:

Oxidation

$$2HNO_2 + O_2 + 2H_2O \rightarrow 2H_3O^+ + 2NO_3^-$$

Reduction

$$SO_2 + 2HNO_2 + H_2O \rightarrow 2NO + H_3O^+ + HSO_4^-$$

$$NH_2OH + HNO_2 \rightarrow NNO + H_2O$$

Nitrogen shows a variety of reasonably stable oxidation states and a number of ions. The powerful reducing agent sodium amalgam (Na/Hg) reduces sodium nitrate to the *hyponitrite* ion $N_2O_2^{2-}$:

$$2NaNO_3 + 8Na/Hg + Hg + 4H_2O \rightarrow Na_2N_2O_2 + 8Na^+ + 8OH^- + 8Hg$$

Crystalline hyponitrites are stabilized by their ionic lattice energy, but solid hyponitrous acid, $H_2N_2O_2$, detonates even when gently rubbed. Some of the reasons for this behavior will be explored in Chapter 8. It is also possible to prepare *oxyhyponitrite*, $N_2O_3^{2-}$, and *nitroxylate*, $N_2O_4^{4-}$ (possibly NO_2^{2-}):

$$C_4H_9ONO_2 + NH_2OH + 2NaOCH_3 \rightarrow Na_2N_2O_3 + C_4H_9OH + 2CH_3OH$$

$$2NaNO_2 + 2Na \xrightarrow{\text{liq. } NH_3} Na_4N_2O_4 \qquad (NaNO_2)$$

Although this section deals in general with oxoanions, nitrogen also forms two oxocations, NO_2^+ and NO^+. These are usually prepared by dissolving the appropriate nitrogen oxide in a strong-acid medium:

$$2N_2O_5 + H_3O^+ClO_4^- \rightarrow NO_2^+ClO_4^- + 3HNO_3$$

$$N_2O_3 + 3H_2SO_4 \rightarrow 2NO^+ + H_3O^+ + 3HSO_4^-$$

NO_2^+, the *nitronium* ion, is linear and has a very short N—O bond (1.10 Å versus 1.22 Å or more in most nitrates), indicating very strong bonding. NO^+, the *nitrosyl* or *nitrosonium* ion, also has a short bond (about 1.14 Å).

Oxygen also forms an "oxocation" and three "oxoanions." The first of these is the dioxygenyl cation O_2^+. This was first prepared by fluorinating platinum in silica apparatus, but is more conveniently made using group Va fluorides:

$$O_2 + \tfrac{1}{2}F_2 + MF_5 \rightarrow O_2^+MF_6^- \qquad (M = P, As, Sb)$$

It is a powerful oxidizing agent, yielding oxygen gas and ozone violently from water. The O—O bond length in O_2^+ is shorter than in O_2 (1.123 Å versus 1.207 Å) even though it has one less electron, a phenomenon that will be considered in Chapter 4.

The oxygen oxoanions are superoxide, O_2^-, peroxide, O_2^{2-}, and ozonide, O_3^-, with bond lengths of 1.28 Å, 1.49 Å, and 1.19 Å (bent), respectively. All are powerful oxidizing agents, though peroxide (as H_2O_2 in aqueous solution) can reduce some other strong oxidizing agents such as permanganate and chlorine gas. They are prepared as follows:

$$2Na + O_2 \rightarrow Na_2O_2$$

$$K + O_2 \rightarrow KO_2 \qquad \text{(also Rb, Cs; Na under high pressure and temperature)}$$

$$5O_3 + 2KOH \rightarrow 5O_2 + 2KO_3 + H_2O$$

Sodium peroxide fusion is sometimes used in organic and organometallic analysis to oxidize all organic species to carbonate. In a somewhat similar application, potassium or sodium superoxide is used in closed-cycle breathing masks to absorb CO_2 and release oxygen:

$$4KO_2 + 2CO_2 \rightarrow 2K_2CO_3 + 3O_2$$

2.7 OXOANIONS—HEAVIER ELEMENTS

The second row of the periodic table is prolific of oxoanions. There are so many silicates and they are of such importance that we shall consider them separately; but in addition phosphorus and sulfur form a great many poly-anions to which an introduction is appropriate. The oxoanions of the first row tend to have small coordination numbers about the central atom—usually two or three, rarely four and never higher. By contrast, in the second row the coordination number is almost always four when an atom is coordinated by oxygens, and tetrahedral coordination geometry is nearly always found, even if some atoms are not oxygens. Silicates are all assembled from SiO_4 tetrahedra, as we shall see, but the many phosphorus oxoanions are all assembled from PO_4 tetrahedra in which one or more oxygens may be replaced by H or P atoms. The coordination number increase is not a steric problem in which the first-row atoms are too small, because CCl_4 is quite stable. Rather, the difference lies in the capacity for pi bonding of the central atom with oxygen. As Chapter 4 will detail, strong pi bonding requires that the overlapping pi-symmetry orbitals have nearly the same size and energy as atomic orbitals. The oxygen atom, because it is very electronegative and has only a $1s^2$ inner core, has quite small pi-symmetry $2p$ orbitals at very low energy. Accordingly, first-row elements, C, N, and O are much better able to pi-bond to O than are the less electronegative, larger atoms Si, P, and S. For first-row atoms with only $2s$ and $2p$ valence orbitals, pi bonding with a p orbital leaves

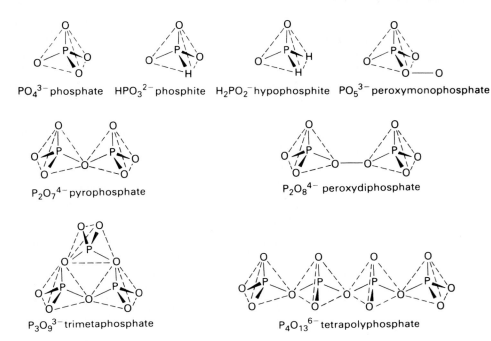

Figure 2.19   Some phosphorus oxoanions.

only sp^2 hybrids for sigma bonding and a maximum coordination number of 3. By contrast, second-row elements will almost always reach greater stability by forming four sigma bonds, normally in tetrahedral geometry. Figure 2.19 shows some of the structures of phosphates; this rule can be seen at work.

Table 2.4 shows some of the many phosphorus oxyanions, grouped by the atoms occupying the four tetrahedral coordination sites. In particular, higher polyphosphates and mixed species can be prepared in considerable variety. Most of these species with only one or two P atoms can be prepared by hydrolysis or oxidation of elemental phosphorus, by hydrolysis of a phosphorus halide or its reaction with another phosphorus acid, or by heating phosphoric acid or acid phosphate salts:

$$P_4 + 4OH^- + 4H_2O \rightarrow 4H_2PO_2^- + 2H_2$$
$$P_2I_4 + 4H_2O \rightarrow H_4P_2O_4 + 4HI \qquad (CS_2)$$
$$2HPO_4^{2-} + H_2PO_4^- \rightarrow P_3O_{10}^{5-} + 2H_2O$$

The table shows both the common names and the IUPAC names of the ions; in general, the common names are convenient for the species shown, but for mixed species and for organic derivatives of the oxoanions the IUPAC names are necessary.

Table 2.4 Phosphorus Oxoanions

Atoms on P:	2H, 2O	1H, 3O	4O	1P, 3O	1H, 1P, 2O
Species:	$H_2PO_2^-$	HPO_3^{2-}	PO_4^{3-}	$O_3PPO_3^{4-}$	$HP(O)_2P(O)_2H^{2-}$
Common:	hypophosphite	phosphite	orthophosphate	hypophosphate	—
IUPAC:	phosphinate	phosphonate	phosphate	hexaoxo-diphosphate(P–P)(4−)	dihydridotetraoxo-diphosphate(P–P)(2−)

4O column (additional species):

$O_3POPO_3^{4-}$
pyrophosphate
μ-oxobis(trioxophosphate)(4−)

$O_3POP(O)_2OPO_3^{5-}$
tripolyphosphate
triphosphate

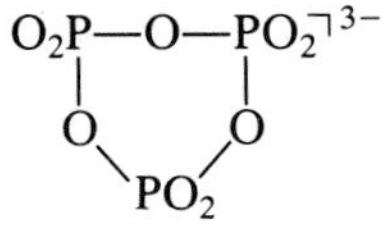

$O_2P\!-\!O\!-\!PO_2^{3-}$

(ring structure with bridging O atoms and PO$_2$ group)

trimetaphosphate
cyclo-triphosphate

1P, 3O column (additional species):

mixed species such as
$HP(O)_2PO_3^{3-}$
hydridopentaoxo-diphosphate(P–P)(3−)

Note: In IUPAC names, μ- indicates a bridging atom and P–P indicates bonded phosphorus atoms in the structure.

Phosphates are extremely important commercially. Enormous quantities are used as "superphosphate" fertilizer, $Ca(H_2PO_4)_2 \cdot H_2O$, mined as fluorapatite, $Ca_5(PO_4)_3F$, and processed using sulfuric acid. Major phosphate rock deposits are found in Florida, North Carolina, Morocco, and Nauru Island in the Pacific, but worldwide agricultural usage threatens to exhaust these resources in a relatively short time. Other orthophosphates are used as flame retardants and as polishing agents in toothpaste. Another major use of ionic phosphates involves sodium tripolyphosphate, which is the most widely used builder (complexing agent for Ca^{2+} and Mg^{2+}) in synthetic detergents. The United States produced 1.3 billion pounds of $Na_5P_3O_{10}$ in 1986, primarily for this use.

Sulfates occur in a variety almost equal to that of phosphates. Their fundamental geometry is also tetrahedral, but there are some sulfur oxoanions in which one of the tetrahedral positions is occupied by a nonbonding electron pair. Also, sulfur–hydrogen bonds, analogous to those in hypophosphorous acid, do not occur in oxoanions, and all sulfur oxoanions have a 2– charge. Figure 2.20 indicates the molecular geometry of several sulfur oxoan-

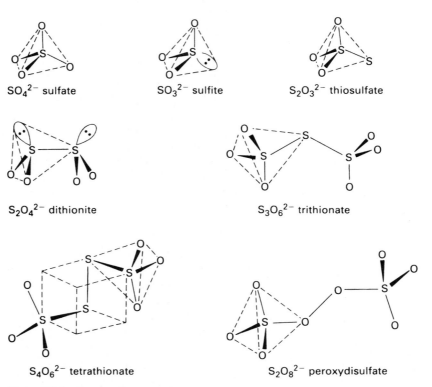

SO_4^{2-} sulfate SO_3^{2-} sulfite $S_2O_3^{2-}$ thiosulfate

$S_2O_4^{2-}$ dithionite $S_3O_6^{2-}$ trithionate

$S_4O_6^{2-}$ tetrathionate $S_2O_8^{2-}$ peroxydisulfate

Figure 2.20   Some sulfur oxoanions.

ions, and Table 2.5—like Table 2.4—gives basic formulas and names. Many more polysulfane sulfonates are known, in addition. Most of the species in the table are prepared by reacting a sulfur oxide with H_2O or H_2S_n, by reacting a sulfite with either an oxidant or a reductant, or simply by heating a sulfate or sulfite:

$$SO_2 + 2H_2O \rightarrow HSO_3^- + H_3O^+$$

$$2HSO_3^- + \text{``}H_2SO_3\text{''} + Zn \rightarrow ZnSO_3 + S_2O_4^{2-} + 2H_2O$$

$$2NaHSO_4 \rightarrow Na_2S_2O_7 + H_2O \qquad \text{(heat)}$$

It should be noted that for both phosphates and sulfates, peroxoanions such as peroxymonosulfate, SO_5^{2-}, from the acid H_2SO_5 (analogous to H_2SO_4) and peroxydiphosphate, $P_4O_8^{4-}$, can be formed, usually by electrochemical oxidation. In these oxoanions the peroxo $-O_2-$ group plays the same structural role as a single $-O-$ oxygen.

Only the simpler sulfur oxoanions have commercial uses. Sulfuric acid is the most common industrial chemical, but its sulfate content is not usually important to the application. Its importance derives from the fact that it is a strong acid, very cheaply made:

$$S + O_2 \rightarrow SO_2$$

$$SO_2 + \tfrac{1}{2}O_2 \rightarrow SO_3 \qquad \text{(V_2O_5 catalyst)}$$

$$SO_3 + H_2SO_4 \rightarrow H_2S_2O_7 \qquad \text{(nonaqueous)}$$

$$H_2S_2O_7 + H_2O \rightarrow 2H_2SO_4$$

Ionic sulfates used in substantial quantities include the following: gypsum, $CaSO_4 \cdot 2H_2O$, used in wallboard and plaster; aluminum sulfate or alum, $NaAl(SO_4)_2 \cdot 12H_2O$, used in water purification, and calcium or magnesium hydrogen sulfite, used in acid–sulfite papermaking to dissolve lignin from wood cellulose. The thiosulfate ion is important both as the fixer in the conventional black-and-white photographic process and as a laboratory reagent in iodimetry, where it reacts quantitatively with elemental iodine to produce the tetrathionate ion:

$$2S_2O_3^{2-} + I_2 \rightarrow S_4O_6^{2-} + 2I^-$$

Most of the sulfur oxoanions in which the sulfur appears in a low formal oxidation state are reducing agents, as in the above reaction. Dithionate is the strongest reducing agent in the group; it reduces Ti^{4+} to Ti^{3+} in aqueous solution, and Cu^+ and Pb^{2+} to the metals. Conversely, peroxydisulfate is one of the strongest convenient laboratory oxidizing agents; it can oxidize aqueous Mn^{2+} to permanganate and Cr^{3+} to dichromate.

Chlorine oxoanions present a simpler picture. There are no binuclear species, and the only known species are the familiar perchlorate, chlorate, chlorite, and hypochlorite. The ClO_4^- ion is essentially an ideal tetrahedron, and both ClO_3^- and ClO_2^- have bond angles near the tetrahedral 109.5° (106°

Table 2.5 Sulfur Oxoanions

Atoms on S:	4O	1S, 3O	1e⁻ pr, 3O	1e⁻ pr, 1S, 2O
Species:	SO_4^{-2}	$O_3SSO_3^{2-}$	SO_3^{2-}	$O_2SSO_2^{2-}$
Common:	sulfate	dithionate	sulfite	dithionite
IUPAC:	tetraoxosulfate(2−)	hexaoxo-disulfate(S–S)(2−)	trioxosulfate(2−)	tetraoxodisulfate(S–S)(2−)
	$O_3SOSO_3^{2-}$	SSO_3^{2-}	*mixed species:*	
	pyrosulfate	thiosulfate	$O_2SSO_3^{2-}$	
	disulfate(2−)	trioxothiosulfate(2−)	disulfite	
		$O_3SSSO_3^{2-}$	pentaoxodisulfate(S–S)(2−)	
		trithionate		
		1,1,3,3,3-hexaoxotrisulfate(2 S–S)(2−)		
		$SSSO_3^{2-}$		
		disulfane monosulfonate		
		1,1,1-trioxotrisulfate(2 S–S)(2−)		

and 111° respectively), suggesting that it is useful to think of the lower-oxidation-state chlorates as tetrahedra in which nonbonding electron pairs occupy one, two, or three positions. We have seen this as a structural possibility for the sulfates above, and by group VII it is preferable to any form of polymerization.

All of the chlorates are strong oxidizing agents, though kinetic factors sometimes make them effectively inert. In dilute aqueous solution, for example, ClO_4^- is almost completely unreactive as an oxidizing agent, but in concentrated solution $HClO_4$ is an extremely treacherous explosive when allowed to contact almost any organic substance. The chlorine oxoanions are prepared as follows:

$$OCl^-: \quad 2Cl_2 + 2HgO \rightarrow HgCl_2 \cdot HgO + Cl_2O$$
$$Cl_2O + 3H_2O \rightarrow 2H_3O^+ + 2OCl^-$$
$$ClO_2^-: \quad 2ClO_3^- + SO_2 \rightarrow 2ClO_2 + SO_4^{2-}$$
$$2ClO_2 + O_2^{2-} \rightarrow 2ClO_2^- + O_2$$
$$ClO_3^-: \quad Cl^- + 3H_2O \rightarrow ClO_3^- + 3H_2$$
(overall cell reaction for brine electrolysis)
$$ClO_4^-: \quad Cl^- + 4H_2O \rightarrow ClO_4^- + 4H_2$$
(overall cell reaction for brine electrolysis)

As oxidizing agents, the ionic salts of the chlorates are widely used both in the laboratory and in industry. Most of the alkali-metal and alkaline-earth salts can be prepared conveniently, although it is apparently impossible to obtain very pure hypochlorites in the solid state. Sodium hypochlorite is used in solution, as a bleach (oxidizing conjugated pi chromophores to absorb at a higher frequency in the ultraviolet), both in industry and in the household. Sodium chlorite is used as a textile bleach in manufacture. Calcium hypochlorite is used in water treatment (oxidizing bacteria). Chlorates are used in matches and pyrotechnics, and ammonium perchlorate is a high-energy oxidizer in solid rocket propellant.

Beyond the second row of the periodic table there are some other important oxoanions, involving both transition metals and nonmetals. Among the transition metals, these include vanadate (VO_4^{3-} and polymers); chromate (CrO_4^{2-}) and dichromate ($Cr_2O_7^{2-}$); molybdate (MoO_4^{2-} and polymers); tungstate (WO_4^{2-} and polymers); and manganate (MnO_4^{2-}) and permanganate (MnO_4^-). As the formulas suggest, these are all tetrahedral as monomers in solution and also in some crystals. Polymers, however, usually form with the metal in octahedral coordination.

A number of the metallates form interesting series of polymers in solution, called *isopolyanions*. These will be explored further in Chapter 6, but Fig. 2.21 suggests the concentration/pH conditions for the isopolyvanadates. They are formed from the VO_4^{3-} ion by two processes:

Protonation $\qquad VO_4^{3-} + H_3O^+ \rightarrow HVO_4^{2-} + H_2O$

Condensation $\qquad 2HVO_4^{2-} \rightarrow V_2O_7^{4-} + H_2O$

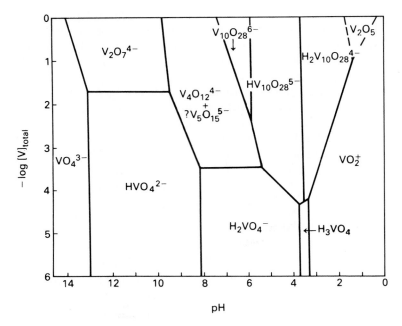

Figure 2.21   Vanadium (V) species predominantly present in aqueous solution. [From M. T. Pope, *Heteropoly and Isopoly Oxometallates,* Springer-Verlag, 1983, Fig. 3.1.]

All of the isopolyanions shown in the figure are believed to be assembled from VO_4 tetrahedra except for the decavanadate, $V_{10}O_{28}^{6-}$, which in crystalline form (and presumably in solution) has edge-sharing VO_6 octahedra (somewhat distorted). None are strongly reactive in either the acid–base or redox sense.

Chromate and dichromate are strong oxidizing agents, of course, and are widely used as such both for analytical purposes (for example, the Fe^{2+} titration) and in synthesis (the oxidation of toluene to benzaldehyde). Since the end product of the reduction of Cr(VI) in chromates is Cr^{3+} and a three-electron transfer is unlikely, the mechanism involves some transient intermediate oxidation states of Cr. Furthermore, the coordination number of the Cr atom must change at some point from 4 in the tetrahedral CrO_4 group to 6 in the octahedral $Cr(H_2O)_6^{3+}$ ion. The mechanism for the Fe^{2+}/CrO_4^{2-} titration, for example, is

$$Fe^{II} + Cr^{VI} \rightleftharpoons Fe^{III} + Cr^{V} \qquad \text{(Rapid equilibrium)}$$

$$Fe^{II} + Cr^{V} \rightarrow Fe^{III} + Cr^{IV} \qquad \text{(Slow, rate-determining)}$$

$$Fe^{II} + Cr^{IV} \rightarrow Fe^{III} + Cr^{III} \qquad \text{(Fast)}$$

The rate-determining step is probably slowed by the necessity to change coordination geometry at that point.

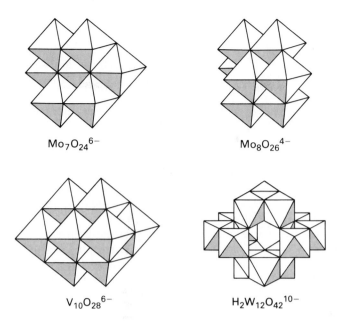

$Mo_7O_{24}{}^{6-}$

$Mo_8O_{26}{}^{4-}$

$V_{10}O_{28}{}^{6-}$

$H_2W_{12}O_{42}{}^{10-}$

Figure 2.22 Some isopolyanion structures.

Molybdate and tungstate ions, like vanadate, form an assortment of isopolyanions in aqueous solution, even though chromate does not (only trichromate and tetrachromate have been characterized, and they are rare). Although it is not difficult to reduce them, they are not strong oxidants as chromates are. When a molybdate solution is acidified, it forms a heptamolybdate $Mo_7O_{24}^{6-}$, then an octamolybdate $Mo_8O_{26}^{4-}$. Similarly, an acidified tungstate solution forms $HW_6O_{21}^{5-}$, then $H_2W_{12}O_{42}^{10-}$ and $H_2W_{12}O_{40}^{6-}$. Like the decavanadate, these have structures consisting of edge-sharing MO_6 octahedra (with some corner sharing in the tungstates). Some of these are shown in Fig. 2.22.

Neither MnO_4^{2-} nor MnO_4^- form polyanions, nor does the rare MnO_4^{3-}. They are tetrahedral ions with no tendency to protonate or condense in aqueous solution as do the vanadates. They are both powerful oxidizing agents, but only permanganate is common in the laboratory because manganate disproportionates in any but the most strongly basic solution:

$$3MnO_4^{2-} + 2H_2O \rightarrow 2MnO_4^- + MnO_2 + 4OH^-$$

Like chromate, permanganate has many laboratory uses in both analytical and synthetic contexts. Unfortunately, permanganate solutions cannot be made up as primary standards because of the tendency of permanganate to decompose water (though slowly):

$$4MnO_4^- + 2H_2O \rightarrow 4MnO_2 + 3O_2 + 4OH^-$$

In spite of this strong oxidizing power, permanganate can be prepared from MnO_2 (through the manganate ion) by air oxidation in fused KOH:

$$2MnO_2 + O_2 + 4OH^- \rightarrow 2MnO_4^{2-} + 2H_2O$$

(disproportionates as above on dilution)

The arsenate (AsO_4^{3-}) and selenate (SeO_4^{2-}) ions are tetrahedral species quite similar to their congeners phosphate and sulfate, but they form many fewer polymeric anions. A trimetaarsenate $As_3O_9^{3-}$ is known in crystalline $K_3As_3O_9$, but it is not certain whether it can survive in solution. Similarly, $Se_2O_7^{2-}$ and $Se_3O_{10}^{2-}$ are presumably comparable to the polysulfates, but little is known about other analogs. Tellurium forms an entirely different species: Telluric acid is $Te(OH)_6$, which is a dibasic acid yielding such salts as $Li_2H_4TeO_6$ with octahedral TeO_6 coordination.

The remaining oxoanions are all strong oxidants: the bromates, iodates, and perxenate (XeO_6^{4-}). All four bromates corresponding to the four chlorates are known as ionic salts, but it is interesting that the perbromate ion, BrO_4^-, was prepared for the first time only in 1968, even though perchlorate and periodate have been common laboratory species since the early nineteenth century.

$$BrO_3^- + F_2 + 2OH^- \rightarrow BrO_4^- + 2F^- + H_2O$$

The delay may be explained by the fact that it is a more powerful oxidizing agent than perchlorate, just as arsenate and selenate are stronger oxidants than phosphate and sulfate. In another parallel with the group VI oxoanions, periodate does not normally have the IO_4^- structure, but rather octahedral IO_6 coordination. Periodic acid in aqueous solution is H_5IO_6, a fairly weak acid sometimes known as *paraperiodic* acid. The ordinary sodium salt is $Na_3H_2IO_6$, but Na_5IO_6 can be prepared by passing dry O_2 over heated NaI and Na_2O, and dilute nitric acid converts the ordinary sodium salt into $NaIO_4$. The IO_4^- ion is known as *metaperiodate*, and there is a brief series of polyperiodates: $I_2O_9^{4-}$, $I_2O_{10}^{6-}$, $I_2O_{11}^{8-}$, and $H_2I_3O_{14}^{5-}$. All of these are composed of linked IO_6 octahedra and decompose to $H_2IO_6^{3-}$ in water.

Although all four of the chlorine oxoanions are known in ionic salts $(ClO^-, ClO_2^-, ClO_3^-, ClO_4^-)$, the heavier halogens are less versatile. No hypoiodites are known as isolated compounds, although IO^- is initially formed when I_2 is dissolved in cold base, like the other halogens:

$$X_2 + 2OH^- \rightarrow X^- + OX^- + H_2O \qquad (X=Cl, Br, I)$$

$$3OX^- \rightarrow X^- + XO_3^- \qquad (X=Br, I)$$

Similarly, no iodites are known, although presumably the IO_2^- ion is an intermediate in the second disproportionation above. Iodate, IO_3^-, is the only other halate besides periodate to polymerize; $I_3O_8^-$ can be isolated as the acid HI_3O_8. Although most iodine oxoanions are octahedral IO_6 units, one I in the $I_3O_8^-$ ion is seven-coordinate, and the salt $NaIO_3$ has the I surrounded by eight O atoms in a square antiprism.

Like the perbromate ion, the perxenate (XeO_6^{4-}) ion is a newcomer to the fairly traditional array of oxoanions. It was first prepared in 1962 by hydrolyzing XeF_6 (itself a novel compound):

$$2XeF_6 + 4Na^+ + 16\,OH^- \rightarrow Na_4XeO_6 + Xe + O_2 + 12F^- + 8H_2O$$

Perxenate is a powerful oxidizing agent; it is reduced in most cases to xenon (VI). XeO_6^{4-} is an ideal octahedron; the Xe—O bond length is quite comparable to that found in IO_6^{5-} and $Te(OH)_6$ (1.86, 1.85, and 1.91 Å respectively), so that the bonding does not appear to be at all unusual in spite of the reputation of the noble gases for chemical unreactivity. Although it can be partially protonated (to $HXeO_6^{3-}$, $H_2XeO_6^{2-}$, and $H_3XeO_6^-$), the free perxenic acid cannot be prepared because in acid solution it decomposes to xenate (VI):

$$H_3XeO_6^- \rightarrow HXeO_4^- + \tfrac{1}{2}O_2 + H_2O$$

2.8 IONIC HALIDES

Although a conservative view of ionic systems (such as that adopted by Shannon in choosing crystals from which to obtain ionic radii) would suggest that only oxides, oxoanions, fluorides, and fluoroanions show predominantly ionic bonding in a crystal, the physical and chemical properties of the alkali and alkaline-earth halides are closely enough related that inorganic chemists usually consider all of these to show similar bonding and structural properties.

Not all halides are ionic, of course—not even all fluorides. When two elements meet, one of them a halogen, they have three thermodynamic choices: to form a compound involving substantial electron transfer and thus substantial ionic bonding; to form a compound involving predominantly electron sharing between atoms, which will thus have covalent or polar covalent bonds and be a molecular compound; or to form no compound at all. The system will choose from among these possibilities the one that minimizes its free energy. All elements form fluorides except He, Ne, and Ar; He, Ne, Ar, and Kr do not form chlorides; no noble gas forms a bromide; and neither the noble gases nor S or Se form stable iodides (although both iodocations and iodoanions of S and Se are well characterized). In all the other cases, a compound forms and a thermodynamic choice must be made as to the most favorable type of bonding. As Fig. 2.5 and the associated discussion have suggested, if the nonhalogen is easy to ionize, the halide will be largely ionic, whereas if it represents a large enthalpy cost the compound can lower its free energy more by adopting a covalent-molecular electronic arrangement.

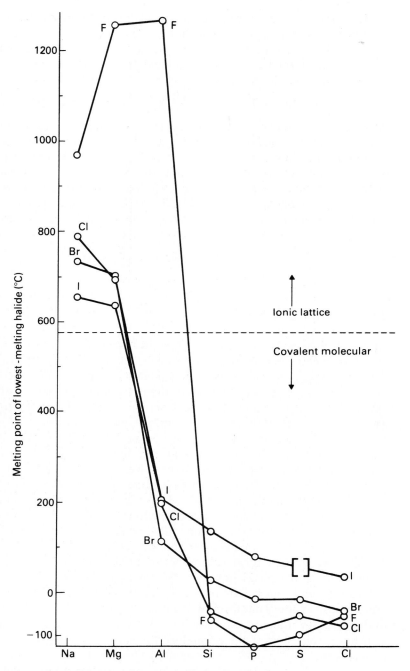

Figure 2.23   Melting points of second-row halides.

If we examine the physical properties of binary halides across a row of the periodic table, a break is apparent that can be correlated with changing bond type (see Fig. 2.23). Since ionic lattices have great thermal stability and molecular lattices (held only by van der Waals forces) have much less, melting point is often taken as an index of ionicity. There is a sharp transition in the melting-point curve for fluorides after AlF_3 and for the other halides after the alkaline-earth salts. Undoubtedly this does correspond to a transition in bonding, but crystal geometry is also extremely important. The reason it must be considered is that electron transfer between two atoms—the degree of ionicity in their bonding—is essentially determined by their electronegativity difference, which changes smoothly, not abruptly, going from Na to Mg to Al to Si for any halide. Presumably gaseous atom-pairs would become smoothly more covalent across this series, but lattice behavior is quite different. Thus all of the silicon halides appear to be about as covalent as the chlorine halides by the melting-point criterion, even though we know that a Si—F bond must be quite polar. The problem is that in an ionic lattice of Si^{4+} $(F^-)_4$, the fluorines would have to have a coordination number of at least two, which would require a silicon coordination number of eight. However, the (hypothetical) silicon ion is too small to occupy a cubic hole in a lattice. The result is tetrahedral molecular SiF_4, which has polar bonds but no net dipole moment. The molecular crystal, in turn, is held together by London dispersion forces between adjacent fluorines. These forces are quite small because the polarizability of a fluorine atom is so low. Arguments of this sort must be applied in interpreting any pattern like that shown in this figure; they will be developed further in Chapter 3.

Fluorine occurs in nature primarily as two ionic salts and as a fluoroanion comparable to the oxoanions we have studied. The largest source is the mineral *fluorspar*, CaF_2. Fluorspar is widely distributed over the earth's surface; over four million tons are mined annually, primarily in Mexico and Europe. In fluorine content an even larger source is *fluorapatite*, $Ca_5(PO_4)_3F$, which is the mineral most often extracted for its phosphate content. For economic reasons, much of the fluorine is discarded as HF even though the total HF discarded in U.S. phosphate plants is equal to our entire annual HF consumption. When the economic availability of fluorspar decreases or environmental pressures increase, recovery of phosphate fluoride will surely become the major source of the element. The third source is natural *cryolite*, Na_3AlF_6, in which the fluorine is present as the fluoroanion AlF_6^{3-}. Cryolite is extremely important as the molten-salt electrolyte in aluminum production, but deposits of it (in Greenland) are not large, and most cryolite is synthetically prepared:

$$Al_2O_3 + 12HF + 6NaOH \rightarrow 2Na_3AlF_6 + 9H_2O$$

Besides the massive use of fluorine by the aluminum industry, a large amount of fluorspar is used in steel production as a flux and liquefying agent for slag in the furnaces. Most fluorine is converted to HF, both for cryolite production and for fluorocarbon manufacture. The latter includes poly-tetrafluoroethylene resins (Teflon) and Halon firefighting liquids, but is pre-dominantly the Freon series of refrigeration fluids and aerosol propellants. The Freon replacements that are coming into production do not use chlorine, but will be net fluorine users at as great a rate as the current Freons. Modest amounts of fluorine are used (both as NaF and as H_2SiF_6) for water fluorida-tion, and sodium fluoride and sodium fluorophosphate are used in tooth-paste. In both cases, the object is to convert dental *hydroxyapatite*, $Ca_5(PO_4)_3OH$, to fluoroapatite, which is much more resistant to chemical attack by mouth acids.

In the laboratory, ionic fluorides are primarily useful for stabilizing high oxidation states of the cation they accompany. For example, the only silver(II) halide is AgF_2 and the only manganese(IV) halide is MnF_4. The reasons for this stabilizing effect are revealed by a Born–Haber cycle focused on the reaction in which the high oxidation state decomposes:

$$MX_n \rightarrow MX_{n-1} + \tfrac{1}{2}X_2$$

We can write the cycle

$$M^{n+}(g) + nX^-(g) \xrightarrow{-IP_n} M^{(n-1)+}(g) + (n-1)X^-(g) + X^-(g)$$

$$\left.-U_n\right\uparrow \qquad\qquad\qquad \left.+U_{n-1}\right\downarrow \qquad \left.-\Delta H_f^0[X^-(g)]\right\downarrow$$

$$MX_n \xrightarrow{\;\;\Delta H^0\;\;} MX_{n-1} \quad + \quad \tfrac{1}{2}X_2$$

ignoring the differences between lattice energies (which are really internal energies) and enthalpy. ΔH^0 for the decomposition reaction is given by

$$\Delta H^0 = U_{n-1} - U_n - IP_n - \Delta H_f^0(X^-(g))$$

Approximating the lattice energies by Kapustinskii's expression,

$$\Delta H^0 = -\frac{1202(n)(n-1)(1)}{r_{M^{(n-1)+}} + r_{X^-}}\left(1 - \frac{0.345}{r_{M^{(n-1)+}} + r_{X^-}}\right)$$

$$+ \frac{1202(n+1)(n)(1)}{r_{M^{n+}} + r_{X^-}}\left(1 - \frac{0.345}{r_{M^{n+}} + r_{X^-}}\right) - IP_n - \Delta H_f^0[X^-(g)]$$

If for simplicity we take the two cation radii to be the same and reduce the factor $1 - 0.345/(r_M + r_X)$ to its value for typical radii, 0.89, this expression reduces to

$$\Delta H^0 = 1070n\left[\frac{n+1}{r_M + r_X} - \frac{n-1}{r_M + r_X}\right] - IP_n - \Delta H_f^0[X^-(g)]$$

$$= \frac{2140n}{r_M + r_X} - IP_n - \Delta H_f^0[X^-(g)]$$

The first of these three terms makes a positive contribution to ΔH^0 (helps prevent decomposition), the second term makes a large negative contribution, and the third term makes a small positive contribution. Decomposition will thus be retarded more by a smaller denominator in the first term—another way of saying that the smallest halide ion will best protect the higher oxidation state. Fluoride is thus the best choice for stabilizing a high oxidation state.

There are a few fluoroanions of main-group elements that are stable in aqueous solution and have properties more or less comparable to similar oxoanions. They show little or no tendency to polymerize in the manner of many oxoanions, essentially because fluorine has such a high electronegativity that it is reluctant to increase its coordination number by donating electrons. The better-known fluoroanions of main-group elements are tetrafluoroborate, BF_4^-, hexafluoroarsenate(III), AsF_6^{3-}, hexafluorosilicate, SiF_6^{2-}, hexafluorophosphate, PF_6^-, hexafluoroarsenate(V), AsF_6^-, and hexafluoroantimonate, SbF_6^-. All of these are at least reasonably stable in water; a number of others can be made in other environments. Among the few polymeric fluoroanions are $As_2F_{11}^-$, $Sb_2F_{11}^-$, $Sb_3F_{16}^-$, $Sb_4F_{16}^{4-}$, and the remarkable species $(XeOF_4)_3F^-$. There are a few fluorocations: NF_4^+, N_2F^+, and NOF_2^+ have been reported.

To an overwhelming extent, chlorine occurs in nature as various ionic chloride salts, or as aqueous chloride ion in the oceans. There are large deposits of rock salt, NaCl, in many locations worldwide; these are the primary source of the element chlorine in commerce. Two other naturally occurring ionic chlorides that are extracted for their potassium content are *sylvite*, KCl, and *carnallite*, $KCl \cdot MgCl_2 \cdot 6H_2O$. Very little chlorine occurs in nature in any form other than as an ionic chloride.

Most chlorine is used as the elemental Cl_2 gas, although substantial amounts of NaCl are used as table salt and both NaCl and $CaCl_2$ are used in large quantities to melt highway ice. Cl_2 is produced primarily by the electrolytic oxidation of aqueous NaCl solutions: $Cl^- \rightarrow \frac{1}{2}Cl_2 + e^-$. At the cathode, water is simultaneously reduced to H_2: $H_2O + e^- \rightarrow OH^- + \frac{1}{2}H_2$. A chlorine plant thus also produces large amounts of NaOH, industrially known as *caustic soda*. The overall process is known as the *chlor-alkali* process.

The first uses of chlorine involved its oxoanions as textile bleaches. This is still a significant use, but over 99% of the chlorine in commerce is now used for other purposes. About two-thirds of it goes into organic chemicals: synthetic intermediates, solvents, plastics, insecticides, refrigerants, and dyes. This is a relatively recent development—industrial chlorine production has increased 300-fold since 1930, while (for comparison) steel production has approximately doubled. Laboratory uses for ionic chlorides involve primarily their high solubility in polar solvents to maintain a constant high ionic strength in solutions, along with the electron-donor properties of the Cl^- ion (to be discussed in Chapters 6, 7, and 12).

Bromine, like chlorine, is essentially found in nature only as the halide anion. However, very few solid minerals contain significant amounts of bromide. Although chemical similarities might lead one to expect bromide substitution in rock salt, most mineral NaCl contains less than 0.04% Br^- because substituting the larger bromide ion in the NaCl lattice is energetically unfavorable. The overwhelming proportion of the earth's bromide ion is found in seawater or brine pools. It is commercially extracted from these by chlorine oxidation to elemental Br_2, which is only slightly soluble in water and can be removed by an air current through the seawater or brine:

$$Cl_2 + 2Br^- \rightarrow Br_2 + 2Cl^-$$

Seawater contains about 65 ppm Br, but some brines (Dead Sea, Michigan, California) contain up to 0.5% Br, simplifying the materials-handling problem.

Bromine is much less abundant than chlorine or fluorine, and it is produced by industry at only about 1% the rate of chlorine. In commerce, its greatest use is in ethylene dibromide, which is a fumigant and was formerly added to leaded gasoline to remove lead from the combustion chamber as the volatile $PbBr_2$. Another major use is as the photochemically active AgBr in photographic film (although here the silver is the critical ingredient rather than the bromine). In the laboratory, bromine is an indispensable reagent for the organic chemist because it combines the electron-acceptor property of the other halogens (in a fairly gentle form) with a convenient liquid state at room temperature. Comparable uses in inorganic synthesis are much rarer, although the electron-donor capability of the bromide ion in metal complexes is of continuing interest in coordination chemistry.

Unlike any of the other halogens, iodine (though a relatively rare element) occurs in nature to a considerable extent as an oxoanion, IO_3^-; the Chilean nitrate deposits contain about 0.1% I as $Ca(IO_3)_2$. Chile thus contributes about half the world's iodine production. The iodate is concentrated in water by repeated crystallization of the sodium nitrate. It is then reduced to I^- and reoxidized to I_2 using more of the iodate solution:

$$2IO_3^- + 6HSO_3^- + 6H_2O \rightarrow 2I^- + 6SO_4^{2-} + 6H_3O^+$$

$$5I^- + IO_3^- + 6H_3O^+ \rightarrow 3I_2 + 9H_2O$$

Iodine also occurs in natural brines along with bromine and is extracted in the same way, although the concentrations are lower (30 ppm).

There are no major industrial uses of iodine or iodides, though it has some key small-scale applications. For example, the catalysts for stereospecific polyolefins use TiI_4, some photographic emulsions use AgI, some cloud-seeding experiments use AgI vapor to form ice nuclei, and I_2 is used as a disinfectant and as a tungsten-vapor scavenger in quartz-iodine lamps. In the laboratory, organic iodides have the same synthetic advantages that organic bromides do, as they react under very mild conditions. Traditionally, a far wider use has been in iodimetric redox titrations for quantitative analysis. The

I^- ion is very easily oxidized to I_2 (for instance, by Cu^{2+} and H_2O_2). In most iodimetric methods, excess iodide is added to a sample having this oxidizing capability, then the iodine produced is titrated with thiosulfate:

$$Cu^{2+} + 2I^- \rightarrow CuI + \tfrac{1}{2}I_2$$

$$I_2 + 2S_2O_3^{2-} \rightarrow S_4O_6^{2-} + 2I^-$$

The endpoint is marked by the disappearance of the intense blue color of the complex formed between I_2 and a small amount of starch indicator.

In later chapters, we shall consider many of the acid–base and redox properties of the halides. Here we will conclude an introductory survey of the halides by summarizing their physical properties as they relate to the degree of ionicity in the compound. Table 2.6 provides a basis for this comparison. In this chapter we have been concerned with the ionic and partly ionic halides; in Chapters 4 and 5 we shall examine the bonding and properties of the covalent molecular halides. The background material already presented, however, should allow a perspective on that class of halides as well.

A final topic in the diverse subject of ionic halides is the group of anions known as *pseudohalides*. The accepted group of pseudohalides includes cyanide, CN^-; azide, N_3^-; cyanate, OCN^-; thiocyanate, SCN^-; selenocyanate, $SeCN^-$; tellurocyanate, $TeCN^-$; and azidothiocarbonate, $SCS(N_3)^-$. The pseudohalide definition depends on the following properties, not all of which are shown by every member of the series:

1. The anions have a single negative charge and an electronegativity (averaged over all atoms) not greatly different from Cl^- and Br^-.

2. They form ionic salts that have most of the properties suggested for ionic halides in Table 2.6.

3. The pseudohalide ion Z^- can be oxidized to a pseudohalogen, Z_2. Of the listed ions, only $TeCN^-$ and N_3^- do not meet this criterion. For example,

$$2AgSeCN + I_2 \rightarrow (SeCN)_2 + 2AgI$$

4. The metal pseudohalides parallel metal halides in their water solubilities. The alkali and alkaline-earth pseudohalides are soluble, but the silver, mercury(I), and lead(II) pseudohalides are only sparingly soluble.

5. The pseudohalogen hydrides are acids, as are the hydrogen halides HX. However, the HZ pseudohalogen acids are weak, with pK values in the 4–10 range.

Alkali salts of the pseudohalides are prepared according to the following reactions:

$$N_2O + NaNH_2 \xrightarrow{\text{Fused}} NaN_3 + H_2O$$

$$CaC_2 + N_2 \xrightarrow{1100°C} CaNCN + C \xrightarrow{+Na_2CO_3} 2NaCN + CaCO_3$$

$$KCN + PbO \longrightarrow Pb + KOCN$$

$$S + NaCN \xrightarrow{\text{Fused}} NaSCN \quad (\text{also } SeCN^-)$$

$$CS_2 + NaN_3 \longrightarrow NaSCSN_3$$

Table 2.6 Physical Properties of MX$_n$ Halides

Property	Ionic halides	Partly ionic halides	Covalent molecular halides
Formed by	Most metals from groups Ia, IIa, IIIa, and transition-metal fluorides.	Transition metals and group IIIb and IVb metals.	Nonmetals and all MX$_n$, where $n > 3$.
Electronegativity difference between M and X	Generally ≥2.0, but as low as ~1.2 for iodides.	Generally between 1.0 and 2.0, but as low as 0.5 for iodides.	Generally between 0 and 1.0.
Bonding description	Electrostatic lattice model; ordered 3-dimension lattices with high coordination numbers for all atoms; lattice energies very well reproduced by Madelung-constant expression.	Electrostatic model or band theory; halide in unsymmetrical environment in lattice, usually chain or layer structures, low coordination numbers frequent; lattice energies from Madelung constant deviate from experimental by 5–20%.	Shared-electron bonds (valence-bond or molecular-orbital models); symmetry of molecules predictable by VSEPR; very weak bonding in solid lattice due to van der Waals forces.
Sublimation energy	← (Increasing) Comparable to bond energies; vapor species frequently polymers of lattice formula.	Vapor polymers less common.	Typically 5–15% of bond energies; vapor species normally molecular unit.
Boiling point	1200°C to 1600°C — ← Generally MF$_n$ > MCl$_n$ > MBr$_n$ > MI$_n$ with order dictated by coulomb forces.	—— (Increasing) ←	−100°C to +300°C Generally MI$_n$ > MBr$_n$ > MCl$_n$ > MF$_n$ with order dictated by polarizabilities.
Melting point	600°C to 1000°C — ← Order similar to the boiling point order.	—— (Increasing) ←	−150°C to +200°C Order similar to the boiling point order.

122

Electrical conductivity:			
Melt	High conductivity.	Relatively low conductivity.	Very low conductivity due to autoionization.
Solid	Low conductivity because of high energy barrier for ion transport.	Varies; some disordered or open lattices permit ion migration; some halides have partial metallic conductance.	Very low conductivity.
Heat of formation per mole of halogen atom	200–400 kJ/mol X $\longrightarrow$ ΔH_f reproduced well by Madelung-constant calculations; variations match those in lattice energy. $MF_n < MCl_n < MBr_n < MI_n$	—— (Increasing) $\longleftarrow$ ΔH_f reproduced poorly by Madelung-constant calculations; deviations increase $MF_n < MCl_n < MBr_n < MI_n$	40–150 kJ/mol X Ionic model unusable; variations of ΔH_f determined by variations in bond energy $M—F < M—Cl < M—Br < M—I$
Solubility	Favored by polar coordinating solvents of high dielectric constant; solubility dictated by balance between lattice energy and solvation energies of ions. Solubility for given cation generally increases $MF_n < MCl_n < MBr_n < MI_n$	Directional bonding stabilizes lattice relative to solution; solubility for given metal generally increases $MI_n < MBr_n < MCl_n < MF_n$	Favored by nonpolar media; solubility in polar or hydrogen-bonding solvents enhanced if MX_n molecule is polar or has H-bonding capability.
Hydrolysis	$\longrightarrow$ (Increasing tendency) $\longrightarrow$		$\longrightarrow$ (Increasing tendency)

Source: Adapted with permission of Pergamon Press Ltd. from Table 21 of A. J. Downs and C. J. Adams, *Chlorine, Bromine, Iodine, and Astatine,* 1975; a reprint of *Comprehensive Inorganic Chemistry* (vol. 2, ch. 26) by J. C. Bailar, Jr. (G. Wilkinson, ed.), Oxford, U.K.: Pergamon Press, 1973.

123

The pseudohalide classification is an interesting but somewhat arbitrary one. It is not entirely clear why, for example, the nitrite ion (which forms N_2O_4) and the hydride ion are not included. Later chapters will take up some of the interesting chemical properties of these ions.

PROBLEMS

A. DESCRIPTIVE

A1. Place the following gaseous atomic species in order of increasing polarizability, with arguments for your ordering:

$$Al^{3+} \; Br^- \; Cl^- \; Mg^0 \; Mg^{2+} \; N^{3-} \; Na^+ \; Ne^0 \; P^{3-} \; S^{2-} \; Sc^{3+} \; Ti^{2+}$$

A2. Suggest which crystal lattice is most likely for each of the following compounds:

$$BN \; CaS \; Cs^+SH^- \; GaP \; KF \; PtS \; TiO \; TlBr$$

A3. Sketch a close-packed layer of spheres in two dimensions. If this layer is the c layer indicated in the chapter's discussion for metal atoms/ions in the defect NiAs structure of Fe_7S_8, mark spheres to be omitted to produce the Fe_7S_8 structure in the most symmetrical way.

A4. The alkali metals do not yield the same product on burning in oxygen gas. Do the charge and size relationships of the oxoanions O^{2-}, O_2^{2-}, and O_2^- explain the pattern of different products? Explain the lattice considerations involved.

A5. Write reasonable Lewis structures for the $N_2O_2^{2-}$ ion, the $N_2O_3^{2-}$ ion, and the $N_2O_4^{4-}$ ion. Which should have least bonding between N atoms?

A6. The Cl—O bond length (internuclear distance) is 1.48 Å for ClO_3^- and 1.44 Å for ClO_4^-. Why is the thermochemical radius of ClO_4^- much greater than that of ClO_3^- (2.22 Å versus 1.86 Å, Table 2.3)?

A7. By analogy with the comparable xenon compound, predict the reaction of IF_5 with water in basic solution.

A8. Write reaction equations for the following oxides dissolving in concentrated aqueous potassium hydroxide: (a) ZrO_2; (b) Li_2O; (c) Al_2O_3; (d) BaO.

A9. Why should it be true for partly ionic halides (Table 2.6) that deviations from ΔH_f^0 increase from MF_n to MI_n when ΔH_f^0 is estimated by a Born–Haber cycle using a calculated lattice energy?

A10. Why should ionic fluorides have a higher boiling point than ionic bromides or iodides, even though covalent molecular halides show just the opposite trend? (See Table 2.6.)

B. NUMERICAL

B1. Calculate the magnitude (in kJ/mol) of the net attraction between the two atoms in a Na^+—Cl^- gaseous ion pair.

B2. Calculate the Madelung constant for a cation in a square-packed layer of ions with alternating charges, a two-dimensional lattice analogous to the one-dimensional lattice in Fig. 2.2. Use the rapid-convergence approach, and include at least three rings of ions around the central cation. How does your most accurate value compare with those for zero-, one-, and three-dimensional lattices?

B3. Calculate the lattice energy for CaH_2 as an ionic crystal, making allowance for appropriate coordination numbers. Use your result to estimate ΔH_f^0 for CaH_2. How good a fit does your calculation yield for the experimental value of -1.89 kJ/mol? Comment in light of the special nature of the hydride ion.

B4. Estimate ΔH_f^0 for ionic ScH_3 as in the previous problem. What is the thermodynamic likelihood of forming this compound? In what way does your answer depend on the polarizability of the hydride ion?

B5. Use a Born–Haber cycle to estimate ΔH^0 for the disproportionation of CaI to Ca^0 and CaI_2. Compare your result with the chapter's ΔH^0 for CaCl. What influence does anion radius have? Is it possible to stabilize Ca^+ in an environment of ions of any size? Explain.

B6. Use data appropriate to the functions of Fig. 2.5 to decide whether SnF_4 might be stable as an ionic lattice. Experimentally, SnF_4 sublimes at about 800°C. Is this consistent with your prediction?

B7. Calculate what the 4-coordinate and 2-coordinate radii of Cu^+ should be, working from the 6-coordinate crystal radius. How do the resulting values compare to the values in Table 2.3? What electronic reasons can you offer for the differences?

B8. Use the metallic radii in Table 4.7 to calculate SPI radii for the isoelectronic ions from K^+ through Cr^{6+}. How do your calculated values compare to those of Table 2.3 and to the crystal radii of the same table?

B9. Show that the limiting radius radio r/R for the CsCl lattice is 0.7321.

B10. In this chapter, we have pointed out that close packing yields the highest possible density of spherical atoms in a lattice. A quantity called the *packing fraction* ϕ measures this density as the ratio of the volume of spheres in a cube of three-dimensional space to the volume of the cube. Calculate the packing fraction for ccp spheres, such as the Cl^- ions in the NaCl lattice of Fig. 2.4 with the Na^+ ions missing. The close-packed atoms touch along the diagonals of the cube faces.

C. EXTENDED REFERENCE

C1. Use a Born–Haber cycle treatment analogous to that for the stabilizing effect of F^- on high oxidation states to indicate the effect of M^+ cation radius on the reaction

$$2MO_2(s) \xrightarrow{\Delta H^0} M_2O_2(s) + O_2(g)$$

[See D. A. Johnson, *Some Thermodynamic Aspects of Inorganic Chemistry* (Cambridge University Press: Cambridge, 1968).]

C2. Just as removing cations from the NiAs lattice leads to the CdI_2 lattice, further removal of cations leads to the BiI_3 lattice and, ultimately, to the UCl_6 lattice. Sketch a symmetrical cation distribution (in the c layer of Fig. 2.14) that would yield the correct stoichiometry for BiI_3 and UCl_6. [See *Acta Crystallographica* (1974), *B 30*, 1481.]

C3. Within the pseudohalides N_3^-, NCS^-, and NCO^- pi bonding is quite extensive in each three-atom system; the pi electrons are readily polarizable. Comment on the relation between this characteristic of the anion structure and the crystal structure of alkali metal pseudohalides. [See Z. Iqbal, *Structure and Bonding* (1972), *10*, 25.]

C4. The salt Cs_2SbCl_6 is known, but it contains half Sb^{III} and half Sb^V, not the Sb^{IV} the formula might suggest. Write a Born–Haber cycle for the disproportionation reaction $2Cs_2Sb^{IV}Cl_6 \rightarrow Cs_2Sb^{III}Cl_6 + Cs_2Sb^VCl_6$, which the experimental results suggest must be exothermic. Assume the salt contains Cs^+ and $SbCl_6^{q-}$ ions, and that the size of the $SbCl_6^{q-}$ ion is determined by the crystal radii of Cl^- and Sb^{n+}. What information will you need to extract ΔH^0 for the disproportionation reaction from your Born–Haber cycle? [See K. Prassides and P. Day, *Inorg. Chem.* (1985), *24*, 1109.]

3

Directional Bond Networks and Solid State Chemistry

The previous chapter has developed the energy relationships that underlie the great stability of ionic crystals, and has surveyed some of the chemical systems that show predominantly ionic bonding. Precisely because of this great stability, these systems include some of the most familiar chemicals—sodium chloride and limestone (calcium carbonate), to name two. But in fact, no chemical system is completely ionic in its bonding. Even cesium fluoride has at least a small shared-electron or covalent contribution to its stability. As we have seen, for the compounds in Chapter 2 this contribution can be neglected to a good approximation because the slight overestimate of electrostatic attraction almost exactly compensates for the neglect of covalent bonding. This leaves us, however, with an obvious question: What changes in solid lattice properties can we expect when covalent bonding becomes an important component of the total lattice stability? In this chapter we will examine these changes and consider some of the important chemical systems and applications that result.

3.1 IONICITY, POLARITY, AND DIRECTIONAL BONDS

"Ionic" compounds form when the electronegativity difference between touching atoms is so great that electron transfer between the two atoms—from the less electronegative to the more electronegative—is overwhelmingly favorable. Under this circumstance, the Madelung calculation of lattice energy will reproduce the Born–Haber cycle value quite well, as we have seen. As a very rough guide, we can expect this to be true whenever the electronegativity difference between touching atoms, $\Delta\chi$, is greater than two electronegativity units. Thus KF has a neighbor-atom $\Delta\chi$ of $4.10(\text{F}) - 0.91(\text{K}) = 3.19$ and would be expected to be ionic to quite a good approximation. Sodium chloride yields $2.83 - 1.01 = 1.82$ and fails to meet this criterion, but not by much, and we can still consider it ionic. Obviously an electronegativity difference of 1.95 is no different from one of 2.05; we have to choose a threshold somewhere, and it's better to have one with only one significant figure in view of the crudeness of the rule.

A major reason for the crudeness of the rule is the fact that the attraction of an atom in a compound for neighboring electrons depends not only on the intrinsic character of the atom (its effective nuclear charge) but on its net charge within the array of atoms. In Chapter 1 we have already seen that a more sensitive characterization of the electron-attracting character of an atom is its differential ionization energy, $I(q)$, which consists of two parameters a_0 and a_1. a_0 characterizes the intrinsic attraction of an electrically neutral atom for electrons, and a_1 measures the rate at which that attraction increases with increasing positive charge. Chapter 1 suggested that electronegativity was more or less proportional to a_0, but it should be clear that an atom with a high net positive charge will have a large contribution to its electronegativity from the a_1 term in $I(q)$. Because the positively charged atom/ion is always the less electronegative one in a neighbor-atom pair, increasing its electronegativity for the net-charge effect reduces $\Delta\chi$; no atom pair in which the formal charge of the cation exceeds about 3 can possibly have $\Delta\chi$ large enough to show predominantly ionic bonding. In an MnO_4^- ion, even though the nominal electronegativity difference between Mn and O is 1.90, the +7 oxidation state of manganese guarantees that the bonding within the MnO_4^- ion will be covalent.

There are also steric considerations that limit the formation of ionic crystals. Consider the coordination number relationships that must hold in an extended ionic crystal. In one mole of A_mB_n crystal, there will be $m \cdot n$ moles of A–B contact points between ions, or an integral multiple of this number. Then the coordination number of each A atom must be $m \cdot n/m$ or n, and by the same reasoning the coordination number of B must be m—or $2n$ and $2m$, or $3n$ and $3m$, or some higher multiple. In Al_2O_3, if Al is three-coordinate O must be two-coordinate, or if Al is six-coordinate O must be four-coordinate, or if Al is nine-coordinate O must be six-coordinate, and so on. Experimentally, Al is six-coordinate and O is four-coordinate in the Al_2O_3 crystal. This does not seem like much of a steric limitation, but it becomes one when the stoichiometry of the compound becomes more lopsided.

In AlF_3 the comparable coordination-number ratio for Al and F would be $3:1$, $6:2$, $9:3$, ... Experimentally the ratio is $6:2$, but the octahedron of F around Al is distorted considerably so that three F are much closer than the other three, which may represent incipient steric problems with the approach of six negatively charged anions around a very small cation. For the comparable compound InF_3, the coordination is again $6:2$, but the larger In^{3+} ion can accommodate six fluorides in a nearly ideal octahedron. In electronegativity terms, there should be very little difference between Al and Si with respect to bonding to F; $\Delta\chi$ is 2.63 and 2.36 respectively. But the 1:4 stoichiometry of SiF_4 requires a coordination number ratio of 4:1, 8:2, 12:3, or a higher multiple. To have an extended lattice, the F coordination number must be at least 2—but then Si must be eight-coordinate, which as we have seen in the previous chapter is an impossible steric burden. Accordingly, SiF_4 forms a species with four-coordinate Si and one-coordinate F, which is an isolated SiF_4 molecule rather than an extended lattice. Of course, the formal charge of 4+ on Si as an ionic species also reduces the effective electronegativity difference between Si and F, which also reduces the ionicity of the bonding for SiF_4.

For most main-group cations, steric saturation will occur at a coordination number between six and eight, as in the NaCl and CsCl lattices. This makes it tempting to assume that when we see a six-coordinate metal in a crystal structure the bonding is predominantly ionic, but this is always unreliable and frequently untrue. For example, the six-coordinate NaCl lattice is adopted by SnAs and the eight-coordinate CsCl lattice is adopted by TlI, and neither of these compounds can possibly be thought predominantly ionic. As Chapter 4 will show, it is perfectly possible to form six or eight covalent bonds to a given atom if it has d valence orbitals. This means that high coordination numbers simply don't provide any information about bond type for a compound. However, the nondirectional character of electrostatic attraction does mean that predominantly ionic compounds will essentially always reach steric saturation with the more highly charged ion (usually the cation) having coordination number six or higher.

A low "cation" coordination number in an extended lattice, however, does indicate that the maximum electrostatic attraction is not occurring. This implies that covalent bonding may well be controlling the "cation" coordination number and that directional bonds probably control the geometry even if electrostatic attraction is an important component of the total lattice stability, as Chapter 2 has suggested.

The conflicting influences of ionicity and steric interference can be seen in Fig. 3.1, which displays the coordination pattern chosen by about 60 binary compounds with A^nB^{8-n} stoichiometry, in which n and $8-n$ refer to the number of valence electrons on each atom. It can be seen that if the electronegativity difference between A and B is large *and* the ion size disparity is not too great, the higher coordination number will always be chosen. On the other hand, if the electronegativity difference is low, four-coordinate geometry (implying four directional polar covalent bonds) will result even if formal ionic radii are not too different, but particularly if they are. If we sketch in a boundary line passing nearly through the three points representing compounds (MgS, MgSe, HgSe) showing both six-coordinate and four-coordinate crystal structures, the line almost perfectly divides the two groups of compounds. The two prominent exceptions are lithium iodide, which apparently does not form covalent bonds strong enough to dictate lattice geometry, and

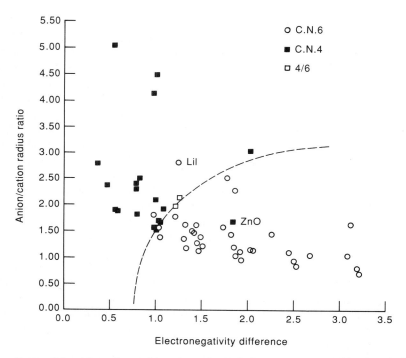

Figure 3.1   Atom size and bond covalence influences on crystal lattice choice.

zinc oxide, which seems to form exceptionally strong polar covalent bonds that limit the zinc coordination number even though the ion sizes are not too disparate.

3.2 POLYMERIC NETWORKS OF POLAR BONDS

As has been suggested before, not all ionic compounds have structures or chemical properties that can be adequately described by the charged-hardsphere model. When Shannon compiled his extensive survey of ionic radii, he used only data for oxides and fluorides to establish cation radii, although he includes other anionic radii from Pauling's table (adjusted for consistency). Sulfides, selenides, iodides, and other such compounds are rarely found in environments that can be thought of as strongly ionic in their bonding— never, for example, in a rutile lattice. It is therefore necessary to include the energetic and geometric requirements of covalent bonding in discussing the structure and properties of all such compounds.

The principal difference between electrostatic attraction of M^+ for X^- and covalent bonding between M and X is that electrostatic attraction is completely isotropic or nondirectional (no angles appear in the coulomb law). Covalent bonding, on the other hand, is highly directional; for stability, it requires good overlap of directional orbitals on the atoms involved. Since covalent-bond energies are comparable to electrostatic lattice energies (per atom pair) and can drop to zero for geometries in which there is no net orbital overlap, it should be clear that there are important energy consequences of alterations in bond angles. For example, in the rutile lattice the trigonal planar coordination of the anion is commonly found in covalent molecules, but the bond angles of 90°, 135°, 135° are quite unusual and do not lead to ideal overlap. This may be a substantial reason for the reluctance of partly ionic systems that are also partly covalent to adopt that lattice.

Another major difference between ionic and covalent bonding is that in ionic lattices, there is no limit on coordination number except that imposed by the repulsions of the many anions crowding about a given cation, or vice versa. Under all but the most unusual circumstances, however, an atom participating in a covalent bond will have its coordination number limited to the number of its valence orbitals. Thus a true sulfide ion could be six-coordinate in the NaCl lattice or even eight-coordinate in the CsCl lattice; but if its coordination number is limited by covalent bonding, it should not exceed four, since sulfur has one $3s$ and three $3p$ valence orbitals. Exceptions to this general rule are possible under circumstances Chapter 4 will explore (such as, for example, in the molecule SF_6), but the rule has important consequences in determining the structure of partly ionic compounds.

The covalent limitation on coordination number has an important effect on the crystal structures of partly ionic compounds beyond the simple preference for, say, the zincblende structure over the NaCl structure. We frequently find experimental evidence for polymeric structure such as chains or layers in the lattices of compounds having significant covalent bonding. We can usefully characterize crystals in terms of the geometric degree to which covalent bonding exists between the basic chemical units of the substance:

1. Three-dimensional polymers: Extended covalent bonding throughout the crystal in all three directions; not always distinguishable from a three-dimensional ionic crystal unless unusual coordination geometries are present.

2. Two-dimensional polymers: Covalently bonded layers of atoms or molecular units are held together by van der Waals forces or by isolated ions.

3. One-dimensional polymers: Commonly seen as chains of atoms or molecular units, but short chains (such as dimers) may form or rings may be present. Also includes multiple cross-linked chains, as long as the number of chains is small compared to the chain length.

4. Zero-dimensional polymers: Monomer units showing no covalent bonding between them, either because the crystal is strongly ionic (Na^+Cl^-) or because it is a molecular crystal (naphthalene).

An interesting example of the effect of changing covalent character on apparent polymerization in crystal structures is provided by the indium halides, which have the general formula InX_3. The indium coordination polyhedra in these crystal structures are shown in Fig. 3.2. The most ionic, InF_3, adopts the ReO_3 lattice (also seen in Fig. 3.20), in which each indium is surrounded by an octahedron of fluoride ions. A way to think of this lattice is as ccp fluoride ions with one-fourth of the ccp positions vacant, containing indium ions in one-fourth of the octahedral sites. Whether one thinks of this as an ionic crystal or as a three-dimensional polymer, there are no unique

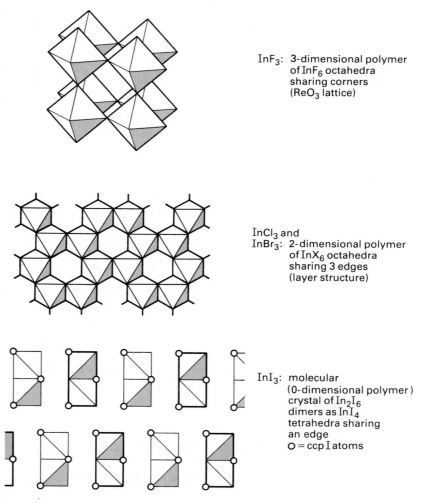

InF_3: 3-dimensional polymer of InF_6 octahedra sharing corners (ReO_3 lattice)

$InCl_3$ and $InBr_3$: 2-dimensional polymer of InX_6 octahedra sharing 3 edges (layer structure)

InI_3: molecular (0-dimensional polymer) crystal of In_2I_6 dimers as InI_4 tetrahedra sharing an edge
O = ccp I atoms

Figure 3.2   Crystal polymers of InX_3 indium halides.

directions within the crystal. $InCl_3$ and $InBr_3$, however, still form octahedral InX_6 coordination polyhedra but have a layer structure analogous to that of CdI_2 (Fig. 2.14). This is a two-dimensional polymer consisting of sheets of octahedra sharing three edges; the sheets are held together by van der Waals forces between the nearly hcp Cl or Br atoms. Finally, InI_3 forms In_2I_6 molecules, each of which consists of two InI_4 coordination tetrahedra sharing one edge. These double-tetrahedron molecules stack in the crystal in such a way as to keep the iodine atoms nearly ccp, even though the only forces between the In_2I_6 molecules are van der Waals forces. If one takes the basic unit of the compound to be an InI_3 monomer, then a linear dimer has formed and is stacked in the crystal appropriately. The thermal properties of these compounds agree with the supposed increase of covalent character from InF_3 to InI_3: the melting points of the compounds are 1170°C for InF_3, 386°C for $InCl_3$, 436°C for $InBr_3$, and 210°C for InI_3. The two with the layer structure sublime well below their melting points.

Although there might seem to be a very limited variety of ways to form a chain-type one-dimensional polymer, there are in fact many examples with quite varied forms of chain linkage. The simplest is that of plastic sulfur and of the stable forms of elemental selenium and tellurium: —Se—Se—Se—Se—Se—, although the bond angles are not 180° and the chain is thus either kinked or helical. The same general kind of chain is formed in compounds such as HgO and AuI, where the atoms alternate along the chain. If the stoichiometry is 2:1 (AB_2), we find structures in which both B atoms are part of the bridging chain:

as in $CuCl_2$ and $PdCl_2$ (with square planar coordination around the metal) and $BeCl_2$ and SiS_2 (with tetrahedral coordination around the metal). However, we also find structures with one bridging and one terminal B atom:

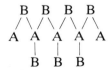

as in SeO_2.

Octahedral coordination is common in chain structures. If only two corners of the octahedron are shared, the overall stoichiometry will be AB_5; if two edges (usually *trans* to each other) are shared, the stoichiometry is AB_4; and if two faces are shared, the stoichiometry is AB_3. Examples of these are CrF_5, NbI_4, and ZrI_3 respectively. An individual chain may have a net charge (normally negative), in which case it will be separated from the next chain in the crystal by counterions to maintain charge neutrality in the bulk crystal.

We have already encountered two related layer structures, each layer of which is really a sandwich: the CdI_2 lattice and the BiI_3 lattice. In both, the center layer of the sandwich is the metal and the outer layers are halides. One further step in this direction is Bi_2Se_3, which is a kind of club sandwich containing three layers of close-packed selenium atoms with bismuth atoms in all the octahedral sites. The bulk crystal is then composed of stacks of these multiple layers. However, there are many simpler layer structures. Graphite is an obvious example of a hexagonal-network layer. Elemental arsenic also has a graphite-type structure, but because the bond angles are less than $120°$, each six-membered ring must be buckled like cyclohexane, so the layers of arsenic atoms are buckled. The 1:1 compounds can have the same structure as graphite (BN, for example) or arsenic (SnS) if the atoms alternate in the layer. We can also imagine layer structures containing square networks instead of hexagonal networks. LiOH forms such a structure with alternating Li and O atoms in which the layers are buckled so that oxygen atoms are above and below the center of the layer and lithium atoms are in the center. HgI_2 has a structure similar to that of LiOH, but half the cation positions must be vacant to maintain the correct stoichiometry. Several of these layer structures are shown in Fig. 3.3.

As we have just noted, a highly symmetrical crystal in which directional bonding produces a three-dimensional polymer is not always geometrically distinguishable from an equivalent strongly ionic system. But the Madelung expression for electrostatic lattice energy yields quite good approximations to the Born–Haber value taken from the enthalpy of formation of the crystal if the bonding is indeed predominantly electrostatic, essentially because the error from ignoring the small covalent contribution is nearly compensated by the slight overestimate of the electrostatic contribution. We saw in Chapter 2 that NaCl has a Born–Haber lattice energy of -787 kJ/mol, while the Madelung calculated value is -757 kJ/mol. By contrast, when a similar comparison is made for a crystal with substantial covalent-network bonding, the Madelung calculation is usually a very poor fit for the Born–Haber cycle value: For AgI, which has an electronegativity difference of less than 0.8 unit and is four-coordinate, the Born–Haber lattice energy is -1059 kJ/mol while the electrostatic approximation yields a lattice energy of only -670 kJ/mol. This is obviously hopelessly deficient, and makes it clear that covalent bonding is thermodynamically important to the AgI crystal.

Two lattices are often seen in three-dimensional polymeric network crystals. The first we have already seen: the ReO_3 lattice, seen in Fig. 3.2 for the InF_3 crystal. We can distinguish it from a truly "electrostatic" lattice because its approximate close-packing of anions has regular vacancies, which leave large cavities in the crystal that could, under the right circumstances, contain other atoms or ions. We shall return to this possibility shortly. Note that

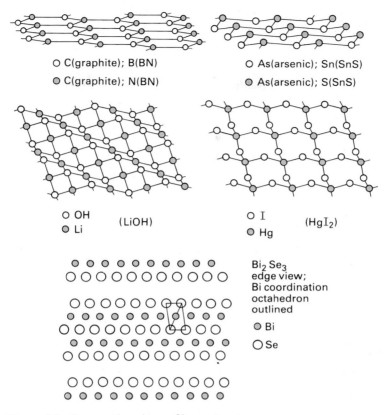

○ C(graphite); B(BN)
◉ C(graphite); N(BN)

○ As(arsenic); Sn(SnS)
◉ As(arsenic); S(SnS)

○ OH   (LiOH)
◉ Li

○ I   (HgI₂)
◉ Hg

Bi₂Se₃
edge view;
Bi coordination
octahedron
outlined

◉ Bi

○ Se

Figure 3.3 Perspective views of layer structures.

because of the 1:3 stoichiometry of the compound the coordination number ratio must be 6:2.

Another three-dimensional polymeric network crystal lattice adopted by over a hundred compounds is the $PbCl_2$ lattice, an idealized version of which is seen in Fig. 3.4. In this lattice the Pb atom is nine-coordinate, approximating the tricapped trigonal prism shown, though with distortion. The three-dimensional polymer is formed by first joining a chain of tricapped trigonal prisms together by sharing capping edges. A second such chain can be placed next to the first, sharing each edge, if the capping Cl atoms from the first chain become prism atoms for the second chain so that the two chains are not at the same height and if the direction of the second chain is reversed relative to the first chain. This edge sharing can be extended both in the plane of the figure and perpendicular to it. In the figure, the triangular prism faces in one chain are shaded for emphasis and its basic direction is indicated by arrows; this chain is slightly higher than the neighbor unshaded chain, leading to a

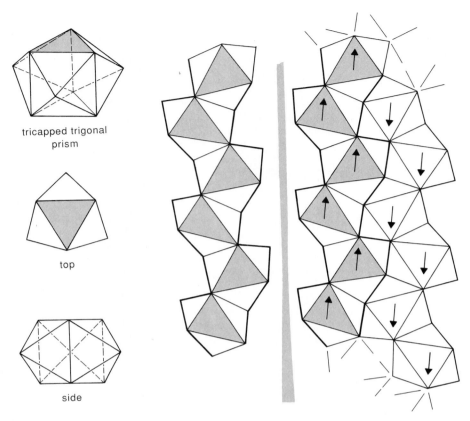

tricapped trigonal
prism

top

side

Figure 3.4 PbCl$_2$ lattice as formed from nine-coordinate polyhedra joined into chains, joined into layers, joined into a three-dimensional network.

corrugated sheet when many chains are joined. The corrugated sheets, in turn, can stack to a true three-dimensional polymer. The AB$_2$ compounds adopting this structure tend to be halides (rarely fluorides), hydrides, sulfides, phosphides, and similar species in which electronegativity differences are not too great and covalent bonding becomes important.

If vacancies are symmetrically arranged in planes within a three-dimensional polymer conceptually built up from layers, we can see two-dimensional layer structures. In Chapter 2 we saw (Fig. 2.14) that the hexagonal-close-packed-anion NiAs structure yields the layered CdI$_2$ structure if all the cations are removed from octahedral holes in alternate layers of those holes. The hcp-anion CdI$_2$ and ccp-anion CdCl$_2$ layered structures are quite common, and in particular the I–Cd–I sandwich layers can be stacked in a

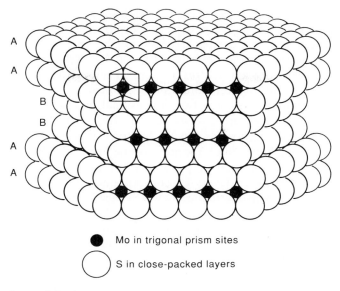

● Mo in trigonal prism sites

○ S in close-packed layers

Figure 3.5 Molybdenite MoS_2 as an example of layer structures bound by van der Waals forces.

number of different symmetries, as we shall see in considering superstructures (Section 3.4).

Regardless of the pattern of network bonds within a given layer of a two-dimensional polymeric lattice, some forces have to hold the layers together in the final crystal. In increasing order of binding strength, these forces can be van der Waals forces between electrically neutral layers (as in CdI_2), hydrogen bonding from a protic hydrogen on one layer to an electron donor atom on the adjacent layer, and electrostatic forces involving charged layers.

In the first case (which represents the largest class of layered compounds), graphite as well as CdI_2 obviously represent layers held together by van der Waals forces, but as we shall see in discussing the silicates, talc $[Mg_3(OH)_2Si_4O_{10}]$ also has a structure consisting of electrically neutral layers (even though electronegativity differences are substantial within the layers and the polarity of the bonds should be high). Because van der Waals forces are so weak, all of these compounds' layers are very easily separated; they are quite soft and often good lubricants. A particularly interesting example is MoS_2 (*molybdenite*), which has the layered structure shown in Fig. 3.5. The electrically neutral sandwich layer is formed by superimposing two sheets of S atoms and placing Mo atoms in sites between S sheets so that each Mo is surrounded by six S at the corners of a trigonal prism and each sulfur is bonded to three Mo atoms. Because the two S sheets are not close-packed the final three-dimensional structure can't be close-packed, but the S-Mo-S layers can pack together in a close-packed fashion; the final sequence of S sheets becomes (using the notation of Fig. 2.10) AA BB AA, and so on. The van der Waals forces between neutral S atoms in the AB plane are weak, and MoS_2 is an excellent high-temperature dry lubricant.

Stronger bonding between two-dimensional layers is provided by hydrogen bonding. Aluminum hydroxide, $Al(OH)_3$, has the same structure (known as the BiI_3 lattice) as the $InCl_3$ shown in Fig. 3.2, with hydrogen bonding occurring between OH groups in one layer and the O atoms in the adjacent layer. An alternative possibility for hydrogen bonding involves water molecules between layers hydrogen-bonding to both layers; *gypsum*, $CaSO_4 \cdot 2H_2O$, has a layer structure in which each layer contains Ca^{2+} coordinated by the oxygens from SO_4^{2-}. Its hydrate water molecules lie between the $CaSO_4$ layers, coordinated through their oxygen atoms to a Ca^{2+} ion in one layer but also hydrogen bonded to a SO_4^{2-} ion in the same layer and another SO_4^{2-} in a neighbor layer.

The strongest forces binding layers together (that can still be distinguished as layer structures) are electrostatic forces, which can arise in either of two ways. The more common way involves layers with a net charge and intercalated counterions with the opposite charge. Thus Na^+ ions bind together layers with overall composition $Si_2O_5^{2-}$ in $Na_2Si_2O_5$, and K^+ ions bind together magnesium aluminosilicate layers in mica, $KMg_3(OH)_2$-Si_3AlO_{10}, a structure we shall return to in considering silicates below. A rarer alternative is a layer structure in which alternate layers have fully developed two-dimensional networks of bonds but also have opposite net charges. A rather elaborate example involves positively charged layers of $Mg(OH)_2$ in which substitution of Al^{3+} for Mg^{2+} gives a net positive charge, crystallizing together with negatively charged layers of $MgCl_2$ in which substitution of Na^+ for Mg^{2+} gives a net negative charge: The overall formula is $[Mg_7Al_4(OH)_{22}]^{4+}[Na_4Mg_2Cl_{12}]^{4-}$.

Layer structures can be planar, as in the case of the CdI_2 lattice, or corrugated. An interesting example of corrugation is As_2O_3, which occurs as a mineral (*claudetite*) in two forms, both of which are corrugated layers but puckered differently. Figure 3.6 shows the simple, nearly planar sheet of close-packed O atoms with one-fourth of the two-dimensional close-packed sites vacant that is the basis for the layers in both forms. Arsenic atoms lie both above and below the O sheet, forming a network of As_6 rings shown as hexagons. The two crystal forms differ in the placement of As atoms above and below the O plane; both yield corrugated layers, but with basically different symmetries.

In addition to three-dimensional and two-dimensional polymeric networks, one-dimensional chain networks are seen for numerous species. These chains can be bound to each other, as for layers, by van der Waals forces,

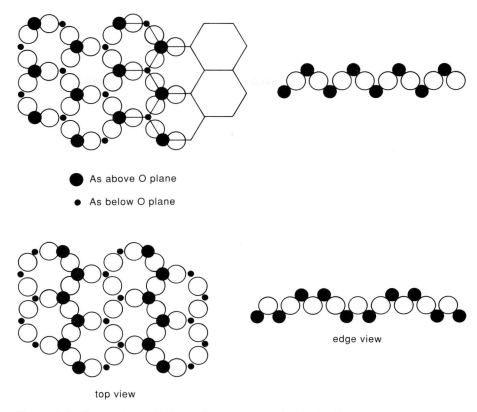

As above O plane

As below O plane

top view

edge view

Figure 3.6 Symmetry variations of layer corrugation in As_2O_3.

hydrogen bonding, or electrostatic forces. Perhaps the simplest chains are found in AuI, whose chains consist of $-Au-I-Au-I-$ units with bond angles of $180°$ at Au and $72°$ at I, resembling the HgO structure of Chapter 2. These chains are bound to each other by the fairly substantial van der Waals forces between the heavy, polarizable Au and I atoms. ZrI_3 also forms chains bound to each other by I–I van der Waals forces, but the chain is formed by ZrI_6 octahedra sharing opposite faces, as in Fig. 3.7. That figure also shows the double-chain structure of $LiOH \cdot H_2O$, for which one double chain is bound to the adjacent double chain by hydrogen bonds between the bridging OH^- ions and hydrated water molecules.

As for layer structures, it is also possible to have charged chains bound electrostatically within the crystal. Most such systems consist of negatively charged chains held together by cations between the chains, as in $CsCuCl_3$ whose basic chain unit is square planar $CuCl_4$ connected at adjacent corners of each square; the chain has the overall formula $[CuCl_3^-]_n$, bound to the next

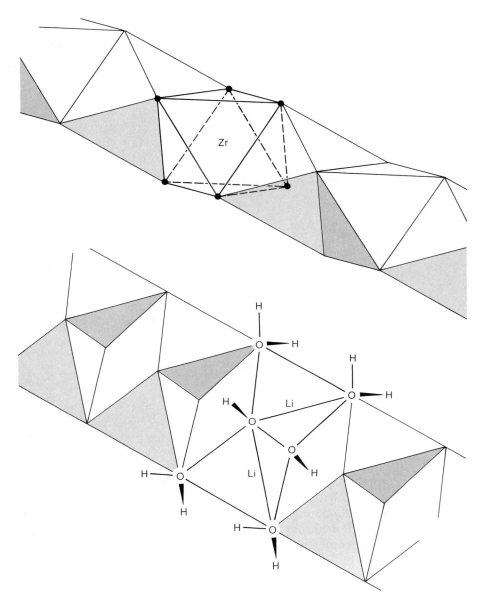

Figure 3.7 ZrI_3 chains bound to each other by van der Waals forces and $LiOH \cdot H_2O$ chains bound to each other by hydrogen bonding.

parallel chain by n Cs$^+$ ions. Rarely, positively charged chains bind directly to negatively charged chains: Borax, Na$_2$[B$_4$O$_5$(OH)$_4$]·8H$_2$O, consists of cationic chains of octahedral Na(H$_2$O)$_6^+$ units sharing two edges, plus anionic chains of B$_4$O$_5$(OH)$_4^{2-}$ units hydrogen-bonded into a one-dimensional network.

There are also what could be considered "zero-dimensional" or finite molecule systems in which covalent bonding is important. As we have noted for three-dimensional networks, it is not always possible to distinguish these from ionic lattices; CaO is essentially an ionic compound with the NaCl lattice, but it has the same symmetry as CaC$_2$, in which the acetylide C$_2^{2-}$ unit lies along the Ca–Ca axis in the crystal. In exactly the same sense, we cannot distinguish CaC$_2$ from the three-dimensional network compounds VC, TiN, or InAs, all of which also have the NaCl lattice. In all of these, directional covalent bonding plays an important role, and it is rather arbitrary to define the C–C bond in CaC$_2$ as defining a finite molecule within the lattice. Still, dimensionality can be a useful formalism in understanding lattice geometries and physical properties, and we shall see it in many applications.

3.3 CRYSTAL DEFECTS AND DEFECT STRUCTURES

Our discussion so far has emphasized the symmetry of perfect crystals. Real crystals, however, will contain defects—sometimes deliberately introduced. We will, in looking at reaction patterns in Part III, distinguish *enthalpy-driven* processes from *entropy-driven* processes. Given that $\Delta G = \Delta H - T\Delta S$, this distinction apportions negative-free-energy (spontaneity) credit to the gaining of energy stability (negative ΔH) or to the creation of randomness (negative $-T\Delta S$) in the system being considered. Occasionally a crystal defect will be enthalpy-driven, as when a heteroatom being doped into a crystal is more strongly bonded into lattice sites than the atom it is replacing. However, it is usually true that the creation of defects is entropy-driven, because in a highly ordered crystal defects can be randomly placed; even though a defect may cost stability, as in the case of a missing atom or ion, the increased randomness drives the creation of the defect. The greater the energy cost per defect, the fewer will be formed before the unfavorable ΔH overtakes the favorable ΔS. However, because ΔS is temperature-weighted in the free-energy definition while ΔH is not, as the temperature is increased the concentration of defects will inevitably increase, with the end result being melting to a disordered liquid if no other decomposition process occurs first.

The simplest crystal defects are *point defects*, which occur at isolated individual lattice sites. If these are very low in concentration within the crystal, they may be nearly impossible to detect. One point defect that does not change the chemical stoichiometry of the crystal is a vacancy representing

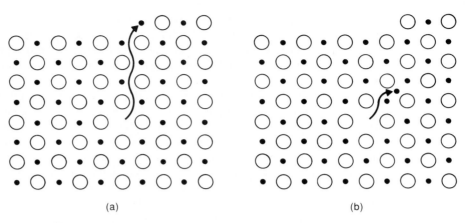

Figure 3.8   (a) Cation movement in forming a Schottky defect. (b) Cation movement in forming a Frenkel defect.

the absence of an atom or ion, known as a *Schottky defect*. Figure 3.8a shows the formation of a Schottky defect; although the missing cation is still bound at the surface of the crystal so that the stoichiometry does not change, there is an energy cost for its creation because lattice bonding is missing around the vacancy. We have created a bubble in the lattice.

Another point defect that does not change the stoichiometry of the crystal is the movement of an atom or ion from one site to a previously vacant interstitial location, known as a *Frenkel defect*. Figure 3.8b shows the formation of a Frenkel defect. This process also requires an energy input, because the new site is a less stable location either through increased repulsions or through reduced site bonding. However, in some cases Frenkel defect formation is easy when size relationships are favorable; in a lattice consisting of close-packed anions with cations in octahedral sites, small cations that fit reasonably well in the smaller tetrahedral sites are fairly likely to migrate there. This is particularly true when formation of four tetrahedrally directed covalent bonds in the defect site can add to the electrostatic lattice energy, which is true for the silver halides. AgF, AgCl, and AgBr adopt the NaCl lattice, but the silver ion can migrate to tetrahedral sites—more readily for the larger Br sites. AgI crystallizes in the four-coordinate wurtzite structure (Fig. 2.13), for which Ag^+ migration could take the cation from a tetrahedral to a larger octahedral site; as a result, cation migration is particularly easy for AgI. This has dramatic significance in photographic chemistry, as we shall see in Chapter 17.

It is also possible to have point defects that alter the chemical composition of the crystal. Substitution of one atom for another with the same net charge or formal oxidation state can yield such a point defect, and any commercially available ionic compound will have at least a small concentration of such defects, as for example a very small concentration of Na^+ ions in crystalline KCl. If substitution involves an atom with a different charge, formal oxidation state, or valence electron count, a point defect will be created but it must be paired with another point defect that compensates for the electrical difference that now exists. Often this compensation is a change in oxidation state for another atom near the defect. We have already noted the continuous change in composition for the Co–Te system between compositions CoTe and $CoTe_2$, with the crystal structure remaining essentially unchanged as Co atoms enter the vacant planes in the CdI_2 lattice of $CoTe_2$ until the NiAs lattice of CoTe is formed (see Fig. 2.14). In $CoTe_2$ we find Te layers contacting each other and oxidation states can reasonably be taken as Co^{2+} and Te_2^{2-}, where the Te_2^{2-} species have one Te atom in each of the two touching Te layers. As Co^{2+} ions are added to the vacant sites between these layers, the two Te atoms with formal oxidation state -1 become telluride ions with oxidation state -2. This sort of compensation is common in solid systems whose composition is changing through the addition of small numbers of substitutional point defects.

Substitution can change the stoichiometry of a crystal without introducing another kind of atom, as in the Co–Te case. If alkali halide crystals are heated in the vapor of the same alkali metal present in the crystal, some alkali metal atoms are added as ions to the crystal lattice. The electrons released in this process diffuse through the crystal until they are trapped in an anion vacancy. At this point the trapped electron is in effect a quantized particle in a box, and it has the appropriate ground state and excited state wave functions. As a result, it can absorb visible light, and the otherwise colorless crystal has a microscopic color center, known generally by the equivalent German term "Farbenzentrum" or *F-center*. Alkali halide crystals containing F-centers have a measurable excess of alkali metal ions and a slightly lower density than a stoichiometric crystal, as might be expected from the presence of vacancies at anion sites.

When numerous point defects are symmetrically arranged they can generate a new phase, as in the case we have already noted of FeS, which has the NiAs lattice but is usually missing some Fe atoms; when one-eighth are missing, the vacancies are symmetrically arranged in layers and constitute the new phase Fe_7S_8. However, point defects can also cluster so that rows of atoms are missing within the crystal, constituting an *extended defect*. Of course, if such rows are regularly arranged this would constitute a new phase, but if there are only a few (and particularly if they are randomly arranged) each can create a *crystallographic shear plane* in which a segment of the crystal slides diagonally past a neighbor segment to collapse together and fill the row of vacancies. Figure 3.9 shows the formation of such a shear plane; part (a) shows the initial top view of a ReO_3 lattice with ReO_6 octahedra

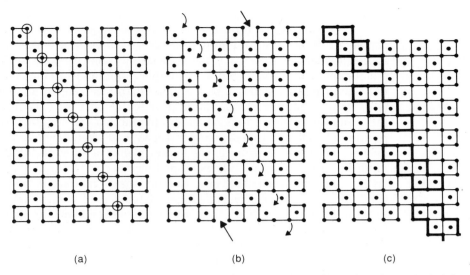

(a)   (b)   (c)

Figure 3.9 Crystallographic shear plane formation. (a) ReO_3 lattice plane; circled O atoms will be removed as an extended defect. (b) Direction of collapse of lattice when O atoms are removed. (c) Blocks of edge-sharing octahedra within the ReO_3-equivalent lattice.

sharing corners, and a row of oxygen atoms is circled for removal. Each missing atom is a point defect, and the row along which they fall, which could be randomly oriented within the crystal, is the shear plane. Part (b) of the figure shows the initial lattice with the shear-plane oxygen atoms removed; the direction of lattice displacement and the individual oxygen atom displacements along the shear plane are indicated. Part (c) shows the resulting collapsed crystal, in which extensive edge-sharing between octahedra (as opposed to the initial sharing of corners) is occurring inside the blocks of octahedra that are heavily outlined.

The shear plane just described for the ReO_3 lattice, if it existed alone in the crystal, would simply be a crystal defect. On each side of the plane, blocks of unchanged ReO_3 lattice remain. But the systematic removal of oxygen atoms has changed the stoichiometry in the vicinity of the plane. If there were parallel shear planes of this sort throughout the crystal, separated by slabs of the original ReO_3 structure that were only a few octahedra thick, the overall stoichiometry of the crystal would be significantly different from the initial 3:1 ratio. Further, the slabs of corner-sharing octahedra would alternate with slabs of edge-sharing octahedra, which would constitute a different (and more complicated) lattice. Many new structural possibilities are opened by this consideration, as the next section will develop.

The oxygen vacancies that we have just described for the ReO_3 lattice, if regularly arranged, could create a new lattice. Because such a lattice could, at least in principle, be prepared by systematically altering the original ReO_3 lattice, the new lattice would be described as a *superstructure* of the ReO_3 lattice. Superstructure lattices can be formed in many ways, but generally these fall in two categories: substitutional structures and defect structures. What we have described above for ReO_3 would constitute a defect superstructure, and we shall return to these. There is a sense in which the CdI_2 lattice is a defect NiAs structure (see Fig. 2.14). However, some simpler cases in what is admittedly a complex area can be obtained by partially substituting one metal for another in a basic lattice, maintaining a symmetrical arrangement.

In Fig. 2.15 we have already seen that compounds with AX_2 stoichiometry frequently adopt either the rutile or the fluorite lattice. Both of these form substitutional superstructures. Given that metal atoms will frequently choose to occupy octahedral sites, or in other terms to be octahedrally coordinated, we might expect to see a number of compounds in which partial substitution of B for A leads to BX_6 coordination octahedra. In such a case, the overall cation:anion ratio of 1:2 can be maintained by having a total of three cations: A_2BX_6. The basic octahedral coordination geometry for 1:2 compound is the rutile lattice, and we see a number of A_2BX_6 compounds that simply adopt the rutile lattice with A and B randomly distributed among the cation sites, such as Mg_2FeF_6. This will only be possible, however, if the charges on the two formal cations are the same and their radii do not differ too much. For fairly strongly ionic systems (mostly fluorides and oxides) that have differing cation charges we often see the *trirutile* lattice, shown in Fig. 3.10a. Remembering from Fig. 2.15 that the octahedra are stacked in the overall rutile structure, what we see in the trirutile lattice is stacks of three octahedra with regularly arranged A and B atoms: AABAABAAB.... This is obviously a substitutional superstructure of the rutile lattice. The trirutile lattice is adopted by Li_2TiF_6, $MgSb_2O_6$, Cr_2WO_6, and even $LiNbWO_6$, among other systems.

In the same way, substitutional superstructures of the fluorite lattice are known in some variety. Figure 3.10b shows a simplified version of the fluorite lattice in which only the cation positions are marked, along with some double- and triple-decked superstructures. These lattices are usually adopted by species containing atoms large enough to reach cation coordination number eight, such as the uranium and strontium compounds shown.

In the same sense in which the layer structure of CdI_2 may be considered a defect structure of NiAs, we often see complex layer structures that are defect structures (in the sense that certain layers of "anions" have no "cation" layers between them) but are also superstructures in the sense that the

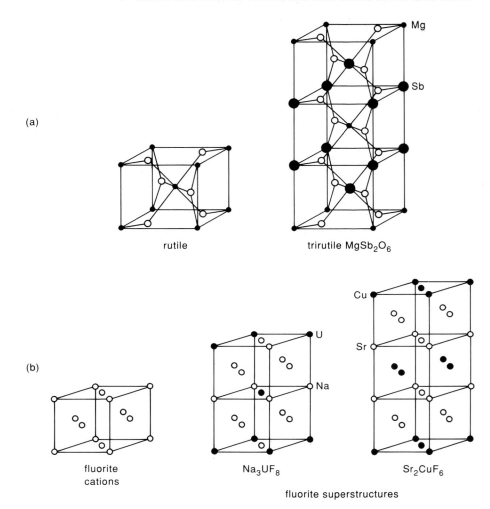

Figure 3.10 AB_2 superstructures for stoichiometric substitution. (a) $MgSb_2O_6$ in the trirutile lattice. (b) Na_3UF_8 and Sr_2CuF_6 in fluorite superstructure lattices.

anion layers do not stack in the hcp fashion ABABAB... or the ccp fashion ABCABCABC... but in a more complicated sequence. Sulfides, selenides, and to some extent the other chalcogenides are particularly likely to adopt such structures, presumably because their modest electronegativity makes covalent bond networks important. For example, Bi_2Se_3 and Bi_2Te_3 form the close-packed layer sequence AcBAcBaCBaCbACb..., where the capital letters denote Se or Te and the lower-case letters Bi positions. This is a sequence of nine layers of chalcogenides and six layers of Bi, leading to the correct 3:2 stoichiometry.

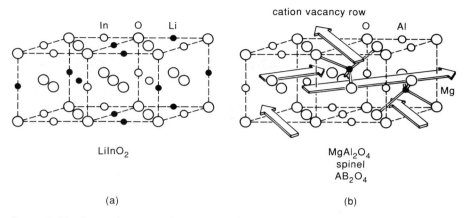

cation vacancy row

LiInO$_2$

(a)

MgAl$_2$O$_4$
spinel
AB$_2$O$_4$

(b)

Figure 3.11 Superstructures based on the NaCl lattice. (a) LiInO$_2$ lattice without vacancy defects. (b) AB$_2$O$_4$ spinel lattice for MgAl$_2$O$_4$ showing both extended vacancy defects and interstitial ions.

Defect structures are common for mixed oxides $A_m B_n O_q$. Because a close-packed lattice has one octahedral site for each close-packed atom, one formula unit of such an oxide in which the oxygens are close-packed will have q octahedral sites; given $m + n$ electropositive atoms, a fraction $m + n/q$ of the octahedral sites will be filled. This does not have to lead to a defect structure: In LiInO$_2$, $m + n = 2$ and with two oxygens in the formula $m + n/q = 1$, and all octahedral sites are filled. LiInO$_2$ crystallizes in a superstructure of the NaCl lattice seen in Fig. 3.11a, which simply shows the unit cell to be two adjacent cubic unit cells of the NaCl lattice, with O atoms at the Cl sites and Li and In atoms symmetrically arranged at all the Na sites. However, defect structures are possible in the same overall framework: In the mineral *spinel*, MgAl$_2$O$_4$, whose structure is shown in Fig. 3.11b, the Al atoms are in half the octahedral sites, with alternate diagonal rows of sites vacant; the Mg atoms are in one-eighth of the tetrahedral sites. This is a very common structure for mixed oxides with the overall stoichiometry AB$_2$O$_4$. In general, the B atoms will be in half the octahedral sites while the A atoms will occupy tetrahedral sites. A variant, the *inverse spinel* structure, has half the B atoms and all the A atoms in octahedral sites, and the other half of the B atoms in tetrahedral sites. The Fe$_3$O$_4$ (magnetite) crystal has the inverse spinel structure, and its seemingly odd oxidation state is accounted for by regarding the crystal as Fe^{2+}(Fe^{3+})$_2$O$_4$. The necessary cation charge of +8 can be achieved in several ways: A^{6+}(B$^+$)$_2$O$_4$, A^{4+}(B^{2+})$_2$O$_4$, and A^{2+}(B^{3+})$_2$O$_4$. All of these combinations of oxidation states are known in compounds with the spinel or inverse spinel structures—Na$_2$WO$_4$, Zn$_2$SnO$_4$, and NiAl$_2$O$_4$, for example. Note that while we have previously mentioned defect structures in which individual atoms or atom-dimensional layers of atoms were missing, the spinel structure has one-dimensional rows of atoms missing. The same "polymerization" possibilities are possible for vacancies as for atoms in these lattices, although it might be better to view them as what's left after the filled fraction of atoms have formed a polymeric network.

For mixed oxides with stoichiometry ABO_3, two structures are commonly seen. If one of the two atoms A or B is large enough, it will fill the cavities in the ReO_3 lattice, in which location it will be 12-coordinate and will in effect be completing the close-packed array of O atoms. Study problem B1 derives the necessary relationship between ionic radii for A, B, and O to achieve a perfect fit:

$$t = \frac{r_{12} + r_O}{r_6 + r_O}\left(\frac{1}{\sqrt{2}}\right)$$

Here $t = 1$ for a perfect fit, r_{12} is the crystal radius of the 12-coordinate atom (corrected for its coordination number), r_6 is the six-coordinate crystal radius of that atom, and r_O is the six-coordinate radius of the oxide ion. As the results of the problem indicate, when t is greater than 1 the system will in general adopt this structure, known as the *perovskite* structure after the mineral $CaTiO_3$, perovskite, and shown in Fig. 3.12a.

If neither electropositive atom in the ABO_3 system is large enough to fill the 12-coordinate site in the ReO_3-lattice symmetry, the commonly seen structure is the *ilmenite* lattice, named for the mineral ilmenite, $FeTiO_3$, and

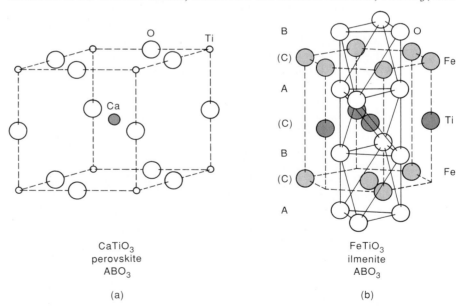

CaTiO₃
perovskite
ABO₃

(a)

FeTiO₃
ilmenite
ABO₃

(b)

Figure 3.12 ABO_3 lattices. (a) Perovskite lattice for large A ion. (b) Ilmenite lattice for smaller A ion.

seen in Fig. 3.12b. Like CdI_2, it is essentially a defect NiAs hcp lattice, but in a layered superstructure in which one layer of cations has A^{q+} in two-thirds of the Ni sites and the next layer of cations also has B^{q+} in two-thirds of the Ni sites, but staggered from the A atoms. The cation symmetry within a layer is that of $InCl_3$ in Fig. 3.2.

The existence of defect structures with general formulas ABO_3 and AB_2O_4 (among others) raises the possibility that even these stoichiometries might have vacancy defects—that is, that there might be nonstoichiometric compounds with related structures and interesting properties. A series of mixed metal oxides called the *bronzes* has been known for many years, the first tungsten bronze having been prepared by Wöhler in 1823. He heated Na_2WO_3 and WO_3 together in a reducing atmosphere of H_2 and obtained brilliantly colored crystals with a metallic luster, having the general formula Na_xWO_3. In these compounds, x can vary over almost the entire range from 0.00 to 1.00 (golden yellow); for x greater than about 0.3 (deep violet), the structure will be the perovskite lattice, with the relatively large Na^+ ion in the 12-coordinate site. However, when the fraction of occupied 12-coordinate sites falls below 0.3 the structure changes to a complicated tetragonal network of WO_6 octahedra linked at their corners as in the ReO_3 lattice, but in which in addition to rings of four octahedra there are rings of three and of five octahedra. The rings of three octahedra yield nine-coordination analogous to that in the $PbCl_2$ lattice in Fig. 3.3, while the rings of five octahedra yield large tunnels through the lattice that can be occupied by a variety of large ions. In addition, there is a hexagonal structure with rings of three and six octahedra, favored by low concentrations of the largest cations such as Rb^+ and Cs^+.

The variety of cation site sizes available in these structures means that sodium is not the only alkali metal forming tungsten bronzes, and all but lithium (presumably too small to stabilize large sites) are seen in variable-composition bronzes. However, many other metals in a high oxidation state also form bronzes. The best-known bronzes are those of tungsten, molybdenum, and vanadium, but Na_xMO_3 can have M = Ti, V, Nb, Ta, Mo, W, Re, and Ru with the brilliant colors and metallic luster seen in tungsten bronzes. The metallic luster and high electrical conductivity indicate that electrons are delocalized through the lattice to some extent, in addition to the electrostatic and covalent-network bonding. This raises the possibility of unusual electronic properties, and many bronzes are being studied as potential high-temperature superconductors.

In this context it may be appropriate to note that oxide superstructures are extremely important in all of the recently discovered high-temperature cuprate superconductors; Chapter 10 will consider these structures in light of the electronic properties of the d valence electrons on the copper atoms, but it will be apparent there that many of these same structural considerations are influencing or controlling those properties.

3.5 LAYERED STRUCTURES AND INTERCALATION

We have seen a number of layered structures, at varying degrees of complexity. These have in general been stoichiometric compounds, although the presence of point defects obviously alters any strict stoichiometric ratio. However, where the layers within these structures are either weakly bound to each other or are spaced apart by occasional cross-links it is often possible to have atoms or molecules fitting between the layers without destroying the layers themselves. These species are in a sense the opposite of point defects— almost "point surpluses." However, they are described as _intercalated_ species. If the intercalated species fill the possible sites between layers in a symmetrical way, so as to produce at least an approximate stoichiometric ratio of intercalated species to layers, the product is usually called a _lamellar compound_. Perhaps the most well-known layered structure is that of graphite, and many intercalated graphitic compounds are known.

Graphite layers are sheets of the familiar benzene rings, though the extended pi systems give each C–C bond region a bond order of 1.33 rather than benzene's value of 1.5. The sheets do not stack with superimposed hexagonal rings, but rather with an ABABAB . . . sequence similar to that in an hcp lattice (an alternate ABCABC . . . rhombohedral lattice is also known for natural graphite but is not seen for synthetic graphite). The layers are held together only by van der Waals attractions between the pi networks, so that other atoms or even molecules can penetrate the region between layers readily if a replacement bonding energy is possible. The pi electrons themselves are available for bonding, and the oldest graphitic compound, graphite oxide, clearly has covalent C–O and C=O bonds. Although graphite oxide was first reported in 1859, its structure is not well established; however, it is formed by oxidizing graphite in mixed nitric and sulfuric acids using either chlorate or permanganate as the primary oxidant. Its formula is not even reliable, but is approximately $C_8O_4H_2$. The intercalated oxygen atoms seem to have formed bridging ether linkages across C_6 rings as well as carbonyl and hydroxyl groups that probably yield a keto/enol equilibrium and a puckering of the graphite layers to a boat or chair configuration for each ring, although the layer structure remains.

Similarly, a fluoride of graphite exists with formula $(CF)_n$, known as carbon monofluoride and prepared by direct fluorination of graphite. The layer structure remains, but each carbon has tetrahedral bonding to the three carbon atoms of the graphite sheet and to one fluorine atom, and each sheet puckers into either the chair or the boat configuration. The product resembles Teflon and has applications as a lubricant. Clearly the covalent bonding of O and F atoms to the graphite layer yields the strongest intercalation bonding, but also disrupts the carbon sheet significantly.

Since little energy is required to separate the graphite layers, we might expect that intercalation could occur with less strongly bound species. Alkali and alkaline-earth metals, but also halogens, will react with graphite—the alkali metals by partially reducing the graphite and the halogens by partially oxidizing it. Atomic size is important, because stable intercalation depends on achieving an approximate fit between the hexagon-center spacing (2.46 Å) and either the ionic diameter of the metal (3.04 Å for K^+ but only 1.46 Å for Li^+) or the X–X bond length for X_2 halogens (2.66 Å for I_2 but only 1.99 Å for Cl_2). Apparently as a result of these size relationships K, Rb, and Cs readily form lamellar compounds with graphite but Li and Na do so much less easily. The heavier metals M form C_8M, but also $C_{24}M$, $C_{36}M$, $C_{48}M$, and $C_{60}M$. In these compounds, the carbon sheets are still planar and the metal atoms are centered over C_6 hexagons in a triangular array for C_8M, hexagonal for higher C_n. Figure 3.13 indicates the stacking sequences of layers that yield these formulas; note that although in graphite the layers are staggered ABAB . . . , in these lamellar compounds layers binding metal atoms always have an eclipsed arrangement AMA or BMB.

Size seems to be particularly important for the halogens (other than F) in forming graphitic compounds. In an unusual pattern of reactivity, F_2 and Br_2 ($r_{Br-Br} = 2.28$ Å versus ring spacing 2.46 Å) react readily, Cl_2 ($r_{Cl-Cl} = 1.99$ Å) only in an unstable fashion, and I_2 ($r_{I-I} = 2.66$ Å) not at all. However, though I_2 and Cl_2 don't react, ICl ($r_{I-Cl} = 2.33$ Å) does. C_8Cl probably forms at dry ice temperatures and decomposes when allowed to warm, but C_8Br and $C_{4n}Br$ to $n = 5$ form readily, with intercalation patterns observed corresponding to the stages seen in Fig. 3.13. None of these lamellar compounds are strongly ionic; the C_8M alkali-metal compounds appear

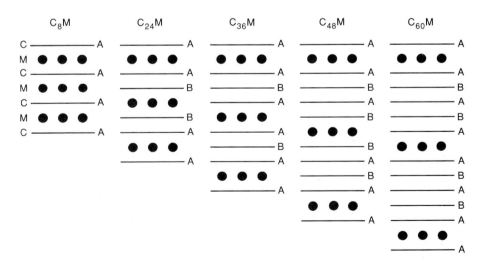

Figure 3.13 Intercalation stages in the formation of alkali-metal/graphite lamellar compounds.

from Mössbauer spectroscopy and chemical inference from hydrolysis reactions to be only about 50% ionized, so that considerable atomic character remains in the intercalated metal atoms. Similarly, intercalated Br_2 appears from conductance measurements and Raman spectra to be only about 2% ionized as Br^-.

Other interhalogen compounds such as IF_5, BrF_3, and IBr can also be intercalated, as can the electronically similar noble gas fluorides such as XeF_4 and KrF_2; the latter are powerful fluorinating agents and usually partially fluorinate the carbon sheets as well as intercalating the original molecules. Perhaps more unusual is the formation of intercalation compounds involving metal halide molecules such as $FeCl_3$. Over sixty different metal halides have been found to intercalate into graphite, from $MgCl_2$ to $HgCl_2$, usually simply by heating the two substances together in a sealed tube. The stoichiometry is usually less precise than for the alkali-metal compounds, and C_8M is rarely observed, although $C:M$ values from 6 to 9 are fairly common, but higher ratios are also seen (up to 70 for WCl_6).

Some of the graphitic intercalation compounds have useful applications: Graphite-XeF_6 is a convenient fluorinating agent for organic molecules, and the $(CF)_n$ fluoride is widely used as a cathode in lithium batteries. Because a number of intercalating molecules substantially increase the electrical conductance of the graphite, there is ongoing interest in their electrochemical and catalytic properties. As an example of the intriguing nature of these compounds, the graphite-AsF_5 intercalation compound has been reported to have a greater electrical conductivity than pure silver, which is the best metallic conductor.

As we have already seen, there is a strong tendency for metal sulfides to form layered crystal structures. Figure 3.5 has shown the layered structure of MoS_2, and in this and in the previous chapter we have noted the layered CdI_2 structure, which is also adopted by many MS_2 sulfides. In these crystals the van der Waals forces binding sulfide layers together are similar to those binding graphite layers. This binding not only makes the sulfides good lubricants, like graphite, but allows intercalation in much the same way as in the graphitic compounds we have been examining. Two sulfides of particular interest in this context are NbS_2, which has the Nb in trigonal prismatic coordination like MoS_2 except that its double layers are stacked ABCABC... rather than ABABAB..., and TaS_2, which has the CdI_2 lattice. Both of these sulfides are superconductors, and their intercalation compounds are as well. In either case, the adjacent S–S layers are close-packed and have octahedral and tetrahedral sites that could accept intercalated metal ions. For small ions such as Li^+ a considerable composition range Li_xNbS_2 $(0 < x < 1)$ is observed, with Li^+ ions occupying symmetrically arranged sites between layers; stages of layer penetration comparable to those in Fig. 3.13 are observed. In these compounds little change in the interlayer spacing is observed when the small ion enters, but there are also sulfide intercalation compounds in which larger molecules, such as hydrated alkali-metal ions or alkyl- or arylammonium ions, are introduced. For these systems large changes in the layer spacing occur, but the layer structure is preserved.

A less obvious but very important layer structure is displayed by a compound known as *β-alumina*, which is actually a sodium aluminate: formally $NaAl_{11}O_{17}$, but usually with a nonstoichiometric sodium content somewhat higher than the formula indicates. Its layer structure is based on the spinel lattice seen in Fig. 3.11, which does not look much like layers. However, note that the missing rows of cations create diagonal directions through the lattice array along which electrostatic attraction is reduced. If the weakly bound oxygen atom at the lower left of the double cube is removed, and similarly the weakly bound oxygens to the upper right of the double cube (not shown) are removed, the result is a slab of spinel lattice four close-packed O layers deep. The slab runs diagonally from lower right to upper left of the drawing; the layers may be counted from upper right to lower left. In β-alumina these four-deep layers, containing only aluminum and oxygen, are joined by oxygen atoms that occupy only one-fourth of the positions in the fifth close-packed O layer. As a result, the four-deep layers are held 11.3 Å apart, which is ample space for intercalated Na^+ ions. Indeed, the spacing is so large that the sodium ions, with a diameter of 2.32 Å, are quite mobile in moving through the lattice plane in which they are bound.

In the sodium–sulfur high-energy battery under development for electric cars, the electrode reactions are $Na^0 \rightarrow Na^+ + e^-$ and $S + 2e^- \rightarrow S^{2-}$, operating at about 300°C so that the sodium metal and sulfur are molten. Obviously it is necessary to physically separate the sodium and sulfur to prevent direct reaction, and a material is necessary to serve the role of a salt bridge to transmit Na^+ ions but not Na^0 atoms. β-Alumina is the preferred material for this purpose; its ionic conductance for Na^+ at 300°C is approximately equal to that of an aqueous NaCl solution. The role of β-alumina as a ceramic ionic conductor is likely to be extremely important as high-energy batteries are developed.

A number of variations on the β-alumina structure are possible. Singly charged cations may be readily substituted for Na^+ simply by soaking the β-alumina in a molten salt of the other cation, such as K^+ from molten KNO_3. In addition, superstructures of the β-alumina lattice are known. β″-Alumina has an alternative stacking pattern: Na(ABCA)Na(CABC) Na(BCAB)NA... as opposed to the β-alumina Na(ABCA)Na(ACBA) Na..., in which successive four-layer blocks are mirror images about the plane containing the sodium ions. The changed electrical symmetry and interlayer spacing in β″-alumina allow intercalation of doubly-charged ions such as Ca^{2+}, Ba^{2+}, Pb^{2+}, and even lanthanide ions with high ionic conductivity, a property that opens many opportunities for solid electrolyte applications.

3.6 SILICATES

The previous chapter has considered oxoanions in general, and it would seem that orthosilicate as SiO_4^{4-} should be more or less equivalent to ClO_4^- or SO_4^{2-}. But we have seen that increased negative charge leads to an increased tendency to polymerize through condensation reactions that eliminate H_2O, and the orthosilicate ion is even more highly charged than the orthophosphate ion considered earlier. The SiO_4^{4-} ion is found in a number of ionic or partly ionic crystals, but that number is tiny compared to the enormous variety of polysilicates that constitute most of the earth's crust or have been synthesized. There are thus many examples within the class of silicate crystals of the patterns of directional bond networks that we have considered.

In all silicate systems, the silicon atoms are arranged in tetrahedral SiO_4 groups, but these normally share corners (O atoms) in a polymeric structure. It is not impossible to achieve coordination number six around Si, and in SiP_2O_7 the silicon is octahedrally surrounded by six oxygen atoms, but only tetrahedral coordination is found in ordinary silicates, even when some substitution for Si by Al or other elements has occurred. Indeed, in these lattices replacement by Al, Ca, or Mg is very common, and these other elements are found with both tetrahedral or octahedral coordination. The substitution often retains structural equivalence, except where additional cations must be added to retain electrical neutrality. For example, SiO_2 occurs in three crystal structures—quartz, tridymite, and cristobalite—and the compound $AlPO_4$ also occurs with all three of these structures, so that one can imagine that half the Si^{4+} has been replaced by Al^{3+} and the other half by P^{5+} with no real lattice symmetry change.

In general, silicates can be characterized by the extent of polymerization of the tetrahedral SiO_4 units present. The simplest are those in which the orthosilicate monomer or a small finite polymer is isolated in the crystal lattice as a discrete anion. One of the simplest of these is *olivine*, which is principally magnesium orthosilicate with some iron(II) and manganese(II) substituted for magnesium. Its lattice consists essentially of close-packed oxygens with Si in one-eighth of the tetrahedral holes and (Mg, Fe, Mn) in half the octahedral holes, although Fig. 3.14 shows the structure in terms of orthosilicate tetrahedra. Other discrete silicate anions include *pyrosilicate*, $Si_2O_7^{6-}$, and the cyclic ions $Si_3O_9^{6-}$ (formed from three SiO_4 tetrahedra) and $Si_6O_{18}^{12-}$ (formed from six tetrahedra). These are also shown in Fig. 3.14. All are rare; probably the best known is *beryl (emerald)*, $Be_3Al_2Si_6O_{18}$.

The next stage in the polymerization of silicates is the class of one-dimensional polymers of the SiO_4 unit, or chain structures. Figure 3.15 shows the geometric possibilities for single and double chains. The two named forms shown, the *pyroxene* chain and the *amphibole* double chain, are particularly important because they commonly occur in the rocks of the earth's crust. Each constitutes a class of minerals in which the cations differ.

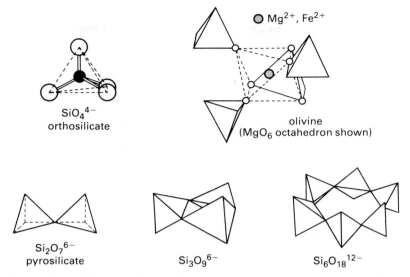

Figure 3.14   Discrete silicate anions and orthosilicate packing in olivine.

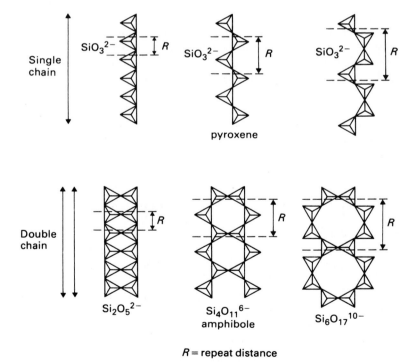

R = repeat distance

Figure 3.15   One-dimensional silicate polymer chains and double chains.

Thus pyroxenes include *enstatite*, $MgSiO_3$, and *diopside*, $CaMg(SiO_3)_2$, and amphiboles include *tremolite*, $Ca_2Mg_5(Si_4O_{11})_2(OH)_2$, and *actinolite*, $Ca_2(Mg, Fe)_5(Si_4O_{11})_2(OH)_2$. Pyroxenes are a major component of basalt rock. Natural asbestos is a mixture of fibrous amphiboles—not all amphiboles are fibrous, because the chains do not always pack together in a parallel fashion for long distances.

The formulas of polysilicates obviously depend on how many of the corners of each SiO_4 tetrahedron are free and how many are shared. If all corners are free, as in the orthosilicate ion, the O : Si ratio is obviously 4. If all corners are shared, each silicon has a half-share of each of the four oxygens surrounding it, and the net number of oxygens per silicon is $4 \times \frac{1}{2} = 2$. For polymeric species, the easiest way to formulate the polymer is to sketch the structure of tetrahedra and draw in lines marking the repeat unit so that the lines do not pass through an atom (that is, neither through the center of any tetrahedron, Si, nor through the corners, O). Counting the atoms within the repeat unit then gives the formula directly. The net charge will be determined by the number of Si^{4+} and the number of O^{2-}.

The chain structures of Fig. 3.15 can link sideways into two-dimensional sheet structures. It should be obvious that pyroxene chains can continue to link as they do in amphiboles, forming sheets of six-membered rings with the formula $Si_2O_5^{2-}$. Perhaps less apparent is the possible linkage between the double chains of eight-membered rings of tetrahedra into sheets; these sheets with the same formula as the sheet of six-membered rings have alternate four- and eight-membered rings of tetrahedra. Sections of these sheets are shown in Fig. 3.16. Although in that figure all of the tetrahedra are shown pointing the same way, they need not. If they alternate pointing up and down, the result is a somewhat puckered sheet. *Petalite*, $LiAlSi_4O_{10}$, is an example of a puckered six-membered sheet and *gillespite*, $BaFeSi_4O_{10}$, is an example of the 4 : 8 sheet.

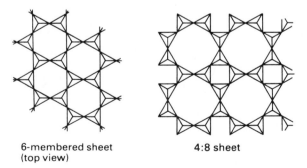

6-membered sheet
(top view)   4:8 sheet

Figure 3.16 Symmetry of two-dimensional silicate polymer layers based on 6-membered rings of tetrahedra and 4- and 8-membered rings of tetrahedra.

Just as layer structures can be thought of as chains joining side-to-side, we can visualize many three-dimensional framework structures as layers joined face-to-face. It is geometrically possible to create three-dimensional frameworks of this sort in which some tetrahedra are not shared between layers and thus have free corners, but these are rare. Much more commonly, three-dimensional framework silicates are formed by having exactly half the tetrahedra point up and the other half down in a symmetrical arrangement in which every tetrahedron is shared at all four corners. The simplest such three-dimensional framework silicates are the various forms of SiO_2; *quartz*, *tridymite*, and *cristobalite*. Figure 3.17 shows two ways in which the sheets of six-membered rings can meet to form a three-dimensional network, yielding the tridymite and cristobalite structures; in the layer sketch, the six-membered rings are emphasized. The symmetry difference between the two

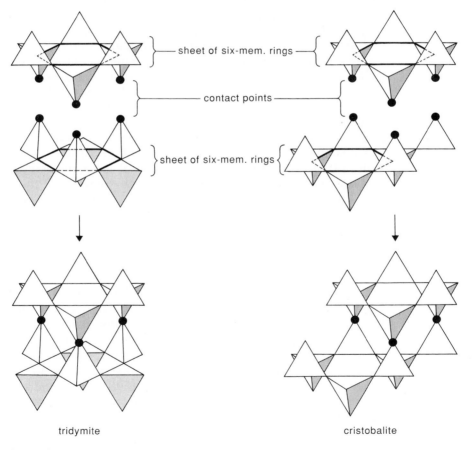

tridymite   cristobalite

Figure 3.17 Symmetries for joining six-membered rings of tetrahedra: tridymite (mirror image rings); cristobalite (identical rings, skewed orientation).

structures is that in the tridymite network each layer is a mirror image of its neighbor, so that the six-membered rings are superimposed. On the other hand, in the cristobalite network each layer has the same orientation, but each layer is displaced sideways enough to bring the three necessary contact points together for a condensation reaction; the rings are no longer superimposed. Quartz, the common low-temperature form of SiO_2, has a more complicated structure in which helical chains of tetrahedra with three tetrahedra per turn are linked to each other at each of the three tetrahedra, somewhat like coil springs in a mattress.

Three-dimensional framework silicates becomes more chemically complex when Al^{3+} substitutes for some of the Si^{4+} in the SiO_4 tetrahedra within the framework. Such substitution produces a net negative charge on the framework, and cations must be accommodated within the framework to achieve electrical neutrality. The commonest category of these—literally the most common chemicals on earth—is that of the *felspars*. In these, either one-fourth or one-half of the Si has been replaced by Al. If the fraction is one-fourth, the framework formula is $AlSi_3O_8^-$ and the framework contains an alkali-metal unipositive ion; if the fraction is one-half, the framework formula is $Al_2Si_2O_8^{2-}$ and the cation is an alkaline-earth dipositive ion. Examples of these are the *plagioclase* felspars *albite*, $NaAlSi_3O_8$, and *anorthite*, $CaAl_2Si_2O_8$. Felspar structures consist of connected sheets of the 4:8 layers shown in Fig. 3.16; adjacent pairs of tetrahedra in each four-membered ring point up and the other two point down. There are multiple structures corresponding to the symmetrical arrangements of these pairs of tetrahedra within a given layer. Felspars form perhaps 60% of the earth's crust. Granite, for example, is a mixture of quartz, felspars, and micas.

If the square of four SiO_4 tetrahedra is formed with all four tetrahedra pointing the same way, they will link in a more complex manner to form three-dimensional polymers containing large tunnels or cavities within the aluminosilicate framework. These systems are the *zeolites*. A few relatively rare naturally occurring zeolites are known, such as *faujasite*, $Na_2Ca(Al_2Si_5O_{14})_2 \cdot 20H_2O$, but most are commercially prepared by crystalization from solutions of sodium silicate and aluminum oxide in strong base. Figure 3.18 shows schematically the way a zeolite crystal is formed: Four SiO_4 tetrahedra unite to form a square of silicon atoms with the tetrahedra pointing the same way from the base of the square; then six of the squares combine to form a cuboctahedron; finally, the cuboctahedra link with other cuboctahedra through either their square or their octahedral faces to form a three-dimensional polymer. The large cavities inside zeolites, with polar sites exposed, are nearly ideal places to isolate molecules from solution or the gas phase and catalyze their reactions.

The survey of symmetries above has given a few examples by way of illustration, but in almost every case the number of naturally occurring examples is very large. Orthosilicates include the $(Fe, Mg)SiO_4$ olivine lattice that has been mentioned, Ca_2SiO_4, *zircon* ($ZrSiO_4$), and *garnet*

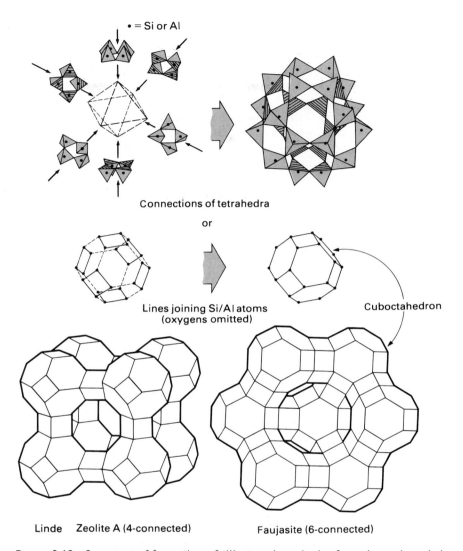

• = Si or Al

Connections of tetrahedra

or

Lines joining Si/Al atoms
(oxygens omitted)

Cuboctahedron

Linde Zeolite A (4-connected)   Faujasite (6-connected)

Figure 3.18 Sequence of formation of silicate cuboctahedra from 4-membered rings of tetrahedra; basic zeolite network structures.

$[M_3^{II}M_2^{III}(SiO_4)_3$ in many variations such as *andradite*, $Ca_3Fe_2(SiO_4)_3]$. Olivine-containing rock is almost the first solid formed when molten basaltic magma from the earth's mantle just below the crust cools on rising to the surface at midocean ridges or volcanic sites. A great deal of it exists, as a result, and it is largely ignored except for the semiprecious stone *peridot*, which is pure Mg_2SiO_4 in transparent deep green (sometimes brown) crystals.

Calcium orthosilicate, Ca_2SiO_4, and the related compound Ca_3SiO_5 are the key ingredients of Portland cement, formed by burning a mixture of limestone and clay to drive off CO_2 and open the silicate layer networks in clays. One form of Ca_2SiO_4 has the olivine lattice and is impervious to water, but the β form has a more open structure and reacts fairly rapidly with water to produce a network lattice that binds the poured concrete into a single monolith.

Zircon, $ZrSiO_4$, is an important ore for zirconium and hafnium; only rarely does it occur in the varicolored transparent crystals that are semiprecious stones. Its structure has Zr^{4+} coordinated by the oxygen atoms from SiO_4^{4-} units as in the olivine structure shown in Figure 3.14, but the Zr is eight-coordinate. The *garnets*, also semiprecious stones that occur as orthosilicates, fall in families depending on which elements are present in the general formula $(Mg, Ca, Fe, Mn)_3(Al, Fe, Cr)_2(SiO_4)_3$. Most are a deep red color, but Cr yields a pale green garnet *uvarovite*; the garnet variety *almandine* is common enough to be used as an inexpensive abrasive in "garnet paper." In all the garnets, the M^{2+} ion is eight-coordinate and the M^{3+} ion is six-coordinate. One particularly important garnet structure is the yttrium aluminum garnet (YAG), which is the host material for a solid-state laser that can be operated continuously at high power levels. The YAG host is doped with about 2% Nd^{3+}, which is the optically active material; the laser is pumped by light at about 800 nm wavelength and emits light at 1060 nm. Neodymium–YAG lasers can operate with continuous power levels of hundreds of watts.

The one-dimensional chain structures mentioned earlier occur in interesting variety, with more variations possible than were recognized for many years. For single chains Fig. 3.15 has shown three symmetry variations with repeat units of one, two, or three tetrahedra (1T, 2T, and 3T respectively). Examples of all three are known, though not always by silicates; 1T is represented by $CuGeO_3$ (GeO_4 tetrahedra), 2T by SO_3 and $LiAsO_3$ but also by the wide variety of pyroxene silicates already noted, and 3T by $NaPO_3$ but also by *wollastonite*, $CaSiO_3$. However, more complex repeat patterns have been observed all the way to 12T, which represents an elaborate superstructure for as simple a system as a chain of tetrahedra.

The double chains of Fig. 3.15 are also known in all three forms, though the four-tetrahedron rings formed from 1T chains constrain the Si–O–Si bond angle to a value smaller than the 145° often seen and thus are relatively rare. The six-tetrahedron rings formed from 2T chains are common as the amphibole forms of asbestos. The fibrous nature of asbestos is an obvious consequence of the one-dimensional chain structure, but asbestos also contains major amounts of chrysotile, a two-dimensional layer rolled up into a microscopic tube as we shall see below. The double chain of eight-tetrahedron rings is seen in *xonotlite*, $Ca_6Si_6O_{17}(OH)_2$.

Intermediate structures are known between the double chain and the infinitely wide ribbon of chains that forms a layer. A triple chain occurs in a mineral with the interesting name *jimthompsonite*, and *chesterite* adopts a superstructure consisting of alternating double and triple chains. Jade occurs both as the pyroxene single-chain *jadeite*, $NaAl(SiO_3)_2$, and as *nephrite*, a calcium magnesium iron silicate with a poorly characterized formula due partly to the fact that it contains single, triple, quadruple, and sextuple chains; the multiple chains have a unique form of defect in which they split apart for several repeat units, then recombine.

An important class of layer structures involves sheets of six-membered rings of tetrahedra in which the tetrahedra all point the same way. The dimensions of the SiO_4 tetrahedra are such that the O—O spacing between oxygens at the peaks of the tetrahedra is very nearly the same as the O—O spacing between adjacent oxygens on an MgO_6 octahedron in crystalline $Mg(OH)_2$ or an AlO_6 octahedron in crystalline $Al(OH)_3$. A layer of MgO_6 octahedra can thus fit right on top of the silicate sheet, as Fig. 3.19 suggests. Because the dimensions are not exactly the same (the sheet of MgO_6 octahedra is slightly larger) this double sheet curls up with the MgO_6 sheet on the outside. The mineral form of this is *chrysotile*, $(OH)_4Mg_3Si_2O_5$, which is the major component of commercial asbestos. The sheets roll up into tubes that, on the bulk scale, appear to be fibers. If the sheet of MO_6 contains Al instead of Mg, the formula becomes $(OH)_4Al_2Si_2O_5$, which is *kaolinite*. Kaolinite is found in kaolin clay, a white clay widely used in making chinaware and also for the coating on glossy paper.

If the layer of MO_6 octahedra is sandwiched between two sheets of SiO_4 tetrahedra in six-membered rings, the resulting formula is $Mg_3(OH)_2(Si_2O_5)_2$, *talc*, or $Al_2(OH)_2(Si_2O_5)_2$, *pyrophyllite*. These are also shown in Fig. 3.19. They are very soft minerals because the sandwich layers are uncharged and are held together only by van der Waals forces.

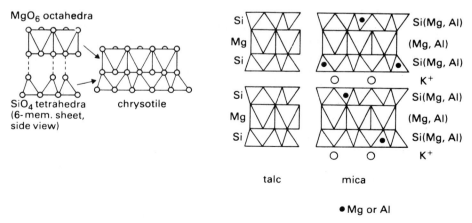

Figure 3.19 Fusion of sheets of MgO_6 octahedra and SiO_4 tetrahedra: chrysotile (asbestos), talc, and mica.

Stronger forces binding the layers together, such as electrostatic attractions, would produce a harder crystal, as Section 3.2 has noted. In the context of this sandwich structure, net negative charges on the sandwich occur if Al^{3+} is substituted for Si^{4+} in the tetrahedra of the silicate sheets. If a stable crystal is to form, the sandwich layers must then be held together by cations between the layers. If one-fourth of the Si atoms are replaced by Al atoms in each layer, the overall stoichiometry of the talc-like sandwich becomes $Mg_3(OH)_2Si_3AlO_{10}^-$. When these layers are interleaved with K^+ ions to maintain electrical neutrality the crystal is *mica* (specifically *phlogopite*, a mica mineral). Mica forms layers on the bulk scale because the crystal easily cleaves along these planes of K^+ ions.

Further Al substitution in the silicate layer (to a formula of $SiAlO_5^{3-}$) increases the negative charge on the sandwich. When the counterions are Ca^{2+} or Mg^{2+} the resulting crystal is known as a "brittle mica"; for example, *margarite*, $CaAl_2(OH)_2(SiAlO_5)_2$. Because of the increased electrostatic attraction of the more highly charged cations and layers, brittle micas are harder than ordinary mica and also do not cleave as readily—hence the name. Finally, there are numerous "hydrated micas" in which the binding ions are hydrates, usually $Mg(H_2O)_6^{2+}$. *Vermiculite*, the familiar expanded mica used as a soil conditioner and packing material, is one of these. It is mined as a fairly dense mineral like mica, but on strong heating it boils off its hydrated water, which in escaping from the crystal pushes the layers apart into expanded mica.

To this point we have considered predominantly naturally occurring silicates, of which there are almost countless variations, depending largely on the conditions within the earth's crust under which they were formed. However, commercially prepared silicates are enormously important in the world economy. The best known such products are those of the glass industry. The characteristic random linkage of SiO_4 tetrahedra in glass, so that the short-range order of the tetrahedron is preserved but no long-range order exists to yield the anisotropy characteristic of a crystal, is found in nature as well as in commercial glass. Many volcanos with low gas content yield molten silicate lava that cools and hardens so rapidly that the high viscosity of the lava prevents rearrangement into an ordered structure. Volcanic glass is called *obsidian*, and there are attractive variants called *snowflake obsidian*, in which small white crystal nuclei have grown to perhaps fingernail size, excluding colored impurity ions into the dark molten lava, before the glass hardens.

Glassy silicate structures can be formed by melting pure silica; the SiO_2 formula requires that all four vertexes of each tetrahedron be shared in the glass as in the original crystal, but this is easily done. However, the melting point of pure silica is quite high (quartz, 1610°C) because of the strong bond network. The technology of the third millennium B.C. could not achieve temperatures above those of a charcoal fire fed by a simple bellows, about 1100°C. Accordingly, the Phoenician discoverers of glass could only prepare it by mixing in another oxide that would lower the melting point of the quartz sand. Sodium carbonate has this effect:

$$Na_2CO_3 \xrightarrow{\text{heat}} Na_2O + CO_2$$

The low charge on Na^+ substantially lowers its lattice binding energy and thus the melting point of the mixed quartz and the softening temperature of the glass.

A mixture of Na_2O and SiO_2 yields a sodium silicate glass, obviously. If the proportion of Na is high, the product is water-soluble and in solution is a widely used commercial adhesive (primarily for corrugated boxes) known as *water glass*; sodium silicates are also used in commercial detergents and as flame retardants. To reduce the solubility and chemical sensitivity, and also to raise the softening temperature somewhat, CaO is added; the Ca^{2+} ion raises the lattice energy of the resulting glass. Ordinary glass, known as *soda lime glass* and which still accounts for perhaps 95% of all glass production, has nearly the same composition it had in Roman times: about 70% SiO_2, 10% CaO, 15% Na_2O.

There are many variations on this composition, yielding glasses with special properties. One of the oldest is *lead glass*, in which about 15% PbO replaces the CaO (though glasses with as high as 90% PbO have been made). The high electron density on the Pb atom raises the refractive index of lead glass, yielding the sparkle desired in cut glass and the high refraction needed for optical prisms and some lenses. *Borosilicate glass*, the original Pyrex brand, eliminates CaO entirely and reduces the Na_2O content to about 5%; the result is 80% SiO_2, about 12% B_2O_3, and the remainder Al_2O_3 and Li_2O.

These glass formulations are all created by molten-oxide reactions in which, at least conceptually, oxide ions are transferred thermally. In recent years a room-temperature solution process known as the *sol-gel process* has permitted some dramatically different formulations. In the sol-gel process alkoxides of silicon and other desired electropositive elements are allowed to polymerize in solution by hydrolysis, then dried. The result is a rigid, brittle glass but one that contains pores where solvent formerly existed. These pores can trap dopant molecules if those molecules are added to the polymerizing solution, including organic species that could not survive traditional molten-glass temperatures. The dopants or *guest species* are in general difficult to leach from the glass, which is to say permanently trapped; they can, however, be reached through the pore structure by other small molecules from the environment of the glass surface, so that chemical reactions can occur. Such glasses can protect laser dyes and photochromic compounds for optical data recording, for instance.

A number of glass applications—more generally, silicate applications—require it to be in the form of a fiber. Familiar applications find glass fiber as a tensile-strength component of fiberglass/plastic composite materials (perhaps best known in Corvette bodies), and as home insulating material, where heat flow is impeded by the many changes from gas to solid phase as molecular motion must transfer heat from one side of an insulated wall to another. Glass has no special properties that make it uniquely suited for this purpose, and molten calcium silicates from the slag of steelmaking blast furnaces are often spun into "rock wool," which is more brittle but equally useful thermally.

A more conceptually elegant application is in the rapidly growing network of optical-fiber communications. For this purpose we are, in effect, shining a light through a block of glass miles thick, and as a result it must be unusually transparent. Intrinsically colored ions such as those of Fe and Mn must be rigorously excluded, and light scattering from the fiber must be minimized. The fiber core, usually fused quartz doped with GeO_2 or P_2O_5, has a high refractive index. It is coated with an outer cladding having a lower refractive index, which has the effect of redirecting scattered light back toward the center of the core fiber. So-called "step-index" optical fiber is made by drawing a fiber as a melt from a quartz rod inside a tube with the cladding composition. A more sophisticated formulation is the "graded-index" fiber, in which the refractive index decreases smoothly from the core to the outside of the fiber. It is made by the chemical vapor deposition (CVD) process, in which a quartz tube is treated on the inside with halide vapors such as BCl_3, $GeCl_4$, and $POCl_3$. This halide coating is oxidized by O_2 at a temperature just below the melting point of the quartz. Repeated deposition with varying compositions (up to 1000 times) allows the building up of a smoothly varying composition and thus refractive index. Finally, the tube is melted and drawn into the final fiber. Light loss from this material is so low that half the entering light will be transmitted through 25 miles of fiber.

There are also "high-tech" thermal fibers. 3M produces an aluminum borosilicate fiber known as Nextel 312, which can be woven into fabrics usable at temperatures up to 1200°C. Pure silica is also used in these applications, and each original insulating tile on the space shuttle was a block of short silica fibers sintered together. These tiles insulated very well but created a weight problem for the shuttle. Newer tiles are a sintered mixture of Nextel 312 and silica fibers, weighing between 9 and 12 pounds per cubic foot (an effective density of only about $0.15 \ g/cm^3$).

Zeolites (see Fig. 3.18) are manufactured at roughly a billion pounds a year for a number of applications. In addition to zeolite minerals, at least 10 different synthetic zeolites are manufactured, varying in both their pore diameters and tunnel connectivity; some have one-dimensional tunnels, while others allow two- or three-dimensional movement of guest molecules. In general, zeolites are prepared in what is called the *hydrothermal process*, by dissolving stoichiometric amounts of NaOH, Na_2SiO_3, and Al_2O_3 in water at high temperature, stripping the water, molding the resulting gel into the desired pellet shape, and heating to complete the cuboctahedron formation reaction and drive off water. Zeolite A is made more simply from kaolin clay by molding the clay into the desired shape, heating, and treating the pellet with NaOH solution. The resulting systems have complex (and only approximate) stoichiometries, Three of the more widely used zeolites are zeolite A, roughly $NaAlSiO_4 \cdot 2H_2O$, with the structure shown in Fig. 3.18; Zeolite Y, roughly $Na_7Al_7Si_{17}O_{48} \cdot 30H_2O$, with the faujasite structure in Fig. 3.18; and ZSM-5, roughly $NaAlSi_{31}O_{64} \cdot 5H_2O$ (but with some replacement of Na^+ by tetraalkylammonium ions). ZSM-5 has a more complex structure including some five-tetrahedra rings, with a three-dimensional network of tunnels.

The porous, polar character of zeolites makes them valuable in several applications. They selectively adsorb molecules of the right size and to some extent the right shape and polarity for their pore entrances. A very common application is water removal; zeolite A, with a pore diameter of 4.1 Å, is commonly used as a *molecular sieve* for drying organic solvents. The small polar water molecule can enter the pore and is tightly bound at a polar site. Thousands of tons of zeolites are used each year in the manufacture of double-glazed windows, to prevent condensation inside the glazing. In a somewhat similar way, zeolites bind cations within their pores at negatively charged sites, and enormous amounts are used in clothes-washing detergents to bind Ca^{2+}, Mg^{2+}, and other metal ions present in hard water that would otherwise precipitate the detergent anions. The major focus of industrial research on zeolites, however, is their usefulness as catalysts for polar organic reactions. To accommodate larger organic molecules large pore sizes are needed, and both zeolite Y with a pore diameter of about 7.4 Å and ZSM-5 with a pore diameter of about 6 Å are widely used. These catalysts are used in isomerization of *n*-alkanes (with Pd or Pt in the channels), in catalytic cracking of crude oil (usually with rare earth ions in the channels), and in methanol-to-gasoline conversion.

3.7 NONSILICATE CERAMICS

We have spent a considerable time on silicate applications; in nearly all of these commercial uses the silicate qualifies as a *ceramic*: a nonmetallic inorganic material prepared or processed at high temperatures. It is the oldest inorganic chemical craft—pottery shards have been dated back 10,000 years. For most of those millennia silicates were the only chemical substances that could be processed as ceramics. However, in this century and particularly in the last 20 years a number of exciting applications of nonsilicate ceramics have emerged. We can consider these by applications, identifying the species chemically as they arise.

Hard materials constitute the first category. Some of their applications require hardness specifically, such as abrasives and metal-cutting tools, while others require strength in high-temperature applications such as firebrick and crucible materials. In these general areas, the ceramics in use are metal borides, carbides, nitrides, and oxides. Particularly for the first two of these it is not clear that the materials are nonmetallic, because the B, C, and to some extent N atoms tend to be dispersed in a lattice that retains much metal–metal contact. These are thus *interstitial* borides, carbides, and nitrides, because a close-packed lattice of metal atoms has interstices large enough to accommodate small heteroatoms of similar electronegativity. The formulas of interstitial compounds do not generally yield recognizable oxidation states, because the upper limit on interstitial atoms is usually the number of tetrahedral sites per close-packed atom, 2. We thus see formulas such as TiB and ThC_2. B, C, and N are small atoms, and the resulting network of polar covalent bonds in all of these yields a very hard substance with a very high melting point. If the melting point exceeds 1500°C the substance is said to be *refractory*, and dozens of refractory substances are in common use.

Metal borides have only limited applications. TiB_2 (mp 2850°C) is very chemically resistant to molten metals, particularly aluminum, and is used as a crucible material for electrometallurgical processes. Metal cutting tools must be hard but must also be abrasion resistant, which is not quite the same property; CrB and CrB_2 are particularly abrasion resistant and are sometimes deposited as surface layers on other hard materials by the chemical vapor deposition process. There is a stoichiometric compound B_4C (actually $B_{12}C_3$) with much the same function; it is produced by heating B_2O_3 with carbon, yielding a compound that is either a boride or a carbide. B_4C is used in making sandblasting nozzles, a tribute to its abrasion resistance, but also in armor plate and in nuclear power reactors as a neutron shield. An interesting variation on the metal borides above is the use of LaB_6 as a thermal electron emitter; it is a refractory boride with a very low photoelectric work function, 2.2 eV. As a result, a heated LaB_6 cathode can emit over 100 A/cm^2 at its highest working temperature, and it is the preferred electron beam source for electron microscopes and electron lithography systems.

Metal carbides are much more widely used, particularly WC and TiC. These can be made by heating a mixture of metal and carbon, but also by heating the metal in a stream of carbon-containing gas such as methane. Because of the very high melting points of these materials (WC, 2720°C; TiC, 2940°C) the product is a finely divided solid, but these particles can be sintered with and partially dissolved in cobalt metal, yielding a *cemented carbide* in which the cobalt metal provides a tough base for the hard (but otherwise brittle) tungsten carbide. Cemented carbide cutting tools are significantly more expensive than an equivalent steel tool, but will last so much longer that they are ultimately cheaper in production applications.

Most transition metals will absorb interstitial carbon at high temperatures; in some cases the carbon remains interstitial, while in others the carbon forms a solid solution replacing metal atoms in the lattice. The composition of either type of carbide is usually variable over wide ranges. By the strictest definition, an interstitial carbide would retain the same lattice as the original metallic element, though perhaps expanded a little to accommodate the carbon atoms. This is usually true at low carbon concentrations, but for compositions approximating M_2C or MC the lattice usually has a different symmetry from the metal, even though the compounds are still described as interstitial. Most transition metals are body-centered cubic (bcc), but most of the phases corresponding to M_2C stoichiometry are hcp and those corresponding to MC are usually face-centered cubic (the ccp NaCl lattice). Usually it is not possible to make a continuous range of compositions within a single phase for a metal-carbon system because of these lattice changes, but the stoichiometry range can still be quite wide. For instance, bcc vanadium dissolves only about 1% carbon, but by adopting the hcp symmetry for the V atoms it can form "V_2C", which actually has a composition range from $VC_{0.37}$ to $VC_{0.50}$. Adding still more carbon yields "VC" in the NaCl lattice but with as wide a range of composition as V_2C.

Although other constituents of steel alloys are important, the steelmaking process consists to a considerable extent of regulating the proportions of *cementite*, Fe_3C, in the iron–carbon mixture. Because of the hardness of the interstitial carbides, steel can be case-hardened for many applications by immersing it in molten NaCN (plus NaCl and Na_2CO_3) or heating it in a reducing atmosphere (H_2, CO) containing CH_4 and NH_3. Either of these procedures yields interstitial iron carbide mixed with interstitial iron nitride, which is also quite hard.

By far the most important ceramic carbide in commercial terms is silicon carbide, SiC, which is produced in an electric resistance furnace (ca. 2400°C), the *Acheson process*:

$$SiO_2 + 3C \rightarrow SiC + 2CO$$

The process is extremely endothermic ($\Delta H = +478$ kJ) and requires high temperatures to enhance the negative-free-energy effect of the positive entropy change associated with the formation of gaseous CO. Under the trademark Carborundum, SiC has become the most widely used abrasive in applications other than ordinary sandpaper. Extensive studies on SiC ceramics in molded applications such as high-temperature gas turbine blades are being carried out, and these uses will be much more widespread in coming years. For these purposes, SiC must be finely ground and hot-pressed into shape (ca. 5000 psi, 2000°C), because sintering SiC is very difficult.

In general, nitrides are less widely used than carbides, mostly because they are not wetted by molten metals as well as the comparable carbides are. As a result, it is much more difficult to manufacture cemented nitrides than cemented carbides, and metal cutting tools are essentially always cemented carbides. However, thin layers of TiN are sometimes deposited on carbide tools by CVD to increase abrasion resistance, as is the case for borides.

Again, the most widely used nitride is silicon nitride, Si_3N_4, made in a process similar to the Acheson process:

$$3SiO_2 + 6C + 2N_2 \rightarrow Si_3N_4 + 6CO$$

or

$$3Si + 2N_2 \rightarrow Si_3N_4$$

or

$$3SiH_4 + 4NH_3 \rightarrow Si_3N_4 + 12H_2$$

If finely divided Si_3N_4 and Al_2O_3 are sintered together, the resulting product is a *sialon* (SiAlON) ceramic. Sialons of various compositions are candidates for high-temperature engine components, whether gas turbine rotors or piston-engine parts such as piston caps, exhaust valve heads, and turbocharger rotors. Given the fundamental thermodynamic limitations on heat-engine efficiency, it is clear that major improvements in fuel mileage can only be achieved by increasing T_{hot} in the efficiency ratio $(T_{hot} - T_{cold})/T_{hot}$, and the necessary high-temperature strength can only be achieved by using ceramic parts at points of greatest thermal stress.

Many oxides are used in ceramic applications. Ordinary clay ceramics are, of course, oxides; however, they are heated only enough to sinter the clay particles together and remain porous. The clay ceramic sintering process is called *vitrification*, and the more thoroughly vitrified a clay ceramic is, the less porous and more translucent it becomes. Common brick and drainpipe ceramics are extremely porous, although they may be glazed by melting a thin layer of a low-melting silicate over the surface to minimize water absorption. Ceramic bathroom fixtures (*whitewares*) are less porous; ironstone and ordinary dinnerware still less porous, although they must still be glazed. Bone china dinnerware is more thoroughly vitrified and is translucent to a strong light. Completely vitrified clay ceramics become in effect a form of glass, though this is not a commercial process. Materials such as Corning's Pyroceram cook-serve-freeze utensils are a glass that has had small amounts of a crystalline oxide such as TiO_2 added to nucleate crystallization of the silicate glass, a process known as *devitrification* although no pores form.

In newer oxide ceramics, the oxide is manufactured, not mined. It must normally be extremely pure, but does not make a plastic molding material like wet clay, and must have a binder such as cellulose or gelatin (even molasses) added in small amounts. These molding materials are heated to perhaps 200°C below the melting point of the oxide for sintering. Unfortunately, the sintering process causes considerable shrinkage, so that the product must be molded oversize and given final dimensions and finish by diamond or boron carbide grinding.

Perhaps the most important oxide ceramic is Al_2O_3, sold as Alundum and used as an abrasive in grinding wheels, as a high-temperature crucible material (mp 2050°C), and as the insulating body material in spark plugs. MgO, *magnesia*, is widely used for firebrick in lining steelmaking furnaces and molten glass tank covers. However, its resistance to thermal shock is improved by mixing in Cr_2O_3 or Al_2O_3, forming a spinel structure MgM_2O_4. The other major pure oxide ceramic is *stabilized zirconia*, ZrO_2, which is nominally a pure oxide in a cubic CaF_2 lattice that is stable only at high temperature but is stabilized at room temperature by the presence of perhaps 15% CaO or MgO. The very high melting point of stabilized zirconia (pure ZrO_2 mp = 2700°C) allows it to be used as a high-temperature resistive heating element and as the nozzle material for continuous casting of steel.

A number of mixed-oxide ceramics are used both as hard materials and as refractories. Firebrick are often made from *mullite*, $Si_2Al_6O_{18}$, which as an aluminosilicate can be prepared from kaolin clay and bauxite. We have noted magnesium spinels above, used not only in steelmaking furnaces but in cement kilns. A final composition that has the advantage of extremely low thermal expansion (and thus resistance to cracking under thermal stress) is Al_2TiO_5, which is being studied as insulating material for engine exhaust systems.

After this extended discussion of hard ceramics, we might close with a soft one. Corning manufactures a ceramic *Macor* that is machinable using ordinary drill bits or lathe tools. It is a devitrified glass (resembling Pyroceram in having crystals embedded in a glass matrix) with a fluorine-rich composition approximating that of the mica crystal phlogopite, but with F replacing OH: $KMg_3F_2Si_3AlO_{10}$. On devitrification, the crystals forming in the glass matrix are microscopic sheet-like fluorophlogopite crystals, whose weak binding forces (like those in mica) allow easy working with ordinary machine tools.

The second category of application of nonsilicate ceramics is that of reinforcing components for composite materials, either as fibers or whiskers. Fiberglass/resin composites have been mentioned above, but more exotic materials are used in the same way as fiberglass for comparable purposes. One of the defects of ceramics as materials of construction is that they are brittle—drop it and you've bought it. Brittleness is the primary reason the ceramics mentioned above have not appeared in common use as, for example, jet engine turbine blades. One of the most promising ways to increase the mechanical toughness of ceramics is to reinforce them internally with fibers of very strong materials. Borosilicate glass, for example can be reinforced as a composite material with graphite fibers or with SiC fibers. Such composites, in which the base material is devitrified lithium aluminosilicate glass (analogous to Pyroceram) reinforced with approximately 50% SiC, have good mechanical toughness up to 1000°C. The SiC fibers are made from $SiCl_2(CH_3)_2$ by polymerizing it to $[Si(CH_3)_2]_n$, heating this polymer to 400°C to rearrange it to $[-SiH(CH_3)-CH_2-]_n$, spinning this polymer to fibers, and heating the fibers to 1250°C to drive off methane and leave SiC in the physical form of the original fiber. Oxide ceramic fibers can be used to reinforce metals, as well. Aluminum oxide fibers dispersed in aluminum metal raise its mechanical strength and heat resistance considerably.

A variation on the use of ceramic fibers is the growth of ceramic whiskers, which are shorter and much smaller diameter than fibers. Whiskers are normally single crystals, while fibers are polycrystalline. This difference can yield as much as a factor of 10 improvement in tensile strength: Where steel wire has a tensile strength of roughly 300,000 psi and SiC fibers are about the same, SiC whiskers can have equivalent values of 4 million psi. It is interesting that, although silicon is not one of the elements thought of as a significant component of plants, SiC whiskers are made by pyrolyzing rice hulls at 2000°C. These whiskers can be used in all the applications of SiC fibers and of some other fibers; in aluminum metal, they raise the tensile strength about 40% and roughly double the stiffness of the metal.

The third category of application of nonsilicate ceramics takes advantage of their specific electrical properties. Insulators are perhaps best known: Not only are spark plug insulators Al_2O_3, but we are all familiar with the large ribbed ceramic insulators on high-voltage power lines and at substations. Aluminum oxide is also the insulating substrate for integrated circuits; it is satisfactory as long as the IC is silicon-based. Gallium arsenide semiconductors, however, operate at higher speeds and require the substrate to have a lower dielectric constant (<2) to reduce dielectric heating by currents in the IC. Possible replacements include ceramics made from strontium silicate or barium silicate, which have lower dielectric constants than alumina.

On the other hand, capacitors need a material with the highest possible dielectric constant between the plates. Barium titanate, $BaTiO_3$, ceramics can have a dielectric constant as high as 12,000, which is roughly 100 times as great as other common materials. As a result, it has found extensive use in capacitor manufacture. The other compounds used in this context (*ferroelectrics*) are the obvious periodic table candidates: $SrTiO_3$ and $BaZrO_3$.

Barium titanate is also piezoelectric, meaning that it develops an electric potential across opposite crystal faces when the crystal is subjected to mechanical stress. It adopts an approximate version of the perovskite lattice (Fig. 3.12), but at room temperature the stable form is distorted along the edge of the cube shown in the figure so that the charge distribution within the unit cell is not centrosymmetric. In these circumstances it is not surprising that pressure on the crystal changes its electric symmetry and thus develops an electrical potential. Such crystals can be used to measure pressure, as for instance in digital laboratory balances where the piezoelectric voltage is nulled by a current applied to a solenoid opposing the weight of the sample; the digital display measures the applied current, but the null condition is the piezoelectric signal. Although $BaTiO_3$ was used in these applications for many years, the most widely used piezoelectric material currently is polycrystalline $Pb_2(TiO_3)(ZrO_3)$, lead zirconate titanate, which also adopts a distorted perovskite lattice. More familiar in this application is the quartz crystal that serves as a frequency reference in quartz watches and in other electronic applications. The natural thermal vibrational modes of the helical quartz crystal yield an extremely stable electrical frequency standard, and the crystals are easily grown by the hydrothermal process, dissolving SiO_2 in aqueous NaOH at 400°C and growing large crystals on small seed crystals.

Yet another electronic application of ceramics is the family of magnetoceramics known as *ferrites*. These are mixed oxides in two structural families. $M^{2+}Fe_2O_4$ spinels (see Fig. 3.11) and another family $M^{2+}Fe_{12}O_{19}$ with the *magnetoplumbite* structure (from the mineral $PbFe_{12}O_{19}$, though that compound is not used commercially). The curious formula of the magnetoplumbites arises from their superstructure, in which the oxygen atoms are close-packed but alternate between cubic and hexagonal layers: BAB*-ABCAC*AC.... In this sequence the layers B* and C* (every fifth layer) have M^{2+} substituted for one-fourth of the O atoms (hence one-twentieth overall), which because cations are in general smaller than anions requires that M^{2+} be a large cation such as Sr^{2+} or Ba^{2+}. These two ferrites, $BaFe_{12}O_{19}$ and $SrFe_{12}O_{19}$, are the most important commercial "hard" ferrites, meaning ferrites that readily acquire and keep a permanent magnetic field. Barium ferrite has a theoretical magnetization of 17,000 gauss, but practical sintered materials are limited to about 4000 gauss, which is still a considerable magnetic field. Hard ferrites are used for permanent magnets, particularly in small permanent-magnet dc electric motors, and for magnetic recording materials such as videotape and floppy disks. On the other hand, the spinel-structure materials are "soft" ferrites, meaning that magnetization is readily induced and removed in them, so that they have no memory of previous magnetic states; they are used for coil and transformer cores, and also for recording heads in both audio and video tape recorders. Mg^{2+} and most of the first-row transition-metal 2+ ions have these properties, and mixed-composition spinel ferrites such as $(Mg, Mn)FeO_4$ are widely used.

These same ferrites are materials for *thermistors*, in that their resistance changes with temperature. $CoFeO_4$ and $NiFeO_4$, but also to some extent the Mn and Mg ferrites, are used for negative-coefficient thermistors (R decreases as T increases). Positive-coefficient thermistors (R increases with T) are usually made from $BaTiO_3$ or $MgAl_2O_4$. A related property is shown by *varistors*, whose resistance decreases sharply above a given applied voltage. Varistors, usually made from doped zinc oxide, are the critical element in surge protectors and lightning arrestors.

The final group of electrical ceramics is the ionic conductors or solid electrolytes. The best known of these is β-alumina, whose properties we have already described. β-Alumina naturally contains mobile Na^+ ions but can have these replaced by other cations, as we have seen. An important alternative ionic conductor would allow anions to move freely; in particular, it would be useful to have a solid electrolyte for air-burning fuel cells that allowed oxide ions to move freely. ZrO_2 doped with small amounts of MgO or Y_2O_3, which yields the stabilized zirconia described earlier, has this property at the high temperatures (ca. 1000°C) at which solid-electrolyte fuel cells are operated. Substituting the lower-charged cations for Zr^{4+} yields charge-compensating vacancies at O^{2-} sites, which allow remaining oxide ions to move through the lattice. Westinghouse has developed air/natural-gas fuel cells using stabilized zirconia electrolyte that may be usable in large arrays to replace gas turbines for electric power generation, with the large increase in thermodynamic efficiency that comes with avoidance of the heat-engine limitation on energy conversion.

3.8 CLUSTER IONS AND CLUSTER LATTICES

In examining polymeric networks and superstructures thus far, we have usually categorized the various examples in terms of their dimensionality: one-, two-, or three-dimensional networks. Thus, for example, we have taken the ReO_3 lattice (Fig. 3.2) as an example of a three-dimensional network. However, within such highly symmetrical systems distortions can yield localized groups of atoms that verge on molecular status. This is particularly likely if the electronegativity differences between neighbor atoms in the crystal are small enough to make the predominant bonding covalent. Figure 3.20 compares the symmetry of ReO_3 against that of $CoAs_3$, a mineral called *skutterudite*. In skutterudite the As atoms, which in the symmetric ReO_3 lattice would

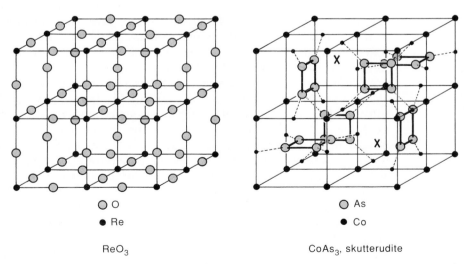

○ O

● Re

ReO_3

○ As

● Co

$CoAs_3$, skutterudite

Figure 3.20 Covalently bound clusters within formal three-dimensional network lattices: $CoAs_3$ as a variation on the ReO_3 lattice.

have been on the Co–Co axes, have collapsed toward the center of the large cavity in the lattice (filled in the perovskite structure by another large cation) to form bonds with each other; in the figure, the dotted lines indicate the directions in which the As atoms have distorted from ideal ReO_3 positions. Although the stoichiometry of the compound suggests some sort of triads of As atoms, the structure in fact consists of square As_4 clusters. Note that the stoichiometry requires that only three-fourths of the eight visible cubes can have As_4 clusters present, so that the cubes whose centers are marked $\times$ are empty. If we consider that the cluster is an ion As_4^{q-}, it becomes difficult to assign q. As_4^{4-} would correspond to Co^{3+}, but represents a rather high net negative charge for an element of modest electronegativity. In the next chapter we will consider cluster bonding and will develop electron-counting rules that would allow a prediction of q, but here we can note that there is a well-characterized Bi_4^{2-} ion. Small cluster ions of post-transition elements such as As and Bi are known as *Zintl ions*. Because they are clearly covalent in their internal bonding we will delay their consideration until Chapter 4. However, there are lattices in which extended networks of comparable directional bonds are responsible for the stability and geometric preference of the lattice, and we can consider those here.

If the Zintl ions are taken as zero-dimensional polymers, we can inquire about one-dimensional chains of clusters in crystal lattices. The earlier transition metals—Sc, Y, the lanthanide metals, Zr and Hf—show a strong tendency to develop cluster structures in their lower oxidation states, for reasons that Chapter 13 will develop. Quite commonly these clusters contain an encapsulated heteroatom, and it is thought that the heteroatom must be present to stabilize most of the known clusters. Figure 3.21 shows an example, the cluster $Sc(Sc_6I_{12}C)$ that constitutes the basic structural element of the compound $Sc_7I_{12}C$. This compound should probably not be thought of as a polymer, but it does show the octahedral cluster that is one of the most characteristic units of chain polymers in comparable compounds. In the crystal lattice, the I atoms are approximately cubic close-packed, as indicated by the ABC layer lettering; however, in each B layer a single I is replaced by the octahedral Sc_6C cluster unit, and an isolated Sc atom lies in an octahedral site between the CA layers. It would be possible to regard this compound as a one-dimensional polymer (Sc_6C clusters joined by ScI_6 octahedra), but the Sc is significantly (0.13 Å) closer to the three C iodine atoms shown than to those in the next A layer.

On the other hand, there is no doubt that Y_4I_5C is a polymer. Figure 3.22 shows the chain of Y_6C octahedra sharing opposite edges. As the polymer is drawn, all exposed edges of the Y_6C octahedra are bridged by I atoms except the equatorial edges; in the true crystal structure these are also bridged by I atoms in a parallel line that also bridge this chain to the next chain. The partially exposed Y atoms at the top and bottom of the octahedra are also weakly bonded to I atoms from adjacent chains, so that there is some degree

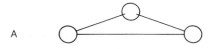

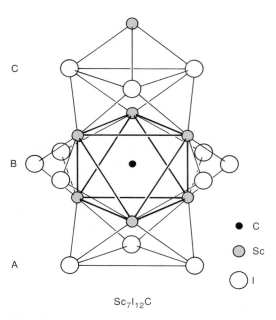

$Sc_7I_{12}C$

Figure 3.21 Sc_6C cluster as a structural replacement for I in nearly close-packed-I lattice for $Sc_7I_{12}C$.

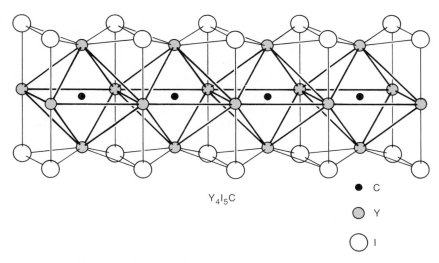

Y_4I_5C

Figure 3.22 Portion of chain of octahedral Y_4I_4C clusters in Y_4I_5C lattice.

of three-dimensional bonding beyond simple van der Waals attractions. This is not a unique structure: Sc_5Cl_8C and Sc_5Cl_8N adopt exactly comparable one-dimensional polymer chains, except that each of the four exposed edges of the polymer chain is bridged to neighbor chains by $(ScCl_4)_n$ chains, which are themselves edge-joined chains of $ScCl_6$ octahedra.

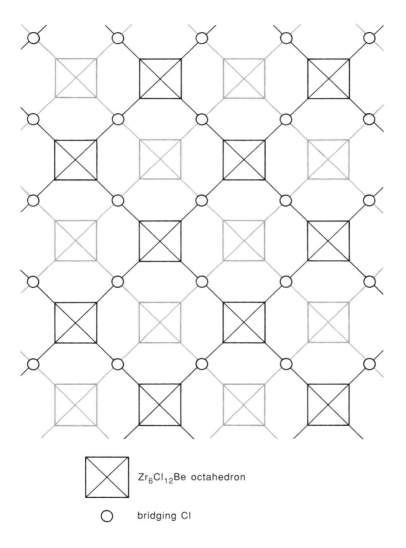

☒ $Zr_6Cl_{12}Be$ octahedron

◯ bridging Cl

gray lines represent second layer

Figure 3.23 Portion of two successive layers of $Zr_6Cl_{12}Be$ octahedra in $Na_4Zr_6Cl_{16}Be$ lattice.

Two-dimensional layer structures have also been established for comparable compounds. $Na_4Zr_6Cl_{16}Be$ adopts the slightly puckered layer structure shown in a simplified form in Fig. 3.23. The Be atom is encapsulated in the Zr_6 octahedron, all 12 of whose Zr–Zr edges are bridged by Cl atoms yielding a $Zr_6Cl_{12}Be$ unit; these form the layer shown, with another Cl bonded out of the octahedron at each Zr. In each layer, the equatorial four of these Cl atoms are shared with adjacent $Zr_6Cl_{12}Be$ octahedra, and the two axial Cl atoms are unshared between layers (though they do provide part of the interstitial site for each Na^+ ion). As the figure indicates, the next layer adopts a staggered arrangement over the first so that the octahedra of the second layer fit in the open spaces of the first.

Other cluster shapes are known in lattices analogous to these, including chains of tetrahedra, double chains of octahedra, chains of nine-atom clusters each of which is a capped square antiprism, and even less symmetrical monomer units in both chains and layers. None as yet have any commercial application, but for example some have metallic electrical conductivity; it seems likely that as our understanding of the electrical and magnetic character of solid materials increases, polymeric networks such as these may well take on importance we cannot now imagine.

PROBLEMS

A. DESCRIPTIVE

A1. In a crystal of K_2CrO_4 what kind of bonding should exist between adjacent atoms? Present arguments based on both electronegativity and the discussion in Section 3.1.

A2. For AB_2 compounds we have discussed the rutile, fluorite, $PbCl_2$, and CdI_2 lattices. Keeping in mind the variation in structure shown by the indium halides, suggest which of the AB_2 lattices is most likely for $SnBr_2$.

A3. Consider chains formed by the fundamental units AB_4 (tetrahedron) or AB_6 (octahedron). What linkages can yield the overall formulas AB_2, AB_3, AB_4, and AB_5? Sketch the possibilities.

A4. How are the structures of HgI and As_2O_3 related? See Figs. 3.3 and 3.6.

A5. Boron nitride BN forms a layer structure similar to graphite. How should the forces between layers compare to those of graphite? What can you predict about the physical properties of the BN compound?

A6. Figure 3.9 shows the formation of a crystallographic shear plane in the ReO_3 lattice. What is the formula of each block of edge-sharing octahedra in the collapsed structure?

A7. V^{2+} and Ta^{5+} differ only modestly in crystal radius. Should VTa_2O_6 adopt the randomly substituted rutile lattice or the trirutile lattice?

A8. What general relationship between r_A and r_B favors the inverse spinel structure over the normal spinel structure for AB_2O_4 compounds?

A9. What arguments can you bring to bear on the question of whether technetium bronzes can be made?

A10. What coordination numbers are shown by the atoms/ions in an ideal ilmenite lattice?

A11. What general structures would you predict for the following silicates?

    (a) Montmorillonite, $Al_2(Si_2O_5)_2(OH)_2$

    (b) Nepheline, $KNa_3Al_4Si_4O_{16}$

    (c) Dioptase, $Cu_6Si_6O_{18} \cdot 6H_2O$

    (d) Spodumene, $LiAl(SiO_3)_2$

    (e) Orthoclase, $KAlSi_3O_8$

A12. Petalite, $LiAlSi_4O_{10}$, has a layer structure. Is the Al substituted for Si in the SiO_4 tetrahedra for each layer, or does it occupy space between the silicate layers? How can you tell?

A13. What should the formula and net charge be on the triple chain of silicate tetrahedra in jimthompsonite?

A14. If Li is substituted for Na in a given glass formulation, what should be the effect on the softening temperature of the glass?

A15. In the manufacture of graded-index optical fibers by CVD, the vapor deposition progressively leaves MO_n oxides inside the quartz tube. If the refractive index must decrease toward the outside of the fiber, should M have a higher or lower atomic number than Si?

A16. In the hard ferrite $BaFe_{12}O_{19}$ (magnetoplumbite structure), Fe^{3+} should occupy octahedral sites between oxide layers, but should avoid the positively charged Ba^{2+} locations in every fifth layer. If no octahedral sites touching the Ba-containing layer are occupied, can 12 Fe be accommodated in the remaining octahedral lattice sites?

B. NUMERICAL

B1. Use the geometry in Fig. 3.12 to derive the relationship

$$\frac{r_{12} + r_O}{r_6 + r_O}\left(\frac{1}{\sqrt{2}}\right) = t = 1$$

for perfect fit of atoms in the perovskite structure, where t is the size tolerance factor. Remembering to adjust tabulated 6-coordinate crystal radii for 12-coordination as in Chapter 2, calculate values of t for $BaTiO_3$ and $CaTiO_3$, which have the perovskite structure, and for $FeTiO_3$, $CoMnO_3$, and $MgMnO_3$, which have the ilmenite structure. Do your results verify the size relationship indicated in the chapter?

B2. Use tabulated electronegativities and radii to locate the following compounds on the graph in Fig. 3.1, and predict their preferred crystal lattices: GaN, KBr, BeO, BaSe, CuCl.

B3. The standard enthalpy of formation of AgI is -62.3 kJ/mol. Use this value in a Born–Haber cycle to obtain a value for the lattice energy U of AgI. Separately, calculate a theoretical value of U for AgI using the Kapustinskii approximation. Use your results to verify the values in Section 3.2.

B4. Use VSEPR principles to predict bond angles in the AuI chain. The experimental Au–I–Au angle is $72°$. Suggest reasons that this angle might be so small.

B5. Pure NaCl has a density of 1.54424 g/cm^3. A sodium chloride crystal heated in sodium vapor develops 1×10^{19} F-centers per cm^3. Calculate the change in density that should occur as a result of this concentration of F-centers.

B6. If the operating temperature of a LaB$_6$ electron emitter cathode is $1800°C$, how does the average thermal energy of an electron at that temperature compare with the work function of LaB$_6$ (2.2 eV)?

B7. The thermal limitations on the metal components of a piston engine mean that T_{hot} for the power stroke of the engine can be only about $1500°C$. Make a reasonable assumption for T_{cold} (the exhaust manifold temperature) and calculate the maximum thermodynamic efficiency of such an engine. If ceramic components allow T_{hot} to equal the flame temperature of about $2200°C$, what change in overall efficiency is possible?

C. EXTENDED REFERENCE

C1. Numerous layer structures have been described in the chapter. How many ways are there to fill a two-dimensional layer with geometrically regular rings (equal sides, equal angles)? How many of the possible layer symmetries are shown by known silicates? [See A. F. Wells, *Structural Inorganic Chemistry*, 5th ed. (Clarendon Press, Oxford, 1984), chapters 3 and 23.]

C2. Corbett [E. Garcia and J. D. Corbett, *Inorg. Chem.* (1990), *29*, 3274] reports a series of interstitial compounds Zr$_5$Sb$_3$Z, where Z (the interstitial atom) is any of 15 elements from C to Ag. The interstitial atom fits inside a Zr$_6$ octahedron. How does the linkage of Zr$_6$ octahedra in these compounds compare to the linkages seen for Y$_6$ and Zr$_6$ octahedra in Section 3.8 of the chapter? Do the bookkeeping on his Figure 1 structural diagram to show that the structure portrayed does indeed add up to the formula Zr$_5$Sb$_3$Z. In his Figure 2 it is clear that interstitial C and O actually shrink the Zr$_6$ cluster, while interstitial P and heavier atoms expand it; in general, across a given row of the periodic table the cluster is expanded less by the atoms toward the right side of the row. Explain the electronic effects that are responsible for these trends in cluster size.

C3. J. D. Johnson et al. [*Inorg. Chem.* (1988), *27*, 1646] report a new vanadium phosphate phase with formula V$_3$P$_4$O$_{15}$. They describe the structure as linked in three dimensions. Copy their structural diagram (Figure 2) and show how their trimer units link to a chain, and the chains into a layer. Is the layer more or less planar or is it corrugated? How can the layers link to a three-dimensional network?

C4. J. M. Thomas et al. [*Angew. Chem. Int. Ed.* (1988), *27*, 1364] report a new layered chlorooxobismuthate with the valuable property of being an efficient catalyst for the oxidative oligomerization of CH_4 to C_2H_4 and C_2H_6, which could be important in preparing synthetic liquids fuels. They report "a layer . . . similar to that in CsCl." Copy their structure diagram and indicate the CsCl structure units. Within the remaining "BiOCl" structure units, what is the coordination number and geometry of the Bi/Ca atoms? If the top layer in their structure diagram were the surface of a catalyst, what electrical symmetry would be shown by the cavities as catalytic sites?

4

Bonding Theory for Covalent Molecules, Clusters, and Crystals ____

The last two chapters dealt with the chemical bonding properties of systems in which the electronegativity difference between neighboring atoms is great enough to make electron transfer (and the resulting electrostatic attraction) the primary binding force for the atoms of the system. As the discussion of the Madelung constant implied, such systems can be described to good accuracy on the basis of classical electrostatics without invoking quantum mechanics. This is because atoms in these systems interact as entire entities; the electrons are attracted to a single center, the nucleus of their own atom, and the theoretical and structural consequences of electron sharing between two or more nuclei need not be explored. We saw, however, that no lattice is completely ionic and that, indeed, many extended arrays of atoms in crystals that might appear ionic actually have properties determined to a considerable degree by the formation of covalent bonds within the crystal.

If we turn now to covalent or predominantly covalent systems of atoms, we face exactly the opposite situation and thus a more complex theoretical problem in describing the bonding. Covalent systems are held together by shared electrons, which experience simultaneous attractions of nearly the same magnitude to two or more nuclei. Parenthetically, it might be noted that the requirement of near-equal attractions to neighboring nuclei is the same as saying that covalent bonding occurs between atoms having similar electronegativities. Such systems necessarily consist of electrons that are moving about two or more centers of nuclear positive charge, not localized on one atom. We must describe the electron distribution in a covalent molecule using quantum-mechanical techniques because the problem is similar to that of describing the distribution of an electron in an atom. The bonding energy for a covalent system, then, becomes the difference between the total energy of the electrons in the multicenter system and the energy those electrons would have if distributed in the same—but isolated—atoms.

4.1 MOLECULAR ORBITAL METHODS

Just as, in general, polyelectronic atoms cannot be treated by mathematically exact quantum-mechanical methods, approximate methods are necessary for all covalent molecular systems of any chemical interest. The only exception is the one-electron molecule H_2^+, which can be solved by exact methods if constant internuclear positions are assumed. The Schrödinger equation can then be solved using elliptical coordinates in which each hydrogen nucleus is at an elliptical focus. The resulting dissociation energy for the H_2^+ ion is an excellent match for the experimental (spectroscopic) value when allowance is made for the vibrational energy the diatomic ion must have. The value of the calculation, however, lies in the fact that it provides a benchmark for comparing approximate methods that can be extended to more interesting molecules.

The appearance of the exact-result wave function (Fig. 4.1) suggests that it could be approximated by the sum of the two one-electron orbitals for the individual hydrogen atoms H_a and H_b: $1s^a + 1s^b$, as shown in Fig. 4.2. This approach amounts to creating a *molecular orbital* to describe the distribution of an electron by taking a linear combination of the atomic orbitals that belong to the atoms in the molecule. Such molecular orbitals are called LCAO-MOs to indicate the general form of the approximate wavefunction. We are thus making the orbital approximation again, this time to describe the electron distribution within a polyatomic molecule.

Since the atomic orbitals (AOs) for a given atom form a mathematically complete set of possible electron distributions, the quality of completeness requires that we produce as many LCAO-MOs as we have AOs to begin with by considering all the possible $+/-$ symmetries of combination of the AOs.

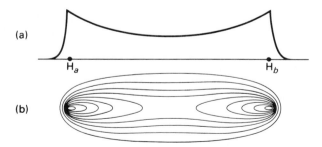

(a)

H_a H_b

(b)

Figure 4.1 Electron distribution in H_2^+ (a) ψ^2 as a function of distance along internuclear axis. (b) Angular dependence of ψ^2 with each equal-probability contour representing 0.9, 0.8, 0.7, and so on, of ψ^2_{max}.

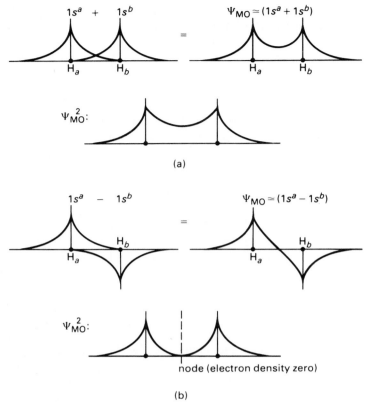

Figure 4.2   Molecular orbitals for H_2^+ under the LCAO approximation. (a) Symmetric bonding MO. (b) Antisymmetric antibonding MO.

For H_2^+, there is only one alternative to $1s^a + 1s^b$: $1s^a - 1s^b$. Figure 4.2 also shows this LCAO-MO. Whereas the first (symmetric) MO tends to place the electron in the middle of the bond where it will be most strongly attracted by the two nuclei, this (antisymmetric) MO has a node at that location and, in effect, forces the electron to spend most of its time outside the H—H bond region. An electron in this region actually tends to pull the nearer nucleus away from the other. This electron distribution is therefore known as an *antibonding orbital*, by contrast with the first, which is said to be a *bonding orbital*.

We can describe the energies of these molecular orbitals by considering the results of inserting the LCAO-MOs into the Schrödinger equation in symbolic fashion. The Schrödinger equation can be rearranged into an operator-eigenfunction expression for the total energy E as the eigenvalue: $\hat{H}\psi = E \cdot \psi$, where $\hat{H}$ represents the total-energy (Hamiltonian) operator.

Since ψ^2 is a probability distribution function for the electron, the probable value or *expectation value* for the total energy can be found by multiplying the Schrödinger equation through by ψ, integrating over all space, and noting that if the wave functions are properly normalized,

$$\int_{\text{all space}} \psi^2 = 1$$

$$\int \psi \hat{H} \psi = E \cdot \int \psi^2 = E \cdot 1$$

If we insert the bonding MO into the integral on the left and expand the products of atomic orbitals that result, we get

$$E = \int 1s^a \hat{H} 1s^a + \int 1s^b \hat{H} 1s^b + \int 1s^a \hat{H} 1s^b + \int 1s^b \hat{H} 1s^a$$

$$= 2 \int 1s^a \hat{H} 1s^a + 2 \int 1s^a \hat{H} 1s^b$$

When the bonding MO is properly normalized, the factors of 2 disappear. The first integral essentially represents the coulomb energy of an electron described by a single atomic orbital; it is usually symbolized α. The second integral represents the mutual-attraction energy of an electron simultaneously described by the AOs of two different atoms and thus the basis of covalent bonding between the atoms; it is usually symbolized β. Figure 4.3 is a molecular-orbital energy-level diagram in which these energies are shown. The figure suggests that the H_2^+ molecule is more stable than a hydrogen atom plus a separate proton by the energy β. In a simple approximation, α, the

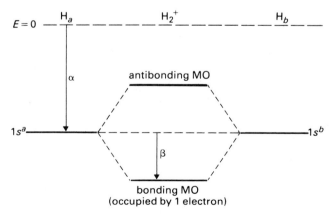

Figure 4.3 MO energy-level diagram for H_2^+.

coulomb energy of an electron described by a single atomic orbital, is equal to the valence-orbital ionization potential for that orbital, as given in Table 1.8. This empirical interpretation saves us the considerable mathematical difficulty of computing the value of the integral. Similarly, we can estimate β to good accuracy in many cases by assuming it to be proportional to the average of the α energies for the two combining AOs, and also to the fractional extent to which the AOs overlap each other in space: $\beta = k \cdot S_{ab}[(\alpha_a + \alpha_b)/2]$. Here S_{ab} represents the integral of the product of the two AOs ϕ_a and ϕ_b over all space: $S_{ab} = \int \phi_a \phi_b \, dv$. This overlap integral is much more easily calculated than the integrals involving the Hamiltonian operator.

For a polyelectronic system (for instance, the two-electron H_2 molecule), we usually assume (as we did for atoms) that the electron distribution can be described by generating one-electron orbitals and populating them with the correct overall number of electrons. For H_2, this yields the energy-level diagram shown in Fig. 4.4. The total bond energy of H_2 should in this approach be 2β, twice that of H_2^+. Experimentally, the dissociation energy of H_2^+ is 2.79 eV and that of H_2 is 4.72 eV, so the approximation seems justifiable in this case.

When we move on to more complex X_2 diatomic molecules in which each X atom has both s and p valence orbitals, the possibilities for AO overlap increase. Because p orbitals are directional, we must establish coordinate axes on each atom, and we normally do this so as to take fullest advantage of the symmetry of the molecule. Figure 4.5 indicates this choice of axes for a diatomic molecule, and also shows the possible overlaps that could lead to molecular-orbital formation. As we shall shortly see, the greatest energy effects are to be expected from the overlap of AOs having similar energies, so the s–s, p_z–p_z, and p_x–p_x/p_y–p_y overlaps will primarily be responsible for MO formation.

Figure 4.5 indicates an important distinction in the symmetry of atomic-orbital overlaps that must be made for inorganic as well as for organic

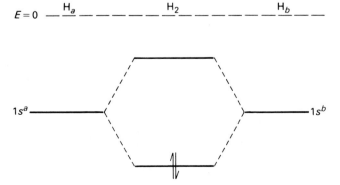

Figure 4.4 MO energy-level diagram for H_2 in the simplest approximation.

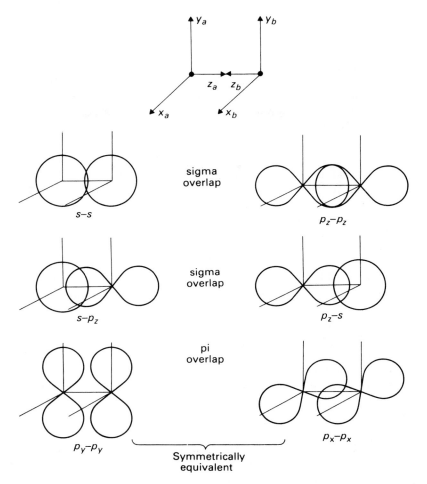

Figure 4.5   Coordinate axes and AO overlaps for an X_2 molecule having s and p valence orbitals.

molecules. Atomic orbitals that overlap in a head-on fashion (without any nodes containing the bond axis) are said to have *sigma* symmetry (σ), by analogy with s atomic orbitals. On the other hand, atomic orbitals that overlap edgewise and have one nodal plane containing the bond axis are said to have *pi* symmetry (π), by analogy with atomic p orbitals, which have one such node. Of course, p AOs can have sigma overlap as well as pi overlap, and d orbitals can have both sigma and pi overlap as well as a face-to-face overlap in which two nodes contain the bond axis (*delta* overlap, δ). We will delay a detailed consideration of this type of bonding until Chapter 13.

Each of the major kinds of overlap mentioned above can lead to a bonding and to an antibonding molecular orbital, with qualitatively the same pattern of energies as that shown in Fig. 4.4 for s–s overlap. However, because there are now several overlaps to consider, we must discuss the factors influencing the relative energies of the MOs. One important distinction is that sigma overlap normally produces greater energy separation between the bonding and antibonding orbitals than pi overlap does. This reflects the fact that edgewise overlap is less effective than head-on overlap. One fairly obvious consequence of the weaker edgewise overlap is that pi bonding is usually quite weak and unstable with respect to rearrangement into a network of sigma bonds for atoms that have large inner cores of nonvalence electrons. Thus we see many examples of classical double bonds consisting of a sigma bond plus a pi bond for first-row elements C, N, and O where the inner core is only $1s^2$, but only rarely double bonds such as Si=Si or P=P (where the inner cores are $1s^2 2s^2 2p^6$). Compounds containing heavy-element double bonds must usually be stabilized with respect to a sigma-forming rearrangement by providing considerable steric bulk around the double-bonded region of the molecule, such as multiple t-butyl groups.

Another consideration is that the lower the energy of the component atomic orbitals, the lower the energy of the MO. Thus a sigma (s–s) antibonding orbital will in general be more stable (lie at lower energy) than a sigma (p–p) bonding orbital even though the overlap might actually be better for the p orbitals. Figure 4.6a gives an energy-level diagram of a typical X_2 molecule in which the X atoms have only s and p valence AOs constructed without considering s–p_z sigma overlap. Note that we use the VOIP values for the atomic orbitals as a guide to placing these orbitals and the resulting MOs. The two kinds of pi overlap shown in Fig. 4.5 are symmetrically equivalent and thus lead to π_x and π_y MOs that have identical energies and are thus said to be degenerate. The pi-bonding energy effect is smaller than the sigma-bonding energy effect for the same type of AOs.

Figure 4.6b shows a slightly more complicated energy level diagram for the same X_2 molecule. In this diagram, s–p interaction is considered. This is necessary whenever the $2s$ and $2p$ (or $n\,s$ and $n\,p$) orbitals are relatively close together in energy. Consideration of the VOIP values in Table 1.8 shows that for molecules such as B_2 and presumably Li_2 (both of which are known only at high temperatures and low pressures in the gas phase), the $2s$–$2p$ separation is fairly small, so the energy levels of Fig. 4.6b should, presumably, be used. On the other hand, the large $2s$–$2p$ energy separation for N_2, O_2, and F_2 means that then s–p interaction will have only a modest energy effect and need not be considered for qualitative MO purposes. For these molecules, we would use the energy levels of Fig. 4.6a.

For our purposes, the essential difference between the two sets of energy levels in Fig. 4.6 is the order in which the σ_z^b and $\pi_{x,y}^b$ MOs occur. If s–p interaction is not important, the natural order is σ_z^b at lower energy, $\pi_{x,y}^b$ at higher energy because of the greater sigma-bonding energy effect. On the other hand, if s–p interaction is important, the σ_s^b and σ_z^b MOs (from s–s and

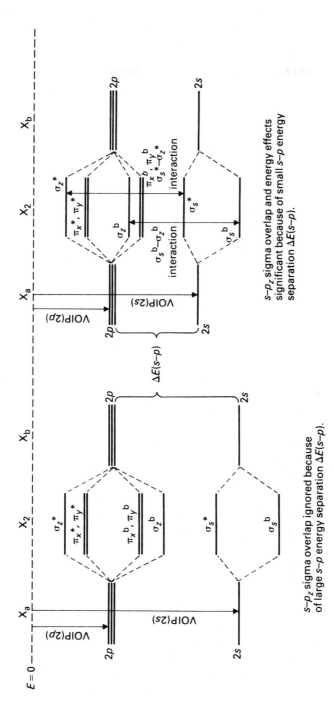

Figure 4.6 Energy-level diagrams for X_2, where X has $2s$ and $2p$ valence orbitals, with and without $s–p$ sigma interaction.

p–p sigma overlap respectively) will combine to yield two new MOs at a lower energy than σ_s^b and at a higher energy than σ_z^b, as indicated by the arrow in Fig. 4.6b. This can lead to an energy inversion of the σ_z^b and $\pi_{x,y}^b$ MOs, because the energy of the π-type MOs is not affected by the s–p interaction.

To use diagrams such as those of Fig. 4.6 to describe the bonding in an X_2 molecule, we must populate the MOs with the correct number of electrons. That number is the sum of the valence electrons of the component atoms, adjusted for any net charge on the molecular species. Thus, for O_2 the correct number would be 12, since each O atom has six valence electrons. On the other hand, for the superoxide ion, O_2^-, the correct number would be 13, because the added electron is considered to be in the valence-electron group. As for polyelectronic atoms, these electrons are placed in the MOs starting with the lowest-energy orbitals. Each MO can accommodate a pair of electrons with opposed spins, except that for degenerate orbitals (as for partially filled sets of atomic orbitals) electrons are placed in separate orbitals with parallel spins as much as possible.

A crucial comparison of the two energy-level diagrams of Fig. 4.6 involves the molecule B_2, which has six valence electrons and thus should fill the σ_s^b, the σ_s^*, and either the σ_z^b MO with a pair of electrons or the π_x^b and π_y^b MOs with one electron each, depending on whether the σ_z^b orbital or the $\pi_{x,z}^b$ orbitals lie at lower energy. If the σ_z^b orbital is filled, all the electrons in the molecule are paired and B_2 would be diamagnetic. Alternatively, if the π_x^b and π_y^b orbitals each contain one electron, the presence of two unpaired electrons would make B_2 paramagnetic. Experiment has shown that B_2 is paramagnetic, requiring the theory to use the energy levels of Fig. 4.6b. This is consistent with the discussion a few paragraphs back.

An important effect for a more familiar molecule occurs for O_2, which has 12 valence electrons. Regardless of which set of energy levels in Fig. 4.6 is chosen, the last two electrons should be placed separately in the π_x^* and π_y^* orbitals. This means that O_2 should be paramagnetic, which in fact is true. This is a natural consequence of the LCAO-MO approach, but is difficult to account for by older bonding models.

Figure 4.7 shows the filled MO energy-level diagrams for the molecules Li_2, N_2, and O_2 on a common energy scale. It can be seen that the energies of the most available (highest-energy) electrons for the three molecules are not as different as the VOIPs for the free atoms would suggest. This is quite generally true, constituting a sort of electronic leveling effect in molecules. In fact, the experimental ionization energy for N_2 is greater than that for O_2 (15.58 versus 12.08 eV), even though the VOIP values for the $2p$ atomic orbitals lie in the reverse order.

We can describe the strength of the bonding in an X_2 molecule by defining the *bond order* as the number of electrons in bonding MOs, minus the number in antibonding MOs, all divided by two (an electron pair is usually taken as a single bond). In the Li_2 molecule of Fig. 4.7, the bond order is $\frac{1}{2}(2-0)$, or 1; in the N_2 molecule of the same figure the bond order is

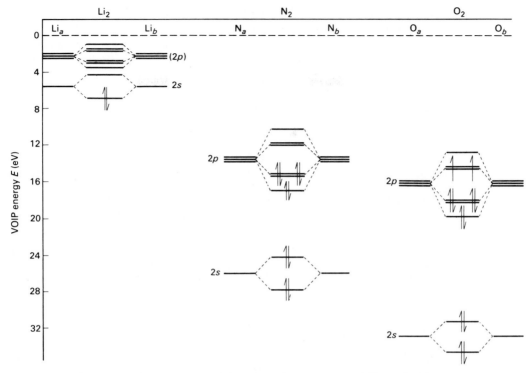

Figure 4.7 MO energy-level diagrams for Li_2, N_2, and O_2 molecules.

$\frac{1}{2}(8-2)$, or 3; and in the O_2 molecule, the bond order is $\frac{1}{2}(8-4)$, or 2. These bond orders all correspond well with our chemical intuition about their bonding, of course. In general, high bond orders correspond to stronger, shorter bonds in thermochemical or structural terms. Pursuing the same examples, Li_2 has an experimental bond energy of 105 kJ/mol and a bond length of 2.672 Å; N_2 has a bond energy of 941.8 kJ/mol and a bond length of 1.098 Å; and O_2 has a bond energy of 494 kJ/mol and a bond length of 1.207 Å. The qualitative correlation of these values with the bond order of the molecule is clear.

All the examples so far have involved homonuclear molecules. In homonuclear molecules, of course, the symmetry of the molecule guaranteed that the coefficients of each atomic orbital in a given MO would be equal in magnitude, though perhaps of opposite sign. The simplest example of a heteronuclear molecule is the HX system (such as LiH or HCl). For this, and for all other heteronuclear molecules, a criterion must be applied to establish the relative magnitudes of the coefficients of overlapping AOs in a given MO. The usual criterion is energy minimization, in which each coefficient is taken as a parameter and varied to give the molecule as a whole the greatest stability, or most negative total electronic energy. There are various procedures for such calculations, yielding theoretical results of varying mathematical sophistication and experimental fidelity. There are some qualitative features common to all calculations, however.

1. When two atomic orbitals overlap, two molecular orbitals will result. One will lie at a lower energy than either of the original AOs, and the other will lie at a higher energy than either AO.

2. In any MO—bonding or antibonding—the AO closer in energy will have the greater coefficient. Since electron-population analyses rely on squared coefficients, this means that the AO closer in energy and the atom on which it is centered will have a greater share of the electrons described by that MO.

3. If one AO on a given atom overlaps two different AOs on a neighbor atom or atoms, three molecular orbitals will result. One will be strongly bonding (lower in energy than any of the three AOs) because the MO electrons are concentrated in the bonding region and there is no $+/-$ sign cancellation in any of the AO overlaps. Another MO will be strongly antibonding (higher in energy than any of the three AOs) because all of the AO overlaps show sign cancellation, yielding nodes between nuclei. The third will be approximately nonbonding, with an energy roughly halfway between that of the bonding and antibonding MOs. It represents favorable overlap by one AO and unfavorable (canceling) overlap by another.

4. For any given type of AO overlap, the bond-energy effect (the lowering of the bonding-MO energy below that of the stablest contributing AO) will be greater the closer the contributing AOs are in energy to each other. If the AO energy gap is very large, the low-energy AOs become essentially nonbonding.

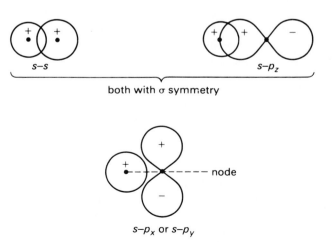

Figure 4.8   Atomic-orbital overlaps for H (1s) and X (s and p) in an HX molecule.

These principles are applied to HX compounds as in Fig. 4.8, which shows the AO overlaps for the 1s orbital of a hydrogen atom and the s and p valence orbitals for any X atom. The H 1s overlaps both the s and p_z AOs from the X atom, so those three yield three MOs—bonding, antibonding, and nonbonding. Because the p_x and p_y have a node along the bond axis, they can have no net overlap with the hydrogen 1s and are thus strictly nonbonding. Figure 4.9 shows the resulting energy-level diagram for the molecular orbitals of the HX molecule under two circumstances: LiH, where the X(Li) AOs lie at higher energy than H, and HF, where the X(F) AOs lie at substantially lower energy than H. From the extent to which the MOs are

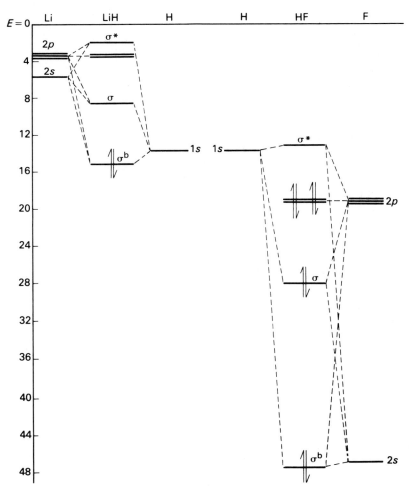

Figure 4.9   Approximate MO energy-level diagrams for LiH and HF.

filled by electrons we can see that in LiH the H atom has a preponderant share of the two electrons and thus a net (fractional) negative charge. In HF, on the other hand, the H atom has only a very small share in the sigma-symmetry MOs and none at all in the pi-symmetry MOs. We expect, therefore, that a population analysis would show the H atom to have a partial positive charge. Although the figure does not show it, one might surmise that the breakeven point for positive charge on the H atom comes when the p orbitals on the X atom have a higher energy than the H $1s$. Under such conditions, electrons will no longer be drained away from the H atom to leave it with a positive charge. The VOIP values are thus a rough guide to the charge distribution to be expected in the molecule.

Heteronuclear diatomic molecules in which both atoms have s and p orbitals, such as CO, CN$^-$, NO$^+$, gaseous LiCl, or a general XY molecule, can be treated by the same qualitative methods. The atomic-orbital overlaps will be essentially the same as those indicated in Fig. 4.5, though no longer symmetrical about the midpoint of the bond axis. Figure 4.10 gives the approximate energy-level diagrams for two XY molecules derived using the

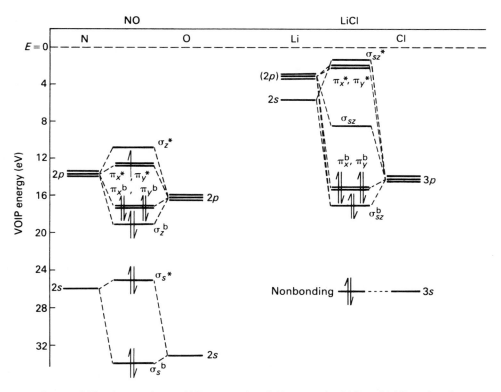

Figure 4.10   Approximate MO energy-level diagrams for NO and LiCl molecules.

guidelines above: NO, in which the two atoms have very similar atomic-orbital energies, and LiCl, in which the energies are quite different. The leveling effect noted before can be seen again here for the highest-energy electrons in these molecules, even though the AO energies are quite dissimilar. The MO energies for NO fall in a pattern not too different from that for homonuclear molecules) (see Fig. 4.6), but those for LiCl are substantially altered by the great energy differences between interacting AOs. In the figure, the Cl $3s$ has been treated as a nonbonding orbital and the three AOs $2s$(Li), $2p_z$(Li), and $3p_z$(Cl) have been allowed to form a bonding, a nonbonding, and an antibonding MO. To a certain extent, the exclusion of the Cl $3s$ is arbitrary, but the qualitative diagram is essentially unaltered if it is included.

The bond order for NO, as suggested by the diagram, is $\frac{1}{2}(8-3)$, or 2.5, which is compatible with traditional chemical intuition. On the other hand, the LiCl bond order appears to be $\frac{1}{2}(6-0)$, or 3, which is unusual. It should be noted however, that the energy of the π_x^b and π_y^b MOs is lowered only slightly from the Cl $3p$ energy, so they would not be expected to make a significant contribution to the overall bonding energy. Experimental results give bond dissociation energies of 626 kJ/mol for NO and 468 kJ/mol for LiCl, which conforms reasonably well with the above argument.

When we move on to triatomic molecules, the qualitative MO approach still works, but it is now necessary to specify the molecular geometry: A typical XY_2(Y-X-Y) molecule can be either linear or bent; numerous examples of each type are known. Limiting ourselves to hydrides for simplicity, we can consider the two cases BeH_2 (known to be linear) and H_2O (bent at 104°). Figure 4.11 gives the coordinate systems for these two molecules and their atomic-orbital overlaps.

In the next few pages we shall develop the simple qualitative MO model for bonding in linear and bent AH_2 molecules, planar and pyramidal AH_3 molecules, and tetrahedral AH_4 molecules. You should already be familiar with using sp, sp^2, and sp^3 hybrid orbitals for direct sigma overlap. It is important to realize, however, that most molecular-orbital treatments do not use hybrid orbitals in quantitative calculations, but rather work directly with Slater atomic orbitals or mathematical simulations of STOs. It is perfectly possible to give a qualitative account of the results of these calculations without imposing hybridization, and that is the approach we shall take. For simplicity, we shall occasionally adopt hybrid orbitals to describe bonding in specific molecules, both in this chapter and later on in the book. In general, however, theoretical inorganic chemists regard hybridization as an unnecessary constraint, and that approach will be presented here.

It can be seen that the changed geometric relationship of the AOs in Fig. 4.11 turns the $s-p_z-s$ overlap from nonbonding in the linear arrangement to bonding–antibonding in the bent geometry. A bit less obvious is the fact that on going from the linear to the bent geometry, the $s-s-s$ bonding MO becomes more stable because the two H $1s$ AOs overlap each other better.

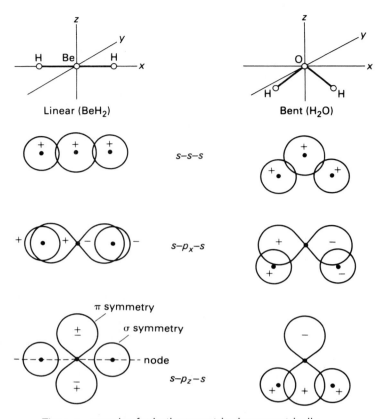

The s-p_y-s overlap for both geometries is symmetrically
equivalent to the nonbonding "sigma-pi" diagram for s-p_z-s
overlap in linear geometry.

Figure 4.11   Coordinates and overlaps for AH_2 triatomic hydrides.

Furthermore, the s-p_x-s bonding MO becomes *less* stable—lies at higher
energy—because the s-p overlap is poorer when the s orbital is not centered on the p axis. The overall results of these energy shifts are shown in
Fig. 4.12.

It is possible to predict AH_2 molecular geometries in a qualitative fashion
on the basis of these energies and the valance electron count of the molecule,
using arguments known as *Walsh's rules*: An AH_2 molecule will be bent if it
contains one or two valence electrons, linear if it contains three or four
valence electrons, and bent if it contains five to eight valence electrons. These
rules rest on the minimization of the molecule's total electronic energy as the
valence electrons fill the MOs from the bottom up. One- or two-electron
systems fill only the s-s-s bonding orbital (with p_z contribution in the bent
form), which prefers the bent geometry. Three- or four-electron systems fill

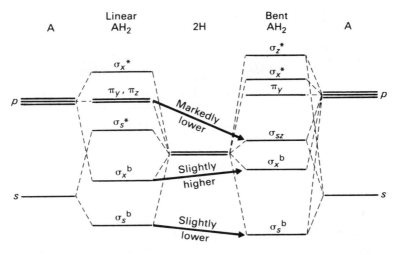

Figure 4.12 MO energy levels for AH_2 in linear and bent geometry.

Table 4.1 Predicted and Experimental Geometries for AH_2 Molecules

No. of val. e^-	Molecule	Predicted H—M—H angle (Walsh)	Experimental H—M—H angle
2	LiH_2^+	Bent slightly	Bent[a]
3	BeH_2^+	Linear	Linear[a]
4	BeH_2	Linear	Linear[a]
	BH_2^+	Linear	Linear[a]
5	BH_2	Bent slightly	131°
	AlH_2	Bent slightly	Bent
6	CH_2	Bent sharply	136° (105° in excited state)
	NH_2^+	Bent sharply	140–150° (115–120° in excited state)
	BH_2^-	Bent sharply	100°[a]
	SiH_2	Bent sharply	97°
7	NH_2	Bent sharply	103°
	PH_2	Bent sharply	92°
8	H_2O	Bent sharply	104°
	H_2S	Bent sharply	92°
	H_2F^+	Bent sharply	135°

[a] Detailed calculation; molecule unknown.

the s–p_x–s bonding orbital as well, which prefers the linear geometry more strongly. Five- through eight-electron systems fill the s–p_z–s MO, which is much more stable in the bent geometry. By way of comparison, Table 4.1 gives some experimentally determined geometries for AH_2 molecules; the qualitative MO model is not flawless, but it does give generally useful predictions.

In a generally similar way, we can construct qualitative MOs for AH_3 molecules such as ammonia or BH_3. The presumed structure for such a hydride has the H atoms arranged symmetrically (equilaterally) about the A atom, but there are still two geometric possibilities: The molecule might be planar or pyramidal. Figure 4.13 shows these geometries with their associated coordinate systems, and also the resulting AO overlaps.

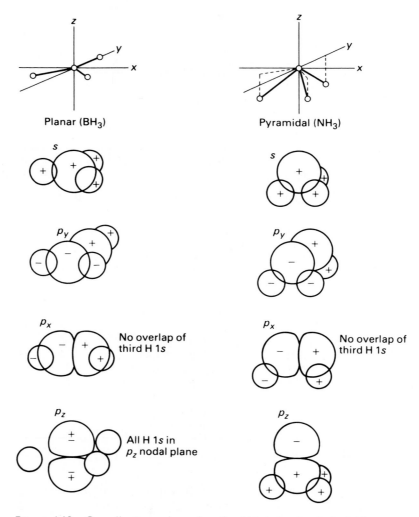

Figure 4.13   Coordinates and overlaps for AH_3 tetraatomic hydrides.

As for the AH_2 hydrides, the overlap sketches in Fig. 4.13 show changes in net overlap that are reflected in the energies of the resulting molecular orbitals. Again the H $1s$ AOs are strictly nonbonding with respect to the p_z orbital on the A atom if the molecular is planar, but have substantial overlap with one lobe of the p_z if the molecule is pyramidal. The MO energies for p_x and p_y overlap change in the same manner as for the AH_2 hydrides. For ideal symmetry, the total overlap is the same for p_y as for p_x even though the individual overlaps appear quite different, so the resulting MO energies are the same for a given geometry. Figure 4.14 shows the results of these energy changes as an AH_3 molecule changes from planar to pyramidal geometry.

Following exactly the same reasoning as for the AH_2 hydrides, we can predict that AH_3 hydrides with one or two electrons (none are known, but the hypothetical H_4^{2+} would be an example) should be pyramidal, those with three through six electrons (such as BH_3) should be planar, and those with eight or more electrons (such as NH_3 or H_3O^+) should be pyramidal. The seven-electron case is ambiguous, since it is not clear how much effect a single electron should have in the lower-energy pyramidal MO, but the methyl radical and the NH_3^+ ion are known to be very nearly planar.

For central atoms that have only s and p valence orbitals, the highest hydride we expect to see is AH_4. The AH_4 hydride is normally tetrahedral. Since the corners of a tetrahedron have the symmetry of alternate corners of a cube, we can draw coordinate axes for the molecule in such a way as to make the three p orbitals equivalent in their total overlap with the hydrogen $1s$ orbitals (see Fig. 4.15). Each molecular orbital thus involves all four H $1s$ AOs, and if the system has eight electrons (which, as the energy levels suggest, is by far the most stable number), there are four net bonds. These,

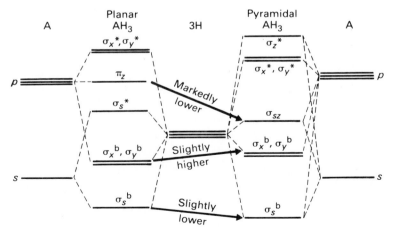

Figure 4.14 MO energy levels for AH_3 in planar and pyramidal geometry.

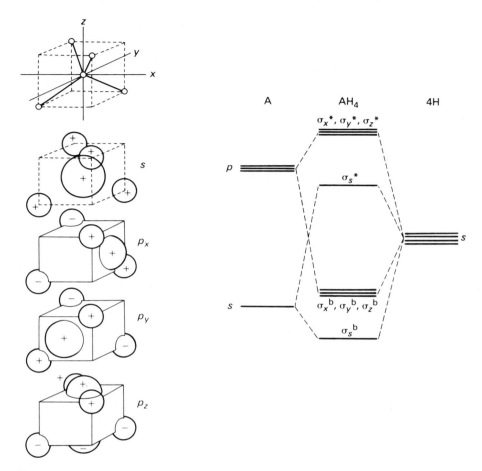

Figure 4.15   Coordinates, overlaps, and MO energy levels for tetrahedral MH_4.

however, are delocalized over the entire molecule. It is possible to transform the molecular orbitals of Fig. 4.15 mathematically to the more familiar localized bonds, but it is perfectly possible to represent the bonding in this generalized way without prior assumptions about sp^3 hybridization.

Although the discussion of the past few paragraphs has dealt with hydrides, it is equally appropriate for other systems in which only sigma bonding need be considered and only one sigma orbital is involved for each outer atom. Thus, for example, the energy levels of Fig. 4.14 are equally useful for BF_3 (planar, bond angle 120°) and NF_3 (pyramidal, bond angle 102.2°). Similarly, the energy levels of Fig. 4.15 can be applied not only to BH_4^-, CH_4, and NH_4^+, but to CCl_4 and BF_4^- as well.

_____ *4.2 MOLECULAR ORBITALS AND POLARITY*

If we perform numerical calculations corresponding to the qualitative MO treatments of the last few pages, the result of the energy minimization principle is an LCAO-MO with specific coefficients. For example, one treatment of the LiH molecule gives coefficients for the lowest-energy bonding MO (as in Fig. 4.9) as follows:

$$\psi(\sigma^b) = 0.700\phi_{1s}(H) + 0.328\phi_{2s}(Li) + 0.204\phi_{2p_z}(Li)$$

If we square this wave function and multiply by the number of electrons occupying it (two), the result will be the electron density associated with the orbital. Squaring a trinomial gives six terms, three squared terms and three cross-multiplied terms:

$$\text{Electron density } \rho = 2 \cdot \psi^2 = 0.980\phi_{1s}^2 + 0.215\phi_{2s}^2 + 0.083\phi_{2p_z}^2$$
$$+ 0.918\phi_{1s}\phi_{2s} + 0.571\phi_{1s}\phi_{2p_z} + 0.268\phi_{2s}\phi_{2p_z}$$

The overall valence-electron density is obviously divided between pure-atomic-orbital electron-density values such as $0.980\phi_{1s}^2$, and what we can call overlap-population values such as $0.918\phi_{1s}\phi_{2s}$ in which the electron is simultaneously described by two atomic orbitals.

Since one of the important considerations in predicting chemical reactivity is the net charge on each atom in a molecule, we often wish to use MO calculations to predict total valence-electron densities for each atom. The predicted net charge is obviously the difference between the total MO valence-electron density and the atom's original complement of valence electrons. We can add the individual atomic-orbital populations easily enough, but some criterion must be adopted to divide the overlap-population electrons. Mulliken proposed multiplying each overlap-population value by the magnitude of the overlap integral S_{ab} for the two AOs (to keep the total counted electron density equal to the total number of valence electrons) and assigning half of the result to each AO. In the calculation being described, the overlap integrals are $S_{1s2s} = 0.469$, $S_{1s2p_z} = 0.506$, and $S_{2s2p_z} = 0$ (AOs on the same atom are orthogonal). So for the total valence-electron population on the H atom in LiH, we have

$$\text{H atom valence-electron population} = [0.980 \text{ (from } \phi_{1s}^2)]$$
$$+ \left[0.918 \times \frac{0.469}{2} \text{ (from } \phi_{1s}\phi_{2s})\right]$$
$$+ \left[0.571 \times \frac{0.506}{2} \text{ (from } \phi_{1s}\phi_{2p_z})\right]$$
$$= 0.980 + 0.215 + 0.145$$
$$= 1.340 \text{ electrons}$$

The predicted charge on the H atom in LiH (gaseous molecule) is -0.340, which coincides well with our intuitive ideas about the atomic polarity in LiH.

We can quantitatively compare the theoretical electron densities with experimental results if we consider the dipole moment of the LiH gaseous molecule, which is defined as the product of the magnitude of the two point charges constituting a dipole and the distance between them. This has been measured as 1.96×10^{-29} C m, or 5.88 debye. If we take the equilibrium internuclear distance in gaseous LiH (1.595 Å) as being the distance between point charges, the theoretical dipole moment based on the present calculation is 2.60 debye. This is not an inspiring match for the measured value, but it does have the right polarity. More sophisticated calculations can usually predict gaseous-molecule dipole moments within about 10%. The calculations are done using the criterion of electronic-energy minimization, so agreement with experimental determinations of electron distribution within the molecule (which are not based on energy measurements) is an important validation of the theory.

At the same time it must be remembered that the dipole moment of an isolated gas-phase molecule will not be the same as its dipole moment in a liquid or solid. As was pointed out in Chapter 1, the presence of a permanent dipole near another molecule will create an induced dipole moment in the second molecule, oriented so as to produce a small attraction between the two. Since the condensed-phase dipole moment is the vector sum of the permanent moment and the induced moment, the result can be significantly different from that measured in (or calculated for) the gas phase.

4.3 POLARITY AND HYDROGEN BONDING

The polarity of X–H bonds is a bonding parameter of particular importance, because such a bond in which the H atom has a significant positive charge can form hydrogen bonds. Hydrogen bonds are interactions between two molecules (or two regions of a single molecule) in which a positively charged hydrogen atom in one molecule attracts nonbonding electrons from an atom in the other molecule so strongly that it establishes a partial chemical bond. This is not unlike the dipole–dipole interactions characteristic of nonideal gases or solutions, but it is a substantially stronger effect, on the order of 5–30 kJ/mol of hydrogen bonds.

The strength of the interaction is not surprising when one considers the unique nature of the hydrogen atom, in which no inner core of electrons shields the positively charge nucleus. In a molecule in which the H atom has a substantial positive charge, the nucleus is nearly naked; the proton, which has only about 10^{-4} the radius of a typical ion core with inner-core electrons, can embed itself in a nonbonding pair of electrons from another atom. The resulting deformation of the nonbonding pair produces an overall attraction substantially greater than the usual dipole–dipole effect.

The preceding argument suggests that we can expect to see hydrogen bonding as a significant feature of intermolecular attractions whenever a hydrogen atom in a single molecule has a substantial positive charge. In our earlier discussion of the HX molecules, it was noted that an approximate breakeven point for positive charge on the H atom in HX comes when the p orbitals of the X atom have about the same energy as the H $1s$. Only if the p orbitals lie at lower energies (have higher VOIP values) than the H $1s$ will the H atom have a significant positive charge. Examination of Table 1.8 reveals that only oxygen and fluorine have p valence orbitals with VOIPs significantly higher than the H $1s$ value of 13.6 V, although N and Cl are close. Accordingly, we should see hydrogen bonding only for molecules with H—F, H—O, H—N, and H—Cl bonds, and the strongest hydrogen bonds should be formed by the first two, F and O. This accords with our experience, but it is also true that the VOIP values for an atom are influenced by its net charge in a given molecule. Some C—H bonds display weak hydrogen bonding, for example, if the other atoms attached to the carbon are strongly electronegative and withdraw electrons from the C—H bond; HCN hydrogen-bonds to other HCN molecules to the extent of about 12 kJ/mol. Table 4.2 gives some of the parameters of commonly seen hydrogen-bonded systems.

The strength of a hydrogen bond is influenced not only by the charge on the hydrogen atom but also by the nature of the electron-donor atom with

Table 4.2 Properties of Some Hydrogen-Bonded Systems

System	ΔH for dissociation of 1 mole bonds (kJ)	$X \cdots X$ distance (Å)	Sum of van der Waals X,X radii (Å)	Bond shortening Δr (Å)
$F \cdots H \cdots F^-$	155	2.27	3.00	0.73
$HF \cdots H—F$	29	2.49	3.00	0.51
$H_2O \cdots H—OH$	21	2.76	3.00	0.24
$CH_3C\overset{O \cdots H—O}{\underset{O—H \cdots O}{\diagup\diagdown}}C—CH_3$	29	2.76	3.00	0.24
$H_3N \cdots H—NH_2$	17	3.38	3.10	−0.28
$Cl \cdots H \cdots Cl^-$	50	3.22	3.50	0.28
$Br \cdots H \cdots Br-$	38	3.35	3.70	0.35
$HCN \cdots H—CN$	12	3.18	3.30	0.12

Value in pm = tabulated value × 100.

which it is interacting. The hydrogen bond will be stronger the more concentrated the nonbonding pair of electrons is; thus, smaller donor atoms form stronger hydrogen bonds. Interestingly, polarizability does not seem to be important. O and F, which are the least polarizable atoms other than the noble gases, form the strongest hydrogen bonds as donors *as well as* providing the most positive hydrogen atoms at the acceptor end (but see study problem C1).

Table 4.2 suggests that hydrogen-bonded atoms are generally significantly closer to each other than the same two atoms would be under conditions of nonbonding contact. The normal distance may be thought of as the sum of the van der Waals nonbonding radii of the atoms; for the strongest hydrogen bonds, this shortening can be over half an angstrom unit. For the very shortest, strongest hydrogen bonds, the H atom is found in the center of the X$\cdots$X bond axis, but for most hydrogen bonds the H atom is unsymmetri-

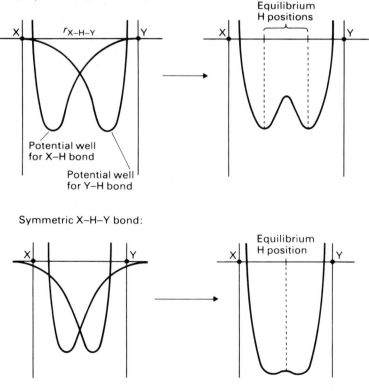

Figure 4.16 Potential energy diagrams for symmetric and unsymmetric hydrogen bonds.

cally placed nearer one atom. Figure 4.16 suggests the potential-energy relationships for the two types of hydrogen bonds, although it is not known whether the shortest bonds have a small energy peak in the center of the potential well or a true single well.

It is possible to construct qualitative molecular-orbital diagrams that account for bonding in symmetrical strongly bonded systems such as FHF^-. Although calculations normally use the s and p valence orbitals of the fluorine atoms, the origin of the bonding is more easily seen if we assume sp hybridization for each F using the $2p$ orbital lying along the bond axis. The resulting overlaps and MO energy levels are shown in Fig. 4.17. The 16 electrons of FHF^- are placed in one sigma bonding orbital, in three sigma nonbonding orbitals (two of which are the outer-directed sp hybrids), and in the four pi nonbonding orbitals (the unhybridized p orbitals of the fluorine atoms, now separated too far to have significant pi overlap). The result is a total bond order of 1.0, or a bond order of 0.5 per F—H atom connection—a stable

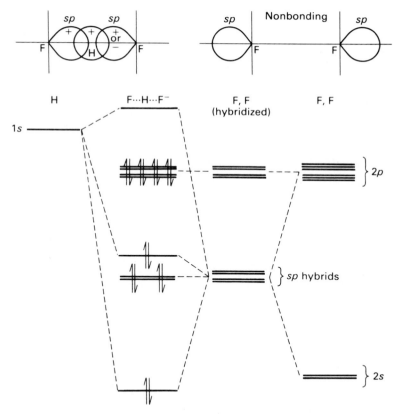

Figure 4.17 Molecular orbitals for the FHF^- ion.

system, even though the isoelectronic F_2^{2-} ion would have zero bond order and is unknown. Within the last 10 years it has become apparent that under the right circumstances a water molecule and a hydroxide ion can hydrogen-bond in the same way as HF and F^-, producing a very stable $H_3O_2^-$ species. $H_3O_2^-$ is a good bridging anion ligand for two transition-metal cations in a crystal, and analytical data suggest that Werner may actually have made such a complex around the turn of the century, although recognition of its structure had to wait until the 1980s.

4.4 MOLECULAR ORBITALS FOR ELECTRON-DEFICIENT MOLECULES

For the simple molecules considered in the first two sections of this chapter, an unspoken rule has been in effect governing bonding stability: The most stable compounds (frequently the only known compounds) are those in which all of the bonding molecular orbitals are filled and all of the antibonding MOs are vacant. The stoichiometry of a given compound will usually arrange itself to this end. However, the mathematical symmetry of the LCAO-MO model limits the number of bonding MOs in which an atom can participate to the number of its valence orbitals (though the coordination number can be greater if delocalized bonding MOs are involved). Some molecules, by the terms of this bonding model, have too many electrons for the number of possible bonding MOs; others have too few. We shall examine these two circumstances in this section and the next.

The first-row elements Li through F, with $2s$ and $2p$ valence orbitals, can presumably form four bonding MOs. The elements Li, Be, and B have too few electrons to contribute one to each of four bonding MOs, although those MOs can frequently be filled by electrons from the other atoms in the molecule. BH_4^-, for example, is as satisfactory as CH_4 to bonding models. However, molecular networks involving catenation—the covalent bonding of an atom to another atom of the same element—provide no such source of extra electrons, or at least no adequate source. Such compounds are relatively rare for Li and Be, but exist in fascinating variety for boron. Boron–hydrogen networks, or *boranes*, are the classical electron-deficient compounds. They are presently known from B_2H_6 (diborane) to $B_{20}H_{16}$. Even larger networks are known for molecules containing heteroatoms with more valence electrons (carboranes, with C substituted for B in the network, and metallaboranes in which a metal atom has been substituted).

Traditional valence rules suggest that boron and hydrogen should form a trigonal-planar BH_3 molecule with three B—H sigma bonds and a nonbonding $2p$ orbital on the boron atom. Although the BH_3 molecule is known spectroscopically as a highly reactive intermediate, the lowest-molecular-weight borane is diborane, whose structure is shown in Fig. 4.18. The struc-

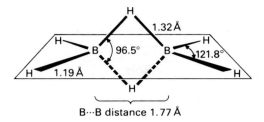

Figure 4.18   The stereochemistry of B_2H_6, diborane.

ture of diborane is markedly different from that of ethane, which has the same stoichiometry. In the molecular-orbital model, the bridging hydrogens are seen as participating in three-center bonds—that is, in bonding overlap involving AOs from three different atoms. Although calculations use only the atomic orbitals of the eight atoms, a qualitative view of these bonds is more easily seen if we take the boron atoms to be sp^3 hybridized, which is reasonably appropriate given the structure. Figure 4.19 shows the bridging overlaps and energy levels for this system. While each of the terminal hydrogen $1s$ orbitals forms a sigma bonding and antibonding MO with the sp^3 hybrid directed toward it, the bridging hydrogen $1s$ orbitals overlap two sp^3 hybrids each and thus form a bonding, a nonbonding, and an antibonding MO. Approximate contours for the bridging MOs are shown at the bottom of the figure.

   In the molecular-orbital mode, the B—H—B bridge bonds are formally similar to the hydrogen bonds of the previous section, since the bonding MO arises from the immersion of a proton with its associated $1s$ orbital in the sigma overlap of two other atoms. However, the situation is not the same viewed as a polar attraction, because there is no electron-pair donor atom. Nonetheless, the unique nature of hydrogen—its lack of inner-core electrons to disrupt a multicenter sigma bond—seems to stabilize boranes as well. All four BX_3 halides are known as monomers, for example, and although dimeric and polymeric halides can be prepared, none have the formula B_2X_6 or a structure that can be interpreted in terms of bridging bonds. (The dimers are planar B_2X_4 molecules with structures entirely analogous to ethylene, but without the pi bond.) However, other electron-deficient molecules do show bridging structures similar to that of the diborane. Trimethylaluminum, for example, is really $Al_2(CH_3)_6$ with the methyl groups arranged in the same locations as the hydrogen atoms in B_2H_6. It might be noted, however, that the bond angles at the bridging carbon atoms in $Al_2(CH_3)_6$ are so small ($75°$) that the bridging is being done not by the carbon nucleus and inner core, but by a carbon orbital (an sp^3 hybrid, perhaps) extending toward the Al–Al axis.

   If we attempt LCAO-MO treatments of the bonding in larger borane molecules, the full treatment becomes correspondingly more complicated but

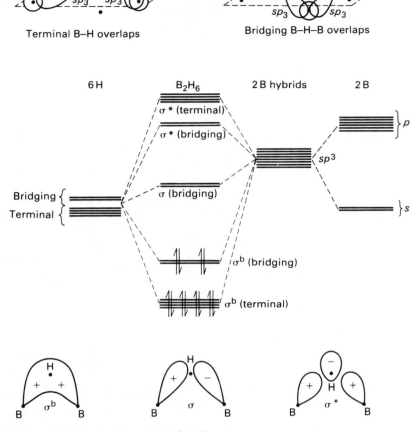

Figure 4.19 Molecular orbitals for diborane.

a new type of multicenter bond appears. For simplicity, let us consider the symmetric hexahydrohexaborate(2−) anion, $B_6H_6^{2-}$, which has the structure shown in Fig. 4.20. The cage of boron atoms is octahedral, and all of the hydrogen atoms are bonded in a terminal rather than a bridging manner. The large number of atomic orbitals involved (30) makes a qualitative MO treatment complex, but the key bonding orbitals can be visualized fairly readily. Figure 4.21 shows the appropriate overlaps. Each boron is assumed to have sp hybridization, with one sp hybrid orbital directed toward the terminal hydrogen atom for sigma bonding, and two unhybridized p orbitals at right angles to that direction. The remaining sp hybrid orbital points toward the

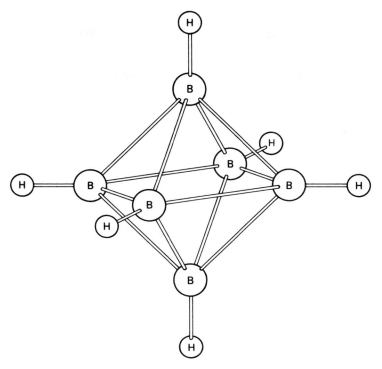

Figure 4.20   The $B_6H_6^{2-}$ anion.

center of the B_6 cage, and one strongly bonding orbital results from the in-phase overlap (that is, overlap with the same algebraic sign) of these six sp hybrids. As the figure suggests, each of the remaining p orbitals can overlap neighbor p orbitals in sigma fashion or (with different neighbors) in pi fashion. There are three such sigma ring MOs around the B_6 cage, with the symmetry of the earth's equator, the 0° meridian, and the 90° meridian. The bonding forms of these have the same energy, by symmetry. Similarly, there are three pi-overlap rings of equivalent energy. Within the B_6 cage, then, there are seven strongly bonding MOs: 1 sp cluster, 3 sigma rings, and 3 pi rings. The unusual multicenter bond is the sp cluster. Such bonds appear to be common in borane cages; perhaps the most common is a three-center cluster of three sigma-symmetry orbitals.

Within the $B_6H_6^{2-}$ cluster, the electrons must be accommodated in 13 bonding MOs—the seven of the B_6 cage and six B—H sigma-bonding MOs. However, six boron atoms and six hydrogen atoms have only 24 electrons, so the molecule should exist as a 2– ion because the two additional electrons fill the set of bonding MOs. Most boranes and carboranes exist as cages or

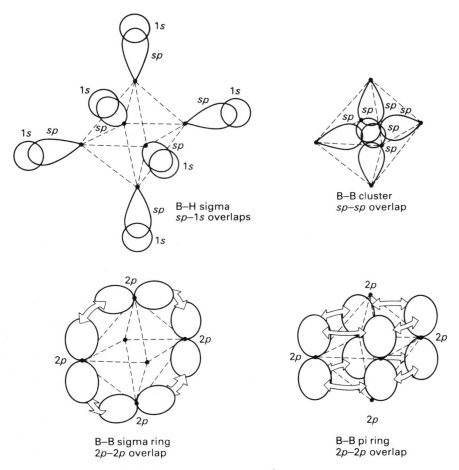

Figure 4.21   Atomic-orbital overlaps for the $B_6H_6^{2-}$ anion.

clusters that should incorporate multicenter bonds, and it is generally true that the number of bonding orbitals is one greater than the number of vertices the cage has (or would have, if it were in its closed, fully symmetric form with only triangular faces). Accordingly, such closed B_nH_n cages normally have the formula $B_nH_n^{2-}$. We shall return to this discussion in a later section on bonding in clusters.

It may be useful at this point to briefly compare the boranes with the alanes (aluminum hydrides). The same sort of bridging bonds are possible using the hydrogen atoms, but where B_2H_6 is a gas, AlH_3 is a polymeric nonvolatile solid. Gaseous AlH, AlH_3, and Al_2H_6 are known, but only as extremely reactive intermediates. There are several crystal forms of $(AlH_3)_n$, the most stable of which has approximately hexagonal-close-packed hydrogen atoms containing aluminum atoms in octahedral holes. All hydrogen atoms form bridging Al—H—Al bonds, and all aluminum atoms are surrounded by six such bridging bonds in nearly ideal octahedral geometry. This marked structural difference between "BH_3" and AlH_3, even though the bridging-hydrogen bonding appears to be similar, presumably arises because aluminum, as a second-row element, has $3s$, $3p$, and $3d$ valence orbitals and can thus participate in a larger number of bonding MOs than boron with only four valence orbitals. The mathematical symmetry of the LCAO-MO approach requires that no atom participate in more bonding MOs than it has valence AOs (though of course it can participate in fewer). If we assume that stable systems in which covalent bonding is dominant must have a bonding MO available for each bonded neighbor atom, the coordination number of first-row atoms (Li to F) cannot exceed four (a more sophisticated statement of the octet rule), whereas the coordination number for second-row atoms could be as large as nine ($s + 3p + 5d$). Normally, steric constraints prevent coordination numbers from increasing beyond about six, but that increased bonding capability can lead to markedly different structures even for systems with identical stoichiometry and similar bond types, as in this case.

We can summarize the MO approach to simple electron-deficient molecules by noting that it relies on the formation of three-center orbitals, each of which has a bonding, nonbonding, and antibonding form. The molecule that, by traditional valence rules, is electron-deficient actually arranges to fill all of its bonding MOs, although its nonbonding MOs are vacant. In the next section we shall take much the same approach to electron-rich molecules, but the surplus electrons are distributed in the nonbonding orbitals.

4.5 MOLECULAR ORBITALS FOR ELECTRON-RICH MOLECULES

As was just suggested, some stable molecules have more valence electrons than can be accommodated in their bonding MOs. It should be clear that the additional electrons do not contribute to the bonding in the molecule because they cannot occupy bonding MOs. In nonbonding MOs they will presumably have little or no effect on the bonding, while in antibonding MOs they will progressively weaken the bond as more electrons are added. The simplest case we can consider is that of the diatomic molecules X_2, where each X atom has both s and p valence orbitals. The MO energy levels for such a system were given in simplified form in Fig. 4.6. It is apparent from that figure and Fig. 4.7 that a system with 10 valence electrons achieves maximum bonding with a bond order of three. The N_2 molecule (bond energy 942 kJ/mol) is the familiar example of this bond pattern. Adding two antibonding electrons (as in O_2) lowers the bond-dissociation energy to 494 kJ/mol, while adding two more antibonding electrons (to the F_2 configuration) fills the π^* orbitals, lowers the bond order to one, and leaves F_2 with a bond-dissociation energy of 158 kJ/mol.

This bond-energy comparison is a bit inexact, since other features of the molecular structure are changing besides the total electron configuration. A comparison that eliminates this complication is that between O_2^+ (dioxygenyl cation), O_2, O_2^-, and O_2^{2-}. These species have, respectively, 11, 12, 13, and 14 electrons in the molecular orbitals of Fig. 4.6, leading to bond orders of 2.5, 2.0, 1.5, and 1.0. If the bond-dissociation energies are calculated for the following four processes (so that one product is always the neutral O atom), the experimental energies are

$$O_2^+ \rightarrow O^+ + O \qquad E_{BE} = 623 \text{ kJ/mol}$$
$$O_2 \rightarrow O + O \qquad\qquad\quad 494$$
$$O_2^- \rightarrow O^- + O \qquad\qquad\ 351$$
$$O_2^{2-} \rightarrow O^{2-} + O \qquad\quad 205$$

As Fig. 4.22 indicates, these bond-dissociation energies are very nearly proportional to the bond order of the MO calculation. For electron-rich diatomic molecules, stability depends only on having a surplus of bonding electrons.

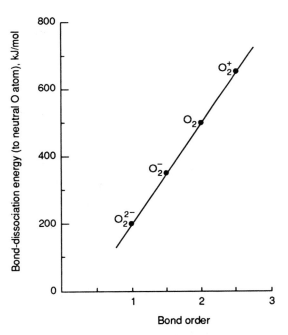

Figure 4.22   Relationship between bond-dissociation energy and MO bond order for dioxygen species.

For polyatomic species, the MO model resembles that for electron-deficient species, as has been suggested. An important difference is that the central atom in a bridge bond uses a *p* orbital to overlap the two terminal atoms, so that the total overlap need not include the region of the central atom's inner-core electrons, where repulsion would be prohibitively high. Since molecules to which this model is applied are usually interhalogens or noble-gas compounds with extremely stable *s* orbitals, the bonding usually involves only *p* overlap. Furthermore, the large inner core of atoms beyond the first row prevents the bonded atoms from approaching closely enough to have significant pi overlap. Accordingly, only *p*–*p* sigma overlap is considered.

An example of this approach is the treatment of the triiodide ion, I_3^-. In crystals in which the cation is large enough to effectively isolate the I_3^- anion,

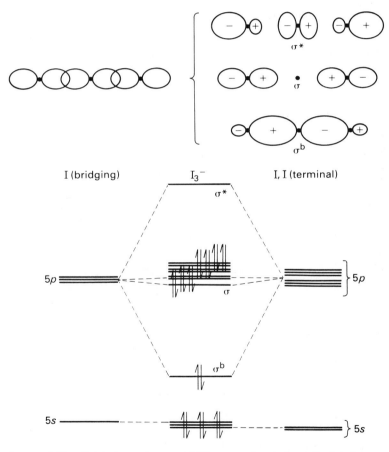

Figure 4.23 Orbital overlaps and MO energy levels for triiodide ion.

the anion is linear and symmetrical, with a bond length significantly greater (2.90 Å) than in the I_2 molecule (2.67 Å). Figure 4.23 shows the p-orbital overlaps and the resulting MO energy-level diagram for such a system. Note that the 22 electrons of the I_3^- system fill all of the bonding and nonbonding MOs, and that there is one net bond for the three-atom system, or (as for FHF⁻) 0.5 bond order per I—I atom connection. Since this is the same total bond order as that calculated for $I^- + I_2$, it is not surprising that the equilibrium constant for the association process in aqueous solution is measurable under ordinary conditions:

$$X^- + X_2 \rightleftharpoons X_3^- \qquad K_{eq} = 710(X = I),\ 16.3(X = Br),\ 0.19(X = Cl)$$

The molecular orbitals of Fig. 4.23 may be applied equally well to the noble-gas compound XeF_2 (which also has a linear geometry) if allowance is made for the VOIP difference between Xe and F. XeF_2 is also a 22-electron system, so the total bonding is the same as for the I_3^- ion: a total bond order of 1.0, or half a bond per bond region in the molecule. This value is consistent with the experimental thermochemistry of XeF_2, which indicates a total bond-dissociation energy (for removing both F atoms) of 268 kJ/mol. This is fairly close to the bond-dissociation energy for the XeF^+ ion (201 kJ/mol), which according to Fig. 4.6 should also have a bond order of 1.0.

4.6 COVALENT RADII

We have assumed in all the preceding discussion of molecular-orbital formation that the bonded atoms come close enough together for their valence orbitals to overlap, but not close enough for the bonding to be disrupted by the repulsion of the inner-core electrons of the bonded atoms. The size of an atom's electron cloud depends on its effective nuclear charge. However, if the bonding is essentially covalent so that there is little net electron transfer between atoms, the effective nuclear charge should be nearly constant for a given atom from one molecule to another, and the size of the inner-core electron cloud should be nearly constant. The possibility thus arises that one could assign a radius to a given atom in such a way that the sum of these radii for two atoms would approximately equal the observed covalent bond length. A corollary of this is that the observed internuclear distance for two given atoms should be the same from one compound to another. Table 4.3 indicates that this is a good approximation, but better for some atoms than others. Even in the poorer cases, however, it is clear that covalent radii could be chosen that would reproduce interatomic distances within about 0.05 Å.

It is possible to establish a set of covalent radii by trial-and-error fitting to observed bond lengths for a variety of molecules. A simpler procedure,

Table 4.3 Covalent Bond-Length Comparisons

Bond	Molecule	Internuclear distance (Å)
C—Cl	$\begin{cases} CF_3Cl \\ CH_3Cl \end{cases}$	1.76 1.78
Sn—I	$\begin{cases} SnI_4 \\ (CH_3)_3SnI \end{cases}$	2.64 2.72
As—F	$\begin{cases} AsF_3 \\ AsF_5 \text{ (axial)} \\ AsF_5 \text{ (equatorial)} \end{cases}$	1.706 1.711 1.656
Sb—Cl	$\begin{cases} SbCl_3 \\ SbCl_5 \text{ (axial)} \\ SbCl_5 \text{ (equatorial)} \end{cases}$	2.36 2.34 2.29

Value in pm = tabulated value × 100.

however, can be used for covalent radii that cannot be used for ionic radii: Take half the X—X bond length in a molecule in which two X atoms are bonded to each other. A complicating factor, however, is that multiple bonds are shorter than single bonds. Accordingly, covalent radii are used only to estimate the lengths of single bonds, except for a few well-established multiple bond lengths such as that for a carbon–carbon double bond (1.34 Å). Covalent radii must be estimated from the X—X distance in a molecule in which no pi bonding is occurring between the X atoms. This involves a theoretical judgment about bonding as well as experimental internuclear distances, and is not always clear-cut.

The effect of multiple bonding on bond lengths for a homonuclear bond can be reproduced fairly well by a simple empirical observation: A double bond is about 0.86 as long as a single bond for first-row elements, and a triple bond is about 0.78 as long as a single bond. For second-row and heavier elements (where multiple bonds are much rarer), the inner-core electrons are less compressible and the factors are 0.91 for a double bond and 0.85 for a triple bond.

Table 4.4 gives single-bond covalent radii for most elements and, for comparison, gives the van der Waals radii wherever they have been established. The van der Waals radii add to give contact distances between nonbonded atoms in touching molecules; they are established from neighbor-molecule distances in crystals and from critical volumes in gases. The van der Waals radii are larger than the corresponding covalent radii because they represent a situation in which the exclusion principle prevents the valence-electron clouds from interpenetrating. Covalent radii, on the other hand, specifically involve valence-orbital overlap and are limited by the need to avoid interpenetration of *core* electron clouds.

Table 4.4 Covalent Radii and van der Waals Radii (Å)

	1	2	3	4	5	6	7	8	9	10	11	12	13	14	15	16	17	18
	H 0.37 / 1.20																	**He** 1.40
	Li 1.34 / 1.82	**Be** 0.90											**B** 0.82	**C** 0.77 / 1.70	**N** 0.75 / 1.55	**O** 0.73 / 1.52	**F** 0.72 / 1.47	**Ne** 1.54
	Na 1.54 / 2.27	**Mg** 1.30 / 1.73											**Al** 1.18	**Si** 1.17 / 2.10	**P** 1.06 / 1.80	**S** 1.02 / 1.80	**Cl** 0.99 / 1.75	**Ar** 1.88
	K 1.96 / 2.75	**Ca** 1.74	**Sc** 1.44	**Ti** 1.32	**V** 1.25	**Cr** 1.27	**Mn** 1.46	**Fe** 1.20	**Co** 1.26	**Ni** 1.20 / 1.63	**Cu** 1.38 / 1.43	**Zn** 1.31 / 1.39	**Ga** 1.26 / 1.87	**Ge** 1.22	**As** 1.20 / 1.85	**Se** 1.16 / 1.90	**Br** 1.14 / 1.85	**Kr** 1.15 / 2.02
	Rb 2.11	**Sr** 1.92	**Y** 1.62	**Zr** 1.48	**Nb** 1.37	**Mo** 1.45	**Te** 1.56	**Ru** 1.26	**Rh** 1.35	**Pd** 1.31 / 1.63	**Ag** 1.53 / 1.72	**Cd** 1.48 / 1.58	**In** 1.44 / 1.93	**Sm** 1.41 / 2.17	**Sb** 1.40 / 1.85	**Te** 1.36 / 2.06	**I** 1.33 / 1.96	**Xe** 1.26 / 2.16
	Cs 2.25	**Ba** 1.98	**La** 1.69	**Hf** 1.49	**Ta** 1.38	**W** 1.46	**Re** 1.59	**Os** 1.28	**Ir** 1.37	**Pt** 1.28 / 1.72	**Au** 1.43 / 1.66	**Hg** 1.51 / 1.55	**Tl** 1.52 / 1.96	**Pb** 1.47 / 2.02	**Bi** 1.46			

r_{cov} : (first value) r_{vdw} : (second value)

Value in pm = tabulated value × 100.

Both covalent radii and van der Waals radii are applicable to systems in which little or no electron transfer has taken place and no ionic attraction is contributing to the overall bonding. If electron transfer does occur, the atoms involved experience changes in their effective nuclear charges and thus in their electron-cloud radii. An atom acquiring electrons will swell, and one losing electrons will shrink. The result of this change and the growing electrostatic attraction between charged atoms is a progressive shortening of the bond length with increased ionic character (electron transfer). If bond lengths are predicted as the sum of two covalent radii without allowing for this effect, the results will often prove substantially too long when compared with experiment.

Because the degree of electron transfer in a bond is roughly proportional to the electronegativity difference between the two atoms involved, Schomaker and Stevenson proposed an empirical correction to the predicted bond length:

$$r_{AB} = r_A + r_B - 0.09(|\chi_A - \chi_B|)$$

where r_A and r_B are covalent radii, χ_A and χ_B are electronegativity values, and the sign of the electronegativity difference is disregarded. A somewhat better fit is provided by adjusting for the square of the electronegativity difference:

$$r_{AB} = r_A + r_B - 0.07(\chi_A - \chi_B)^2$$

Again it should be noted that this predicts only single (sigma) bond lengths. When observed bond lengths are shorter than the predicted values in covalent molecules, it is usually assumed that some degree of pi bonding is involved. Some progress has been made in developing a single set of radii—one per atom—that could be used with corrections for the atom's bonding environment to predict internuclear distances in either an ionic or a covalent system. However, most chemists still use separate tables of ionic (crystal) radii and covalent radii for different bonding situations.

As an example of the use of covalent radii, we might predict the length of an As—F bond. The covalent radius of arsenic is taken as 1.20 Å from the two molecules shown in Fig. 4.24, and the radius of fluorine is 0.72 Å from the bond length of F_2, 1.43 Å. The sum of these two radii is 1.92 Å, but there is a substantial electronegativity difference to be considered:

$$
\begin{aligned}
r_{As-F} &= r_{As} + r_F - 0.07(\chi_{As} - \chi_F)^2 \\
&= 1.20 + 0.72 - 0.07(2.20 - 4.10)^2 \\
&= 1.92 - 0.253 \\
&= 1.667 \quad \text{or} \quad 1.67 \text{ Å}
\end{aligned}
$$

This value can be compared with the various values observed, as given in Table 4.3.

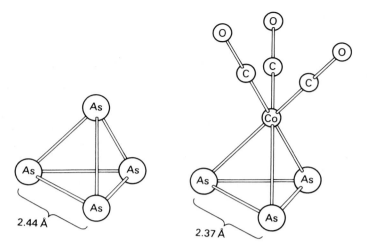

Figure 4.24   Arsenic–arsenic bond lengths.

4.7 COVALENT BOND ENERGIES

If bond lengths are nearly constant in a covalent bond between two atoms, it follows that the degree of overlap of the valence orbitals is nearly constant and that the bond-energy integral β for those two atoms should also be nearly constant. Even if the electronegativity difference between the two atoms causes some electron transfer and a partial ionic attraction is a component of the total bond energy, that component should be nearly constant for two given atoms—although, as the differential-ionization-energy quantity suggests, the electronegativity of an atom is affected by its net charge. This near constancy for a given charge suggests that it should be possible to derive a set of covalent bond-energy values that could be summed over all the bonds in a molecule to approximate the total electronic energy of the molecule. Of course, the total electronic energy of a molecule is not experimentally measurable, but changes in that energy during a chemical reaction are measurable by the usual calorimetric techniques. That is, we can approximate the energy change during a chemical reaction as the difference between the total bond energy of the products and the total bond energy of the reactants:

$$\Delta E = \sum_{\text{Products}} \text{BE} - \sum_{\text{Reactants}} \text{BE}$$

where ΔE is the internal energy change for the reaction and BE is an individual bond energy.

This expression is not unlike the usual Hess's law expression by which enthalpy changes are calculated from enthalpies of formation,

$$\Delta H^0 = \sum_{\text{Products}} \Delta H_f^0 - \sum_{\text{Reactants}} \Delta H_f^0$$

but there is an important difference. Enthalpies of formation are thermodynamic constructs that are useful and accurate to within the experimental error of thermochemical measuring techniques because enthalpy is a state function. Hess's law for enthalpy changes from enthalpies of formation is exact. On the other hand, covalent bond-energy tables represent a kind of average evaluation for a given bond over several compounds; in one compound it may represent almost complete covalence, while in another compound the rest of the molecular environment of the bond may give it a significant ionic component and a different total energy. So the form of Hess's law that involves these bond energies is only an approximation, not a thermodynamic state relationship.

The approximation is sometimes quite good and sometimes rather crude. For an example of each, we can calculate the enthalpies of formation of ammonia and hydrazine by the bond-energy approach, assuming the internal-energy change to be essentially equal to the enthalpy change.

$$N_2 + 3H_2 \rightarrow 2NH_3 \qquad \Delta H = 2(3 \times BE_{N-H}) - 3 \times BE_{N-H} - BE_{N \equiv N}$$

Two comments are important here. First, the N_2 molecule has a bond order of three, so we must use the triple-bond value for the NN bond energy. In general, it is necessary to establish the bond order for each bond in a molecule from a simple sketch of a Lewis electron-dot structure before beginning a calculation. Second, the numerical values of bond energies in Table 4.5 are all thermodynamically negative quantities, since they represent the energy *released* on formation of the bond. With these two cautions in mind, we write (from Table 4.5)

$$\Delta H = 2[3 \times (-390.8)] - 3 \times (-432.2) - (-941.4)$$
$$= -106.8 \text{ kJ/mol reaction}, \qquad \text{or} \quad -53.4 \text{ kJ/mol } NH_3$$

Since the experimental enthalpy of formation of ammonia is -46.2 kJ/mol, this represents an excellent approximation. For hydrazine, the numerical accuracy is not as good:

$$N_2 + 2H_2 \rightarrow N_2H_4 \qquad \Delta H = BE_{N-N} + 4 \times BE_{N-H} - 2 \times BE_{H-H} - BE_{N \equiv N}$$
$$= -251 + 4 \times (-390.8) - 2 \times (-432.2) - (-941.4)$$
$$= -8 \text{ kJ/mol } N_2H_4$$

In this case, the experimental value is $+50.4$ kJ/mol and the bond-energy calculation is clearly less reliable.

Table 4.5   Covalent Bond Energies (kJ/mol)

	H	Li	Be	B	C	N	O	F	Cl	Br	I	S
H	432	243	222	293	411	391	459	561	428	362	295	364
Li		105			238		351	573	464	423	347	
Be								632	456	372	289	
B			289	372		385	536	615	456	377		
C					346 —	305 —	358 —	485	326	285	213	272 —
					610 =	615 =	799 =					573 =
					835 ≡	890 ≡	1071 ≡					
N						251 —	201 —	285	301	155		
						418 =	607 =					
						941 ≡						
O							213 —	188	251	234	201	
							494 =					
F								155	251	230	272	285
Cl									238	218	209	255
Br										188	176	218
I											151	
S												226

The term "bond energy" is used in two senses that can be quite different. In one sense it is a *bond-dissociation energy*, referring to breaking a specific bond in a specific molecule without disrupting the rest of the molecule. For example, the bond-dissociation energy of a carbon–oxygen double bond might be taken as the energy of the following process:

$$CO_2 \rightarrow CO + O \qquad \Delta H_{dissoc} = 531 \text{ kJ/mol}$$

However, it is also possible to think of the carbon–oxygen double-bond energy as half of the *atomization energy* of CO_2:

$$CO_2 \rightarrow C + O + O \qquad \tfrac{1}{2}\Delta H_{at} = \tfrac{1}{2}(1602) = 801 \text{ kJ/mol}$$

These numbers are strikingly different because they refer to different processes. In the first, breaking one C=O bond does *not* leave the other unchanged. The CO molecule has a bond order of 3.0, so its stability has actually increased as the other bond was breaking. In the second, both bonds are being treated equally. We might expect the average bond energy taken from the atomization energy to more generally applicable, but it should not be surprising that when we use tabulated values to estimate ΔH for reactions, the dissociation of a particular bond in the reaction often does not require the same energy as the tabulated average.

In our earlier discussion of molecular-orbital bond-energy effects, it was suggested that the β bond-energy integral for pi overlap is geometrically less favorable in pi bonding. Since we always assume that a single bond is sigma and multiple bonds arise from additional pi overlap, instinct suggests that double bonds should be less than twice as strong as single bonds and triple bonds less than three times as strong. However, Table 4.5 suggests that bond strengths are fairly closely proportional to bond order (as Fig. 4.22 indicated), even when taken from different molecules. Multiple bonds are shorter than single bonds. This shortening when a pi bond forms usually improves the sigma overlap and presumably strengthens that bond as well as allowing the pi overlap. The result is that some double bonds are even stronger than two single bonds would be.

4.8 BONDING IN ELEMENTS

Of necessity, the bonding in all elements that occur in polyatomic forms is covalent or metallic (or both). All of the elements except the noble gases do occur in polyatomic forms, but the tendency to extended bonding is quite strong; all of the elements except the noble gases form infinite chains, layers, or three-dimensional networks except N, O, S, and the halogens. The formation of chains of atoms of the same element in a finite compound is called *catenation*, and it is much rarer than the infinite networks of the bulk elements. Carbon, of course, forms infinite networks in diamond and graphite, but it also forms an enormous number of catenated compounds that are quite stable with respect to the infinite networks. Elemental silicon, on the other hand, has the diamond structure and is about as stable toward ordinary chemical attack as diamond, but has a much smaller capability for chain formation. Although Si–Si chains are well known, it is difficult to prepare and isolate silanes (Si_nH_{2n+2}, analogous to saturated hydrocarbons) beyond about $n = 8$; very long polymers protected by organic groups are stable, however. Other elements that form reasonably long atom chains in compounds include sulfur (in sulfanes that have been characterized up to HS_8H), selenium and tellurium (in alkali-metal polyselenides and polytellurides, up to K_2Se_4), germanium (in germanes, to Ge_5H_{12}), and phosphorus (in the triphosphine P_3H_5). Most of the other nonmetals and metalloids can form X—X bonds but no chains longer than two atoms. Boron forms a very extensive set of borane cage compounds, but these do not in general have a chain structure and will be considered separately. As we shall see in later sections, many nonmetallic elements display what is known as *pseudocatenation*, in which very long chains of alternating atoms are formed with very nearly covalent bonds between them: —X—Y—X—Y—.

After examining the short list of elements forming significant chains, we can formulate criteria for catenation in terms of our bonding theory. To form reasonably stable chains, an element must have valence electrons nearly equal in number to the number of its valence orbitals. Such an atom can contribute enough electrons to fill all the bonding MOs it can form. One with fewer electrons remains too good an electron acceptor in a catenated compound, while one with more electrons has too many donor nonbonding electron pairs. Furthermore, each element forming catenated compounds must have reasonably small valence orbitals to allow good overlap and the resulting stable bonding.

A few elements do occur as small molecules in their most stable form, and others can be prepared as analogous molecules even though they slowly revert to polymerized forms. The only stable molecule in which the halogens occur is the diatomic X_2 (although the triiodide ion has already been discussed, and other polyhalide ions are also known: Cl_3^-, Br_3^-, I_5^-, I_7^-, I_8^{2-}, and I_9^-). The X_2 molecule is also, of course, the most stable form for hydrogen, nitrogen, and oxygen, but in the vapor phase other diatomic molecules can be observed: Li_2 and the other alkali metal M_2 molecules, Co_2, Ni_2, Cu_2, Ag_2, As_2, Au_2, C_2, P_2, S_2, Se_2, Te_2, and Po_2. Ions such as C_2^{2-}, S_2^{2-}, O_2^+, O_2^- and O_2^{2-} are also known.

Triatomic species of a single element are quite rare. Only one neutral molecule, ozone, is known, but besides the O_3 molecule the O_3^- ozonide ion is reasonably stable. Other triatomic ions include S_3^{2-}, Se_3^{2-}, the N_3^- azide ion, and the trihalide ions Cl_3^-, Br_3^-, and I_3^-. Of these species, the ones in which each atom has an *even* atomic number are bent, with bond angles at the central atom of 116.8° (O_3), ~100° (O_3^-), and ~103° (S_3^{2-}). The species in which each atom has an odd atomic number for each atom are linear (N_3^-, Cl_3^-, Br_3^-, and I_3^-). The congener of O_3, S_3, is known only at very low pressures and high temperatures in the gas phase.

Tetraatomic molecules of an element are almost equally rare; the tetrahedral P_4 molecule is the classic example. Arsenic forms a similar As_4 molecule in the gas phase, but it is not certain whether antimony forms Sb_4 or not. As_4^+, Sb_4^+, and Bi_4^+ have all been observed in a mass spectrometer, but these are gaseous ions at extremely low pressures. Although crystals containing the X_4 molecules can be obtained for P and As, neither is the most stable form of the element. Sulfur, which normally exists as the cyclooctasulfur S_8 molecule, has been observed as S_4 (apparently an open chain) in the vapor phase (at least to a small extent). In the proper solvents, both open-chain S_4^{2-} and cyclic square-planar S_4^{2+} have been prepared, as have the analogous Se_4^{2+} and Te_4^{2+}. This is an appropriate place to mention *Zintl ions*, which will be discussed in the next section; a number of these cluster ions exist in the *p* block, particularly groups IV and V. For example, Sb_4^{2-} and Bi_4^{2-} are known, and other Zintl ions have been characterized at least up to As_{11}^{3-}. The bonding in these is an interesting extension of our qualitative MO model, as we shall see.

Elemental sulfur has a particularly complicated structural chemistry. The only stable form at room temperature is cyclooctasulfur in an orthorhombic crystal structure; the rings are puckered into a crown configuration with a bond angle S—S—S of 107°48′. Above 95.4°C, the S_8 molecules adopt a monoclinic crystal packing that can persist at room temperature for as long as a month if the monoclinic crystal has been annealed at 100°C. Other modifications are possible, however. For example, cyclohexasulfur, S_6, is prepared by allowing H_2S_2 to react with S_4Cl_2. It is stable in crystals, but it is extremely reactive by comparison with S_8 and decomposes in visible light. Other sulfur molecules, all cyclic, include S_7, S_9, S_{10}, and a remarkably stable S_{12} (mp 145°C, compared to the 119.3°C of S_8). Most of the higher cyclopolysulfur molecules are prepared by variations on an interesting reaction:

$$(C_2H_5)_2TiS_5 + S_2Cl_2 \rightarrow (C_2H_5)_2TiCl_2 + S_7$$

The structures of this titanium–sulfur compound and of some elemental sulfur species are shown in Fig. 4.25.

As Fig. 4.25 indicates, sulfur can form a helical chain polymer in addition to its molecular species. When ordinary sulfur (S_8) is melted, the liquid is very mobile and a pale yellow color, but very abruptly at 159°C it changes to a dark red, extremely viscous material through a heat-induced polymerization. If this polymer is quenched in water and stretched, the resulting *plastic sulfur* is quite elastic. In its stretched, more or less crystalline form it has the helical structure shown, known as polycatenasulfur. Both selenium and tellurium form similar helical polymers. These polymers are the stablest forms of those elements.

In addition to these one-dimensional polymer chains, elements sometimes occur in one of several two-dimensional polymers. The best known is

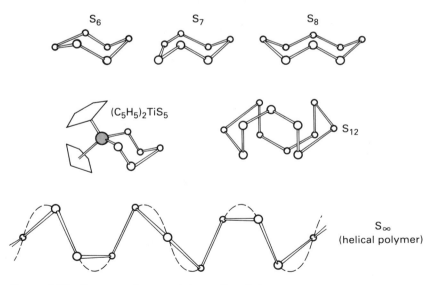

Figure 4.25   Some species containing sulfur chains.

graphite, the most thermodynamically stable form of carbon, whose compounds we have already studied. Its layers of carbon atoms are planar, with an extended hexagonal network in which continuous pi bonding is presumably the stabilizing factor. However, phosphorus, arsenic, antimony, and bismuth also form puckered layer structures, as Fig. 4.26 indicates. These are the stablest forms of these elements. It can be seen that the layers contain six-membered rings like those of graphite, but the absence of effective pi overlap and the presence of a nonbonding electron pair cause the rings to pucker into the cyclohexane chair configuration.

In addition to three-dimensional metallic bonding, which we shall examine in a later section, a number of nonmetals or metalloids form three-dimensional polymeric lattices. The best known of these lattices, of course, is the diamond lattice of carbon, in which each C atom is tetrahedrally surrounded by four other C atoms in a rigid but rather open lattice also adopted by silicon, germanium, and gray tin, as well as a number of compounds in which covalent bonding or hydrogen bonding is important, such as water (ice), ZnS (zincblende), and AgI. (The diamond lattice was shown in Fig. 2.13.)

A more complicated pattern of three-dimensional polymer formation is unique to boron, which has three well-characterized crystal forms. Each of

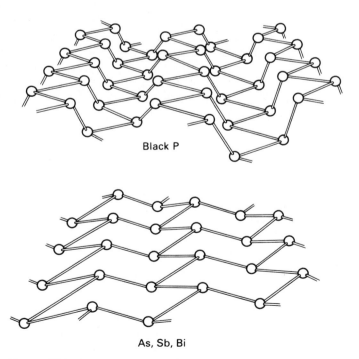

Black P

As, Sb, Bi

Figure 4.26   Layer structures of group V elements.

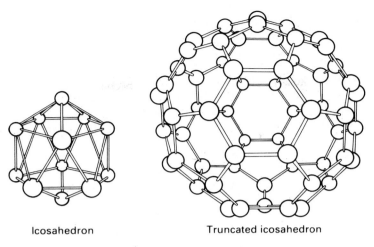

Icosahedron   Truncated icosahedron

Figure 4.27 Icosahedral cages in elemental boron crystals.

these involves a B_{12} icosahedron or a truncated icosahedron, as shown in Fig. 4.27. In α-rhombohedral boron, the structure consists of almost-cubic-close-packed icosahedra with three-center two-electron bonds between icosahedra within a plane of icosahedra and with conventional two-center bonds between layers to form six-membered chair-conformation rings of icosahedra. In α-tetragonal boron, B_{12} icosahedra are arranged in a tetrahedron about an isolated B atom to which they are bonded, but the icosahedra are also linked by direct two-center bonds. In the polymeric structure, each icosahedron is bonded to two isolated B atoms at opposite ends of the icosahedron and to 10 other icosahedra (one per B atom). Finally, in β-rhombohedral boron, each B atom in an icosahedron is bonded to a B atom in another neighbor icosahedron. The geometric requirements of this bonding fuse the neighbor icosahedra in groups of three (each pair sharing a triangular face). If only the closest half of each neighbor icosahedron is considered, the sixty surface atoms lie at the vertices of a truncated icosahedron centered on the original icosahedron. This nearly spherical structure, whose bonds have the same symmetry as the seams on a soccer ball, is also seen in the novel C_{60} *buckminsterfullerene*, and obviously lends considerable stability to networks of small atoms.

4.9 BONDING IN CLUSTERS

In general, our qualitative molecular orbital model has given us a good picture of the bonding in covalent systems. We saw in earlier sections that essentially the same pattern of AO overlap to give bonding, nonbonding, and antibonding MOs yields a satisfactory explanation for the stoichiometry and stability of electron-deficient and electron-rich compounds. The p block of the periodic table provides a fairly rigorous test of the model in this context, since the p block spans electronic behavior from the classic electron-deficient compounds of boron to the electron-rich halogens. A simple model has emerged that describes quite well, if not flawlessly, the geometries and net charges of cluster ions across the p block. We can consider the model and its molecular-orbital origins here; later we will see that both the MO treatment and the model itself can be extended to transition-metal clusters in the d block.

We see for the boranes, for boron halide clusters, and for Zintl ions a common pattern of cluster structures. Nearly all of these adopt geometries consisting of polyhedra all of whose faces are equilateral triangles, or these polyhedra with one or two vertexes missing. Such polyhedra, which for instance include the octahedron but not the cube, are often called *deltahedra* because the triangular faces resemble an upper-case Greek delta. The deltahedra with four to 12 vertexes are shown in Figure 4.28; a few such as the tetrahedron and octahedron are familiar, but most seem unusual to the casual viewer. However, there are clearly electronic advantages to deltahedral structures for these species.

The model for these cluster structures is variously known as the polyhedral skeletal electron pair theory (*PSEPT*) or as *Wade's rules*. It assumes deltahedral geometry for all cluster species; mathematical graph theory shows that the triangular network in a deltahedron makes most efficient use of a limited electron supply for bonding. However, not all clusters will appear as closed deltahedra. Those that do are described as having *closo* geometry, but if one vertex is missing from the deltahedron the structure is *nido* and if two vertexes are missing the structure is *arachno*.

How does the cluster decide among these possibilities? In the discussion associated with Fig. 4.21 we demonstrated that there are seven cluster-

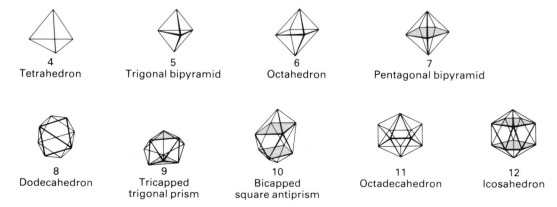

Figure 4.28   Closed deltahedra with four to 12 vertexes.

bonding MOs for an octahedral B_6H_6 cluster. It can be shown in general that there will be $n + 1$ bonding MOs for a deltahedron M_n with n vertexes. In each case there are sets of n bonding MOs that lie essentially in the more-or-less spherical surface of the deltahedron; the extra bonding MO is the M–M cluster overlap in which there is an inward radial overlap of an sp orbital from each M atom (as in Fig. 4.21). This extra bonding MO persists for a deltahedron even if a vertex is missing (*nido*) or two vertexes are missing (*arachno*). Accordingly, to decide on the geometric structure of a p-block cluster it is only necessary to know how many cluster-bonding electrons must be accommodated in bonding orbitals. The number of pairs of such electrons must equal the number of bonding MOs, which in turn defines the number of vertexes the appropriate deltahedron must have. If there are, for example, 14 cluster-bonding electrons in a given cluster there are obviously seven pairs, which must be accommodated in seven bonding MOs. Because of the $n + 1$ pattern, a deltahedron with six vertexes (the octahedron) will yield seven bonding MOs. Therefore the basic geometry of this cluster will be that of an octahedron *even if it does not have six cluster atoms*. If there are only five cluster atoms, the geometry of those atoms will be that of an octahedron missing one vertex, or a square pyramid. If there are only four cluster atoms, their geometry will be that of an octahedron missing two vertexes: either an open square or two triangles sharing an edge but folded along that edge, called *butterfly* geometry. So we can move immediately from valence electron count to a geometric structure prediction.

We have reduced the problem of cluster structure prediction to electron counting, and it only remains to define the cluster electron count. For a given atom, valence electrons are apportioned by assuming the sp hybridization seen for the boron atoms of $B_6H_6^{2-}$ in Figure 4.21. The overlaps seen in that figure are generally applicable to other clusters except for the B–H sigma overlap; not all clusters have radial outer atoms. However, Wade's rules assume that the outward-pointing sp hybrid will be filled by electrons before counting up cluster-bonding electrons. If there is a radial outer atom, it is assumed that an electron-pair sigma bond is formed using one electron from the outer atom and one electron from the cluster atom. If there is no outer atom, the outward-pointing sp hybrid will be filled by two valence electrons from the cluster atom. The electrons available for cluster bonding by a given atom are thus the valence electrons minus one, $v - 1$, if there is an outer atom, but $v - 2$ if the cluster atom is naked. Table 4.6 summarizes both the available deltahedra and the cluster valence electrons donated by various elements.

An example of the use of Wade's rules will be useful. What geometries should be adopted by the two nine-atom clusters $B_9H_9^{2-}$ and Ge_9^{2-}? In $B_9H_9^{2-}$ each boron has a terminal H atom, so in Table 4.6 we check the number of cluster electrons per B atom under the EH heading; a BH unit will contribute two electrons to its cluster. Nine of these BH units will provide 18 electrons to

Table 4.6 Cluster Structure Electron Counting Rules

Element	Cluster electrons per atom E		Total cluster e⁻ incl. net charge	Deltahedron vertexes needed	Deltahedron
	E	EH or EX			
Li, Na	−1	0	8	3	triangle
Be, Mg	0	1	10	4	tetrahedron
B, Al, Ga, In, Tl	1	2	12	5	trigonal bipyramid
C, Si, Ge, Sn, Pb	2	3	14	6	octahedron
N, P, As, Sb, Bi	3	4	16	7	pentagonal bipyramid
O, S, Se, Te	4	5	18	8	dodecahedron
F, Cl, Br, I	5	6	20	9	tricapped trigonal prism
			22	10	bicapped square antiprism
			24	11	octadecahedron
			26	12	icosahedron

the cluster, plus two additional electrons as the net charge equals 20 cluster electrons. A 20-electron cluster has 10 pairs, which can be obtained from a nine-vertex deltahedron—a tricapped trigonal prism. For the germanium cluster each Ge atom is bare with respect to outer atoms, and we look in Table 4.6 under the E heading. Because a bare Ge atom contributes two electrons to its cluster, we again see a 20-electron cluster when the net charge is included, and again we expect a tricapped trigonal prism. Both predicted structures are correct; it is interesting that there is an isoelectronic principle at work for these clusters.

Clusters will not always adopt the *closo* form, of course. If we work through Wade's rules for the Zintl anion Sn_9^{4-}, the cluster electron count is 22 so that the basic cluster structure is that of a bicapped square antiprism. But there are only nine, not 10, atoms in the cluster, so the structure is the *nido* form shown in Fig. 4.29a, in which one of the capping atoms is missing. It is also possible to predict structures for heteronuclear clusters. It is easy to show that $TlSn_8^{3-}$ will have the same cluster electron count and the same geometry as Sn_9^{4-}; perhaps less obvious is the $Tl_2Te_2^{2-}$ cluster shown in Fig. 4.29b. All four cluster atoms are bare, so two Tl contribute one each, or two; and two Te contribute four each, or eight. The cluster electron count (with net charge) is 12 electrons, requiring a trigonal bipyramid. But there are only four atoms, so the deltahedron must be *nido*; the butterfly geometry in the figure is consistent if we assume that one equatorial atom has been removed.

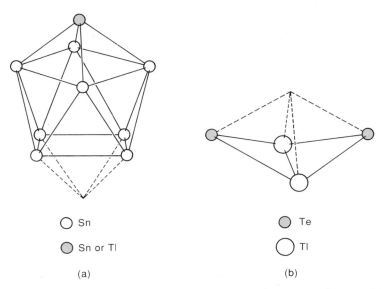

○ Sn

◉ Sn or Tl

(a)

◉ Te

○ Tl

(b)

Figure 4.29 *p*-Block cluster geometries. (a) Sn_9^{4-} and $TlSn_8^{3-}$ (*nido* bicapped square antiprism). (b) $Tl_2Te_2^{2-}$ (*nido* trigonal bipyramid).

Many such clusters are known, and in Chapter 5 we will survey a number of the known Zintl ions. Perhaps even more important, in Chapter 13 we will see that many of these ideas can be extended to *d*-block clusters, which because of their potential catalytic applications are a particularly exciting area of current research.

4.10 BAND THEORY AND METALLIC BONDING

Most of the elements in the periodic table are metals, which are characterized by the familiar metallic luster, high electrical and thermal conductivity, varying (but usually substantial) malleability, and relatively low ionization energies. More theoretically, the metallic elements have *s* electrons that contribute to the bonding with the bulk element and relatively low effective nuclear charges for the valence electrons. The *s* orbitals are nondirectional in their overlap and, because of the low Z_{eff}, fairly diffuse. To the extent the *s–s* overlap is responsible for conventional covalent bonding inside the metallic crystal, that bonding is nondirectional—which accords with the malleable character of the crystals. However, metals have too few electrons for this type of bonding to be a satisfactory explanation of their properties. In general, metals have fewer valence electrons than valence orbitals, and as most metals crystallize in close-packed structures with 12 nearest neighbors, the number of electrons is far too small for the number of two-electron bonds required.

The diffuse nature of the valence orbitals in metals is the key to understanding metallic bonding. The observed internuclear distances in metals are such that the valence orbitals place substantial electron density near the nuclei of all the nearest-neighbor atoms. It is thus necessary to think of the bonding in terms of delocalized molecular orbitals extending over the entire crystal. If each metal atom contributes one *s* orbital to the general overlap (as would be the case for any alkali metal), the number of MOs formed is equal to the number of atoms in the crystal—on the order of Avogadro's number. However, the energy span between the lowest-energy bonding orbital and the highest-energy antibonding orbital is limited by the number of neighbors overlapping a single *s* orbital, because that determines the total overlap and thus the magnitude of the bond-energy effect. It follows that the MOs are very closely packed energetically, the spacing much smaller than kT thermal energy at room temperature. Figure 4.30 suggests the progressively closer packing as the number of overlapping atomic orbitals increases. Figure 4.31 shows the effect on MO energies of changing internuclear distance (with internuclear repulsion and inner-core repulsions added at short range). In Fig. 4.31b the resulting diagram of energy-level bands is shown for sodium metal. The observed internuclear distance r_0 is well within the overlap distance for the 3*s* and 3*p* orbitals on each Na, and is almost small enough for the 2*s* orbitals to overlap.

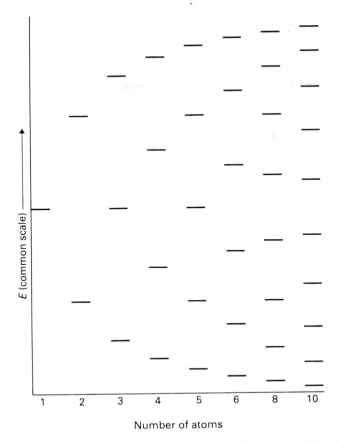

Figure 4.30 Density of MO energy-level diagrams as a function of increasing number of overlapping AOs in a one-dimensional array of atoms.

The appearance of Fig. 4.31b should make it clear why this approach to metallic bonding is called the *band theory*. Many of the properties of metals (and, indeed, of other solids with delocalized bonding) can be readily interpreted from the extent to which such bands are filled by electrons. In sodium, for example, the 2s band is, in effect, only N_0 degenerate nonbonding orbitals, since r_{Na-Na} is too large for significant overlap. This band is filled by electrons, but the 3s band is only half filled, since the N_0 molecular orbitals only contain N_0 electrons. The band covers a span of energy substantially greater than the pairing energy for sodium 3s electrons, so most of the electrons are paired in the bottom half of the band. This accounts for the fact that in alkali metals, the paramagnetism is only about 1% as great as it would be if each atom present had one unpaired electron. Only a narrow strip kT

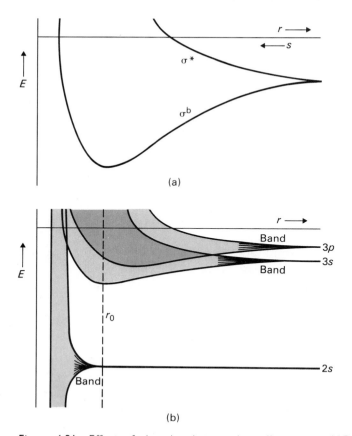

Figure 4.31   Effect of changing internuclear distance on MO energies (repulsions included). (a) Effect on two overlapping orbitals. (b) Effect on valence and inner-core overlap for N_0 atoms of Na metal in bcc crystal lattice.

wide at the top of the filled part of the band contains unpaired electrons, and kT is only about 1% as large as the filled width of the band.

The good electrical conductivity of metals comes from the existence of a partly filled band of orbitals. Since the spacing between orbitals is small compared to kT, there is no significant energy barrier to placing an additional electron in one end of a metal crystal, at which point it occupies a delocalized MO extending to the other end of the crystal. These electrons that are free to move within the crystal also account for the high thermal conductivity of the metallic elements, because when they are present their large numbers and high velocities make them much more efficient means of transferring heat energy than the atomic vibrations that are the only other mechanism. Still further, the presence of the conduction electrons as a sort of free-electron gas within the metal crystal allows them to interact with the oscillating electromagnetic field of visible light in such a mobile fashion that the light is totally reflected, giving the familiar metallic luster. There is, however, a characteristic plasma frequency (usually in the ultraviolet) above which the electron gas or plasma can no longer respond to the rapidly changing electromagnetic field and absorbs the high-frequency radiation. In a few particularly polarizable metals, this plasma frequency is in the blue end of the visible spectrum, which accounts for the characteristic yellow colors of copper, gold, and to a slight extent cesium.

The foregoing discussion rests on the possibility of having unpaired electrons and very closely spaced energy levels within a band. If, on the other hand, the band is filled (as in the alkaline-earth metals with two s electrons), its properties should change drastically since strict pairing would be enforced. However, as Fig. 4.31b suggests, valence-electron bands frequently overlap in energy at the observed internuclear distance. This is the case for the alkaline-earth metals, and they show fully metallic properties because there are vacant p-band orbitals adjacent to filled s-band orbitals. On the other hand, it is sometimes true that filled bands do not quite overlap empty ones. If the energy gap between the top of the filled band and the bottom of the empty one is comparable to kT, a few thermally excited electrons will be found in the otherwise empty band and a few electron vacancies or holes in the filled band. Such a solid does not have the full electrical properties of a metal, but is an *intrinsic semiconductor*. Silicon and germanium are examples of elements showing this property. Chemical modification of the lattice, either by altering the band positions or by changing the valence-electron density per atom, can produce semiconductor characteristics in other lattices. As an example of the first kind of modification, gallium is a metal with an unfilled band; but gallium arsenide, GaAs, has the same crystal lattice as silicon and is a semiconductor. Gallium has only three valence electrons, but arsenic has five; the average of four electrons per atom gives the same pattern of band filling as in silicon. Figure 4.32 indicates the pattern of band filling for semiconductors (in which the bands do not quite overlap) and for the alkaline-earth metals (in which they do). In a semiconductor crystal such as pure Si, the number of excited electrons in the empty band is necessarily equal to the number of holes in the lower filled band, but chemical substitution in very small amounts—typically on the order or 0.1 ppm—by elements having more or fewer valence electrons than Si can create a slight surplus of electrons or a slight surplus of holes, as Fig. 4.32 also suggests. In the former case, electric charge is carried through the crystal predominantly by the surplus of negatively charged electrons and the crystal is said to be a negative or *n-type* semiconductor; if there is a surplus of electron vacancies or positively charged holes, the crystal is a *p-type* semiconductor.

The preceding discussion has assumed a generalized overlap of atomic orbitals that increases as internuclear distance decreases. This is true, but

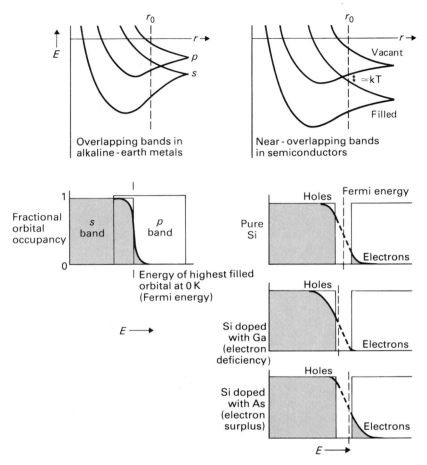

Figure 4.32 Energy bands and band-filling patterns for overlapping bands (alkaline-earth metals) and near-overlapping bands (semiconductors).

when p or d orbitals are involved in the overlap, it can be quite directional and energy bands can be split by the crystal lattice symmetry. When the overlap is strongly directional it is to a considerable extent destroyed by melting the crystal to a disorderly liquid; in consequence, the melting points of metals are a guide to the extent of d–d overlap in the lattice. Figure 4.33 indicates the systematic increase in directional bonding with increasing d-electron (and p-electron) bonding across the periodic table, as reflected in the elements' melting points. The constraints of lattice symmetry allow only about six bonding MOs per atom, which will be filled when each atom donates six valence electrons to the bonding pool. Greater numbers of electrons per atom require that some antibonding orbitals be filled, which progressively

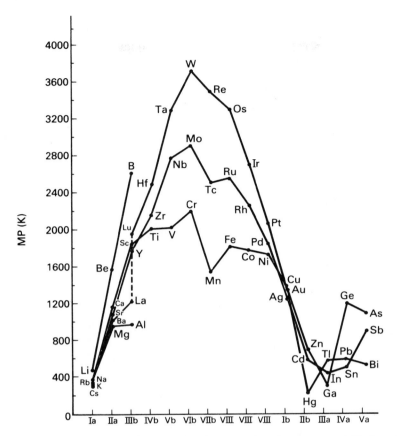

Figure 4.33 Melting points of the elements showing some metallic bonding.

reduces the total directional bonding effect and thus the melting point. By the time Zn, Cd, and Hg are reached, the *d* electrons are so tightly held that they are unavailable for bonding; therefore, very little thermal energy is required to melt these elements, which are bound only by the filled *s* band.

The increase in directional bonding seen going across a row of the periodic table is accompanied, of course, by an increase in the electronegativity and a decrease in the size of the metal atoms involved. Although the interatomic distances can readily be determined by x-ray diffraction studies, they are not always simple to compare because the crystal lattices differ. As has been suggested, most metals adopt a close-packed lattice, either hcp or ccp; in either of these a given metal atom has 12 neighbors. However, a significant number of metals adopt a body-centered-cubic lattice (bcc) like that of CsCl (Fig. 2.12), in which the coordination number is eight. As for

Table 4.7 Metallic Radii for Atoms in 12-Coordinate Lattices

Li 1.57	**Be** 1.12													
Na 1.91	**Mg** 1.60											**Al** 1.43		
K 2.35	**Ca** 1.97	**Sc** 1.64	**Ti** 1.47	**V** 1.35	**Cr** 1.29	**Mn** 1.37	**Fe** 1.26	**Co** 1.25	**Ni** 1.25	**Cu** 1.28	**Zn** 1.37	**Ga** 1.53	**Ge** 1.39	
Rb 2.50	**Sr** 2.15	**Y** 1.82	**Zr** 1.60	**Nb** 1.47	**Mo** 1.40	**Tc** 1.35	**Ru** 1.34	**Rh** 1.34	**Pd** 1.37	**Ag** 1.44	**Cd** 1.52	**In** 1.67	**Sn** 1.58	**Sb** 1.61
Cs 2.72	**Ba** 2.24	**La** 1.88	**Hf** 1.59	**Ta** 1.47	**W** 1.41	**Re** 1.37	**Os** 1.35	**Ir** 1.36	**Pt** 1.39	**Au** 1.44	**Hg** 1.55	**Tl** 1.71	**Pb** 1.75	**Bi** 1.82

Ce 1.82 — **Lu** 1.72

Lanthanides: (Eu 2.06, Yb 1.94, others smooth interpolation)

Actinides:

Th 1.80	**Pa** 1.63	**U** 1.56	**Np** 1.56	**Pu** 1.64

Value in pm = tabulated value × 100.

ionic radii, it is likely that metallic radii are influenced by coordination number. From experimental studies on a number of metals showing more than one crystal lattice, Goldschmidt estimated that the radius of an eight-coordinate metal is about 0.97 of the radius of that same metal in a 12-coordinate environment, while the radius of a six-coordinate metal atom is about 0.96 that of the 12-coordinate atom. Table 4.7 gives a set of 12-coordinate radii for metal atoms, which can be compared with the covalent radii and van der Waals radii in Table 4.4. In general, metallic radii are larger than the corresponding covalent radii for the lower-melting metals of Fig. 4.33, but nearly the same as the covalent radii for the high-melting metals in the center of the transition-metal sequence. This observation reinforces the view that the high melting points for those metals are caused by the directional overlap of d orbitals in something approximating conventional covalent bonding.

PROBLEMS _____

A. DESCRIPTIVE

A1. Sketch AO overlaps for a square planar AH_4 molecule lying in the xy plane. Which AO on the A atom cannot overlap *any* combination of H orbital signs? Which combination of H $1s$ orbital signs has no net overlap with any A orbital? Set up a qualitative energy-level diagram. Comment on the reasons why a species such as SiH_4 prefers the tetrahedral geometry.

A2. Use the heteronuclear diatomic MOs for the XY molecule (as shown in Fig. 4.10 for NO) to set up energy-level diagrams for linear H—X—Y and bent H—X—Y triatomic molecules. Assume that in XY, σ_z^b and π_x^b, π_y^b are very close in energy. Sketch the H $1s$ overlaps with the XY MOs and show that, for the linear geometry, four XY MOs overlap the H $1s$ to produce five MOs for that set in the triatomic system, whereas for the bent geometry, five XY MOs overlap the H $1s$, yielding six MOs. Construct the two appropriate energy-level diagrams and use them to explain—without recourse to hybridization—why HCN should be linear and HOCl bent.

A3. The $B_3H_8^-$ ion is an equilateral triangle of BH_2 units (analogous to cyclopropane) with bridging hydrogen atoms on two sides. Sketch AO overlaps like those of Fig. 4.19 and set up the corresponding MO energy-level diagram for $B_3H_8^-$. Show that $B_3H_8^-$ fills all the resulting bonding MOs, and that a total of three bonds holds the triangle together.

A4. $TeBr_2$ forms complexes with thiourea, $SC(NH_2)_2$, that have a square-planar $TeBr_2S_2$ coordination unit rather than the more obvious tetrahedral arrangement. Explain the square-planar geometry by extending the arguments accompanying Fig. 4.23.

A5. Why should O_3^- and S_3^{2-} have bent geometry, while Cl_3^-, Br_3^-, and I_3^- have linear geometry?

A6. The paramagnetism of solid potassium metal is less than that for sodium. Rubidium is still less paramagnetic. Suggest electronic reasons for this in the context of band theory.

A7. It is easy to dissolve sodium metal in liquid mercury to form sodium amalgam, but iron metal is essentially insoluble in mercury. What reasons can you offer for this difference?

A8. Why is tungsten carbide harder than tungsten metal?

A9. Use the X_2 molecular orbital energy level diagram to estimate the ionization energy of the acetylide ion, C_2^{2-}. See Fig. 4.7 and Table 1.9 for reference energies. Should the acetylide ion be harder or easier to ionize than a neutral C_2 gaseous molecule?

A10. Use the qualitative MO energy level diagram of Fig. 4.15 to account for the nearly nonpolar nature of C–H bonds in CH_4.

A11. Should the VOIP values for AOs on a given atom change if the net charge on that atom in a molecule is nonzero? If so, in what way?

A12. Digallane, Ga_2H_6, has recently been shown to have the same molecular geometry in the gas phase as diborane. However, the bridging Ga–H–Ga angle is $97.9°$ as opposed to the B–H–B angle of $83.0°$. What electronic differences between B and Ga might open up the bridging bond angle for Ga?

A13. Going across the p block from B to F, or from Ga to Br, the covalent radius for each element decreases to the right in a given row. Relate this change to periodic trends in atomic properties.

A14. Use Wade's rules to predict the geometry of the Zintl ions Pb_5^{2-} and Bi_8^{2+}. How many isomeric possibilities are consistent with the rules for the bismuth cluster?

A15. If the structure of the Sn_8Sb^{q-} ion is consistent with that of the $TlSn_8^{3-}$ ion in Fig. 4.29, what should q be? What should q be if the Sn_8Sb^{q-} deltahedron is *closo*?

B. NUMERICAL

B1. Calculate bond order for diatomic X_2 molecules across the second-row elements from Na_2 to Cl_2. Compare the results for P and S with the simplest qualitative bond-order estimate for the observed P_4 and S_8 species. Why are the larger species more stable? What electronic factors prevent N and O from forming the analogous molecules N_4 and O_8?

B2. Assuming Pauling's electronegativity value for C, calculate the Pauling electronegativity value for F from data in Table 4.5.

B3. Use the bond energies in Table 4.5 to estimate ΔH^0 for the reaction

$$BCl_3 + NH_3 \rightarrow Cl_2BNH_2 + HCl$$

What is the dominant thermodynamic driving force for the reaction?

B4. Estimate an ionic crystal radius for NH_4^+ by comparing it with other ions in Table 2.3. Use your result to calculate a theoretical lattice energy for $NH_4^+BF_4^-$. Calculate a hypothetical lattice energy for ionic $B^{3+}N^{3-}$ (4-coordinate). Use these values to estimate ΔH^0 for the reaction

$$4BF_3 + 4NH_3 \rightarrow 3NH_4BF_4 + BN$$

Compare your answer to that of problem B3 above.

B5. The mass magnetic susceptibility of metallic sodium is $8.8 \times 10^{-9} \, m^3kg^{-1}$ at 300 K. At that same temperature, a mole of free electrons in sodium would have a mass susceptibility of $6.88 \times 10^{-7} \, m^3kg^{-1}$. How wide is the valence band of sodium in kilojoules?

C. EXTENDED REFERENCE

C1. A simple "classical" scheme for estimating hydrogen-bond dissociation energies can be developed for the $Y \cdots H—X$ system by considering the bond to be a monopole/induced-dipole attraction. The potential energy expression for this attraction is $q_H^2 \alpha_Y / 2r^4$ (see Table 1.11). For the molecules NH_3, H_2O, and HF, nine separate hydrogen-bonding combinations are possible ($H_3N \cdots HNH_2$, $H_3N \cdots HOH$, and so on). Given the experimental polarizabilities of these three molecules ($\alpha_{NH_3} = 2.26 \, Å^3$; $\alpha_{H2O} = 1.48 \, Å^3$; $\alpha_{HF} = 0.80 \, Å^3$), calculate the bond-dissociation energy for each of the nine structures, assuming (a) that q_H^2 is proportional to the VOIP difference between the H $1s$ and the sp^3 hybrid VOIP on the atom to which the H is covalently bonded (the X atom), and (b) that all r values are equal. Calibrate your values by setting the $H_2O \cdots HOH$ value equal to the experimental 5.1 kcal/mol. Plot your calculated results against the MO results calculated by J. D. Dill et al. [*J. Amer. Chem. Soc.*(1975), *97*, 7220]. How good a fit does the classical expression yield?

C2. Use the three schemes described in this chapter to estimate the covalent-bond lengths of all the bonds in B_2F_4 (B—B and B—F), P_2I_4 (P—P and P—I), SCl_2, SF_2, and H_2Se. Compare your three predictions for each bond against experimental values in A. F. Wells, *Structural Inorganic Chemistry*, 5th ed. (Clarendon Press: Oxford, 1984).

C3. Describe the bonding in $I_3Cl_2^+$ in simple MO terms. Its structure has been determined by T. Birchall and R. D. Myers [*Inorg. Chem* (1982), *21*, 213].

C4. M. Ardon and A. Bino report the crystal structures of two metal completes containing coordinated $H_3O_2^-$ in *Inorg. Chem.* (1985), *24*, 1343. What O—O—H bond angle do they report for the terminal hydrogen atoms? What hybridization does this suggest for the oxygen atom in a qualitative MO treatment of the $H_3O_2^-$ ion? Use the appropriate hybrid orbitals to set up an MO energy level diagram for $H_3O_2^-$ analogous to that developed in the chapter for FHF^-.

5

Covalent Molecules and Crystals ———

The variety of ionic compounds is limited to some extent by the necessity of having very different electronegativities on adjacent atoms within the ionic system. Only elements from the left of the periodic table can be ionically bound to elements at the far upper right of the periodic table. As a result, it is difficult to describe any compound of antimony (for instance) as having ionic bonding involving the Sb atom. Covalence has no such limitations: Any element can form covalent bonds if the neighbor atom is properly chosen, save only the lighter noble gases. As a result, the array of covalent compounds is enormous. Chapter 4 has provided the theoretical background for understanding the bonding in these systems, whether they form isolated molecules, cluster molecules, or extended networks in crystals. In this chapter we need to survey the wide variety of covalent main-group species, focusing on the structures and preparation of each. We will consider their patterns of reactivity separately in Part III of the text, by looking not at specific reactions but at broad categories of reaction.

Our organizational principle in this chapter will be the grouping of species into categories with similar chemical and physical properties. Because covalence requires similar electronegativities of neighbor atoms, the simplest pattern will be to categorize covalent compounds by their overall approximate electronegativity. We can begin with intermetallic compounds, in which all atoms have low electronegativities, and systematically move to covalent systems with higher electronegativity.

——— 5.1 INTERMETALLIC COMPOUNDS, ALLOYS, AND GLASSES

Most metals are soluble in each other when molten, and in the solid phase a bewildering variety of species and structures is known. It is impossible to survey these with any thoroughness here, but we can indicate the circumstances under which certain general kinds of structures form.

Conceptually, the simplest structure is the solid solution, in which the two metals form an orderly crystal structure—frequently a close-packed lattice—in which the two kinds of atoms are randomly arranged. There is thus a sort of chemical disorder in these materials, even though the system is physically ordered. All metals miscible in the liquid phase form such random solid solutions if they are quenched rapidly enough. However, slow cooling or subsequent heat treatment can in many cases produce partial ordering in which, for example, atoms of one metal may cluster together in the lattice, forming a superstructure within the overall lattice. This process is sometimes known as *precipitation.* It normally hardens the alloy if it proceeds far enough to provide microscopic grains of one metal within the lattice of the other.

The tendency to form solid solutions is influenced by the electronegativities, sizes, valence electron count, and natural crystal structure of the atoms involved. Most metallic elements are at least slightly soluble in other molten metals, but if the electronegativity difference between the two elements is at all large—even a few tenths of a unit—the delocalized molecular orbitals in the valence bands of the solid become localized around the more electronegative atoms, and compound formation with partial electron transfer is preferable to solid solution formation. Similarly, even if the two elements have nearly the same electronegativity, the atoms must be nearly the same size if a solid solution is to form. If random replacement is to occur in the lattice, significant size differences will distort the structure and lower the lattice's stability. Pairs of metallic elements in which the metallic radii differ by more than about 15% do not form solid solutions with more than a very small percentage of the solute element, no matter how similar they may be chemically. Even if size and electronegativity are a good fit for two metals, a difference of more than about one electron per atom in valence electrons usually prevents solubility, because the element with more electrons achieves greater stability by remaining in its own phase with greater bonding. The natural-crystal-structure-fit requirement is related to the electron count because the band theory of solids yields three-dimensional energy bands for each crystal symmetry, called *Brillouin zones;* just as we expect stable molecules to fill all bonding orbitals but leave antibonding orbitals vacant, a stable solid solution of metals will have its low-energy Brillouin zones filled and its high-energy zones vacant, at least approximately.

A graphic way of combining these principles is seen in a *Darken–Gurry* plot; Fig. 5.1 shows a Darken–Gurry plot for magnesium. The circle, with a radius of about 0.5 electronegativity unit and 0.25 Å (16% of r_{Mg}), approximates the first two rules above, and it can be seen that although Zr is almost a perfect fit for Mg in size and electronegativity, its four valence electrons drastically limit its solubility in the two-electron Mg matrix, even though both metals are hcp when pure.

There is one more useful generalization about metal–metal solid solutions that has a less clear interpretation in structural terms: Solid-solution concentration limits are not reciprocal. Thus Zn will dissolve up to 6% Ag, but Ag will

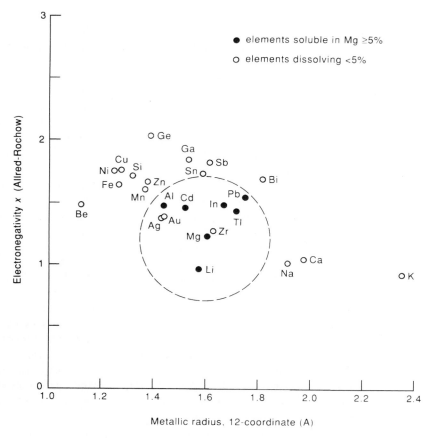

Figure 5.1   Darken–Gurry plot for solubility of metals in magnesium.

dissolve up to 38% Zn. Fairly generally the metal with the lower number of valence electrons will be the better solvent.

Alloys are sometimes found with solid-solution structures, but there are some characteristic structures that show high symmetry even though the resulting stoichiometry does not correspond to any of the usual ideas of valence or oxidation state. For compositions approximating 1:1 stoichiometry, a common structure is that of CsCl (Fig. 2.12) and a variant, the NaTl structure, in which the atoms are arranged as in CsCl but each atom has four neighbors of its own sort and four of the other component. Besides these fairly conventional A_1B_1 structures, there are also numerous lattices for higher B:A ratios in which a three-dimensional network of B atoms forms with such a symmetry that there are regularly spaced large holes occupied by A atoms. Examples of the ideal stoichiometry corresponding to such structures are AB_5, AB_{11}, and AB_{13}. An interesting special case is that of the *Laves phases* AB_2 ($MgNi_2$, $MgCu_2$, $MgZn_2$), for which the transition-metal atoms form tetrahedral four-atom clusters. These tetrahedra are linked into chains sharing a face or into an open three-dimensional network like that of cristobalite (SiO_2, Fig. 3.17). In either case, there are large truncated tetrahedral holes in the lattice that accommodate a Mg atom (or other A atom with a metallic radius roughly 1.225 times that of the B atom in the tetrahedra) and give it a coordination number of 12.

A surprising recent development in alloys is the discovery of *metallic glasses.* A glass, of course, is a supercooled liquid with a rigid structure but without long-range order. For the traditional silicate glasses, the rigid structure is provided by the random linking of silicate tetrahedra. It has already been noted that all miscible metals form solid solutions if they are quenched rapidly enough, and that in solid solutions, physical order is maintained even though the crystal is chemically random. For a number of alloy compositions, it has been proved possible to quench a molten mixture rapidly enough to achieve physical randomness—a glassy structure—as well. The glassy-metal compositions tend to involve either transition metals and metalloids ($Ni_{63}Pd_{17}P_{20}$ and $Pd_{78}Cu_6Si_{16}$, for example) or an early transition metal with a late transition metal ($Nb_{40-60}Ni_{60-40}$, for example). Presumably some directional bonding is involved, but it is not clear that any model like that for silicate glasses can be applied. A quenching rate of 10^{6}°C/sec is necessary, which is usually achieved by directing a stream of the molten metal upon a spinning copper roller that is internally cooled. The product is a ribbon having much the appearance of an ordinary alloy. The glassy metal is extremely strong, apparently because it lacks the grain boundaries found in all normal microcrystalline alloys. One metal glass shows more than three times the tensile strength of stainless steel, for example. The glasses also show good corrosion resistance and are "soft" magnets (easily magnetized and demagnetized), properties that also seem to be related to the absence of grain boundaries.

5.2 MAIN-GROUP ORGANOMETALLIC COMPOUNDS

Strictly speaking, an organometallic compound is one in which a metal-to-carbon bond is formed. However, the term is often also applied to compounds in which an organic molecule (sometimes with a net negative charge) is bonded to a metal atom through an oxygen or nitrogen atom, as in sodium acetylacetonate and beryllium bipyridyl:

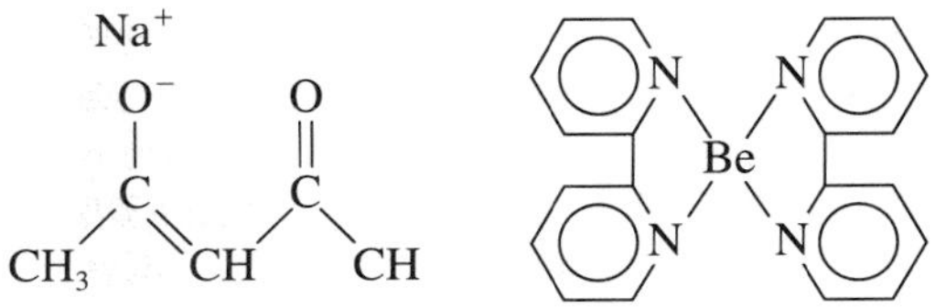

The first of these is predominantly an ionic compound. In the main-group metals, there are numerous such compounds among the heavier alkali metals and alkaline-earth metals. Even some hydrocarbon-based organometallics can be ionic if the hydrocarbon has an electronic structure that makes the formation of an anion particularly favorable. The best-known such compound is sodium cyclopentadienide, in which the hydrocarbon (cyclopentadiene, C_5H_6) becomes aromatic by losing a proton to become the $C_5H_5^-$ cyclopentadienide ion. An interesting special case of such anion formation is the formation of paramagnetic radical anions when sodium or potassium metal reacts with a large aromatic molecule such as naphthalene or anthracene, transferring a single electron to a relatively stable antibonding MO of the organic molecule:

Much more generally, however, the bond between a metal atom and a carbon atom is very nearly covalent. It is in this context that we should examine the structures of organometallic species. To a good approximation, main-group metals form only sigma bonds with organic ligands. As we shall see later, a major influence in the stability of transition-metal organometallic species is d-electron pi bonding, but this is impossible for main-group metals. The p-orbital pi overlap that one might expect, particularly for the p-block metals such as gallium, tin, and antimony, does not observably affect the overall bonding. When we consider silicon compounds, we shall look at possible exceptions to this generalization, but it is universally true for elements on the left side of the periodic table.

Considering first the metals in groups Ia, IIa, and IIIa, we find compounds corresponding to the traditional valences: RLi, R_2Be, R_3B, and their congeners. The more common methods of preparing these compounds are outlined in Table 5.1 together with their principal reaction types. Further comment on their structures is appropriate here, however. There is in general no difficulty in satisfying the bonding requirements of the carbon atoms involved in these species, because a metal atom simply replaces a hydrogen atom in the carbon coordination sphere. However, the stoichiometry is such that not all the possible bonding MOs centered on a given metal atom can be filled by the available metal and carbon bonding electrons. Thus, the compounds in the simple stoichiometries indicated above are electron-deficient, like the boranes mentioned earlier. The response of the molecules is to form clusters or polymers in which bridging bonds or metal–metal bonding can occur. Methyllithium and methylsodium, for example, form $M_4(CH_3)_4$ tetramers in crystals and in polar solvents, in which the metal atoms form a compact tetrahedron with a methyl centered over each face of the tetrahedron as in Fig. 5.2. A similar crystal structure is also found for ethyllithium. However, when EtLi dissolves in nonpolar solvents, it forms a hexamer, presumably the octahedral structure also shown in Fig. 5.2. The lithium salt of the enolate of methyl t-butyl ketone forms the same hexamer; an interesting feature of the structure is that the twelve Li and O atoms form an almost perfect hexagonal prism. By way of contrast, methylpotassium adopts the nickel arsenide crystal

Table 5.I. Organometallic Species of Groups Ia, IIa, IIIa

Ia: MR	IIa: MR$_2$	IIIa: MR$_3$

Preparation

Ia: MR	IIa: MR$_2$	IIIa: MR$_3$
1. R—Br + 2Li $\xrightarrow{Ar}$ R—Li + Li$^+$Br$^-$	1. BeCl$_2$ + 2 n-BuLi $\longrightarrow$ Be(n-Bu)$_2$ + 2Li$^+$Cl$^-$	1. BF$_3$ + 3EtMgBr $\longrightarrow$ B(Et)$_3$ + 3MgFBr
2. 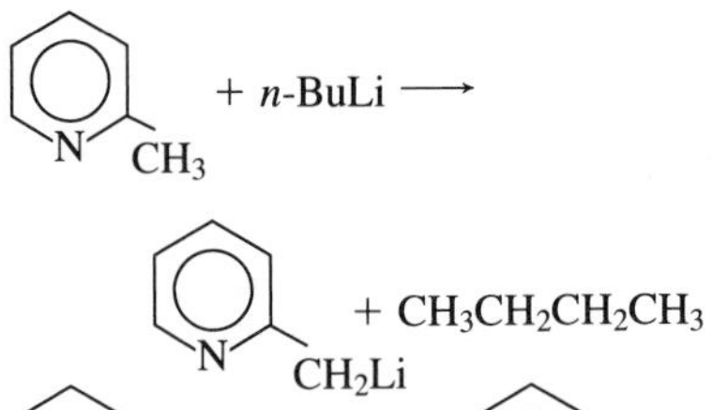 [2-methylpyridine] + n-BuLi $\longrightarrow$ [pyridine–CH$_2$Li] + CH$_3$CH$_2$CH$_2$CH$_3$	2. R$_2$Hg + Be $\longrightarrow$ R$_2$Be + Hg°	2. 3R$_2$Hg + 2Al $\longrightarrow$ 2R$_3$Al + 3Hg°
3. 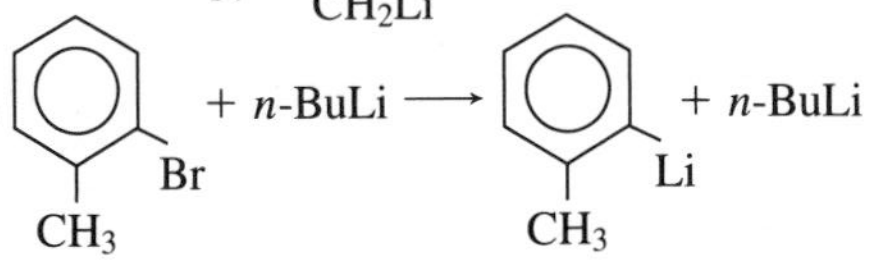 [bromo-methylbenzene] + n-BuLi $\longrightarrow$ [lithio-methylbenzene] + n-BuLi	3. RX + Mg $\longrightarrow$ RMgX	3. Al + $\frac{3}{2}$H$_2$ + 2Al(Et)$_3$ $\longrightarrow$ 3Al(Et)$_2$H $\xrightarrow{C_2H_4}$ 3Al(Et)$_3$
4. CH$_3$—C≡CH + Li $\xrightarrow{Ar}$ CH$_3$—C≡C—Li + H$_2$		
5. R$_2$Hg + 2Li $\xrightarrow{Ar}$ 2R—Li + Hg°		

Reactions

Ia: MR	IIa: MR$_2$	IIIa: MR$_3$
1. 2R—Li + CO$_2$ $\longrightarrow$ R—C(=O)—C—R + Li$_2$O	1. R$_2$Be + H$_2$O $\longrightarrow$ 2RH + Be^2 + 2OH$^-$	1. Al(Et)$_3$ + LiEt $\longrightarrow$ LiAl(Et)$_4$
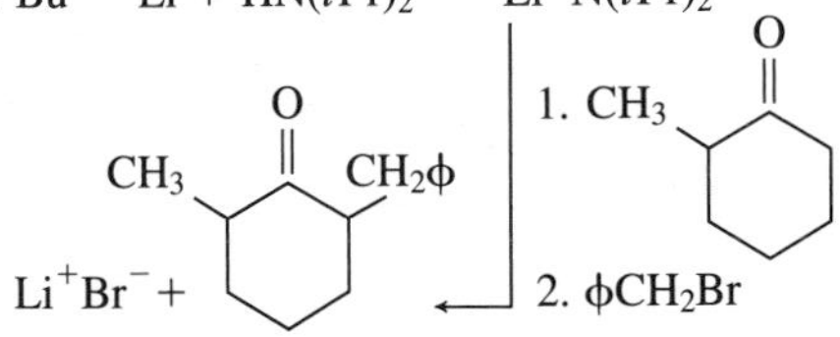 2. Bu—Li + HN(iPr)$_2$ $\longrightarrow$ Li$^+$N(iPr)$_2^-$ [then with methyl-cyclohexanone: 1. CH$_3$; 2. ϕCH$_2$Br $\longrightarrow$ Li$^+$Br$^-$ + (methyl,benzyl-cyclohexanone)]	2. n(i=Pr)$_2$Be $\xrightarrow{Heat}$ [(i-Pr)BeH]$_n$ + nCH$_3$—CH=CH$_2$	2. Al(Et)$_3$ + C$_2$H$_2$ $\longrightarrow$ Al(Et)$_2$—CH$_2$CH$_2$Et Polyolefin catalyst
3. Industrial anionic polymerization initiator (n-BuLi)	3. Organic Grignard reactions	3. Al(Et)$_3$ + CrCl$_3$ + 6CO $\longrightarrow$ Cr(CO)$_6$ + AlCl$_3$
4. 2Na$^+$ CH$_3$C$_5$H$_4^-$ + MnCl$_2$ $\longrightarrow$ Mn(C$_5$H$_4$CH$_3$)$_2$ + Na$^+$Cl$^-$		4. R—C≡N + Al(i-Bu)$_2$ $\longrightarrow$ RCH=N—Al(Bu)$_2$ $\xrightarrow{H_2O^+}$ RCHO + NH$_4^+$ + Al^{3+} + Isobutane

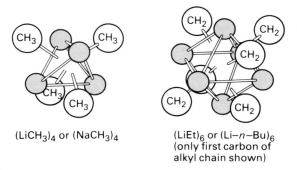

(LiCH$_3$)$_4$ or (NaCH$_3$)$_4$ (LiEt)$_6$ or (Li–n–Bu)$_6$
                                       (only first carbon of
                                       alkyl chain shown)

Figure 5.2 Organolithium polymeric clusters (shaded atoms represent metal).

structure (Fig. 2.14), perhaps as a result of its greater ionic character or the larger radius of the potassium atom.

Beryllium alkyls form chains with bridging alkyl groups (Fig. 5.3), thereby achieving tetrahedral coordination of the Be atom. Bulky alkyl groups such as t-butyl, however, prevent the BeR$_2$ monomer from associating. In Li$^+$Be(t-Bu)$_3^-$ the three R groups give Be trigonal planar geometry, though Li should perhaps be included in the Be coordination; the Li–Be distance is equal to the sum of the Li

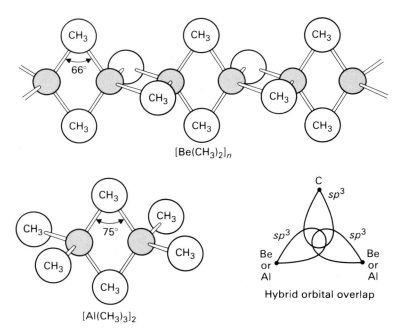

[Be(CH$_3$)$_2$]$_n$

[Al(CH$_3$)$_3$]$_2$

Hybrid orbital overlap

Figure 5.3 Polymeric structures for beryllium and aluminum alkyls.

and Be covalent radii. Bridging similar to that of the Be chains occurs in aluminum alkyls, but the stoichiometry allows full tetrahedral coordination of the Al simply by forming a dimer, seen also in Fig. 5.3. The same structure is seen for the isoelectronic species $Mg_2(C_2H_5)_6^{2-}$. The M—C—M bond angles in these bridged compounds are remarkably narrow. They presumably result from the three-center overlap, which has the property that although sigma overlap is occurring, its directional character is not greatest along the internuclear axis.

Metal-atom size is apparently quite important in forming stable clusters or polymers. The magnesium alkyls are all extensively polymerized, even those involving alkyl groups bulky enough to prevent the polymerization of the corresponding beryllium compound. However, the tetrahedral [Mg(*o*-phenylene)]$_4$·4THF has been shown to have the same structure as the methyllithium tetramer in Fig. 5.2. In the same size relationship, most aluminum alkyls are dimers, but the much smaller boron atom forms only planar BR_3 monomers even for trimethylboron. Bridging is possible for boron only when the bridging atom is as small as a hydrogen atom.

The conventional long form of the periodic table is often criticized for giving the impression that hydrogen is an alkali metal. However, a small group of compounds known as perlithio organics suggests a greater resemblance than we usually expect between lithium and hydrogen. For example, the following reaction proceeds smoothly:

$$H_3C—C{\equiv}CH + 4n\text{-}BuLi \rightarrow 4CH_3CH_2CH_2CH_3 + [Li_3C—C{\equiv}CLi]$$
$$\downarrow$$
$$Li_2C{=}C{=}CLi_2$$

Few other lithiocarbons are known, but in favorable cases, polylithiation of acetylenes and aromatic systems is possible up to at least three-fourths of the hydrogens present.

The metals in groups IIb, IIIa, and IVa of the periodic table also form a wide variety of organometallic compounds, some with important uses in organic synthesis. The first synthetic organometallic compound (Frankland, 1852) was diethylzinc, prepared by the pyrolysis of ethylzinc iodide, which in turn was obtained by simply heating zinc metal in ethyl iodide:

$$Zn + C_2H_5I \rightarrow C_2H_5ZnI$$
$$2C_2H_5ZnI \rightarrow (C_2H_5)_2Zn + ZnI_2$$

For many of these b-group metals, the latter reaction is an equilibrium skewed predominantly toward the mixed alkyl halide, as it is for Grignard reagents. However, in this case diethylzinc can be distilled off because of its low boiling point (117°C). Preparative methods and common reactions of these b-group organometallics are given in Table 5.2.

The b-group organometallics are less electron-deficient than the a-group systems, and show less tendency to form clusters or chains than the compounds of

Table 5.2 Organometallic Species of Groups IIb, IIIa, IVa

IIb: MR_2	IIIa: MR_3	IVa: MR_4
Preparation		
1. $C_2H_5I + Zn/Cu \longrightarrow C_2H_5ZnI$ $\quad\quad\quad\quad\quad\quad\quad\;\downarrow$ Heat $\quad ZnI_2 + (C_2H_5)_2Zn$	1. $GaCl_3 + 3RMgBr \longrightarrow GaR_3 + MgBrCl$	1. $GeCl_4 + nRMgCl \longrightarrow R_n GeCl_{4-n}$
2. $ZnCl_2 + 2LiR \longrightarrow ZnR_2 + 2LiCl$	2. $GaBr_3 + Al(Et)_3 \longrightarrow Ga(Et)_3 + AlCl_3$	2. $Ge + 2CH_3Cl \xrightarrow{Cu} (CH_3)_2GeCl_2$
3. $CdCl_2 + 2RMgBr \longrightarrow CdR_2 + MgBrCl$	3. $CH_3Br + Mg/In \longrightarrow In(CH_3)_3 + MgBr_2$	3. $\phi_3GeLi + RCH{=}CH_2 \longrightarrow \phi_3GeCH_2CHLiR$ $\quad LiOH + \phi_3GeCH_2CH_2R \xleftarrow{\;\;} {}\;\;\downarrow H_2O$
4. $2Hg + 2RBr \longrightarrow HgR_2 + HgBr_2$	4. $2LiCH_3 + CH_3I + TlI \longrightarrow$ $\quad\quad Tl(CH_3)_3 + 2LiI$	4. $CH_3Cl + Sn \xrightarrow{175°C}$ $\quad CH_3SnCl_3 + (CH_3)_2SnCl_2 + (CH_3)_2SnCl$
5. $HgBr_2 + R_2C{=}CR_2 \longrightarrow R_2C{-}CR_2$ $\quad\quad\quad\quad\quad\quad\quad\quad\;\; \overset{\mid}{Br}\;\; \overset{\mid}{HgBr}$	5. $TlOH + C_5H_6 \longrightarrow Tl(C_5H_5) + H_2O$ $\quad\quad\quad\quad\; \text{Cyclopentadiene}$	5. $4EtCl + Na/Pb \longrightarrow Pb(Et)_4 + 3Pb + 4NaCl$
Reactions		
1. $Zn\phi_2 + \phi CN \longrightarrow \phi CN \cdot Zn\phi_2$ $\quad\quad\quad\quad\quad\quad\quad\;\downarrow$ Heat $\quad\quad \phi_2C{=}NZn$	1. $Ga\phi_3 + HCl \longrightarrow \phi_2GaCl + C_6H_6$	1. $R_4Ge + Br_2 \longrightarrow R_3GeBr + RBr$
2. $ZnR_2 + R'CHO \longrightarrow RR'CHOZnR$ $\quad Zn^{2+} + RR'CHOH \xleftarrow{\;\;} {}\;\;\downarrow H_2O$	2. $GaCl_3 + Sn(CH_3)_4 \longrightarrow$ $\quad CH_3GaCl_2 + (CH_3)_3SnCl$	2. $nR_4Sn + (4-n)SnX_4 \longrightarrow 4R_nSnX_{4-n}$
3. $CdR_2 + 2R'COCl \longrightarrow 2R'COR + CdCl_2$	3. $In(Et)_3 + EtOH \longrightarrow (Et)_2InOEt + C_2H_6$	3. $R_4Sn + BCl_3 \longrightarrow RBCl_2 + R_3SnCl$
4. $HgR_2 + M \longrightarrow MR/MR_2/MR_3$ $\quad\quad\quad\quad$ Active main-group metal	4. $CH_3CH_2CCH_3 + TlOEt \longrightarrow$	4. $n(CH_3)_2SnCl_2 + 2n\,Na \xrightarrow{NH_3}$ $\quad [Sn(CH_3)_2]_n + 2n\,NaCl$
		5. $R_4Pb + SOCl_2 \longrightarrow R_2PbCl_2 + R_3PbCl$

beryllium and aluminum. Normally the alkyls are monomers with a linear, trigonal-planar, or tetrahedral structure dictated by the stoichiometry. A few clusters, however, are known for alkylmetal halides (for instance, EtZnCl, Fig. 5.4), and a number of organotin clusters are known with oxygenated organic species analogous to the lithium enolate cluster described above; exactly the same cluster symmetry is seen in a hexamer $[PhSn(O)O_2CCy]_6$, formed from triphenyltin cyclohexanecarboxylate. Chains of up to 20 tin atoms can be produced by the sodium reduction of $(CH_3)_2SnCl_2$ (see Table 5.2), and in much the same way chains of 24 Si atoms can be prepared by the reductive coupling of $Me(SiMe_2)_6Cl$ and $Cl(SiMe_2)_6Cl$ with Na/K. The geometric requirements for the organometallic bonds are apparently quite specific. For example, R_2Hg compounds must have a linear C—Hg—C geometry. Sodium amalgam and 1,2-dibromobenzene yield o-phenylenemercury, but the latter forms a hexamer like that in Fig. 5.5 or a planar trimer rather than the anthracene analog

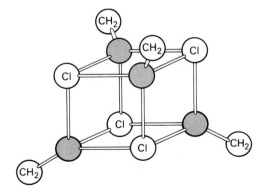

in order to allow the 180° bond angle at Hg.

The chemical reactivities of the b-group organometallics are quite varied, even within a single group. Zinc alkyls burn spontaneously in air; the corresponding mercury alkyls are not affected by air or water. The same pattern applies to the group IIIa metals, particularly to the mixed alkylmetal halides: R_2TlX compounds are extremely stable toward practically all kinds of attack. For the group IVa metals the pattern is less clear. Tetramethylsilane, the common nmr reference, and tetraethyllead, the gasoline antiknock additive, are about equally resistant to hydrolysis and most forms of chemical attack. Germanium alkyls are somewhat more reactive than those of silicon. Some experiments show that metal–carbon

Figure 5.4 Ethylzinc chloride tetramer (in nonpolar solvents). Only C coordinated to Zn in shown.

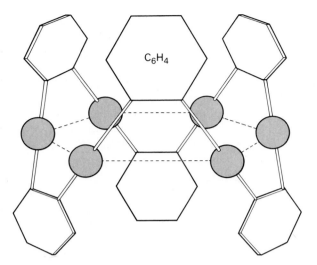

Figure 5.5   *o*-Phenylenemercury hexamer.

cleavage rates (by acid) increase uniformly from Si to Pb by $1:10^8$, a trend opposite to that shown by the alkyls of groups IIb and IIIa.

Commercial applications of the b-group organometallics were once quite extensive but are now diminishing. The most obvious application is the use of
tetraethyllead for octane enhancement in motor fuel, which is rapidly shrinking
as more nations ban it because of the accumulation of environmental lead and the
resulting health problems that arise from heavy-metal poisoning. In somewhat the
same way organotin compounds, particularly R_3SnX variations, have been extensively used because of their toxic qualities in applications such as fungicides,
wood preservatives, and marine antifouling paint. However, the very success of
these compounds as biocides has led to their removal from marine paints because
the amount leaching from the paint into seawater was damaging crustaceans and
other species in bays and estuaries.

5.3 BORANES

We have considered intermetallic systems and organometallic systems in our
progress to covalent systems of higher overall electronegativity. The metalloid elements with electronegativities near 2 are next beyond metals, and their simplest
covalent compounds are hydrides. Nearly all the *p*-block elements form covalent
molecular hydrides, and we shall consider them together as a class in the next section. However, the molecular hydrides of boron are different from essentially all
the other molecular hydrides in showing the unique bonding patterns that arise for
electron-deficient compounds. They form clusters, and these clusters have been a

paradigm for the theory of such compounds; we can look at these compounds and the associated theory in more depth in this section.

The tendency to clustering in boranes is so strong that the monomer, BH_3, cannot be prepared in the condensed phase, or even in an undiluted gas at reasonable pressures. Instead, the dimer B_2H_6, whose structure we have already considered, is the simplest stable borane. However, derivatives of monoborane can be prepared in some variety, since BH_3 is an excellent electron acceptor; its four valence orbitals contain only six electrons (3 from $B + 3 \times 1$ from H). Many species with a pair of nonbonding electrons form donor–acceptor compounds with BH_3: $H_3B \leftarrow CO$, $H_3B \leftarrow PF_3$, $H_3B \leftarrow N(CH_3)_3$, and so on. The strongest such bond is formed by the hydride ion, which yields BH_4^- (tetrahydroborate) ion in which all B—H bonds are equivalent.

Alkali-metal compounds with the BH_4^- ion (still frequently called the borohydride ion) are essentially ionic. $LiBH_4$ and $NaBH_4$ are conveniently prepared by allowing diborane to react with the metal hydride in an ethereal solvent:

$$2LiH + B_2H_6 \rightarrow 2LiBH_4$$

Although the hydride ion is both a very strong base and a powerful reducing agent, the ionic borohydrides are much gentler; for example, $NaBH_4$ does not hydrolyze in water solution at room temperature above pH 9. Some typical reactions include

$$R_3NHCl + LiBH_4 \rightarrow R_3N \cdot BH_3 + LiCl + H_2$$
$$PCl_3 + 3LiBH_4 \rightarrow PH_3 + 3LiCl + \tfrac{3}{2}B_2H_6$$
$$Ph_3AsCl_2 + 2LiBH_4 \rightarrow Ph_3As + B_2H_6 + H_2 + 2LiCl$$
$$Fe^{3+} + 3BH_4^- \rightarrow Fe^0 + \tfrac{3}{2}B_2H_6 + \tfrac{3}{2}H_2$$

The heavier group II metals also form more or less ionic $M(BH_4)_2$ compounds, but $Be(BH_4)_2$ and its neighbor $Al(BH_4)_3$ are volatile, presumably covalent compounds whose current reported molecular structures in the vapor phase are shown in Fig. 5.6. Since Be and Al are both neighbors of B in the periodic table, it is possible to regard these compounds as heteronuclear cluster compounds with bridging H atoms analogous to those in diborane. The compounds are formed by heating the chlorides with $LiBH_4$ in the absence of solvent:

$$BeCl_2 + 2LiBH_4 \rightarrow BeB_2H_8 + 2LiCl$$

These clusters are particularly mobile in their structures, and at least six different structures have been reported at one time or another for BeB_2H_8, mostly differing in the number of bridging hydrogens but also including triangular BeB_2 arrangements with bridging hydrogens in varying numbers. Unfortunately for simplicity, more than one may be correct.

Borane clusters are known from B_2H_6 to $B_{20}H_{16}$. The formulas of the better-characterized compounds are given in Table 5.3, and the stereochemistry of some

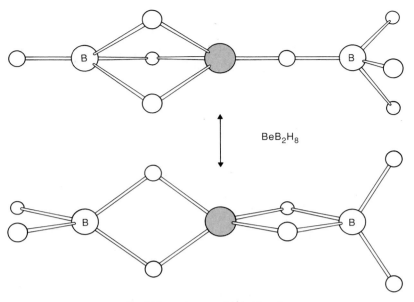

BeB_2H_8

(rapid interchange of H bridges)

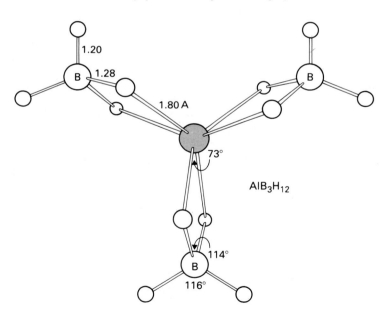

AlB_3H_{12}

Figure 5.6 Beryllium and aluminum tetrahydroborates.

Table 5.3 Some Polyboranes and Borane Anions

Formula	Preparation	Key reactions
B_2H_6	$NaBH_4 + H_2PO_3F \xrightarrow{} \frac{1}{2}B_2H_6 + H_2 + NaHPO_3F$	Pyrolysis to many higher polyboranes
$[B_3H_8]^-$	$NaBH_4 + B_2H_6 \xrightarrow{Ethers} NaB_3H_8 + H_2$	
B_4H_{10}	$2B_2H_6 \xrightarrow{High\ P} B_4H_{10} + H_2$	$B_4H_{10} + C_2H_2 \xrightarrow{100°C} C_2B_4H_6$ (plus other carboranes and H_2)
B_5H_9	$\frac{5}{2}B_2H_6 \xrightarrow{250°C} B_5H_9 + 3H_2$	$B_5H_9 + C_2H_2 \xrightarrow[discharge]{Electric} C_2B_{3-5}H_{5-7} + H_2$
B_5H_{11}	$2B_4H_{10} + B_2H_6 \rightleftharpoons 2B_5H_{11} + 2H_2B_6H_{10}$	
B_6H_{10}	$B_5H_{11} \xrightarrow{Diglyme} B_6H_{10}\ (+B_2H_6, B_4H_{10}, B_5H_9)$	
B_6H_{12}	$2B_5H_{11} \xrightarrow{} B_6H_{12} + B_4H_{10}$	
$[B_6H_6]^{2-}$	$2BH_4^- + 2B_2H_6 \xrightarrow{100°C} B_6H_6^{2-} + 7H_2$	
$[B_7H_7]^{2-}$	$B_9H_9^{2-} \xrightarrow{Air} B_8H_8^{2-} \xrightarrow{} B_7H_7^{2-} \xrightarrow{} B_6H_6^{2-}$	
B_8H_{12}	$2B_9H_{15} \rightleftharpoons 2B_8H_{12} + B_2H_6$	
B_8H_{14}	$B_8H_{12} + NaH \xrightarrow{Me_4N^+Cl} Me_4N^+B_8H_{12}^- \xrightarrow{HCl} B_8H_{14}$	
$[B_8H_8]^{2-}$	see $B_7H_7^{2-}$	
B_9H_{15}	$2B_5H_{11} \xrightarrow{Hexamethylenetetramine(L)} B_9H_{15} + L \cdot BH_3 + 2H_2$	
$[B_9H_9]^{2-}$	$20CsB_3H_8 \xrightarrow{230°} 2Cs_2B_9H_9 + 2Cs_2B_{10}H_{10} + Cs_2B_{12}H_{12} + 10CsBH_4 + 35H_2$	Oxidative degradation to lower dianions
$B_{10}H_{14}$	$B_2H_6 \xrightarrow{Me_2O,\ 150°C} B_{10}H_{14} + H_2\ (+\ other\ boranes)$	$B_{10}H_{14} + C_2H_2 \xrightarrow{Base} C_2B_{10}H_{12} + 2H_2$
$B_{10}H_{16}$	$2B_5H_9 \xrightarrow{Electric\ discharge} B_{10}H_{16} + H_2\ (+\ other\ boranes)$	
$[B_{10}H_{10}]^{2-}$	$B_{10}H_{14} + 2Et_3N \xrightarrow{} Et_3NH^+_2B_{10}H_{10}^{2-} + H_2$	
$[B_{11}H_{11}]^{2-}$	$B_{10}H_{14} + BH_4^- \xrightarrow{OH^-} B_{11}H_{14}^{2-} \xrightarrow{250°} B_{11}H_{11}^{2-}$	
$[B_{12}H_{12}]^{2-}$	$2NaBH_4 + B_{10}H_{14} \xrightarrow{} Na_2B_{12}H_{12} + 5H_2$	
$B_{16}H_{20}$	$B_9H_{15} \xrightarrow{Me_2S} Me_2S \cdot B_9H_{13} \xrightarrow{Pyrolysis} B_{16}H_{20}\ (+\ other\ boranes)$	
$B_{18}H_{22}$	$2B_{10}H_{10}^{2-} + 2Fe^{3+} \xrightarrow{} 2Fe^{2+} + B_{20}H_{18}^{2-} \xrightarrow{H_3O^+} B_{18}H_{22}$	
$B_{20}H_{16}$	$2B_{10}H_{14} \xrightarrow{Electric\ discharge} B_{20}H_{16} + 6H_2$	

of these compounds is shown in Figure 5.7. There is also a family of stable dianions $B_nH_n^{2-}$ from $n = 6$ to 12, all of which have deltahedral structures of boron atoms as shown in Fig. 4.28, with a radial H from each B. Boranes are named by using the Greek prefix for the number of boron atoms, the root "-borane," and an Arabic numeral in parentheses for the number of hydrogen atoms. Using this system, $B_{10}H_{14}$ is decaborane(14), for example. In addition, inspection of Fig. 5.7 reveals that many boranes or their anions form deltahedra with one or two vertexes of the deltahedron missing. The *closo, nido-* ("nest"), and *arachno-* ("cobweb") prefixes can be added to the stoichiometric name to describe structures. In general, *closo-* structures are adopted only by the $B_nH_n^{2-}$ anions (and isoelectronic carboranes, to be considered later), whereas *nido-* structures are found for B_nH_{n+4} boranes and *arachno-* structures are found for B_nH_{n+6} boranes. In addition, a few boranes have structures consisting of linked polyhedral fragments; these are given the prefix *conjuncto-*.

The cluster bonding in boranes has been interpreted in Chapter 4 in terms of MOs overlapping over the surface of the cluster and an additional core MO at the center of the cluster. The similarities shown by the MOs can be summarized as the cluster structure rules in Table 4.6 (Wade's rules). In that discussion, however, we

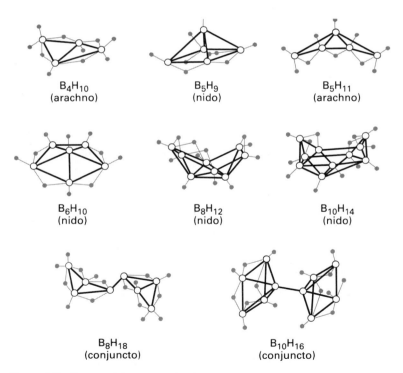

B_4H_{10}
(arachno)

B_5H_9
(nido)

B_5H_{11}
(arachno)

B_6H_{10}
(nido)

B_8H_{12}
(nido)

$B_{10}H_{14}$
(nido)

B_8H_{18}
(conjuncto)

$B_{10}H_{16}$
(conjuncto)

Figure 5.7   Some polyborane molecular structures.

did not mention the *nido* and *arachno* boranes. Quite generally it is found that B_nH_{n+4} boranes adopt *nido* structures, in which n of the hydrogens are radially arranged from the n borons and the extra four hydrogens are found bridging B–B edges of the open face of the *nido* deltahedron. Similarly, B_nH_{n+6} boranes are normally *arachno* fragments of deltahedra from which two adjacent vertexes have been removed; the six extra hydrogens bridge edges of the open face. These structures are consistent with Wade's rules if the electron from each bridging H is considered part of the cluster-bonding pool. Thus B_5H_9 has a cluster electron count of $(5 \times 2) + (4 \times 1) = 14$, which from Table 4.6 predicts an octahedron as the basic structure; the five B atoms are seen to form a square pyramid that can obviously be seen as an octahedron missing one vertex; and the four extra hydrogens bridge the four edges of the square open face.

Sometimes more than one *nido* or *arachno* isomer will be possible for a given deltahedron. Boranes almost always choose the isomer with the largest possible open face, meaning that the B atom on the *closo* deltahedron with the largest coordination number will be removed. This allows the electrons from the bridging hydrogens the largest possible ring in which to circulate, and reduces repulsion. The locations of the bridging hydrogens can be predicted from the form of the HOMO (highest occupied MO); no bridging H will be found over a B–B edge that has a node in the HOMO. The H nuclei (protons) will choose to locate themselves over edges with positive electron overlap populations in the HOMO.

As the last chapter has suggested, we can tinker with the cluster electron count by inserting heteroatoms with different numbers of valence electrons. For example, one or two B atoms can be replaced by carbons. Such species are known as *carboranes,* named as polycarba-polyboranes with carbon positions indicated by numbering the borane skeleton. $C_2B_4H_6$, with the structure shown in Fig. 5.8, is thus 1,2-dicarba-*closo*-hexaborane(6). *Nido-, arachno-,* and *closo*-structures are known with up to four carbon atoms substituted, though the vast majority have two carbons regardless of the size of the boron cage. This feature of the stoichiometry might result only from the synthetic route chosen, but also might occur because *closo*- structures are the most stable. Since the stoichiometry is determined by the electron count and a C with four electrons has the same electron count as B^-, the *closo*- $B_nH_n^{2-}$ ions are electronically equivalent to $C_2B_{n-2}H_n$ uncharged molecules.

Small carboranes are frequently made by reacting small boranes with alkynes. Frequently the reaction will yield *nido*- products at low temperatures and *closo*-products at high temperatures:

$$B_5H_9 + C_2H_2 \xrightarrow{215°C} 2,3\text{-}C_2B_4H_8 + 2\text{-}CB_5H_9CH_3$$

$$\xrightarrow[450°C]{} 1,6\text{-}C_2B_4H_6 + \text{other } closo\text{- products}$$

This behavior reflects the increased stability of the *closo*- structures. The best-known carborane is 1,2-*closo*-$C_2B_{10}H_{12}$ (also shown in Fig. 5.8), which is prepared by a similar reaction using a Lewis-base catalyst:

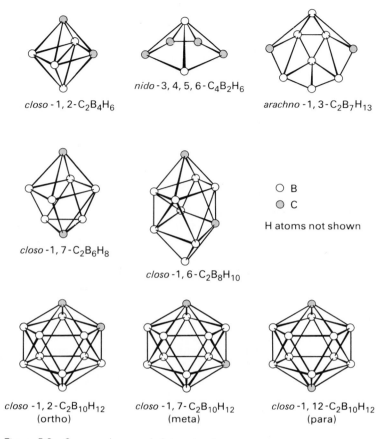

$nido$ - 3, 4, 5, 6 - $C_4B_2H_6$

$closo$ - 1, 2 - $C_2B_4H_6$

$arachno$ - 1, 3 - $C_2B_7H_{13}$

$closo$ - 1, 7 - $C_2B_6H_8$

$closo$ - 1, 6 - $C_2B_8H_{10}$

○ B
◐ C

H atoms not shown

$closo$ - 1, 2 - $C_2B_{10}H_{12}$
(ortho)

$closo$ - 1, 7 - $C_2B_{10}H_{12}$
(meta)

$closo$ - 1, 12 - $C_2B_{10}H_{12}$
(para)

Figure 5.8   Some carborane skeleton structures.

$$B_{10}H_{14} + C_2H_2 \xrightarrow{Et_2S} C_2B_{10}H_{12} + 2H_2$$

There are three isomers of this carborane, 1,2-, 1,7-, and 1,12-. The 1,7- isomer is prepared by heating the 1,2- at 470°C, and the 1,12- by flash pyrolysis of the 1,7- at 700°C for a few seconds. Other intermediate-sized carboranes (B_{6-9}) are frequently made by partially degrading 1,2- or 1,7-$C_2B_{10}H_{12}$ in strong base:

$$C_2B_{10}H_{12} + CH_3O^- + 2CH_3OH \longrightarrow C_2B_9H_{12}^- + H_2 + B(OCH_3)_3$$

$$C_2B_9H_{12}^- \xrightarrow{H^+} C_2B_9H_{13} \xrightarrow{Heat} 2,3\text{-}C_2B_9H_{11} + H_2$$

While most small boranes are extremely unstable toward air oxidation, the $closo$- carboranes, particularly the three icosahedral isomers of $C_2B_{10}H_{12}$, are quite resistant both to oxidation and pyrolysis, as the preparative temperatures will suggest. Because the two C atoms can be readily metalated and substituted, some extremely heat-stable polymers can be made in which $C_2B_{10}H_{10}$ icosahedra alternate with organic or siloxane groups in polymeric chains.

We have already suggested that most metals can be considered electron-deficient with respect to covalent bond formation. From this we might expect that metals could be substituted in borane cages subject only, perhaps, to the electron-counting rules. This is indeed true, both for main-group metals and for transition metals, with appropriate adjustments made for d electrons. Chapter 13 will specifically compare transition-metal clusters and borane clusters and develop the electron-counting rules further at that point. Here we can briefly indicate the scope of formation of mixed clusters, the *metallaboranes*. Their chemistry is quite varied: Over 40 elements other than boron have been incorporated into borane cages.

As this variety might suggest, there are several synthetic techniques for incorporating heteroatoms into a borane cage. The most frequently used borane is $B_{10}H_{14}$. It can be either allowed to react directly with a basic heteroatom (such as S^{2-}) or converted into an anion by a strong base such as NaH and allowed to react with an acidic species such as PCl_3 or $NiBr_2$. H_2 or a hydrocarbon is often released in the acid–base reaction:

$$B_{10}H_{14} + S^{2-} + 4H_2O \longrightarrow B_9H_{12}S^- + B(OH)_4^- + 3H_2$$

$$CsB_9H_{12}S \xrightarrow{200°C} CsB_{10}H_{11}S + \tfrac{3}{2}H_2 + ?$$

$$B_{10}H_{11}S^- \xrightarrow[\text{2. FeCl}_2]{\text{1. OH}^-} [Fe(B_{10}H_{10}S)]^{2-}$$

$$Na_3B_{10}H_{10}CH·2THF + PCl_3 \rightarrow 1,2\text{-}B_{10}H_{10}CHP \xrightarrow{485°C} 1,7\text{-}B_{10}H_{10}CHP$$

$$1,7\text{-}B_{10}H_{10}CHP + C_5H_{10}NH \rightarrow [C_5H_{10}NH_2]^+[B_9H_{10}CHP]^-$$

$$[B_9H_{10}CHP]^- \xrightarrow[\text{2. FeCl}_2]{\text{1. NaH}} [Fe(B_9H_9CHP)_2]^{2-} \xrightarrow{CH_3I} Fe(B_9H_9CHPCH_3)_2$$

$$B_{10}H_{14} + NaH \rightarrow B_{10}H_{13}^- + Na^+ + H_2$$

$$B_{10}H_{13}^- \xrightarrow[\text{2. Me}_4\text{N}^+\text{Cl}^-]{\text{1. NiBr}_2(\text{PPh}_3)_2} [Me_4N]_2^+[Ni(B_{10}H_{12})_2]^{2-}$$

$$B_{10}H_{14} + Me_3N·AlH_3 \xrightarrow{Et_2O} [Me_3NH]^+[B_{10}H_{10}AlH_2]^-·2Et_2O = H_2$$

$$B_{10}H_{14} + CdEt_2 \xrightarrow{Et_2O} CdB_{10}H_{12}·2Et_2O + C_2H_6 \xrightarrow{H_2O} [Cd(B_{10}H_{12})_2]^{2-}$$

For boranes smaller than $B_{10}H_{14}$, there is an interesting thermodynamic effect of metal substitution: All neutral boranes with fewer than 10 boron atoms are at least somewhat air-sensitive (some of them extremely so), but in many cases metal substitution stabilizes the borane—now a metallaborane—toward air and moisture. For example, Fig. 5.9 shows two metallaboranes that form air-stable crystals even though the corresponding boranes are spectacularly unstable. The preparation is analogous to that of the larger systems; it starts with the sodium salt of pentaborane-9, $Na^+B_5H_8^-$, and adds a metal halide:

$$B_5H_8^- + CoCl_2 + C_5H_5^- \rightarrow (C_5H_5)CoB_4H_8$$

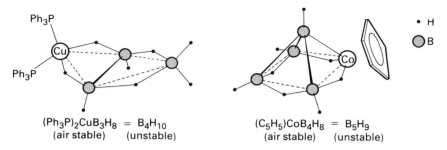

$(Ph_3P)_2CuB_3H_8$ = B_4H_{10} $(C_5H_5)CoB_4H_8$ = B_5H_9
  (air stable)   (unstable)   (air stable)   (unstable)

Figure 5.9   Metallaborane structures.

Multiple metal substitution is also possible, to the point that the product becomes perhaps as much a metal cluster as a metallaborane cage. The same synthesis also yields $(C_5H_5)_3Co_3B_4H_4$ with three Co atoms and four B atoms in the cage.

The stoichiometric flexibility being shown in these compounds can be extended to some remarkable structures. It is possible to imagine a compound like the cobalt cluster of Fig. 5.9 in which the borane fragment is planar like the cyclopentadienide ring shown; these can be stacked into triple- and quadruple-decker sandwich clusters. Figure 5.10 shows the structures in a reaction in which an iron(III) compound with a structure analogous to the earlier cobalt compound is reduced, then allowed to react with $NiBr_2$ and a planar-carborane cobalt cluster, yielding two products, one featuring an iron–nickel–carbon–boron eight-membered cluster and the other showing four stacked five-membered rings. At this

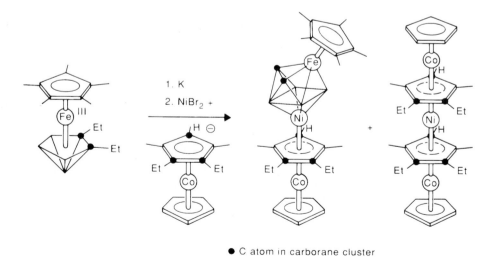

● C atom in carborane cluster

Figure 5.10   Stacked rings and clusters in organometallacarborane systems.

point the boron content is relatively low, and we may delay further discussion of these structures till Chapter 13.

5.4 SILANES AND MOLECULAR HYDRIDES

A special kind of cluster formation is shown by elements that form chains or rings of atoms, like carbon in traditional organic compounds. This process is called catenation; Chapter 4 has suggested bonding criteria for catenation, in which a chain of atoms must fill all its possible bonding MOs but no antibonding MOs. Straight-chain hydrocarbons meet this condition, of course; the uncatenated bonding orbitals are shared with hydrogen atoms. An isoelectronic chain can be constructed from sulfur atoms with no hydrogens, assuming sp^3 hybridization at each S. Many catenated-sulfur molecules are known, and we have seen that the observed bond angles of about $103–113°$ are consistent with the tetrahedral sp^3 angle of $109°27'$. However, although we might also expect that the other group IVa elements Si, Ge, Sn, and Pb would form catenated hydrides analogous to the hydrocarbons, this is only partially true.

Another condition for stable catenation is good orbital overlap of neighbor atoms, which requires that inner-core electrons not form so large a cloud as to prevent close approach of the atoms. As one moves down the column of group IVa elements, the inner core expands from $1s^2$ for C to $1s^2 2s^2 2p^6 3s^2 3p^6 3d^{10} 4s^2 4p^6 4d^{10} 4f^{14} 5s^2 5p^6 5d^{10}$ for Pb. The result is that catenated systems are much less stable for the lower members of the group. Thus whereas linear hydrocarbons can in principle be extended to any desired length, the longest well-characterized linear silicon hydride polymer or *silane* is Si_8H_{18}. For Ge the longest *polygermane* is Ge_9H_{20}, but for Sn no polystannanes are known beyond Sn_2H_6, *distannane,* and Pb forms only the PbH_4 monomer. We have already seen, however, in Table 5.2 and the associated discussion that extended chains of Sn atoms are possible if the chain is stabilized by CH_3 groups on each Sn atom. The same is true for Si and Ge; not only is it possible to cleanly prepare $Me(SiMe_2)_{22}Me$, but if the protective organic bulk is increased still further, as with alternating methyl and hexyl groups, Si chain polymers up to molecular weights of 830,000 can be prepared, corresponding to roughly 6000 Si atoms in a chain. Even lead can be induced to form three-atom chains by providing organic protective groups, as in the compound $Pb[Pb(C_6H_5)_3]_4$.

Silane (or monosilane), SiH_4, was originally prepared by adding magnesium silicide (formed by reducing SiO_2 with Mg metal) to aqueous acid, a process that also produced most of the known polysilanes in decreasing quantity with increasing chain length. However, a more convenient preparation involves $LiAlH_4$ acting on either finely divided dry SiO_2 at about $170°C$ or $SiCl_4$ in ether solutions:

$$2SiO_2 + 2LiBH_4 \rightarrow 2SiH_4 + Li_2O + Al_2O_3$$
$$SiCl_4 + LiAlH_4 \rightarrow SiH_4 + LiCl + AlCl_3$$

Silane and disilane are gases at room temperature, but all higher polysilanes are liquids that decompose slowly into solid polymeric silicon hydrides and H_2. The silanes are much less stable than the corresponding alkanes; indeed, the thermal decomposition of SiH_4 is used to provide controlled deposition of pure silicon at semiconductor junctions:

$$SiH_4 \xrightarrow{500°C} Si° + 2H_2$$

The silanes spontaneously burst into flames when exposed to air, are immediately hydrolyzed by water, and even decompose in methanol:

$$SiH_4 + 2O_2 \rightarrow SiO_2 + 2H_2O$$
$$SiH_4 + 4ROH \rightarrow Si(OR)_4 + 4H_2$$

The vigor, even violence, of these and other similar reactions contrasts sharply with the kinetic inertness of methane. Presumably in either case the reaction intermediate would be a five-coordinate C or Si, necessarily involving the participation of d orbitals in any model involving localized bonds. A number of Si compounds are known in which the Si atom has not only five, but even six directional bonds: The SiF_6^{2-} ion is an obvious case, but Figure 5.11 shows two additional such species; there is obviously no four-orbital limitation on the Si atom. Since C has no valence d orbitals but Si does, it is not surprising that the reaction intermediate is energetically more accessible for silane and the rate correspondingly more rapid.

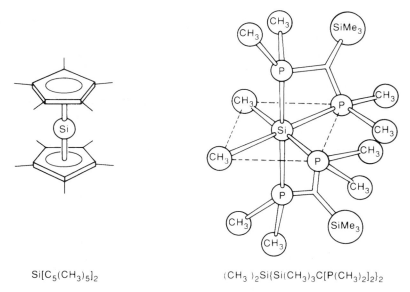

$$Si[C_5(CH_3)_5]_2 \qquad\qquad (CH_3)_2Si\{Si(CH_3)_3C[P(CH_3)_2]_2\}_2$$

Figure 5.11   Unusual coordination numbers for Si in organosilicon compounds.

The comparison between methane and the other first-row molecular hydrides is instructive in a different sense. Molecular hydrides XH_n have, by definition, covalent bonding and are held in a solid lattice only by van der Waals forces; they are therefore all gases or volatile liquids at room temperature. Of the first-row elements, lithium is sufficiently electropositive relative to hydrogen that its hydride is strongly polar and is bound in a lattice by significantly ionic attraction. Therefore, LiH cannot be vaporized without decomposition. BeH_2 is also a solid, though it is not thought to be ionic. It is so electron deficient that it forms a cluster extending over the entire crystal, including hydrogen bridging bonds analogous to those of the boranes. The nature of BH_3 and the boranes has already been described, and CH_4 is the natural model of a small molecule with all possible bonding MOs formed and occupied, but no nonbonding or antibonding MOs occupied.

Adding one more electron to the central atom, as in the nitrogen atom, causes the atom to have five valence electrons but only four valence orbitals. The mathematical symmetry of the LCAO model requires that an atom cannot take part in more bonding MOs than it has valence orbitals to provide for the linear combination. The model is faithful to experiment in this respect, because in the familiar compound NH_3, ammonia, nitrogen places two electrons in a nonbonding orbital and uses the other three possible bonding orbitals. Ammonia is a gas at room temperature, but the presence of a nonbonding pair of electrons and at least a slight positive charge on the H atoms makes weak hydrogen bonding possible in addition to the van der Waals forces that condense methane. This means that slightly more thermal energy will be required to boil ammonia, so its boiling point is higher than would be expected for a comparable molecule having no polarity or hydrogen-bonding property. Figure 5.12 compares the boiling points of the simple molecular hydrides against those of the corresponding isoelectronic noble gases. It can be seen that the three first-row hydrides NH_3, H_2O, and HF show very great deviations from what might be considered the periodic trends. The deviations can be interpreted directly in terms of hydrogen bonding, since the effect for water, which can form two hydrogen bonds per molecule, is approximately twice as great as that for NH_3 or HF, which can form only one hydrogen bond per molecule.

Ammonia is prepared (to the extent of some 17,000,000 tons in the United States annually) by the familiar Haber reaction:

$$N_2 + 3H_2 \rightarrow 2NH_3 \quad \Delta H^0 = -92.2 \text{ kJ/mol reaction}$$

Although the negative $\Delta H°$ favors the reaction and the equilibrium constant is favorable ($K = 6 \times 10^5$ at 298 K), even the most effective industrial catalysts require a temperature of about 400°C to promote the reaction at a satisfactory rate. At the higher temperature, the equilibrium is less favorable because entropy favors the reactants over the products and is a more important component of the free energy change at the higher temperature. However, high pressures (150–400 atm) restore the yield since $K_p \cdot P_{tot}^2 = K_c$ is the equilibrium constant written in terms of concentrations.

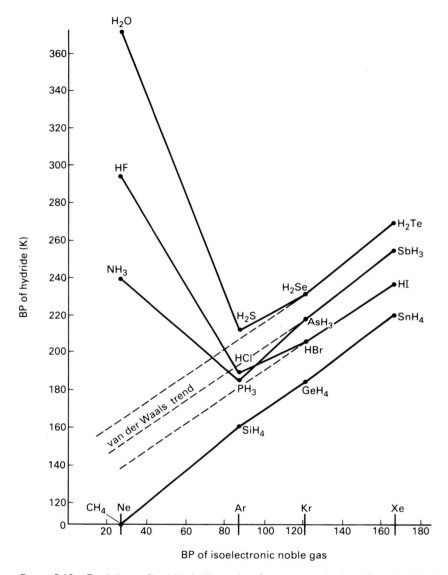

Figure 5.12 Deviations of hydride boiling points from expected values if van der Waals forces were only lattice binding force.

Aqueous ammonia, which is usually called ammonium hydroxide even though little of the NH_3 has hydrolyzed to NH_4^+ and OH^-, is one of the most familiar laboratory reagents. The fact that N is less electronegative than O makes the pair of nonbonding electrons on NH_3 more readily available for bonding to electron-acceptor species than are the nonbonding electron pairs on H_2O. As Chapter 6 will indicate in more detail, this makes NH_3 a base in water solution. Its polarity and hydrogen-bonding capability make NH_3, as the anhydrous liquid, a good solvent for ions and polar molecules for the same reasons water is. Liquid ammonia is the preferred solvent for many organic reactions in which charged, strongly basic species (usually carbanions) are to be made.

Water, of course, is not prepared but simply purified. Ordinary distillation reduces impurities to about 1 ppm; 20 ppb is readily attainable by deionization, along with activated-carbon removal of dissolved organics, followed by reverse osmosis to still further remove any remaining dissolved ions and organics. Pure water in a condensed phase has a uniquely great ability to form three-dimensional hydrogen-bond networks, because the number of positively charged hydrogen atoms equals the number of nonbonding electron pairs (although anhydrous H_2SO_4 has the same proton count, it has many more potential donor electron pairs). The resulting extensive hydrogen bonding is responsible for the open crystal structure of ice I (the common form—eight other lattices are stable at various combinations of low temperature and high pressure) and for the relatively ordered structure of liquid water at and below room temperature. The ice I crystal structure is that of diamond (Fig. 2.13), with O substituted for C and asymmetric hydrogen bonds ($O—H \cdots O$) substituted for $C—C$ bonds. There is evidence to suggest that this ordered structure persists in liquid water, though disrupted often enough by disordered interstitial water molecules that no long-range order can exist, and with the regions of order constantly shifting as disordered individual water molecules are attached to the lattice and others break away.

Water is a powerful solvent for ionic or strongly polar substances. This subject will be explored in detail in Chapter 6. Here we might note that its prevalence in our surroundings has made it the reference solvent for most inorganic solubility studies and the reference proton-transfer material for Brønsted acid–base models. The nonbonding electrons on H_2O are readily donated, and many electron-deficient inorganic species are destroyed in water solution because they accept electrons from the solvent molecules.

Hydrogen fluoride, HF, is quite similar to ammonia and water in many of its properties, as Table 5.4 suggests. It is a less familiar laboratory reagent or solvent because it is highly toxic and causes severe flesh burns, besides being corrosive. It is prepared by heating CaF_2 (fluorite or fluorspar) in sulfuric acid and trapping the gaseous product in anhydrous sulfuric acid for redistillation:

$$CaF_2 + H_2SO_4 \rightarrow 2HF + CaSO_4$$

Because of the increased electronegativity of F over O or N, the nonbonding electrons on the HF fluorine atom are less readily available for donation than those from H_2O or NH_3, and HF is acidic rather than basic in water solution. It is a relatively weak acid in water ($K_a = 6.46 \times 10^{-4}$) but extremely acidic as a pure liquid solvent. The difference is presumably due to the differing extent to which hydrogen bonding is possible in H_2O and HF liquids.

Hydrogen bonding is extremely important to any discussion of the properties of HF. In the solid phase, HF forms linear polymers in which succeeding F atoms

Table 5.4 Physical Properties of NH_3, H_2O, and HF

Property	NH_3	H_2O	HF
Solid			
Melting point (°C)	−77.74	0	−83.55
Density, solid (g/cm^3 at mp)	0.817	0.91671	1.653
$\Delta H^0{}_{fus}$ (J/mol at mp)	5,655	6,008	3,929
K_f, cryoscopic const. (deg · kg solvent/mol solute)	0.9567	1.86	1.55
Liquid			
Boiling point (°C)	−33.42	100	19.51
Density, liquid (g/cm^3 at bp)	0.682	0.9584	0.952
$\Delta H^0{}_{vap}$ (kJ/mol at bp)	23.35	40.66	7.49
Viscosity (cp)	0.254 (−33°)	1.0019 (20°)	0.256 (0°)
Surface tension (erg/cm^2 at mp + 25°C)	38.4	71.97	14.0
Dielectric const. (at mp)	23	87.74	175
Autoprotolysis K_{eq}	~10^{-27}	$1.008 \cdot 10^{-14}$	$\leq 2 \cdot 10^{-12}$
K_b, ebullioscopic const. (deg kg solvent/mol solute)	0.3487	0.512	1.9
Gas			
ΔH^0_f (298 K) (kJ/mol)	−46.11	−241.82	−271.5
S^0 (298 K) (J/mol · deg)	192.3	188.72	173.6
Dipole moment (debye)	1.46	1.84	1.83

are held together by F—H···F hydrogen bonds, with F—F—F angles of 120°C. Both the liquid and the vapor seem to be composed predominantly of H_6F_6 cyclic hexamers, and the rather large decrease in density of HF on melting (see Table 5.4) is presumably due to this structural rearrangement. In Chapter 6 we shall examine the effects of this hydrogen bonding on the chemical properties of HF.

Nitrogen and oxygen—but not fluorine—form catenated hydrides. Hydrazine, N_2H_4, and hydrogen peroxide, H_2O_2, are both manufactured on a fairly large scale. Hydrazine is made by oxidizing ammonia by aqueous hypochlorite,

$$2NH_3 + OCl^- \rightarrow N_2H_4 + Cl^- + H_2O$$

a process in which the initial product is chloramine, $ClNH_2$. Hydrogen peroxide is manufactured predominantly by the electrochemical oxidation of HSO_4^- and hydrolysis of the peroxydisulfate formed:

$$2HSO_4^- + 2H_2O \rightarrow O_3SOOSO_3^{2-} + 2H_3O^+ + 2e^-$$

$$S_2O_8^{2-} + 2H_2O \rightarrow 2HSO_4^- + H_2O_2$$

The N_2H_4 and H_2O_2 molecules form an interesting structural comparison. Figure 5.13 shows the possible conformers of the two; both are apparently stablest in the *gauche* form. Rotation barriers through the *trans* position are 15.9 kJ/mol for N_2H_4 and 4.60 kJ/mol for H_2O_2, and through the *cis* or eclipsed position are 49.7 kJ/mol for N_2H_4 and 29.3 kJ/mol for H_2O_2. These numbers reflect the modest repulsion of the nonbonding electrons for a neighbor bond pair and the significantly greater repulsion of the nonbonding pairs for each other in the *cis* arrangement.

Hydrazine and hydrogen peroxide also have parallel acid–base properties. If we think of them as ammonia and water with a hydrogen replaced by —NH$_2$ and —OH respectively, it is clear that in each case the added electronegative element will withdraw electrons from the rest of the molecule and make them less available for donation. Accordingly, N$_2$H$_4$ should be less basic than ammonia, and H$_2$O$_2$ should be less basic (more acidic) than water: $K_b = 1.8 \times 10^{-5}$ for NH$_3$, but $K_b = 8.5 \times 10^{-7}$ for N$_2$H$_4$; $K_a = 1.0 \times 10^{-14}$ for H$_2$O, but $K_a = 1.4 \times 10^{-12}$ for H$_2$O$_2$. Like most hydrides, N$_2$H$_4$ is a strong reducing agent and H$_2$O$_2$ at least a weak one, though the extreme electronegativity of the oxygen atom makes it a better electron acceptor and oxidizing agent. By contrast, ammonia and water have very weak reducing properties except at high temperatures, apparently for kinetic reasons.

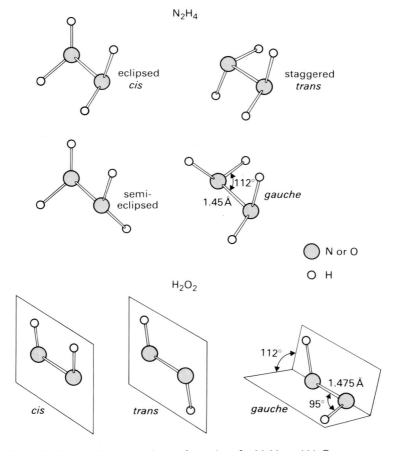

Figure 5.13 Possible molecular conformations for N$_2$H$_4$ and H$_2$O$_2$.

The main-group elements in groups III–VII all form molecular hydrides, although the properties of the hydrides differ and all of the heavier elements' hydrides are unstable or unknown. Table 5.5 indicates preparation methods for these hydrides. Those not listed cannot be prepared in a condensed phase or even at significant pressures as a gas. Aluminum hydride (alane) is a white polymeric solid with the AlF_3 lattice that cannot be heated to provide monomeric AlH_3 vapor without complete decomposition, although ion–molecule reactions similar to those shown for the noble gases in Table 5.5 have yielded very low pressures of the monomer. Similarly, GaH_3 (gallane) is stable only below $-30°C$, as a dimer with the diborane structure mixed with an uncharacterized polymer; at room temperature gallane decomposes into the elements Ga and H_2. These hydrides are vigorous reducing agents; some organic reductions are done with AlH_3 or the milder reducing agents $RAlH_2$ and R_2AlH. All other hydrides in Table 5.5 are gases at room temperature, as Fig. 5.12 has indicated. The weak intermolecular attractions implied by this property are the result of the low electron density at the surface of the hydride molecules and the consequent weak induced-dipole/induced-dipole force between the molecules. Although some of the molecular hydrides are moderately polar, this does not seem to make a large contribution to the total intermolecular attraction. Likewise, hydrogen bonding is usually insignificant for any of the hydrides except possibly HCl.

5.5 ORGANOMETALLOIDS

In a previous section we have already mentioned organometallic compounds of the main-group elements from groups Ia through IVa. Since hydrogen and carbon have very similar electronegativities, we can extend the discussion of molecular hydrides to include the organometallic compounds of the metalloid elements from groups IV through VI. Of course, most of the elements in these groups are not metals, but the term "organometallic" is fairly generally used to refer to organo-substituted compounds of at least the heavier members.

The most important single category of such compounds is that of the *silicones,* which are polymers of alternating silicon and oxygen atoms, with alkyl or aryl groups on each silicon: —O—SiR_2—O—SiR_2—O—. Such a structure meets the valence-electron/valence-orbital requirements mentioned earlier for catenated systems, but isolates the O atoms to prevent the facile elimination of O_2. Silicones are prepared by the hydrolysis and subsequent condensation of organosilicon chlorides, which are themselves prepared by adding organolithium or Grignard-reagent to $SiCl_4$ or by directly combining silicon with alkyl chlorides at elevated temperatures:

$$2CH_3Li + SiCl_4 \xrightarrow{\text{Ether}} (CH_3)_2SiCl_2 + 2LiCl$$

$$2CH_3Cl + Si^0 \xrightarrow{300°C} (CH_3)_2SiCl_2 \quad (70\% \text{ plus other } (CH_3)_nSiCl_{4-n})$$

The chlorides hydrolyze readily:

$$(CH_3)_2SiCl_2 + 2H_2O \rightarrow (CH_3)_2Si(OH)_2 + 2HCl$$

Table 5.5 Preparative Reactions for Molecular Hydrides

Group III	Group IV	Group V
$3\text{LiH} + \text{AlCl}_3 \xrightarrow{\text{Ether}} 3\text{LiCl} + \text{AlH}_3$	$\text{GeO}_2 + \text{NaBH}_4 \xrightarrow{1M \text{ HBr}} \text{GeH}_4 + \text{H}_2 + \text{NaBr}$	$\text{Ca}_3\text{P}_2 + \text{H}_2\text{O} \longrightarrow \text{PH}_3 + \text{Ca}^{2+} + \text{OH}^-$
$\text{GaCl}_3 + 4\text{LiH} \longrightarrow \text{LiGaH}_4 + 3\text{LiCl}$	$\text{LiAlH}_4 + \text{SnCl}_4 \longrightarrow \text{SnH}_4 + \text{LiCl} + \text{AlCl}_3$	$\text{As}_2\text{O}_3 + \text{BH}_4^- \xrightarrow{\text{OH}^-} \text{AsH}_3 + \text{B(OH)}_4^-$
$2\text{LiGaH}_4 + (\text{H}_2\text{GaCl})_2 \longrightarrow \frac{3}{n}(\text{GaH}_3)_n + 2\text{LiCl}$	$\text{Pb} \cdot \text{Mg} + \text{H}_3\text{O}^+ \longrightarrow \text{Mg}^{2+} + \text{PbH}_4$ Alloy	$\left.\begin{array}{l}\text{SbCl}_3 \\ \text{BiCl}_3\end{array}\right\} + \text{LiAlH}_4 \xrightarrow{90°} \left\{\begin{array}{l}\text{SbH}_3 \\ \text{BiH}_3\end{array}\right. + \text{LiCl} + \text{AlCl}_3$

Group VI	Group VII	Group O
$\text{Na}_2\text{S} + n\text{S} \longrightarrow \text{Na}_2\text{S}_n \xrightarrow{\text{H}_3\text{O}^+} \text{H}_2\text{S}_n$	$\text{NaCl} + \text{H}_2\text{SO}_4 \longrightarrow \text{HCl} + \text{NaHSO}_4$	$\left.\begin{array}{l}\text{Ar}^+ \\ \text{Kr}^+ \\ \text{Xe}^+\end{array}\right\} + \text{H}_2 \xrightarrow[\text{spectrometer}]{\text{Mass}} \left\{\begin{array}{l}\text{ArH}^+ \\ \text{KrH}^+ + \text{H} \\ \text{XeH}^+\end{array}\right.$
$\text{FeS} + \text{H}_3\text{O}^+ \longrightarrow \text{H}_2\text{S} + \text{Fe}^{2+}$	$\left.\begin{array}{l}\text{NaBr} \\ \text{NaI}\end{array}\right\} + \text{H}_3\text{PO}_4 \longrightarrow \left\{\begin{array}{l}\text{HBr} \\ \text{HI}\end{array}\right. + \text{NaH}_2\text{PO}_4$	No neutral hydrides
$\text{H}_2 + \text{Te} \xrightarrow{650°C} \text{H}_2\text{Te}$	$3\text{I}_2 + 2\text{P} + 6\text{H}_2\text{O} \longrightarrow 6\text{HI} + 2\text{H}_3\text{PO}_3$	

This product, a silanol, is a solid that condenses with itself on gentle heating to yield a mixture of a cyclic trimer (sometimes known as D_3) and tetramer (D_4), along with linear polymers:

$$3(CH_3)_2Si(OH)_2 \longrightarrow \qquad \text{(for the trimer)}$$

This product is a siloxane. If it (or the tetramer) is mixed with a small quantity of the trimethylsiloxane $(CH_3)_3SiOSi(CH_3)_3$ and shaken with H_2SO_4, the Si—O bonds rearrange into linear polymers in which the end groups are $(CH_3)_3Si$— and the average molecular weight is determined by the proportions of trimethyl and dimethyl siloxanes in the starting mixture. The linear polymers are still siloxanes, but the mixture is usually called silicone oil. (The name is misleading, because no "silanone" $R_2Si{=}O$ ketone-like double bond has ever been observed. Only sigma bonding seems to have been demonstrated for these two aroms.)

As Chapter 4 has noted, second-row atoms and third-row atoms do not readily form double bonds involving pi overlap. However, the systems for which these bonds have been demonstrated are almost always organometallics, with protecting bulky alkyl groups. We may digress here briefly to consider the stabilities of these double bonds. The original effort to make "silicones" was intended to yield small silanone molecules with $Si{=}O$ bonding. The observed rearrangement to oligomers and ultimately to polymers is the thermodynamic result of the weaker pi overlap that occurs for Si because of the enlarged inner core of Si relative to the first row elements, $1s^2 2s^2 2p^6$ versus $1s^2$. Even though the O atom has the smaller inner core, the necessarily greater internuclear distance for Si–O, which periodic trends in radii suggest would have to be about 0.25 Å greater than C–O, reduces the pi overlap so much that rearrangement to a tetrahedral network of sigma bonds is overwhelmingly favorable. However, that rearrangement must be accomplished through some reaction mechanism in which a fourth atom forms a bond to Si (or to another second- or third-row element). For second- and third-row elements E, bulky protecting groups can prevent such attack and stabilize an $E{=}E$ or $E{=}X$ bond.

Photolysis of $Me_3Si{-}Si(Ms)_2{-}SiMe_3$, where *Ms* is mesityl (2,4,6-trimethylphenyl), yielded the first $Si{=}Si$ bond in 1984:

$$Me_3SiSi(Ms)_2SiMe_3 \xrightarrow{h\nu} Ms_2Si{=}SiMs_2 + Me_3SiSiMe_3$$

This product was the first *silene;* it is obviously sterically protected by the very bulky mesityl groups. If the methyl groups on the protecting mesityl group are replaced by the still bulkier *t*-butyl groups (2, 4, 6-$(Me_3C)_3C_6H_2$ or *BPh*), the P=P bond can be stabilized as BPh–P=P–BPh. In a similar fashion the luxuriant foliage of the tris(trimethylsilyl)methyl group, $(Me_3Si)_3C$, can stabilize the third-row As=As bond: $(Me_3Si)_3C$–As=As–$C(SiMe_3)_3$. Similarly, $[(Me_3Si)_2CH]_2$ Sn=Sn$[CH(SiMe_3)_2]_2$ has been prepared. Heteronuclear double bonds are also possible: $[(Me_3Si)_2CH]_2Sn$=P–BPh. Obviously, all of these E=E species have been protected by bulky organic groups; even though the Si=N double bond has been stabilized in $(tBu)_2Si$=N–$Si(tBu)_3$, the totally unprotected nature of the O atom in Si=O makes it unlikely that it can be stabilized short of an almost encapsulating group on the Si. The Si=O group has been characterized in very short-lived species trapped on solid Ar at low temperatures, a technique called *matrix isolation*. But we have come a long way since the original quest for a "silicone."

If very pure dimethylsiloxanes are used in the preparation of what we now know to be silicone polymers, with no trimethylsilyl endgroups, the molecular weight can become very high—up to perhaps 10^7—and the product is an elastomer gum rather than an oil. Dimensional stability is provided by adding very finely divided SiO_2 filler to the polymerizing mixture; and vulcanization or cross-linking is achieved by adding benzoyl peroxide to the reaction mixture and heating the silicone when it is molded as a product. Such silicone rubbers have high tensile strengths and, like the silicone oils, have unusually great thermal stability and undergo unusually small thermal changes in their mechanical properties.

For silicon and the rest of the group IV elements, the formation of bonds to carbon is accompanied by very little electron transfer. The bonding is best thought of as involving sp^3 hybrid orbitals on the central atom, and the stoichiometry is almost always MR_4 as a result. Although the +2 oxidation state is increasingly stable going down the group from Si to Pb, very few MR_2 organometallic compounds are known for any of the elements because further M—C bond formation is a highly exothermic process (Si—CH_3: 314 kJ/mol bond; Ge—CH_3: 247 kJ/mol bond; Sn—C: 192 kJ/mol bond; Pb—C: 130 kJ/mol bond).

The group V elements, with five valence electrons, might be expected to form MR_5 organosubstituted compounds. Indeed, some are known, particularly where R is an aryl rather than an alkyl group. However, they require the use of five valence orbitals on the central atom if localized bonding orbitals are assumed, which makes them impossible for N with only the $2s$ and $2p$ valence orbitals. Their existence for the heavier elements (for instance, pentaphenylphosphorus, pentamethylantimony) appears to require that the $3d, 4d, \ldots$ orbitals participate in the bonding. Since most such compounds are trigonal bipyramidal, the appropriate hybrid orbitals are dsp^3. An attractive alternative (particularly for the lighter elements having d orbitals at higher energies) is to use only the tetrahedral sp^3 hybrids and retain a nonbonding pair of electrons in one hybrid. The resulting MR_3 compounds are much more common than the MR_5 stoichiometry, and since none of the atoms involved is strongly electronegative, the nonbonding electron pair is readily donated as in the ammonia molecule or the comparable amines. For examples, trialkylphosphines and arsines react readily with alkyl halides to form quaternary phosphonium and arsonium salts:

$$(CH_3)_3As + CH_3Br \rightarrow (CH_3)_4As^+Br^-$$

The use of d orbitals for bonding in P and the heavier elements makes phosphine and arsine oxides, R_3PO and R_3AsO, much more stable than the corresponding amine oxides R_3NO. Phosphorus and arsenic can form a partial double bond with the oxygen through d–p pi overlap, while no such stabilization is possible for the N—O bond if the nitrogen is already four-coordinate.

Similarly, in group VI, essentially all organosubstituted compounds are MR_2. The S, Se, or Te atoms use sp^3 hybrid orbitals, two of which are filled by non-bonding pairs of electrons. Only a very few MR_4 compounds are known, all of which involve Te, and no MR_6 compounds have been reported. Localized bonds for the TeR_4 molecules require the use of $5d$ orbitals by Te: There must be four bonding MOs and one nonbonding MO to accommodate the remaining pair of electrons; five valence orbitals are thus required. This is energetically more favorable for the heavier elements, because the energy gap between $n\,s$, $n\,p$, and $n\,d$ orbitals decreases as n increases.

Some preparative methods and typical reactions for group V and VI organometallic compounds are shown in Table 5.6.

Table 5.6   Preparation and Typical Reactions of Group V and VI Organometallics

Group V	Group VI
Preparations	
$PBr_3 + 3RMgBr \longrightarrow R_3P + 3MgBr_2$	$Na_2Se + RBr \longrightarrow R_2Se + 2NaBr$
$BiCl_3 + Al_2(C_2H_5)_6 \longrightarrow 2Bi(C_2H_5)_3 + 2AlCl_3$	$Hg\phi_2 + Se \longrightarrow \phi_2Se + Hg$
$AsCl_3 + 3\phi Cl + 3Na \longrightarrow \phi_3As + 3NaCl$	$\phi_2TeCl_2 + 2\phi Li \longrightarrow \phi_4Te + 2LiCl$
$SbCl_5 + 5\phi Li \longrightarrow \phi_5Sb + 5LiCl$	$Se + C_2H_2 \xrightarrow{350°C}$ (ring structure with Se)
Reactions	
$\phi_3P + CH_3Br \longrightarrow \phi_3PCH_3^+Br^- \xrightarrow{BuLi}$	$R_2Se + Cl_2 \longrightarrow RSeCl + RCl$
$\phi_3P=CH_2$	$\downarrow H_2O$
$\phi_3P=O + R_2C=CH_2 \longleftarrow \overset{R_2C=O}{\rule{0pt}{0pt}}$	$RSeCl_3 \xleftarrow{HCl} RSeOOH$
$\phi AsI_2 \xrightarrow{Hg} \phi IAs—AsI\phi \longrightarrow (\phi As)_5 \text{(cyclic)}$	$R_2Se + R'I \longrightarrow R_2R'Se^+I^-$
$\phi_5Sb + \phi Li \longrightarrow Li^+Sb\phi_6^-$	
$\phi_5Sb + 2Cl_2 \longrightarrow \phi_3SbCl_2 + 2\phi Cl$	

5.6 METALLOID CLUSTERS

An unusual category of covalent metalloid compounds is the group of clustered metalloid atoms E_n, for which we have already laid a theoretical basis in Chapter 4. Many of these are homonuclear clusters (of only one element) and carry a net charge; these are the Zintl ions that we introduced earlier.

As Chapter 8 will develop in more detail, the alkali metals will dissolve in liquid ammonia yielding what may be termed ammoniated electrons—a very strongly reducing environment. In 1891 Johannis dissolved metallic lead in such a solution and obtained a solid compound with the approximate empirical formula $NaPb_2$. Over some 40 years occasional investigation refined this formula and explored other systems, yielding formulas Pb_9^{4-}, Sb_7^{3-}, and Te_4^{2-} for some of the electrolyte solution species. Zintl, beginning in 1931, investigated many of these solution species (subsequently known as Zintl ions) and also numerous similar solid intermetallic compounds (Zintl phases). On electrochemical grounds he suggested the solution species Sn_9^{4-}, Pb_7^{4-}, Pb_9^{4-}, As_3^{3-}, As_5^{3-}, As_7^{3-}, Sb_3^{3-}, Sb_7^{3-}, Bi_3^{3-}, Bi_5^{3-}, Se_2^{2-}, Se_3^{2-}, Se_4^{2-}, Se_6^{2-}, Te_2^{2-}, Te_3^{2-}, and Te_4^{2-}. The solutions are brilliantly colored: Pb_9^{4-} is green, As_3^{3-} is yellow, and Bi_3^{3-} is deep violet. Many of these were obtained by dissolving binary alloys such as $NaPb_{2.25}$, or Na_4Pb_9, into liquid ammonia. However, subsequent x-ray crystallography studies have shown that the solid Zintl phases rarely contain the same cluster structures as the dissolved Zintl ions.

The Zintl ions themselves have been difficult to secure in crystalline form from their solutions. Essentially the problem has been that the sodium ion was not large enough to isolate the cluster ions from each other in a lattice. However, the introduction of the macrocyclic *crypt* ligands to coordinate the alkali metal cation in the early 1970s allowed crystallization of a number of Zintl anions from solution, and reliable x-ray structures are now known for Ge_9^{2-}, Ge_9^{4-}, Sn_9^{3-}, Sn_9^{4-}, Sn_5^{2-}, Pb_5^{2-}, Sb_4^{2-}, Bi_4^{2-}, As_{11}^{3-}, and Sb_7^{3-}. Comparable cations are also known: Bi_5^{3+}, Bi_8^{2+}, Bi_9^{5+}, Se_4^{2+}, Se_8^{2+}, Se_{10}^{2+}, Te_4^{2+}, Te_6^{2+}, and Te_6^{4+}, usually isolated with large anions such as $AlCl_4^-$. In general these ions have geometric structures consistent with Wade's rules for cluster electron count, though those rules fail for a few of the nine-atom clusters.

There are also heteronuclear Zintl ions, with comparable geometric and electronic structures in most cases. Structures are known for $Sn_2Bi_2^{2-}$, $Pb_2Sb_2^{2-}$, $Tl_2Te_2^{2-}$, $TlSn_8^{3-}$, $TlSn_9^{3-}$, $As_2Se_6^{3-}$, $As_2Te_6^{3-}$, $Sn_2Te_6^{4-}$, $Hg_4Te_{12}^{4-}$, and some others—particularly tellurides—that have more or less classical electron-pair bonding rather than the delocalized form seen in the cluster MOs. Figure 5.14 shows a few of these cluster structures with their electron counts; it is interesting to attempt to fit these to the skeletal electron count rules.

As we have noted, in addition to the molecular Zintl ions there are also solid Zintl phases, which are usually taken to be binary solids involving an alkali or alkaline-earth element and a group IV, V, or VI element, particularly if clustering

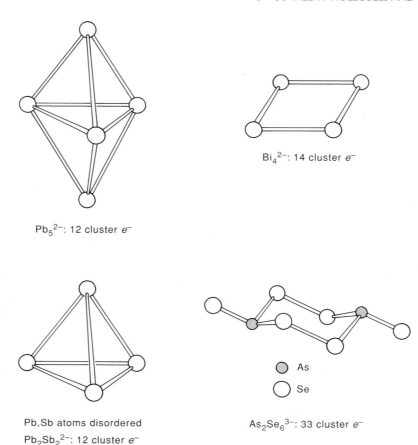

Pb_5^{2-}: 12 cluster e^-

Bi_4^{2-}: 14 cluster e^-

Pb,Sb atoms disordered
$Pb_2Sb_2^{2-}$: 12 cluster e^-

As

Se

$As_2Se_6^{3-}$: 33 cluster e^-

Figure 5.14   Structures of some homonuclear and heteronuclear Zintl anions.

of the more electronegative element occurs in the lattice. Some of these seem to yield to conventional localized electron-pair bond models, such as Ba_4GeAs_4, in which the $GeAs_4$ unit as $GeAs_4^{8-}$ is formally equivalent to the orthosilicate ion. Similarly, in $CaGa_2$ an assumed Ga_2^{2-} is electronically equivalent to a neutral group IV system, and the system displays the graphite lattice structure. On the other hand, Li_9Ge_4 shows Ge_2 dimers, $Li_{12}Si_7$ has planar Si_5 rings and trigonal planar Si_4 stars, KGa_3 has Ga_8 dodecahedra bridged into a three-dimensional network by single Ga atoms, and $Na_{22}Ga_{39}$ has Ga_{12} icosahedra bonded to each other directly and by single Ga atoms. All of the Zintl phases have nonmetallic properties in spite of their constituent elements; they are brittle, have fixed stoichiometry, are often brightly colored, and are usually semiconductors. An interesting parallel exists between the Zintl phase clusters of main-group metals or metalloids and a comparable set of transition-metal compounds known as Chevrel phases, featuring Mo_6S_8 clusters; we will consider these in Chapter 13.

_____ *5.7 NONMETAL OXIDES*

All elements except the lighter noble gases form oxides. Although oxygen is extremely electronegative and substantial charge transfer occurs in most cases, the oxides of the nonmetals have structures determined by the presence of extensive directional covalent bonding. We shall consider them here in the context of their covalent bonding. These oxides are found in three basic structural forms: discrete monomeric molecules, small finite polymers, and extended polymeric lattices. Table 5.7 shows the distribution of these forms across the nonmetals in the periodic table (some of the less-stable or less-well-characterized oxides are not shown).

It can be seen from Table 5.7 that isolated monomeric molecules of the nonmetal oxides are largely limited to the halogens and the top row of the periodic table. The difference in physical properties between compounds with the same stoichiometric formula can be as dramatic as the contrast between gaseous CO_2 and quartz SiO_2. The atomic property responsible for this difference is the possibility of pi bonding between the central atom and the surrounding oxygens, coupled with the greater M—O electronegativity difference for the polymeric systems (compare Table 5.7 with Fig. 1.17). The elements C, N, and O can all form strong pi bonds with each other as a result of their partially filled pi-symmetry $2p$ orbitals and the small inner core that allows the atoms to approach closely for good pi overlap. As a result, the structures shown in Fig. 5.15 all seem to involve significant pi bonding (except for OF_2 and Cl_2O) and represent a greater stabilization of the MO_n systems than an extended sigma-bonded lattice could provide. By contrast, the larger inner cores of Si, As, Se, and the other heavier atoms require internuclear distances so great that pi overlap is poor and M=O double bonding is rarely if ever observed. As has already been mentioned, the use of vacant $n\,d$ valence orbitals by elements with $n\,s$ and $n\,p$ valence electrons can be important in some compounds, but if those orbitals are vacant, they will be too diffuse to have good pi overlap with the compact O $2p$. The increased electronegativity of Cl (that is, its increased effective nuclear charge) shrinks the Cl $3d$ orbitals enough to make pi overlap possible with the oxygen $2p$ orbitals. The chlorine oxides show spectroscopic evidence of pi bonding, but this is not the case for bromine or iodine oxides, presumably because their $4d$ and $5d$ orbitals are too big.

Common preparative reactions for the nonmetal oxides of Table 5.7 are given in Table 5.8. In general, laboratory syntheses intended to produce small quantities have been given, but a few significant industrial processes are also included. Of course, a number of the gaseous species such as O_2 and CO_2 occur naturally in the atmosphere in large quantities and can be obtained by cryogenic separation. Because of the oxidizing nature of the earth's atmosphere, several

Table 5.7 Nonmetal Oxides

B	C	N		O	F
⟨B_2O_3⟩	CO	N_2O	NO_2	O_2	OF_2
	CO_2	NO	N_2O_4	O_3	O_2F_2
		N_2O_3	⟨N_2O_5⟩		

Si	P	S	Cl	
⟨SiO_2⟩	[P_2O_3]$_2$?	SO_2	Cl_2O	Cl_2O_6
	⟨[P_2O_5]⟩$_2$	⟨[SO_3]⟩$_3$*	Cl_2O_2	Cl_2O_7
			ClO_2	

Ge	As	Se	Br
GeO?	⟨[As_2O_3]⟩$_2$	⟨SeO_2⟩	[Br_2O]
⟨GeO_2⟩	⟨As_2O_5⟩?	[SeO_3]$_4$	[BrO_2]$_2$

Sn	Sb	Te	I	Xe
⟨SnO⟩	⟨[Sb_2O_3]⟩$_2$	⟨TeO_2⟩	⟨I_2O_4⟩?	XeO_3
⟨SnO_2⟩	⟨Sb_2O_5⟩	⟨TeO_3⟩	⟨I_2O_5⟩	XeO_4

Pb	Bi
⟨PbO⟩	⟨Bi_2O_3⟩
⟨Pb_3O_4⟩	
⟨PbO_2⟩	

Key: No border = monomer   ⟨◯⟩ = extended polymeric lattice

[   ]$_n$ = small polymer, *n*-fold   * = monomer also known

? = structure not definitively established

other nonmetal oxides are produced in industrial and fuel-based reactions in quantities great enough to be a major focus of concern over air pollution: CO, NO, NO_2, SO_2, SO_3, O_3. We shall briefly examine their roles in atmospheric chemistry after first considering the structural characteristics of the nonmetal oxides in general.

The molecular oxides of Fig. 5.15 have structures that are generally consistent with simple electron-pair repulsion theory. However, some of the bond lengths and implied bond orders are unusual and have been the subject of theoretical controversy. In one of the more settled areas, all MO treatments of CO agree that the bond order in that molecule is 3.0; the C—O bond length is only 1.128 Å compared to 1.163 Å for CO_2. Since CO is isoelectronic with N_2 and CN^-, the triple bond is not in any sense unusual.

When molecular oxides are studied, the ready possibility of pi bonding sometimes makes it difficult to establish what constitutes a single sigma bond and what its length should be. This is particularly true for the nitrogen oxides, several of which contain N—N bonds. Presumably the N—N bond in hydrazine (H_2N—NH_2), which is 1.45 Å long, constitutes a single bond. However, the N—N bond in N_2O_3 ($ONNO_2$) is 1.86 Å and that in N_2O_4 (O_2NNO_2) is 1.75 Å, both of which are much longer than the hydrazine bond and thus are presumably weaker. N_2O_4 is isoelectronic with the $C_2O_4^{2-}$ oxalate ion, which also shows a central bond slightly longer than a standard C—C single bond, but the difference is much smaller (0.03 Å) in the oxalate ion. Some MO calculations have suggested that there is a net pi anti-bonding overlap in the N—N region, which would be consistent with experiment since it leads to a bond order less than 1.0.

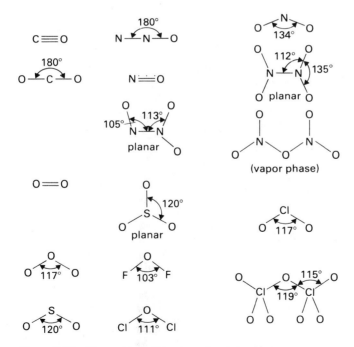

Figure 5.15 Geometries of simple molecular oxides.

Table 5.8 Preparation of Nonmetal Oxides

B and Group IVa

$H_3BO_3 \xrightarrow{\text{Heat}} B_2O_3$ (normally glass)

$C + H_2O \rightleftharpoons CO + H_2$ (water gas)

$CaCO_3 + 2H_3O^+ \longrightarrow Ca^{2+} + CO_2 + 3H_2O$

$SiCl_4 + 2H_2O \longrightarrow SiO_2 + 4HCl$

$Ge + GeO_2 \longrightarrow 2GeO$

$GeCl_4 + 2H_2O \longrightarrow GeO_2 + 4HCl$

$SnCl_4 \cdot 2H_2O + OH^- \longrightarrow SnO \cdot nH_2O \xrightarrow{\text{Heat}} SnO$

$Pb^{2+} + 2NH_3 + H_2O \longrightarrow PbO + 2NH_4^+$

$Pb^{2+} + OCl^- + 3H_2O \longrightarrow PbO_2 + Cl^- + 2H_3O$

$PbCO_3 \xrightarrow{\text{Heat, }O_2} Pb_3O_4 + CO_2$

Group Va

$NH_4NO_3 \xrightarrow{250°C} N_2O + 2H_2O$

$2NO_2^- + 2I^- + 4H_3O^+ \longrightarrow 2NO + I_2 + 6H_2O$

(Industrial: $4NH_3 + 5O_2 \xrightarrow{\text{Pt/Rh}} 4NO + 6H_2O$)

$2NO + N_2O_4 \xrightarrow{-20°C} 2N_2O_3$

$2NO + O_2 \longrightarrow 2NO_2$

$2NO_2 \xrightarrow{0°C} N_2O_4$

$4HNO_3 + P_4O_{10} \xrightarrow{-10°C} 2N_2O_5 + 4HPO_3$

$P_4 + 3O_2 \xrightarrow{\text{90 torr, 75\% }O_2} P_4O_6$

$P_4 + 5O_2 \xrightarrow{\text{Excess }O_2} P_4O_{10}$

$2MCl_3 + 3H_2O \longrightarrow M_2O_3 + 6HCl$ (M = As, Sb, Bi)

$As_2O_3 + HNO_3 \longrightarrow As_2O_5$

$2SbCl_5 + 5H_2O \longrightarrow Sb_2O_5 + 10HCl$

Group VIa

$3O_2 \xrightarrow{\text{Silent electric discharge}} 2O_3$

$\left. \begin{array}{l} \text{S} + O_2 \longrightarrow SO_2 \\ SO_2 + \tfrac{1}{2}O_2 \xrightarrow{V_2O_5} SO_3 \end{array} \right\}$ Industrial

$SO_3^{2-} + 2H_3O^+ \longrightarrow SO_2 + 3H_2O$

$Fe_2(SO_4)_3 \xrightarrow{\text{Heat}} Fe_2O_3 + 3SO_3$

$Se(Te) + 4HNO_3 \xrightarrow{\text{Evap.}} SeO_2 (TeO_2) + 4NO_2 + 2H_2O$

$K_2SeO_4 + SO_3 \xrightarrow{SO_3 \text{ solvent}} K_2SO_4 + SeO_3$

$Te(OH)_6 \xrightarrow{300°C} TeO_3 + 3H_2O$

Group VIIa

Warning: Halogen oxides are violent and unpredictable explosives!

$2F_2 + H_2O \xrightarrow{\text{Solid KF}} OF_2 + 2HF$

$2Cl_2 + 2HgO \longrightarrow HgCl_2 \cdot HgO + Cl_2O$

$2ClO_3^- + H_2C_2O_4 + 2H_3O^+ \longrightarrow 2ClO_2 + 2CO_2 + 4H_2O$

$12HClO_4 + P_4O_{10} \longrightarrow 6Cl_2O_7 + 4H_3PO_4$

$Br_2 + 4O_3 \xrightarrow{-78°C} 2BrO_2 + 4O_2$

$BrO_2 \xrightarrow{-60°C} Br_2O + Br_2O_5?$

$2HIO_3 \xrightarrow{200°C} I_2O_5 + H_2O$

$4HIO_3 \xrightarrow{H_2SO_4} 2I_2O_4 + O_2 + 2H_2O$

Presumably the situation is similar in N_2O_3, since the structures are closely related.

The nonmetal oxides that form small polymers include some interesting structures. The most familiar are the oxides of phosphorus, P_2O_3 and P_2O_5, which actually exist as dimers P_4O_6 and P_4O_{10} with the four phosphorus atoms in the same tetrahedral arrangement as in the white phosphorus P_4 structure. Figure 5.16 shows the related structures of these three phosphorus species, along with the related CaF_2 lattice. If the P_4O_6 structure is idealized to the 109° tetrahedral angle at P, it is equivalent to the CaF_2 in terms of the P(Ca) positions and the coordination, although five of the eight atoms coordinating Ca are missing. In fact, the O—P—O angle is only 99°, which tends to pull the O atoms in from the cube faces to isolate the P_4O_6 molecule. It should not be surprising, however, that in addition to As_4O_6 and Sb_4O_6, both of which have the P_4O_6 structure, there is a Bi_2O_3 extended lattice equivalent to the CaF_2 cube, but with the center and corner O(F) atoms missing. This is similar to the comparison in Chapter 3 between the ReO_3 lattice and its distorted variation $CoAs_3$. As the discussion of indium halides in that chapter suggested, there is often only a very subtle difference between discrete molecules and polymeric lattices.

The other small-polymer oxides of interest are SO_3 and SeO_3. Both form small rings, as indicated in Fig. 5.17. Although the molecules are a trimer and tetramer, respectively, the rings that form are six-membered and eight-membered. The SO_3 trimer is similar to the $Si_3O_9^{6-}$ silicate ring of Fig. 3.14, except that the terminal O atoms are not all equivalent. Three are axial, with an S—O bond length of 1.37 Å, and three are equatorial, with a bond length of 1.43 Å. The SeO_3 tetramer has a similar nonplanar structure, with a preferred chair conformation, for its Se—O—Se ring. We shall encounter six- and eight-membered rings in other systems. It is interesting to note that each monomer unit in such a ring will have 24 electrons, as in the three systems just mentioned and in the isoelectronic cyclic trimetaphosphate ion $P_3O_9^{3-}$.

The covalent oxides with extended polymeric lattices—generally the heavier elements from groups IV–VII—occur as chains, layers, and three-dimensional networks. The group IV oxides occur as three-dimensional networks of MO_4 tetrahedra (Si, Ge) or of MO_6 octahedra (SnO_2 and PbO_2 adopt the rutile structure). SnO and PbO have a layer structure of flattened MO_4 pyramids, while Pb_3O_4 has a three-dimensional network of PbO_6 octahedra linked by PbO_3 flattened pyramids. TeO_3 and probably As_2O_5 and Sb_2O_5 also form three-dimensional networks. P_2O_5 has a layer structure and a three-dimensional network of PO_4 tetrahedra in addition to its P_4O_{10} dimer. As_2O_3 and TeO_2 have layers of distorted MO_4 units, and Sb_2O_3 and SeO_2 have chains of MO_3 pyramids. Beyond group IV, oxides in less than the maximum oxidation state tend to form distorted MO_3 or MO_4 groups in chains or layers, since the remaining nonbonding electrons have a substantial effect on the overall M coordination geometry. The unique polymeric structure in this group is that of N_2O_5, which in the solid phase

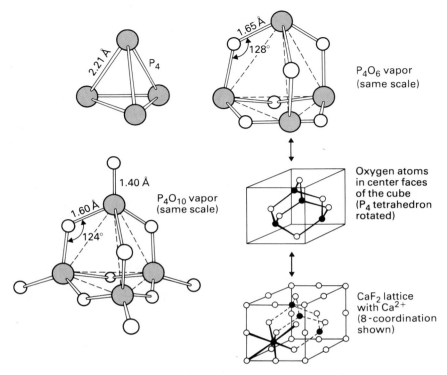

Figure 5.16   Phosphorus oxides and their relation to the cubic CaF_2 lattice.

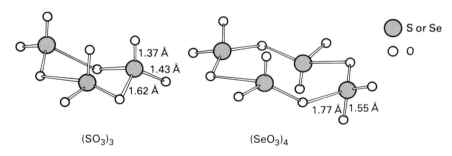

Figure 5.17 MO_3 ring structures.

near room temperature forms the ionic $NO_2^+NO_3^-$, in which NO_2^+ is linear and NO_3^- planar.

The atmosphere normally contains very large amounts of N_2 and O_2, but very little of any nitrogen oxide. At ordinary atmosphere temperatures, the reaction between N_2 and O_2 is extremely unfavorable:

$$N_2 + O_2 \rightleftharpoons NO \qquad \Delta G_{298}^0 = +173.4 \text{ kJ/mol reaction}$$

However, at the high temperatures of furnace or internal-combustion engine chambers, the positive entropy change ($\Delta S^0 = +24.7$ J/mol rn deg) for the above reaction begins to make NO stable, though this reaction is not the mechanism for its production. In a lean combustion mixture with a fairly long high-temperature residence time, as in fossil-fuel power plants, most NO is produced by the *Zel'-dovich mechanism:*

$$O + N_2 \rightarrow NO + N; \qquad N + O_2 \rightarrow NO + O$$

The activation energy for the first reaction is roughly 300 kJ/mol rn, which explains the need for high temperatures. In lean mixtures at lower temperatures, an alternative mechanism is

$$O + N_2 + M \rightarrow N_2O + M; \qquad O + N_2O \rightarrow NO + NO; \qquad H + N_2O \rightarrow NH + NO$$

In richer mixtures a faster mechanism operates, the *prompt NO mechanism:*

$$CH + N_2 \rightarrow HCN + N; \qquad HCN + O \rightarrow NCO + H; \qquad NCO + H \rightarrow NH + CO$$
$$NH + H \rightarrow N + H_2; \qquad N + O_2 \rightarrow NO + O$$

Although this appears more complicated, the activation energies are lower and it dominates NO production in combustion mixtures when the CH hydrocarbon radical is available, as it normally is in automobile engine operation. Given the composition of air, any combustion mixes roughly four times as much N_2 as O_2 in the combustion mixture, so significant amounts of NO are produced.

Cooling the exhaust gases should allow the equilibrium to readjust itself to N_2 and O_2, but in fact cooling occurs so rapidly that the mixture is quenched with most of the NO still present. Although NO is a free radical, it does not readily dimerize. However, it does react readily with O_2, which is also a paramagnetic diradical:

$$2NO + O_2 \rightarrow 2NO_2$$

This reaction is kinetically second-order in NO, so that the half-life for the oxidation depends on the NO concentration ($t_{1/2} = 1/k[NO]_0$). At the low NO concentrations in even polluted air (very roughly 0.1 ppm), the oxidation is very slow. However, under the conditions that promote photochemical smog a much more rapid mechanism takes over. These conditions are small concentrations of naturally occurring NO_2, sunlight to provide $h\nu$ activation energy for photochemical reactions, and hydrocarbon emissions present in the atmosphere.

Sunlight causes the following reaction of the free radical NO_2:

$$NO_2 + hv \rightarrow NO + O$$

The highly reactive O atom reacts promptly with O_2 to yield ozone if another molecule M is present to carry away the energy released by bond formation:

$$O + O_2 + M \rightarrow O_3 + M$$

Ozone attacks hydrocarbons with many possible oxidation products. One possible sequence is

$$RCH{=}CHR + O_3 \rightarrow RCHO + RO\cdot + HCO\cdot$$
$$RCHO + O + hv \rightarrow RC(O)\cdot + \cdot OH$$
$$RC(O)\cdot + O_2 \rightarrow RC(O)OO\cdot$$
$$RC(O)OO\cdot + NO_2 \rightarrow RC(O)OONO_2$$

The last product is a peroxyacyl nitrate, which is an extremely reactive oxidant and hydrolyzes to nitric acid—obviously an unpleasant pollutant. Ozone will also oxidize NO directly to prevent depletion of the initial NO_2:

$$NO + O_3 \rightarrow NO_2 + O_2$$

The other free radicals produced in hydrocarbon oxidation can also rapidly oxidize NO:

$$\cdot OH + CO \rightarrow CO_2 + \cdot H$$
$$\cdot H + O_2 + M \rightarrow HO_2\cdot + M$$
$$HO_2\cdot + NO \rightarrow NO_2 + \cdot OH$$

The net result of this cycle is the oxidation of both NO and CO:

$$O_2 + NO + CO \rightarrow NO_2 + CO_2$$

Since these or other equivalent processes produce NO_2 rapidly, the concentrations of NO and NO_2 in photochemical smog are roughly equal, and the aggregate is usually referred to as NO_x. Photochemical smog is characterized by high concentrations of NO_x, O_3, and photochemical oxidants such as the peroxynitrates. Because of the critical role of hydrocarbons, it is characteristic of cities with a high density of automotive traffic.

An entirely different set of nonmetal oxides is responsible for a medically dangerous smog that has sometimes been quite severe in major coal-burning areas. Where coal combustion is the major use of fossil fuel, SO_2 and SO_3 (collectively SO_x) are of greatest concern. All fossil fuels contain sulfur, but it is present in only a very small concentration in gasoline because it is largely removed during refining. Coal, on the other hand, is usually burned essentially without treatment. Sulfur and sulfur-containing compounds are burned to SO_2:

$$S + O_2 \rightarrow SO_2 \qquad \Delta G^0_{298} = -300 \text{ kJ}/\text{mol reaction}$$

SO_2 is thermodynamically unstable with respect to SO_3,

$$2SO_2 + O_2 \rightleftharpoons 2SO_3 \qquad \Delta G^0_{298} = -140 \text{ kJ/mol reaction}$$

but the reaction does not occur at the high combustion temperatures because of its negative entropy change. Like the NO oxidation, it is kinetically unfavorable at low temperatures. In sulfuric acid manufacture, the SO_2 oxidation is catalyzed by V_2O_5, but in ordinary fuel combustion nearly all the sulfur is emitted as SO_2. Photochemical oxidation of SO_2 is also quite slow under conditions approximating those in the atmosphere at sea level, but dissolved SO_2 in droplets of mist is oxidized from SO_3^{2-} to SO_4^{2-} more rapidly, particularly if catalyzed by traces of transition metal ions. SO_3 formed in the atmosphere has an extremely short lifetime, because as the anhydride of sulfuric acid it is quite hygroscopic. It dissolves as SO_4^{2-} in water that it attracts from the atmosphere, forming mist droplets of aqueous sulfuric acid that are soon removed from the atmosphere by precipitation as acid rain. The average lifetime of SO_4^{2-} in the atmosphere is thus only about a day. On the other hand, the average lifetime of an SO_2 molecule (which is much less hygroscopic than SO_3) is about a month. The adverse bronchial effects of sulfuric acid mist are obvious, but when weather conditions raise the concentration of SO_2 it also has an irritating effect on mucous membranes and a corrosive effect on building materials.

5.8 NONMETAL HALIDES AND OXOHALIDES

Halides are the most widespread type of compound in the periodic table. Only helium, neon, and argon fail to form a halide of any sort. Many of the halides, however, have structures and chemical reactivities determined by the requirements of electrostatic attraction in a predominantly ionic lattice. This is a consequence of the high electronegativity of the halogens, particularly fluorine. Here we shall examine only the halides of those elements for which directional covalent bonding is the primary stabilizing effect, which mostly limits us to the nonmetals. Table 5.9 indicates the stoichiometries and (in a very general way) the structures of the nonmetal halides. In this section we shall consider some of the trends in stability and bonding represented in this table and in Table 5.12 for oxohalides.

Some of the methods of preparing the nonmetal halides appear in Table 5.10. In general, many of the halides can be made by directly combining the elements, either in an inert solvent or by passing the gaseous halogen over the other element. However, this tends to produce the highest nonmetal oxidation state found for that halogen (for instance, $Ge + Cl_2$ yields $GeCl_4$, not $GeCl_2$). When a lower halide is desired, common techniques involve the nonmetallic element plus the hydrogen halide or a metal halide, particularly for fluorides.

The stability of covalent halides depends to a considerable extent on the strength of the nonmetal-to-halogen bond. In the simple LCAO-MO picture, that

Table 5.9 Nonmetal Halides

B	C	N	O	F	Ne
BX_3	CX_4	NF_3 NCl_3	(see Table 4.14)	(below)	
B_2X_4	C_2X_6 (etc.)	NBr_3 $NI_3 \cdot NH_3$ (circled)			
$\boxed{BY}$ 4,7,8,9,10,12		N_2F_4			

Si	P	S	Cl	Ar
SiX_4	PF_4 PCl_5(circled)* PBr_5(circled)	SF_6 SCl_4	ClF_5	
Si_nCl_{2n+2}	PX_3	SF_4 SCl_2	ClF_3	
Si_nF_{2n+2}	P_2Cl_4 P_2I_4	SF_2	ClF	
	P_2F_4	FSSF SSF_2 S_nCl_2 S_nBr_2	(and below)	
		S_2F_{10} $[S_2I_2][SI_2]$ (boxed)		

Ge	As	Se	Br	Kr
GeX_4	AsF_5	SeF_6 $\boxed{SeCl_4}_4$ $SeBr_4$?	BrF_5	KrF_2
GeX_2 (circled)	AsX_3	SeF_4 Se_2Br_2	BrF_3	
Ge_2Cl_6	As_2I_4	Se_2Cl_2	BrF $BrCl$	
			(and below)	

Sn	Sb	Te	I	Xe
SnX_4 SnF_4(circled)	$SbCl_5$ SbF_5	TeF_6 $\boxed{TeX_4}_4$	IF_7 $\boxed{ICl_3}_2$	XeF_6
SnX_2 (circled)	SbX_3	$TeCl_2$ $TeBr_2$ IF_3	IF_5 IBr (circled)	XeF_4
		Te_2F_{10} Te_2I_2 IF	ICl (circled)	XeF_2 $XeCl_2$

Pb	Bi
$PbCl_4$	BiF_5 (circled)
PbF_4 (circled)	BiX_3 (circled)
PbX_2 (circled)	

Key:

X = all halogens
Y = Cl, Br, I only
no border = monomer
$\boxed{n}$ = small polymer, n-fold

[ ] = species unstable at room temperature
(○) = extended polymeric lattice
* = monomer also known
? = structure not definitively established

Table 5.10 Preparation of Nonmetal Halides

B and Group IVa	Group Va
F $Na_2B_4O_7 + 12HF \longrightarrow Na_2O(BF_3)_4 + 6H_2O$	$NH_4^+HF_4^- \xrightarrow{Electrolysis} NF_3$
$4BF_3 + 2NaHSO_4 + H_2O \xleftarrow{H_2SO_4}$	$NH_3 + F_2 \xrightarrow{Cu} N_2F_4 + NF_3$
$SiO_2 + 2CaF_2 + 2H_2SO_4 \longrightarrow SiF_4 + 2CaSO_4 + H_2O$	$2PF_2I + 2Hg \longrightarrow P_2F_4 + Hg_2I_2$
$GeO_2 + BaCl_2 \xrightarrow{Aq. HF} BaGeF_6 \xrightarrow{Heat} GeF_4$	$2PCl_3 + 3CaF_2 \longrightarrow 2PF_3 + 3CaCl_2$
$SnCl_4 + 4HF \longrightarrow SnF_4 + 4HCl$	$2PCl_5 + 5CaF_2 \longrightarrow 2PF_5 + 5CaCl_2$
$SnO \xrightarrow{Evap. aq. HF} SnF_2$	$M_2O_3 + 6HF \longrightarrow 2MF_3 + 3H_2O$ (M = As, Sb, Bi)
	$M + \frac{5}{2}F_2 \longrightarrow MF_5$ (M = As, Sb, Bi)
Cl $B_2O_3 + 3C + 3Cl_2 \longrightarrow 6CO + 2BCl_3$	$NH_4Cl + 3Cl_2 \longrightarrow NCl_3 + 4HCl$
$M + 2Cl_2 \xrightarrow{400°C} MCl_4$ (all Group IVa)	$P_4 + 6Cl_2 \xrightarrow{PCl_3 \text{ soln.}} 4PCl_3$ (also for As, Sb, Bi)
$GeO_2 \xrightarrow{Aq. HCl} GeCl_4$	$PCl_3(SbCl_3) + Cl_2 \longrightarrow PCl_5(SbCl_5)$
$GeCl_4 + Ge \xrightarrow{300°C} 2GeCl_2$	
$M + 2HCl \longrightarrow MCl_2 + H_2$ (M = Ge, Sn, Pb)	
Br $B_2O_3 + 3C + 3Br_2 \longrightarrow 6CO + 2BBr_3$	$[(CH_3)_3Si]_2NBr + 2BrCl \longrightarrow NBr_3 + 2(CH_3)_3SiCl$
$M + 2Br_2 \xrightarrow{500°C} MBr_4$ (all Group IVa)	$P_4 + 6Br_2 \xrightarrow{CCl_4} 4PBr_3$ (also for As, Sb, Bi)
$GeBr_4 + Zn \longrightarrow GeBr_2 + ZnBr_2$	$PBr_3 + Br_2 \xrightarrow{CS_2} PBr_5$
I $LiBH_4 + 3I_2 \longrightarrow BI_3 + LiI + 2HI + H_2$	$3I_2 + 5NH_3 \longrightarrow NI_3 \cdot NH_3 + 3NH_4I$
$M + 2I_2 \xrightarrow{Heat} MI_4$ (M = Si, Ge, Sn, Pb)	$P_4 + 6I_2 \xrightarrow{CS_2} 4PI_3$ (also for As, Sb, Bi)
$GeO_2 \xrightarrow{H_3PO_2} Ge(OH)_2 \xrightarrow{HI} GeI_2$	$2As + 2I_2 \xrightarrow{260°C} As_2I_4$
$Sn + I_2 \xrightarrow{Aq. HCl} SnI_2$	

Group VIa	Group VIIa
F $M + 3F_2 \longrightarrow MF_6$ (M = S, Se, Te)	$\left.\begin{array}{l} X_2 + F_2 \longrightarrow 2XF \\ X_2 + 3F_2 \longrightarrow 2XF_3 \end{array}\right\}$ (X = Cl, Br, I)
$3SCl_2 + 4NaF \longrightarrow S_2Cl_2 + SF_4 + 4NaCl$	
$3S + 2AgF \xrightarrow{125°C} S_2F_2 + Ag_2S$	$ClF_3 + F_2 \xrightarrow{h\nu} ClF_5$
$SF_4 + SeO_2 \xrightarrow{200°C} SeF_4 + SO_2$	$3I_2 + 5AgF \longrightarrow 5AgI + IF_5$
$Te + 2TeF_6 \xrightarrow{180°C} 3TeF_4$	$NaCl + 3F_2 \xrightarrow{100°C} ClF_5 + NaF$
	$I_2 + 7F_2 \xrightarrow{250°C} 2IF_7$
Cl $2S + Cl_2 \longrightarrow S_2Cl_2 \xrightarrow{Cl_2} SCl_2$	$X_2 + Cl_2 \longrightarrow 2XCl$ (X = Br, I)
$Se(Te) + 2Cl_2 \longrightarrow SeCl_4 (TeCl_4)$	$I_2 + 5Cl^- + ClO_3^- + 6H_3O^+ \longrightarrow I_2Cl_6 + 9H_2O$
$2Se + Cl_2 \longrightarrow Se_2Cl_2$	
$Te + CCl_2F_2 \xrightarrow{450°C} TeCl_2 + ?$	
Br $2S + Br_2 \longrightarrow S_2Br_2$	$I_2 + Br_2 \longrightarrow 2IBr$
$Se(Te) + 2Br_2 \xrightarrow{CS_2} SeBr_4(TeBr_4)$	
I $Te^{4+} + 4HI + 4H_2O \longrightarrow TeI_4 + 4H_3O^+$	(See above)

strength is proportional to the β integral, which in turn is proportional to the overlap integral and to the average of the valence-orbital ionization energies for the two atoms. As the covalent radius of the halogen increases, both of these quantities decrease. As Fig. 5.18 indicates, covalent bond energies do decrease quite uniformly from M—F to M—I. The values plotted in Fig. 5.18 are experimental enthalpies of atomization for specific nonmetal-halide molecules, divided by the number of halogen atoms produced in the atomization. From our earlier discussion of bond-dissociation energies versus atomization energies, it should be clear that these numerical values might be quite different for a particular bond dissociation or even for atomization of a different halide (SF_4 versus SF_6, for instance). Quite generally, however, nonmetal iodides are more unusual and less stable than chlorides or fluorides because of the lower bond energies.

One consequence of the progressively smaller bond energies for the heavier halogens is that group oxidation states for a nonmetal central atom in a molecule are stable only for the lighter halogens. Thus silicon forms SiF_4, $SiCl_4$, $SiBr_4$, and SiI_4; phosphorus forms PF_5, PCl_5, and PBr_5, but not PI_5; and sulfur forms SF_6, but none of the other SX_6 species. Chlorine forms no ClX_7 molecules at all, while ClF_5 is known but no other ClX_5. Consider the Born–Haber cycle for the loss of a halogen molecule by MX_6:

$$MX_4(g, \text{sq. planar}) + 2X(g) \xrightarrow{-E_{\text{rearr}}} MX_4(g, \text{trig.bipyr.}) + 2X(g)$$

$$+2E_{\text{mx}} \uparrow \qquad\qquad\qquad\qquad \downarrow -E_{\text{xx}}$$

$$MX_6(g, \text{octahedral}) \xrightarrow{\Delta H} MX_4(g, \text{trig.bipyr.}) + X_2(g)$$

If two *trans-* X atoms are removed from the original MX_6, a square planar MX_4 remains. This species can gain stability ($-E_{\text{rearr}}$) by puckering to the trigonal bipyramidal electron geometry, as in Fig. 5.19. The net gain of sigma overlap by the p orbital on the M atom is very roughly equal to the energy of one M—X bond. Finally, the overall system gains stability by allowing the two X atoms to form a single bonded X_2 molecule. ΔH for the total reaction becomes

$$\Delta H = +2(E_{\text{M-X}}) - E_{\text{rearr}} - E_{\text{X-X}}$$

or, approximately,

$$\Delta H = +2(E_{\text{M-X}}) - E_{\text{M-X}} - E_{\text{X-X}} \simeq E_{\text{M-X}} - E_{\text{X-X}}$$

Table 5.11 shows the result of this treatment for the sulfur, selenium, and tellurium MX_6 compounds, of which only the fluorides are known. When the effect of entropy on the reaction (which generates gas) is considered, it is not surprising that only the fluorides are thermodynamically stable.

There is also a vertical trend in the stability of oxidation states for any given column of nonmetals in the periodic table. An element with n valence electrons will, of course, have a maximum formal oxidation state of $n+$ (except for O, F,

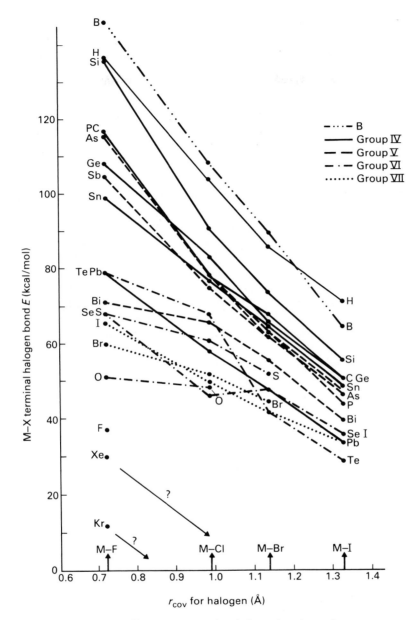

Figure 5.18 Size effects on nonmetal-to-halogen bond energies.

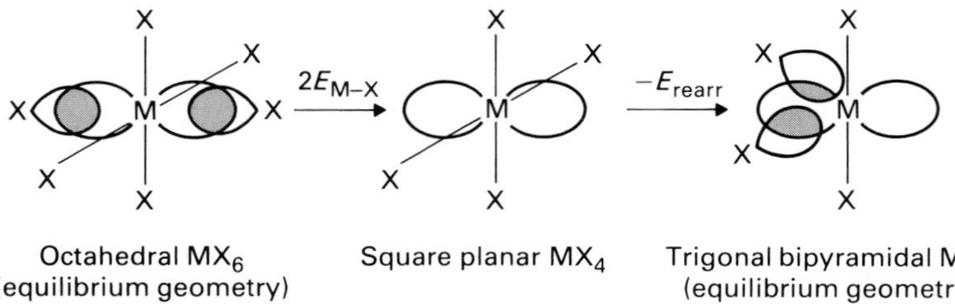

Octahedral MX_6 Square planar MX_4 Trigonal bipyramidal MX_4
(equilibrium geometry)   (equilibrium geometry)

Figure 5.19 Orbital overlap change for $MX_6 \rightarrow MX_4$.

and the noble gases, which are so electronegative that they do not readily delocalize valence electrons out into bonds). For the nonmetals, which are p-block elements with valence electron configurations $s^2 p^{n-2}$, the oxidation state $(n-2)+$ is also observed. This corresponds to retaining the more stable s electrons while releasing the p electrons into bonds. In any particular periodic table group, however, the $(n-2)+$ oxidation state becomes more stable relative to the $n+$ state for the heavier elements near the bottom of the group. For example, silicon(II) compounds are known—SiF_2, SiO, SiS, and a few others—but they are so unstable with respect to silicon(IV) that most of them spontaneously ignite when exposed to moist air. For practical purposes, the chemistry of silicon is thus the chemistry of Si^{4+}.

On the other hand, germanium(II) is much more stable, so that most germanium compounds can be made with Ge in either the 2+ or 4+ state, and the Ge^{2+} compounds are only moderate reducing agents. Tin is almost as stable in the 2+ as in the 4+ state, and vigorous redox reactions occur only when Sn^{2+} is in a

Table 5.11   Energies for the Reaction $MX_6 \rightarrow MX_4 + X_2$

MX_6	$E_{M-X}(kJ/mol)$	$E_{X-X}(kJ/mol)$	$\Delta H(\approx E_{MX} - E_{XX})$
SF_6	285	155	+130
SCl_6	255	238	+17
SBr_6	218	188	+30
SeF_6	285	155	+130
$SeCl_6$	192	238	−46
$SeBr_6$	201	188	+13
SeI_6	151	151	0
TeF_6	331	155	+176
$TeCl_6$	310	238	+72
$TeBr_6$	176	188	−12
TeI_6	121	151	−30

crystal with a strongly oxidizing anion such as NO_3^-; tin(II) is a very gentle reducing agent in most chemical environments.

Finally, lead is familiar to us primarily as Pb^{2+} compounds, and the few stable lead(IV) compounds are all strong oxidizing agents. This behavior is not due to greater stability of the s electrons in the heavier elements; in fact, the total ionization energy for the Pb s electrons is actually less than that for the Si s electrons: 71.1 eV versus 78.6 eV (the sum of IP_3 and IP_4). Rather, as Fig. 5.20 suggests and in a fashion reminiscent of electronic reasons for the failure of heavy elements to pi bond, it is because the inner core for the heavier elements is so large that a neighbor-atom orbital cannot effectively overlap the nondirectional Pb $6s$ orbital. Similar behavior is seen for the metals immediately to the left of Sn and Pb in the periodic table: Although Al is always 3+, Ga to some extent and In and Tl prominently display the 1+ oxidation state. Mercury is unusually inert as the metal in the 0 oxidation state compared to the electropositive zinc, which forms only the 2+ ion.

We have now mentioned four categories of covalent compounds of the nonmetal and metalloid elements: hydrides, organometalloids, oxides, and halides. It should not be surprising that, since all of these depend for their stability on directional covalent bonds, a variety of mixed substitution is possible. We have already

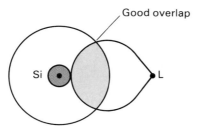

Si inner core occupies small fraction of $3s$ volume.

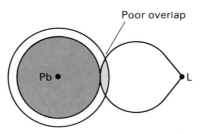

Pb inner core occupies large fraction of $6s$ volume.

Figure 5.20   LCAO overlap for nondirectional s orbitals.

mentioned some organosubstituted hydrides, but all the other permutations are possible. The Rochow direct process for making dimethyldichlorosilane from silicon and chloromethane is very important commercially, since it provides precursors for silicone polymers:

$$Si + 2CH_3Cl \rightarrow (CH_3)_2SiCl_2 \qquad [70\%, \text{ plus } (CH_3)_nSiCl_{4-n}]$$

Similarly, HCl will react with Si^0 or with SiH_4 to form chlorosilanes, and reducing agents will often convert alkylmetalloid halides to the hydride:

$$2RAsCl_2 + LiAlH_4 \rightarrow 2RAsH_2 + LiCl + AlCl_3$$

Most of the group IV–VI elements, and xenon among the noble gases, form oxohalides. A few, such as NOCl, nitrosyl chloride, and $SOCl_2$, thionyl chloride, are useful synthetic reagents. Table 5.12 shows the known oxohalides; note that it

Table 5.12   Nonmetal Oxohalides

B	C	N	O	F	
—	$COX_2[\neq I]$	$NOX[\neq I]$	—	—	
	$XC(O)C(O)X$	$NO_2X[\neq I]$			
	$[\neq I]$	NOF_3			
Si	**P**	**S**	**Cl**		
$X_3SiOSiX_3[Cl, Br]$	$POX_3[\neq I]$	$SOX_2[\neq I]$	ClO_2F		
$Si_3O_2Cl_8$	POX_nY_{3-n}	$SO_2X_2[F,Cl]$	$ClOF_3$		
$Si_4O_3Cl_{10}$	$XYP(O)OP(O)XY$		ClO_3F		
$(SiBr_2O)_4$	$[X, Y=F, Cl]$		O_3ClOF		
	$(PO_2Cl)_3$		ClO_2F_3		
Ge	**As**	**Se**	**Br**		
$Cl_3GeOGeCl_3$	$AsOF$	$SeOX_2[\neq I]$	BrO_2F		
	$\boxed{AsOF_3}$	SeO_2F_2	BrO_3F		
		F_5SeOF			
Sn	**Sb**	**Te**	**I**	**Xe**	
$\boxed{X_2SnO}$	$\boxed{SbOX}$	$Te_3O_2F_{14}$	IO_2F	$XeOF_2$	
			IOF_3	$XeOF_4$	
$[F, Cl, Br, I]$	$\boxed{SbOF_3}$		IO_3F	XeO_2F_2	
	$\boxed{SbO_2F}$		IO_2F_3	XeO_3F_2	
			IOF_5		
Pb	**Bi**				
—	$\boxed{BiOX} [\neq F]$				

is often true that the lighter elements will not form an oxoiodide, but that bismuth, at least, does not form an oxofluoride. Oxohalides are often made by allowing the halogen to react with the nonmetal oxide or with the element:

$$2NO + X_2 \rightarrow 2XNO$$

$$P_4 + 4SO_2 + 10Cl_2 \rightarrow 4OSCl_2 + 4OPCl_3$$

The M—O unit in a molecule usually has a trivial name ending in a -yl. $NOCl$ is nitrosyl chloride, NO_2Cl is nitryl chloride, $POCl_3$ is phosphoryl chloride, $SOCl_2$ is thionyl chloride, SO_2Cl_2 is sulfuryl chloride, and so on.

5.9 INTERHALOGENS

An interesting special case of the nonmetal halides is the group of interhalogen compounds and ions. The stoichiometries of the well-established interhalogen compounds were shown in Table 5.9. All of the diatomic XY molecules are known, although it is extremely difficult to purify them because of the ease with which they disproportionate to X_2 and Y_2. The larger neutral molecules uniformly consist of an atom of the more electropositive halogen bonded to an odd number of atoms of the more electronegative halogen. The central atom in the molecule is always the more electropositive. This is in accord with the differential-ionization-energy treatment of partial charges in Chapter 1, in which we saw that positive charge builds up on the central atom in proportion to the number of outer atoms. The odd number of ligand atoms, of course, allows complete electron pairing with the seven central-atom electrons. Valence-shell electron-pair repulsion theory accurately predicts the shapes of the gaseous molecules: XY_3 molecules are T-shaped, XF_5 molecules are square pyramids, and IF_7 is a slightly distorted pentagonal bipyramid. In IF_7 and the pentafluorides the F—F distance is much less than the sum of two van der Waals radii for fluorine, so some F—F bonding is presumably occurring in addition in addition to X—F bonding.

The halogen fluorides illustrate vividly an important principle mentioned in an earlier section: Covalent bond energies depend on the compound in which the bond occurs. Figure 5.21 shows the experimental bond energies for the halogen fluorides as a function of the electronegativity of the central atom. It can be seen that the energy of the Cl—F bond varies from over 250 kJ/mol for ClF to less than 170 kJ/mol for ClF_5. A curious feature of the bonding in these compounds is that the usual relationship between bond length and bond strength does not seem to apply. We usually assume that the shorter a bond is between two given atoms, the stronger it is, but the bonds in IF_7 are significantly shorter than in IF (1.786 Å axial, 1.858 Å equatorial for IF_7 versus 1.909 Å in IF), even though the bond is stronger in IF.

The interhalogens are prepared by mixing the elements under various conditions, as Table 5.11 has indicated. The ease with which these reactions are reversed makes the interhalogens vigorous fluorinating or chlorinating agents:

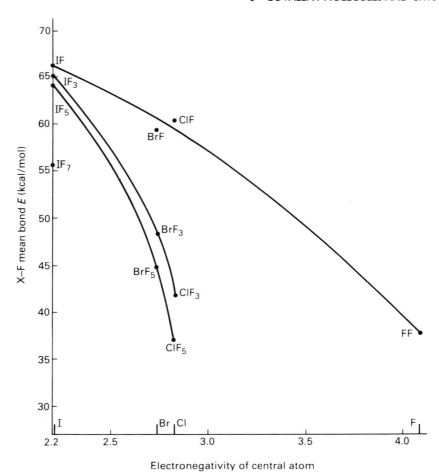

Figure 5.21   Interhalogen fluoride-bond energies as influenced by electronegativity and stoichiometry.

$$ClF + SF_4 \longrightarrow SF_5Cl$$

$$NOCl + SnF_4 \xrightarrow{BrF_3} (NO)_2SnF_6$$

$$N_2O_4 + Sb_2O_3 \xrightarrow{BrF_3} (NO_2)SbF_6$$

A variant of the last reaction gives an interesting product whose formula emphasizes the fact that if an interhalogen is to serve as a fluoride donor, a cation must also be formed:

$$6Sb_2O_3 + 32BrF_3 \rightarrow 12[BrF_2]^+[SbF_6]^- + 10Br_2 + 9O_2$$

This fluoride donor capability is observed in the pure liquid interhalogens as well, though not to the same degree in all. It has been shown through conductance and vibrational-spectroscopy studies that BrF_3, the most commonly used interhalogen solvent, undergoes the following autoionization reaction:

$$2BrF_3 \rightleftharpoons BrF_2^+ + BrF_4^- \quad \left(\text{Specific conductance} = 8 \times 10^{-3} \text{ ohm}^{-1}\text{cm}^{-1}\right)$$

Halide-ion transfer is quite mobile in most liquid interhalogens, and autoionization sometimes takes more complicated forms involving disproportionation of the species initially formed:

$$6ICl \rightleftharpoons I_3^+ + ICl_2^+ + 2ICl_2^-$$

Although the last equilibrium is still speculative to some extent, a large number of interhalogen cations and anions (the latter usually called polyhalides) has been characterized. Table 5.13 categorizes the ions for which crystalline salts

Table 5.13 Interhalogen Cations and Anions

Total no. of atoms in ion	Central halogen						
	Cl		Br		I		
2			Br_2^+		I_2^+		
3	Cl_3^+	Cl_3^-	Br_3^+	Br_3^-	I_3^+	I_3^-	
	ClF_2^+	ClF_2^-	BrF_2^+	BrF_2^-	IF_2^+	IF_2^-	
	Cl_2F^+			$BrCl_2^-$	ICl_2^+	ICl_2^-	
				Br_2Cl^-		IBr_2^-	
						I_2Br^-	
						I_2Cl^-	
						$IBrF^-$	
						$IBrCl^-$	
4					I_4^{2+}		
5		ClF_4^-	BrF_4^+	BrF_4^-	I_5^+	I_5^-	I_4Cl^-
					IF_4^+	IF_4^-	I_4Br^-
					$I_3Cl_2^+$	ICl_4^-	ICl_3F^-
						$I_2Cl_3^-$	$IBrCl_3^-$
						$I_2Br_3^-$	$I_2BrCl_2^-$
							$I_2Br_2Cl^-$
7	ClF_6^+	ClF_6^-		BrF_6^-		I_7^-	
				Br_6Cl^-		IF_6^-	
						I_6Br^-	
8					I_8^{2-}		
9					I_9^-		
					IF_8^-		

have been isolated; there is a surprising number. The anions are usually prepared by simply mixing the appropriate halide ion or polyhalide ion with a halogen or interhalogen in solution in an inert solvent. Equilibrium constants for some of these reactions are shown in Table 5.14 for water solvent, where it can be seen that most are only moderately stable. The cations are more difficult to prepare, because the combination of several electronegative atoms and a positive charge makes each cation a powerful electron acceptor. Most of the cations are prepared by using an extremely electronegative solvent molecule such as HSO_3F or AsF_5. Either a halogen molecule is oxidized by peroxydisulfuryl difluoride (FSO_2OOSO_2F) or an interhalogen molecule is mixed with a good fluoride acceptor such as SbF_5:

$$3I_2 + S_2O_6F_2 \rightarrow 2I_3^+ + 2SO_3F^-$$

$$IF_3 + SbF_5 \rightarrow IF_2^+ SbF_6^-$$

$$Cl_2 + ClF + AsF_5 \rightarrow Cl_3^+ AsF_6^-$$

The bonding in interhalogens, both molecules and ions, can be described reasonably well by the simple MOs formed only from the p orbitals of the halogen atoms that we developed earlier for electron-rich species. Figure 4.23 showed the orbital overlaps and MO energies for the I_3^- ion; here we can compare the energy-level diagrams and bond orders for ICl, ICl_2^+, and ICl_2^-. Valence-shell electron-pair repulsion theory correctly predicts the molecular geometries shown in Fig. 5.22. The central iodine atom in each case has five p electrons and each Cl contributes one sigma electron in its one sigma-symmetry orbital, so ICl has six electrons to be placed in the MOs, ICl_2^+ also has six, and ICl_2^- has eight. The overall bond order is 1.0 for ICl, 2.0 for ICl_2^+, and 1.0 for ICl_2^-, which corresponds well to the observed bond lengths. The bond order per bond region is 1.0 for ICl, 1.0 for ICl_2^+, but only 0.5 for ICl_2^-; the experimental bond lengths are 2.321 Å for ICl, 2.28 − 2.31 Å for ICl_2^+, but 2.47 − 2.55 Å for ICl_2^-. We find in general that bond lengths are longer for polyhalide anions than in the neutral interhalogen molecules, and (for the few cases in which x-ray data are available) that interhalogen cation bond lengths are slightly shorter than those in the most similar neutral molecules.

Table 5.14 Aqueous Stability Constants for Polyhalide Ions

Reaction	Equilibrium constant (aq, 25°C)
$Cl^- + Cl_2 \rightleftharpoons Cl_3^-$	0.2
$I^- + Cl_2 \rightleftharpoons ICl_2^-$	1×10^{20}
$Cl^- + ICl \rightleftharpoons ICl_2^-$	125
$I^- + ICl \rightleftharpoons I_2Cl^-$	3×10^8
$Cl^- + I_2 \rightleftharpoons I_2Cl^-$	1.7
$I^- + I_2 \rightleftharpoons I_3^-$	750

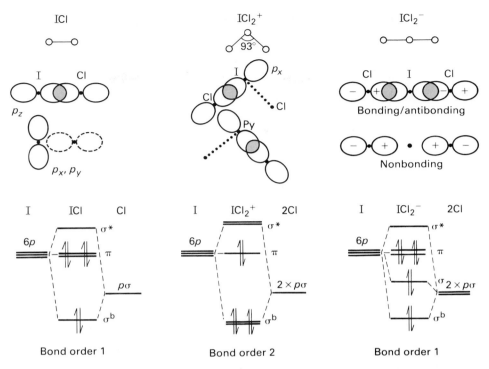

Figure 5.22 Qualitative MOs for ICl, ICl_2^+, and ICl_2^-.

5.10 NOBLE-GAS COMPOUNDS

The noble gases He, Ne, Ar, Kr, Xe, and Rn were considered totally unreactive chemically by their discoverers, and this impression gained theoretical stature when G. N. Lewis formulated the electron-pair model for the covalent bond and the octet rule. However, as understanding of the nature of the chemical bond matured a few chemists, notably Linus Pauling, suggested that a powerful oxidizing agent such as F_2 might be able to form covalent fluorides with the heavier and more polarizable noble gases such as Xe and Rn. The synthesizing experiment was tried, but not under the proper conditions, and the idea of noble-gas bonding was abandoned for some 25 years. In 1962 Neil Bartlett, investigating the properties of the powerful oxidizing agent PtF_6, prepared the platinum oxyfluoride PtO_2F_6, which proved to be $O_2^+PtF_6^-$. The similarity between the ionization potentials of O_2 (12.08 eV) and Xe (12.1 eV) led Bartlett to the successful preparation of $XePtF_6$, causing a surge of interest in noble-gas chemistry. Within about two years the compounds KrF_2, $KrF_2 \cdot 2SbF_5$, XeF_2, XeF_4, XeF_6, XeO_3, XeO_4, $XeOF_2$, XeO_2F_2, and a number of compounds with very electronegative ligand groups such as $-OSO_2F$ and $-OClO_3$ had been prepared. Ionic species such as

XeF_5^+, XeF_7^-, XeF_8^{2-}, XeO_6^{4-}, and $Xe_2F_{11}^+$ are also known.

In retrospect (always the best view) it should have been possible to predict the thermodynamic stability of the xenon fluorides. Consider Fig. 5.23, in which trends of M—X bond energies are plotted as a function of group number. For each of the four trend lines, the bond energy at each experimental point is that of the comparable electron configuration—one bonding pair or two nonbonding pairs. No data exist for two of the necessary compounds (AsF and ICl_5), and the nearest

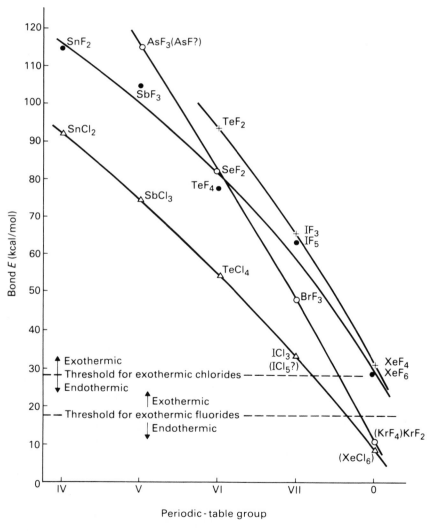

Figure 5.23 Trends in bond energies and in the stability of noble-gas halides.

equivalents have been used. If we write the reaction for forming one mole of Xe—F bonds,

$$\tfrac{1}{n}Xe(g) + \tfrac{1}{2}F_2(g) \rightarrow \tfrac{1}{n}XeF_n(s)$$

the entropy change is clearly unfavorable, but the reaction may still be favorable if the formation of one mole of Xe—F bonds is associated with a more negative enthalpy change than the formation of half a mole of F_2 (or Cl_2, in that case). Accordingly, Fig. 5.23 shows the energy thresholds for the exothermic formation of the noble-gas fluorides and chlorides. Even if the xenon compounds were not known, it can be seen that the formation of the two fluorides ought to be exothermic, and quite possibly thermodynamically favored. On the other hand, because of the generally lower M—Cl bond energies and the higher Cl—Cl energy, the unknown $XeCl_6$ ought to be strongly endothermic (though $XeCl_2$ has probably been prepared). Equally, because of the sharply decreasing M—F bond energies from AsF_3 to BrF_3, KrF_4 should not be stable; it has never been prepared.

The noble-gas fluorides are all prepared from mixtures of the noble gas and fluorine. All other compounds are prepared by starting with the appropriate fluoride and hydrolyzing it or allowing it to react with a very electronegative anhydrous acid:

$$Kr + F_2 \xrightarrow[-196°C]{\text{Electric discharge}} KrF_2$$

$$Xe + nF_2 \xrightarrow{400°C} XeF_{2n} \qquad \text{(Primary product determined by Xe:}F_2 \text{ ratio and pressure)}$$

$$XeF_6 + H_2O \rightarrow XeOF_4 + 2HF$$

$$2XeF_6 + 4Na^+ + 16OH^- \rightarrow Na_4XeO_6 + Xe + O_2 + 12F^- + 8H_2O$$

$$XeF_2 + HOSO_2F \rightarrow FXeOSO_2F + HF$$

$$3XeOF_4 + CsF \rightarrow Cs^+[(XeOF_4)_3F]^-$$

The last remarkable compound above has a three-coordinate fluorine atom, which is exceptionally rare.

The molecular geometry and bonding properties of the noble-gas compounds are of interest both because of the electron-rich nature of the compounds and because of the reputed stability of the noble-gas electron configuration. We shall consider the bonding first, because it exactly parallels the MO treatment we developed for the polyhalide ions. The MOs of Fig. 5.22 are equally applicable to XeF^+, the unknown XeF_2^{2+}, and neutral XeF_2 (linear), as these species are isoelectronic with the iodine chlorides in the figure. As in the polyhalide case, XeF^+ has a much stronger bond (200 kJ/mol) than either of the bonds in XeF_2 (total bond energy 268 kJ/mol XeF_2). The general tendency of xenon fluorides toward bond angles of 90° or 180° agrees well with the assumption that only p orbitals are involved in forming the MOs. However, XeF_6 represents an exception; its crystal structure consists of XeF_5^+ units with bridging fluorides such that two-thirds of the Xe atoms are seven-coordinate and one-third are eight-coordinate. The p-only MO

Table 5.15 VSEPR Predictions of Xenon-Compound Molecular Geometries

Molecule	Xe electron-pair coordination	Predicted geometry	Observed geometry (r in Å)
XeF_2	5 pairs, 3 nonbonding; trigonal bipyramid	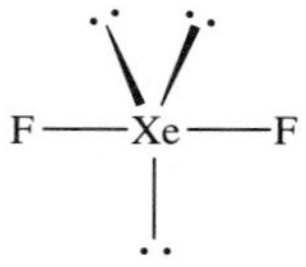	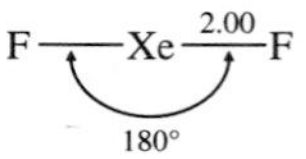
XeF_4	6 pairs, 2 nonbonding; octahedron	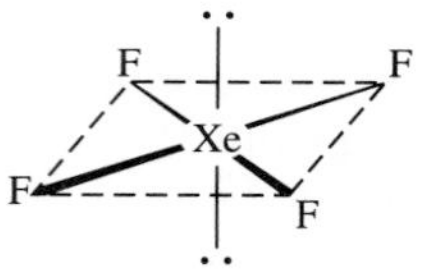	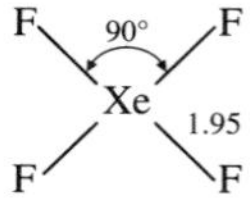
XeF_6	7 pairs, 1 nonbonding; pentagonal bipyramid or capped octahedron	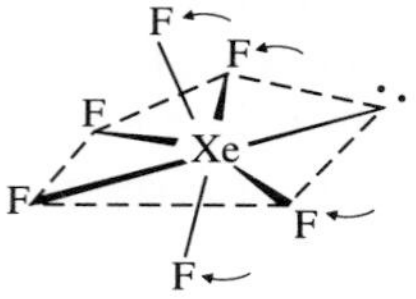	Gaseous XeF_6 nonoctahedral; $\frac{2}{3}$ of Xe atoms in crystal have capped octahedral geometry
XeF_5^+	6 pairs, 1 nonbonding; octahedron	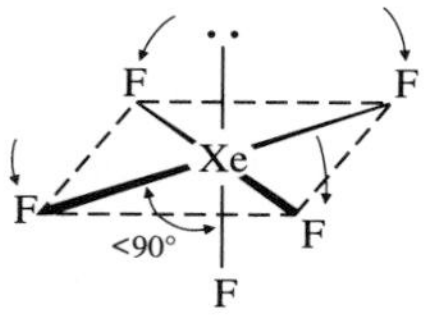	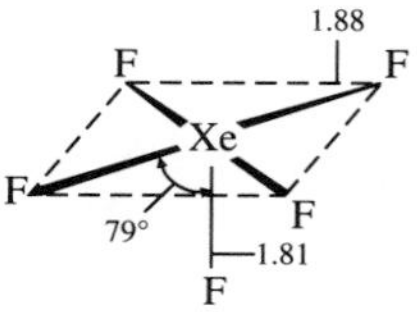
XeO_2F_2	7 pairs, 4 in 2 double bonds, 1 nonbonding; trigonal bipyramid	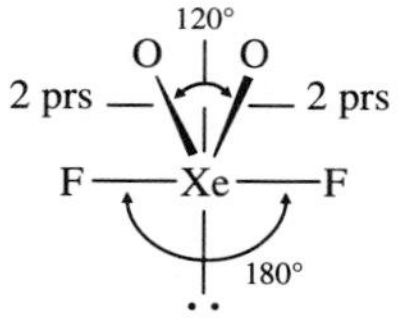	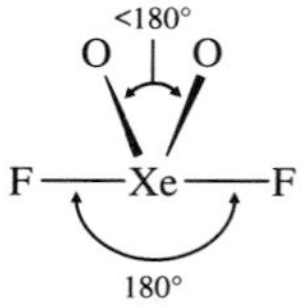
XeO_4	8 pairs, all in 4 double bonds (0 nonbonding); tetrahedron	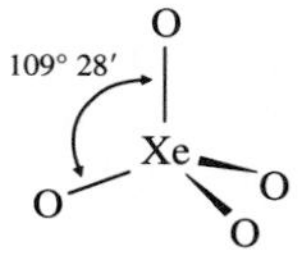	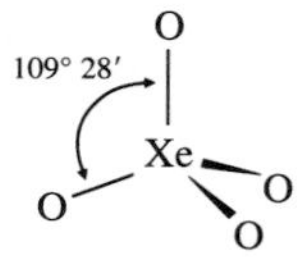

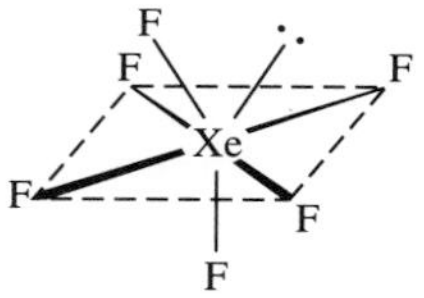

model predicts a simple octahedral XeF_6 structure, and it is necessary to consider more atomic orbitals to account for the less-symmetric structure. If not only the Xe $5s$ and $5p$ but also the $5d$ orbitals are included, even a qualitative treatment shows that the seven-coordinate structure has a lower electronic energy than the octahedral structure. However, the orbitals are now less readily visualized, and the structure prediction correspondingly more difficult.

A qualitative prediction of the molecular geometries of the noble-gas compounds is most readily made using valence-shell electron-pair repulsion theory (VSEPR). This technique, which should be familiar in its rudiments from previous courses but is reviewed in Appendix B, assumes perfect pairing of valence-shell electrons and minimization of repulsion between electron pairs. It further assumes that a nonbonding pair has a greater steric repulsion effect for neighbor pairs than a bonding pair, but about the same effect as two pairs in a M=O double bond. Table 5.15 summarizes the VSEPR predictions for several xenon compounds. All are experimentally correct, including that of nonoctahedral XeF_6, though the gas-phase geometry of the latter has not been fully characterized.

5.11 NONMETAL SULFIDES, NITRIDES, AND POLYMERS

In terms of stoichiometry and structure, the covalent compounds in this group are among the most unusual known to inorganic chemists. In particular, they show a strong tendency toward chain, ring, and cluster formation, frequently in defiance not only of ordinary valence rules but also of the cluster electron-counting rules we have developed. We have discussed earlier the electronic requirements for catenation. Here it may be useful to define a related phenomenon, *pseudocatenation,* in which chains form from two elements X and Y; —X—Y—X—Y—. In pseudocatenation, the electron-to-valence-orbital ratio is averaged over the X and Y neighbor atoms, and the resulting structures are frequently similar to true catenated systems. A remarkable example of this is boron nitride, which has the empirical formula BN. It is equivalent electronically to elemental carbon, since B has one fewer electron than C, and N has one more. BN occurs in two polymeric forms. One of these has the alternating atoms arranged in sheets of six-membered rings exactly like graphite (though the stacking of sheets differs slightly). The other form, obtained at high pressures—remember that graphite is the more stable form of carbon at 1 atm—has the diamond lattice, and is essentially equivalent to diamond in hardness.

With pseudocatenation in mind, we first consider nonmetal sulfides, for which preparative reactions are shown in Table 5.16. Note that for elements after the first row of the periodic table, the synthesis usually consists of simply heating the element and sulfur in suitable proportions. B_2S_3 can also be prepared in this way, but the very high temperature required causes the reactants to attack the reaction container. When H_2S is used, as in the table, the two initial cyclic products both have the empirical formula HSBS. We will later want to compare this 16-electron unit

Table 5.16 Preparation of Nonmetal Sulfides

B	N	Si
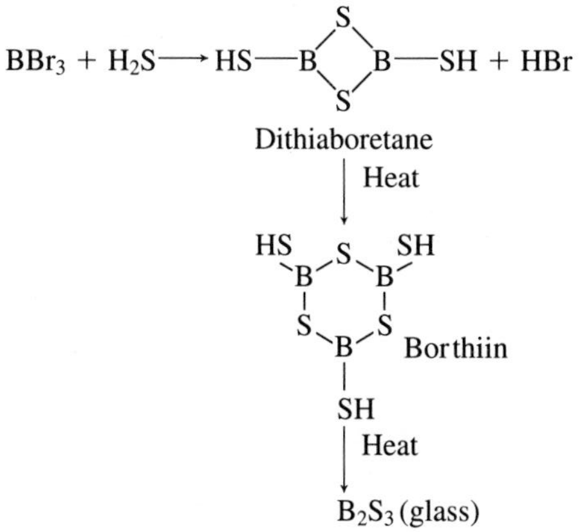 $BBr_3 + H_2S \longrightarrow HS\!-\!B\overset{S}{\underset{S}{\diamond}}B\!-\!SH + HBr$ Dithiaboretane Heat ↓ borthiin (ring with SH groups) SH Heat ↓ B_2S_3 (glass)	$6S_2Cl_2 + 4NH_4Cl \longrightarrow S_4N_4 + S_8 + 16HCl$ $3S_4N_4 + 2S_2Cl_2 \longrightarrow 4S_4N_3^+Cl^-$ $S_4N_4 \xrightarrow{Ag\ 220°C} S_2N_2$ (cond. $-190°C$) $\xrightarrow{room\ T}$ (SN)x $S_4N_4 \xrightarrow{Cl_2,\ CCl_4} (NSCl)_3$ $S_4N_4 \xrightarrow{AgF_2} (NSF)_4$	$Si + S_2 \xrightarrow{1000°C} SiS_2$

P	As	Sb
$8P_4 + 3S_8 \xrightarrow{180°C} 8P_4S_3$ $4P_4S_3 + S_8 \xrightarrow[hv]{I_2CS_2} 4P_4S_5$ $4P_4 + 3S_8 \xrightarrow{Heat} 4P_4S_7$ $4P_4 + 5S_8 \xrightarrow{300°C} 4P_4S_{10}$ $PSBr_3 + Mg \longrightarrow (PS)_x$	$8As + S_8 \xrightarrow{Heat} 2As_4S_4$ (Realgar) $32As + 3S_8 \xrightarrow{Heat} 8As_4S_3$ $16As + 3S_8 \xrightarrow{Heat} 4As_4S_6$	$16Sb_2O_3 + 9S_8 \xrightarrow{Heat} 16Sb_2S_3 + 24SO_2$ (Stibnite)

with some other species. Four-membered rings, as in dithiaboretane, are rare in inorganic chemistry; those that exist frequently involve sulfur, perhaps because of its great tendency to bond to itself. Heating B_2S_3 and S_8 together at low pressure yields the remarkable compound B_8S_{16}, which has the planar molecular symmetry of a porphine: B_2S_3 rings linked at the B atoms by S bridges.

The nitrogen sulfides (sometimes called sulfur nitrides) are remarkable compounds because of their bonding and geometry. Figure 5.24 shows the structures of these species and raises a few points that deserve comment. The conformation of the N_4S_4 ring has a square of N atoms concentric with a tetrahedron of S atoms. The dashed lines in the figure indicate that the opposed sulfur atoms are much closer together than the sum of their van der Waals radii, and so presumably have some degree of bonding, but the nitrogen atoms on one side of the square are also closer than the sum of *their* van der Waals radii. The fact that all bond distances are effectively equal indicates that considerable delocalization must be occurring. The 1.62 Å bond length is much shorter than the sum of the N and S covalent radii

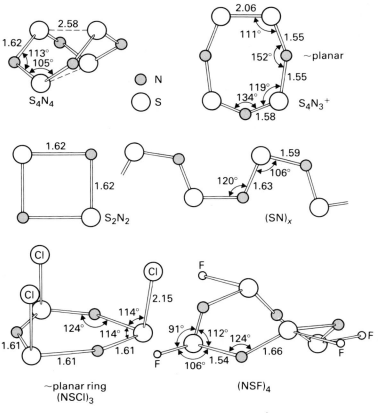

Figure 5.24   Some nitrogen sulfide structures (r_{N-S} in Å).

(1.77 Å), even after it has been corrected for the electronegativity difference (1.74 Å). This suggests significant multiple bonding. The $S_4N_3^+$ ring is a planar system with 10 pi electrons (2 per S, 1 per N, and a + charge overall), so that in the Hückel sense it is aromatic. This is consistent with the extremely short, equal bonds, even though some of the bond angles are rather far removed from the usual sp^2 hybrid angle of 120°. If N_4S_4 vapor is passed through silver wool at about 220°C and immediately condensed at liquid-nitrogen temperatures, the low-temperature product is the square-planar N_2S_2 (a dimer of the NS· radical, analogous to NO). On warming to room temperature, N_2S_2 rapidly isomerizes to the linear polymer $(SN)_x$. Single crystals of $(SN)_x$ have the remarkable property of being one-dimensional metals. The crystal is fibrous like asbestos, and the sides have a golden metallic reflectance while the ends (looking at the ends of the fiber cluster) are more or less flat black. This nonmetallic crystal displays electrical conductance comparable to that of metallic mercury along the fiber axis, but not perpendicular to it. At sufficiently low temperatures, it is even a superconductor. Finally, note the difference between the chlorination and fluorination products of S_4N_4: Chlorination yields a six-membered, nearly planar ring with equal bond lengths and, presumably, aromatic pi delocalization (though the odd positioning of the Cl atoms would seem to require that each S use a $3d$ orbital for pi overlap). Fluorination, on the other hand, only spreads and flattens the N_4S_4 eight-membered ring, leaving alternative N—S bond lengths that indicate a localization of charge into "single" and "double" bonds.

The formation of fibers and clusters is continued by the heavier nonmetal sulfides. SiS_2 forms fibrous, asbestos-like crystals in which SiS_4 tetrahedra (analogous to SiO_4 tetrahedra) share edges as in Fig. 5.25. Note again the presence of four-membered rings containing sulfur. With one possible exception, the phosphorus sulfides form clusters rather than chains. The P_4 tetrahedron of white phosphorous is retained in P_4S_3, P_4S_5, P_4S_7, and P_4S_{10}; but, as in the oxides, the phosphorus atoms are spread so far apart that P—P bonding disappears along most of the tetrahedron edges. As the number of S atoms increases, of course, the P—P bond disruption becomes greater and greater. P_4S_3 and P_4S_5 are relatively straightforward (if unsymmetric) structures, as shown in Fig. 5.25. In that figure, P_4S_7 is drawn with the P_4 tetrahedron on its side to emphasize the relation between its structure and that of S_4N_4 (Fig. 5.24), which is even more pronounced if the comparison is made to the $S_4N_5^-$ ion, which has a bridging N above the S_4 tetrahedron just as P_4S_7 has a bridging S above the P_4 tetrahedron. Otherwise the structure is intermediate between that of P_4S_5 and P_4S_{10}, which is the same as that of P_4O_{10} (Fig. 5.16). Finally, the As_4S_4 structure (found in nature as the mineral *realgar*) is the same as N_4S_4, except that now the sulfur atoms form the square and the arsenic atoms the tetrahedron. It is interesting that this structure is so prevalent for compounds between group V and group VI elements, since the structure occurs essentially for no other systems. An interesting related structure is that of $Fe_4S_3(NO)_7^-$, which has Fe and S atoms in the same locations as the P and S atoms in P_4S_3; we will return to transition metal clusters in Chapter 13.

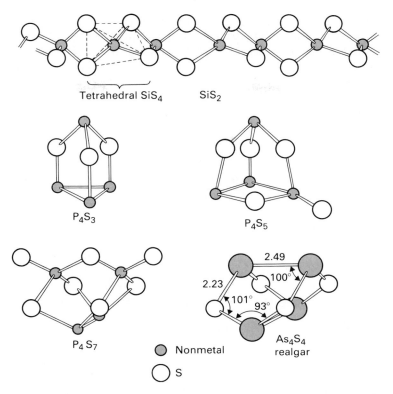

Figure 5.25   Polymeric structures of some sulfides of Si, P, and As (r_{M-S} in Å).

When we turn to the nonmetal nitrides, we find fewer cluster systems but a number of species that can shed light on different modes of covalent bonding. Considering boron first, perhaps the simplest B—N bond is formed by mixing the electron-deficient $B(CH_3)_3$ with $N(CH_3)_3$, which has a nonbonding pair of electrons:

$$(CH_3)_3B + :N(CH_3)_3 \rightarrow (CH_3)_3B \leftarrow N(CH_3)_3$$

The ← arrow is normally used to symbolize a bond in which one atom has provided both electrons, but the bond is not different from other sigma bonds except that negative charge tends to pile up on the electron-acceptor atom. A very large number of amine–borane adducts of this sort have been prepared. The electron-donor ability of the amine nitrogen atom correlates well with the various inductive and steric effects that might be expected. If both the boron and nitrogen have a hydrogen attached, heating the amine-borane adduct yields a different kind of B—N bond:

$$R_2HN \rightarrow BHR'_2 \xrightarrow{\text{Heat}} R_2NBR'_2 + H_2$$

In this aminoborane product, the N and B presumably share a sigma pair consisting of one electron from each atom, as in a hydrocarbon C—C bond. However, the N still has a nonbonding pair, and it can form a partial pi bond superimposed on the sigma bond. In general, physical properties such as vibrational stretching force constants indicate that the B—N bond in aminoboranes is stronger than a single sigma bond would be. Alternatively, the nonbonding pair can be donated to a neighbor aminoborane molecule. Both dimers and trimers are known:

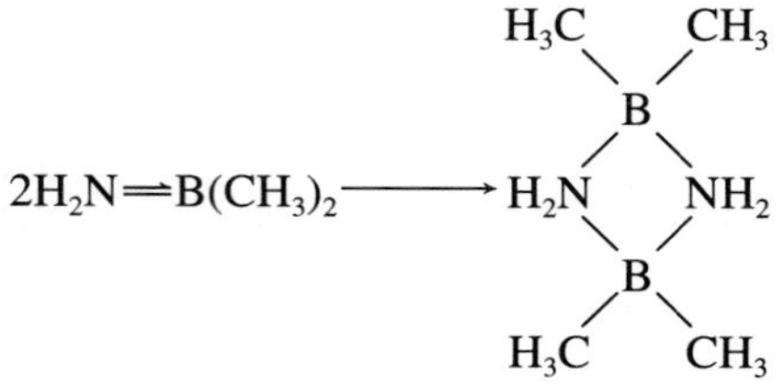

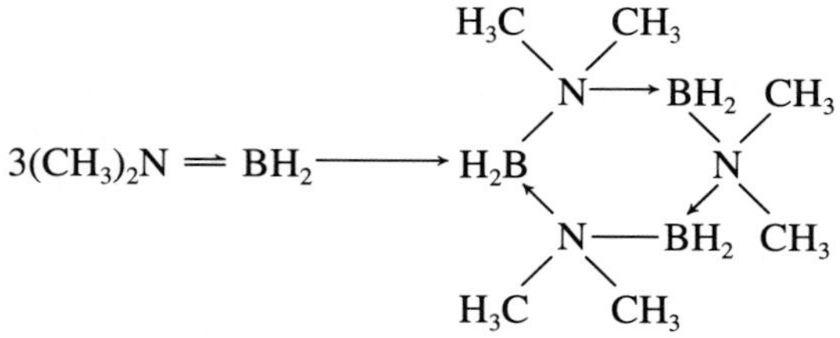

$$2H_2N{=}B(CH_3)_2 \longrightarrow$$

$$3(CH_3)_2N \rightleftharpoons BH_2 \longrightarrow$$

The trimer compound is known as a triborazane; it exists in the cyclohexane chair conformation. The equivalent unsaturated compound, borazine, is prepared by heating BCl_3 with NH_4Cl and reducing the product with $NaBH_4$:

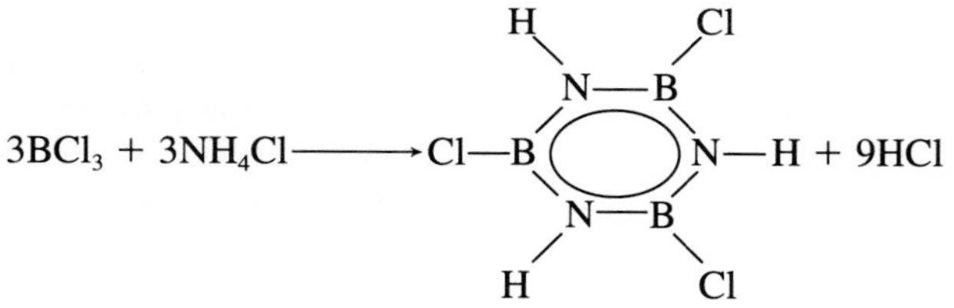

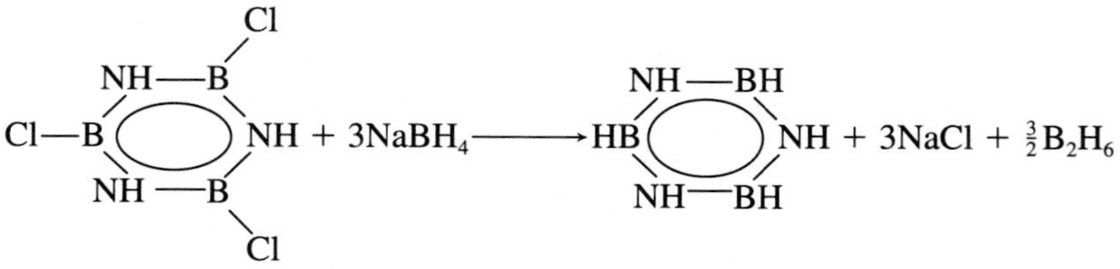

$$3BCl_3 + 3NH_4Cl \longrightarrow$$

$$+ \; 9HCl$$

$$+ \; 3NaBH_4 \longrightarrow$$

$$+ \; 3NaCl + \tfrac{3}{2}B_2H_6$$

Borazine strongly resembles benzene in its physical properties, and as a planar regular hexagon it is structurally equivalent. However, the alternating electronegativities of the atoms around the ring prevent complete pi delocalization, so borazine does not have fully aromatic chemical properties. We have seen that pi bonding is even more difficult for third-row than for second-row elements, which must be protected by sterically bulky substituents; in the present context we may note the existence of "alumazine" [more properly a tris(N-alkyliminoalane)]. The planar six-membered Al_3N_3 ring has only a methyl group on each Al, but the N atoms are from 2,6-diisopropylpyridine, which provides a sheltering cloud of isopropyl groups on each side of the ring. The Al–N distance is 1.78 Å, which compares to 1.89–1.96 Å in four-coordinate Al–N systems; the shortening suggests that some pi bonding is occurring in spite of the substantial polarity of the Al–N bonds.

If the borazine synthesis is carried out at 750°C in an atmosphere of ammonia, the product is hexagonal boron nitride, $(BN)_x$, with the graphite-like structure we mentioned earlier. The bonding resembles that in graphite in the same sense that borazine resembles benzene; however, graphite's fairly high electrical conductivity is caused by the motion of the delocalized pi electrons, and the localization of pi electrons at the electronegative N atoms in $(BN)_x$ makes it a very poor conductor. Borazon, the high-pressure form of $(BN)_x$, has the previously mentioned diamond structure with only sigma bonds. Since all bond lengths are equivalent, no unique electron donor–acceptor bond can be defined at each B atom, and the bonding is most accurately described using band theory.

Two silicon–nitrogen compounds are of interest: silicon nitride, Si_3N_4, and trisilylamine, $(H_3Si)_3N$. Si_3N_4 is comparable to quartz and other silicates in that it is a three-dimensional polymer of tetrahedral SiN units. However, the overall stoichiometry of the compound requires that each nitrogen atom be three-coordinate. Although we might expect each N to be pyramidal like NH_3, it is in fact trigonal planar to a good approximation. This change in geometry, which suggests a change in bonding, might be merely an artifact of the crystal packing. However, trisilylamine has the same planar coordination geometry around the nitrogen atom, even as an isolated monomer. Figure 5.26 contrasts the geometries of $N(CH_3)_3$ and $N(SiH_3)_3$ and shows the d–p pi bonding that has been proposed to account for the difference. Although silicon has no $3d$ electrons, its valence electrons are $3s$ and $3p$, so the $3d$ orbitals must be of comparable size and not too high in energy. In the planar geometry, the two N electrons left over after forming the three sigma bonds can be pi-delocalized over all four atoms in the molecular framework much as the nitrogen atom serves as a pi donor in aminoboranes—except, of course, that the acceptor orbital is a d rather than a p orbital. This type of bonding is impossible for trimethylamine, because carbon has no valence d orbitals. As a consequence, trimethylamine is a good electron-pair donor because of the nonbonding pair on the N atom, whereas trisilylamine is a very weak donor, having already engaged those electrons in bonding.

These two compounds are not the only examples of trigonal planar Si coordination. We have already noted the Zintl phase compound $Li_{12}Si_7$ with a trigonal planar Si_4 star species. A more versatile group of compounds with this pattern of Si coordination is the *silatranes* and azasilatranes, examples of which are seen in Fig. 5.27. The planar SiO_3 group in a silatrane is formed by allowing

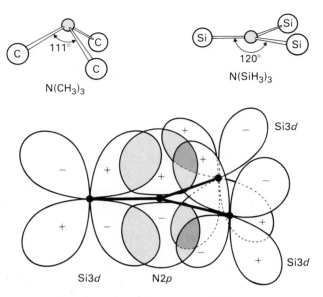

Figure 5.26 Pi bonding in trisilylamine involving $d-p$ overlap.

triethanolamine or an equivalent to condense with a trialkoxysilane; if the tetrahedral arrangement around Si were maintained the molecule should simply form
a cage with Si at the top, N at the bottom. However, the N donates an electron
pair to the Si, forcing overall trigonal bipyramidal geometry. Azasilatranes are
similarly prepared by the reaction of a substituted triaminosilane with a tetramine
compound. Apart from the interesting aspects of their bonding, it might be noted

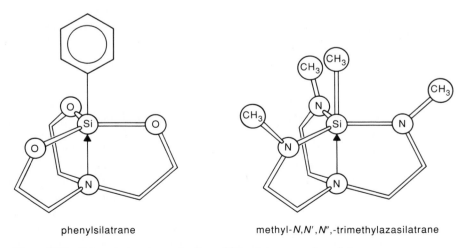

phenylsilatrane   methyl-N,N',N'',-trimethylazasilatrane

Figure 5.27 Trigonal planar coordination of Si in silatranes and azasilatranes.

that Russian chemists have reported remarkable physiological properties of sila-tranes from hair growth to increased testicular mass; the present author is not pre-pared to endorse these uses, however.

Phosphorus forms a very large number of compounds with nitrogen. Most of these are derivatives of phosphorous acid or phosphoric acid in which one or more of the —OH groups have been replaced by —NH_2 groups (or =O by =NH). Our purpose here is to look at the phosphonitrilic polymers $(PNCl_2)_n$. They are pre-pared by reacting PCl_5 with NH_4Cl in a chlorinated inert solvent:

$$n\ PCl_5 + n\ NH_4Cl \xrightarrow{\text{CHCl}_2\text{CHCl}_2,\ 140°C} (PNCl_2)_n + 4n\ HCl$$

Cyclic polymers from $n = 3$ to $n = 8$ are known, of which the trimer and the tetramer are most important. In addition a linear polymer with n near 15,000 is formed. If the P—Cl bonds (which are unstable with respect to hydrolysis) are re-placed by various organic groups, further heating produces a cross-linked polymer that, depending on the side group, can be a tough thermoplastic film or a rubbery elastomer with outstanding low-temperature flexibility. The bonding in the rings is presumably related to the bonding in the high polymer, since heating the rings to about 250°C causes rearrangement to the polymer. The structures of the trimer, tetramer, and pentamer are shown in Fig. 5.28. Some inferences can be drawn from these structures—but others should not. First, some pi bonding is certainly occurring, because all P—N bond lengths are significantly less than the sum of the covalent radii (1.81 Å). The pi bonding is delocalized, because there is no sig-nificant alternation of bond lengths. However, even though the six-electron ring (the trimer) and the 10-electron ring are planar and the eight-electron ring is not, one should not assume that the Hückel rules for aromaticity apply to these sys-tems. Because the P atoms are four-coordinate, they can only participate in pi bonding through the sort of d–p overlap shown for trisilylamine in Fig. 5.26. If the ring is taken to lie in the xy plane, the d_{xz} and d_{yz} orbitals can be chosen to provide good pi overlap; but since each has a nodal plane containing the z axis, the delocalized pi cloud that results will be interrupted at each P atom, as Fig. 5.28 also shows. Furthermore, the wider the ring bond angles at N become, the more the N hybridization can be described as sp; both remaining $2p$ orbitals can then engage in pi overlap with the P $3d$, even in the plane of the ring. Finally, ring planarity is not a criterion for d–p pi delocalization, because the greater number of d orbitals in a given set and the greater number of nodes mean that pi overlap can occur almost regardless of bond angles or puckering.

If the phosphonitrilic amide $[NP(NH_2)_2]_n$ is prepared and pyrolyzed, the product is phospham, $(HNPN)_n$, an amorphous polymeric material with a basic bonding unit of 16 electrons $(1 + 5 + 5 + 5)$. This is the same number found in $(HSBS)_n$, HN_3, H_2NCN (cyanamide), CO_2, NO_2^+, HSCN, BrCN, and many other systems of three heavy atoms. By sketching atomic-orbital overlaps for a three-atom system in which each atom has s and p valence orbitals, we can readily de-velop an MO energy-level diagram with exactly eight bonding and nonbonding or-bitals. Sixteen electrons will, of course, be the optimum number for such a system.

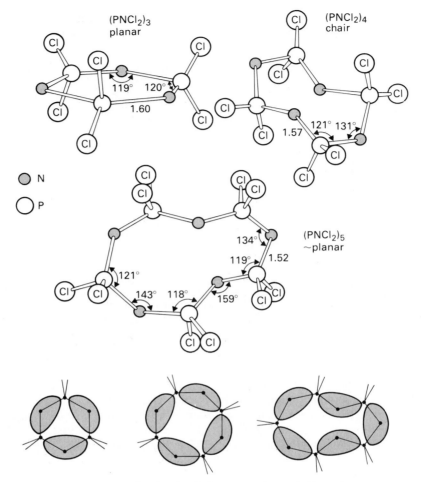

Partially delocalized π clouds (interrupted at each P atom)

Figure 5.28   Phosphonitrilic chloride structure and bonding (r_{P-N} in Å).

PROBLEMS

A. DESCRIPTIVE

A1. What formulas and general structures should organoscandium compounds have? How much clustering or polymerization should be expected?

A2. In nonpolar solvents $(CH_3)_2GaBr$ is a dimer. Should the AXY_2 dimer have a structure resembling that of ethane or that of diborane? Why?

A3. Suggest a preparative reaction for $(C_6H_5)GaCl_2$.

A4. What geometry do Wade's rules suggest for the $B_9H_{12}^-$ ion? Give as much detail as you can.

A5. In Table 5.4 it can be seen that liquid ammonia is much less dense than either liquid water or liquid HF. How can you account for the trend in density seen for these liquids?

A6. Table 5.5 quotes different preparative reactions for HCl and HBr. What chemical property of the molecular hydrides might make HBr incompatible with H_2SO_4? Write a balanced reaction equation for the decomposition to be expected.

A7. For group IV elements E the E–C bond energy decreases steadily from C–C (346 kJ/mol bond) to Pb–C (130 kJ/mol bond). Why should this be true, in electronic terms?

A8. The preparation of BCl_3 and BBr_3 from B_2O_3 requires elemental carbon. What is the thermodynamic reason for carbon's participation?

A9. Suggest a preparative reaction for tetravinyltin.

A10. How could fluorobenzene be converted to benzyne (⬡) by an organometallic reaction?

A11. What geometry should the $B_{10}H_{14}$ borane cluster have according to Wade's rules? Use the deltahedral structures shown in Fig. 4.28 as a basis for choice. When $B_{10}H_{14}$ reacts with $Me_3 \cdot AlH_3$ to form $[B_{10}H_{10}AlH_2]$, what geometry should the new cluster have (ignoring the solvated ether molecules also attached)?

A12. Use simple VSEPR arguments to account for the *gauche* conformation of hydrazine.

A13. How should the basicity of hydroxylamine, H_2NOH, compare with that of ammonia? Of hydrazine?

A14. The molecular oxide SO is known only as a ligand in the compound $SFe_3(CO)_9(SO)$, but it is formally similar to O_2 and S_2, both of which are more stable. Suggest an exothermic reaction for decomposing gaseous SO that is not feasible for O_2 or S_2, and explain why neither of these species can undergo that reaction.

A15. Suggest a preparative reaction for $[NP(NH_2)_2]_3$.

B. NUMERICAL

B1. Use data from Tables 1.10 and 4.7 to estimate which metals are likely to be able to form a solid solution with titanium.

B2. Calculate the value of K_p at 500°C for the Haber ammonia synthesis reaction equation given in the chapter, assuming that $\Delta H°$ is constant with temperature. What is the value of K_c at 500° if the total pressure is 200 atm?

B3. Johannis's original preparation of what we now think of as the Zintl anion Pb_9^{4-} yielded a sodium salt that he took to be $NaPb_2$. How much do the elemental analyses of "$NaPb_2$" and Na_4Pb_9 differ in weight percent Na?

B4. How does the Al–N internuclear distance in the Al_3N_3 "alumazine" ring compare with what might be expected for an Al=N double bond?

B5. The I—F bond energy is essentially the same in IF_3 and IF_5, but it is lower in IF_7 (see Fig. 5.21). Show by appropriate bond-energy calculations that this implies that IF_3 gas should disproportionate to $I_2(g)$ and $IF_5(g)$, but that IF_5 gas should be stable with respect to disproportionation.

B6. Show trigonometrically that the nitrogen atoms in S_4N_4 are closer together than the sum of their van der Waals radii.

C. EXTENDED REFERENCE

C1. A molecule containing the Si$=$C bond is called a *silene*. The silene

$$(Me_3Si)_2Si{=}C(tBu)OSiMe_3$$

reacts with propenal in a $[2 + 4]$ cycloaddition:

$$(Me_3Si)_2Si = C(tBu)OSiMe_3 + CH_2 = CH\text{-}CH = O \rightarrow$$

```
     (Me₃Si)₂Si—C(tBu)OSiMe₃              (Me₃Si)₂Si—C(tBu)OSiMe₃
            /   \                                /   \
         H₂C     O              +              O     CH₂
            \   /                                \   /
          HC = CH      A                        HC = CH      B
```

Should isomer **A** or isomer **B** predominate? Why? [See A. G. Brooks *et al., Organometallics* (1991), *10,* 2752.]

C2. What should the structure of the $B_{11}H_{13}^{2-}$ ion be? If $Na_2B_{11}H_{13}$ is allowed to react with $Al(CH_3)_3$, what product should result? Write a balanced reaction equation. [See T. D. Getman and S. G. Shore, *Inorg. Chem.* (1988), *27,* 3439.]

C3. R. C. Haushalter et al. have reported the synthesis and crystal structure of the $SnAs_{14}^{4-}$ Zintl anion ([*J. Chem. Soc., Chem. Commun.* (1988), 1027]). The As atoms are in two seven-atom clusters, each with the geometry of the P_4S_3 molecule (Fig. 5.25), linked by the Sn atom. How many electrons must the tin atom contribute to the overall cluster bonding if each seven-atom cluster is to be isoelectronic with P_4S_3? Does the electron count help to explain the curious bonding geometry of the tin atom?

C4. The chapter has described the formation of siloxanes by the hydrolysis of R_2SiCl_2 species. If $HSiCl_3$ is hydrolyzed under "scarce-water" conditions, one significant product is $H_8Si_8O_{12}$. ^{1}H NMR shows that all eight hydrogen atoms in this molecule are symmetrically equivalent. Propose a structure for the Si_8 product. [See P. A. Agaskar, *Inorg. Chem. (1991), 30,* 2707.]

6

Acids, Bases, and Solvents ━━━━━━━━━

We have now examined the basics of the current theories of the structure of the main-group elements and their more common compounds. The vast majority of the discussion has dealt with either the solid state (crystal structures and bonding) or the gaseous state (isolated molecular structures). The experimental fact, however, is that most inorganic reactions are carried out in liquid solution. We therefore need to examine the structures of those solutions, as well as the energy relationships accompanying dissolution and the solvated condition.

Liquids are more like solids than like gases under ordinary conditions (that is, liquids are much closer to their freezing point than to their critical point). Internuclear distances between touching nonbonded atoms are only slightly greater than in the corresponding crystalline solid, and around any individual molecule a reasonable degree of short-range (nearest-neighbor) order is maintained. The principal difference is that some atoms or molecules in the liquid have a smaller coordination number than in the solid, and the space left by the missing neighbors is spread out irregularly through the liquid "lattice." At random intervals during the normal vibrational motion of the molecules, the open space appears as molecule-sized holes into which a neighbor molecule can move. Diffusion thus occurs much more rapidly in liquids than in solids, and reactants can be brought together efficiently.

If one begins with the fully ordered solid, creating the holes in the liquid lattice requires an energy input since neighbor–neighbor attractions are being destroyed. This is the reason why all substances have a positive heat of fusion. However, the heat of fusion is always much less than the heat of vaporization, which corresponds to the energy input required to remove *all* neighbors. Thinking back to the discussion in Chapter 1 of intermolecular forces, it is clear that neighbor–neighbor attractions will be quite different in different liquids. Some liquids will have only relatively weak van der Waals forces binding them; others will have the forces between strongly polar molecules or hydrogen-bonded molecules; and molten salts will have the very strong monopole–monopole attractive

Table 6.1   Intermolecular Forces in Solvent Liquids (J/mL liquid)

Solvent	ΔH_{fus} (J/mL liquid)	Solvent	ΔH_{fus} (J/mL liquid)
NaCl	803	SO_2	193
$AlCl_3$	335	H_2SO_4	185
H_2O	320	$POCl_3$	142
BrF_3	246	C_2H_5OH	86
NH_3	227	CH_3OH	78
HF	226	CS_2	73
DMSO	203	CCl_4	35

Note: Values are ΔH_{fus} at mp per gram, multiplied by the liquid density at the temperature at which it is commonly used.

forces. A very rough way of comparing these forces is to look at their enthalpies of fusion per milliliter of liquid formed, on the crude assumption that melting produces the same degree of disruption in all liquids. Such data are given in Table 6.1 for some liquids commonly used as solvents. It can be seen that on this assumption, the forces in NaCl (and other molten salts with singly charged ions) are some 25 times as great as the forces in CCl_4, in which presumably only London dispersion forces are at work. Although the order in which the liquids fall in the table could not have been predicted in detail, it is not surprising in light of what we know about the relative magnitudes of intermolecular forces.

When we dissolve a solute in a liquid, we expect to see some change in the structure of the liquid. Some of the intermolecular contacts in the solution are now between different molecular species, and different forces of attraction may come into play. There are thus inevitable energetic differences between a pure liquid (in which each molecule is surrounded by other molecules identical to itself) and a solution (in which the solute molecules are predominantly surrounded by solvent molecules). The smaller the energy difference is, the more nearly ideal the solution is. Molecular size, on the other hand, has very little effect on solubility or solution properties, essentially because the disordered liquid–solvent structure has enough free space to accommodate atoms of different sizes.

6.1 SOLVATION AND ELECTRON DONATION

The formation of a solution from a liquid and a potential solute (ionic or molecular) is a process whose spontaneity is influenced both by the solvent–solute energy relationship mentioned above (ΔH_{soln}) and by the degree of randomness and enhanced probability introduced by dissolving the solute in the solvent (ΔS_{soln}).

For a solid, we would ordinarily expect ΔS_{soln} to be quite favorable, because of the random nature of a solute compared to its crystalline form:

$$Na^+Cl^-(crystal) \xrightarrow{H_2O} Na^+(aq) + Cl^-(aq) \qquad \Delta S^0 = +43.1 \text{ J/deg} \cdot \text{mol rn}$$

This will be true in general for covalent molecular solids that dissolve in molecular form, and to some extent for ionic solids, as above, although 40 J/K is perhaps not as big as might have been expected. As we shall see shortly, ions in solution in a polar solvent normally have an ordering effect on the solvent molecules surrounding them, and it is quite possible for this ordering effect to exceed the disorder induced by the dissolution of the crystal. The ordering effect depends on the magnitude of the charge on each ion, and in water solvent most salts with a +2 or −2 ion (or greater charge) have a negative entropy of solution. In such cases, of course, spontaneous dissolution requires a negative enthalpy change so that ΔG ($= \Delta H - T\Delta S$) can be negative.

One more consideration with respect to the entropy of solution is that under ordinary conditions, it is not very large. Since T is about 300 K for most solution preparation, an entropy change of +40 J/deg · mol rn leads to a $-T\Delta S$ free energy contribution of only about 12 kJ/mol rn. If ΔH were exactly zero, this would lead to a solubility equilibrium constant of a little over 100:

$$\begin{aligned} K_{eq} &= e^{-\Delta G^0/RT} = e^{-\Delta H^0/RT} \cdot e^{\Delta S^0/R} \\ &= e^{-0/RT} \cdot e^{\Delta S^0/R} = 1 \cdot e^{\Delta S^0/R} \\ &= e^{+40/8.3} = e^{4.8} = 124 \end{aligned}$$

This would certainly represent a spontaneous process. However, many dissolution processes have enthalpy changes much greater than 12 kJ/mol rn. If the enthalpy change dominates the overall free energy change (and it frequently does at the relatively low thermodynamic temperatures at which we work), solubility and solution properties will be determined by the energy relationships accompanying the process of dissolution.

When a solution involving a given solute and solvent is formed, three energy relationships must be considered: (1) The lattice energy or other cohesive energy of the solute is being destroyed. (2) Some of the cohesive energy of the liquid solvent is being lost, since the solute is intruding between solvent molecules that would otherwise be neighbors. (3) Attractions are being created between solute molecules (or ions) and solvent molecules (or ions, in the case of molten salts). To break even on the enthalpy change for the overall process, the solvent–solute attractions must equal the loss of solute–solute attractions and solvent–solvent attractions. We can consider a few cases. If the solvent has only weak van der Waals attractions, such as CCl_4, that loss will not be important. A CCl_4 molecule, on the other hand, is uncharged, nonpolar, and cannot participate in hydrogen bonding. It is very unlikely to interact with a solute molecule in any way that will yield strong attractions. Accordingly, CCl_4 cannot dissolve any solute that has strong solute–solute attractions, such as hydrogen-bonding or strong dipole–dipole attractions, let alone any ionic species. You can't dissolve salt in carbon tetrachloride.

At the other extreme, water solvent is strongly hydrogen-bonded and water molecules are quite polar. There are thus substantial solvent–solvent attractions to be overcome, but also the opportunity for substantial solvent–solute attraction through hydrogen-bonding, dipole–dipole attraction, and even ion–dipole attraction. Water is thus a good solvent for all solutes except those incapable of any of these interactions—that is, nonpolar solutes having only London dispersion forces between molecules. Such solutes cannot yield enough solute–solvent attraction to pay the price of disrupting the solvent–solvent attractions, and ΔH_{soln} becomes quite unfavorable. You can't dissolve oil in water.

There is an extremely important kind of interaction between solvent and solute that we have not yet mentioned. It has been tacitly accepted that the molecules in a pure liquid solvent cannot react chemically with each other in the sense of forming new bonds. One assumes that the bonding MOs of the solvent molecules are filled with paired electrons, so that even if nonbonding electrons are present, a neighbor molecule cannot accept (share) them in a low-energy orbital. However, most solvent molecules do have nonbonding electrons, and many solute molecules or ions have vacant acceptor orbitals. We frequently find that solvent molecules serve as electron-pair donors and solute ions or molecules as electron-pair acceptors, in the sense that they share an electron pair originating on the solvent molecule in a covalent bond. Since covalent bond energies can be as large as ionic lattice energies, this donor–acceptor interaction is an extremely important influence on the enthalpy of solution. The electron donor–acceptor interaction is the basis of the Lewis definition of acids and bases, and we shall return to it after examining the origins of the large interaction energy between ions and polar solvent molecules.

Consider a positive ion dissolved in a polar solvent and surrounded by an octahedral array of six solvent-molecule dipoles as in Fig. 6.1. The total potential energy of this array (the equivalent of the lattice energy of the same ion in a crystal) can be accounted for by five terms:

$$U_1 = -\frac{6q\mu_0}{r^2} - \frac{6q\mu_i}{r^2} + \frac{6(1.19)(\mu_0 + \mu_i)^2}{r^3} + \frac{6q\mu_i^2}{2\alpha} + \frac{6b}{r^9}$$

$$(1) \qquad (2) \qquad\qquad (3) \qquad\qquad (4) \quad\ (5)$$

The first term represents the attraction of an ion with net charge q for six dipoles with a permanent dipole moment μ_0, arranged so that the center of each dipole is r units away from the center of the ion. The second term is the additional attraction of the ion for the six induced dipole moments μ_i that are created in the solvent molecules as a result of being placed next to the point-charge ion. The third term represents the total repulsion of the six dipoles for each other, lumping the permanent and induced dipole moments together; the 1.19 quantity is a geometric factor analogous to the Madelung constant for a crystal. The fourth term represents the

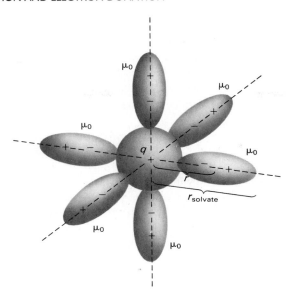

Figure 6.1 A cation in an octahedral array of dipoles.

energy cost of creating the six induced dipoles in molecules each of which has a polarizability α. Finally, the fifth term is the Born equivalent of the van der Waals repulsion of the interpenetrating electron clouds of the ion and the neighbor dipoles. In solving the potential-energy equation to obtain a numerical value for the overall ion–dipole attraction, this term (or at least the b parameter) is eliminated as we eliminated the corresponding term for the lattice energy in Chapter 2—by differentiating the energy equation with respect to internuclear distance and setting the result equal to zero at the equilibrium distance.

We can use this potential-energy expression to estimate the magnitude of these ion–dipole attractions if we substitute numbers typical of common ions and solvents: $r = 3$ Å, $\mu_0 = 1$ debye $= 3.3 \times 10^{-30}$ C · m, $\alpha = 3 \times 10^{-30}$ m^3, and $q = 1.6 \times 10^{-19}$ C (for a +1 ion). If we solve for μ_i by substituting into $\mu_i = \alpha \mathscr{E}$ the electrical field $\mathscr{E}$ due to the ion and the other five dipoles, the result is $\mu_i = 1.06$ debye. This indicates that induced-dipole effects will be nearly as important as permanent-dipole effects, even after allowing for the energy expended in inducing the dipoles. Specifically, if we examine the calculated energies term by term, we find that the ion–permanent-dipole attraction is -193 kJ/mol ion; the ion–induced-dipole attraction is -204 kJ; the total dipole–dipole repulsion is $+67$ kJ; the energy required to induce the μ_i dipoles is $+67$ kJ; and the van der Waals repulsion is $+109$ kJ. The overall potential energy of the ion is -153 kJ/mol ion, which is a very substantial contribution to the stability of the ion in solution.

This is not the only energy effect on dissolving an ion in a polar solvent, however. The solvated ion we have now created will attract other polar solvent molecules in exactly the same manner that it attracted the first layer, but we need not go through the same detailed calculation. For ordinary solvent molecules, the solvate ion will be large enough relative to the individual solvent molecules that one can treat the solvate ion as a charged sphere immersed in a continuous dielectric medium with a dielectric constant equal to that of the pure solvent liquid. Under this approximation, we can calculate a second contribution to the potential energy of the ion in the polar solvent:

$$U_2 = -\frac{q^2}{2r_{solvate}}\left(1-\frac{1}{\epsilon}\right) \qquad r_{solvate} = r_{cation} + 2r_{solvent}$$

Here $r_{solvate}$ is the radius of the ion with the first layer of solvent molecules around it and ϵ is the dielectric constant of the solvent. The dielectric constant is quite low for covalent-molecule solvents and for molten salts (on the order of 1.5–5). For strongly polar and hydrogen-bonding solvents, on the other hand, it is much higher (on the order of 20–100). Water has the best-known value, 78.54. If we continue the sample calculation by assuming that this is a +1 ion with a solvated radius of 5 Å in a solvent with a dielectric constant of 20, U_2 will be 263 kJ/mol solvate. This is, in effect, the energy difference between the solvated ion in a vacuum and that same ion in the dielectric solution.

It can be seen that the presence of a polar, dielectric solvent has a substantial stabilizing effect on an ion: $U_1 + U_2 = (-153) + (-263) = -416$ kJ/mol. If the ion had originally been in a salt M^+X^-, the ion of opposite charge would also be dissolved and solvated. If its solvation energy were comparable to that just calculated, the total solvation energy for the salt would be about 800 kJ/mol, which is fully comparable with ionic-lattice energies for such salts. However, the energy yield of the solute–solvent interaction (solvation energy) must compensate not only for the loss of solute–solute interaction (lattice energy), but also for the loss of solvent–solvent interaction. In our hypothetical case the solvent has had a cavity formed in it for each ion present. The solvent molecules formerly in the cavity have "evaporated," which costs $7 \times \Delta H_{vap}$ for the solvent (six for the solvating molecules plus one to make room for the ion itself). However, the extra one solvent molecule recondenses elsewhere, and the six solvated molecules (which have not actually entered the vapor phase) regain most of their stability from the pure liquid nearby by reorienting neighbor solvent molecules in solution. The result is that the net solvent–solvent energy cost for forming an octahedral solvated ion in solution is only about $2 \times \Delta H_{vap}$ for the pure liquid. Since ΔH_{vap} for most common solvents (such as those in Table 6.1) is about 35 kJ/mol, there is a net solvent–solvent energy cost of perhaps 70 kJ/mol of dissolved ions. For M^+X^-, the cost is about 150 kJ/mol, although cations and anions do not have exactly the same orienting effect on neighbor solvent molecules in solution, and thus have slightly different overall enthalpies of solvation. This leaves us with an overall solvation energy of about $-830 + 150 = -680$ kJ/mol MX, which is almost exactly equal to the lattice energy of alkali halides ($J_{NaCl} = -778$ kJ; $U_{KCl} = -707$ kJ).

This result leads to the interesting conclusion that in spite of the very great lattice energies of ionic salts, there is likely to be only a very small heat effect when they are dissolved in polar solvents. Furthermore, that effect could be either endothermic or exothermic. However, not all ionic salts will be soluble. Solubility requires a negative ΔG, and neither ΔH nor ΔS is as favorable for highly charged species such as $CaCO_3$ and other $+2/-2$ salts. We have already seen that ΔS is usually negative (unfavorable) for highly charged salts. Examining the equation for the U_1 component of the solvation energy reveals that the attraction between the ion and the permanent dipoles increases only in proportion to the charge on the ion, whereas the lattice energy increases in proportion to the product of the charges on the two ions. In general, the solvation energy of an ionic compound does not increase as rapidly with charge as the lattice energy does, and ΔH_{soln} becomes less and less favorable for highly charged salts. Inevitably, ΔG_{soln} also becomes less favorable, and solubility is reduced.

For ions that form well-defined stoichiometric solvates, such as $Mg(H_2O)_6^{2+}$, it is equally possible to account for the U_1 portion of the solvation energy by molecular-orbital methods. Each solvent molecule is assumed to have a sigma orbital containing two electrons directed toward the central ion, which has vacant s, p, and d orbitals that can overlap the solvent-ligand sigma orbitals. The electrons are ultimately accommodated in sigma bonding MOs; the lower overall electronic energy of the bound system is the equivalent of the U_1 ion–dipole attraction. Such calculations are quite successful, but require extensive computer facilities and are therefore less convenient than the classical approach just presented. However, they do directly consider the internal atomic structure of the solvent molecules and explain stability directly in terms of electron distributions within the molecule. In this model, the solvating properties of polar solvents depend directly on their electron-donor ability. Of course, species in solution other than the solvent molecule can serve as donors of nonbonding electrons, forming a bond with an electron-acceptor ion (or other species). We shall next examine such electron donors and acceptors as participants in the general reaction between acids and bases.

6.2 ELECTRON ACIDS AND BASES

There are many definitions of the terms "acid" and "base." Most inorganic chemists shift from one definition to another, depending primarily on the solvent system being used. As this chapter progresses, we shall examine several of the more common definitions, each in its solvent context. The definition most nearly independent of the solvent system is that of G. N. Lewis:

An acid is an electron-pair acceptor.
A base is an electron-pair donor.

This definition is, of course, exactly parallel to the model of solvation just described. A cation in solution is serving as a Lewis acid, and the solvent is serving as a Lewis base. Not all solutes are Lewis acids, and not all solvents are Lewis bases; the terms are definitions of convenience, which cover a certain kind of chemical reactivity. Any acid–base definition defines a particular chemical reaction, even though it appears to be a structural description. As soon as we choose a reaction type, we automatically choose an acid–base definition *if* there is any sense in which the reaction has the complementary character of acids and bases generally. We can construct a general principle of acid–base reactivity that is independent of the definition we chose, but that sets permissible limits for acid–base definitions:

> *The strongest acid present in a reaction mixture will react with the strongest base present to reduce the availability of the characteristic acid–base material, without changing the formal oxidation state of any atom.*

Within this very broad convention (which, for example, excludes redox reactions) we can choose any acid–base definition that represents a convenient way to categorize reactions of interest. If the Lewis definition is substituted into this reactivity principle, it says that the strongest electron pair donor in a mixture will react with the strongest electron pair acceptor to reduce the availability of nonbonding electrons. The question of the strength of acids and bases will be deferred until Chapter 7, where we deal with patterns of reactivity of acids and bases. Here we shall limit the discussion to the structural features of acids and bases.

Since a Lewis acid must accept a pair of electrons, it must have a vacant low-energy orbital. Furthermore, since the donating base must not lose its share in the electrons (which would change the oxidation state of the donating atom and make the process a redox reaction), the Lewis acid must have a vacant coordination site so it can form a new bond with the Lewis base. The obvious Lewis-acid candidate from first-row elements is boron in various BX_3 compounds (where X can be halogens or organic groups). BX_3 has only six bonding electrons in boron-based orbitals, but eight can be accommodated in the four possible sigma bonding orbitals (see Fig. 4.15), and boron can have a coordination number of four without undue steric hindrance. Pursuing the same reasoning, we find other good Lewis acids are AlX_3 and SbX_5. In each case the central atom is one short of its usual maximum coordination number (though Al is found to have coordination up to C.N. = 6) and is two short of the maximum number of electrons that can be accommodated in the bonding orbitals for the higher coordination number.

A Lewis base must have a pair of sigma-symmetry nonbonding electrons, although one lobe of a filled, weakly bound pi bonding orbital can also serve as a donor (see Fig. 6.2). This is true in general of elements in groups IV–VII of the periodic table that are in less than their most positive oxidation state. Thus carbon in CO can serve as a Lewis base, as can NH_3 and other NR_3 amines, H_2O, and the X^- halide ions. Heavier elements in these groups can of course function in the same way, as for example SnR_2 and PR_3. Since the Lewis-base atom is also increasing its coordination number, an effective Lewis base must have a donor atom whose coordination number in the base molecule is less than the maximum possible value for that atom—generally, less than the number of valence orbitals for that atom.

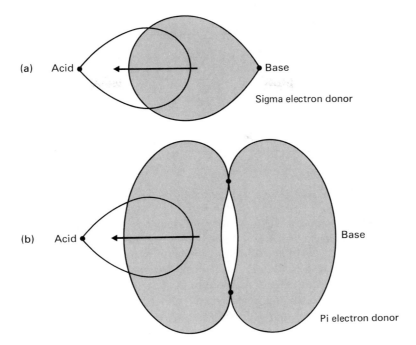

Figure 6.2   Possible orbital overlaps for Lewis acid–base interaction. (Shaded area represents initial base electron density.)

Lewis acids and bases display definite preferences in reacting with each other. It has been found in general that the relative affinity of two bases for a given acid depends strongly on what the acid is. Ahrland, Chatt, and Davies used the gradual accumulation of thermodynamic data for Lewis acid–base reactions, such as stability constants for adducts,

$$A + B \rightleftharpoons AB \qquad K_{stab} = \frac{[AB]}{[A][B]}$$

to group a large number of Lewis acids into two classes, (*a*) and (*b*), depending on their relative base preferences. Class (*a*) acids prefer group V bases in the order $N \gg P > As > Sb$, group VI bases in the order $O \gg S > Se > Te$, and group VII bases in the order $F \gg Cl > Br > I$. Conversely, class (*b*) acids prefer group V bases in the order $N \ll P > As > Sb$, group VI bases in the order $O \ll S \approx Se \approx Te$, and group VII bases in the order $F \ll Cl < Br < I$. For example, Table 6.2 gives the logarithm of the stability constant for some metal-ion acceptors (acids) and halide donors (bases). It is clear that Fe^{3+} and Zn^{2+} are fundamentally different from Hg^{2+} and Pb^{2+}, and many more examples could be given. What features of electronic structure are common to class (*a*) acids and different from class (*b*) acids? Without reproducing all of the data, we can summarize as follows: Class

Table 6.2   Relative Stabilities of Metal-Ion Halide Complexes

		log K_1 of substituent X			
	M^{n+}	F^-	Cl^-	Br^-	I^-
Class (a)	Fe^{3+}	6.04	1.41	0.49	—
	Zn^{2+}	0.77	−0.19	−0.60	−1.3
Class (b)	Hg^{2+}	1.03	6.74	8.94	12.87
	Pb^{2+}	0.3	0.96	1.11	1.26

Note: $M^{n+} + X^- \overset{K_1}{\rightleftharpoons} MX^{(n-1)+}$

(a) acids include all the ions from the s-block of the periodic table, cations with a charge greater than +3, and the lighter transition-metal ions with a charge greater than +1. Class (b) acids include cations from the p-block with less than a +3 charge and the heavier transition-metal ions with less than a +3 charge.

The common electronic feature of class (a) acceptors is that, regardless of their specific electronic structure, the remaining electrons are exposed to a large effective nuclear charge, either because they are inner-core electrons (as in Na^+) or because they are in a highly charged ion such as Sn^{4+}. They are thus tightly bound and not readily deformable by a neighboring atom. These acids form more stable complexes in solution with bases that have high electronegativity: N, O, and F in preference to P, S, and Cl. By contrast, class (b) acceptors have a relatively low charge and larger radii, and in many cases have valence electrons remaining before a base donates more. These acids prefer the less electronegative bases P, S, and Cl (or even I) to N, O, and F. Pearson and Busch proposed the term "soft" to apply to class (b) acids and to the less electronegative bases they prefer, and "hard" to apply to class (a) acids and the more electronegative bases. Pearson equated the terms to polarizability, since the class (a) acids and their preferred electronegative bases both have strongly bound, only slightly deformable electrons, and are thus only slightly polarizable or hard, while the converse is true for soft acids and bases. The descriptive rule for the stability of an acid in solution in the presence of several possible bases, or vice versa, becomes

Hard acids prefer to bind to hard bases.
Soft acids prefer to bind to soft bases.

A common misconception is to equate hardness with strength, which is completely without foundation. It is possible to have strong hard bases in solution, such as OH^- in water, but it is equally possible to have strong soft bases such as S^{2-} in water. The same is true of weak bases and both strong and weak acids. We shall return to this distinction in Chapter 7.

We are at this point considering Lèwis acid/base characteristics in the context of the deformability of the electron cloud of the atom directly involved as an acceptor or donor. One way to think about this deformability is to measure it by the upward curvature of the graph of atomic energy versus net charge for the atom (Fig. 1.15)—the more rapidly the energy required to remove charge increases with charge, the more resistant to deformation the atom is. Now, this curvature for any given atom is given by the a_1 parameter of its differential ionization energy. We have already suggested that the a_0 parameter could reasonably be taken as the absolute electronegativity of the atom; here we may suggest that a_1 represents a value of the *absolute hardness* of the atom. In the discussion above, we have referred to hard bases as those donating through an electronegative atom such as F or O; this is a shorthand way of recognizing that a_0 and a_1 usually increase together across the periodic table. However, it is useful to have a definition of absolute hardness for an atom independent of a particular Lewis acid–base reaction in which it may be involved.

Table 6.3 gives a fairly extensive list of hard, soft, and intermediate acids and bases. From these, one can predict a wide range of reactions in solution. Note that the acids listed include many electron-pair acceptors that are not ions or even metals, which is irrelevant as long as they meet the conditions for a Lewis acid species. For an example of the qualitative use of the hard/soft acid/base principle (HSAB), consider the question: Which of the following ions can form bromo complexes in water solution: Ag^+, As^{3+}, Ni^{2+}, Pt^{2+}, Te^{4+}, Ti^{4+}? The Br^- base is intermediate in its hard/soft properties; the competing base, H_2O, is hard because electron pair donation is from the electronegative O atom. Of the six cations listed, Ag^+, Pt^{2+}, and Te^{4+} are soft acids and will certainly prefer the intermediate base Br^- to the hard base H_2O; they form $AgBr_2^-$, $PtBr_4^{2-}$, and $TeBr_5^-$ respectively. As an intermediate acid, Ni^{2+} will form the complex $NiBr_4^{2-}$ with the intermediate base Br^-. In water solution, however, a mixture of species $NiBr_n(H_2O)_{4-n}^{(n-2)-}$ is formed, indicating a competition between the bases Br^- and H_2O. The hard acids As^{3+} and Ti^{4+} form only hydrated species or oxoanions in water solution (that is, they prefer the hard base H_2O to the intermediate base Br^-). There is no evidence of As—Br bonding or Ti—Br bonding in water, though of course species such as $AsBr_3$ and $TiBr_4$ can be prepared in other solvents.

It is important to note that hardness or softness is not a permanent intrinsic property of a given element, even though we can define an absolute hardness for an isolated atom. For acceptors (acids) with variable oxidation states, the net positive charge on the ion has a strong influence on the hardness. The greater the positive charge, the harder the species. Thus Co^{3+} is classified as a hard acid, Co^{2+} as an intermediate acid, and Co^0 (neutral metal atom) as a soft acid. Similarly, Sn^{4+} is hard but Sn^{2+} is intermediate. There are also inductive effects on hardness for both acids and bases. A hard acid that is not *too* hard can be softened enough by the presence of several soft-base ligands that it will prefer to add another soft base rather than another hard base. The moderately hard Co^{3+} forms both $Co(NH_3)_5X^{2-}$ and $Co(CN)_5X^{3-}$ complexes, where NH_3 is a hard base, CN^- a soft base, and X is a halide-ion base. When five hard NH_3 bases are already present, the complex with $X = F^-$ is much more stable than the one with $X = I^-$, which is consistent with the

Table 6.3 Hard and Soft Acids and Bases

	Acids		

Hard	Intermediate	Soft
H^+, Li^+, Na^+, K^+ $(Rb^+$, $Cs^+)$	Fe^{2+}, Co^{2+}, Ni^{2+}, Cu^{2+}, Zn^{2+}	$Co(CN)_5^{3-}$, Pd^{2+}, Pt^{2+}, Pt^{4+}
Be^{2+}, $Be(CH_3)_2$, Mg^{2+}, Ca^{2+}, Sr^{2+} (Ba^{2+})	Rh^{3+}, Ir^{3+}, Ru^{3+}, Os^{2+}	Cu^+, Ag^+, Au^+, Cd^{2+}, Hg^+, Hg^{2+}, CH_3Hg^+
Sc^{3+}, La^{3+}, Ce^{4+}, Gd^{3+}, Lu^{3+}, Th^{4+}, U^{4+}, UO_2^{2+}, Pu^{4+}	$B(CH_3)_3$, GaH_3	BH_3, $Ga(CH_3)_3$, $GaCl_3$, $GaBr_3$, GaI_3,
Ti^{4+}, Zr^{4+}, Hf^{4+}, VO^{2+}, Cr^{3+}, Cr^{6+}, MoO^{3+}, WO^{4+}, Mn^{2+},	R_3C^+, $C_6H_5^+$, Sn^{2+}, Pb^{2+}	Tl^+, $Tl(CH_3)_3$
Mn^{7+}, Fe^{3+}, Co^{3+}	NO^+, Sb^{3+}, Bi^{3+}	CH_2, carbenes
BF_3, BCl_3, $B(OR)_3$, Al^{3+}, $Al(CH_3)_3$, $AlCl_3$, AlH_3, Ga^{3+}, In^{3+}	SO_2	HO^+, RO^+, RS^+, RSe^+, Te^{4+}, RTe^+
CO_2, RCO^+, NC^+, Si^{4+}, Sn^{4+}, CH_3Sn^{3+}, $(CH_3)_2Sn^{2+}$		Br_2, Br^+, I_2, I^+, ICN, etc.
N^{3+}, RPO_2^+, $ROPO_2^+$, As^{3+}		O, Cl, Br, I, N, RO, RO_2.
SO_3, RSO_2^+, $ROSO_2^+$		M^0 (metal atoms) and bulk metals
Cl^{3+}, Cl^{7+}, I^{5+}, I^{7+}		
HX (hydrogen-bonding molecules)		

	Bases		

Hard	Intermediate	Soft
NH_3, RNH_2, N_2H_4	$C_6H_5NH_2$, C_5H_5N, N_3^-, N_2	H^-
H_2O, OH^-, O^{2-}, ROH, RO^-, R_2O	NO_2^-, SO_3^{2-}	R^-, C_2H_4, C_6H_6, CN^-, RNC, CO
CH_3COO^-, CO_3^{2-}, NO_3^-, PO_4^{3-}, SO_4^{2-}, ClO_4^-	Br^-	SCN^-, R_3P, $(RO)_3P$, R_3As
F^- (Cl^-)		R_2S, RSH, RS^-, $S_2O_3^{2-}$
		I^-

fact that F^- is harder than I^-. But when five soft CN^- bases are already present, the complex $Co(CN)_5I^{3-}$ is the most stable of the series and the equivalent with $X = F^-$ is not even known. One can even prepare $Co(CN)_5H^{3-}$ with the very soft base H^- as the sixth ligand. Presumably the soft CN^- base molecules have softened the Co^{3+} considerably with respect to further interaction with bases. Similarly, although alkyl amines RNH_2 are moderately hard bases (because of the electronegativity of the N donor atom), the presence of polarizable pi electrons in aniline and pyridine make those molecules only intermediate in hardness, even though the N atom is still the donor.

The HSAB principle is useful in making qualitative estimates of the solubility of ionic salts in water and to some extent in other solvents, though not many other solvents yield solvation energies large enough to dissolve many ionic salts. In water solutions, the O atom in H_2O is the electron donor. Since oxygen is strongly electronegative, water is a hard base. However, it is not as hard as F^-, though it is harder than the other halide ions. The result is that the solubility of fluorides in water is frequently quite different from that of other halides, and in a way that is consistent with the HSAB principle. Consider the data in Table 6.4 for the solubility of lithium and silver halides. In the solid state, all eight compounds have the metal surrounded symmetrically by halide ions (viewing them for simplicity as ionic compounds). To dissolve any of the compounds in water, the halide bases must be replaced by H_2O base. Lithium, which is a quite hard acid, will prefer the hard base F^- to the less hard H_2O. However, it will prefer the relatively hard H_2O to the softer Cl^-, Br^-, and I^-. One can thus qualitatively rationalize the low solubility of LiF as compared to the high solubilities of LiCl, LiBr, and LiI. On the other hand, the soft acid Ag^+ will prefer the soft bases Cl^-, Br^-, and I^- to the hard base H_2O, but will prefer dissolving in H_2O to remaining in the lattice of even harder F^- ions. In the softer-base solvent SO_2 the Li^+ (hard acid) compounds are less soluble, but the Ag^+ (soft acid) compounds are more soluble, except for AgF.

In another interesting comparison, we can examine the two compounds TlBr and $TlBr_3$, which differ only in the oxidation state of the metal. TlBr is soluble in water to the extent of 0.24 g/L, whereas $TlBr_3$ dissolves 322 g/L. Since Tl^+ is quite a soft acid, it is not surprising that it prefers the soft Br^- lattice to the hard H_2O solvation sphere. Conversely, Tl^{3+} is much harder (roughly comparable to Ga^{3+} and In^{3+} in Table 6.3) and prefers hydration to the Br^- bases in the lattice.

Table 6.4 Solubilities of Lithium and Silver Halides (g salt/100 mL solvent)

Ion	F^-	Cl^-	Br^-	I^-
Li^+ (in H_2O)	0.27	64	145	165
Ag^+ (in H_2O)	182	10^{-4}	10^{-5}	10^{-7}
Li^+ (in SO_2)	0.06	0.012	0.05	20
Ag^+ (in SO_2)	—	0.29	10^{-3}	0.016

6.3 PROTON ACIDS AND BASES AND PROTIC SOLVENTS

Pearson's original description of the HSAB principle compared bases in their behavior toward the extremely hard acid H^+, the proton, which has no electrons at all. The proton represents a very special case in acid–base behavior, for two reasons. First, as it is an elementary particle with no accompanying inner-core electrons, it is easy to move. Even if we think of acids and bases in the Lewis sense, we can readily visualize a proton being transferred from a bonding pair in one molecule to a nonbonding pair in another. This changes our focus on the change occurring in the acid–base system. Second, the proton is the potential Lewis acid in the liquid-water system, which is overwhelmingly the most important solvent in chemical use. Just as the oxygen atom with two nonbonding electron pairs can serve as a Lewis base, the polar O—H bond in water leaves the hydrogen with a high enough positive charge (about 0.15+, according to some quantum mechanical calculations) that it can serve as an electron acceptor or Lewis acid. If two water molecules come together, the proton in one O—H bond can transfer to the previously nonbonding electron pair on the other water molecule's oxygen atom. This is the familiar K_W or *autoprotolysis* reaction:

$$H_2O + H_2O \rightleftharpoons H_3O^+ + OH^-$$

Once our attention is fixed on the transfer of a proton from one electron pair to another, the acid–base reaction represented by this transfer is more conveniently described in terms of the Brønsted acid/base definition:

An acid is a proton donor.
A base is a proton acceptor.

Even within the context of O–H proton attraction for a neighbor electron pair, it is not always necessary to have proton transfer. Kamlet and Taft have proposed an acid/base definition that is similar to the Brønsted definition but focuses on this behavior without transfer of the proton:

An acid donates a proton within a hydrogen bond that is formed.
A base donates an electron pair to a proton within a hydrogen bond that is formed.

Because we so often see proton transfer in, for example, aqueous reactions, the Brønsted definition is more generally useful. However, one specific important application of the Kamlet–Taft definition is in the stabilization of solvated anions (Kamlet–Taft bases) by protic solvent molecules (Kamlet–Taft acids) in solution, an interaction that goes beyond the simple electrostatic solvation the previous section has described.

The Brønsted definition, substituted into the general acid–base reactivity principle, says that the strongest donor in a mixture will react with the strongest proton acceptor to reduce the availability of protons. Thus, for example, the hydrogen sulfite ion will react with carbonate to produce hydrogen carbonate ion, which is a weaker acid than hydrogen sulfate and thus has less proton availability:

$$HSO_4^- + CO_3^{2-} \rightarrow SO_4^{2-} + HCO_3^-$$

If these ions are in aqueous solution, the water molecules represent both a competing acid (against HSO_4^-) and a competing base (against CO_3^{2-}). However, water is both a weaker acid than HSO_4^- and a weaker base than CO_3^{2-}, so it does not take any direct part in the reaction that occurs. Because of the very great number of known hydrides and oxyacids, a wide variety of proton-transfer reactions can be carried out in water solution. Many of these should be familiar from the weak-acid/weak-base equilibria studied in introductory chemistry courses. Several general categories are possible:

Strong acid/water: $HX + H_2O \rightarrow X^- + H_3O^+$
Weak acid/water: $HX + H_2O \rightleftharpoons X^- + H_3O^+$
Strong base/water: $B + H_2O \rightarrow BH^+ + OH^-$
Weak base/water: $B + H_2O \rightleftharpoons BH^+ + OH^-$
Strong acid/weak base: $H_3O^+ + B \rightleftharpoons BH^+ + H_2O$
Weak acid/strong base: $HX + OH^- \rightleftharpoons X^- + H_2O$
Weak acid/weak base: $HX + B \rightleftharpoons BH^+ + X^-$
Strong acid/strong base: $H_3O^+ + OH^- \rightleftharpoons H_2O + H_2O$

The Brønsted definition of acids and bases is most useful in describing many important industrial and biochemical reactions carried out in water solutions. However, several other solvents are known in which comparable reaction equations can be written, since proton transfer is possible from one solvent molecule to another. Such solvents (said to be *protic*) are molecules in which hydrogen is bonded to an atom or group that is electronegative enough to give the hydrogen atom a significant positive charge. Obvious candidates are the hydride neighbors of water in the periodic table, NH_3 and HF. These are in fact widely used (particularly NH_3), but there are also other convenient liquids whose molecules contain protic hydrogens: acetic acid, H_2SO_4, and even HCN. Just as the characteristic acidic and basic species in water are H_3O^+ and OH^-, each of these solvents is capable of autoprotolysis. Table 6.5 shows these characteristic species and gives the equilibrium constant value for the autoprotolysis reaction.

Table 6.5 Some Autoprotolysis Reactions and Equilibrium Constants

		Acid	Base	K_{eq} (25°C)
NH_3	$+ NH_3$	$\rightleftharpoons NH_4^+$	$+ NH_2^-$	5×10^{-27}
HF	$+ HF$	$\rightleftharpoons H_2F^+$	$+ F^-$	2×10^{-12}
$HOAc$	$+ HOAc$	$\rightleftharpoons H_2OAc^+$	$+ OAc^-$	10^{-14}
H_2SO_4	$+ H_2SO_4$	$\rightleftharpoons H_3SO_4^+$	$+ HSO_4^-$	3×10^{-4}

In each of these solvents there is a characteristic acid ion, as Table 6.5 indicates, and a characteristic base ion. It follows that each of the categories of aqueous acid–base reactions has a counterpart in a nonaqueous protic solvent. In liquid ammonia, for example, urea is a weak acid and can be titrated by amide ion (the characteristic strong base of ammonia):

$$\underset{\substack{\| \\ \text{(O)}}}{H_2N-C-NH_2} + NH_2^- \rightarrow \underset{\substack{\| \\ \text{(O)}}}{H_2N-C-NH^-} + NH_3$$

We do not usually think of urea as an acid, but this only reveals the extent to which we are conditioned to think of acids and bases in terms of their behavior in water (where urea is a base). The lower electronegativity of N relative to O makes the nonbonding electrons on NH_3 more readily available than those on H_2O. As a better electron donor, NH_3 is a fundamentally more basic solvent in which molecules such as urea with no measurable acidic properties in water behave as weak acids, and many aqueous weak acids become strong acids:

$$HOAc + NH_3 \rightarrow OAc^- + NH_4^+$$

Conversely, some strong bases in water become weak bases in liquid ammonia,

$$OCH_3^- + NH_3 \rightleftharpoons CH_3OH + NH_2^-$$

and only the very strongest bases in our common experience are also strong bases in ammonia:

$$H^- + NH_3 \rightarrow H_2 + NH_2^-$$

The other direct analog of water is liquid HF solvent. Since F is even more electronegative than O, the HF molecule is a weaker electron donor than H_2O, is thus a weaker Lewis base, and therefore is a more acidic solvent in either the Lewis or Brønsted sense. As Chapter 5 suggested, hydrogen bonding is particularly strong in liquid HF (solvent HF will be a particularly strong Kamlet–Taft acid). The autoprotolysis equilibrium forms hydrogen-bonded polymeric species that are at least as complex as the following:

$$3HF(\text{liquid HF}) \rightleftharpoons H_2F^+(\text{liquid HF}) + HF_2^-(\text{liquid HF})$$

In Fig. 4.17 we have already presented molecular-orbital energy levels for the symmetrical FHF^- ion. The bent HFH^+ ion is isoelectronic with H_2O and can be described by the same MO energy-level diagram. The chain and ring polymers of HF in the liquid state also stabilize these ions. In water, the ionic mobilities of H_3O^+ and OH^- are unusually high because proton transfer through a hydrogen bond can create the electrical effect of ionic motion without actual diffusion (see Fig. 6.3). In liquid HF, the mobilities of H_2F^+ and HF_2^- are much higher than other ions for the same reason. In liquid ammonia, however, the weaker hydrogen bonds

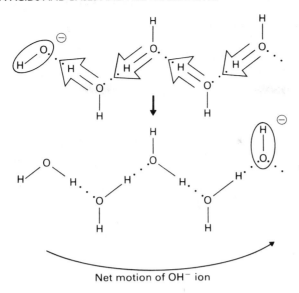

Net motion of OH⁻ ion

Figure 6.3 OH⁻ mobility in aqueous solution through hydrogen-bond proton transfer.

formed have a higher energy barrier in the middle of the potential-energy well (see Fig. 4.16). Proton transfer is thus impeded, and the NH_4^+ and NH_2^- ions have mobilities only modestly higher than those of other ions in that solvent.

In the relatively acidic liquid HF solvent, substances that are weak bases in water become strong bases:

$$NH_3 + 2HF \rightarrow NH_4^+ + HF_2^-$$

Water itself is a moderately strong base:

$$H_2O + 2HF \rightleftharpoons H_3O^+ + HF_2^- \qquad K_b = 2 \times 10^{-1}$$

Even acetic acid is a strong base:

$$CH_3COOH + 2HF \rightarrow CH_3C(OH)_2^+ + HF_2^-$$

Very few species serve as acids in liquid HF. Only $HClO_4$ and HSO_3F donate protons directly when added to liquid HF, but the more electronegative fluorides serve as Lewis acids to the fluoride end of the HF molecule and have the same effect:

$$HClO_4 + HF \rightarrow H_2F^+ + ClO_4^-$$

$$SbF_5 + 2HF \rightarrow H_2F^+ + SbF_6^-$$

This is perhaps a good place to point out that one acid/base definition emphasizes the solvent dependence of acid/base behavior:

An acid dissolves in a solvent or reacts with it to produce the characteristic cation of that solvent.

A base dissolves in a solvent or reacts with it to produce the characteristic anion of that solvent.

This solvent-system definition of acidity is compatible with the Brønsted definition if the cation is defined as the protonated solvent molecule and the anion as the deprotonated solvent molecule. Thus $HClO_4$ and SbF_5 are equally acids in the HF solvent system because both dissolve to produce the H_2F^+ cation characteristic of that solvent. As we shall see later, this definition can be extended to some aprotic solvents in which cations and anions arise through different mechanisms.

It is particularly clear from the discussion of hydrogen bonding and solvent polymerization in liquid HF—and perhaps even from the solvent-system definitions—that solvation energies play a very large role in determining the Brønsted acidity or basicity of species in solution. A somewhat purer view of the Brønsted acid/base properties of individual molecules independent of the solvent's influence is provided by the gas-phase *proton affinity* of the molecule (or ion). The proton affinity, analogous to the electron affinity, is the energy released when a proton is added to an isolated atom or molecule:

$$X(g) + H^+(g) \xrightarrow{\text{PA}} XH^+(g)$$

Proton affinities for bases are measured on a relative basis by comparing the equilibrium distribution of protons between two competing gas-phase bases in a technique called *ion–cyclotron resonance*. In ion–cyclotron resonance, molecular ions are produced by electron impact as in mass spectrometry and allowed to drift through a moderate vacuum ($\approx 10^{-5}$ torr) under the combined influences of an electric and a magnetic field, which is scanned. At the pressures used, molecular collisions occur with possible proton transfer; the product species are detected at different values of the magnetic field.

With respect to the potential Brønsted acidity of various hydrides, it is helpful to look at the proton affinity (PA) of XH_{n-1}^-, which is the conjugate base of the acid XH_n. The greater the proton affinity, the more tightly the proton in XH_n is held, and the weaker XH_n is as a gas-phase Brønsted acid. We can use a Born–Haber cycle to put PA in the context of the other energy quantities with which we are already familiar:

$$
\begin{array}{c}
X^0(g) \;+\; H^0(g) \\
{}^{-EA}\nearrow \quad {}^{-IP_H}\nearrow \quad \searrow {}^{-D_{X-H}} \\
X^-(g) \;+\; H^+(g) \xrightarrow{\text{PA}} XH(g)
\end{array}
$$

Here EA is the electron affinity of the neutral X atom or molecular fragment, IP is the ionization energy of hydrogen atoms (1312 kJ/mol), and D_{X-H} is the bond-dissociation energy of the X—H bond in the XH molecule. Since IP is the same for any proton acid, we can concentrate on the periodic trends in the other two quantities, the covalent-bond energy and the electron affinity. Table 6.6 gives these values for a number of the molecular hydrides and their anions. Considering

Table 6.6 Proton Affinities and Related Energy Quantities (kJ/mol)

	CH_4	NH_3	H_2O	HF
HA	444	535	598	761
IP_X	1230	979	1218	1523
PA	528	866	686	548
	CH_3^-	NH_2^-	OH^-	F^-
D_{X-H}	435	456	497	569
EA	≤50	71	176	331
PA	≥1695	1695	1632	1548

	SiH_4	PH_3	H_2S	HCl
HA	≤439	427	405	506
IP_X	1138	962	1004	1230
PA	≤611	774	711	585
	SiH_3^-	PH_2^-	HS^-	Cl^-
D_{X-H}	335	351	377	431
EA	100	121	222	347
PA	1544	1540	1464	1393

	AsH_3	H_2Se	HBr
HA	389	355	402
IP_X	967	962	1121
PA	732	711	590
	AsH_2^-	HSe^-	Br^-
D_{X-H}	301	318	368
EA	121	209	326
PA	1490	1418	1351

	HI
HA	297
IP_X	1000
PA	607
	I^-
D_{X-H}	297
EA	297
PA	1310

the acid properties first by looking at the anion data, we see at once that the proton affinity is primarily influenced by the charge on the species. The lowest PA for an anion (the least basic X^- anion in the gas phase) is 1310 kJ/mol, whereas the highest PA for a neutral species (the most basic XH molecule in the gas phase) is only 866 kJ/mol. This says that no neutral hydride should be able to serve as a proton acid in any molecular-hydride solvent in the absence of solvation energies! For HCl in water, for example, which in the liquid phase is a strong acid,

$$HCl(g) + H_2O(g) \rightarrow H_3O^+(g) + Cl^-(g) \qquad \Delta E = 1393 - 686 = +707 \text{ kJ/mol rn}$$

In general, the trends in PA are consistent with our instinctive judgments about the acidity of these hydrides, in spite of this surprising observation. CH_4 and NH_3 are extremely weak acids, H_2O is only slightly stronger, and HF is only slightly stronger still. Other horizontal trends are similarly appropriate, but it is again surprising to note that SiH_4 has essentially the same PA as HF, and that in the gas phase, phosphine is a stronger proton acid than water. The vertical trends are also in accord with solution behavior. For example, HF is the weakest of the hydrogen halide acids.

To consider the base properties of these hydrides in the gas phase, we can examine the neutral molecules in the context of the following Born–Haber cycle:

$$
\begin{array}{ccc}
& X^+(g) \quad + \quad H^0(g) & \\
IP_X \nearrow & -IP_H \nearrow & \searrow HA_{(=-D_{X^+-H})} \\
X^0(g) \quad + & H^+(g) \xrightarrow{\;PA\;} & XH^+(g)
\end{array}
$$

Again we need not consider IP_H, but IP_X is the atomic or molecular ionization energy of the X species and HA is the neutral hydrogen-atom affinity of the X^+ cation, which is just the negative of the X^+—H bond dissociation energy. (Note that in general D_{X^+-H} is not the same as D_{X-H}, since the charges are different.) These values are also given in Table 6.6. Since a low PA is characteristic of a weak Brønsted base in the gas phase, we can see from the table that CH_4 is, not surprisingly, a weak base relative to NH_3, which is in turn stronger than H_2O and HF. In keeping with the electronegativities, HF is a slightly weaker base than the other hydrogen halides, which are about equal. On the other hand, NH_3 is significantly stronger as a base than PH_3 or AsH_3. Curiously, there is not much difference between H_2O, H_2S, and H_2Se as bases.

When we look at the energy components of the two Born–Haber cycles, we can see some of the origins of these trends. For example, HF is less acidic than HCl—HI because the bond dissociation energy D_{X-H} is so high for fluorine, which is largely a reflection of the small size of the F atom. A similar trend in D_{X-H} accounts for the acidity trend of H_2O, H_2S, and H_2Se. Although the same trend exists for the HA bond energy in these compounds seen as bases, it is compensated for by an equal trend in the molecular ionization energy; H_2Se is thus more acidic than H_2O, but it is also slightly more basic. CH_4 is a much weaker base than NH_3 essentially because CH_4 has a much greater molecular-ionization energy, which reflects the fact that sigma-bonding electrons must be ionized in CH_4, but nonbonding electrons can be ionized in NH_3.

6.4   WATER AND AQUEOUS SOLUTION SYSTEMS

The proton-affinity energy relationships have emphasized again the primary importance of solvation energies in establishing acid/base properties. Although the formation of H_3O^+ and Cl^- is energetically quite unfavorable in the gas phase, HCl is a strong acid in liquid water precisely because water solvates (hydrates) the ions so strongly. Because of the importance of water as a solvent, we need to look in greater detail at its structure and solvating properties.

Perhaps the most important thing to keep in mind about the structure of liquid water is that there is a lot of structure present by comparison to other liquids. The open tetrahedral structure of ice I, seen in Fig. 6.4, is maintained by strong hydrogen bonds that limit the coordination number of any O atom to four. This is a striking contrast to the isoelectronic Ne atom, which in the solid state is close-packed and thus has twelve neighbors. The openness of the ice I structure is reflected in the relative densities of ice at its melting point (0.92 g/mL) and solid neon at its melting point (1.44 g/mL). Melting the ice to liquid water changes the situation surprisingly little. Liquid water at its freezing point has a density of 1.00 g/mL, whereas liquid neon has a corresponding density of 1.25 g/mL. This evidence of openness in the liquid lattice of water is reinforced by x-ray diffraction results. The radial distribution function from x-ray studies of liquid water,

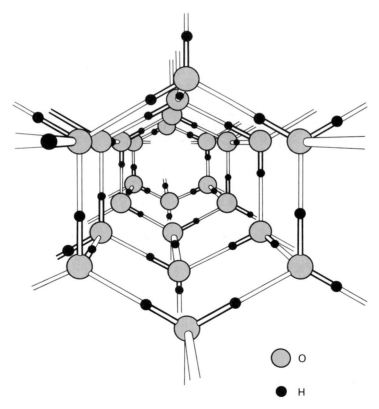

O

H

Figure 6.4 Crystal lattice of ice I, viewed down open shaft of cavities with threefold symmetry (see also Fig. 2.13 diamond lattice).

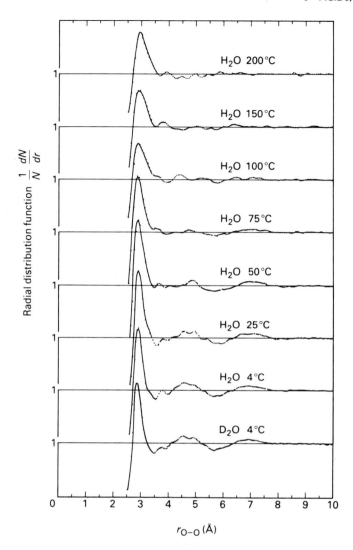

Figure 6.5 Radial distribution function for O atoms in H_2O molecules surrounding a given water molecule in the liquid state. [From A. H. Narten *et al., Disc. Farad. Soc.* (1967), *43*, 97. By permission from The Royal Society of Chemistry, London.]

reproduced in Fig. 6.5, shows a strong peak at about 2.9 Å (the O···O separation), which when integrated indicates that there are on average 4.4 neighbor water molecules in the liquid. The equivalent data for liquid neon indicate the presence of 8.6 neighbors. The inference is that the ice-like structure is maintained to a considerable extent in liquid water, but that some hydrogen bonds are broken or substantially deformed so that the long-range order breaks down. There is a small peak in the distribution function at about 3.5 Å that is not consistent with the tetrahedral structure; this probably represents a few molecules enclosed in the linked cavities or open shafts visible in the ice structure of Fig. 6.4. Near room temperature, water also retains ordering to the extent of having small peaks in the radial distribution function at about 4.5 and 7 Å, though that ordering disappears with a gradual increase in temperature. Thermodynamic data suggest that near room temperature about 50% of the hydrogen bonds of ice are retained in liquid water, though different models yield varied interpretations. Diffusion studies indicate that a given water molecule in a site in the liquid lattice vibrates 100–1000 times before diffusing to a new site, and most theoretical models of liquid water consider vibrationally averaged positions of the molecules. Presumably there are low-energy clusters of water molecules with ice-like structures and nearly maximum hydrogen bonding, and other higher-energy disordered molecules with fewer hydrogen bonds. The ordered clusters change their boundaries as rapidly as molecules can diffuse away and others approach. There have even been suggestions that large regions of a cluster (perhaps 20 or so molecules) relax their hydrogen bonding simultaneously, in a cooperative fashion, to become disordered.

When this curious liquid with an open structure, strong hydrogen bonding, and substantial dipole moment and dielectric constant serves as a solvent for ions, it can accommodate ions within the open structure without very much disruption of the hydrogen bonding of the liquid lattice. The solvation-energy effects (called *hydration energies* for water solvent) are substantial. Calculating them by the approach of the earlier section on solvation, using experimental values for the dipole moment and polarizability of water, yields the results in Table 6.7.

Table 6.7 Hydration Energies for Some Spherical Singly Charged Ions (kJ/mol)

Ion	U_1	U_2	U_3	Theoretical ion hydration energy	"Experimental" hydration energy
Li^+	−391	−202	+63	−531	−561
Na^+	−323	−179	+63	−439	−446
K^+	−267	−164	+63	−368	−362
Rb^+	−247	−159	+63	−343	−341
Cs^+	−228	−153	+63	−319	−317
F^-	−341	−185	+63	−463	−465
Cl^-	−226	−153	+63	−316	−323
Br^-	−209	−148	+63	−294	−296
I^-	−185	−141	+63	−262	−255

Notes: U_1 = ion–dipole net attraction; U_2 = hydrate immersion in dielectric; U_3 = loss of stability from solvent disruption.

The theoretical ion-hydration energy is the sum of U_1, U_2, and U_3. The "experimental" hydration energy assumes $\Delta H_{hyd}(H^+) = -1131$ kJ/mol.

Li^+ and F^- are assumed to be 4-coordinate; all other ions are assumed to be 6-coordinate.

The calculations summarized in Table 6.7 call for some comment. For each ion, the overall attractions have already been estimated as comparable to those in an ionic crystal lattice, so the radii can reasonably be taken as the Shannon and Prewitt crystal radii for the appropriate coordination number. Li^+ and F^- in the first row are assumed to be four-coordinate; all other ions are assumed to be six-coordinate. The radius of a water molecule coordinated to a cation (which, added to the ionic radius, produces r in Fig. 6.1) is assumed to be 1.22 Å, the O^{2-} radius. However, while coordinated to an anion the radius of the water molecule is taken as 1.38 Å, half the $O \cdots O$ distance in ice (shortened to the $F \cdots H \cdots F^-$ distance for $O—H \cdots F^-$). The outer radius (added to produce $r_{solvate}$ in Fig. 6.1) is taken as 1.44 Å, half the $O \cdots O$ distance in liquid water. The calculated values match the "experimental" values fairly well; the deviations are on the order of ±2%.

The quotation marks around "experimental" are apt because it is not possible to make measurements on a single ion. The following Born–Haber cycle shows the energy relationships:

$$M^+(g) \quad + \quad X^-(g)$$

$$U \nearrow \quad \searrow \Delta H_{hyd}(M^+) \quad \searrow \Delta H_{hyd}(X^-)$$

$$M^+X^-(s) \xrightarrow{\Delta H_{hyd}} M^+(aq) \quad + \quad X^-(aq)$$

If one assumes that the lattice energy U can be calculated accurately, then the *sum* of the hydration energies of the cation and anion can be obtained from the cycle

$$\Delta H_{hyd}(M^+) + \Delta H_{hyd}(X^-) = -U_{MX} + \Delta H_{soln}(MX)$$

This presents a problem analogous to that of galvanic-cell potentials, in which no single half-cell potential can be measured. The approach here is the same as for cell potentials: The hydration energy of the proton is arbitrarily taken as zero, so that the measured enthalpy of solution of gaseous HX is, on this scale, equal to the hydration energy (enthalpy) of X^-:

$$\Delta H_{soln}(HX) = \Delta H_{hyd}(H^+) + \Delta H_{hyd}(X^-) \equiv 0 + \Delta H_{hyd}^{rel}(X^-)$$

Relative hydration energies of cations can then be established by subtracting in the previous equation. We could immediately establish absolute hydration energies if we had an experimental value for the hydration energy of the proton. One interesting feature of the solvation-energy theory helps us in this quest: None of the terms in the U_1 and U_2 equations depend on the sign of the charge on the ion. At this level of theory, therefore, a cation and an anion of equal size should have equal hydration energies.

Using the definition of relative hydration energies given above, it is possible to express the absolute hydration energy of an anion in terms of its relative hydration energy and the absolute hydration energy of H^+:

$$\Delta H^{rel}(X^-) = \Delta H_{soln}(HX) = \Delta H_{hyd}(H^+) + \Delta H_{hyd}(X^-)$$

$$\Delta H_{hyd}(X^-) = \Delta H^{rel}(X^-) - \Delta H_{hyd}(H^+) \tag{6.1}$$

An equivalent expression for the cation M^+ can be derived from the enthalpy of solution of the salt MX:

$$\Delta H_{soln}(MX) = \Delta H_{hyd}(M^+) + \Delta H_{hyd}(X^-) = \Delta H^{rel}(M^+) + \Delta H^{rel}(X^-)$$

(canceling the lattice energy from both sides). Rearranging,

$$\begin{aligned}
\Delta H_{hyd}(M^+) &= \Delta H^{rel}(M^+) + \Delta H^{rel}(X^-) - \Delta H_{hyd}(X^-) \\
&= \Delta H^{rel}(M^+) + \Delta H^{rel}(X^-) - [\Delta H^{rel}(X^-) - \Delta H_{hyd}(H^+)] \\
&= \Delta H^{rel}(M^+) - \Delta H_{hyd}(H^+) \qquad\qquad (6.2)
\end{aligned}$$

Subtracting Eq. (6.1) from Eq. (6.2),

$$\Delta H_{hyd}(M^+) - \Delta H_{hyd}(X^-) = \Delta H^{rel}(M^+) - \Delta H^{rel}(X^-) + 2\Delta H_{hyd}(H^+)$$

Since at our level of theory, the left side of this equation should be equal to zero for ions of equal size, we can immediately obtain—for that condition—an expression for $\Delta H_{hyd}(H^+)$ in terms of the arbitrarily assigned relative hydration energies of M^+ and X^-:

$$2\Delta H_{hyd}(H^+) = \Delta H^{rel}(X^-) - \Delta H^{rel}(M^+)$$

$$\Delta H_{hyd}(H^+) = \tfrac{1}{2}[\Delta H^{rel}(X^-) - \Delta H^{rel}(M^+)]$$

Figure 6.6 plots the experimental relative hydration energies of alkali-metal ions and halide ions against their ionic radii. The spherical-potential radii (see Table 2.3 and associated discussion) are used because the characteristic disorder of the liquid approximates the spherical potential on which SPI radii are based. It can be seen that the relative hydration energies do fall on smooth curves and that ΔE, the difference in hydration energies between cation and anion at a constant radius, is indeed very nearly constant for SPI ionic radii between 1.1 Å and 1.8 Å. Since the average of these values is 2262 kJ/mol ion, it follows that the absolute hydration energy of the proton at this level of theory should be half that value, or 1131 kJ/mol. More sophisticated theory yields a $\Delta H_{hyd}(H^+)$ of -1103 kJ/mol and, by including ion interactions with water as an electric quadrupole, gives better agreement with experimental hydration energies.

The large hydration energies imply that the hydrated ion has a strong ordering effect on the surrounding water molecules. Even though the large hydration energies make ΔH favorable for dissolving many ionic salts in water, the ordering effect (as we have already noted) makes ΔS unfavorable for small or highly charged ions. Table 6.8 classifies some molar entropies of solution in water by ion charges. It can be seen that beyond 1+/1− salts, only the bulkiest ions have positive entropies of solution. The ordering effect of highly charged ions can control the solubility of some salts. $CaSO_4$, which is quite insoluble in water, actually has a negative enthalpy of solution of about 17 kJ/mol, but its high negative entropy of solution (-140 J/mol K) makes its free energy of solution positive. Many alkaline-earth carbonates and sulfates show this pattern.

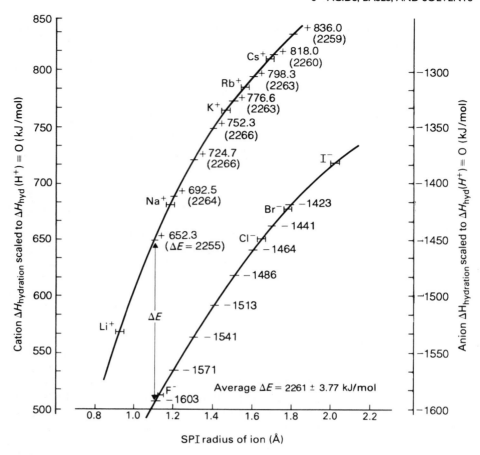

Figure 6.6   Relative ionic hydration energies for alkali metals and halides as a function of ionic radius.

Table 6.8   Entropies of Solution of Ionic Compounds (J/mol K)

1+/1− charge		2+/1− charge		3+/1− charge		2+/2− charge		2+/3− charge	
NaCl	+43	$MgCl_2$	−115	$AlCl_3$	−263	$CaSO_4$	−140	$Ca_3(PO_4)_2$	−838
$NaNO_3$	+89	$CaCl_2$	−45						
NaOAc	+23	$Ca(NO_3)_2$	+46					3+/2− charge	
NH_4Cl	+75								
NH_4NO_3	+109	1+/2− charge						$Al_2(SO_4)_3$	−823
		Na_2SO_4	−11						
		$(NH_4)_2SO_4$	+25						

6.5 ACIDS AND BASES IN WATER

For the simple hydrides, transfer of protons to water in the gas phase was extremely unfavorable thermodynamically. The very large proton-hydration energy that has been revealed by our most recent considerations, however, changes the situation dramatically. We saw that in the gas phase the transfer of a proton from HCl to H_2O was endothermic by some 707 kJ/mol. However, as Chapter 7 will show in the process of comparing acid strengths in water, the huge hydration energy of the proton and the substantial hydration energy of the chloride ion more than overcome this, so that the experimental enthalpy of proton transfer for HCl dissolving in liquid water is −63 kJ/mol HCl. Even though the entropy of solution of gaseous HCl is quite negative (−130 J/mol K) because of the strong ordering effect of the hydrated proton, the favorable enthalpy change is large enough to make the overall free-energy change negative and hence favorable for complete proton transfer. Thus, the crucial quantity in this analysis of why HCl is a strong acid in water is the very large hydration energy of the proton.

Not all hydrides are acids in water, even though all would presumably benefit from the large proton-hydration energy. As the HCl example shows, there is a rather delicate balance of energies that can be affected by several other terms. In particular, there must be a large electron affinity for the hydride's anion fragment. Although we will delay a discussion of acid strengths until the next chapter, the electron-affinity requirement means that, in general, acids in water solution must contain a hydrogen bound to a halogen or some other electronegative element. If a compound's enthalpy of proton transfer to water is to be large enough to yield a substantial negative free energy and thus make it a strong acid, it must contain a hydrogen bound either to oxygen or to a halogen. Not even all of these are strong acids. A strong acid, of course, transfers essentially all of its protons to water; the original acid species, such as HCl, is said to be *leveled* to the acid cation characteristic of the solvent, or H_3O^+ for water solvent. Table 6.9 lists the better-known strong proton acids in aqueous solution. Except for HCl, HBr, and HI, all are protonated oxoanions or fluoroanions that contain several O or F atoms to raise the overall electronegativity of the anion.

Ammonia and other nitrogen hydrides, such as hydrazine and the alkylamines, are the only hydrides that serve as bases in water solution. In the next chapter,

Table 6.9 Strong Proton Acids in Water

HBF_4	$HMnO_4$	HNO_3	H_2SO_4	$HClO_4$	HCl
	$H_2Cr_2O_7$	HPO_2F_2	$H_2S_2O_7$	$HClO_3$	HBr
		HPF_6	HSO_3F	$HBrO_4$	HI
			HSO_3Cl	$HBrO_3$	
			HSO_3NH_2	HIO_3	
			H_2SeO_4		

again, we shall compare the acid/base behavior of NH_3 against that of HCl by way of a Born–Haber cycle; there we shall see that only a fairly delicate thermodynamic balance favors proton acceptance over proton donation for ammonia. More than anything else, the high proton affinity of ammonia shifts the balance to proton-base behavior. However, although only protonated Lewis acids can, in a direct sense, serve as Brønsted acids in water, most Lewis bases can serve as Brønsted (proton acceptor) bases. Only the softest bases such as CO show no tendency to bind a proton (H^+ is a very hard acid). Very strong bases in water will be leveled to the OH^- ion characteristic of H_2O:

$$NH_2^- + H_2O \rightarrow NH_3 + OH^-$$

Because OH^- has such a high proton affinity itself, only species with very high proton affinities will be able to strip protons away from water quantitatively, thereby being leveled. As Table 6.10 indicates, only anions have this property. Since a strong proton base must have a high X—H bond energy, the specific electron-donor atom in the base ion must be small. In fact, it is usually a first-row atom.

Because the oxide ion appears in Table 6.10 as a strong base, and because oxides are perhaps the most familiar compounds of nearly all elements, it is important to make distinctions among the elements according to the acid–base behavior of their oxides. Any metallic element sufficiently electropositive to form an ionic oxide will form a strongly basic solution when its oxide dissolves in water, because the oxide ion hydrolyzes to OH^- but no compensating hydrolysis of the cation occurs. Such an oxide is termed a *basic oxide*. However, more electronegative elements form covalent molecular oxides with high overall molecular electronegativities. A water molecule will serve as an electron-donor base toward these oxides, transferring a proton to another water molecule in solution to create H_3O^+, as in SO_3:

$$SO_3 + 2H_2O \rightarrow HSO_4^- + H_3O^+$$

SO_3 is thus an *acidic oxide*. Not all acidic or basic oxides are water soluble, either for thermodynamic or for kinetic reasons relating to their lattice energies. Basic oxides, however, are usually soluble in concentrated strong acids and acidic oxides in strong bases:

$$MgO + 2H_3O^+ \rightarrow Mg^{2+} + 3H_2O$$

$$Sb_2O_5 + 2OH^- + 5H_2O \rightarrow 2Sb(OH)_6^-$$

Table 6.10 Strong Proton Bases in Water

H^-	CH_3^-	N^{3-}	O^{2-}	BO_3^{3-}
		NH^{2-}	OCH_3^-	PO_4^{3-}
		NH_2^-	S^{2-}	
		P^{3-}		

Table 6.11 Amphoteric Oxide Reactions

Reactions as acid	Oxide	Reactions as base
$Be(OH)_4^{2-} \leftarrow H_2O + 2OH^- +$	BeO	$+ 2H_3O^+ + H_2O \rightarrow Be(H_2O)_4^{2+}$
$2Al(OH)_4^- \leftarrow 3H_2O + 2OH^- +$ (or polymers)	Al_2O_3	$+ 6H_3O^+ + 3H_2O \rightarrow 2Al(H_2O)_6^{3+}$
$Zn(OH)_4^{2-} \leftarrow H_2O + 2OH^- +$	ZnO	$+ 2H_3O^+ + 3H_2O \rightarrow Zn(H_2O)_6^{2+}$
$Sn(OH)_3^- \leftarrow H_2O + OH^- +$	SnO	$+ 2H_3O^+ + H_2O \rightarrow Sn(H_2O)_4^{2+}$ (and $Sn_3(OH)_4(H_2O)_3^{2+}$)
$Sn(OH)_6^- \leftarrow 2H_2O + 2OH^- +$	SnO_2	$+ 4H_3O^+ \rightarrow Sn(H_2O)_6^{4+}$
$Pb(OH)_3^- \leftarrow H_2O + OH^- +$ (and $Pb_6(OH)_8^{4+}$)	PbO	$+ 2H_3O^+ + H_2O \rightarrow Pb(H_2O)_4^{2+}$

Some oxides react with both strong acid and strong base and thereby show both basic and acidic properties (though they usually do not react with water at all). These are the *amphoteric oxides.* The metal atom in an amphoteric oxide must be fairly electropositive to give the oxygen sufficient negative charge to strip a proton from a neighboring H_3O^+. However, the metal ion must also be electronegative enough to serve as an electron acceptor from a neighboring OH^-. Most metals with electronegativities between about 1.5 and 1.8 show amphoteric behavior to some extent, but the best known amphoteric oxides are BeO, Al_2O_3, ZnO, SnO, SnO_2, and PbO (see Table 6.11). Some oxides, called variously *inert* or *neutral oxides,* react with neither acid nor base. Most of these (CO, N_2O, NO) are nonmetal covalent oxides that do not react with a proton because the electron-pair donor atom is too soft and do not react with bases because they already have nonbonding pairs of electrons. The list, however, also includes MnO_2, which does react with acids and bases, but only in a redox fashion.

Since the acidity or basicity of an oxide is in effect determined by the electronegativity of the central atom, it follows that the acidity is also related to the formal oxidation state of the central atom. Although the formal oxidation state should not be confused with the true charge on an atom in a molecule or lattice, it does show the same trend as the net charge of an element that has several stable oxidation states. A positively charged atom attracts electrons much more strongly than a neutral one and thus appears more electronegative, as the differential-ionization-energy expression of Chapter 1 suggests. For this reason, the oxides of elements in high formal oxidation states are more acidic than the corresponding oxides in low oxidation states. Chromium, for example, forms three oxides: CrO is a basic oxide, Cr_2O_3 is amphoteric, and CrO_3 is acidic. Making allowances for this effect, Fig. 6.7 shows the distribution of acidic and basic oxides across the periodic table.

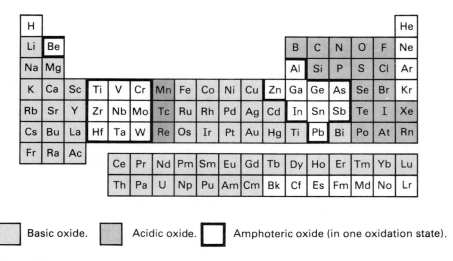

Basic oxide.	Acidic oxide.	Amphoteric oxide (in one oxidation state).

Figure 6.7 Acid/base properties of oxides.

6.6 HYDROXIDES, HYDROUS OXIDES, OXOCATIONS, AND POLYANIONS

The characteristic cations and anions of most solvents are rather elusive in any environment other than that of the solvent itself. Usually a protonated solvent cation will be stable in its own solvent liquid or in more acidic solvent liquids; a deprotonated solvent anion will be stable in its own solvent liquid or in more basic solvent liquids. The cations of basic solvents, such as NH_4^+ (from NH_3), form a number of stable salts; so do the anions of acidic solvents, such as HSO_4^- (from H_2SO_4). The conjugate ions NH_2^- and $H_3SO_4^+$, however, are by far most stable in the solvents themselves. Water is no different; the oxonium or hydronium cation is known only in a few salts such as $H_3O^+ClO_4^-$, which is usually written $HClO_4 \cdot H_2O$. The hydroxide ion, despite its familiarity, is stable in only a few ionic crystals. If the absence of polymeric structures is taken as a criterion for ionicity, only K^+ and the heavier alkali-metal ions, Sr^{2+} and Ba^{2+}, and La^{3+} and most of the lanthanide rare earths form ionic hydroxide crystals. When hydroxides of the other metal cations are prepared from aqueous solution, the product either contains layer structures in which directional covalent bonding is at work in the polymer sheets, small polymer ions in which directional bonding produces rings or clusters, or nonstoichiometric materials called *hydrous oxides* containing hydroxide (and possibly oxide) ions coordinated to a metal cation, along with a variable amount of water hydrogen-bonded into the structure as well as coordinated to the cation.

Layer structures are formed by $LiOH$, $NaOH$, $Mg(OH)_2$, $Ca(OH)_2$, $Mn(OH)_2$, $Fe(OH)_2$, $Co(OH)_2$, $Ni(OH)_2$, $Cu(OH)_2$, $Al(OH)_3$, and $Cd(OH)_2$. In addition, Zn, Sc, and In have crystalline hydroxides in which otherwise symmetrical ionic con-

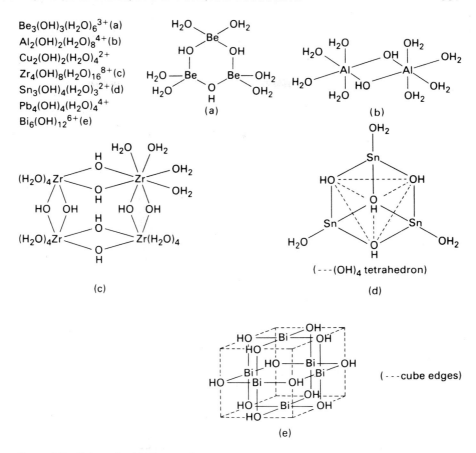

Figure 6.8 Polymerized hydroxycations.

figurations are severely distorted by hydrogen bonding between OH groups. Small polymers are formed by the hydroxides of a number of metals. Some are listed, along with a few structures, in Fig. 6.8. It is likely that more such structures will be discovered as techniques for investigating the structure of solution species improve. Such structures depend on the ability of the OH^- ion to serve as a Lewis base two or even three times in bridging metal ions, since it has three pairs of nonbonding electrons. Such polymers can form in water solution from most metal ions that are electropositive enough to be distinct cations but that have a large enough charge to be good Lewis acids.

Most of the cations that form small hydroxide polymers will also form hydrous oxides, which are gelatinous, flocculent, rather slimy semisolids resembling dilute catsup. To understand these, consider a fairly highly charged ion such as Al^{3+} which in acidic solution will be hydrated by six water molecules as in

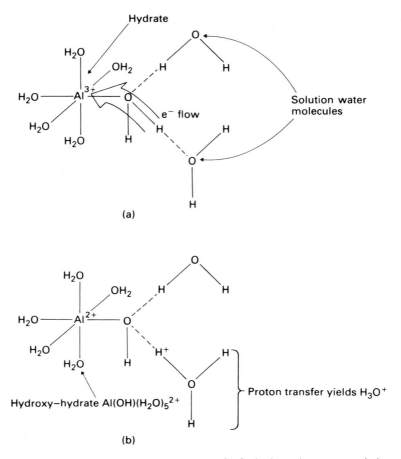

Figure 6.9   Electron flow and proton transfer for hydrates in aqueous solution.

Fig. 6.9a. Each hydrated water molecule will be hydrogen-bonded to other water molecules, as shown. If base is added to the solution, hydronium ions are removed through the K_w equilibrium, and the positively charged hydrogen atoms on the hydrate water molecules find it progressively more advantageous to release electrons to the Al^{3+} and transfer through the hydrogen bond to a neighbor H_2O (or OH^-) to form either H_3O^+ to be neutralized or H_2O directly. The hydrated Al^{3+} ion thus acts as a weak acid being neutralized. When three protons have been stripped from the hydrated ion in this way, it no longer has any net charge. The absence of monopole electrostatic repulsion allows the $Al(OH)_3(H_2O)_3$ molecules to agglomerate through shared OH units and hydrogen bonding, which can include intermediate water molecules as suggested in Fig. 6.10. Such a structure obviously does not have the symmetry of a crystalline hydroxide. Since the hydrogen bonds are much weaker than normal covalent bonds or ionic attraction, the agglomerate is ex-

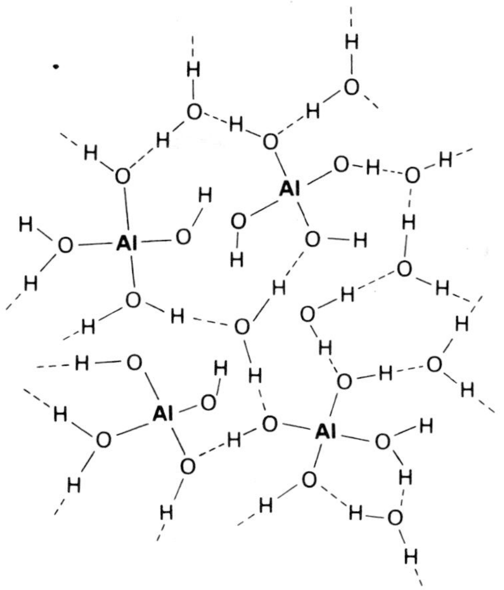

Figure 6.10   Hydrogen bonding in a hydrous oxide.

tremely soft and ill-defined. If it is filtered from the solution and dried by anything less than rigorous means, it will contain the poorly defined, nonstoichiometric amount of water characteristic of a hydrous oxide. The most electropositive elements, such as the alkali metals and alkaline earths, do not form hydrous oxides because they are too large or have insufficient charge to withdraw electrons from their hydrated water molecules, thereby serving as a proton source. On the other hand, the most electronegative elements also do not form hydrous oxides. If they are dissolved in a positive oxidation state, they withdraw electrons so strongly that they make the hypothetical hydrated species a strong acid that will exist in solution as an anion. Addition of base to such an anion either has no effect or, if protons remain, increases the negative charge and prevents agglomeration. An example of this behavior is S^{6+}, which could form the hypothetical hydrate $S(H_2O)_4^{6+}$. Actually, of course, the positive charge on such a species would be removed instantly by proton transfer, leaving the neutral molecule $SO_2(OH)_2$ or H_2SO_4. In solution, the strong acid forms HSO_4^- or SO_4^{2-}.

An electropositive element in solution in a high oxidation state is likely to attract electrons from hydrate water molecules so strongly that it will deprotonate one or two oxygens entirely and form strong multiple bonds with the oxygens. In the best-known examples, V^{4+} forms VO^{2+}, the vanadyl ion, and U^{6+} forms UO_2^{2+}, the uranyl ion. These units (and a few others listed in Table 6.12) are stable enough to persist in a variety of compounds. Since at least a few of them have been characterized in what must be nearly ionic solids [$UO_2(OH)_2$ and UO_2F_2] the class is referred to as *oxocations*. In general, the M—O bond distances are quite short (1.7–1.8 Å for UO_2^{2+}, 1.54–1.62 Å for VO^{2+}), which is the principal basis

Table 6.12   Metal Oxocations

(TiO^{2+})?	VO^{3+}	CrO^{4+}	ReO^{5+}	RuO^{4+}	UO_2^{2+}
(ZrO^{2+})?	VO_2^+	CrO_2^{2+}	ReO_2^{3+}	RuO_2^{2+}	UO_2^+
	VO^{2+}	CrO^{3+}	ReO_3^+	RuO_2^{2+}	(and other
	NbO^{3+}	MoO^{4+}	ReO^{4+}	OsO_3^{2+}	early actinides)
	NbO_2^+	MoO_2^{2+}	ReO_2^{2+}	OsO^{5+}	
	TaO^{3+}	MoO^{3+}	ReO^{3+}	OsO^{4+}	
	TaO_2^+	MoO_2^+	ReO_2^+	OsO_2^{2+}	
		WO^{4+}			
		WO_2^{2+}			
		WO^{3+}			
		WO_2^+			

for suggesting multiple bonding. Since directional bonding with substantial covalent character is occurring in oxocations, the MO or MO₂ group can also occur in molecular compounds. For example, the complex $VOCl_2 \cdot 2N(CH_3)_3$ is soluble in benzene.

Both of the processes just mentioned—the formation of hydrous oxides and the formation of oxocations—are acid–base reactions in water that reduce the net charge on the solute species. Both deal with highly charged, strongly polarizing cations. There is a mirror-image acid–base reaction that reduces the high net negative charge on some anions to form *polyanions*. Since the initial anion species have a charge opposite to those just considered, the charge-reduction process occurs when acid is added, rather than base. Specifically, additional H_3O^+ reduces negative charge on a MO_4^{n-} or MO_6^{n-} oxoanion by stripping an oxide ion out as water. The oxoanion than maintains its coordination number of 4 or 6 by forming a bridging M—O—M bond with another oxoanion. Perhaps the most familiar example is the formation of dichromate ion from chromate in acid solution:

$$2CrO_4^{2-} + 2H_3O^+ \rightleftharpoons Cr_2O_7^{2-} + 3H_2O$$

The reaction occurs by protonating CrO_4^{2-} to $CrO_3(OH)^-$ or $HCrO_4^-$, two of which hydrogen-bond to each other, then form a bridging oxygen bond by eliminating a water molecule. The resulting structure consists of two CrO_4 tetrahedra sharing a corner.

Polyanions are formed by the elements indicated in Fig. 6.11. There is a good deal of resemblance between elements that form amphoteric oxides (Fig. 6.7) and those that form polyanions. However, they are not usually the same oxidation state of the element; polyanions normally involve an element in an oxidation state that corresponds to an acidic oxide. In Chapter 2 we have already identified some of the polyanion species, with structural diagrams in Fig. 2.22. There are usually complex relationships between the degree of polymerization of an anion, the concentration of the solution, and the pH of the solution as Fig. 2.21 indicates for

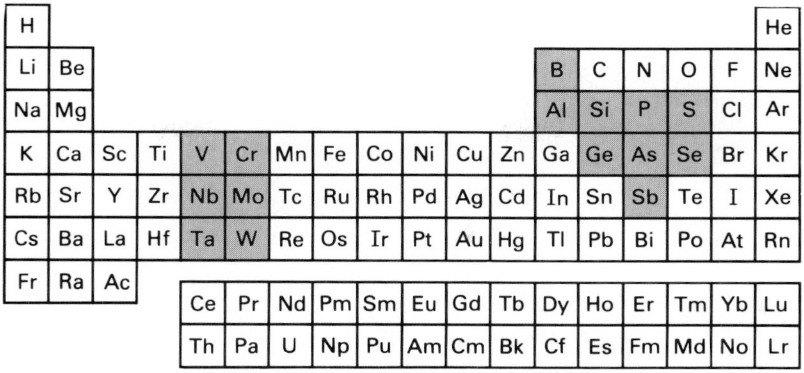

H																	He
Li	Be											B	C	N	O	F	Ne
Na	Mg											Al	Si	P	S	Cl	Ar
K	Ca	Sc	Ti	V	Cr	Mn	Fe	Co	Ni	Cu	Zn	Ga	Ge	As	Se	Br	Kr
Rb	Sr	Y	Zr	Nb	Mo	Tc	Ru	Rh	Pd	Ag	Cd	In	Sn	Sb	Te	I	Xe
Cs	Ba	La	Hf	Ta	W	Re	Os	Ir	Pt	Au	Hg	Tl	Pb	Bi	Po	At	Rn
Fr	Ra	Ac															

Ce	Pr	Nd	Pm	Sm	Eu	Gd	Tb	Dy	Ho	Er	Tm	Yb	Lu
Th	Pa	U	Np	Pu	Am	Cm	Bk	Cf	Es	Fm	Md	No	Lr

Figure 6.11   Elements forming isopolyanions in aqueous solution.

polyvanadates. With the sole exception of polyborates, polyanions are composed of tetrahedra sharing corners or octahedra sharing edges and corners. Polyborates contain some tetrahedral BO_4 groups, but they are composed primarily of planar BO_3 groups. BO_3 and BO_4 groups tend to form primarily rings and linear polymers, whereas MO_6 octahedra tend to form three-dimensional polymers, as the earlier figures suggest. Table 6.13 and the earlier Tables 2.4 and 2.5 give overall formulas for some—but by no means all—of the polyanions formed in this way.

All of the species so far identified have been *isopolyanions* in which the polymerizing oxoanions involve only one element other than oxygen. In a solution containing more than one oxoanion, however, it is often possible to form *heteropolyanions* by condensing the different oxoanions together. Most heteropolyanions involve three-dimensional, roughly spherical cages of MoO_6 or WO_6 octahedra with a foreign ion coordinated inside the cage. Occasionally a second

Table 6.13   Composition of Some Isopolyanions

$V_2O_7^{4-}$	$Cr_2O_7^{2-}$	$B_2O_5^{4-}$
$V_{10}O_{28}^{4-}$	$Mo_7O_{24}^{6-}$	$B_3O_6^{3-}$
$Nb_6O_{19}^{8-}$	$Mo_8O_{26}^{4-}$	$(BO_2)_n^{n-}$
$Ta_6O_{19}^{8-}$	$Mo_{36}O_{112}^{8-}$	$Si_2O_7^{6-}$
	$W_2O_8^{4-}$	$Ge_3O_9^{6-}$
	$W_4O_{16}^{8-}$	$As_3O_{10}^{5-}$
	$W_6O_{22}^{4-}$	$Se_2O_7^{2-}$
	$W_{10}O_{32}^{4-}$	$Se_3O_{10}^{2-}$
	$W_{12}O_{42}^{12-}$	
	$W_{12}O_{40}^{8-}$	

Note: Most of the more highly charged species are at least partly protonated in solution.

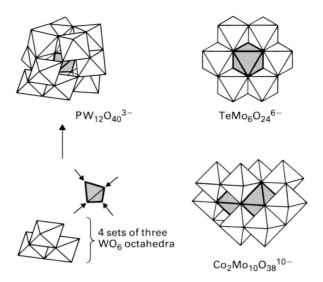

$PW_{12}O_{40}{}^{3-}$

$TeMo_6O_{24}{}^{6-}$

4 sets of three
WO$_6$ octahedra

$Co_2Mo_{10}O_{38}{}^{10-}$

Figure 6.12 Some heteropolyanion cage structures. (Shaded areas represent heteroatom coordination polyhedron.)

foreign ion can be accommodated. A wide variety of foreign ions form such hetero-poly ions: P^{5+}, Si^{4+}, Te^{6+}, Sn^{4+}, Co^{3+}, I^{7+}, Ce^{4+}, Mn^{4+}, and others. The inner atom's coordination can be tetrahedral ($PW_{12}O_{40}^{3-}$), octahedral ($MnMo_9O_{32}^{6-}$), or even icosahedral ($CeMo_{12}O_{42}^{8-}$). (An icosahedron is shown in Fig. 4.28.) The best-known heteropoly ions are the phosphotungstates, silicotungstates, and phosphomo-lybdates. The latter can be reduced to yield a blue color used as a sensitive phos-phate test in water analysis. Elements other than Mo and W can form host cages (for example $MnNb_{12}O_{38}^{12-}$), but such examples are unusual. Figure 6.12, which should be compared to Figure 2.22, shows a few structures of heteropolyanions.

In particular, the $PW_{12}O_{40}$ structure shown at the left of Fig. 6.12 is very commonly found in a variety of systems, probably because of the flexible size of the central hole. Such anions are said to have the *Keggin structure,* and include $PMo_{12}O_{40}^{3-}$, $SiW_{12}O_{40}^{4-}$, and $GeMo_{12}O_{40}^{4-}$. There are also isomers of the Keggin structure in which one or several of the triangles of MO_6 octahedra have been ro-tated to change their linkages to the other triangles and thus the symmetry of the overall anion. "Polymerization" of a sort can also occur for Keggin units; if one triangle of octahedra is removed, the remaining polyanion is cup-shaped with an open face containing the central atom. Two of these cup-shaped units can join at the open face of the cages. This brings the central atoms close together, which is acceptable as long as they do not have lone pairs of electrons facing each other; the $SiW_9O_{34}^{10-}$ open cage can dimerize, but not the $AsW_9O_{34}^{11-}$, for this reason. Even these systems can be linked into "clamshells" linked at one side and bridged across the other side by units such as Hg_2^{2+}, or into square units of four cage monomers that coordinate other cations inside the central opening.

_____ *6.7 NONAQUEOUS PROTIC SOLVENTS*

As long as we stay within the general confines of protic solvents, the Brønsted acid/base definition is appropriate even when we leave the water system. We have already looked at some general properties of the simple water analogs NH_3 and HF. Because N is less electronegative than O, the nonbonding electrons on NH_3 are more readily available than those on H_2O, which makes NH_3 a better Lewis base and liquid ammonia a more basic solvent even when the definition is cast in proton-transfer terms. By an equivalent argument, of course, HF is a more acidic solvent.

We can begin a short survey of commonly used nonaqueous solvents by considering the properties of liquid ammonia in more detail. Ammonia is not as powerful a solvent as water for most simple ionic salts, as Table 6.14 suggests, but it is an unusually good solvent for a few specific cations. The differences in the molecular properties of water and ammonia are summarized as follows: Ammonia has a lower dipole moment (1.46 D for ammonia versus 1.84 D for water), a higher polarizability (2.21 $Å^3$ for ammonia versus 1.48 $Å^3$ for water), and a lower dielectric constant (23 at bp for ammonia versus 78 at 25°C for water). Ammonia forms weaker hydrogen bonds than water and has a higher potential barrier for proton transfer within the hydrogen bond. In solvating an ordinary ion such as K^+, the permanent-dipole attraction of NH_3 molecules will obviously be less, but the induced-dipole attraction will be greater because of the increased polarizability of the NH_3 molecule. Since the NH_3 molecule is about the same size as H_2O, ammonia-solvation energies will be nearly the same as hydration energies for single ions. For example, K^+ has a net ion–dipole attraction of −230 kJ/mol (calculated on the same basis as Table 6.7), a solvate-immersion energy of −152 kJ/mol, a solvent-disruption energy of only about 35 kJ/mol (because of the weaker hydrogen bonds), and a total calculated solvation energy of −344 kJ/mol—nearly the same as that calculated for water solvent (−368 kJ/mol). Individual ions are thus about as stable in ammonia as in water.

Table 6.14 Solubilities of Some Simple Salts in NH_3 and H_2O at 0°C (g/100 g solvent)

Salt	NH_3	H_2O	Salt	NH_3	H_2O
LiCl	1.4	63.7	NH_4Cl	66.4	29.7
LiI	~7	151.	NH_4I	335.	154.2
$LiNO_3$	138.	53.4	NH_4NO_3	274.	118.3
KCl	0.1	27.6	$CaCl_2$	0.0	59.5
KI	184.2	127.5	CaI_2	4.0	181.9
KNO_3	10.7	13.3	$Ca(NO_3)_2$	84.1	102.0
CsCl	0.4	162.2	AgCl	0.3	0.0
CsI	151.8	44.0	AgI	84.2	0.0
$CsNO_3$	—	9.2	$AgNO_3$	~80	122.

However, ion pairing is a much greater problem in ammonia than in water. That is, a solvated ion, once formed, does not spontaneously separate from its neighbor counterion. The ion-pairing potential energy is given simply by Coulomb's law:

$$U_{\text{ion-pr}} = -\frac{q_1 q_2}{\epsilon r_{12}}$$

For singly charged ions of average radius separated by one solvent molecule, $U_{\text{ion-pr}}$ is about 3 kJ/mol when the solvent is water with its high dielectric constant. Since the room-temperature thermal energy (kT) is about 2.5 kJ/mol, ion pairing in water is not extensive. For ammonia, however, $U_{\text{ion-pr}}$ is about 10 kJ/mol. Since this is about five times kT at the bp of ammonia, there is a strong tendency for ions in ammonia to remain together in an effectively uncharged ion pair or complex, behaving as a nonelectrolyte.

One interesting result of ion pairing in ammonia is that the acid-leveling effect takes a curious form. If, recognizing ion pairing, we write the proton-transfer reactions in two stages,

$$\text{HA} + \text{NH}_3 \rightleftharpoons (\text{NH}_4^+)(\text{A}^-) \qquad K_{\text{prot}}$$

$$(\text{NH}_4^+)(\text{A}^-) \rightleftharpoons \text{NH}_4^+ + \text{A}^- \qquad K_{\text{sep}}$$

it is clear that no matter how thorough proton transfer may be (K_{prot} very large) separation of the ion pair will not be very extensive (K_{sep} small). If measured by experimental techniques that detect independent ions, K_a may appear to be small even when proton transfer has been essentially complete. HCl and $HC_2H_3O_2$ appear to have nearly the same K_a (which is appropriate, since both have been leveled to the anion), but that K_a value is about 10^{-4}. For acetic acid, this is somewhat stronger than the value in water (1.8×10^{-5}), but HCl appears much weaker.

A related effect of the low dielectric constant of ammonia is its inability to insulate highly charged ions from each other in order to dissolve solids containing multiply charged ions. Although water dissolves many 2+/2− compounds (particularly if the cation is already hydrated) liquid ammonia dissolves essentially only singly charged ions.

The forms of hydrolysis by which hydrated water molecules transfer protons to other solvent water molecules are mirrored by *ammonolysis* reactions. However, these have yet to be thoroughly investigated:

$$\text{Co(NH}_3)_6^{3+} + \text{NH}_3 \rightleftharpoons \text{Co(NH}_3)_5(\text{NH}_2)^{2+} + \text{NH}_4^+$$

$$\text{BCl}_3 + 6\text{NH}_3 \rightarrow \text{B(NH}_2)_3 + 3\text{NH}_4^+ + \text{Cl}^-$$

Some amides are amphoteric, just as some hydroxides are. $AlCl_3$ dissolved in water precipitates the hydrous oxide when OH^- is added, but the solid dissolves in excess OH^- as the aluminate $Al(OH)_4^-$. $IrBr_3$ (yielding a softer 3+ ion) dissolves in liquid ammonia, precipitates $Ir(NH_2)_3$ on the addition of KNH_2, then redissolves in excess NH_2^- as $Ir(NH_2)_6^{3-}$.

The increased polarizability of NH_3, which makes it a softer Lewis base, makes it a better solvent not only for soft-acid cations such as Ag^+ and Zn^{2+}, but also for more polarizable anions such as I^-. In both cases, the London dispersion attraction, which depends on the product of the polarizabilities of the two neighbors, is enhanced in a solvent of greater polarizability. This effect is visible in Table 6.14. The formation of complexes in which nitrogen atoms donate electron pairs to a transition-metal ion will be explored in later chapters.

Although liquid ammonia is the most commonly used proton-base solvent, some studies have been made of alkylamines, which have a more convenient liquid temperature range than ammonia. Amines are not widely used as solvents, however, because they have only a weak solvating ability, which prevents most ionic species from dissolving. Although the Lewis-base electron pair is as readily available as in ammonia, the alkyl groups on the amines increase the bulk of the molecule and, to some extent, impose steric requirements on solvates.

The primary alcohols constitute a group of protic solvents that are neither acidic nor basic, but that have reduced capabilities (relative to water) for acid/base behavior. Methanol and ethanol are quite polar ($\mu = 1.71$ and 1.68 D, respectively), have dielectric constants comparable to ammonia (32 and 24 at 25°C), and form strong hydrogen bonds. However, their autoprotolysis constants are lower than water (10^{-17} and 10^{-19}), and most weak acids and bases in water solution appear even weaker in alcohols (see Table 6.15). The same steric problems that reduce the solvating ability of alkylamines affect these alcohols, so that low solubility is often a problem for ionic species. The low solvating ability does make the alcohols much more nearly inert solvents for Lewis acid–base reactions, and they are often used in this application.

An important analytical application for anhydrous methanol solvent is the determination of small quantities of water in solids or liquids that are methanol-soluble by the *Karl Fischer titration*. The sample is dissolved in anhydrous methanol and titrated with a methanol solution of SO_2 and I_2, dissolved as pyridine adducts. Methanol reacts with SO_2 to form a methyl sulfite, which is oxidized to methylsulfonate by I_2 but only in the presence of stoichiometric amounts of water:

Table 6.15   Values of pK_a for Weak Acids and of pK_b for Bases

	pK_a in	Water	Methanol	Ethanol
Acid	$HC_2H_3O_2$	4.8	9.5	10.3
	C_6H_5OH	10.0	14.0	
	pK_b in	Water	Methanol	Ethanol
Base	NH_3	4.8	6.3	
	$C_6H_5NH_2$	9.4	11.0	13.3

$$CH_3OH + SO_2 + py \rightarrow pyH^+ + CH_3OSO_2^-$$

$$CH_3SO_3^- + I_2 + H_2O + 2py \rightarrow 2pyH^+ + CH_3OSO_3^- + 2I^-$$

Although the titration can be done by using an iodometric indicator, it is now usually done by generating the I_2 coulometrically from iodide and measuring the time to generate excess iodine. This is, of course, an application in which the alcohol is not only solvent but reactant.

Proton-acid solvents are used in a variety of applications. The two commonest are glacial acetic acid and sulfuric acid. The names are unfortunate in that they describe the behavior of these compounds in aqueous solution, whereas acetic "acid" actually serves as a strong base in H_2SO_4 solvent; but there are no good alternatives.

Acetic acid, the less acidic of the two solvents, has a convenient liquid temperature range ($16-118°C$), is quite polar ($\mu = 1.74$ D in the gas phase), and forms strong hydrogen bonds. Ionic solubilities are not very high, however, because acetic acid has a very low dielectric constant ($\epsilon = 6.2$ at $25°C$). Ion pairing is quite extensive for the same reason. Substances that are bases in water are normally bases in acetic acid, and weak bases are much stronger. Even some carboxylic-acid species serve as bases; potassium hydrogen phthalate (KHP) is a strong base in acetic acid, since it is entirely converted to acetate and phthalic acid:

$$K^+OOCC_6H_4COOH^- + HC_2H_3O_2 \rightarrow K^+C_2H_3O_2^- + HOOCC_6H_4COOH$$

Anhydrous perchloric acid and HBr appear to be strong acids in acetic acid solution (though with extensive ion pairing), but the other common mineral acids, H_2SO_4, HCl, and HNO_3, are present in solution at least partly in molecular form and must be considered weak acids. Some metal species are amphoteric, as the following reaction sequence indicates:

$$ZnCl_2 + 2C_2H_3O_2^- \rightarrow Cl^- + Zn(C_2H_3O_2)_2(s)$$

$$Zn(C_2H_3O_2)_2 + 2C_2H_3O_2^- \rightarrow Zn(C_2H_3O_2)_4^{2-}$$

This sequence should be compared with that found for $IrBr_3$ in liquid ammonia. Acetic acid does not form as many solvates as water (at least partly because of the steric requirements of the —COOH group), but several are known including a disolvate of the $ZnCl_2$ above: $ZnCl_2 \cdot 2HC_2H_3O_2$. The behavior of $ZnCl_2$ in acetic acid thus parallels almost exactly that of amphoteric metal-ion hydrates in water.

A great deal of attention has been given to H_2SO_4 solvent because of the unusual chemical species that are stabilized by its extremely acidic nature. Sulfuric acid is a powerful solvent because of its high dielectric constant ($\epsilon = 100$ at $25°C$), its very high polarity, and its strong hydrogen bonding. It has a remarkably high autoprotolysis constant (about 10^{-4}), which emphasizes its proton-transfer capability. Because it is so acidic, most solutes serve as bases yielding the HSO_4^- ion:

$$H_2O + H_2SO_4 \rightarrow H_3O^+ + HSO_4^-$$

$$NH_4ClO_4 + H_2SO_4 \rightarrow NH_4^+ + HClO_4 + HSO_4^-$$

Note that the formation of H_3O^+ by water does not make the solution acidic; in sulfuric acid solvent, H_3O^+ is just another cation. Note also that the formation of undissociated $HClO_4$ in the solvolysis of NH_4ClO_4 indicates that strong acids are quite rare in sulfuric acid solvent. The only strong acid is made by allowing SO_3 to dehydrate boric acid in sulfuric acid solution:

$$B(OH)_3 + 3SO_3 + 2H_2SO_4 \rightarrow B(OSO_3H)_4^- + H_3SO_4^+$$

Because of the extensive self-ionization, most solutes solvolyze to bind the HSO_4^- ion (as B^{3+} does above). Tin(IV) compounds not only solvolyze but also bind solvated H_2SO_4 molecules. The resulting solvate is a weak acid:

$$SnCl_4 + 6H_2SO_4 \rightarrow Sn(HSO_4)_4(H_2SO_4)_2 + 4HCl$$

$$Sn(HSO_4)_4(H_2SO_4)_2 + H_2SO_4 \rightleftharpoons H_3SO_4^+ + Sn(HSO_4)_5(H_2SO_4)^-$$

This is exactly comparable to the behavior of $AlCl_3$ in water, where the hydrated Al^{3+} ion is a weak acid comparable to acetic acid.

The strong hydrogen bonds in liquid H_2SO_4 make proton transfer and autoprotolysis particularly easy. They also greatly enhance the electrical conductance of the $H_3SO_4^+$ and HSO_4^- ions by a mechanism like that of Fig. 6.3. Since the multiple hydrogen bonds also make H_2SO_4 quite viscous, the diffusion and ionic mobility of other charged species are restricted so much that the electrical conductance of a sulfuric acid solution is essentially due to the characteristic solvent ions alone, no matter what else is present. Most soluble species are electrolytes in H_2SO_4 because of the great reactivity and potential for solvolysis of the solvent. The few soluble nonelectrolytes (such as CH_3SO_2F and SO_2Cl_2) actually depress the conductivity of their solutions below that of the free solvent, apparently because they solvate the autoprotolysis ion HSO_4^- and reduce its free concentration. Acid–base titrations in sulfuric acid solvent can be carried out by monitoring the conductance of the solution. The conductance will drop sharply as a strong acid, $HB(HSO_4)_4$, is added to a solution containing the base HSO_4^-, then rise after the equivalence point is passed as excess $H_3SO_4^+$ is added to the solution. If the acid or base is weak rather than strong, the titration curve will be curved and displaced from the equivalence point by a factor that can be calculated from the K_a or K_b values. Conductance and freezing-point data are frequently used to interpret the progress of reactions in sulfuric acid medium. Conductance gives the number of moles of HSO_4^- (or occasionally $H_3SO_4^+$) formed by a mole of solute; freezing-point depression gives the total number of solute particles formed per mole of initial solute. Freezing-point depression measurements are convenient because H_2SO_4 has a relatively large K_f (6.12 deg · kg/mol). Thus when HNO_3 is dissolved in H_2SO_4, four particles are formed per HNO_3 molecule, two of which conduct electricity significantly. From this we can infer the following reaction:

$$HNO_3 + 2H_2SO_4 \rightarrow NO_2^+ + H_3O^+ + 2HSO_4^-$$

As the nitronium ion in the last reaction equation suggests, H_2SO_4 is a good host medium for electronegative cations and ions in high oxidation states, because such species are strong Lewis acids and H_2SO_4 has only very weak Lewis-base properties that might destroy the species. For this reason, it is often used as a solvent for oxidation reactions producing such species. Conversely, liquid ammonia (a good Lewis base) is a good host medium for other strong Lewis bases such as carbanions; it is used to carry out many reduction reactions that produce unusual species that are strong Lewis bases. We shall examine some uses of these and other nonaqueous solvents in Chapter 8.

6.8 APROTIC SOLVENTS

A very wide range of liquids can serve as aprotic solvents for inorganic species and reactions. The useful solvents include CS_2 and hexane at the bottom end of the polarity scale, from which internal electrical forces escalate all the way to molten salts at the other end of the scale. We can usefully group the solvents in terms of the dominant intermolecular forces present in the liquid. A few solvents such as CS_2, CCl_4, hexane, and benzene offer essentially only the very weak London dispersion forces of attraction. We can group these as *van der Waals solvents*. The next group, with stronger interactions, has polar molecules that usually have substantial Lewis-base electron-donor capability and many have Lewis-acid acceptor ability as well under appropriate circumstances. There are many of these, including the protic solvents just discussed. Two of the better-known aprotic solvents in this category are acetonitrile and dimethylsulfoxide. Extremely polar solvents can be arbitrarily separated from the *Lewis-base solvents* by their ability to promote ion-transfer reactions between solute species. *Ion-transfer solvents* include such polar molecules as $POCl_3$ and BrF_5 and also molten salts such as LiCl–KCl eutectic and Na_2CO_3–K_2CO_3 eutectic. We shall briefly consider each of these categories in turn.

The van der Waals solvents have only the very weakest forces between solvent molecules. Therefore, in forming solutions there is essentially no energy cost for the disruption of solvent–solvent attractions. However, they are equally unable to generate solute–solvent attractions except for the London dispersion force (which makes them better solvents for highly polarizable solutes containing heavy atoms). This prevents them from dissolving ionic salts or even strongly polar molecular compounds. Since there can be little enthalpy yield for the process of dissolution, entropy effects dominate the free-energy change for the process. The entropy of solution is invariably favorable for nonpolar solutes, since they are unable to orient solvent molecules around them as an ion does water molecules. Most of the simple nonpolar molecules are therefore reasonably soluble in these solvents. As examples of the use of these solvents, the following reactions are fairly typical:

$$(C_2H_5)_6Al_2 + Sb_2O_3 \xrightarrow{\text{Hexane}} 2(C_2H_5)_3Sb + Al_2O_3$$

$$2Br_2 + 2HgO \xrightarrow{\text{CCl}_4} Br_2O + HgBr_2 \cdot HgO$$

$$Se_8 + 4Cl_2 \xrightarrow{\text{CS}_2} 4Se_2Cl_2$$

Lewis-base solvents (also called coordinating solvents) are the group that most closely resembles water in solvent properties. Typically, a Lewis-base solvent will contain an oxygen (sometimes a nitrogen) atom bound in a sterically prominent position in a small organic molecule. The hydrocarbon portions of the molecule enhance the donor ability of the oxygen atom through their low electronegativity, and the sterically prominent position of the donor atom ensures a substantial dipole moment and minimizes steric hindrance in a solvated species. The more commonly used Lewis-base solvents have substantial dielectric constants (above 20), which increase their solvent ability for ionic solutes and reduce ion pairing. Table 6.16 lists some of these solvents, arranged in order of their dielectric constants. All have substantial dipole moments as well, so that if the solvent molecules are small enough, the ion-dipole component of the solvation energy should be substantial. This ability to specifically coordinate cations is the common property of Lewis-base solvents. The N atoms that serve as donors in acetonitrile and pyridine are somewhat softer bases than the O donors in the other solvent molecules. This affects specific solubilities, but cation–solvent interactions are similar in general.

These solvents, however, differ markedly in their behavior toward anions. This can have a strong influence on the nature of the stable solute species. For example, each of the solvents in Table 6.16 coordinates Fe^{3+} by solvating $FeCl_3$. All but pyridine form $FeCl_2S_4^+$ (where S is a solvent molecule) and sometimes other cationic species. Pyridine forms $FeCl_3 \cdot py$, a neutral species that is consistent with the low dielectric constant of pyridine. However, there is a sharp distinction between the anionic species present. If no specific coordination of an anionic species is possible, a large anion can be solvated away from its cation more easily than a small one of the same charge; thus, $FeCl_4^-$ is the stable anion in MeCN, $MeNO_2$, acetone, DMA, and py. On the other hand, the small Cl^- is the stable ion in NMA and DMSO (and methanol). These solvents can coordinate a Cl^- by serving as

Table 6.16   Some Lewis-Base Coordinating Solvents

Name (common designation)	mp (°C)	bp (°C)	μ (D)	ϵ
N-methylacetamide (NMA)	28	206	3.73	165
Propylene carbonate (PC)	−55	240	4.98	69
Dimethyl sulfoxide (DMSO)	18	189	3.96	45
Acetonitrile (MeCN)	−48	82	3.92	39
Dimethylacetamide (DMA)	−20	165	3.81	38
Dimethylformamide (DMF)	−61	153	3.82	37
Nitromethane ($MeNO_2$)	−29	101	3.50	36
Hexamethylphosphoramide (HMPA) (CARCINOGEN)	7	232	5.37	30
Acetone	−95	56	2.84	20
Pyridine (py)	−42	115	2.19	12

electron acceptors or Lewis acids; NMA, which strictly speaking should be considered a protic solvent, can hydrogen bond from its N—H to the Cl^- and is thus a Kamlet–Taft acid. The same is true of methanol. DMSO has vacant d orbitals on the sulfur atom at the positive end of the dipole that can accept electrons from the Cl^-. To the extent that these Lewis or Kamlet–Taft acid–base interactions can be viewed as ion–dipole attractions, the specific coordination of the anion by the solvent dipole will be favored by the small anion.

Ion-transfer solvents also serve as Lewis bases toward cations in solution and are similar in this respect to the solvents of Table 6.16. However, they have the additional property of *autoionization,* analogous to the autoprotolysis of protic solvents. In the autoionization reaction, a small electronegative anion such as Cl^-, F^-, or O^{2-} is transferred from one solvent molecule to another:

$$POCl_3 + n\ POCl_3 \rightleftharpoons POCl_2^+ + Cl(POCl_3)_n^- \qquad \kappa = 2 \times 10^{-8}$$

$$SbCl_3 + SbCl_3 \rightleftharpoons SbCl_2^+ + SbCl_4^- \qquad \kappa = 8 \times 10^{-7}$$

$$BrF_3 + BrF_3 \rightleftharpoons BrF_2^+ + BrF_4^- \qquad \kappa = 8 \times 10^{-3}$$

$$IF_5 + IF_5 \rightleftharpoons IF_4^+ + IF_6^- \qquad \kappa = 5 \times 10^{-6}$$

The κ values given for each reaction are the specific conductances of the pure liquids (in $ohm^{-1}\ cm^{-1}$), presumably caused by the indicated ions. The chloride- and fluoride-transfer properties implied by these autoionization reactions make the liquid solvents strong chlorinating and fluorinating agents. In fact, any substance soluble in BrF_3 will be converted into the corresponding fluoride.

$$KCl + Ta^0 \xrightarrow{BrF_3} KTaF_6$$

$$Au^0 \xrightarrow{BrF_3} AuF_3$$

$$N_2O_4 + Sb_2O_3 \xrightarrow{BrF_3} (NO_2^+)(SbF_6^-)$$

Other solvents from the list above serve sometimes as chlorinating agents,

$$SbCl_5 + POCl_3 \rightarrow POCl_2^+ + SbCl_6^-$$

and sometimes simply as coordinating solvents,

$$TiCl_4 + 2POCl_3 \rightarrow TiCl_4(OPCl_3)_2$$

$$TiCl_4(OPCl_3)_2 + POCl_3 \rightleftharpoons [TiCl_3(OPCl_3)_3]^+ + Cl^-$$

$$TiCl_4(OPCl_3)_2 + Cl^- \rightleftharpoons [TiCl_5(OPCl_3)]^- + POCl_3$$

$$[TiCl_5(OPCl_3)]^- + Cl^- \rightleftharpoons TiCl_6^{2-} + POCl_3$$

although it should be noted that even in the latter case, the solvent's primary role is to facilitate ion transfer.

One of the most thoroughly studied ion-transfer solvents is liquid SO_2, which does not autoionize. A great deal of solution chemistry in liquid SO_2 has been rationalized in terms of the supposed oxide-transfer reaction

$$2SO_2 \rightleftharpoons SO^{2+} + SO_3^{2-}$$

but it seems clear that no such reaction occurs. Isotopically labeled sulfur does not exchange between SO and $SOCl_2$—a reaction that should occur rapidly if the SO^{2+} ion exists in even the smallest concentrations, since isotopically labeled chloride does exchange rapidly between Cl^- and $SOCl_2$ in liquid SO_2.

In spite of its failure to autoionize, liquid SO_2 promotes both oxide- and chloride-ion transfer among solutes. For example,

$$Et_2Zn + SO_2 \longrightarrow ZnO + Et_2SO$$

$$PCl_5 + SO_2 \longrightarrow POCl_3 + SOCl_2$$

$$2\ UCl_5 \xrightarrow{\text{Liq. } SO_2} UCl_4 + UCl_6$$

$$AsCl_3 + 3\ KF \xrightarrow{\text{Liq. } SO_2} AsF_3 + 3\ KCl$$

If autoionization did occur, the solvent-system acid/base definitions would make the hypothetical SO^{2+} the acid in liquid SO_2 and the real SO_3^{2-} the base. A number of metathetical reactions are known in which these species do not actually react, but the products are those that would be expected from such an acid–base reaction:

$$Cs_2SO_3 + SOCl_2 \xrightarrow{\text{Liq. } SO_2} 2CsCl + 2SO_2$$

Such a reaction can be carried out as a conductimetric titration with a reasonably clean equivalence point. In the context of this acid/base system, it is even possible to show amphoterism, as in the following reaction sequence (compare with previous amphoteric reactions in other solvents):

$$2AlCl_3 + 3SO_3^{2-} \xrightarrow{\text{Liq. } SO_2} Al_2(SO_3)_3(s) + 6Cl^-$$

$$Al_2(SO_3)_3(s) + 3SO_3^{2-} \xrightarrow{\text{Liq. } SO_2} 2Al(SO_3)_3^{3-}$$

$$2Al(SO_3)_3^{3-} + 3SOCl_2 \xrightarrow{\text{Liq. } SO_2} Al_2(SO_3)_3(s) + 6Cl^- + 6SO_2$$

The parallel to the behavior of $AlCl_3$ in water is so exact that one can reasonably use the term amphoterism even though the acid cation SO^{2+} apparently does not exist.

Molten salts are a logical extension of the molecular ion-transfer solvents, as suggested by the series of autoionization reactions:

$$AsCl_3 + AsCl_3 \rightleftharpoons AsCl_2^+ + AsCl_4^-$$

$$HgCl_2 + HgCl_2 \rightleftharpoons HgCl^+ + HgCl_3^-$$

$$Na^+Cl^- \rightarrow Na^+ + Cl^-$$

$HgCl_2$, the intermediate case, is traditionally termed a molten salt even though its electrical conductivity is quite low compared to that of, say, NaCl in the molten state (10^{-3} ohm^{-1} cm^{-1} versus 800 ohm^{-1} cm^{-1} in the middle of their liquid ranges). The boundary is somewhat arbitrary, obviously. The commonest molten-salt media are the alkali halides, which are often used in eutectic low-melting mixtures to bring the liquid-temperature range closer to the usual laboratory conditions. The LiCl–KCl eutectic mixture melts at 450°C, the KCl–ZnCl$_2$ eutectic at 262°C, and the KCl–AlCl$_3$ eutectic at only 128°C. The lower temperatures of the Zn and Al systems are associated with the formation of bulky chlorometallate anions such as AlCl$_4^-$ that have lowered lattice energies.

Metals and metalloids are often soluble in their own molten halides, forming solutions of two fundamentally different types. Very electropositive metals such as the alkali and alkaline-earth metals form colored solutions in their halides in which there is little interaction between metal and solvent. The conductivity of the solution increases markedly over that of the molten halide; in such solutions, as much as 99% of the conductance can be electronic as opposed to ionic, even though the solution may be only about 10 mole percent metal. On the other hand, less electronegative metals such as Cd and metalloids such as Bi form solutions in their molten halides in which there is strong interaction between solute and solvent and little or no increase in conductivity. Solubility in this sort of solution usually depends on the possibility of forming a *subhalide* of the metal. Most subhalides involve small clusters of metal atoms in a charged species, including but not limited to the Zintl anions. Perhaps the simplest is Hg$_2^{2+}$, formed when mercury is dissolved in HgCl$_2$. Most elaborately, Bi dissolves in BiCl$_3$ to give Bi$_9^{5+}$, BiCl$_5^{2-}$, and Bi$_2$Cl$_8^{2-}$.

Molten salts are excellent solvents for ion-transfer reactions. NaCl and PbCl$_2$ react in molten HgCl$_2$ to yield Na$_2$PbCl$_4$, and the unusual oxidation state Cu^{3+} is formed in a 3:1 KCl–CuCl$_2$ melt on fluorination:

$$CuCl_2 + 3KCl + 3F_2 \rightarrow K_3CuF_6 + \tfrac{5}{2}Cl_2$$

Mixed chloride–fluoride melts such as KF–KCl–ZnCl$_2$ allow the convenient fluorination of a number of molecular chlorides or oxychlorides:

$$Me_3SiCl + NaF \rightarrow Me_3SiF + NaCl$$

$$SOCl_2 + 2NaF \rightarrow SOF_2 + 2NaCl$$

In a fully dissociated molten salt such as NaCl, the characteristic-cation and characteristic-anion definitions of acid and base are not particularly helpful, since they imply that the melt is at all times full of acidic and basic species that do not react with each other. However, melts involving oxoanions can show slight oxide dissociation as in molten NaNO$_3$:

$$NO_3^- \rightleftharpoons NO_2^+ + O^{2-}$$

This implies a capability for oxide-transfer reactions, which in fact occur:

$$Cr_2O_7^{2-} + CO_3^{2-} \rightarrow 2CrO_4^{2-} + CO_2$$

The above reaction has been characterized in molten nitrates, but the wide variety of possible oxoanion melts—particularly those involving silicate and borate glasses—makes it convenient to define acids and bases for those solvents in terms of their potential for oxide-transfer. The *Lux–Flood acid/base* definitions represent, again, the Lewis definitions made specific to these solvents:

An acid is an oxide acceptor.
A base is an oxide donor.

In a molten oxoanion salt, the strongest oxide donor will react with the strongest oxide acceptor to reduce the availability of O^{2-}. In the reaction above, $Cr_2O_7^{2-}$ serves as an acid and CO_3^{2-} as a base. Although NO_3^- can serve as an oxide donor, it is a weaker oxide donor than CO_3^{2-} and thus takes no part in the reaction. The Lux–Flood definitions make it easy to understand the role of limestone in steelmaking, where the production of a ton of pig iron in a blast furnace requires about a ton of coke as a reducing agent, but also about 800 lb of $CaCO_3$. The carbon (coke) reduces Fe_2O_3 but not the associated SiO_2. The $CaCO_3$ serves as an oxide-donor base so that the oxide acceptor SiO_2 will dissolve in the molten-silicate slag:

$$CaCO_3 \rightarrow CaO + CO_2$$
$$CaO + SiO_2 \rightarrow CaSiO_3$$

Many of the polymerization and depolymerization reactions of oxoanions, particularly those occurring in glasses or molten silicate minerals, can be interpreted conveniently in terms of the Lux–Flood acid/base model.

6.9 SUPERACIDS

We have looked briefly at a wide variety of solvents and the possible patterns of acid/base behavior in them. Particularly in the category of protic solvents, we have seen that some solvents can be intrinsically more acidic or basic than others. Although we have stayed away from the relative strengths of acids and bases, it is possible to use the various definitions of acid/base reactivity to design the most acidic possible solvent. The resulting solvent, which has been investigated extensively by Gillespie, Olah, and others, is called a *superacid* medium.

If we wish the strongest possible Brønsted acid as a pure liquid, we seek a hydroxylic compound with the smallest possible number of protons and the most electronegative possible anion. These properties maximize electron withdrawal from the O—H bond and make the H as protic as possible. One possibility is $HClO_4$, but as a pure liquid it is a treacherously explosive oxidizing agent. A less vigorous oxidizing agent is sulfuric acid, which can be rendered even more electronegative and acidic by replacing one —OH group with either —F to form fluorosulfuric acid (HSO_3F) or —CF_3 to form trifluoromethylsulfonic acid (HSO_3CF_3). Both of these are acids with strength comparable to that of perchloric acid in media such as acetic acid. Liquid HSO_3F, which has been more thoroughly investigated than HSO_3CF_3, is thus a first approach to a superacid. It has a convenient liquid range ($-89°C$ to $163°C$) and an autoprotolysis about 1% as extensive as that of sulfuric acid. It is also less viscous than H_2SO_4 because of the reduced number of hydrogen bonds per molecule.

We can increase the acidity of HSO_3F by adding to it the strongest possible Lewis acid. The Lewis acid must accept an electron pair without net electron transfer (since that would reduce the Lewis-acid molecule), so the strongest possible Lewis acid will be the most electronegative molecule that (*a*) has a vacant coordination site on its central atom and (*b*) is two electrons short of filling the bonding MOs for the full coordination geometry. Such a molecule accepts a neighbor in its inner coordination sphere that donates two electrons to the central atom, and it exerts the strongest possible attraction on those electrons. Since the highest common coordination number is 6, we choose a pentafluoride with no nonbonding electrons in the sixth coordination site, which is to say a group V pentafluoride: SbF_5 or AsF_5. SbF_5 dissolved in HSO_3F does behave as a weak acid; it can be titrated conductometrically by the base SO_3F^-. SbF_5 apparently accepts a pair of electrons from an oxygen atom in HSO_3F, thereby withdrawing electrons even more strongly from the O—H bond in the coordinated HSO_3F.

An alternative approach would be to mix a strong Lux–Flood acid with HSO_3F. SO_3, which is chemically compatible with HSO_3F, is such an acid, but it does not strip an oxide ion from HSO_3F to leave HSO_2F^{2+}, which would presumably be a very strong acid. Instead, it forms HSO_3OSO_2F, in which SO_3 accepts a pair of electrons from HSO_3F but in which the O—H has moved from the —SO_2F sulfur atom to the SO_3 sulfur atom. The resulting HS_2O_6F is not acidic enough to donate protons to the HSO_3F solvent, so SO_3 does not increase the acidity of the HSO_3F solvent.

However, when SO_3 *and* SbF_5 are dissolved in HSO_3F, the SO_3 sulfur serves as an F^- acceptor, forming the solvent's characteristic anion SO_3F^-, which then coordinates the Sb atom in the former fluoride site. This continues in the following sequence, in which the final neutral molecule $H[SbF_2(SO_3F)_4]$ serves as a strong acid, even in HSO_3F:

$$SbF_5 + SO_3 \rightleftharpoons SbF_4(SO_3F)$$

$$SbF_4(SO_3F) + SO_3 \rightleftharpoons SbF_3(SO_3F)_2$$

$$SbF_3(SO_3F)_2 + SO_3 \rightleftharpoons SbF_3(SO_3F)_3$$

$$SbF_2(SO_3F)_3 + HSO_3F \rightleftharpoons H[SbF_2(SO_3F)_4]$$

$$H[SbF_2(SO_3F)_4] + HSO_3F \rightarrow H_2SO_3F^+ + SbF_2(SO_3F)_4^-$$

SO_3 thus forms a strong acid in the SbF_5–HSO_3F superacid solvent, being leveled to the characteristic cation $H_2SO_3F^+$.

As one might imagine, proton bases in SbF_5–HSO_3F solution are not difficult to find. Organic amines, of course, are strong bases, but so is nitrobenzene:

$$\phi NO_2 + HSO_3F \rightarrow \phi NO(OH)^+ + SO_3F^-$$

Sulfuric acid, in fact, serves as a weak base:

$$H_2SO_4 + HSO_3F \rightleftharpoons H_3SO_4^+ + SO_3F^-$$

The perchlorate ion serves as a strong base. The molecular $HClO_4$ formed by solvolysis shows no proton-donor capability:

$$ClO_4^- + HSO_3F \rightarrow HClO_4 + SO_3F^-$$

Sulfur dioxide, in keeping with the nearly nonexistent Lewis-base qualities it shows as a solvent, is not protonated in SbF_5–HSO_3F solution. Since it breaks up the hydrogen-bond network without absorbing protons, it reduces the viscosity of the solution without changing its acidity. For this reason, SO_2 is often added to a superacid solvent to simplify laboratory manipulations.

Because any available nonbonding electrons in a superacid solution have already been leveled to the extremely weak base SO_3F^-, the solvent is a good host for unusual and unstable cations. Organic carbonyl and carboxyl groups can be protonated:

$$\underset{\text{CH}_3\text{CCH}_3}{\overset{\overset{\displaystyle O}{\|}}{}} + HSO_3F \rightarrow \underset{\text{CH}_3\text{CCH}_3}{\overset{\overset{\displaystyle OH^+}{|}}{}} + SO_3F^-$$

$$\underset{\text{OH}}{\overset{\displaystyle O}{\text{CH}_3\text{C}}} + HSO_3F \rightarrow \underset{\text{OH}}{\overset{\displaystyle OH^+}{\text{CH}_3\text{C}}} + SO_3F^-$$

Even hydrocarbons can be protonated under favorable circumstances:

$$(CH_3)_4C + HSO_3F \rightarrow [(CH_3)_3C(CH_4^+)] \rightarrow (CH_3)_3C^+ + CH_4$$

The oxidizing agent peroxydisulfuryl difluoride, FSO_2OOSO_2F or $S_2O_6F_2$, oxidizes the less electronegative nonmetal elements to unusual cations:

$$2I_2 + S_2O_6F_2 \rightarrow 2I_2^+ + 2SO_3F^-$$
$$3I_2 + S_2O_6F_2 \rightarrow 2I_3^+ + 2SO_3F^-$$
$$Se_8 + S_2O_6F_2 \rightarrow Se_8^{2+} + 2SO_3F^-$$
$$Se_8^{2+} + S_2O_6F_2 \rightarrow 2Se_4^{2+} + 2SO_3F^-$$

As is frequently the case with fluorinated ion-transfer solvents, the SbF_5–HSO_3F medium is a strong fluorinating agent as well as a superacid. Most nonmetal and metalloid oxides and oxoanions will be partially or completely fluorinated by the medium, particularly at high temperatures. Boric acid yields BF_3, As_2O_3 yields AsF_3, P_4O_{10} yields POF_3, selenate ion yields SeO_2F_2, and permanganate and perchlorate ions yield MnO_3F and ClO_3F. These species, most of which would be instantly hydrolyzed by water, are stabilized by the acidic environment and can usually be distilled out of the superacid solvent without decomposition.

PROBLEMS

A. DESCRIPTIVE

A1. What do the values in Table 6.1 (intermolecular forces for liquids) suggest about the mutual solubility of those liquids?

A2. Classify the following as Lewis acids or Lewis bases with respect to the less-electronegative atom in each: PH_3, PCl_5, BeH_2, $BeCl_2$, CO_2, CO, $SnCl_2$, $SnCl_4$.

A3. Use the HSAB principle to decide which of the following should be readily soluble in water: $CaI_2(s)$, $CaF_2(s)$, $PbCl_2(s)$, $PbCl_4(l)$, $CuCN(s)$, $Cu(CN)_2(s)$, $ZnSO_4(s)$, $ZnSO_3(s)$.

A4. Consider the autoprotolysis reactions of Table 6.5. For each solvent, produce a "pH" scale equivalent to that of water. Write the definitions of the standard acidic and basic solutions, and relate "pH" to "pOH."

A5. Why is the viscosity of liquid water at its boiling point more than twice as great as that of the isoelectronic liquid neon at its boiling point?

A6. Why is the entropy of solution of the $+1/-1$ salt NH_4NO_3 more than twice as positive as the entropy of solution of the $+1/-1$ salt NaCl?

A7. Why are the oxyacids H_3PO_4 and HIO_4 *not* strong acids in water?

A8. For the elements of group Va, write balanced reactions for the hydrolysis of the M_2O_3 oxides, of which N and P are acidic, As and Sb are amphoteric, and Bi is basic.

A9. Of the first-row elements Li through Ne, only one forms a hydrous oxide. Which element is it? In making your prediction, be sure your answer is consistent with the discussion of oxoanions in Chapter 2.

A10. Can you suggest a structural reason why many polyanions have formulas containing six or more metal atoms, often with no smaller polyanions detectable?

A11. Why is LiI much less soluble in liquid ammonia than in water, even though CsI is much more soluble in ammonia? See Table 6.14.

A12. What reaction would you expect if the strong-acid species $HB(OSO_3H)_4$, prepared in H_2SO_4 solvent, is dissolved in water?

A13. Suggest the general type of aprotic solvent (van der Waals, Lewis-base, ion-transfer) best suited as host to each of the following reactions:

(a) $CuCl + AlCl_3 \rightarrow CuAlCl_4$

(b) $2 LiCH_3 + LiI + InCl_3 \rightarrow (CH_3)_2InI + LiCl$

(c) $2AgSCN + I_2 \rightarrow 2AgI + (SCN)_2$

(d) $Et_4Pb + HCl \rightarrow Et_3PbCl + C_2H_6$

(e) $N_2F_4 + AsF_5 \rightarrow N_2F_3^+AsF_6^-$

(f) $TiI_4 + 4N_2O_4 \rightarrow Ti(NO_3)_4 + 4NO + 2I_2$

(g) $SPCl_3 + 3NH_4SCN \rightarrow SP(NCS)_3 + 3NH_4Cl$

(h) $B_{10}H_{14} + 2Et_3N \rightarrow (Et_3NH)_2B_{10}H_{10} + H_2$

(i) $6S_2Cl_2 + 16NH_3 \rightarrow S_4N_4 + S_8 + 12NH_4Cl$

(j) $KF + SO_2 \rightarrow KSO_2F$

A14. In glassmaking, glass sand (approximately SiO_2) is melted with sodium carbonate and limestone (calcium carbonate). The melt evolves CO_2 gas. Interpret the process in terms of the Lux–Flood acid/base model.

B. NUMERICAL

B1. Develop an algebraic expression for the first-layer solvation energy U_1 of a cation with charge q^+ in a molten-salt solvent, surrounded by an octahedral array of spherical anions X^-. Let the internuclear distance be R, and assume the X^- ions have polarizability α. Work out the numerical value of the geometric constant for anion–anion repulsion.

B2. Consider the two solvents A and B. Solvent A has a dielectric constant of 100, whereas that of B is only 10. For simplicity, the solvents are assumed to be otherwise identical in their molecular properties (not actually possible). In solvent A, the cation M^+ has a total solvation energy of 400 kJ/mol M^+, and U_1 and U_2 are equally important. Calculate the solvation energy of M^+ in solvent B. Discuss the probable thermodynamic significance of the changed dielectric constant.

B3. For the two solvents of problem B2, calculate the ion-pairing potentials for the salt M^+X^-. Both solvent molecules have a radius of 2.0 Å and both ions have a radius of 1.0 Å. Calculate the fraction of ion pairs (separated by a solvent molecule) in each solvent that have sufficient thermal energy under a Boltzmann distribution ($e^{-U_{\text{ion-pr}}/RT}$) to dissociate at 25°C. Comment on the probable thermodynamic significance of the changed dielectric constant.

B4. Use the proton affinities of Table 6.6 to calculate ΔH^0 values ($\approx \Delta E$) for each of the following gas-phase proton-transfer reactions:

$$NH_3 + HCl \rightarrow NH_4^+ + Cl^-$$

$$NH_3 + Cl^- \rightarrow NH_2^- + HCl$$

What reaction should occur in aqueous solution? How can this result be reconciled with your calculated ΔH^0 values?

B5. Use data from this chapter and Chapter 2 to show that there is virtually no net heat flow when a mole of NaCl is dissolved in water.

B6. Use the Kapustinskii treatment to estimate the lattice energy for $AlCl_3$ as $Al^{3+}(Cl^-)_3$. Combine this value with the hydration energy of Cl^- (Table 6.7) and the experimental heat of solution of $AlCl_3$ (-338 kJ/mol $AlCl_3$) to yield a value for the hydration energy of Al^{3+}. Compare your result with the tabulated hydration energy of the isoelectronic Na^+, and discuss the contrasting values, suggesting reasons for the widely differing values.

C. EXTENDED REFERENCE

C1. Plot ΔH_{hyd}^{rel} values for the alkali-metal ions and halide ions against Shannon crystal radii to yield a graph comparable to Fig. 6.6. Calculate several values of $\Delta H_{hyd}(H^+)$ from your graph, covering the ion-radius range form 1.2 Å to 1.8 Å. Radii are tabulated in Chapter 2, and relative hydration energies are given in J. O'M. Bockris and A. K. N. Reddy, *Modern Electrochemistry* (New York: Plenum, 1970). List the assumptions you make, and discuss the relative constancy of the resulting proton hydration energy.

C2. Explain how the electrical conductance of a superacid solution can be used to indicate the formation of the strong acid $H[SbF_2(SO_3F)_4]$ as SO_3 is added to SbF_5-HSO_3F. [See R. J. Gillespie, *Acc. Chem. Res.* (1968), *1*, 202.]

C3. B. S. Ault [*Inorg. Chem.* (1981), *20*, 2817] has reported the formation of $SiF_4 \cdot NH_3$ when SiF_4 and ammonia are codeposited on a solid argon surface. However, he was unable to form any comparable complex with SiF_4 and PH_3. Using gas-phase proton affinities as a guide, show from this result that ammonia should be the only simple nonmetal hydride that can form a complex with SiF_4.

C4. J. B. Lambert *et al.* have prepared and studied the trimethylsilyl cation Me_3Si^+ and some related organosilyl cations in nonaqueous solvents [*Organometallics* (1991), *10*, 2578]. In Lewis-base solvents several of these species have high electrical conductivity, indicating reasonably clean ionic behavior. In acetonitrile, the $PhMe_2Si^+ClO_4^-$ system has an equivalent conductance of 836 ohm cm^2 equiv^{-1}, while the $PhMe_2Si^+BPh_4^-$ (tetraphenylborate) system has an equivalent conductance of 2120 ohm cm^2 equiv^{-1}. Why is the conductance so much higher for the tetraphenylborate solution? In CH_2Cl_2 solvent, the perchlorate has a conductance of only 0.38 ohm cm_2 equiv^{-1}. Why should it be so much lower in methylene chloride?

Part III

Main-Group Reactions

7

Enthalpy-Driven Reactions I:
Acid–Base Reactions ━━━━━━━━

Although the previous chapter dealt extensively with acid/base chemistry, it did so primarily from a structural perspective. Primary emphasis was laid on the nature of liquid solutions and the solvent–solute interactions that stabilize them. Since in many cases these conform to an acid/base definition, it seemed appropriate to deal with solvents and solution species in that context. In this and the next two chapters, however, we shall change our emphasis. On the basis of the structures we have surveyed in all the chapters of Part II, we shall look at the potential for chemical reaction among these species. This chapter begins by examining the strength of acid–base interactions, the origins of the thermodynamic driving force for acid–base reactions, and the range of reaction types that are usually taken as having an acid–base nature.

━━━━━━ ## 7.1 SPONTANEITY AND THERMODYNAMICS

The strength of an acid–base interaction is an assessment of its thermodynamic driving force, quantified as the free-energy change ΔG for the acid–base reaction. We have already noted that at ordinary laboratory temperatures, the entropy change associated with most inorganic reactions is less important than the enthalpy change, because $-T\,\Delta S$ is usually smaller for $T = 300$ K than ΔH in the free-energy definition $\Delta G = \Delta H - T\,\Delta S$. For the specific case of acid–base reactions, we also observe that most acid–base reactions involve a combination of an acid species and a base species: $A + B \rightarrow AB$. Even for proton-transfer reactions in solution, where the total number of molecular species does not change, many reactions increase the number of charged ions in solution and thereby increase the extent of solvation: $AH + B \rightarrow HB^+ + A^-$. In either case, the system becomes more ordered as the reaction progresses and consequently has a negative entropy change. To have a negative free-energy change (that is, to be spontaneous), the acid–base reaction must have a negative enthalpy change, which means that the total electronic energy of the system must become more negative (more stable). We shall shortly look at the origins of this increased electronic stability for different acid/base definitions.

It may be appropriate to detour briefly to look at the circumstances under which an acid–base reaction can have a positive ΔS, since that adds to the overall thermodynamic stability of the reaction products. In the simplest case, for instance, proton transfer in a Brønsted acid–base reaction often reduces the net charge on the original acid and base species. Smaller net charges will have a weaker ordering effect on the surrounding solvent:

$$H_3O^+(aq) + S^{2-}(aq) \rightarrow HS^-(aq) + H_2O(l) \qquad \Delta S = +109 \text{ J/mol K}$$

$$NH_4^+(aq) + PO_4^{3-}(aq) \rightarrow NH_3(aq) + HPO_4^{2-}(aq) \qquad \Delta S = +185 \text{ J/mol K}$$

There is an important kind of Lewis acid–base reaction that characteristically has a positive entropy change. Some Lewis-base molecules have more than one pair of nonbonding electrons that are available for donation both in an electrostatic and a steric sense. Molecules that can serve as multiple electron-pair donors through several different atoms to a single acceptor atom or ion are called *chelating agents*. When a solvated cation in solution forms a chelated acid–base compound with such a chelating agent, the total number of particles present increases, and with it the entropy of the system:

$$Ca(H_2O)_6^{2+}(aq) + Y^{4-}(aq) \rightarrow CaY^{2-}(aq) + 6H_2O(l) \qquad \Delta S = +118 \text{ J/mol K}$$

In this equation, Y^{4-} represents the ethylenediaminetetraacetate ion

(commonly abbreviated EDTA), which serves as a Lewis base through each of the two N atoms and one O atom on each of the four carboxyl groups, forming an extensively chelated octahedral coordination pattern around the Ca^{2+} acceptor ion. This ΔS value represents a large contribution to the stability of the calcium–EDTA complex. Since ΔH for the same reaction is -27.0 kJ/mol rn, the $-T\,\Delta S$ contribution of -35.0 kJ/mol rn (at 25°C) to the overall free-energy change of -62.0 kJ/mol rn is actually more important to the stability of the complex than the enthalpy change. A similar (though not necessarily as large) effect will occur whenever more than one single-electron-pair Lewis base is replaced by a chelating Lewis base on a given Lewis acid. This is the *chelate effect*, which simply says that a Lewis acid–base complex involving a chelating donor will be more stable than the equivalent complex involving single-pair donor molecules with the same donor atoms. We will see a number of examples of the chelate effect in considering transition-metal complexes in Chapter 12.

More generally, however, we find that entropy effects do not dominate Lewis acid–base reactions—rather, the reactions proceed because there is a substantial favorable (negative) ΔH for the acid–base combination. The origin, of course, of this added electronic stability is the fact that two electrons in a nonbonding (or only weakly bonding) orbital redistribute themselves to occupy a strongly bonding orbital and are thus attracted by an increased number of nuclei. If we think in molecular-orbital terms about the magnitude of the energy change associated with this redistribution, we can isolate no fewer than five factors that affect the energy change and hence ΔH for the acid–base reaction.

The first of these factors is the valence-orbital ionization potential (VOIP) of the acceptor atom or acid, which should be considered in terms of the energy-level diagrams in Figs. 4.3 and 7.1. The binding energy quantity β is usually taken to be proportional to the average of the VOIPs of the two atoms forming the bond. Since the acid is the "new" atom, the greater its VOIP, the greater β becomes. The second factor is the VOIP of the base atom. This is true even though the donor–acceptor electrons have been on the base all along, so that that aspect of their environment is not changing. The reason for the $VOIP_{base}$ influence is that the difference between $VOIP_{acid}$ and $VOIP_{base}$, $\Delta VOIP$, influences β. The greater $\Delta VOIP$ is, the smaller the β quantity will be, essentially because a very large $\Delta VOIP$ corresponds to a situation in which the bonding electrons spend so little time near the nucleus with the high-energy orbital that it can have little binding effect on them.

A third factor influencing β is the degree of AO overlap, S_{ab}, between the donor and acceptor orbitals. The greater the overlap, the greater the binding energy in most MO treatments. S_{ab} and β are smaller both if the donor and acceptor orbitals are quite different in size and if the donor orbital is not very directional (compared to,

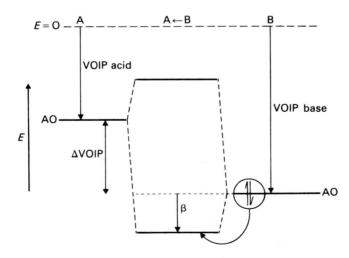

Figure 7.1 MO energy quantities for Lewis acid–base interaction.

for example, a highly directional hybrid orbital). This can result in what is called *back-strain* in Lewis acid–base interactions—that is, a reduction of donor–acceptor overlap because one or both molecules have internal repulsions that reduce favorable hybridization. For example, simple amines like trimethylamine, with a bond angle C—N—C of 110.9 °, can be described using sp^3 hybrid orbitals so that the donor orbital is a directional sp^3 hybrid. Very bulky amines, such as triphenylamine with a C—N—C bond angle of 116°, presumably use orbitals close to sp^2 hybridization to bond the organic groups, leaving a more or less unhybridized p orbital to serve as donor. The p orbital cannot overlap an acceptor as well as an sp^3 hybrid, so the donor–acceptor bond energy is lower. Although one cannot usually isolate a single effect, it is at least consistent that the proton affinity of trimethylamine (958 kJ/mol) is significantly greater than that of triphenylamine (~925 kJ/mol).

Still another factor influencing ΔH for acid–base reactions is a van der Waals repulsion between acid substituents and base substituents—steric hindrance, in other words. Even if the β binding-energy quantity is essentially the same for a given acid and two different bases, increased steric repulsions can significantly affect the overall βH. For instance, quinuclidine and triethylamine have the same proton affinity within 1 kJ/mol (which is not surprising since they have essentially the same structure except that quinuclidine has its ears pinned back). Since a proton has no inner core and no substituents, it cannot encounter any steric repulsions. However, when the same two amines form Lewis acid–base compounds with trimethylboron, ΔH for the formation of the triethylamine compound is about 8kJ less favorable than ΔH for the quinuclidine compound. In the context of these acid–base interactions, this sort of steric repulsion is called *front-strain*. The front-strain in this case is the van der Waals repulsion between the methyl groups of the triethylamine and those of the trimethylboron.

Finally, ΔH for a Lewis acid–base interaction can be affected by geometric changes that occur within either the acid or the base when the donor–acceptor bond is formed. If, say, the free base molecule has optimum stability with a certain set of bond angles, those bonds may be weakened if the angles are deformed when the base coordinates with an acid. The amine manxine is an example.

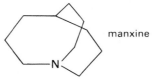

manxine

As a free molecule, it takes on a nearly planar bridgehead C and N geometry. Its low molecular ionization potential should give it a quite high proton affinity (see Chapter 6 and Table 6.6). Its experimental proton affinity, however, is slightly lower than that of tripropylamine (which it resembles) because the N atom is forced to pucker toward tetrahedral geometry (∡ C—N—C = 115°) when it forms the N—H bond. This produces a large internal strain energy relative to the free amine, which lowers the stability of the ammonium ion and thus the proton affinity. This is an example of a third category of strain, *internal strain,* that has a steric effect on ΔH for acid–base interaction.

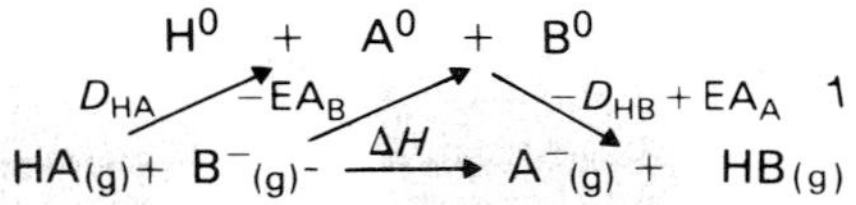

$$H^0 \quad + \quad A^0 \quad + \quad B^0$$

$$D_{HA} \nearrow -EA_B \qquad -D_{HB} + EA_A \quad \uparrow$$

$$HA_{(g)} + B^-_{(g)} \xrightarrow{\ \Delta H\ } A^-_{(g)} + HB_{(g)}$$

$$\Delta H = D_{HA} - EA_B - D_{HB} + EA_A = \Delta D_{(HA-HB)} + \Delta EA_{(A-B)}$$

HA	B⁻	A⁻	HB	$\Delta D_{(HA-HB)}$	$\Delta EA_{(A-B)}$	ΔH(kJ/mol)
(1) HF + OH⁻		→ F⁻ + HOH		(569 − 498)	(−331 + 176)	
				+71	−155	−84
(2) HI + OH⁻		→ I⁻ + HOH		(297 − 498)	(−297 + 176)	
				−201	−121	−322
(3) HI + F⁻		→ I⁻ + HF		(297 − 569)	(−297 + 331)	
				−272	+34	−238

Figure 7.2 Born–Haber cycle data for gas-phase Brønsted acid–base reactions.

For Brønsted acid–base reactions, the electronic origin of the favorable ΔH is not as simple as forming bonds from nonbonding electrons. In the reaction $B^-: + H—A \rightarrow B—H + :A^-$, a pair of nonbonding electrons on B is forming a bond, but a pair of bonding electrons on A is becoming nonbonding. To appreciate the molecular origins of ΔH for this reaction, consider the Born–Haber cycle in Fig. 7.2, in which the imaginary intermediates are the neutral atoms H, A, and B. The three real reactions tabulated in the figure actually occur in both the gas phase and in solution, but for different reasons. The enthalpy change is equal to the difference in the bond-dissociation energies of HA and HB, plus the difference in the electron affinities of A^0 and B^0. For the first reaction, in which HF serves as the acid and OH⁻ as the base, the bond-energy difference is unfavorable but is outweighed by the much more favorable electron-affinity difference. For the third reaction, in which HI is the acid and F⁻ the base, the situation is reversed: The electron-affinity difference is unfavorable, but it is outweighed by the very high H—F bond energy, which makes the bond-energy difference favorable. And in the strongly driven case of the second reaction, both energy differences are favorable. The data given in the figure are taken from Table 6.6, which refers to unsolvated gaseous species. Solvation energies will substantially affect any ordinary laboratory reaction. Even so, it is clear that a Brønsted acid–base reaction is favored if a very strong H—B bond is formed with the potential base (usually more important if there is a substantial size difference between the competing base atoms—see Fig. 5.18), or if an extremely electronegative A⁻ anion is formed in which the A^0 atom has a large electron affinity.

7.2 SPONTANEITY, MECHANISMS, AND RATES

Some nonthermodynamic considerations can affect the structure and, by inference, the reactivity of acids and bases. For example, it should be possible to form

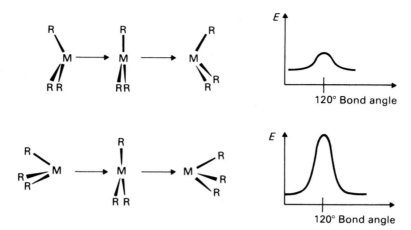

Figure 7.3 Energies for inverting MR$_3$ species.

an optically active (chiral) Lewis acid–base compound between an electron-acceptor acid A and an unsymmetrical base :NRR′R″, since the tetrahedral N in the donor-acceptor compound would then have four dissimilar ligands. Such chiral coordination compounds are known. However, the unsymmetrical amine NRR′R″ itself should also be chiral, since the nonbonding electron pair itself represents a "dissimilar ligand." Such amines, however, cannot be experimentally resolved (unless the amine is polycyclic, such as manxine) because the activation-energy barrier for inversion of amines is so low (20–30 kJ/mol reaction) that a D- or L-conformation racemizes faster than the resolution can be carried out. In general, the activation-energy barrier will be higher the more the bond angles in the free base deviate from the planar 120° (see Fig. 7.3). An N donor atom in a small ring or at a bridgehead obviously raises the inversion energy barrier substantially, but substituted phosphines and arsines also have smaller angles and much higher inversion barriers than the corresponding amines (Table 7.1).

Inversion of the sort we have been discussing can occur during an acid–base reaction in which a new base substitutes for another in a Lewis acid–base adduct.

Table 7.1 Geometries and Inversion Barriers for MR$_3$

Molecules	R—M—R angle (degree)	Inversion barrier (kJ/mol)
NH$_3$	106.6	25
NMe$_3$	110.9	33
PH$_3$	93.8	113
PMe$_3$	98.9	134
AsH$_3$	91.8	
AsMe$_3$	96	

Figure 7.4 suggests the course of the reaction under two circumstances: one in which the incoming base and the departing one are both extremely electronegative, and another in which they are not. The first inverts the molecule's conformation; the second leads to retention. Both lead to stereospecific products. We can approach the difference between the two in terms of valence-shell electron-pair repulsion (VSEPR) theory, according to which a trigonal bipyramidal system, such as the intermediate for the mechanisms shown in Fig. 7.4, will tend to place its most electronegative ligands at the axial positions of the trigonal bipyramid. This happens because the axial and equatorial positions are not symmetrically equivalent, and the axial ligands (which suffer the greatest repulsion) are most stable if they are so electronegative that they can withdraw the bonding electrons from the repelling equatorial pairs. Consequently, if the entering and departing base are electronegative, they will tend to occupy the axial positions in the reaction intermediate and inversion will accompany the substitution. On the other hand, if the remaining ligands are more electronegative, the departing base will tend to occupy an equatorial position, and the substitution will retain its original configuration. Most such substitution reactions are stereospecific for second-row and heavier elements, which suggests that they have associative mechanisms involving a reaction intermediate having a larger coordination number than the original molecule. If the mechanism were dissociative, so that a four-coordinate molecule $R_3M:B$ became a three-coordinate intermediate R_3M^+ by losing B^-, the new base would presumably enter at random, showing no stereospecificity.

An important consideration in running acid–base reactions involving a solvated cation is the rate at which the solvated solvent molecules leave the inner coordination sphere around the cation, since in general that rate limits the rate at which the new base can enter the coordination sphere. Water molecule replacement

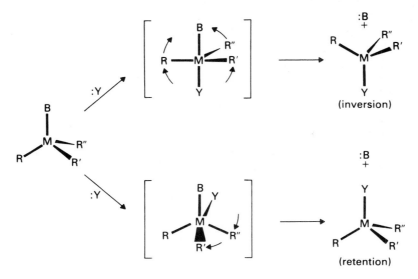

Figure 7.4 Stereospecific mechanisms for MR_3B substitution.

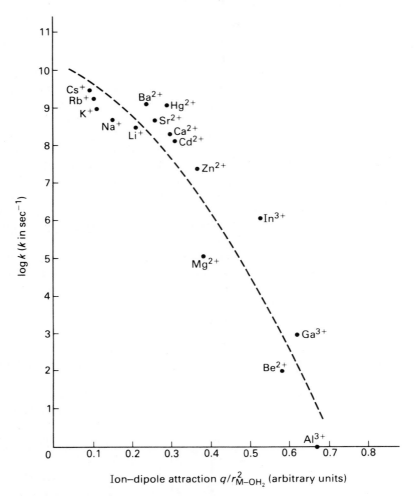

Figure 7.5   Rate constants for the exchange of hydrated H_2O molecules as a function of electrostatic attraction.

rates in a number of aqueous hydrates have been studied, with results shown in Fig. 7.5 for main-group metal ions. (Transition-metal ions, which have also been well studied, show strikingly different behavior. We will examine this behavior in Chapter 14.) The upper limit to the rate of substitution is given by the rate at which water molecules can diffuse toward the ion through the solution. This limit is on the order of 10^{10} sec^{-1}. Substitution rates approach this limit for larger, less highly charged ions without a large ion–dipole attraction for the water molecule present in the coordination sphere. The rate of replacement decreases as the ion–dipole attraction increases, whether because the ion becomes smaller (such as Li^+) or the charge increases (from Na^+ to Mg^{2+} to Al^{3+}). The substantial scatter of the experimental points both reflects the difficulty of measuring very fast reaction rates and indicates that factors other than simple ion–dipole electrostatic attraction are involved in binding the hydrate water molecules. As a very rough guide, we may note that hard acids such as Be^{2+} and Al^{3+} tend to bind the H_2O base so strongly that they exchange it very slowly (below the line approximating the functional dependence in Fig. 7.5). By comparison, most of the softer acids such as In^{3+} and Hg^{2+} exchange the weakly bound hard base H_2O quite rapidly. For the most highly charged ions, the rate of water substitution is apparently influenced by the proton-acceptor ability of the anions present, which suggests that hydrolysis of the cation hydrate may be important in the mechanism:

$$M\text{—}OH_2^{n+} + B^- \rightarrow M\text{—}OH^{(n-1)-} + HB$$

The extent of hydrolysis of cations and the strength of other Brønsted acids is an area that has received a great deal of attention, and we shall turn to it next.

7.3 PROTIC ACID/BASE STRENGTH IN WATER

We have already seen in Section 7.1 above that the favorable negative ΔH for a gas-phase proton-transfer process is influenced both by the competition of electron affinities and by the bond energies to hydrogen of the potential bases. In solution, as Chapter 6 indicated, solvation energies will also be important—particularly because, as we shall see, the energy cycles yield an enthalpy change for proton transfer that is the difference between two large numbers. The relative strengths of acids or bases in this type of reaction will be determined primarily by the enthalpy change, because the entropy change will be nearly constant for a given class of reactions: an electrically neutral acid reacting with water to yield H_3O^+ and a singly charged anion, for example. Accordingly, in what follows we shall concentrate on enthalpy cycles for proton transfer.

The simplest systems to begin with are the volatile hydrides that resemble water in structure: NH_3, H_2S, and HX. Everyone who has studied descriptive chemistry knows that ammonia is a base in water and the other compounds are acids—but why? Consider first the Born–Haber cycle below for the reaction in which the gaseous hydride dissolves in water to produce H_3O^+ and the deprotonated anion:

$$
\begin{array}{ccc}
H(g) + X(g) + H_2O(l) \xrightarrow{IP_n + EA_x} & X^-(g) & + \; H^+(g) + H_2O(l) \\
D_{HX} \uparrow & \downarrow \Delta H_{hyd}(X^-) & \downarrow \Delta H_{hyd}(H^+) \\
HX(g) \;\; + \; H_2O(l) \xrightarrow{\Delta H} & X^-(aq) & + \quad H_3O^+(aq)
\end{array}
$$

For this cycle, $\Delta H = D_{HX} + IP_H + EA_X + \Delta H_{hyd}(H^+) + \Delta H_{hyd}(X^-)$. If we calculate ΔH for the two hydrides NH_3 and HCl, we have

NH_3: $\Delta H = 389 + 1310 - 71 - 1130 - 460 = +38$ kJ/mol reaction
HCl: $\Delta H = 427 + 1310 - 347 - 1130 - 322 = -67$ kJ/mol reaction

Assuming a modest negative ΔS (because gas is lost and charged species in solution are formed) it is clear that NH_3 can never behave as an acid in water, but HCl can. The primary differences are that Cl has a much greater electron affinity than NH_2, and that NH_2^- (presumably comparable to OH^-) has a much greater hydration energy than Cl^-, but not enough greater to stabilize the amide ion in water solution.

By way of comparison, we can write a cycle for gaseous ammonia reacting with water to serve as a base:

$$NH_3(g) + H(g) + OH(g) \xrightarrow{IP_H + EA_{OH}} NH_3(g) + H^+(g) + OH^-(g)$$

with D_{H-OH} (up) on the left and PA (down) on the right:

$$NH_3(g) + H_2O(g) \qquad\qquad NH_4^+(g) + OH^-(g)$$

with ΔH_{vap} (up) on the left, $\Delta H_{hyd}(NH_4^+)$ (down) and $\Delta H_{hyd}(O^{H^-})$ (down):

$$NH_3(g) + H_2O(l) \xrightarrow{\Delta H} NH_4^+(aq) + OH^-(aq)$$

Here $\Delta H = \Delta H_{vap}(H_2O) + D_{H-OH} + IP_H + EA_{OH} + PA_{NH_3} + \Delta H_{hyd}(NH_4^+) + \Delta H_{hyd}(OH^-)$, yielding $\Delta H = 42 + 498 + 1310 - 176 - 866 - 377 - 460 = -29$ kJ/mol reaction. The principal differences between ammonia as an acid and ammonia as a base are that (*a*) the base cycle has a much smaller hydration enthalpy (for NH_4^+) than the acid cycle (for H^+), and (*b*) the base cycle has a large negative contribution from the proton affinity of NH_3, which is not present at all in the acid cycle. The result is that ammonia has a modest negative enthalpy change for the base reaction, and does indeed behave as a weak base in water.

It is not possible to calculate ΔH for HCl behaving as a base in water as we just did for NH_3, because too many values are not known for the H_2Cl^+ species that would presumably be formed. However, we can examine the base properties of the Cl^- ion through a very similar cycle:

$$Cl^-(g) + H(g) + OH(g) \xrightarrow{IP_n + EA_{OH}} Cl^-(g) + H^+(g) + OH^-(g)$$

with D_{H-OH} (up) on the left and PA (down) on the right:

$$Cl^-(g) + H_2O(g) \qquad\qquad HCl(g) + OH^-(g)$$

with $-\Delta H_{hyd}$ (up) and ΔH_{vap} (up) on the left, $\Delta H_{soln}(NH_4^+)$ (down) and $\Delta H_{hyd}(OH^-)$ (down):

$$Cl^-(aq) + H_2O(l) \xrightarrow{\Delta H} HCl(aq) + OH^-(aq)$$

Here HCl(aq) is the hypothetical dissolved molecule. Its enthalpy of solution is assumed to be the same as that of its neighbor H_2S in the periodic table. In this cycle,

$$\Delta H = -\Delta H_{hyd}(Cl^-) + \Delta H_{vap}(H_2O) + D_{H-OH} + IP_H + EA_{OH} + PA_{Cl^-}$$
$$+ \Delta H_{soln}(HCl) + \Delta H_{hyd}(OH^-)$$

In numerical terms, $\Delta H = 322 + 42 + 498 + 1310 - 176 - 1393 - 21 - 460 = +122$ kJ/mol reaction. This is substantially unfavorable, so chloride ion should show essentially no proton-base properties in water solution (though it could serve as a Lewis base to other acids). The reason is that the proton affinity of Cl^- is not quite large enough to overcome both the large H—OH bond energy and the large ionization energy of the H atom. An interesting comparison can be made between the singly charged anions of the volatile hydrides, when they are treated as bases in the same manner as Cl^- (see Table 7.2). Neither Br^- nor I^- have any proton-base properties in water; F^- and HS^- might possibly be weak bases; OH^- is simply exchanging protons; and NH_2^- is a strong base. These values accord with our experience, of course, but the various energy terms in the table yield a better understanding of the molecular reasons for the base behavior of these anions.

The hydrohalic acids present an interesting and (at first sight) unusual pattern of reactivity in water as acids. HF is a weak acid; all the others are strong acids. This is implied by the ΔH values in Table 7.2, since a strong Brønsted acid is a conjugate to a weak base. We can see it more directly, however, from a cycle written in free-energy terms for the aqueous acids. The reason for using free energies here—apart from the fact that they correlate directly with K_a ($\Delta G^0 = -RT \ln K_a$)—is that the entropy effect of HF in water solution is significantly different from that of the other HX acids because of its very strong hydrogen bonding. For the cycle

$$
\begin{array}{ccccc}
H(g) + X(g) + H_2O(l) & \xrightarrow{IP_H + EA_{OH}} & X^-(g) + H^+(g) + H_2O(l) \\
\uparrow {\scriptstyle D_{H-X} - \frac{3}{2}RT} & & \downarrow {\scriptstyle \Delta G_{hyd}(X^-)} \quad \downarrow {\scriptstyle \Delta G_{hyd}(H^+)} \\
HX(g) \quad + H_2O(l) & & \\
\uparrow {\scriptstyle -\Delta G_{soln}} & & \\
HX(aq) \quad + H_2O(l) & \xrightarrow{\Delta G(K_a)} & X^-(aq) \quad + \quad H_3O^+(aq)
\end{array}
$$

we have

$$\Delta G = -\Delta G_{soln}(HX) + D_{H-X} - \tfrac{3}{2}RT + IP_h + EA_x + \Delta G_{hyd}(H^+) + \Delta G_{hyd}(X^-)$$

The appropriate values for this cycle are given in Table 7.3. Even though we might expect the very electronegative F atom to withdraw electrons from its H—F bond and be a strong acid, the high H—F bond energy and surprisingly low electron affinity make HF a weak acid. Furthermore, the free energy of hydration of the F^- ion is considerably less favorable than its enthalpy of hydration, because the water molecules undergo extensive ordering around the F^- as it forms strong hydrogen bonds. For the other HX acids, of course, K_a on the order of 10^{10} simply indicates that the acid has been completely leveled to H_3O^+. In nonaqueous solvents, HCl behaves as a slightly weaker acid than HBr or HI, as the theoretical K_a values imply.

Table 7.3 also shows the cycle data for water and its congeners serving as acids. For these compounds, X^- is OH^-, SH^-, and so on. The H_2Y compounds become more acidic going down the column of the periodic table, even though this

Table 7.2 Born–Haber Cycle Data for Anions as Bases in Water (kJ/mol reaction)

	$-\Delta H_{hyd}(X^-)$	$+ \Delta H_{vap}(H_2O)$	$+ D_{H-OH}$	$+ IP_H$	$+ EA_{OH}$	$+ PA_{X^-}$	$+ \Delta H_{soln}(HX)$	$+ \Delta H_{hyd}(OH^-)$	$= \Delta H$
I⁻	+255	+42	+498	+1310	-176	-1310	-8	-460	+151
Br⁻	+297	+42	+498	+1310	-176	-1351	-8	-460	+152
Cl⁻	+322	+42	+498	+1310	-176	-1393	-21	-460	+122
F⁻	+464	+42	+498	+1310	-176	-1548	-50	-460	+80
SH⁻	+322	+42	+498	+1310	-176	-1464	-21	-460	+51
OH⁻	+460	+42	+498	+1310	-176	-1632	-42	-460	0
NH₂⁻	+460	+42	+498	+1310	-176	-1695	-29	-460	-50

Table 7.3 Born–Haber Cycle Data for HX Acids in Water (kJ/mol reaction)

	$-\Delta G_{soln}(HX)$	$+ (D_{H-X} - \frac{3}{2}RT)$	$+ IP_H$	$+ EA_X$	$+ \Delta G_{hyd}(H^+)$	$+ \Delta G_{hyd}(X^-)$	$= \Delta G^0$	K_a
HF	+25	+536	+1310	-331	-1100	-423	+17	10^{-3}
HCl	-4	+402	+1310	-347	-1100	-305	-44	10^8
HBr	-4	+339	+1310	-326	-1100	-276	-57	10^{10}
HI	-4	+272	+1310	-297	-1100	-238	-57	10^{10}
H₂O	+8	+464	+1310	-176	-1100	-427	+79	10^{-14}
H₂S	-4	+351	+1310	-222	-1100	-297	+38	10^{-7}
H₂Se	-4	+289	+1310	-209	-1100	-264	+22	10^{-4}
H₂Te	-4	+255	+1310	-213	-1100	-234	+14	10^{-3}

trend is opposite to the trend in bond polarity. Again we see that the OH radical has an unusually small electron affinity, and that the H—YH bond energy drops off more rapidly than the YH$^-$ hydration free-energy does. Since all the electron affinities are relatively low, none of the H_2Y compounds is a strong acid.

The wide variety of oxyacids H_mXO_n shows a striking range of acidities in water. The pK_a values range from about -11 (HSO$_3$F) to $+12$ (HPO$_4^{2-}$ and others). Table 7.4 gives some experimental pK_a values for oxyacids; the differences are apparent. A number of electronic factors influence the strength of any given oxyacid; when due allowance is made for each, we can calculate the pK_a to within

Table 7.4 pK_a Values for Inorganic Oxyacids

Acid	pK_1	pK_2	pK_3	pK_4
$H_3BO_3(HB(OH)_4)$	9.2			
H_2CO_3	3.6^a	10.3		
H_4SiO_4	9.8			
H_4GeO_4	9.0	12.3		
HNO_3	-1.4			
HNO_2	3.3			
H_3PO_4	2.1	7.2	12.4	
$H_4P_2O_7$	1.0	2.0	5.6	9.4
$H_3PO_3((HO)_2PHO)$	2.0	6.6		
H_3AsO_4	2.0	6.9	11.6	
H_3AsO_3	9.3			
H_2SO_4	-3.0	1.9		
HSO_3F	-10.8			
HSO_3Cl	-10.4			
HSO_3NH_2	1.0			
$H_2S_2O_3 ((HO)_2SSO)$	0.6	1.7		
H_2SO_3	1.9	7.2		
H_2SeO_4		1.9		
H_2SeO_3	2.6	8.0		
$H_6TeO_6(Te(OH)_6)$	7.7	11.0		
$HClO_4$	-7.3			
$HClO_3$	-2.7			
$HClO_2$	2.0			
$HClO(HOCl)$	7.3			
H_5IO_6	3.3	6.7		
HIO_3	0.8			
H_4XeO_6	<0	4.3	10.8	

^{a}Refers to molecular H_2CO_3, not to dissolved CO_2.

about one order of magnitude (in K_a). The oxyacid will be a stronger proton donor the more electron-withdrawing the rest of the molecule is, and a little reflection will suggest what factors are involved:

1. a high formal oxidation state on the central atom
2. a large number of terminal oxygen atoms or other electronegative ligand atoms such as F or Cl

On the other hand, the oxyacid will be a weaker proton donor the more available electrons are from the oxyanion:

3. extra hydroxylic hydrogen atoms
4. nonbonding electron pairs
5. less-electronegative atoms bonded within the oxyanion
6. fewer electronegative ligand atoms (than the commonly found H_mXO_4)
7. a net negative charge

We can write an empirical expression that takes these factors into account:

pK_a = 3(no. of OH)
          − (no. of terminal O, F, Cl, CF$_3$)
          − (oxidation state of central atom)
          + 2(no. of pairs of nonbonding electrons or electrons binding atoms no
              more electronegative than the central atom)
          + 2(4 − central-atom coordination no. including nonbonding pairs)
          − 4(net charge)

For example, for nitrous acid, HO—N—O, we calculate

$$pK_a = 3(1) - (1) - (3) + 2(1) + 2(4 - 3) - 4(0) = 3$$

For phosphorous acid, $(HO)_2OPH$, in which the P—H hydrogen is hydridic and the P is formally 5+,

$$pK_a = 3(2) - (1) - (5) + 2(1) + 2(4 - 4) - 4(0) = 2$$

For anions, the structure of the neutral molecule is assumed, but the ion charge is included at the end, as for $H_2AsO_4^-$:

$$pK_a = 3(3) - (1) - (5) + 2(0) + 2(4 - 4) - 4(-1) = 7$$

Although this empirical expression is fairly accurate, it is awkward. An equation developed by Ricci is nearly as accurate, and more convenient:

$$pK_a = 8 - 9 \text{ (formal charge on central atom)} + 4(n - m)$$

Here the formal charge on the central atom is calculated by assuming that a terminal oxygen carries a 1− charge to be balanced by the central atom in the neutral molecule, and n and m refer to the formula H_mXO_n. In any case, a strong acid will typically have no net charge, few hydrogens, and many oxygens, which accounts for the general form of the strong acids in Table 6.9.

Another important area of proton acid/base chemistry in water is the hydrolysis of species in solution. Anions can hydrolyze by serving as Lewis bases for a proton Lewis acid on a water molecule:

$$S^{2-} + HOH \rightleftharpoons HS^- + OH^-$$

Cations hydrolyze in a slightly more complicated way. In Fig. 6.9 and the associated discussion of hydrous oxides, we indicate how the transfer of a proton through a hydrogen bond from a hydrated water molecule to a nearby free water molecule promotes the formation of hydrogen-bonded polymeric hydrous oxides. Even without added OH^- ion, such hydrated cations serve as weak Brønsted acids in water. Hydrated aluminum ion is about as strong an acid as acetic acid:

$$Al(H_2O)_6^{3+} + H_2O \overset{K_a}{\rightleftharpoons} Al(OH)(H_2O)_5^{2+} + H_3O^+ \qquad K_a = 1.07 \times 10^{-5}$$

The acid strength of the metal ions in this hydrolytic reaction varies a great deal, depending mostly on how strongly the charged atom attracts the electrons binding the protons to the hydrate water molecules. The stronger this attraction, the more positively charged the hydrate-water hydrogen atoms become, and the more readily they are transferred by a reaction like that of hydrated aluminum ion. Figure 7.6 gives pK_a values for a number of metal ions and shows their functional dependence on the coulomb attraction of the metal ion for electrons at the distance of a hydrate-water oxygen nucleus. Baes and Mesmer pointed out that the numerous data points tend to fall along four reasonably parallel straight lines. Each line connects acids of quite different *strength* that have comparable *hardness*. Thus the uppermost line, connecting ions such as Ca^{2+}, Al^{3+}, and Th^{4+}, represents the hardest acids (those that least prefer the softer OH^- to the harder H_2O). The lowest line, on the other hand, connects the soft acids Sn^{2+}, Hg^{2+}, and Pd^{2+}. As a rough guide to aqueous solution behavior, it can be noted that since hydrous oxides form at a pH only one or two units higher than the onset of hydrolysis by the reaction given in Fig. 7.6, the pK_a values given there are about the highest pHs at which the hydrous oxide or hydroxide will not precipitate. Thus a solution of Fe^{3+} in water cannot have a concentration higher than about 10^{-5} M if the pH is above about 2, and a solution of Al^{3+} cannot have a concentration higher than 10^{-5} M above pH 5.

These pK_a values refer only to the first stage of hydrolysis, in which the solution species produced has only one metal atom and one OH. Such K_a values might more specifically be termed K_{11} values, since further steps in hydrolysis often occur. More protons can be removed,

$$Al(H_2O)_6^{3+} + 2H_2O \rightleftharpoons Al(OH)_2(H_2O)_4^+ + 2H_3O^+ \qquad K_{12}$$

or a sort of condensation polymerization can occur:

$$13Al(H_2O)_6^{3+} \rightleftharpoons Al_{13}O_4(OH)_{24}(H_2O)_{12}^{7+} + 32H_3O^+ + 6H_2O \qquad K_{13,32}$$

Figure 7.7 shows how the various hydrolysis species contribute to the total dissolved Al^{3+} concentration as the pH changes. Note that the amphoteric $Al(OH)_3$

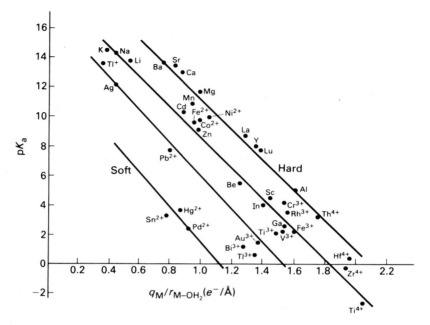

Figure 7.6 pK_a for metal-ion M^{q+} hydrolysis as a function of metal-ion attraction for proton-binding electrons on $M-OH_2$.

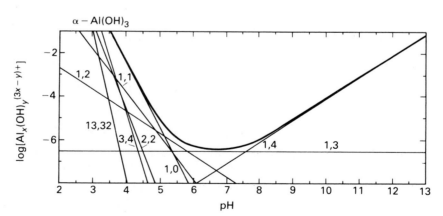

Note: The labels 1, 1; 1, 2; and so on, refer to $Al_x(OH)_y$ species with $x = 1$ and $y = 2$, and so on. The heavy line shows the total concentration of all Al^{3+} species.

Figure 7.7 Al^{3+} aqueous solution species as a function of pH. [From C. F. Baes and R. E. Mesmer, *The Hydrolysis of Cations*, John Wiley, New York, 1977. By permission from John Wiley & Sons, Inc.]

redissolves as $Al(OH)_4^-$ in strongly basic solution. (The structure of the Al_{13} species is related to that of $PW_{12}O_{40}^{3-}$, the Keggin structure, in Fig. 6.12.) A number of such polymerized hydroxycations are known, as Chapter 6 indicated (see Fig. 6.8). This tends to complicate the acid–base behavior of these ions.

7.4 PROTIC ACID/BASE STRENGTH IN NONAQUEOUS SOLVENTS

Although water is the most important solvent to the inorganic chemist, it is sometimes inconvenient that it hydrolyzes—or even levels—many protic species to H_3O^+ or OH^-. As Chapter 6 indicated, other protic solvents have comparable reactions. To the extent, however, that electronegativities and electron-donor properties are different in other solvent molecules, leveling will occur at different absolute levels of acidity or basicity. If the solvent molecule is HA, acid leveling will occur when the solvent acquires an extra proton: $HA + H^+ \rightarrow H_2A^+$. The pK_a for the hydronium ion corresponds to a pH of 0 in water. Similarly, base leveling will occur when the solvent loses a proton to a solute particle: $HA \rightarrow A^- + H^+$. The pK_a for water (forming OH^-) corresponds to a pH of 14 in water. These two pK_a values are related through the autoprotolysis constant:

$$H_2O + H_2O = H_3O^+ + OH^- \qquad K_W \equiv K_{autoprot} = 1 \times 10^{-14} = [H_3O^+][OH^-]$$

$$pK_{autoprot} = p(H_3O^+) + p(OH^-)$$

For a general solvent HA, the corresponding result is

$$pK_{autoprot} = p(H_2A^+) + p(A^-)$$

In any given solvent, two acids or bases can be discriminated only if at least one has its pK_a within the range from $p(H_2A^+)$ to $p(A^-)$. Thus pH titrations in water can distinguish acetic acid (pK_a 4.74) from perchloric acid (pK_a −7.3), since the acetic acid pK_a is between the solvent limits of 0 and 14. It follows that the farther apart the solvent protonation pKs are, the more acids or more bases can be discriminated. But since these are linked by the autoprotolysis constant, the result is that more acids or bases can be discriminated in a solvent with a small autoprotolysis constant.

Consider 10^{-3} M solutions of a strong acid and strong base in water. The first will have a pH of 3, the second a pH of 11. Titrating one against the other yields an endpoint break of eight pH units. This means that weak acids or bases can be discriminated over a range of about 8 in pK_a. In ethanol, on the other hand, autoprotolysis yields

$$EtOH + EtOH \rightleftharpoons EtOH_2^+ + EtO^-$$

$$pK_{autoprot} = [EtOH_2^+][EtO^-] = 10^{-19}$$

If 10^{-3} M solutions of a strong acid (yielding $EtOH_2^+$) and a strong base (yielding EtO^-) are used, the $p(EtOH_2^+)$—which corresponds to $p(H_3O^+)$ in water or pH—

will be 3 for the acid solution. The $p(EtO^-)$ for the base solution will also be 3, but its "pH" [actually, $p(EtOH_2^+)$] will be

$$pK_{autoprot} \simeq 19 = p(EtOH_2^+) + p(EtO^-)$$

$$19 = p(EtOH_2^+) + 3$$

$$p(EtOH_2^+) = 9 - 3 = 16$$

The pK_a range in which acids or bases can be discriminated is thus 13 units, from the $p(EtOH_2^+)$ of 3 for the acid solution to the $p(EtOH_2^+)$ of 16 for the base solution—substantially greater than the eight units in water. The smaller degree of autoprotolysis—which yields a numerically larger $pK_{autoprot}$—is responsible for the greater range.

By contrast, H_2SO_4 has an autoprotolysis constant of 3×10^{-4}, which produces an acid-to-base range so short that very few acids or bases can be discriminated. By an extension of the argument above, the greatest possible range for H_2SO_4 is only about four units on a pK_a scale, and its practical range for reasonably dilute solutions is only about one unit. Most compounds that can accept a proton in any manner serve as strong bases

$$EtOH + 2H_2SO_4 \rightarrow EtOSO_3H + H_3O^+ + HSO_4^-$$

$$HNO_3 + 2H_2SO_4 \rightarrow NO_2^+ + H_3O^+ + 2HSO_4^-$$

A few very weak electron donors, such as nitro-compounds and nitriles, are not completely protonated and thus are weak bases in H_2SO_4. Most common strong acids, such as HNO_3 above, show no acidic properties whatsoever in H_2SO_4. HSO_3F and $H_2S_2O_7$, however, are weak acids, and we have already noted the strong acid $HB(HSO_4)_4$.

A useful way to summarize the pK_a discrimination ranges and intrinsic acidities of different solvents is the chart in Fig. 7.8. The vertical axis is the pK_a for a given substance in water; the vertical bars for different solvents indicate the pK_a discrimination ranges for each. Acidity increases toward the bottom of the chart, basicity toward the top. For any given solvent, only the "acid" species appearing above its bar will be stable in solution, because the corresponding "base" species will be leveled to the solvent anion. Conversely, only the "base" species appearing below a solvent's bar will be stable in that solvent, because the "acid" species are leveled.

Perhaps the most unusual protic solvent in Fig. 7.8 is methyl isobutyl ketone (4-methyl-2-pentanone), which has an extremely long range in which acids or bases of different strengths can be discriminated. The reason for the long range is the absence of —OH groups, which effectively prevents autoprotolysis. At the acid end, very strong acids can protonate the carbonyl $=O$; at the basic end, extremely strong bases can presumably deprotonate a C—H bond, though this has not yet been experimentally proven. Since methyl isobutyl ketone has only a modest dielectric constant (about 15) it has substantial ion pairing and does not have

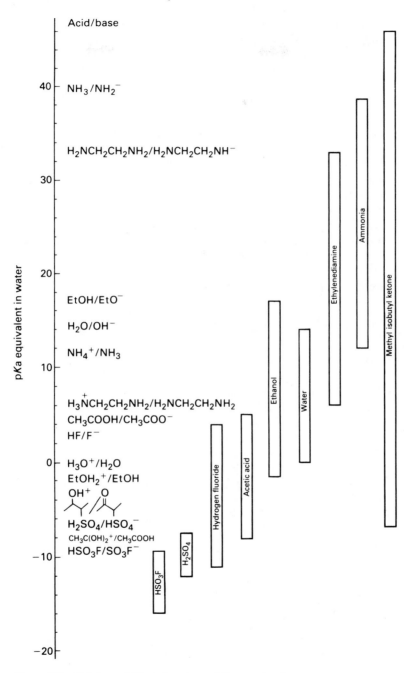

Figure 7.8 Relative acidities of proton acid/base solvents.

strong solvent properties. However, if a base titrant such as tetrabutylammonium hydroxide is used, one can discriminate acids in mixtures over a very wide range of pK_a values by potentiometric titration. For example, separate sharp potential breaks can be obtained in a single titration of $HClO_4$, HCl, salicylic acid, acetic acid, and phenol ($pK_a = 10$).

In general, acids or bases retain their intrinsic proton donor or acceptor capability in different solvents. Therefore, the order of the pK_a values in tables such as Table 7.4 will remain more or less unchanged in solvents other than water, even though the apparent pK_a values change markedly. However, minor changes in the pK_a order do occur, essentially because solvation affects acids and bases with different structures differently. Charge effects, in particular, are important. The pK_a of a cation-acid/neutral-base conjugate pair can be shifted significantly relative to that of a neutral-acid/anion-base pair. For example, in water the anilinium ion (phenylammonium) has very nearly the same pK_a as acetic acid (4.60 versus 4.76), but in dimethylsulfoxide (DMSO), anilinium's apparent pK_a changes only to 3.6, while that of acetic acid changes to 12.3. Benzoic acid, which has the same charge relationship as acetic acid, changes from 4.21 in water to 11.1 in DMSO, a change nearly parallel to that of acetic acid.

7.5 ELECTRONIC ACID/BASE STRENGTH: METAL-ION LEWIS ACIDS

Protic acid–base reactions are an important special case of the more general Lewis acid–base interaction. Since nearly all molecules other than saturated hydrocarbons and boranes have nonbonding or weakly bonding electron pairs, and since nearly all cations and many neutral molecules have vacant low-energy orbitals, an immense variety of Lewis acid–base reactions are possible. We can categorize them as follows: metal-ion-acceptor/nonmetal-donor, electron-deficient metalloid-acceptor/nonmetal-donor, and nonmetal-acceptor/nonmetal-donor. It is also possible to see metal atoms serving as Lewis bases, but we shall delay that discussion until the chemistry of transition metals is considered. Each of the general areas above covers a large number of donor–acceptor compounds. We consider them in order.

The largest single category is the first, which involves metal ions behaving as acceptors. An extraordinary number of such Lewis acid–base compounds have been prepared; most involve transition-metal ions. Because of their d electrons and vacant d orbitals, however, they show different kinds of bonding than are seen in compounds involving main-group metals. Accordingly, we shall limit ourselves here to s-block metal ions (and a few p-block ions) as acceptors. We shall examine transition-metal complexes in Chapter 12 in some detail.

We may begin by considering Lewis acid–base adducts involving an alkali-metal ion as the Lewis acid. Because the alkali-metal cations have only a 1+ charge, and because (except for lithium) they are relatively large, the overall electrostatic attraction is not very great for an electron pair on a donor atom. Since in addition the alkali metals are quite low in electronegativity, strong covalent bonds cannot form with a donor. In general, therefore, very few electron donor–acceptor compounds of these cations have been characterized. There is an outstanding exception, however. In the mid-1960s, Pedersen synthesized the first *crown ether* complexes of alkali cations, using cyclic polyethers like the one shown in Fig. 7.9. Crown ethers received their name because the uncomplexed ring resembles a cartoon crown with oxygen atoms at the raised points. The polymer is $(-CH_2-CH_2-O-)_n$. It is given a convenient trivial name by first giving the exocyclic substituents, the number of atoms in the ring, the word "crown", and the number of oxygen electron-donor atoms in the ring. By this system, Figure 7.9 shows dibenzo-18-crown-6.

If a space-filling model of dibenzo-18-crown-6 is arranged with the six O atoms coplanar and pointed toward the center of the ring, there is a hole of approximately 3.1 Å in the center. Since the ionic radius of K^+ is 1.52 Å when it is 6-coordinate, its ionic diameter of 3.04 Å is almost a perfect fit for the hole. The stereochemistry of the crown ether thus forces six donor–acceptor bonds to form at once, and the potassium ion is substantially stabilized in the crown-ether complex. Since the size fit is important to the stability of the complex, it is not surprising that crown ethers are fairly selective in forming complexes with alkali cations. A "crown-4" ether with a smaller ring and only four oxygen donor atoms prefers Li^+, whereas a crown-5 ether prefers the larger Na^+ and crown-6 ethers prefer the still-larger K^+.

Although the stability constants for the formation of alkali-crown complexes in water tend to be only about 2, because of the strong competition from the small polar water molecules, the complexes are extremely stable in less polar solvents. The complexes have three important applications: to minimize ion pairing in low-dielectric-constant solvents, to dissolve ionic solids in nonpolar solvents (*phase-transfer catalysts*), and to increase the diameter of alkali cations to assist in crystallizing large anions. In all three cases, the experimenter takes advantage of the large diameter of the crown ether and the nonpolar nature of its outside atoms. As

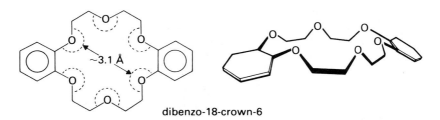

dibenzo-18-crown-6

Figure 7.9 Stereochemistry of crown ethers.

a phase-transfer catalyst, crown ethers can, for example, dissolve potassium hydroxide in toluene. Since it is virtually impossible for toluene to solvate the OH^- ion, KOH behaves as a much stronger base than in water or alcohols. Similarly, $KMnO_4$ can be dissolved to give "purple benzene," in which clean, specific oxidations of organic substances can be carried out. As a size-adjustment tool, the crown ethers allow the convenient crystallization of Zintl anions for x-ray structural studies.

Three-dimensional equivalents of the two-dimensional crown ethers can be prepared. Figure 7.10 shows a *"cryptand"* called 2,2,2-crypt, together with its Na^+ complex. As might be expected, these complexes are even more stable and size-specific than crown-ether complexes are. The most striking application of crypt complexes is the preparation of stable crystals containing Na^- anion (sometimes called *sodide*, but more properly *natride*) and other alkali metallides. This is done by simply dissolving sodium metal in cold ethylamine, from which golden crystals with a metallic luster crystallize:

$$2Na^0 + C \rightarrow [Na^+ \cdot C][Na^-]$$

Here C is the 2,2,2-crypt molecule previously dissolved in the ethylamine. The $Na^+ \cdot$ C complex is shown in Fig. 7.10. In the crystal, a Na^- ion is positioned over each of the six open segments where the inner Na^+ is visible through the crypt

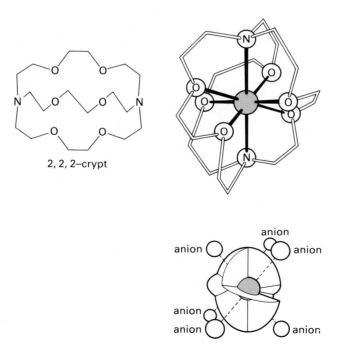

2, 2, 2–crypt

Figure 7.10   Stereochemistry of 2,2,2-crypt in its Na^+ complex.

molecule. Crystallographic data show that the Na^- ion has a radius somewhat greater than that of I^-, perhaps about 2.17 Å. This is consistent with the trend from an ionic radius of 1.16 Å for Na^+ to a covalent radius of 1.54 Å for Na, and is also consistent with a van der Waals radius of 2.27 Å for Na^0.

The alkaline-earth metal cations, with their high charge and smaller radius, form donor–acceptor compounds in which the electrostatic attraction for the donor pair is much stronger than in a similar complex with the neighboring alkali-metal cation. For example, although Na^+ and the heavier alkali cations are undoubtedly hydrated in aqueous solution, it is very difficult to crystallize the hydrated cations. The waters escape into solution, and the more compact anhydrous ionic crystal forms. On the other hand, not only are crystalline hydrates of the alkaline-earth cations such as $[Mg(H_2O)_6]^{2+}2Cl^-$ and $[Ca(H_2O)_4^{2+}(NO_3^-)_2$, well known, but many anhydrous alkaline-earth compounds such as $MgSO_4$, $CaCl_2$ and $Mg(Cl)_4)_2$ are commonly used laboratory drying agents precisely because they form strongly bonded donor–acceptor compounds between the cation acceptor and the H_2O donor.

In fact, the alkaline-earth cation hydrates are so stable that it is difficult to form any other donor–acceptor compound in water solution. Because Ca^{2+} is so common in the earth's crust, ground water usually contains significant concentrations of hydrated Ca^{2+}, the compound principally responsible for water hardness. To prevent the precipitation of calcium soaps, which are water-insoluble, a more effective donor must complex the Ca^{2+} if soap or synthetic detergents are to be effective. In general, donors strong enough to displace water from the hydrated Ca^{2+} ion must take advantage of the chelate effect mentioned earlier. EDTA forms a very stable donor–acceptor complex with Ca^{2+} in water solution ($K_{stab} = 5 \times 10^{10}$); thus, hydrated Ca^{2+} can be conveniently titrated by EDTA. Synthetic detergents use chelating polyphosphates, principally sodium tripolyphosphate, as electron donors to keep Ca^{2+} in solution (see the related structure in Fig. 2.19). Such detergent additives are called "builders." Unfortunately, because phosphates are also essential nutrients for algae, lakes receiving major amounts of urban wastewater containing large quantities of detergent-builder phosphate tend to suffer eutrophication. Other builders (Lewis bases for the Ca^{2+} acid) have been substituted in some metropolitan areas, but for a variety of reasons they are generally less satisfactory than tripolyphospate. The best known potential replacement was nitrilotriacetic acid (NTA):

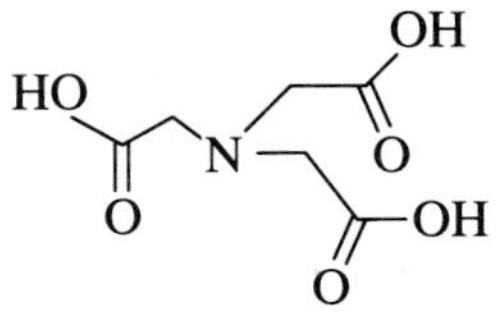

which forms a four-coordinate complex with Ca^{2+} and obviously benefits from the chelate effect. However, because of its carcinogenic properties, its use has been banned in the United States.

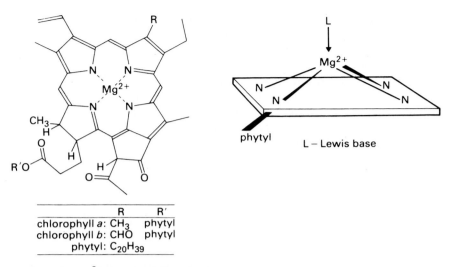

	R	R'
chlorophyll *a*:	CH_3	phytyl
chlorophyll *b*:	CHO	phytyl
phytyl:	$C_{20}H_{39}$	

Figure 7.11 Mg^{2+} in chlorophyll molecules.

Another critically important series of alkaline-earth cation donor–acceptor compounds is that of the cholorophylls (Fig. 7.11). The organic portion of a chlorophyll molecule is a porphyrin macrocycle with four N donor atoms. The whole molecule, except for the phytyl side chain, is nearly planar, though the individual five-membered rings are tipped slightly. The magnesium ion, which is shown as 4-coordinate in the center of the macrocycle ring, lies a few tenths of an angstrom unit above the plane of the four N atoms and has a fifth Lewis-base ligand, with overall square-pyramidal geometry. The fifth electron donor can be water or other small molecule. It can also be a carbonyl oxygen atom from a neighboring chlorophyll molecule, promoting a kind of coordination dimerization or even polymerization. The role of chlorophylls in biological systems, of course, is photochemical. We shall consider these reactions in Chapter 17.

The group Ib and IIb metals—Cu, Zn, and their congeners—belong in this discussion, even though they represent the completion of the *d*-block elements. Copper, silver, and gold in the +1 oxidation state, and zinc, cadmium, and mercury have completely filled sets of *d* orbitals in all their compounds. The Lewis-acid properties of these species thus depend exclusively on their *s* and *p* orbitals, which in that respect makes them comparable to the Ia and IIa metals. However, the relatively loosely held *d* electrons and the low net charge make these ions quite soft acids that strongly prefer soft bases such as N, S, Br, or I to O or F. Thus, although AgCl is insoluble in water, it dissolves readily in aqueous ammonia to yield the $Ag(NH_3)_2^+$ ion, in which each ammonia molecule donates a pair of electrons through its N atom to the Ag^+ acceptor. Because these atoms are higher in electronegativity than the alkali and alkaline-earth metals, there should be a considerable degree of covalent bonding in the donor–acceptor compounds. Accordingly, the coordination number of each metal atom should be limited to four, attainable using sp^3 hybrid orbitals. This is usually true, and coordination numbers 2 and 3 are not unusual. The much less common coordination numbers of 5 and 6 normally involve very electronegative hard bases such as NO_3^-, which may be assumed to involve primarily electrostatic bonding rather than covalent bonding and thus are not limited by the number of valence orbitals.

Copper(I) and silver(I), as very soft acids, can form donor–acceptor compounds with organic pi systems. Silver forms a compound with ethylene in which the $C=C$ pi electrons are donated as in Fig. 6.2b; the pi electrons are now delocalized over three nuclei, which substantially increases their stability. Of course, since hydrocarbon pi systems represent extremely soft bases, it is impossible to form compounds of this sort with the group Ia and IIa metals, whose ions are much harder acids. Similarly, copper(I) forms donor–acceptor compounds with butadiene that can be used to separate butadiene from the other components of a hydrocarbon mixture. In a donor–acceptor reaction that follows exactly the same pattern but is more unusual in appearance, $AgClO_4$ dissolves in benzene to form a compound that crystallizes into columns of sharply angled benzene molecules with a $C=C$ bond region donating to a silver ion on each side. Each column thus has the composition

$$Ag^+ - C_6H_6 - Ag^+ - C_6H_6 -$$

as in Fig. 7.12. Copper(I) in $CuAlCl_4$ also forms such a compound, but the benzene molecule only donates to a single Cu^+—and in a curiously asymmetric fashion, as the figure suggests. In both cases, the hexagonal symmetry of the benzene ring has been deformed in such a way as to increase the electron density in the pi-donor region.

7.6 ELECTRONIC ACID/BASE STRENGTH: METALLOID LEWIS ACIDS

The most commonly studied donor–acceptor compounds (other than those of the main-group and transition-metal ions) are those in which a metalloid atom, usually in a neutral compound, serves as the Lewis acid. The most extensively studied of these involve boron compounds from group III and phosphorus and its congeners from group V, particularly antimony. Boron's three valence electrons give it a stoichiometry BX_3 in simple compounds. This leaves one coordination site and one bonding orbital vacant with respect to a potential tetrahedral geometry. Similarly, the group V elements' MX_5 compounds can accept an electron pair from a Lewis base into an octahedral geometry.

Considering boron first, we have already touched on some of the steric influences (back strain, front strain, and internal strain) on the acidity of BX_3 compounds toward specific bases. For most simple bases, boron's acid strength lies in the order $BBr_3 > BCl_3 > BF_3 \simeq BH_3 > BMe_3$. The reasons for this ordering are not immediately apparent, since one might assume that the most electronegative

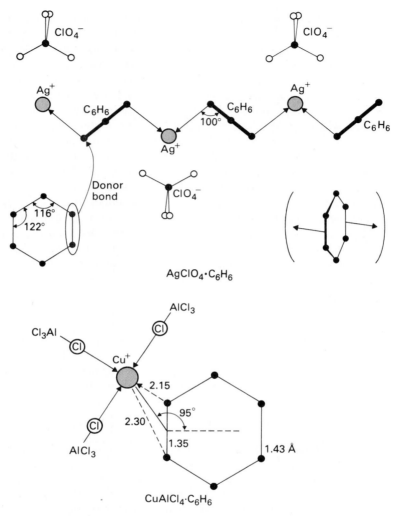

Figure 7.12 Ag$^+$ and Cu$^+$ donor–acceptor compounds with benzene.

X atoms in BX$_3$ would make the boron the strongest electron acceptor. Remember, however, that BX$_3$ compounds are uniformly planar. Therefore, the bond angles must change substantially when X$_3$B:L (with a more or less tetrahedral geometry) is formed. To the extent that B can engage in pi bonding with the X atoms in the planar BX$_3$ molecule, there will be a *reorganization energy* cost when that bonding is lost in becoming tetrahedral. Whenever the X atom is N, O, or F, the good match of valence $2p$ orbital sizes means that pi bonding will be quite favorable in the planar geometry and the change to tetrahedral Lewis-acid geometry correspondingly less favorable, which accounts for the fact that BBr_3 is the strongest boron Lewis acid.

BF_3 is a gas at room temperature; but because it is a Lewis acid, it is easy to store in ether solution as the donor–acceptor complex $F_3B{:}OEt_2$. If this ethereal solution is added to phenylmagnesium bromide, the product is not simply triphenylborane, but a uniquely symmetrical donor–acceptor species, the tetraphenylborate ion

$$F_3B{:}OEt_2 + 4PhMgBr \rightarrow BPh_4^- + 3MgFBr + MgBr^+ + Et_2O$$

The tetraphenylborate ion is useful because it is unable to form hydrogen bonds with any solution or lattice species. It is also useful in precipitating very bulky singly charged cations, since it can be demonstrated from the Madelung equation that lattice energies—and hence insolubility—are greatest when the lattice ions are of approximately equal sizes. This is the anion equivalent of the role of crown ethers in bulking up alkali metal cations to precipitate bulky anions. For this purpose, other useful large inert cations include tetraalkylammonium ions such as tetrabutylammonium, and other useful large inert anions include PF_6^- and to a lesser extent ClO_4^- and BF_4^-.

A final note on BX_3 donor–acceptor compounds is that B_2H_6 usually does not behave as two BH_3s in reacting with Lewis bases. More commonly, the diborane molecule undergoes unsymmetrical cleavage, forming the tetrahydridoborate anion

$$B_2H_6 + 2NH_3 \rightarrow [H_2B(NH_3)_2]^+ BH_4^-$$

The cation is an amine-boronium ion in which both ammonia molecules serve as donors to a single B acceptor atom. The hydride transfer from the hypothetical $H_3B{:}NH_3$ maintains the same number of each type of bond, but adds electrostatic attraction to the stability of the compound.

Group V elements are also strong Lewis acids; some estimates based on enthalpy of reaction place $SbCl_5$ above BCl_3 in acceptor strength. The valence-electron count and stoichiometry of the MX_5 species make them ideal for forming $X_5M{:}L$ octahedral Lewis acid–base compounds with a wide variety of electron-pair donors. Here, L is the Lewis base. In general, this bonding and geometry will occur when L is an oxygen atom in a ketone, ether, amide, sulfoxide, sulfone, or a P=O species such as a phosphine oxide, organophosphate, or $POCl_3$. It will also occur when L is a nitrogen atom in pyridine, some tertiary amines, and a few nitriles. The situation is complicated, however, by two factors: First, the competition between bridging structures and ion-transfer structures ($X_5M{:}X$—R versus $R^+MX_6^-$) can lead to entirely different structures in different solvents; and second, amphoterism can take place in which PCl_5, for instance, accepts an electron pair from pyridine ($Cl_5P{:}py$) but donates a Cl^- to $AlCl_3$ ($PCl_4^+AlCl_4^-$).

As an example of the results of ion transfer in these complexes, gaseous PCl_5 consists of isolated trigonal bipyramidal molecules. On condensing, these molecules presumably form momentary dimers in which a Cl serves as donor to the P in its neighbor molecule: Cl_4P—$Cl:PCl_5$. In the solid state, however, chloride transfer occurs instead: $PCl_4^+PCl_6^-$. In somewhat the same fashion, ICl and PCl_5 combine to form a solid complex $ICl\cdot PCl_5$ that dissolves readily in acetonitrile, nitrobenzene, and chloroform to yield PCl_4^+ and ICl_2^- ions. This complex, however, dissolves only sparingly in benzene and carbon tetrachloride to yield a non-conducting solution in which significant dissociation into ICl and PCl_5 has occurred. ICl_3 and $SbCl_5$ combine to form chain polymers (with the overall formula $ICl_3\cdot SbCl_5$) that have the structure shown in Fig. 7.13. In this structure, the bridging Cl atoms are about 0.6 Å farther from the I than the terminal Cl atoms are, which suggests a formulation as $ICl_2^+SbCl_6^-$. On the other hand, the bridging Cl atoms are also about 0.12 Å farther from the Sb than its terminal Cl atoms are, which suggests $SbCl_4^+ICl_4^-$. Although PCl_5 spontaneously forms PCl_6^- on condensing, the hexachlorophosphate ion is very rarely formed in any other acid–base complex. Almost any Cl^- donor can also serve as a Cl^- acceptor, and the system becomes $PCl_4^+MCl_{n+1}^-$, as in $PCl_4^+AlCl_4^-$, $PCl_4^+TlCl_4^-$, and $PCl_4^+SO_3Cl^-$. PBr_5, in which the Br atoms are more crowded than Cl atoms would be, dissociates readily into $PBr_3 + Br_2$, ionizes in the solid as $PBr_4^+Br^-$, and never serves as a Lewis-acid electron acceptor, even though it forms a number of Lewis acid–base complexes.

It is curious that $AsCl_5$ has never been prepared, since both PCl_5 and $SbCl_5$ are quite stable. However, when Cl_2 is passed into a mixture of $AsCl_3$ and $AlCl_3$, the compound $AsCl_4^+ AlCl_4^-$ forms; if trimethylphosphine oxide (Me_3PO) (which cannot serve as a chloride acceptor) is substituted for $AlCl_3$, the product is $AsCl_5\cdot OPMe_3$. If a strong chloride donor such as $Et_4N^+Cl^-$ is substituted, the compound $Et_4N^+AsCl_6^-$ forms—even though the parent $AsCl_5$ is unknown! Clearly, judiciously chosen Lewis acid–base coordination can play a major role in stabilizing unusual species.

Even when halide transfer is not possible, Lewis acid–base complexes can have a major effect on the structure of the starting molecules. Normally the Lewis acid, the coordination number of which is increasing for the central atom, changes its geometry significantly (usually from trigonal bipyramidal to octahedral).

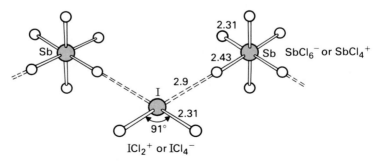

Figure 7.13   The $SbCl_5$—ICl_3 complex.

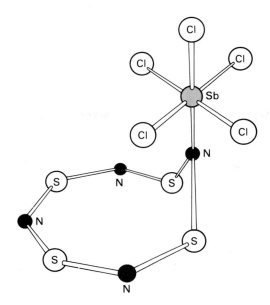

Figure 7.14 The $SbCl_5$—S_4N_4 complex.

The Lewis base, on the other hand, changes little except perhaps for a slight stretching of the bond to the donor atom. However, the S_4N_4 molecule (Fig. 5.24) can serve as a Lewis base to $SbCl_5$ through a N atom. The resulting structure (Fig. 7.14) has a nearly planar ring except for the donor N atom—almost as if a coordinated N^- had been inserted in the $S_4N_3^+$ ion of Fig. 5.24. Relatively subtle changes in the rather mobile electron density of the S_4N_4 molecule can apparently cause major changes in the equilibrium geometry of the molecule.

The acceptor strength of the MX_5 molecules increases toward the bottom of the group; although PCl_5 normally functions as a Cl^- donor to form PCl_4^+, no unequivocal example of $SbCl_4^+$ is known. By this reasoning, BiX_5 molecules should be the most powerful Lewis acid in the group. Against this tendency, however, must be balanced the fact that the maximum oxidation state is increasingly unstable toward the bottom of the group. Only BiF_5 of the four possible BiX_5 molecules has been prepared, and it is a powerful oxidizing agent and fluorinating agent. In the few cases that have been studied, it does indeed serve as a strong acceptor. For example, it forms the BiF_6^- ion with alkali fluorides.

Occasionally an MX_5 molecule can serve as a dibasic Lewis acid by accepting electron pairs from two donors. Although for the transition metals Nb and Ta this results in 7-coordinate complexes such as TaF_7^{2-}, in group Va octahedral geometry is maintained:

$$2SbCl_5 + 2MeCN \rightarrow SbCl_4(NCMe)_2^+ + SbCl_6^-$$

Of course, this amounts to ion transfer exactly like that in $PCl_4^+PCl_6^-$, but it also suggests that the group IVa halides MX_4, with which $SbCl_4^+$ is isoelectronic, might show dibasic Lewis-acid behavior. These systems do form a wide variety of MX_4L_2 complexes such as $SnCl_4(NCMe)_2$ (isoelectronic to the antimony complex just mentioned), but some monobasic 5-coordinate complexes, such as $Me_3SnCl_3 \cdot py$, are also known. As for the neighboring MX_5 molecules, Sn is a stronger acceptor than Ge or Si, and the PbX_4 molecules (X = F, Cl) are such strong oxidizing agents that few coordination compounds are stable.

The MX_4 compounds of the group VIa elements are quite different from those of group IVa in their Lewis acid–base behavior because of the nonbonding electron pair on the central atom. These compounds show true Lewis amphoterism; the MX_4 molecule can either donate an electron pair (usually by donating a halide ion) or accept a pair. SF_4, for example, forms $SF_4 \cdot SbF_5$, in which SF_4 serves as a base, since the structure is $SF_3^+SbF_6^-$; but it also forms $SF_4 \cdot py$ and $Cs^+SF_5^-$ in which the SF_4 serves as an acid. Similarly, $TeCl_4$ forms 1:1 complexes with both $AlCl_3$ and PCl_5. The first of these is $TeCl_3^+ AlCl_4^-$; the second is $PCl_4^+TeCl_5^-$. Some species can also serve as dibasic acids in a manner analogous to $SbCl_5$ in acetonitrile:

$$2TeF_4 + 2py \rightarrow TeF_3(py)_2^+ + TeF_5^-$$

7.7 ELECTRONIC ACID/BASE STRENGTH: NONMETAL ACIDS AND CHARGE-TRANSFER COMPLEXES

In the preceding discussion of Lewis acid/base compounds involving metals and metalloids as acids, it was fairly clear in each case that the acid was electron-deficient in some sense—that is, that it had (at least potentially) a vacant bonding orbital. In the nonmetals, most atoms form molecules with all possible bonding orbitals filled. Even so, there are specific stoichiometric interactions in which these nonmetals combine with electron-donor molecules in a Lewis acid/base sense. The interactions are all relatively weak in that ΔH is small for the acid–base combination, and many seem to be only weak solvent interactions for a nonmetal-molecule solute, but many others can be crystallized. Generally, the complexes form between a molecule with high electron affinity (the acid) and a molecule with a low ionization energy (the base). Either sigma-symmetry nonbonding electrons or pi bonding electrons are donated (Fig. 6.2), and the acceptor orbital is either a vacant sigma or a vacant pi orbital. Table 7.5 gives a few of the more common acids and bases found in these categories. For any molecular complex of this sort, the electrons binding the acid and base species are stabilized in part by polarization (analogous to strong van der Waals attraction) and in part by transfer from the base to the acid and the resulting electrostatic attraction. For this reason, these complexes are often called *charge-transfer complexes*. Although the bond-

Table 7.5   Some Common Acids and Bases in Charge-Transfer Complex Formation

Acids		Bases	
σ acceptors	π acceptors	σ donors	π donors
I_2		RNH_2	Benezene
Br_2		Pyridine	Naphthalene
Cl_2		R_2O	Anthracene
ICl		$R_2C=O$	(Fused aromatics)
SO_2		R_2S	Pyridine
CHI_3		X^- (X = Cl, Br, I)	
$IC≡CI$			
$(BF_3, SbCl_5,$ etc.)			

ing can be relatively weak, there is usually a very strong spectroscopic electron transition in the visible or near-UV range that corresponds to the return of an electron to the donor. This is called a *charge-transfer transition.*

As the appearance of the sigma-acceptor BF_3 and the sigma-donor pyridine in Table 7.5 will suggest, there is no clear distinction between these charge-transfer complexes and the more classical donor–acceptor complexes we have been considering. What is different is that the acids are often electron-rich, requiring a bond analogous to that in I_3^- (Fig. 4.23). I_3^- itself can be considered a charge-transfer complex between the acid I_2 and the base I^-, but the bonding follows the same pattern in less symmetrical σ–σ complexes. Figure 7.15 gives a qualitative MO scheme for the Br_2-acetone charge-transfer complex that is fully analogous to Fig. 4.23. The overlaps are a bit more complex, however, and it may be worth working through the MO energy levels. We shall assume that the bonding involves only sigma overlap of p orbitals on the Br atoms and sp^2 hybrids on the acetone O atoms (because of the bond angles). Each sp^2 hybrid contains an electron pair, whereas the Br p orbitals are assumed to be the ones that would provide the single net sigma bond in free Br_2, containing one electron each. There are thus six electrons to be accommodated in the four MOs to be formed from the four basis AOs. There are three individual overlap regions, shown shaded in the AOs sketched in Fig. 7.15. The AO signs are arranged relative to one arbitrarily

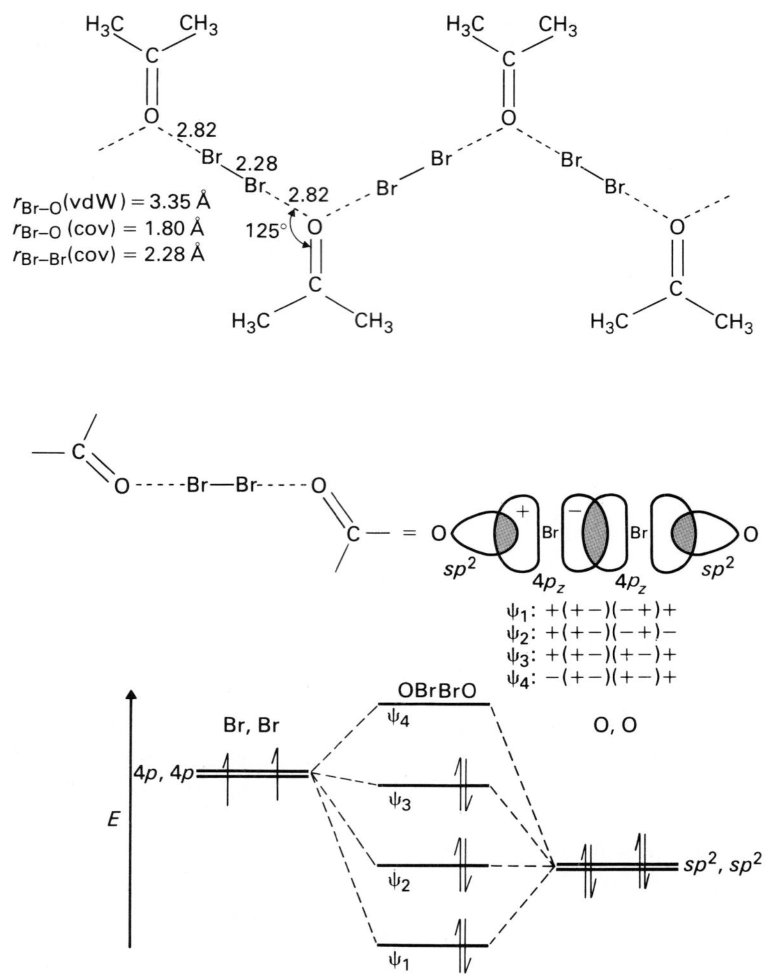

Figure 7.15　Structure and MO energies for the Br_2–acetone charge-transfer complex.

chosen AO whose sign is held constant so that the four MOs will provide the following overlaps:

Ψ_1: (bonding)(bonding)(bonding)　　Strongly bonding MO
Ψ_2: (bonding)(bonding)(node)　　Slightly bonding MO (net $\frac{1}{3}$)
Ψ_3: (bonding)(node)(node)　　Slightly antibonding MO (net $\frac{1}{3}$)
Ψ_4: (node)(node)(node)　　Strongly antibonding MO

These yield the energy levels shown in the figure. The first three of these are filled by the six electrons. In a net bonding sense, Ψ_2 and Ψ_3 cancel each other out. However, Ψ_1 has three favorable AO overlaps, whereas the p–p sigma overlap in free

Br_2 would only have one possible favorable overlap. So although the overall bond order in the Br_2–acetone complex is still only 1, it is a significantly stronger bond, which accounts for the enthalpy-driven formation of the charge-transfer complex. However, because the total bonding is relatively weak, the bond formed is usually significantly longer (and weaker) than the sum of the covalent radii for the atoms, though much less than the sum of the nonbonded van der Waals radii of the atoms (see Fig. 7.15). Usually, neither the acid nor the base changes its geometry much in forming the complex, also because of the weakness of the interaction.

It is sometimes possible to form a charge-transfer complex between a σ acceptor and a π donor. The best-known examples of this bonding are the complexes between the X_2 halogen molecules (except F_2) and benzene or other aromatic hydrocarbons. In crystalline form, the complex is a hexagonal column of parallel benzene rings with an X_2 molecule lying on the common sixfold axis between each pair of benzene rings. The X atom at each end of the X_2 molecule is presumably accepting a pair of electrons from the ring of pi-electron density on its neighbor benzene molecule. This sort of chain formation is common when both the acceptor and donor have two available sites. The crystal structure is frequently dictated by the desirability of chain formation; in solution, the complexes may well have quite different geometry.

The last type of charge-transfer complex is one in which a planar pi donor is coupled to a planar pi acceptor. The donor has a low ionization energy; the acceptor has electronegative substituents within the pi system that stabilize vacant antibonding pi MOs. Such complexes always form stacks of alternating donor and acceptor molecules. One example is shown in Fig. 7.16 with tetramethyl-*p*-phenylenediamine as donor and tetracyanoquinodimethane (TCNQ) as acceptor. Apparently the two molecules stack so that the electron-rich regions of the donor overlap as much as possible with electron-poor regions of the acceptor. However, the bonding is weak enough that the stacking is sometimes modified to

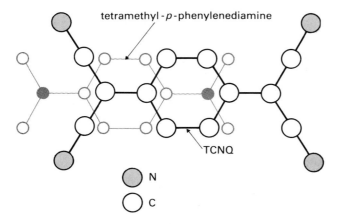

Figure 7.16   Stacking pattern in a π–π charge-transfer complex.

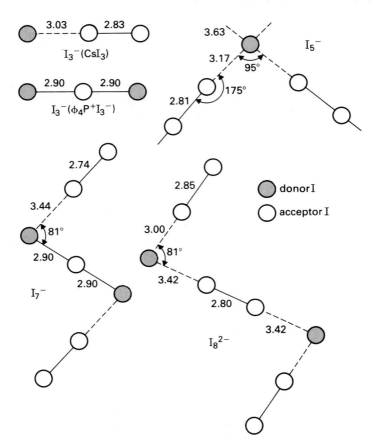

Figure 7.17   Polyiodide donor–acceptor complexes.

accommodate large London dispersion forces, which require neighboring regions of high polarizability.

The two most familiar examples of charge-transfer complexes are the triiodide ion and solutions of I_2 in water and other donor solvents. Iodine vapor is violet colored; its solutions in nondonor solvents are also violet or red-violet. In water, alcohol, and a number of other Lewis-base solvents, however, solutions of I_2 are brown. The brown color is caused by a very intense charge-transfer transition at around 2500 Å in the near ultraviolet, which in turn is caused by the transfer of an electron within the I_2–solvent charge-transfer complex. Different donor solvents give different values of λ_{max}, and the absorption frequency ν_{max} is in a roughly linear relation to the donor ionization energy for a given category of donor molecule (amines, for example). The I_2–solvent complexes have 1:1 stoichiometry and involve bonding like that suggested for the Br_2–acetone complex, though the solution species are presumably not polymerized.

The polyiodide ions, as characterized by x-ray diffraction in crystals, represent an interesting series of variations on the I_2-acceptor/I^--donor model for the triiodide ion. Figure 7.17 shows a number of these structures. In crystals with small cations, an iodide ion can serve as a weak donor to an I_2 molecule, producing an unsymmetrical I_3^-. On the other hand, in crystals with large cations, the I_3^- is isolated from other interactions such as halide contact or hydrogen bonding, and the ion becomes symmetrical. In I_5^-, a single iodide ion serves as a weak donor to two I_2 acceptors. The bond angle is determined by VSEPR considerations, even though only a very weak bond is being formed. In I_7^-, a symmetrical triiodide ion serves as a very weak donor from both terminal I atoms to two I_2 acceptors. (Note the extremely long bond length, not much shorter than the non-bonded van der Waals internuclear distance of 3.92 Å.) Finally, in I_8^{2-}, two iodide ions each donate strongly to a terminal I_2 and weakly to a central I_2. Bond angles in all these systems are about 90° at a donor atom and 180° at an acceptor atom. The strength of these interactions varies not only with the stoichiometry, as in this discussion, but also with the nature of the interhalide species being formed; in Table 5.14, we have seen some values of stability constants for a few species. In general the heavier, more polarizable halogens form the stronger charge-transfer bonds.

7.8 THERMODYNAMICS OF ELECTRONIC ACID–BASE INTERACTION: $E_A E_B + C_A C_B$

In our discussion of the factors affecting the strength of a Lewis acid–base interaction, we isolated the individual influences on the MO energy-quantity β. This quantity represents electron stabilization caused by atomic-orbital overlap or electron sharing by multiple nuclei. As such, it is a measure of covalent attraction and has a major influence on ΔH for acid–base interaction. However, in addition to the electron-sharing energy, there is a possible—and sometimes large—contribution to ΔH from electron-transfer energy. Figure 7.18 gives a simplified, extreme situation in which the base electron pair in a very high-energy orbital is relocated to an MO closely related to the low-energy acceptor orbital on the acid. Because ΔVOIP is so large for the two overlapping orbitals, β is quite small and there is little covalent contribution to the acid–base binding energy. Still, the electron pair has been greatly stabilized, because it was transferred almost completely to the low-energy acceptor orbital. There is thus a large electron-transfer or electrostatic contribution to the acid–base binding energy. To numerically predict ΔH for an acid–base interaction, we must allow for two contributions to the bonding: an electron-transfer or electrostatic contribution, and an electron-sharing or covalent contribution. ΔVOIP affects both of these, increasing the electron-transfer contribution as it decreases the electron-sharing contribution (although the first of these is a fairly direct relationship and the second is affected by many other molecular-structure factors). Since ΔVOIP involves both the donor and the

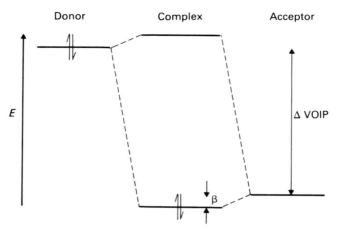

Figure 7.18 MO energies for a donor–acceptor interaction with a large electron-transfer energy.

acceptor, we may expect that ΔH could be predicted using an electrostatic parameter and a covalent parameter for the donor, and similar parameters for the acceptors. An acid with a very large electrostatic parameter should prefer to bind with a base having a large electrostatic parameter, since that combination would yield a large electrostatic contribution to the energy of interaction. Similarly, acids and bases with large covalent parameters should combine preferentially. This is an alternative way of thinking about the rule of thumb that hard acids prefer hard bases and soft acids soft bases: Hard acids and bases are just those with large electron-transfer energies; and soft acids and bases are those for which we expect largely covalent bonding.

Drago has proposed such an algebraic relationship for the enthalpy of acid–base interactions:

$$-\Delta H = E_A E_B + C_A C_B$$

Here E_A and C_A are the electrostatic and covalent parameters for the acid, and E_B and C_B are the corresponding parameters for the base. ΔH is the enthalpy change for the acid–base interaction when carried out under conditions free from significant lattice or solvation energies (that is, in the gas phase or in an inert solvent). The four E and C quantities for any given reaction are empirical parameters rather than any kind of theoretical construct, so a single reference acid or base is chosen with arbitrary E and C parameters. Other values are then scaled to that compound to fit observed enthalpies. Drago chose I_2 as the reference acid and set its E_A and C_A parameters to 1.00. These values are truly arbitrary—they do *not* imply that covalent and electrostatic bonding are equally important for the I_2 acid. At least one value must also be chosen for a base. Drago initially made E_B proportional to the dipole moment of each of a series of amines and C_B proportional to the total distortion polarization of the amine's nonbonding electron pair on forming the bond. For ammonia, this led to $E_B = 1.34$ and $C_B = 3.42$ and a fit of better than 0.1 kcal/mol rn error for the I_2/NH_3 reaction. In a later least-squares fit of a large number of enthalpy data, the arbitrary base parameters were taken to be $E_B = 1.32$ for the base dimethylacetamide and $C_B = 7.40$ for the base diethyl sulfide. Table 7.6 gives Drago's recommended values for about 30 acids and

Table 7.6 E and C Parameters for Lewis Acids and Bases, as Scaled to yield ΔH (kcal/mol reaction)

Acid	E_A	C_A
Iodine	1.00	1.00
Iodine monochloride	5.10	0.830
Iodine monobromide	2.41	1.56
Thiophenol	0.987	0.198
p-tert-Butylphenol	4.06	0.387
p-Methylphenol	4.18	0.404
Phenol	4.33	0.442
p-Fluorophenol	4.17	0.446
p-Chlorophenol	4.34	0.478
m-Fluorophenol	4.42	0.506
m-Trifluoromethylphenol	4.48	0.530
tert-Butyl alcohol	2.04	0.300
Trifluoroethanol	4.00	0.434
Hexafluoroisopropyl alcohol	5.56	0.509
Pyrrole (C_4H_4NH)	2.54	0.295
Isocyanic acid (HNCO)	3.22	0.258
Isothiocyanic acid (HNCS)	5.30	0.227
Boron trifluoride	7.96	3.08
Boron trifluoride (g)	9.88	1.62
Boron trimethyl	6.14	1.70
Trimethylaluminum	16.9	1.43
Triethylaluminum	12.5	2.04
Trimethylgallium	13.3	0.881
Triethylgallium	12.6	0.593
Trimethylindium	15.3	0.654
Trimethyltin chloride	5.76	0.0296
Sulfur dioxide	0.920	0.808
Bis(hexafluoroacetylacetonate)copper(II)	3.39	1.40
Antimony pentachloride	7.38	5.13
Chloroform	3.31	0.150
1-Hydroperfluoroheptane [$CF_3(CF_2)_6H$]	2.45	0.226

(continues)

Table 7.6 (*Continued*)

Base	E_B	C_B
Pyridine	1.17	6.40
Ammonia	1.36	3.46
Methylamine	1.30	5.88
Dimethylamine	1.09	8.73
Trimethylamine	0.808	11.54
Ethylamine	1.37	6.02
Diethylamine	0.866	8.83
Triethylamine	0.991	11.09
Acetonitrile	0.886	1.34
Chloroacetonitrile	0.940	0.530
Dimethylcyanamide	1.10	1.81
Dimethylformamide	1.23	2.48
Dimethylacetamide	1.32	2.58
Ethyl acetate	0.975	1.74
Methyl acetate	0.903	1.61
Acetone	0.987	2.33
Diethyl ether	0.963	3.25
Isopropyl ether	1.11	3.19
n-Butyl ether	1.06	3.30
p-Dioxane [$(CH_2)_4O_2$]	1.09	2.38
Tetrahydrofuran [$(CH_2)_4O$]	0.978	4.27
Tetrahydropyran	0.949	3.91
Dimethyl sulfoxide	1.34	2.85
Tetramethylene sulfoxide [$(CH_2)_4SO$]	1.38	3.16
Dimethyl sulfide	0.343	7.46
Diethyl sulfide	0.339	7.40
Trimethylene sulfide [$(CH_2)_3S$]	0.352	6.84
Tetramethylene sulfide	0.341	7.90
Pentamethylene sulfide	0.375	7.40
Pyridine *N*-oxide	1.34	4.52
4-Methylpyridine *N*-oxide	1.36	4.99
4-Methoxypyridine *N*-oxide	1.37	5.77
Tetramethylurea	1.20	3.10
Trimethylphosphine	0.838	6.55
Benzene	0.486	0.707
p-Xylene	0.416	1.78
Mesitylene	0.574	2.19
Quinuclidine [$HC(C_2H_4)_3N$]	0.704	13.2
Hexamethylphosphoramide	1.73	1.33

Source: Reprinted with permission from R. S. Drago *et al.*, *J. Amer. Chem. Soc.* (1971), *93*, 6014. Copyright 1976 American Chemical Society.

Value in $(kJ/mol)^{1/2}$ = (tabulated value)$^{1/2} \times 2.045$.

30 bases; values in the table have been left in kilocalories (kcal) units rather than the SI kilojoules (kJ) values to emphasize the arbitrary values of 1.00 for the I_2 acid. Calculated ΔH values can, of course, be converted to kJ simply by multiplying the kcal value by 4.184. From the values in the table, about a thousand ΔH values can be calculated, most within 0.1 kcal of the experimental value, and all but a half dozen within 0.3 kcal or about 1 kJ. The exceptions deserve some consideration.

One prominent exception is the reaction between $B(CH_3)_3$ acid and $N(CH_3)_3$ base, which has a measured ΔH of -17.6 kcal/mol rn but a quite different calculated value:

$$-\Delta H = (6.14)(0.808) + (1.70)(11.54) = 24.6 \text{ kcal/mol rn}$$

This is chemically and stereochemically very similar to a case we considered earlier in the chapter: $B(CH_3)_3$ acid and $N(C_2H_5)_3$ base. In the previous example, front-strain, or steric hindrance between acid and base, lowered the bond energy and thus, presumably, the enthalpy of combination. Since front-strain is a property of the *combined* acid and base and not of either alone, it is not surprising that the E and C parameters do not include or predict it. Presumably the energy magnitude of the steric hindrance is the difference between the calculated and experimental values of ΔH: 7.0 kcal or 29 kJ in this case.

On the other hand, the reasons for the striking difference observed between the calculated and experimental values for BF_3 acid and $S(C_2H_5)_2$ base are not at all clear: $\Delta H_{\text{expt}} = -2.9$ kcal, $\Delta H_{\text{calc}} = -15.3$ kcal. Front-strain should be very modest because of the small size of the fluorine atoms. Perhaps the reorganization energy involved in changing BF_3 from its planar geometry in the free molecule to the more-or-less tetrahedral geometry in the acid–base complex differs enough from that for other BF_3 complexes to account for the difference. It is in any event surprising. Perhaps one of the virtues of the E and C parametric approach is that it focuses attention on the truly unusual cases for experimental investigation.

It is tempting—but misleading—to compare the hardness or softness of acids or bases by comparing E values or C values. The difficulty is that the E and C parameters deal not only with hard/soft behavior, but also with strength and weakness. Diethylamine is not a softer base than diethyl sulfide just because its C_B value is larger (8.83 versus 7.40), and triethylgallium is not a harder acid than BF_3 even though its E_A value is larger (12.6 versus 9.88). Not even ratios are reliable: One might take E/C as a measure of hardness, but IBr is not a harder acid than $SbCl_5$ even though its E_A/C_A is larger (1.54 versus 1.44). What one *can* say is that if, in comparing two acids or two bases, both E and C for one acid (or base) are greater than E and C for the other, then the first acid (base) is stronger in combining with any base (acid). If $E_1 > E_2$ but $C_1 < C_2$, there will be some situations in which acid 1 appears stronger and some in which acid 2 appears stronger, depending on the base. The same is true for base comparisons. In short, Drago's four parameters appear to be the minimum number we can use to describe the strength of interaction of a wide variety of acids and bases, but they cannot be readily resolved into judgments as to hardness or softness. The HSAB principle is useful, because it is so convenient and easily remembered, but by its qualitative nature it is unable to predict the strength of acid–base interactions and so is subject to occasional reversals of its predictions.

The E and C parameters are extremely useful for the prediction of enthalpies of acid–base interaction for uncharged molecular species as acids or bases, whether in solution or in the gas phase. It would be useful to have an extended version of the E and C system that would similarly allow for the interaction of charged species. Drago has produced such an extension:

$$\Delta H = e_A e_B + c_A c_B + t_A t_B$$

where the e and c parameters have the same physical significance as before, but with different numerical values because the t term has been added. The t parameters for the acid and base allow for actual electron transfer between acid and base and for the Coulomb attraction that results. The resulting equation is a more powerful representation, because it now covers a larger field of acids and bases, but it is somewhat more cumbersome. For our purposes the two-parameter expression will be adequate.

7.9 OPTICAL BASICITY

Much of our recent discussion, for both Brønsted and Lewis acid/base systems, has focused (with some success) on predicting the intensity or spontaneity of a specific acid–base reaction. There is a different sense, however, in which we might wish to measure intrinsic acidity or basicity, namely: What is the intrinsic acidity of a liquid or solid medium that serves as host matrix for a solute species capable of acid/base interaction? If we restrict our discussion to the specific case of a metal ion in solution, the metal ion serves as a Lewis acid and the question becomes one of predicting the intrinsic basicity of the solvent or lattice in which the metal ion is dissolved or implanted. Duffy and Ingram have proposed a simple spectrophotometric probe in which the UV-absorption frequency of dissolved Pb^{2+} is scaled against the frequency of the same transition in media of differing electron basicity. Although in principle the test can be used in a wide variety of media, it is most frequently applied to media in which an oxygen atom is the electron donor. We shall restrict our discussion to those systems.

The Pb^{2+} ion does not absorb light in the visible range of the spectrum, but it has a transition in the near-UV region in which one of its $6s$ electrons is excited to a $6p$ state (a $^3P_1 \leftarrow {}^1S_0$ transition). In a strongly basic (electron-rich) environment, the added electron density near the Pb nucleus reduces the effective nuclear charge attracting an s electron more than it does that attracting a p electron, which makes it easier to excite the s electron to a p distribution and reduces the frequency of the s-to-p transition. Using a scale of electron-donor ability that fits the spectra of many transition-metal ions, we can extrapolate the Pb^{2+} transition frequency to a hypothetical condition of zero electron-donor ability or zero basicity. Under zero basicity, the transition would occur at 60,700 cm^{-1} (near the gaseous free-ion transition frequency of 64,400 cm^{-1}). By contrast, in a pure ionic-oxide medium (Pb^{2+} doped into $Ca^{2+}O^{2-}$ crystals in small concentrations) the Pb^{2+} transition occurs at 29,700 cm^{-1}. This very large frequency shift is caused by the strongly basic oxide-ion environment of the CaO. For a given medium, then, we can dissolve a small concentration of Pb^{2+}, measure its spectrum, and characterize the oxide basicity of the medium by setting up an *optical basicity* ratio Λ:

$$\Lambda_{medium} = \frac{\nu_{free\ ion} - \nu_{medium}}{\nu_{free\ ion} - \nu_{CaO}} = \frac{60,700 - \nu_{medium}}{31,000}$$

The Λ quantity will be near zero for a highly acidic medium and near 1.00 for a highly basic medium. For example, Λ is 0.332 in 97% H_2SO_4; 0.404 in 100% H_3PO_4; 0.439 in B_2O_3; and 0.680 in $Na_4B_2O_5$. It should be clear that the Pb^{2+} probe is as convenient for complex mixed media, glasses, and molten salts as it is for the simplest solutions.

Any oxide medium has counterions present, whether it is nearly ionic or strongly covalent. These counterions strongly influence the base capability of the oxides. Extensive experimental studies on mixed-oxide glasses show that Λ depends on the fraction of total oxide charge neutralized by each counterion and on the identity of each counterion. By "fraction of charge neutralized" is meant, for example, that in $Ca_3(PO_4)_2$ the Ca^{2+} ions, which have a total charge (or formal oxidation state) of 6+, are neutralizing three-eighths of the total oxide charge of 16−, while the P^{5+} ions are neutralizing five-eighths of the total oxide charge. For each element, Ca and P, its oxide-charge fraction is multiplied by a weighting factor that assesses its ability to moderate the donor capability of a neighbor oxide. Specifically, we have

$$\nu_{free\ ion} - \nu_{Ca_3(PO_4)_2} = \nu_{free\ ion} - \nu_{CaO}\left[\frac{3}{8}\left(1-\frac{1}{\gamma_{Ca}}\right) + \frac{5}{8}\left(1-\frac{1}{\gamma_P}\right)\right]$$

where γ_{Ca} and γ_P are the *basicity-moderating parameters* characteristic of the elements Ca and P (see Table 7.7). These γ values are very similar to the electronegativities of the elements, which is reasonable: A very electronegative element will do more to reduce the electron-donating power of a nearby oxide than a relatively electropositive element will.

The optical basicity Λ, as measured by the spectrum of a Pb^{2+} probe ion, reflects the average basicity of all the various oxide environments in a liquid, glass, or crystal. It can be reproduced to a good approximation by a rearrangement of the foregoing equation:

$$\Lambda = 1 - \left[f_A\left(1-\frac{1}{\gamma_A}\right) + f_B\left(1-\frac{1}{\gamma_B}\right) + \cdots\right]$$

where f_A represents the fraction of oxide charge neutralized by cationic element A. We can thus calculate, to a good approximation, the optical basicity of a given

Table 7.7 Basicity-Moderating Parameters (γ)

H							
2.50							
Li	Be		B	C	N		
1.00	(1.65)		2.36	3.04	3.73		
Na	Mg		Al	Si	P	S	Cl
0.87	1.28		1.65	2.09	2.50	3.04	3.73
K	Ca	Zn	Ga	Ge	As	Se	Br
0.73	1.00	1.82	(2.12)	(2.39)	(2.36)	(3.02)	(3.37)
Rb	Sr						I
0.73	(0.99)						(3.04)
Cs							
0.60							

Note: Parenthesized values are calculated from a linear relation of measured values to electronegativity.

ionic environment from its stoichiometry and from the tabulated γ values of the elements present.

In view of the reasonably effective semitheoretical expression we have developed for γ, we can extend the optical basicity concept to describe the basicity of microscopic systems—oxygen electron-donor atoms on individual molecules or ions—in terms of an analogous quantity λ, the *microscopic optical basicity* of a given oxygen atom:

$$\lambda = 1 - \left[f_A\left(1 - \tfrac{1}{\gamma_A}\right) + f_B\left(1 - \tfrac{1}{\gamma_B}\right) + \cdots \right]$$

Here f_A is the fraction of that oxygen atom's charge that is neutralized by its specific neighbor atom A (in a positive oxidation state). For example, consider the calculated basicities of the oxygen atoms on a carbonate ion and a nitrate ion:

CO_3^{2-}	NO_3^-
$f_C = \tfrac{4}{6} = \tfrac{2}{3}$	$f_N = \tfrac{5}{6}$
$\lambda = 1 - \left[\tfrac{2}{3}\left(1 - \tfrac{1}{3.04}\right)\right]$	$\lambda = 1 - \left[\tfrac{5}{6}\left(1 - \tfrac{1}{3.73}\right)\right]$
$= 0.552$	$= 0.390$

Remembering that a high λ corresponds to a strongly basic system, we interpret these results to mean that a carbonate O atom is much more basic than a nitrate O atom. This, of course, is consistent with our chemical intuition.

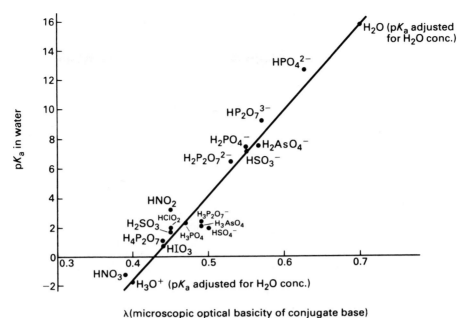

Figure 7.19   Aqueous Brønsted pK_a values as a function of optical basicity.

If λ really represents the basicity of a given ion, it should be possible to correlate the λ values of a variety of oxyanions with the pK_a values in water of those ions' conjugate acids. Figure 7.19 shows that there is in fact a good linear relationship for a variety of ions, regardless of their net charge or degree of protonation. Not all the oxyacids of Table 7.4 fit the line in Fig. 7.19, but the large number that do provides an interesting correlation between optical basicity and an entirely different experimental acid/base measurement. Since the optical-basicity technique was developed to describe the Lewis basicity of nonprotic media such as glasses and silicate minerals, its applicability to Brønsted systems is an unusual confirmation of the unity of acid/base chemistry.

PROBLEMS ———————————————————————————————————————

A. DESCRIPTIVE

A1. The ΔH^0 values for the reactions $I_2 + NR_3 \rightarrow I\!-\!I\!:\!NR_3$ are given below, along with the average R—N—R bond angle for the amine. Which of the five ΔH influences described in the chapter is primarily responsible for the observed trend in ΔH^0?

NR_3	$\measuredangle$ R—N—R (°)	ΔH^0 (kJ/mol rn)
NH_3	106.0	−20
NH_2CH_3	110.0	−30
$NH(CH_3)_2$	109.7	−41
$N(CH_3)_3$	110.9	−51

A2. Predict whether the reactions below, in which a new base substitutes into an acid–base adduct, should occur with retention or inversion of the adduct's configuration.

$$F_3B—SEt_2 + PEt_3 \rightarrow F_3B—PEt_3 + Et_2S$$

$$Br_3B—OEt_2 + MeCN \rightarrow Br_3B—NCMe + Et_2O$$

A3. Estimate the ligand-exchange rate constant for hydrated Pb^{2+} by comparison with Fig. 7.5, taking into account the charge, size, and hardness of the cation.

A4. On Fig. 7.6, pencil in a vertical line separating +1 ions from +2 ions, another separating +2 ions from +3 ions, and a third separating +3 ions from +4 ions. Note the hard/soft acid characteristics within each charge group. Estimate the hydrolysis pK_a for the following ions *not* shown on the figure: Au^+, Ra^{2+}, Pt^{2+}, Mn^{3+}, Mn^{4+}. Give your reason for each value.

A5. What is the most probable reaction when H_3PO_4 is dissolved in H_2SO_4 solvent? When HSO_3F is dissolved in H_2SO_4? When the two resulting solutions are mixed?

A6. Estimate the pK_a of NH_4^+ ion in DMSO solvent, given that in water its pK_a is 9.25.

A7. Detergent builders such as tripolyphosphate, $P_3O_{10}^{5-}$, coordinate Ca^{2+} in hard water to form a negatively charged species that does not precipitate the detergent anion. One of the alternatives to $P_3O_{10}^{5-}$ is the carbonate ion, CO_3^{2-}. What kind of bonding would stabilize a carbonate complex of Ca^{2+} in the presence of large amounts of the hard base H_2O?

A8. Why does toluene form a more stable complex with Ag^+ than benzene does?

A9. Pyridine is a weaker base in water than 2,6-lutidine (dimethylpyridine), but the pyridine-BF_3 complex is more stable than the comparable lutidine-BF_3 complex. What is responsible for this inversion of stabilities?

A10. Besides the complexes $LSbCl_5$, in which $SbCl_5$ is a strong Lewis acid and L is a Lewis base such as $OP(CH_3)_3$, there are complexes $LSbCl_3$ and L_2SbCl_3 and L_2SbCl_3, in which $SbCl_3$ is the Lewis acid. Why aren't there L_3SbCl_3 complexes? On the other hand, the solid compound $[Co(NH_3)_6]^{3+}[SbCl_6]^{3-}$ is known: What makes it more stable than the other L_3SbCl_3 complexes?

A11. Following the *p*-only approach of Figs. 4.23 and 7.15, set up qualitative AO overlaps and MO energy levels for a charge-transfer complex between a Br atom on CBr_4 and pyridine.

A12. Many crystalline ionic polyhalides contain Lewis-base solvent molecules that seem to be essential to the stability of the crystal, since they cannot be removed without decomposing the polyhalide—for example, $K^+I_3^-·H_2O$. What is the probable bonding mode of the solvent molecule in the crystal?

B. NUMERICAL

B1. The entropy changes for the successive protonations of aqueous phosphate ion are given below. Propose a functional relationship between ΔS and ionic charge on a reasonable physical basis.

$$H_3O^+ + PO_4^{3-} \rightarrow HPO_4^{2-} + H_2O \qquad \Delta S^0 = +188 \text{ J/mol·K}$$

$$H_3O^+ + HPO_4^{2-} \rightarrow H_2PO_4^- + H_2O \qquad\qquad +124 \text{ J/mol·K}$$

$$H_3O^+ + H_2PO_4^- \rightarrow H_3PO_4 + H_2O \qquad\qquad +68 \text{ J/mol·K}$$

B2. The reaction $HCl + OH^- \rightarrow Cl^- + H_2O$ is thermodynamically favorable both in the gas phase and in solution, like the similar reactions in Fig. 7.2. Set up a Born–Haber cycle for the HCl reaction to predict the gas-phase ΔH^0. Which of the three reactions in Fig. 7.2 does the HCl reaction most strongly resemble in terms of its driving force? What fundamental difference is there between the driving force of the HCl reaction and that of the most closely comparable reaction?

B3. Use the appropriate Born–Haber cycles from the chapter discussion to show that PH_3 should have a positive ΔH for reacting with water either as an acid or as a base, and hence should show no acid–base properties in water. Take data from the chapter calculations and from Table 6.6; estimate the hydration energies of PH_4^+ and PH_2^- from their radii, each of which is about 0.4 Å greater than that of the comparable nitrogen ion (NH_4^+/NH_2^-).

B4. Use the two equations given in the chapter for estimating pK_a to predict pK_a for the following acids: $HClO_4$, HSO_3F, H_2SO_4, H_3PO_4, H_3PO_3, H_4SiO_4. For which acids do the equations yield the poorest fit? Suggest some electronic reasons for the equations' failure.

B5. Use the E and C parameters in Table 7.6 to estimate ΔH^0 for the acid–base interaction of $Ga(C_2H_5)_3$ with $N(CH_3)_3$. Compare the result to the experimental value of -17.0 kcal/mol rn. Why is this a much better fit than that of the calculation in the chapter for $B(CH_3)_3$ and $N(CH_3)_3$?

B6. Calculate the microscopic optical basicity λ of an oxygen atom on a free orthosilicate ion SiO_4^{4-}. Should it be possible to observe SiO_4^{4-} in aqueous solution? Industrial uses of sodium orthosilicate require it to be more basic than sodium metasilicate (SiO_3^{2-}), which, in turn, is more basic than sodium carbonate. Do optical basicity values bear this out?

C. EXTENDED REFERENCE

C1. In methanol solvent, the conductance of KCl steadily decreases as dicyclohexyl[18]-crown-6 is added until 1:1 stoichiometry is reached. On the other hand, when the same conductimetric titration is carried out in 90%-chloroform/10%-methanol, the conductance steadily *increases* until the equimolar concentration is reached. Why is the conductance behavior reversed when the solvent is changed? [See C. J. Pedersen and H. K. Frensdorff, *Angew. Chem., Int. Ed.* (1972), *11*, 16.]

C2. U. Schindewolf and H. Schwab [*J. Phys. Chem.* (1981), *85*, 2707] report K_b for ammonia reacting with water in liquid-ammonia solution:

$$NH_3(liq) + H_2O(amm) \rightleftharpoons NH_4^+(amm) + OH^-(amm) \qquad K_b = 6 \times 10^{-23}$$

For the same reaction in water, $K_b = 3 \times 10^{-7}$ (both K_b values include the molar concentration of the liquid solvent). For the liquid-ammonia reaction, $\Delta G^0 = +92$ kJ/mol, $\Delta H^0 = +21$ kJ/mol, and $\Delta S^0 = -303$ J/mol K. Using thermodynamic data for aqueous solutions from NBS Technical Note 270 or the *Handbook of Chemistry and Physics,* calculate ΔH^0 and ΔS^0 for the aqueous reaction. Compare these values with those for liquid ammonia. Is the striking change in K_b from water to ammonia solution primarily an enthalpy effect, or is it an entropy effect? Explain your answer in terms of liquid and solution structures.

C3. R. A. Kovar *et al.* [*Inorg. Chem.* (1980), *19*, 3264], have studied acid–base adducts of mixed chloro/butyl gallium compounds, $GaCl_x(n\text{-Bu})_{3-x}$, with methylamines, $(CH_3)_zNH_{3-z}$. For a given gallium compound, the tendency of amines to react is NH_3 (most reactive) $\geq MeNH_2 > Me_2NH > Me_3N$. For a given amine, the gallium compounds react most readily in the order $GaCl_3 > GaCl_2Bu > GaClBu_2 > GaBu_3$. Discuss the probable structural electronic influences responsible for these trends. What experimental technique was used for the study?

8

Enthalpy-Driven Reactions II: Redox Reactions

We have now devoted a good deal of attention to acid–base reactions seen in various guises. Perhaps the most fundamental acid/base definition (the Lewis definition) treats an acid as an electron acceptor and a base as an electron donor. In this chapter we shall consider reducing agents as electron donors and oxidizing agents as electron acceptors. It is important to distinguish between acid/base reactions and redox reactions because the terminology is so similar. In Chapter 6 we stated that the distinguishing characteristic of an acid–base reaction was that no atom changed its formal oxidation state. A redox reaction also occurs between an electron donor and an acceptor, but the electron transfer causes one atom to increase its oxidation state and another to decrease its oxidation state. Atom transfer frequently occurs at the same time, the atom carrying bonding electrons along with it:

$$X—O + Y \rightarrow X + O—Y$$

This is not a necessary characteristic of a redox reaction, however. Electron transfer is the key feature. A base donates electrons but retains partial ownership of them in its valence orbitals, whereas a reducing agent transfers its electrons entirely, with or without atom transfer. In a mechanistic sense, an electron-transfer reaction almost always occurs as a sequence of one-electron or two-electron transfer steps, regardless of the total number of electrons ultimately transferred in the stoichiometric reaction. One-electron transfer steps rarely involve atom transfer and the resulting major changes in the stereochemistry of the reacting species. Two-electron transfer steps, on the other hand, often proceed by atom transfer and almost always show major changes in coordination number or at least in the molecular geometry of the donor and acceptor species. We shall begin by looking at these processes from both a thermodynamic and a kinetic point of view.

8.1 THE MOLECULAR BASIS FOR ELECTRON TRANSFER

As the chapter title suggests, redox reactions are usually enthalpy-driven rather than entropy-driven. Electron transfer yields a more energetically favorable overall electronic arrangement. Just as we investigated the sources of the favorable enthalpy change in acid–base reaction, we inquire here into the sources of redox enthalpy changes. We must first distinguish, however, between two different electron-transfer mechanisms, because the principal energy sources are quite different for the two.

These are called the *inner-sphere* mechanism and the *outer-sphere* mechanism. In an inner-sphere mechanism, the atom being oxidized and the atom being reduced form directed bonds to a common atom or small group, which then serves as a bridge for electron transfer:

$$AX + B(H_2O) \rightleftharpoons AXB + H_2O \qquad \text{(in water)}$$

$$AXB \rightleftharpoons AXB^* \qquad \text{(activated complex)}$$

$$AXB^* \rightleftharpoons A^- \!\!-\!\! X \!\!-\!\! B^+$$

$$A^- \!\!-\!\! X \!\!-\!\! B^+ + H_2O \nearrow A(H_2O)^- + BX^+ \qquad \text{(atom transfer)}$$
$$\searrow AX^- + B(H_2O)^+ \qquad \text{(no transfer)}$$

The bridging group itself does not undergo any redox reaction. Oxygen atoms and halogen atoms are common bridging atoms, but some small molecular groups such as CN^- can also serve that function.

In an outer-sphere mechanism, the atom being oxidized and the atom being reduced sometimes meet directly (for example, when sodium metal reacts with chlorine gas). In solution, however, they more commonly meet only in the sense that their solvation spheres or coordination spheres touch, without the formation of any new directed bonds. If we use $\|$ to represent contact between the valence-electron shells of two atoms without electron sharing or bond formation, an outer-sphere mechanism would proceed as follows:

$$A + B \rightleftharpoons A\|B$$

$$A\|B \rightleftharpoons A\|B^*$$

$$A\|B^* \rightleftharpoons A^-\|B^+$$

$$A^-\|B^+ \rightleftharpoons A^- + B^+$$

This notation is equally appropriate whether A and B are individual atoms or fully coordinated molecular species.

To gain some insight into the thermodynamic driving force of a simple outer-sphere reaction, we can consider the reaction already referred to:

$$Na(s) + \tfrac{1}{2}Cl_2(g) \rightarrow NaCl(s)$$

For this very spontaneous reaction, we have the following thermodynamic data:

$$\Delta G^0 = -384 \text{ kJ/mol reaction} \quad \text{(spontaneous)}$$
$$\Delta H^0 = -411 \text{ kJ/mol reaction} \quad \text{(favorable)}$$
$$\Delta S^0 = -90 \text{ J/deg} \cdot \text{mol reaction} \quad \text{(unfavorable)}$$
$$\mathscr{E}^0 = +3.98 \text{ V}$$

It is clear that the reaction is enthalpy driven. Where does the highly favorable enthalpy change come from? If we form a Born–Haber cycle for the reaction (with all energy quantities expressed in kJ/mol) we see—as was indicated at the beginning of Chapter 2—that lattice or environmental effects are crucial:

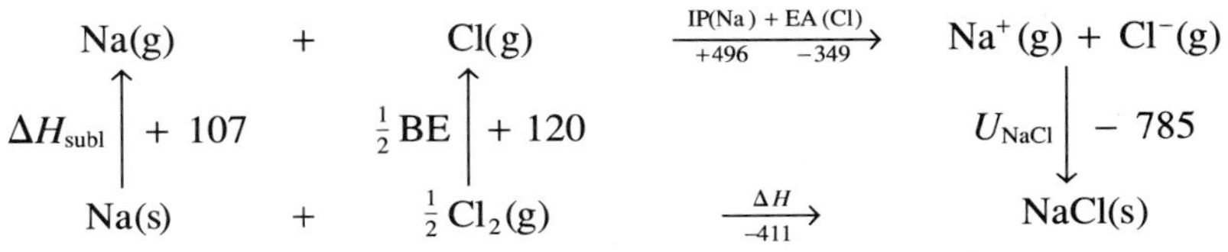

The top line of this cycle, which represents the intrinsic electron-attracting qualities of a neutral chlorine atom and a sodium cation, is clearly unfavorable energetically. It is only the tremendous difference in lattice stability between the ionic NaCl and the less strongly bound Na^0 and Cl_2 that makes the overall reaction so favorable.

Some outer-sphere reactions, however, do not show strong or even significant environmental contributions to the overall enthalpy change. Consider the reaction

$$Fe(CN)_6^{4-}(aq) + Mo(CN)_8^{3-}(aq) \rightarrow Fe(CN)_6^{3-}(aq) + Mo(CN)_8^{4-}(aq)$$

for which $\Delta G^0 = -36$ kJ/mol reaction and $\mathscr{E}^0 = +0.37$ V. Although no entropy data are available for the reaction, the symmetry of the charges and sizes of reactant and products suggests that the entropy change must be small because the overall ordering effect on surrounding water molecules cannot change significantly, and the geometry of the reactant and product molecules also changes very little. Focusing our attention on the enthalpy change, we find that the same symmetry considerations indicate that while there can be a considerable hydration-energy difference between a 4- ion and a 3- ion, the fact that all four ions in the reaction are of comparable radius and symmetrical charge requires that the overall solvation energy change be extremely small. Presumably in this case the favorable enthalpy change arises from the intrinsic difference in electron affinity between Mo(V) and Fe(III).

For inner-sphere reaction mechanisms, the energy source is quite different. Such mechanisms are often accompanied by atom transfer, as the general mechanism above has indicated. One such reaction is that between sulfite and chlorate:

$$SO_3^{2-}(aq) + ClO_3^-(aq) \rightarrow SO_4^{2-}(aq) + ClO_2^-(aq)$$

This reaction is also quite spontaneous (sulfite is a mild reducing agent and chlorate is a strong oxidizing agent), with the following thermodynamic data:

$$\Delta G^0 = -224 \text{ kJ/mol reaction} \qquad \text{(spontaneous)}$$
$$\Delta H^0 = -251 \text{ kJ/mol reaction} \qquad \text{(favorable)}$$
$$\Delta S^0 = -88 \text{ kJ/mol reaction} \qquad \text{(unfavorable)}$$
$$\mathscr{E}^0 = +1.16 \text{ V}$$

Again, this reaction is clearly enthalpy driven. Some of the Born–Haber cycle data are less reliable than those for the familiar NaCl formation. Even so, the cycle for this reaction suggests that environmental effects—solvation energies—are less important than the difference in bond energies for the atom being transferred:

$$
\begin{array}{ccccccc}
SO_3^{2-}(g) & + & ClO_3^-(g) & \xrightarrow[]{\substack{BE(S-O)\ -\ BE(Cl-O) \\ +435 \qquad -289}} & SO_4^{2-}(g) & + & ClO_2^-(g) \\
-\Delta H_{hyp} \Big\uparrow +1142 & & -\Delta H_{hyd} \Big\uparrow +289 & & -\Delta H_{hyd} \Big\downarrow -1109 & & \Delta H_{hyd} \Big\downarrow -389 \\
SO_3^{2-}(aq) & + & ClO_3^-(aq) & \xrightarrow[-251]{\Delta H^0} & SO_4^{2-}(aq) & + & ClO_2^-(aq)
\end{array}
$$

In effect, this reaction proceeds because the sulfur–oxygen bond is so much stronger than the chlorine–oxygen bond. This is no surprise, of course. However, it is worth noting that a change in bond energies is quite different from a change in lattice or solvation energies, or even from a change in individual atomic-electron affinities, both of which are enthalpy sources for outer-sphere electron-transfer mechanisms.

For a general redox reaction occurring in solution, then, we have identified several major contributions to the reaction's favorable enthalpy change: atomic or molecular ionization energies and electron affinities (as measured in the gas phase); solvation and lattice energies for condensed-phase reactants and products; and bond-energy differences where old bonds are broken and new bonds formed. A gas-phase ionization potential, such as

$$Fe^{2+}(g) \xrightarrow{IP_3} Fe^{3+} + e^-(g)$$

is different from the half-cell potential for what is apparently the same reaction,

$$Fe^{2+}(aq) \xrightarrow{\mathscr{E}(oxidation)} Fe^{3+}(aq) + e^-$$

for two reasons. One is the obvious difference that must be accounted for between the hydration energies of the Fe^{2+} and Fe^{3+} ions, in addition to the ionization potential. The other reason is the unspecified environment of the "free" electron in the aqueous half-reaction. Since free electrons are unstable in condensed media, a reference electronic environment must be chosen, and the gas-phase ionization potential must be corrected for the energy difference between a gaseous electron and the same electron in the reference environment. As general-chemistry texts point out, the standard redox electronic environment is the aqueous hydrogen half-cell:

$$H_3O^+(aq, 1\ M) + e^- \xrightarrow{\mathscr{E}(reduction)\ =\ 0\ V} \tfrac{1}{2}H_2(g, 1\ atm) + H_2O(l)$$

Since this reference redox environment is chosen by convention, solution half-cell potentials are meaningful only within that convention (that is, as potential *differences* against a hydrogen half-cell), whereas gas-phase ionization potentials are measurable on an individual basis.

8.2 REDUCTION POTENTIALS AND FREE ENERGIES

Although chemists are accustomed to using free-energy changes to predict or describe the spontaneity of reactions, free energy cannot be measured directly—not even free-energy *changes* can be measured directly. Since the electric potential necessary to remove an electron from an atom is a direct guide to the spontaneity of electron transfer from that atom, that potential is presumably proportional to ΔG^0 for that process. However, to be dimensionally consistent with energy, electric potential must be multiplied by charge. The free-energy change for electron transfer is equal to the potential necessary times the total charge transferred. If we wish to use ΔG^0 in electronvolts, the total charge is simply the number of electrons transferred per half-reaction:

$$\Delta G = -n\mathscr{E}$$

where the negative sign simply aligns the positive potential with the negative free-energy change for spontaneity. Since the electronvolt is not a particularly convenient unit, we more often insert the faraday, $\mathscr{F}$, as a unit-conversion factor:

$$\mathscr{F} = 23.0607\frac{kcal/mol}{eV} = 96.486\frac{kJ/mol}{eV}$$

Using the faraday conversion factor gives us the more familiar expression

$$\Delta G = -n\mathscr{F}\mathscr{E} \qquad or \quad \Delta G^0 = -n\mathscr{F}\mathscr{E}^0$$

We can then use a table of standard reduction potentials such as Table 8.1 in conjunction with this expression to find an immediate measure of the free-energy change for any redox reaction whose half-reactions are included in the table. Note that in this table many metals appear as halometallates or as other complexes, and that many simple aqueous half-reactions are missing, such as the Fe^{3+}/Fe^{2+} couple. A large number of the latter appear in Table 8.2, of which more shortly; the two tables should be used together for redox reaction prediction.

In fact, we can combine half-reactions and their potentials from any table such as 8.1 to yield the potential of a new, unlisted half-reaction. This is only the familiar process of using Hess's law on a thermodynamic state function to

Table 8.1   Standard Reduction Potentials in Aqueous Solution

Acid solution	$\mathscr{E}^0$ (V)
$F_2(g) + 2H^+ + 2e^- = 2HF(aq)$	+3.06
$F_2(g) + 2e^- = 2F^-$	+2.87
$H_4XeO_6 + 2H^+ + 2e^- = XeO_3 + 3H_2O$	+2.3
$F_2O + 2H^+ + 4e^- = 2F^- + H_2O$	+2.15
$XeO_3 + 6H^+ + 6e^- = Xe + 3H_2O$	+1.8
$PdCl_6^{2-} + 2e^- = PdCl_4^{2-} + 2Cl^-$	+1.288
$S_2Cl_2 + 2e^- = 2S + 2Cl^-$	+1.23
$ICl_2^- + e^- = 2Cl^- + \frac{1}{2}I_2$	+1.056
$IrCl_6^{2-} + e^- = IrCl_6^{3-}$	+1.017
$AuCl_4^- + 3e^- = Au + 4Cl^-$	+1.00
$AuBr_2^- + e^- = Au + 2Br^-$	+0.956
$AuBr_4^- + 3e^- = Au + 4Br^-$	+0.87 (60°C)
$IrCl_6^{3-} + 3e^- = Ir + 6\ Cl^-$	+0.77
$(CNS)_2 + 2e^- = 2CNS^-$	+0.77
$C_2H_2(g) + 2H^+ + 2e^- = C_2H_4(g)$	+0.731
$PtCl_4^{2-} + 2e^- = Pt + 4Cl^-$	+0.73
$C_6H_4O_2 + 2H^+ + 2e^- = C_6H_4(OH)_2$	+0.6994
$PtCl_6^{2-} + 2e^- = PtCl_4^{2-} + 2Cl^-$	+0.68
$Ag_2SO_4 + 2e^- = Ag + SO_4^{2-}$	+0.654
$Cu^{2+} + Br^- + 2e^- = CuBr$	+0.640
$PdCl_4^{2-} + 2e^- = Pd + 4Cl^-$	+0.62
$Hg_2SO_4 + 2e^- = 2Hg + SO_4^{2-}$	+0.6151
$RuCl_5^{2-} + 3e^- = Ru + 5Cl^-$	+0.601
$PdBr_4^{2-} + 2e^- = Pd + 4Br^-$	+0.60
$PtBr_4^{2-} + 2e^- = Pt + 4Br^-$	+0.581
$C_2H_4(g) + 2H^+ + 2e^- = C_2H_6(g)$	+0.52
$Ag_2CrO_4 + 2e^- = 2Ag + CrO_4^{2-}$	+0.464
$RhCl_6^{3-} + 3e^- = Rh + 6Cl^-$	+0.431
$\frac{1}{2}C_2N_2(g) + H^+ + e^- = HCN(aq)$	+0.373
$Fe(CN)_6^{3-} + e^- = Fe(CN)_6^{4-}$	+0.36
$AgIO_3 + e^- = Ag + IO_3^-$	+0.354
$Hg_2Cl_2 + 2e^- = 2Hg + 2Cl^-$	+0.2676
$AgCl + e^- = Ag + Cl^-$	+0.2222
$CuCl + e^- = Cu + Cl^-$	+0.137
$CuBr + e^- = Cu + Br^-$	+0.033
$Ag(S_2O_3)_2^{3-} + e^- = Ag + 2S_2O_3^{2-}$	+0.017
$2H^+ + 2e^- (SHE) = H_2$	+0.0000
$HgI_4^{2-} + 2e^- = Hg + 4I^-$	−0.038
$Hg_2I_2 + 2e^- = 2Hg + 2I^-$	−0.0405

Table 8.1 (*Continued*)

Acid solution	$\mathscr{E}^0$ (V)
$WO_3(s) + 6H^+ + 6e^- = W + 3H_2O$	−0.090
$AgI + e^- = Ag + I^-$	−0.1518
$CuI + e^- = Cu + I^-$	−0.1852
$SnF_6^{2-} + 4e^- = Sn + 6F^-$	−0.25
$PbCl_2 + 2e^- = Pb + 2Cl^-$	−0.268
$PbBr_2 + 2e^- = Pb + 2Br^-$	−0.284
$PbSO_4 + 2e^- = Pb + SO_4^{2-}$	−0.3588
$PbI_2 + 2e^- = Pb + SO_4^{2-}$	−0.365
$TlCl + e^- = Tl + Cl^-$	−0.5568
$TlBr + e^- = Tl + Br^-$	−0.658
$Zn^{2+} + 2e^- = Zn$	−0.7628
$SiO_2(quartz) + 4H^+ + 4e^- = Si + 2H_2O$	−0.857
$H_3BO_3(aq) + 3H^+ + 4e^- = B + 3H_2O$	−0.698
$TiF_6^{2-} + 4e^- = Ti + 6F^-$	−1.191
$SiF_6^{2-} + 4e^- = Si + 6F^-$	−1.24
$Al^{3+} + 3e^- = Al$	−1.662
$U^{3+} + 3e^- = U$	−1.789
$Be^{2+} + 2e^- = Be$	−1.847
$AlF_6^{3-} + 3e^- = Al + 6F^-$	−2.069
$\frac{1}{2}H_2 + e^- = H^-$	−2.25
$Mg^{2+} + 2e^- = Mg$	−2.363
$La^{3+} + 3e^- = La$	−2.522
$Na^+ + e^- = Na$	−2.714
$Ca^{2+} + 2e^- = Ca$	−2.866
$Sr^{2+} + 2e^- = Sr$	−2.888
$Ba^{2+} + 2e^- = Ba$	−2.906
$Cs^+ + e^- = Cs$	−2.923
$Rb^+ + e^- = Rb$	−2.925
$K^+ + e^- = K$	−2.925
$Li^+ + e^- = Li$	−3.045

Base solution	$\mathscr{E}^0$ (V)
$O_3(g) + H_2 + 2e^- = O_2 + 2OH^-$	+1.24
$Cu^{2+} + 2CN^- + e^- = Cu(CN)_2^-$	+1.103
$HXeO_6^{3-} + 2H_2O + 2e^- = HXeO_4^- + 4OH^-$	+0.9
$HXeO_4^- + 3H_2O + 6e^- = Xe + 7OH^-$	+0.9
$ClO^- + H_2O + 2e^- = Cl^- + 2OH^-$	+0.89
$BrO^- + H_2O + 2e^- = Br^- + 2OH^-$	+0.761

(*continues*)

Table 8.1   (*Continued*)

Base solution	$\mathscr{E}^0$ (V)
$Ag(NH_3)_2^+ + e^- = Ag + 2NH_3$	+0.373
$Co(NH_3)_6^{3+} + e^- = Co(NH_3)_6^{2+}$	+0.108
$Cu(NH_3)_2^+ + e^- = Cu + 2NH_3$	−0.12
$Ag(CN)_2^- + e^- = Ag + 2CN^-$	−0.31
$Hg(CN)_4^{2-} + 2e^- = Hg + 4CN^-$	−0.37
$Cu(CN)_2^- + e^- = Cu + 2CN^-$	−0.429
$Ni(NH_3)_6^{2+} + 2e^- = Ni + 6NH_3(aq)$	−0.476
$HgS(black) + 2e^- = Hg + S^{2-}$	−0.69
$NiS(\alpha) + 2e^- = Ni + S^{2-}$	−0.830
$SnS + 2e^- = Sn + S^{2-}$	−0.87
$Cu_2S + 2e^- = 2Cu + S^{2-}$	−0.89
$PbS + 2e^- = Pb + S^{2-}$	−0.93
$CNO^- + H_2O + 2e^- = CN^- + 2OH^-$	−0.970
$Cd(CN)_4^{2-} + 2e^- = Cd + 4CN^-$	−1.028
$NiS(\gamma) + 2e^- = Ni + S^{2-}$	−1.04
$Zn(NH_3)_4^{2+} + 2e^- = Zn + 4NH_3(aq)$	−1.04
$HV_6O_{17}^{3-} + 16H_2O + 30e^- = 6V + 33OH^-$	−1.154
$CdS + 2e^- = Cd + S^{2-}$	−1.175
$Zn(OH)_2 + 2e^- = Zn + 2OH^-$	−1.245
$Zn(CN)_4^{2-} + 2e^- = Zn + 4CN^-$	−1.26
$ZnS(wurtzite) + 2e^- = Zn + S^{2-}$	−1.405
$SiO_3^{2-} + 3H_2O + 4e^- = Si + 6OH^-$	−1.697
$H_2BO_3^- + H_2O + 3e^- = B + 4OH^-$	−1.79
$Al(OH)_3 + 3e^- = Al + 3OH^-$	−2.30
$H_2AlO_3^- + H_2O + 3e^- = Al + 4OH^-$	−2.33
$UO_2 + 2H_2O + 4e^- = U + 4OH^-$	−2.39
$BeO + H_2O + 2e^- = Be + 2OH^-$	−2.613
$Mg(OH)_2 + 2e^- = Mg + 2OH^-$	−2.690
$Ce(OH)_3 + 3e^- = Ce + 3OH^-$	−2.87
$Sr(OH)_2 + 2e^- = Sr + 2OH^-$	−2.88
$Ba(OH)_2{\cdot}8H_2O + 2e^- = Ba + 2OH^- + 8H_2O$	−2.99
$Ca(OH)_2 + 2e^- = Ca + 2OH^-$	−3.02

yield a new value for the function. However, it is important to realize that electric potential $\mathscr{E}$ is *not* a state function and its values cannot be combined directly through Hess's law. We must convert the potential for any given half-reaction to its equivalent free-energy change, which *is* a state function, by multiplying the potential by the number of electrons transferred in the half-reaction. These free-energy changes are then combined, and the resulting new ΔG^0 is reconverted to a potential by dividing out the number of electrons transferred in the new half-reaction.

For example, suppose we have the two tabulated half-reactions.

$$ClO_3^- + 6H_3O^+ + 5\,e^- \rightarrow \tfrac{1}{2}Cl_2 + 9H_2O \quad \mathscr{E}^0 = +1.47 \text{ V}$$

$$\tfrac{1}{2}Cl_2 + e^- \rightarrow Cl^- \quad \mathscr{E}^0 = +1.36 \text{ V}$$

and we wish to obtain the potential for the half-reaction

$$ClO_3^- + 6H_3O^+ + 6e^- \rightarrow Cl^- + 9H_2O \quad \mathscr{E}^0 = ?$$

To apply Hess's law to the first two half-reactions (which obviously add to give the desired half-reaction), we must transform the potentials into free energies:

	$\mathscr{E}^0$ $\quad n\mathscr{E}^0 = -\Delta G^0$
$ClO_3^- + 6H_3O^+ + 5e^- \rightarrow \tfrac{1}{2}Cl_2 + 9H_2O$	$+1.47(\times 5 =)$ 7.35 (eV)
$\tfrac{1}{2}Cl_2 + e^- \rightarrow Cl^-$	$+1.36(\times 1 =)$ 1.36
$ClO_3^- + 6H_3O^+ + 6e^- \rightarrow Cl^- + 9H_2O$	$+1.45 \xleftarrow{\;+6\;} 8.71$

The potential for the new half-reaction is 1.45 V, quite different from the result that would be obtained if the potentials were simply added with the half-reactions (an erroneous 2.83 V).

Of course, it is not necessary to go through the conversion to free energies when one simply combines two half-reactions to a complete balanced cell reaction and seeks the cell voltage. One simply subtracts the less-positive potential from the more-positive potential and notes that the half-reaction with the less-positive potential will be driven in the reverse direction. The difference in half-reaction or half-cell potentials will be the standard cell potential. The reason conversion to free energies is not necessary is precisely that the overall cell reaction is balanced—one half-reaction accepts exactly as many electrons as the other donates. Since the number of electrons transferred is the same for the two half-reactions, and is thus also the same for the full-cell reaction, that number cancels out of any conversion like that above, and we need not involve ourselves in the conversion at all.

In Chapter 6 we established a rule for acid–base reactivity that can be paralleled here for mixtures of components that are capable of redox reaction. The rule is most conveniently applied using tables in which half-reactions are arranged in order of their associated potentials, such as Table 8.1:

The strongest oxidant present in a reaction mixture (the reactant in a reduction-potential half-reaction that has the most positive potential) will react with the strongest reductant present (the product of the reduction-potential half-reaction that has the least positive potential) to reduce the availability of electrons for transfer in the mixture.

In thermodynamic terms, it is not hard to understand why this rule works.

If we combine the half-reactions that have the most positive and the least positive potentials, the resulting overall potential is the greatest obtainable in that reaction mixture, and the corresponding free-energy change is also the greatest (most negative) obtainable.

Most redox reactions are run in solution. Since solvents themselves are generally oxidizable or reducible, the reactivity rule establishes a limited range of $\mathscr{E}^0$ values within which oxidants or reductants can be dissolved without decomposing the solvent. This is entirely comparable to the acid–base behavior of solvents, which level solute acids or bases that are too strong for the range of acidities allowed by the solvent's autoprotolysis. Here a kind of redox "leveling" is occurring. For example, aqueous solutions of elemental fluorine cannot be prepared because of the relationship of the two half-reactions

$$F_2 + 2e^- \rightarrow 2F^- \qquad \mathscr{E} = +2.87 \text{ V}$$

$$O_2 + 4H_3O^+ + 4e^- \rightarrow 6H_2O \qquad \mathscr{E} = +1.229 \text{ V}$$

The presence of the oxidant F_2 (reactant in most positive half-reaction) and the reductant H_2O (product in least positive half-reaction) means that the F_2 will be entirely converted to F^- and the solution will liberate O_2. Similarly, stable solutions of Ti^{2+} in water cannot be prepared because of the relationship between the two half-reactions

$$Ti^{3+} + e^- \rightarrow Ti^{2+} \qquad \mathscr{E}^0 = -0.369 \text{ V}$$

$$2H_3O^+ + 2e^- \rightarrow H_2 + 2H_2O \qquad \mathscr{E}^0 = 0 \text{ (defined)}$$

Here the presence of the oxidant H_3O^+ (the solvent's characteristic acid species, present at 1 M if the potential is 0 V) and the reductant Ti^{2+} means that the Ti^{2+} will be converted to Ti^{3+} and the solution will liberate H_2.

As might be expected, different solvents have different patterns of redox behavior, even though behavior analogous to that just described for water is seen for nearly all solvents under appropriate circumstances (which may be extreme). Since bases and reductants both serve as electron donors, it should not be surprising that basic solvents such as liquid ammonia are usually congenial to strong reducing agents and thus are often chosen for reduction reactions. Conversely, acidic solvents such as liquid HF, H_2SO_4, and superacids are congenial to strong oxidizing agents and are often chosen for oxidation reactions. We shall explore this pattern in more detail in a later section.

The intrinsic electron-transfer capability described by the reduction potential $\mathscr{E}^0$ is modified in practice by two additional factors. The first is thermodynamic: All $\mathscr{E}^0$ values refer to solutions in which ions and other solute species are present at unit activity (ideal 1 M) and gases at unit fugacity (ideal 1 atm). Real solutions are always nonideal and rarely have 1 M concentrations. The standard $\mathscr{E}^0$ is modified to an effective $\mathscr{E}$ by the Nernst equation:

$$\mathscr{E} = \mathscr{E} - \frac{RT}{n\mathscr{F}} \ln Q$$

where Q is the activity quotient in which product concentrations divided by reactant concentrations (each raised to its stoichiometric power) have the same form as in an equilibrium constant.

Because balancing redox-reaction equations in aqueous solution usually requires the participation of the water-based species H_3O^+ or OH^-, these species appear in the Nernst equation for such reactions or half-reactions. This produces a significant pH dependence for the potential associated with the reaction, particularly since the 1 M standard state for solutions means that $\mathscr{E}^0$ is quoted for pH 0 or 14. For example, consider the half-reaction for electron transfer between chlorate and chlorine:

$$ClO_3^- + 6H_3O^+ + 5e^- \rightarrow \tfrac{1}{2}Cl_2 + 9H_2O \qquad \mathscr{E} = +1.47 \text{ V}$$

The Nernst equation for this reduction half-reaction becomes

$$\mathscr{E} = \mathscr{E} - \frac{RT}{n\mathscr{F}} \ln \frac{P_{Cl_2}^{1/2}}{[ClO_3^-][H_3O^+]^6}$$

$$= 1.47 - \frac{0.05915}{5} \log \frac{P_{Cl_2}^{1/2}}{[ClO_3^-][H_3O^+]^6}$$

$$= 1.47 - 0.005915 P_{Cl_2} + 0.01183 \log [ClO_3] - 0.07098 \text{ pH}$$

At first glance, this pH dependence does not seem too significant. But suppose the chlorine species are maintained at unit activity and the pH is varied. At pH 7 we have

$$\mathscr{E} = 1.47 - (0.07098)(7) = +0.973 \text{ V}$$

and at pH 14 we have

$$\mathscr{E} = 1.47 - (0.07098)(14) = +0.476 \text{ V}$$

The chlorate ion is obviously a much weaker oxidizing agent in neutral or basic solution than it is in strongly acidic solution. The large stoichiometric coefficient of H_3O^+ in many redox reactions gives it considerable leverage in the Nernst equation; such reactions show strong pH dependence. This is true for most oxoanion reactions, and in particular for those half-reactions in which the number of oxygen atoms attached to the element being reduced changes sharply. Accordingly, chlorate, perchlorate, and nitrate are strong oxidizing agents only in strongly acidic solutions, and are nearly inert in more or less neutral solution. Thus perchloric acid explodes on contact with organic solvents (frequently, if unpredictably), but tetrabutylammonium perchlorate is widely used as a supporting electrolyte for electrochemical studies in those same solvents—with complete safety.

The linear pH dependence of $\mathscr{E}$ for oxoanion and oxocation half-reactions makes it convenient to present the pH influence on the redox stability of water solvent in graphical form, as in Fig. 8.1. The solid line in Fig. 8.1 from +1.229 V at

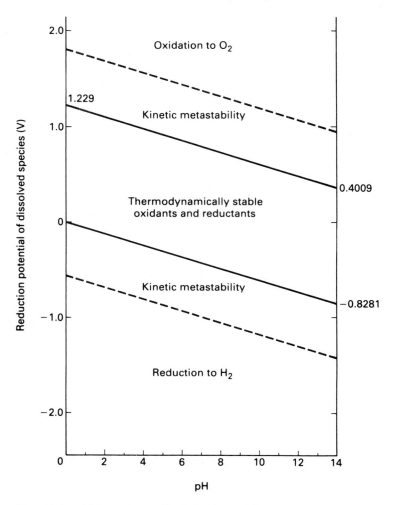

Figure 8.1   pH dependence of water redox stability.

pH 0 to +0.4009 V at pH 14 represents the pH dependence of the half-reaction in which water is oxidized to O_2. Written as a reduction,

$$O_2 + 4H_3O^+ + 4e^- \rightarrow 6H_2O \qquad \mathscr{E} = +1.229 \text{ V}$$

$$\mathscr{E} = \mathscr{E}^0 - \frac{0.05915}{4} \log \frac{1}{P_{O_2} \cdot [H_3O^+]^4}$$

$$= 1.229 + 0.01479 \log P_{O_2} - 0.05915 \text{ pH}$$

If a dissolved aqueous species has a strong tendency to be reduced, it will strip electrons from nearby water molecules, reducing itself and oxidizing water by reversing the above half-reaction. The solid line in Fig. 8.1 that corresponds to the above Nernst equation thus represents the highest reduction potentials that can be associated with thermodynamically stable oxidants in water solution.

Similarly, a powerful reductant will reduce water to H_2 and will itself be oxidized. The half-reaction for the reduction of water is

$$2H_3O^+ + 2e^- \rightarrow H_2 + 2H_2O \qquad E^0 = 0 \text{ V}$$

$$\mathscr{E} = \mathscr{E}^0 - \frac{0.05915}{2} \log \frac{P_{H_2}}{[H_3O^+]^2}$$

$$= 0 - 0.02958 \log P_{H_2} - 0.05915 \text{ pH}$$

This represents the lower solid line in Fig. 8.1. Since the electron affinity of water molecules (and particularly the oxonium ions) is substantial, either can strip electrons from a strong reductant. The potential represented by the lower solid line is thus the lowest (most negative) reduction potential that can be associated with thermodynamically stable reductants in water solution. Note that in basic solution, water becomes more hospitable to reducing agents and less hospitable to oxidizing agents. For example, the dithionite ion, $S_2O_4^{2-}$, is a stable strong reducing agent in basic solution, but neither the ion nor its conjugate acid $H_2S_2O_4$ (dithionous acid) can be prepared in acid solution.

Figure 8.1 also shows the effect of the other important factor in redox spontaneity, a sort of "activation potential" required for kinetic reasons if the reaction is to proceed. We commonly observe that about 0.5 to 0.6 volt more than would be required thermodynamically is needed to observe the redox decomposition of water, whether it is being reduced or oxidized. Since an electronvolt corresponds to 96 kJ/mol, this additional potential is equivalent to an activation energy on the order of 50 kJ/mol for a one-electron transfer, which is a common range for solution reactions. Because of this kinetic effect, a number of systems are stable in water that would not be if strict thermodynamic limits were the only factor involved. For example, in acid solution, H_3PO_2 should reduce water to H_2 ($\mathscr{E} = -0.499$ V), and so should tin metal ($\mathscr{E} = -0.136$ V). In fact, H_3PO_2 is stable, and tin dissolves only very slowly in dilute acid. Similarly, CN^- should be oxidized in base to CNO^- ($\mathscr{E} = -0.970$ V) and SO_3^{2-} to SO_4^{2-} ($\mathscr{E} = -0.93$ V), but neither reaction is observed. In acid solution, dichromate ion, $Cr_2O_7^{2-}$, should oxidize water to O_2, and so should PbO_2 ($\mathscr{E} = +1.33$ V and $+1.455$ V respectively), but both species are stable even in very strong acid. Many such examples can be found. Accordingly, in Fig. 8.1 dashed lines are used to indicate regions of apparent stability or metastability that reflect this kinetic inertness.

As a final note on the thermodynamic spontaneity of redox reactions, we shall look at the relative magnitudes of standard potentials and equilibrium constants. Since ΔG^0 equals both $-n\mathscr{F}\mathscr{E}^0$ and $-RT(\ln K_{eq})$,

$$\ln K_{eq} = \frac{n\mathscr{F}\mathscr{E}}{RT} \qquad \text{or (25°C)} \qquad \log K_{eq} = \frac{n}{0.05915}\mathscr{E}$$

The appearance of K_{eq} as a logarithm ensures that it will change dramatically with potential. To see this, we calculate the K_{eq} values for the two aqueous redox reactions

$$ClO_4^- + ClO_2^- \rightarrow 2ClO_3^-$$

and

$$S_2O_8^{2-} + 2Cr^{2+} \rightarrow 2Cr^{3+} + 2SO_4^{2-}$$

For the perchlorate reaction, the half-reactions in basic solution and potentials are

$$ClO_4^- + H_2O + 2e^- \rightarrow ClO_3^- + 2OH^- \qquad \Delta\mathscr{E}^0 = 0.36 \text{ V}$$

$$ClO_3^- + H_2O + 2e^- \rightarrow ClO_2^- + 2OH^- \qquad \mathscr{E}^0 = 0.33 \text{ V}$$

The potential for the overall reaction is thus only about 0.03 V, or 30 millivolts. The corresponding K_{eq} is 10.3—a rather small value for an equilibrium constant.

By contrast, the peroxydisulfate reaction involves the half-reactions and potentials

$$S_2O_8^{2-} + 2e^- \rightarrow 2SO_4^{2-} \qquad \mathscr{E}^0 = 2.01 \text{ V}$$

$$Cr^{3+} + e^- \rightarrow Cr^{2+} \qquad \mathscr{E}^0 = -0.408 \text{ V}$$

This overall reaction has an associated potential of 2.418 V. This is a substantial potential, but it is even more impressive expressed as an equilibrium constant: 6×10^{81}. We therefore can generalize that a redox reaction will proceed in good yield if its potential is about 1 V, and may even be violent near a potential of 2 V, depending on reaction conditions and kinetic factors.

8.3 REDOX REACTIONS IN AQUEOUS SYSTEMS

Much of our discussion of redox reactions has involved aqueous solutions. We can go still farther toward predicting redox reactions and choosing oxidants and reductants if we address aqueous solutions specifically. One of the most convenient tools for predicting redox reactions in aqueous solution is a table of *Latimer diagrams* for the elements, shown here as Table 8.2. In a Latimer diagram, the reduction potentials for an element in its various oxidation states are compiled into a single sequence of reduction reactions, omitting solvent species and listing only the solute molecule or ion present for each oxidation state. The reduction potential associated with a given half-reaction is indicated over the arrow (in volts). Each half-reaction can of course be balanced using the water species H_2O, H_3O^+, and OH^-. The table is given in alphabetical order by element symbol.

In Table 8.2, the most highly oxidized species for each element is given at the left, and each successive half-reaction proceeds as a reduction. You will note that, in general, the reduction potentials become less positive moving to the right in a diagram. This reflects the fact that as an atom of a given element acquires more

Table 8.2 Latimer Diagrams for Common Elements

Acid solution

$AgO^+ \xrightarrow{2.1} Ag^{2+} \xrightarrow{1.980} Ag^+ \xrightarrow{0.799} Ag^0$

$H_3AsO_4 \xrightarrow{0.560} H_3AsO_3 \xrightarrow{0.248} As^0 \xrightarrow{-0.607} AsH_3$

$Bi_2O_5 \xrightarrow{1.59} BiO^+ \xrightarrow{0.320} Bi^0$

$BrO_4^- \xrightarrow{1.763} BrO_3^- \xrightarrow{1.505} BrOH \xrightarrow{1.595} Br_2 \xrightarrow{1.065} Br^-$
$\qquad\qquad\qquad\qquad\overset{1.52}{\underset{\longrightarrow}{}}$

$ClO_4^- \xrightarrow{1.230} ClO_3^- \xrightarrow{1.21} HClO_2 \xrightarrow{1.645} ClOH \xrightarrow{1.63} Cl_2 \xrightarrow{1.360} Cl^-$
$\qquad\qquad\qquad \overset{1.145}{\longrightarrow} ClO_2 \overset{1.275}{\longrightarrow}$
$\qquad\qquad\qquad\qquad\quad \underset{1.468}{}$

$Co(H_2O)_6^{3+} \xrightarrow{1.808} Co(H_2O)_6^{2+} \xrightarrow{-0.277} Co^0$

$Cr_2O_7^{2-} \xrightarrow{1.33} Cr(H_2O)_6^{3+} \xrightarrow{-0.408} Cr(H_2O)_6^{2+} \xrightarrow{-0.912} Cr^0$
$\qquad\qquad\qquad\qquad\underset{-0.744}{}$

$Cu^{3+}(?) \xrightarrow{1.8} Cu(H_2O)_6^{2+} \xrightarrow{-0.153} Cu(H_2O)_4 \xrightarrow{-0.521} Cu^0$
$\qquad\qquad\qquad\qquad\underset{0.337}{}$

$FeO_4^{2-} \xrightarrow{2.20} Fe(H_2O)_6^{3+} \xrightarrow{0.771} Fe(H_2O)_6^{2+} \xrightarrow{-0.440} Fe^0$
$\qquad\qquad\qquad\qquad\underset{-0.036}{}$

$Hg(H_2O)_4^{2+} \xrightarrow{0.920} Hg_2(H_2O)_4^{2+} \xrightarrow{0.788} Hg^0$
$\qquad\qquad\underset{0.854}{}$

$H_5IO_6 \xrightarrow{1.644} IO_3^- \xrightarrow{1.133} IOH \xrightarrow{1.45} I_2 \xrightarrow{0.536} I^-$
$\qquad\qquad\qquad\underset{1.196}{}$

$In(H_2O)_6^{3+} \xrightarrow{-0.49} In(H_2O)_6^{2+} \xrightarrow{-0.40} In(H_2O)_4^+ \xrightarrow{-0.14} In^0$
$\qquad\qquad\qquad\underset{-0.343}{}$

$\qquad\qquad\qquad\qquad\qquad\qquad\overset{1.51}{}$
$MnO_4^- \xrightarrow{0.564} MnO_4^{2-} \xrightarrow{2.261} MnO_2 \xrightarrow{0.95} Mn(H_2O)_6^+ \xrightarrow{1.51} Mn(H_2O)_6^{2+} \xrightarrow{-1.180} Mn^0$
$\qquad\quad\underset{1.695}{} \qquad\qquad \underset{1.23}{}$

$\qquad\qquad\qquad\qquad\overset{1.45}{} \qquad\qquad \overset{-0.23}{}$
$NO_3^- \xrightarrow{0.803} N_2O_4 \xrightarrow{1.07} HNO_2 \xrightarrow{1.00} NO \xrightarrow{1.59} N_2O \xrightarrow{1.77} N_2 \xrightarrow{-3.09} HN_3 \xrightarrow{0.34} N_2H_5^+ \xrightarrow{1.275} NH_4^+$
$\qquad\underset{0.94}{} \qquad\qquad \underset{1.29}{} \qquad\qquad\qquad\qquad\qquad \underset{1.96}{}$
$\qquad\qquad\qquad\qquad\qquad\qquad\qquad\qquad\qquad\qquad\qquad\underset{0.27}{}$

$NiO_2 \xrightarrow{1.678} Ni(H_2O)_6^{2+} \xrightarrow{-0.250} Ni^0$

(continues)

Table 8.2   (*Continued*)

Acid solution

$$O_3 \xrightarrow{2.07} O_2 \xrightarrow{0.682} H_2O_2 \xrightarrow{1.776} H_2O$$
$$O_2 \xrightarrow{1.229} H_2O$$

$$H_3PO_4 \xrightarrow{-0.276} H_3PO_3 \xrightarrow{-0.449} H_3PO_2 \xrightarrow{-0.508} P_4 \xrightarrow{-0.063} PH_3$$
(−0.283 from H₃PO₃ to PH₃; −0.174 from H₃PO₂ to PH₃)

$$PbO_2 \xrightarrow{1.455} Pb(H_2O)_6^{2+} \xrightarrow{-0.126} Pb^0$$

$$PdO_3 \xrightarrow{\sim 2} Pd(H_2O)_6^{4+} \xrightarrow{\sim 1.6} Pd(H_2O)_4^{2+} \xrightarrow{0.987} Pd^0$$

$$ReO_4^- \xrightarrow{0.73} ReO_3 \xrightarrow{0.40} ReO_2 \xrightarrow{0.251} Re^0 \xrightarrow{-0.4} Re^-$$
(0.362 from ReO₄⁻ to Re⁰; 0.510 from ReO₄⁻ to ReO₂)

$$RuO_4 \xrightarrow{0.9} RuO_4^- \xrightarrow{1.6} RuO_4^{2-} \xrightarrow{1.3} Ru(H_2O)_6^{2+} \xrightarrow{0.45} Ru^0$$

$$S_2O_8^{2-} \xrightarrow{2.01} SO_4^{2-} \xrightarrow{0.172} SO_2 \xrightarrow{0.51} S_4O_6^{2-} \xrightarrow{0.08} S_2O_3^{2-} \xrightarrow{0.50} S_8 \xrightarrow{0.142} H_2S$$
(0.450 from SO₂ to S₂O₃²⁻; −0.082 from SO₂ to S₂O₄²⁻; S₂O₄²⁻ →0.88→)

$$Sb_2O_5 \xrightarrow{0.581} SbO^+ \xrightarrow{0.152} Sb^0 \xrightarrow{-0.510} SbH_3$$

$$SeO_4^{2-} \xrightarrow{1.15} H_2SeO_3 \xrightarrow{0.740} Se^0 \xrightarrow{-0.399} H_2Se$$

$$Sn(H_2O)_6^{4+} \xrightarrow{0.15} Sn(H_2O)_6^{2+} \xrightarrow{-0.136} Sn^0$$

$$H_6TeO_6 \xrightarrow{1.02} H_2TeO_3 \xrightarrow{0.529} Te^0 \xrightarrow{-0.739} H_2Te$$

$$TiO(H_2O)_5^{2+} \xrightarrow{0.099} Ti(H_2O)_6^{3+} \xrightarrow{-0.369} Ti(H_2O)_6^{2+} \xrightarrow{-1.628} Ti^0$$
(−0.882 from TiO(H₂O)₅²⁺ to Ti⁰)

$$Tl(H_2O)_6^{3+} \xrightarrow{1.25} Tl(H_2O)_4^+ \xrightarrow{-0.336} Tl^0$$

$$VO_2(H_2O)_4^+ \xrightarrow{1.00} VO(H_2O)_5^{2+} \xrightarrow{0.359} V(H_2O)_6^{3+} \xrightarrow{-0.256} V(H_2O)_6^{2+} \xrightarrow{1.186} V^0$$
(−0.254)

$$H_4XeO_6 \xrightarrow{2.3} XeO_3 \xrightarrow{1.8} Xe^0$$

Basic solution

$$Ag_2O_3 \xrightarrow{0.739} AgO \xrightarrow{0.607} Ag_2O \xrightarrow{0.345} Ag^0$$

$$AsO_4^{3-} \xrightarrow{-0.68} H_2AsO_3^- \xrightarrow{-0.675} As^0 \xrightarrow{-1.21} AsH_3$$

$$Bi_2O_5 \xrightarrow{\sim 0.6} Bi_2O_3^- \xrightarrow{-0.46} Bi^0$$

Table 8.2 (*Continued*)

Basic solution

$$BrO_4^- \xrightarrow{0.99} BrO_3^- \xrightarrow{0.54} BrO^- \xrightarrow{0.45} Br_2 \xrightarrow{1.07} Br^-$$

(0.61 spanning BrO_3^- to Br^-; 0.761 spanning BrO^- to Br^-)

$$ClO_4^- \xrightarrow{0.36} ClO_3^- \xrightarrow{0.33} ClO_2^- \xrightarrow{0.66} ClO^- \xrightarrow{0.40} Cl_2 \xrightarrow{1.360} Cl^-$$

(0.56 spanning ClO_3^- to Cl^-; 0.50 spanning ClO_3^- to ClO^-; 0.89 spanning ClO^- to Cl^-)

$$Co(OH)_3 \xrightarrow{0.17} Co(OH)_2 \xrightarrow{-0.73} Co^0$$

$$CrO_4^{2-} \xrightarrow{-0.13} Cr(OH)_3 \xrightarrow{-1.1} Cr(OH)_2 \xrightarrow{-1.4} Cr^0$$

(−1.34 spanning $Cr(OH)_3$ to Cr^0)

$$Cu(OH)_2 \xrightarrow{-0.08} Cu_2O \xrightarrow{-0.358} Cu^0$$

$$FeO_4^{2-} \xrightarrow{0.72} Fe(OH)_3 \xrightarrow{-0.56} Fe(OH)_2 \xrightarrow{-0.877} Fe^0$$

$$IO_4^- \xrightarrow{0.7} IO_3^- \xrightarrow{0.14} IO^- \xrightarrow{0.45} I_2 \xrightarrow{0.54} I^-$$

(0.37 spanning IO_3^- to I^-; 0.26 spanning IO_3^- to IO^-)

$$MnO_4^- \xrightarrow{0.558} MnO_4^{2-} \xrightarrow{0.603} MnO_2 \xrightarrow{-0.25} Mn(OH)_3 \xrightarrow{0.15} Mn(OH)_2 \xrightarrow{-1.55} Mn^0$$

(0.588 spanning MnO_4^- to MnO_2; −0.05 spanning MnO_2 to $Mn(OH)_2$)

$$NO_3^- \xrightarrow{-0.86} N_2O_4 \xrightarrow{0.88} NO_2^- \xrightarrow{-0.46} NO \xrightarrow{0.76} N_2O \xrightarrow{0.94} N_2 \xrightarrow{-3.04} NH_2OH \xrightarrow{0.73} N_2H_4 \xrightarrow{0.11} NH_3$$

(0.41 spanning NO_2^- to N_2; 0.01 spanning NO_3^- to NO_2^-; 0.15 spanning NO_2^- to NO; −1.05 spanning NO to N_2; 0.42 spanning NH_2OH to NH_3)

$$NiO_2 \xrightarrow{0.490} Ni(OH)_2 \xrightarrow{-0.72} Ni^0$$

$$O_3 \xrightarrow{1.24} O_2 \xrightarrow{0.365} O_2^- \xrightarrow{-0.517} HO_2^- \xrightarrow{0.878} OH^-$$

(0.401 spanning O_2 to OH^-; −0.076 spanning O_2 to HO_2^-; 0.413 spanning O_2^- to OH^-)

$$PO_4^{3-} \xrightarrow{-1.12} HPO_3^{2-} \xrightarrow{-1.565} H_2PO_2^- \xrightarrow{-2.05} P_4 \xrightarrow{-0.89} PH_3$$

(−1.31 spanning HPO_3^{2-} to P_4; −1.18 spanning $H_2PO_2^-$ to PH_3)

$$PbO_2 \xrightarrow{0.247} PbO \xrightarrow{-0.580} Pb^0$$

$$Pd(OH)_4 \xrightarrow{\sim 0.73} Pd(OH)_2 \xrightarrow{0.07} Pd^0$$

$$ReO_4^- \xrightarrow{\sim 0.7} ReO_4^{2-} \xrightarrow{\sim 0.4} ReO_2 \xrightarrow{-0.577} Re^0 \xrightarrow{-0.4} Re^-$$

(−0.594 spanning ReO_4^- to ReO_2; −0.584 spanning ReO_4^- to Re^0)

(*continues*)

Table 8.2   (*Continued*)

Basic solution

$$SO_4^{2-} \xrightarrow{-0.93} SO_3^{2-} \xrightarrow{-0.80} S_4O_6^{2-} \xrightarrow{0.08} S_2O_3^{2-} \xrightarrow{-0.74} S_8 \xrightarrow{-0.447} S^{2-}$$

with $SO_3^{2-} \xrightarrow{-0.66} S_2O_3^{2-}$ and $SO_3^{2-} \xrightarrow{-1.12} S_2O_4^{2-} \xrightarrow{-0.04} S_4O_6^{2-}$

$$Sb(OH)_6^- \xrightarrow{-0.4} H_2SbO_3^- \xrightarrow{-0.66} Sb^0 \xrightarrow{-1.34} SbH_3$$

$$SeO_4^{2-} \xrightarrow{0.05} SeO_3^{2-} \xrightarrow{-0.366} Se^0 \xrightarrow{-0.92} Se^{2-}$$

$$Sn(OH)_6^{2-} \xrightarrow{-0.93} Sn(OH)_3^- \xrightarrow{-0.909} Sn^0$$

$$TeO_2(OH)_4^{2-} \xrightarrow{\sim0.4} TeO_3^{2-} \xrightarrow{-0.57} Te^0 \xrightarrow{-1.143} Te^{2-}$$

$$Tl(OH)_3 \xrightarrow{-0.05} TlOH \xrightarrow{-0.343} Tl^0$$

$$HXeO_6^{3-} \xrightarrow{0.9} HXeO_4^- \xrightarrow{0.9} Xe^0$$

electrons, its effective nuclear charge decreases, and along with it the atom's attraction for electrons. However, there are exceptions. Choosing a portion of the copper diagram as an example, we have

$$Cu(H_2O)_6^{2+} \xrightarrow{0.153} Cu(H_2O)_4^+ \xrightarrow{0.521} Cu^0$$

Suppose we had a solution of 1 M Cu^+. In the solution, one Cu^+ could react with a neighbor Cu^+ to produce Cu^{2+} and copper metal. The oxidation half-reaction would have an associated potential of -0.153 V, while the reduction reaction would have a potential of $+0.521$ V. The overall reaction $2Cu(H_2O)_4^+ \rightarrow Cu^0 + Cu(H_2O)_6^{2+} + 2H_2O$ would have a positive potential ($0.521 - 0.153 = +0.368$ V) and would be thermodynamically spontaneous. This is a *disproportionation* reaction in which an element in a given oxidation state spontaneously reacts to form a higher and a lower oxidation state. Disproportionation reactions are easy to predict using Latimer diagrams, since one should occur whenever a half-reaction potential is more positive than the one to its left. In such cases, the intermediate oxidation state can always move toward more-negative free energy by driving the half-reaction with the less positive potential in reverse, yielding a net positive potential and negative free-energy change. In this case, the result is that a 1 M solution of Cu^+ cannot be prepared by using water as a Lewis base to coordinate the copper.

The environment of the intermediate oxidation state can make a substantial difference, however. Consider the following Latimer diagram for copper in a solution containing 1 M Cl^-:

$$Cu(H_2O)_6^{2+} + Cl^- \xrightarrow{0.538} CuCl(s) \xrightarrow{0.137} Cu^0 + Cl^-$$

The Cl^- ion, which is a softer base than water, stabilizes the Cu^+ (a softer acid than Cu^{2+}) more than it does the Cu^{2+}, and the potentials no longer fall in the order that makes disproportionation spontaneous. Solid CuCl is thus quite stable, even in the presence of Cu^{2+} solutions or Cu^0.

Still another complicating factor is that some species that are thermodynamically unstable with respect to disproportionation are so kinetically inert that they can be prepared and stored in aqueous solution for extended periods. For example, the Latimer diagram for phosphorus indicates that white phosphorus (P_4) should disproportionate into phosphine (PH_3) and hypophosphorous acid (H_3PO_2), which should in turn disproportionate into phosphorous acid and more phosphine. In fact, white phosphorus is stored under water, and aqueous solutions of H_3PO_2 are stable for extended periods at room temperature, although the latter will both disproportionate and reduce water when heated.

It is straightforward to predict redox reactions with Latimer diagrams. One simply inspects the diagrams for the two elements involved to see which will be oxidized and which reduced. If either species is at one end of its diagram, it can only react in the remaining direction, forcing the other species to react in the opposite sense. That is, dichromate ion $Cr_2O_7^{2-}$ cannot be oxidized, so it must be reduced if it is to react at all. Likewise, whatever it reacts with must be oxidized if it is to react at all. Suppose dichromate is mixed with a solution containing Fe^{2+}. Will any redox reaction occur? Yes, because the reduction potential for $Cr_2O_7^{2-}$ (+1.33 V) is greater than the potential that must be overcome in order to oxidize Fe^{2+} to Fe^{3+} (+0.771 V). Further oxidation will not occur, even with excess dichromate, because the potential to produce ferrate ion, FeO_4^{2-}, is too high (2.20 V). On the other hand, dichromate cannot oxidize MnO_2 because both of the possible products, MnO_4^{2-} and MnO_4^-, could only be formed by overcoming a higher potential than the 1.33 V available from dichromate (2.261 V and 1.695 V, respectively).

If both potentially reacting species are in intermediate oxidation states, we must decide which will be oxidized and which reduced. For example, suppose solutions of bromate, BrO_3^-, and nitrous acid, HNO_2, are mixed. One possibility is that bromate will be oxidized to perbromate, BrO_4^-, and nitrous acid will be reduced to NO, N_2O, or N_2. Alternatively, bromate might be reduced to bromine or to hypobromous acid, BrOH, and nitrous acid might be oxidized to N_2O_4 or nitrate ion. We rule out the oxidation of bromate because it requires a potential greater than that available from any of the possible reduction half-reactions of HNO_2 (−1.763 V versus at most 1.45 V). In this mixture, therefore, BrO_3^- can only serve as an oxidizing agent. It can oxidize HNO_2 because the most positive potential for its reduction (+1.52 V to Br_2) is greater than the potential required to oxidize HNO_2 to either N_2O_4 (−1.07 V) or NO_3^- (−0.94 V). In general, the overall reaction with the most positive potential will proceed, so the products should be Br_2 and NO_3^-. In fact, because the potentials are not far apart, some N_2O_4 is initially formed, but it is readily oxidized to NO_3^- by excess BrO_3^-. In the same sense, the potential for forming BrOH is very close to that for forming Br_2, but we will see no BrOH because that species would disproportionate if it were formed.

The ultimate products of a redox reaction must be compatible with the reactant that is present in excess at the end of the reaction. In this example, if the BrO_3^-

solution is added dropwise to the HNO_2 solution, so that excess HNO_2 is always present, the Br_2 formed can react to oxidize HNO_2 to NO_3^- (-0.94 V), in the process being itself reduced to Br^- ($+0.165$ V). In such a situation, the products will be NO_3^- and Br^-, rather than the NO_3^- and Br_2 that result if the order of addition is reversed. Finally, bubbles of a colorless gas (as opposed to the blue-green N_2O_4) slowly form in the HNO_2 solution, because HNO_2 is unstable with respect to disproportionation into N_2 and NO_3^-, as the diagram predicts.

The compilation of potentials in Tables 8.1 and 8.2, which deals specifically with aqueous solution species, is not limited to species stable in water solution. That is, although the potential -2.714 V applies to the reduction of an aqueous Na^+ ion, we should not assume that the Na metal product can actually be formed in water! Like sodium metal, many of the species in Tables 8.1 and 8.2 lie so far outside the stability boundaries indicated in Fig. 8.1 that they will spontaneously oxidize water to O_2 or reduce it to H_2. It may be useful here to list the more common laboratory oxidizing and reducing agents for aqueous solutions and to comment briefly on their reactions.

In Table 8.3, these species are generally given in the form in which they occur in acid solution. The division into "strong" and "gentle" is rather arbitrary, but reflects the commoner laboratory uses. As has already been pointed out, the strength of a given oxidizing or reducing agent can be profoundly influenced by pH, for both thermodynamic and kinetic reasons. A strong oxidant in acid solution can become quite gentle or even inert in base. This effect, of course, is caused by the Nernst-equation dependence of $\mathscr{E}$ on $[H_3O^+]$ or $[OH^-]$.

Of the strong oxidizing agents listed, perhaps the strongest is peroxodisulfate ion, $S_2O_8^{2-}$, which, for example, oxidizes Mn^{2+} to MnO_4^-. It is prepared by electrolyzing cold sulfuric-acid solutions and forms at the anode: $2SO_4^{2-} = S_2O_8^{2-} + 2e^-$. Its reactions are often slow even though they are overwhelmingly thermodynamically favorable. This explains why its aqueous solutions are reasonably stable even in dilute acid, although the potential lies far outside the boundaries of stability in-

Table 8.3   Common Aqueous Oxidizing and Reducing Agents

Oxidizing agents		Reducing agents	
Strong	Gentle	Strong	Gentle
$S_2O_8^{2-}$	H_2O_2	Zn amalgam	HSO_3^-
MnO_4^-	NO_3^-	$Cr(H_2O)_6^{2+}$	$HS_2O_3^-$
$Cr_2O_7^{2-}$	O_2	$Ti(H_2O)_6^{3+}$	HNO_2
$ClO_4^-, IO_4^-(H_5IO_6)$	I_2	$HS_2O_4^-$	I^-
ClO_3^-, BrO_3^-, IO_3^-	$Cu(H_2O)_4^{2+}$	BH_4^-	$Sn(H_2O)_4^{2+}$
$Ce(H_2O)_7OH^{3+}$	$Fe(H_2O)_6^{3+}$	(basic solution)	$As(OH)_3$
			HOC_6H_4OH
			(hydroquinone)

dicated in Fig. 8.1 for water. It slowly hydrolyzes to peroxomonosulfuric acid, H_2SO_5, an even stronger oxidizing agent that hydrolyzes in turn to H_2O_2. The halates and perhalates are also extremely strong oxidants (H_5IO_6 also oxidizes Mn^{2+} to MnO_4^-). Of the two groups, the halates are more often used. Bromate, for example, oxidizes PH_3 to H_3PO_4 and any sulfur species to SO_4^{2-}. It is not clear why the perbromate potential is greater than that of either perchlorate or periodate, but the difference is significant (1.76 V versus 1.23 and 1.64 V respectively).

Permanganate and dichromate are traditionally used as quantitative oxidizing agents in many redox-titration methods of quantitative analysis. Titrations of Fe^{2+}, NO_3^-, H_2O_2, and U(IV) are widely described in analytical texts. Permanganate solutions, however, slowly decompose water and the organic material present in most distilled water, and solutions must be restandardized frequently. Cerium(IV) is almost as useful in analytical work, and is much more stable in solution. Although the exact products of reaction for most of the strong oxidizing agents depend on the experimental conditions (as the discussion of Latimer diagrams has indicated), one can say that $S_2O_8^{2-}$ normally forms SO_4^{2-}, MnO_4^- usually forms $Mn(H_2O)_6^{2+}$ in acid solution and MnO_2 in base, and $Cr_2O_7^{2-}$ forms $Cr(H_2O)_6^{3+}$ or $Cr(OH)_3$.

The strongest laboratory reducing agent that does not reduce water to H_2 is zinc amalgam (or zinc–mercury solution), which can be used to prepare all of the other strong reducing agents in Table 8.3 except BH_4^-. A convenient laboratory device for using amalgamated zinc is the *Jones reductor,* a short chromatographic column filled with granular zinc metal that has been amalgamated by washing with dilute acidic Hg^{2+}. The column is kept filled with water, and the material to be reduced is eluted through the Jones reductor by dilute acid (though not nitric acid, since it is reduced to hydroxylammonium ion).

Dithionite ion, $S_2O_4^{2-}$ (sometimes misleadingly called hydrosulfite), is prepared by reducing sulfite with zinc amalgam. As the Latimer diagram for sulfur indicates, $S_2O_4^{2-}$ is unstable with respect to disproportionation, but solutions decompose only slowly and are widely used as strong reductants. It is used industrially in reducing vat dyes to soluble forms, in bleaching paper stock, and in removing the red color due to Fe(III) in china clay.

The BH_4^- ion (tetrahydroborate or borohydride) is actually too strong a reductant to exist in water. It hydrolyzes rapidly in acidic solution:

$$BH_4^- + H_3O^+ + 2H_2O \rightarrow H_3BO_3 + 4H_2$$

However, above pH 9 this reaction is very slow and solutions are reasonably stable; hydrates such as $NaBH_4 \cdot 2H_2O$ can be crystallized. In water solution, transition metals are rapidly reduced: V(V) to V(IV), Cr(VI) to Cr(III), Mo(VI) to Mo(V) or Mo(III), Fe(III) to Fe(II), and Co(II) to Co_2B (an interstitial boride comparable to the interstitial carbides of Chapter 3).

Most soft-acid transition-metal ions in low oxidation states can be reduced to the metal [for instance, Pd(II) and Pt(II)]. An interesting modification of this last reaction involves a polymeric amine-borane resin in which BH_3 groups are coordinated to a polyamine. The result is an ion-exchange resin with a strong, selective capacity for reducing precious metals from even very dilute solution. Gold, platinum, palladium, silver, rhodium, iridium, and mercury are all reduced to the metal within the resin, while lighter transition metals—and even such species as Sn(IV) and Pb(II)—are not. The capacity of the resin is approximately one gram of metal per gram of resin, and the metal is recovered by roasting the polymer to gaseous products, leaving the metal behind.

8.4 REDUCTION POTENTIALS IN NONAQUEOUS SOLVENTS

Once we move away from the familiar aqueous environment for dissolved species, many new possibilities for redox reactions appear. In a nonaqueous solvent, there are two factors we must consider before we can describe or predict redox reactions: First, what is the stability range of the solvent toward oxidants and reductants, both in thermodynamic and kinetic terms? Second, how do the electron-transfer potentials for redox couples change from those given in Tables 8.1 and 8.2, which were compiled for aqueous species?

The range of redox stability of solvent molecules toward dissolved species is more or less proportional to the electronegativity difference between the atoms that must be oxidized or reduced in the solvent. For water, as we saw in Fig. 8.1, the thermodynamic stability range is 1.229 V, which reflects the ease with which the extremely electronegative O atom in O_2 is reduced to H_2O, compared to the modest difficulty of reducing H_3O^+ to H_2. If we go to the closely related solvent liquid ammonia, we find that the characteristic reduction potentials are much closer together:

$$NH_4^+ + e^- = NH_3 + \tfrac{1}{2}H_2 \qquad \mathscr{E}^0 = 0 \text{ V (defined for } NH_3)$$
$$3NH_4^+ + \tfrac{1}{2}N_2 + 3e^- = 4NH_3 \qquad \mathscr{E}^0 = +0.04 \text{ V}$$

This range of only 40 millivolts is so narrow that essentially no oxidants or reductants are thermodynamically stable in liquid ammonia. However, the range of kinetic metastability—often called overvoltage—is even greater in ammonia than in water, about 1 volt for both oxidation to N_2 and reduction to H_2, as Fig. 8.2 suggests. Note that the very limited autoprotolysis in NH_3 allows a very wide range of "pH" variation, and that consequently very strong reducing agents are stable (or metastable) in basic NH_3 solutions (that is, solutions that contain significant NH_2^-).

On the other hand, in the molten-salt solvent $LiF_{(1)}$, the Li^+ must be reduced by a solute species, or the F^- must be oxidized (or both). The great difference in the electronegativities of Li and F makes these potentials quite different:

$$\tfrac{1}{2}F_2 + e^- = F^- \qquad \mathscr{E}^0 = 0 \text{ V (defined for molten LiF)}$$
$$Li^+ + e^- = Li^0 \qquad \mathscr{E}^0 = -5.564 \text{ V (500°C)}$$

It is thus essentially impossible to cause a redox decomposition of molten LiF solvent with a chemical solute.

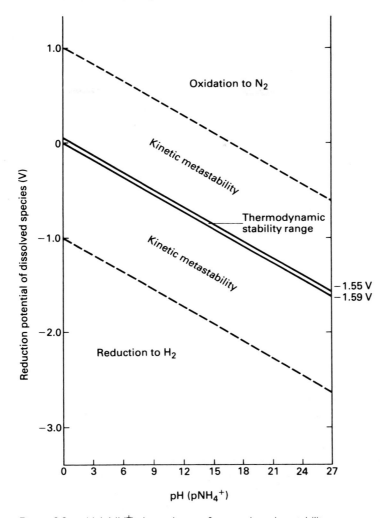

Figure 8.2 pH (pNH$_4^+$) dependence of ammonia redox stability.

When a nonaqueous solvent is chosen for a redox reaction, the reduction potential for a given couple will usually differ from the corresponding potential in water. Table 8.4 illustrates this pattern for a number of couples. In general, the values in different solvents parallel each other fairly well and lie in the approximate order of the electronegativities of the elements being reduced. However, there are some striking differences, such as the Hg^{2+}/Hg^0 values of +0.85 V in water, +0.39 V in CH_3CN, and −0.5 V in molten LiCl–KCl. As the solvent discussion of Chapter 6 suggested, the differences arise from the changes in the stability of the dissolved ions because of (1) the coordinating (Lewis acid/base)

Table 8.4 Comparative Reduction Potentials in Different Solvents at 25°C (V)

Redox couple	H_2O	CH_3OH	CH_3CN	HCOOH	NH_3	LiCl–KCl (450°C)
Cl_2/Cl^-	1.36	1.09	0.72	1.24	1.04	0.32
Br_2/Br^-	1.07	0.86	0.61	0.99	0.84	0.18
Hg^{2+}/Hg^0	0.85	—	0.39	—	−0.24	−0.5
Ag^+/Ag^0	0.80	0.73	0.37	0.64	−0.16	−0.74
I_2/I^-	0.54	0.33	0.21	0.44	0.46	−0.20
Cu^+/Cu^0	0.52	—	−0.14	—	−0.58	−0.96
H^+/H_2	0	−0.03	0.14	0.47	−0.99	—
Pb^{2+}/Pb^0	−0.13	−0.23	0.02	−0.25	−0.67	−1.10
Cd^{2+}/Cd^0	−0.40	−0.46	−0.33	−0.28	−1.19	−1.32
Zn^{2+}/Zn^0	−0.76	−0.77	−0.60	−0.58	−1.52	−1.57
Na^+/Na^0	−2.71	−2.76	−2.73	−2.95	−2.84	—
Ca^{2+}/Ca^0	−2.76	—	−2.61	−2.73	−2.73	−3.17
K^+/K^0	−2.92	—	−3.02	−2.89	−2.97	—
Cs^+/Cs^0	−2.92	—	−3.02	−2.97	−2.94	—
Li^+/Li^0	−2.96	−3.13	−3.09	−3.01	−3.23	−3.30

properties of the neighboring solvent molecules, and (2) the dielectric constant of the solvent. Of these two influences, the solvent's coordinating ability is by far the more important.

All of the couples shown in Table 8.4 involve an uncharged element (usually as a separate phase) and its characteristic ion. When the characteristic ion is a cation, the reduction potential will be more negative the more the cation is stabilized by interaction with the Lewis-base solvent. In particular, soft-acid cations such as Ag^+ and Hg^{2+} will be stabilized a great deal more by interacting with a soft-base solvent such as NH_3 or molten Cl^- than by interacting with the hard base H_2O. These are the couples that show the greatest change in $\mathscr{E}^0$ from water to molten-chloride solvent.

The situation is reversed for anions, where solvent coordination *increases* $\mathscr{E}^0$, but requires that the solvent show Lewis-acid properties (usually through hydrogen bonding). In the Cl_2/Cl^- couple, for example, water, methanol, formic acid, and ammonia all stabilize the Cl^- ion through hydrogen bonding and have relatively high reduction potentials (1.36, 1.09, 1.24, and 1.04 V, respectively). Acetonitrile and the molten-chloride solvent, on the other hand, have no electron-acceptor hydrogen-bonding capability, stabilize the Cl^- ion less, and show lower potentials (0.72 and 0.32 V, respectively). It is interesting to note that for any couple, cationic or anionic, the net effect of changing from water to a nonaqueous solvent is usually to reduce its reduction potential.

8.5 OXIDIZING SOLVENT SYSTEMS

Just as we grouped solvents in Chapter 6 by their acid–base properties, we can group solvents here by their characteristic redox properties. For acid–base systems, some solvents are intrinsically acidic and tend to react with solutes in that manner, some are intrinsically basic and react as bases with many solutes, and others have essentially no acid–base reactivity of their own but serve as polar host-media. Similarly, some nonaqueous solvents are used primarily because they oxidize solutes much more vigorously than would be possible in water, others reduce solutes—or permit their reduction—much more than water would permit, and still others are extremely unreactive toward solute oxidants and reductants and serve as polar host-media for redox reactions. As has already been pointed out, these groups tend to be the same as the acid–base groupings: Acidic solvents frequently serve as oxidants or as hosts for oxidation reactions, basic solvents are often used for reductions, and the Lewis-base coordinating solvents (Table 6.16), which are primarily polar host-media for acid–base reactions, are frequently used similarly for redox reactions. Here we shall consider solvents that are themselves strong oxidants or that promote oxidation reactions. Later sections will deal with the other groups.

Perhaps the most characteristic group of molecules that serve as oxidants and (at least in some experiments) as oxidizing solvents is the halogen oxides and oxyfluorides. The stoichiometry and preparation of these species have been described in Tables 5.9, 5.10, and 5.12. The most thoroughly studied of these compounds are chloryl fluoride, ClO_2F, and perchloryl fluoride, ClO_3F. Both are strong oxidants, as are all the other members of this group. ClO_3F also shows a substantial kinetic inertness due to its high symmetry (quasi-tetrahedral) and low dipole moment. Some reactions showing the oxidizing character of these species (either pure or diluted with the redox-inert $CFCl_3$) are

$$3ClO_2F + AsF_3 \rightarrow ClO_2^+ \, AsF_6^- + 2ClO_2$$
$$ClO_2F + SF_4 \rightarrow SF_6, \, SOF_4, \, SO_2F_2$$
$$3ClO_3F + 4H_2S \rightarrow 4SO_2 + 3HF + 3HCl + H_2O$$
$$ClO_3F + 7HCl \rightarrow HF + 4Cl_2 + 3H_2O$$

It is worth noting that even though the methods by which these strongly oxidizing species are prepared are not in general redox reactions, they tend to rely on highly acidic solvents to stabilize both reactants and products against redox reaction with the solvent.

One such solvent is sulfuric acid, which has been used in many experiments in which a neutral molecule is to be oxidized to a cation. Perhaps the best known such solution is that of I_2, which dissolves in oleum (sulfuric acid containing excess SO_3, usually as $H_2S_2O_7$) to give a solution whose deep blue color is primarily due to the I_2^+ ion formed when I_2 reduces SO_3:

$$2I_2 + 2SO_3 + H_2SO_4 \rightarrow 2I_2^+ + SO_2 + 2HSO_4^-$$

If oxidizing agents such as IO_3^- or $S_2O_8^{2-}$ are used instead of SO_3, the I_2 is oxidized to IO^+, I_3^+ and I_5^+. Similar reactions were described at the end of Chapter 6 for the superacid solvent HSO_3F-SbF_5 and for the oxidant $S_2O_6F_2$, which in addition to yielding the ions above can also give the dication I_4^{2+}.

Sulfuric acid has also been used extensively in electron-spin resonance studies on aromatic-hydrocarbon radical cations. As an extremely acidic protic solvent, it first protonates the hydrocarbon, then oxidizes it:

$$\text{[anthracene]} + H_2SO_4 \longrightarrow \text{[protonated anthracene]}^+ + HSO_4^-$$

$$\text{[protonated anthracene]}^+ + \text{[anthracene]} \longrightarrow \text{[anthracene]}^0 + \text{[anthracene]}^+$$

$$\text{[anthracene]}^0 + 2H_2SO_4 \longrightarrow \text{[anthracene]}^+ + 2H_2O + SO_2 + HSO_4^-$$

If we represent anthracene (or a general aromatic molecule) by A, the overall reaction is thus

$$2A + 3H_2SO_4 \rightarrow 2A^+ + SO_2 + 2H_2O + 2HSO_4^-$$

Other Lewis-acid molecules that are good solvents for oxidizing reactions include IF_5 and SbF_5 (which oxidize I_2 to I_2^+), dilute $SbCl_5-CH_2Cl_2$ solutions (which produce radical cations as readily as H_2SO_4), and liquid SO_2 (which has been used for electrolytic oxidations). As an example of the last of these, complexes of Cu(II) can be oxidized electrolytically to comparable complexes of the unusual oxidation state Cu(III) in liquid SO_2.

8.6 INERT ORGANIC-SOLVENT SYSTEMS

Many common organic solvents have a greater resistance than water to oxidation or reduction by solute species. In addition to the obvious advantage this confers on the organic solvents, it is often true that dissolved species would react with water in a protic acid/base sense to decompose. The major difficulty organic solvents present is that ionic or highly polar species are frequently insoluble in them because they offer only a modest solvation energy compared to the substantial lattice energy of the pure substance (or compared to the substantial hydration energy of an aqueous solution). However, several organic solvents have been developed that are composed of fairly small molecules, are quite polar, and have high dielectric constants. These are fairly good solvents for polar species. The top half-dozen solvents in Table 6.16 are widely used: *N*-methylacetamide, propylene carbonate, dimethyl sulfoxide, acetonitrile, dimethylacetamide, and dimethylformamide. Acetonitrile and propylene carbonate are the most popular of these, especially for electrochemical reactions.

The following are examples of the use of these popular organic solvents for inorganic redox reactions:

Acetonitrile

$$2B_{10}H_{10} \xrightarrow{\text{Electrolysis}} B_{20}H_{18}^{2-} + H_2 + H^+ + 3e^-$$

$$NiCl_2 + H_2O_4 \longrightarrow NOCl + Ni(NO_3)_2 \cdot 3CH_3CN \xrightarrow{\Delta} Ni(NO_3)_2$$

Nitromethane

$$U + N_2O_4 \rightarrow UO_2(NO_3)_2 \cdot N_2O_4 \xrightarrow{\Delta} UO_2(NO_3)_2$$

$$S_4N_4 + 2SOCl_2 \rightarrow S_3N_2O_2 + 2Cl_2 + S_2N_2 + S$$

Ethyl carbonate

$$2LiSnPh_3 + (EtO)_2CO \rightarrow Sn_2Ph_6 + 2LiOEt + CO$$

Ethyl ether

$$2LiAlH_4 + SbCl_3 \xrightarrow{-90°C} SbH_3 + 3AlH_3 + 3LiCl$$

Propyl ether

$$B_{10}H_{14} + 2Et_2S + C_2H_2 \rightarrow B_{10}C_2H_{12} + 2H_2 + 2Et_2S$$

Dimethoxyethane

$$Na + GeH_4 \rightarrow NaGeH_3 + \tfrac{1}{2}H_2$$

In addition to these preparative reactions, the extended voltage range of redox stability for solvents such as acetonitrile allows the polarographic reduction of alkali and alkaline-earth metal ions to the metal, which of course is not possible in aqueous solution.

Alternatively, if the redox reactants are predominantly covalent or only slightly polar, solutions can usually be prepared in the more traditional organic solvents, such as CH_2Cl_2, $CHCl_3$, benzene, hexane, and the like. All of these molecules are inert toward even the stronger common oxidizing and reducing agents; in fact, they are often used to dilute oxidizing solvent media such as $SbCl_5$ and even H_2SO_4. A few such reactions are

Carbon tetrachloride

$$NH_3 + S + 4AgF_2 \rightarrow \underset{\substack{\text{Thiazyl} \\ \text{fluoride}}}{NSF} + 3HF + 4AgF$$

Carbon disulfide

$$PI_3 + S \rightarrow PSI_3$$

Toluene

$$(EtO)_3PO + 3RCl + 3Na^+AlH_2(OCH_2CH_2OCH_3)_2^-$$
$$\rightarrow R_3PO + 3H_2 + 3\ NaCl + 3Al(OEt)(OCH_2CH_2OCH_3)_2$$

Two strong reducing agents frequently used in organic solvents are lithium aluminum hydride, $LiAlH_4$, and sodium borohydride, $NaBH_4$. $LiAlH_4$ is familiar from its many uses in organic reductions, but it has been almost as thoroughly investigated in its reactions with inorganic molecules, particularly halides. It decomposes violently in water:

$$LiAlH_4 + 4H_2O \rightarrow LiOH + Al(OH)_3 + 4H_2$$

It is probably more helpful to think of this as an acid–base reaction between the weak acid water and the very strong base H^- than as a redox reaction. Similarly, many of the reactions shown by $LiAlH_4$ in organic solvents are acid–base reactions of the H^- base rather than true reductions, although they are frequently called reductions:

$$LiAlH_4 + 4CuI \xrightarrow{\text{Pyridine}} 4CuH + LiI + AlI_3$$

$$LiAlH_4 + SiCl_4 \xrightarrow{\text{Et}_2O} SiH_4 + LiCl + AlCl_3$$

Some true reductions (with changing oxidation states) do occur, however:

$$3LiAlH_4 + 4PBr_3 \xrightarrow{\text{Et}_2O,\ -30°C} \tfrac{4}{n}(PH)_n + 4H_2 + 3LiBr + 3AlBr_3$$

$$LiAlH_4 + TiCl_4 \xrightarrow{-110°C} [Ti(AlH_4)_4] \xrightarrow{-85°C} Ti^0 + 4Al^0 + 8H_2$$

Ether is frequently used as a solvent for $LiAlH_4$ reactions because it dissolves the hydride to the extent of 29 g/100 g ether, forming a convenient reaction medium. A useful variant on $LiAlH_4$ is the sodium bis(2–methoxyethoxy)aluminum hydride mentioned above, $Na^+AlH_2(OCH_2CH_2OCH_3)_2^-$, which is soluble even in nonpolar solvents such as toluene. (It apparently encrypts the Na^+ in the methoxy groups, rather like a crown ether.)

Sodium borohydride shows many of the same reactions as $LiAlH_4$, though in general it is a slightly weaker reducing agent:

$$NaBH_4 + 4I_2 \xrightarrow{\text{Cyclohexane}} BI_3 + 4HI + NaI$$

$$2LiBH_4 + PhBiBr_2 \longrightarrow \tfrac{1}{n}(PhBi)_n + 2LiBr + B_2H_6 + H_2$$

$$2LiBH_4 + Ph_3AsCl_2 \longrightarrow Ph_3As + B_2H_6 + H_2 + 2LiCl$$

(Lithium borohydride is very similar to $NaBH_4$, but it is more soluble in solvents such as ethyl ether and tetrahydrofuran). $NaBH_4$ is more resistant to hydrolysis than the aluminohydride ion. We have already noted the uses of BH_4^- as a reducing agent in aqueous solution.

8.7 INERT MOLTEN-SALT SOLVENT SYSTEMS

Molten ionic salts constitute another category of inert solvents appropriate for certain kinds of redox reactions. As our previous discussion indicated, the emf stability range of simple ionic compounds in the liquid state is quite high because of the large electronegativity difference between the cation and anion. It is very difficult, for example, to reduce Na^+ or to oxidize Cl^-, so a very wide voltage range of redox reactivity is possible in a solvent consisting of molten Na^+Cl^-. This would appear to make the molten-salt solvent systems particularly suitable for preparing unusual oxidation states, for example. However, the relatively high temperatures required to keep ionic salts liquid (at least simple ones with monatomic ions), together with the corrosive nature of the molten salts, have made experimental studies difficult.

While molten salts are usually inert with respect to redox reactions, we noted in Chapter 6 that they are usually quite reactive in an acid–base sense, particularly with respect to anion transfer. For this reason, many promising oxidants or reductants are unstable in molten-salt solvents. Most redox chemistry done in molten salts involves either elements as reducing agents,

$$Cd^0 + CdCl_2 \xrightarrow{Na^+AlCl_4^-} Cd_2Cl_2$$

$$44Bi^0 + 28BiCl_3(l) \longrightarrow 3Bi_{24}Cl_{28} \qquad ([Bi_9^{5+}]_2[BiCl_5^{2-}]_4[Bi_2Cl_8^{2-}])$$

or electrochemical reactions

$$2CdCl_2 + 2e^- \rightarrow Cd_2^{2+} + 4Cl^-$$

$$PbCl_2 + e^- \rightarrow Pb^+ \text{ [or } Pb_n^{(2n-1)+}] + 2Cl^-$$

Perhaps the best-known example of the latter is the electrolytic production of aluminum metal in molten cryolite, Na_3AlF_6:

$$2Al_2O_3 + 12e^- \xrightarrow{\hspace{3cm}} 4Al^0$$

$$\xrightarrow{Na_3AlF_6-AlF_3-CaF_2} 12e^- + 3O_2 \xrightarrow[\text{Electrode}]{\text{Carbon}} 3CO_2$$

A later section will examine metal smelting and electrowinning in more detail.

The relative convenience of electrochemical reactions in molten salts and the very wide voltage range of stability for these systems make it possible to design batteries with molten-salt electrolytes that offer very high energy density and power density. For example, consider a cell in which the electrolyte is molten LiCl just above its melting point (608°C), and the electrode reactions are

Cathode

$$Cl_2 + 2e^- = 2Cl^-$$

Anode

$$Li^0 = Li^+ + e^-$$

From Table 8.4, we can estimate the potential of such a cell in the closely related LiCl–KCl electrolyte as 3.62 V. Actually, at the higher temperature in pure LiCl, the open-circuit voltage is about 3.49 V for a single cell. This high voltage (relative to the 2.2 V of a lead–acid cell, for instance) raises the energy density of the cell, since energy density in J/(kg battery weight) is proportional to cell voltage:

$$\text{Energy density} = \mathscr{E}(V) \times \mathscr{F}\frac{\text{coul}}{\text{mol } e^-} \times \frac{\text{mol } e^-}{\text{mol rn}} \div \frac{\text{kg electroactives}}{\text{mol rn}}$$

$$= \text{J/kg electroactives} \quad (\text{or kJ/kg})$$

In this case, the cell reaction is $2\,\text{Li} + \text{Cl}_2 = 2\text{LiCl}$, in which two moles of electrons pass per mole reaction and the mass per mol rn is $2(6.94) + 2(35.45)$, or 84.78, g (0.08478 kg/mol rn). The energy density is thus

$$\text{Energy density} = 3.49 \times 96{,}486 \times 2 \div .08478 = 7.944 \times 10^6 \text{ J/kg}$$

$$= 7944 \text{ kJ/kg}$$

Compare this to the energy density to the familiar lead–acid battery calculated on the same basis, which is 388 kJ/kg. It should be apparent that the high voltage and low formula weights both contribute to the high energy density.

Besides the high cell potentials that are possible, molten-salt electrolytes have such a high conductance (on the order of 1 $\text{ohm}^{-1}\text{ cm}^{-1}$) that the molten-salt cell will have a very low internal resistance. The high temperature at which such cells must be operated, the simple structure of the molten-salt species, and the absence of inert solvent all contribute to a high exchange current, or electron turnover rate at the electrode surface. This allows very high current densities to be drawn from the cell without reducing the open-circuit voltage very much. The lithium–chlorine cell can be operated at current densities of 40 A/cm^2 without significant polarization at the electrode surfaces. This raises the power density of the cell, in watts or kW per kg, to much higher levels than conventional batteries.

Another interesting molten-salt cell that has been the subject of considerable developmental work involves sodium and sulfur:

Cathode

$$\text{S} + 2e^- = \text{S}^{2-}$$

Anode

$$\text{Na}^0 = \text{Na}^+ + e^-$$

This cell, using molten Na_2S as electrolyte, has a theoretical energy density of 4300 kJ/kg and an open-circuit voltage of 2.08 V, which are promising qualities. Unfortunately, sulfur readily dissolves in molten Na_2S to yield various polysulfide ions, so a separator is needed to prevent the molten sodium from reacting directly with polysulfide ions. The separator material is the layered intercalated material we have already considered in Chapter 3—β-alumina, or sodium aluminate, $\text{NaAl}_{11}\text{O}_{17}$, with a crystal structure consisting of layers of close-packed oxygen atoms four deep containing tetrahedrally coordinated Al atoms but no Na^+ ions. The Na^+ ions are in a very loosely packed, open layer between the close-packed aluminate layers. At molten-salt temperatures, sodium ions readily penetrate the open layer and are quite mobile in two dimensions through the crystal. In an electrical cell, the beta-alumina separator allows Na^+ ions, but not Na^0 atoms or any sulfur species, to pass through into the electrolyte.

Two major difficulties have slowed the development of molten-salt high-energy batteries: the need for a high operating temperature and the highly corrosive nature of the electrode materials. While the high temperature of the melts raises the power density of the cells, it also increases the corrosive nature of the electrolyte and requires supplemental heating when the battery is not in use. It is often possible to freeze a molten-salt battery and remelt it without damage, but this is not practical if the battery is to be used in an electric car, for example. The corrosive quality of the chemical species in a molten-salt cell requires extensive materials study and careful design of the cell components, but development continues and the successful application of these batteries seems more likely than ever in the fairly near future.

One approach to reducing the corrosive character of the molten salt solvent is to combine it with an inert organic solvent that will also allow ionic conduction. A promising approach to this is the broad field of *polymer electrolytes,* in which small alkali metal cations are coordinated by a polyether polymer much as a crown ether coordinates isolated M^+ ions. The straight-chain polymer polyethylene oxide, $(-CH_2CH_2O-)_n$, is the simplest candidate; it can wrap around Na^+ ions to coordinate them, but under the influence of an applied voltage can flex so as to allow the Na^+ to move along the chain to a new coordination site. Neighbor chains can also engage in both the coordination and the flexural ion transport. It helps to have a loose, random polymer structure that can flex readily, and polyethylene oxide is too crystalline to do this very well. Other effective polymers include a phosphonitrilic polymer (Section 5.11) in which each P atom has two pendant $-O(CH_2CH_2O)_3CH_3$ groups, which can both wrap around the M^+ ion and flex readily to transport it. In a limited sense, the M^+ ion is mobile under the influence of an applied potential just as it is in a molten salt. Of course, there must be an anion as counterion for each M^+, and to permit maximum mobility the anion needs to form the loosest possible ion pair with M^+. Large singly charged anions are usually used: I^-, AsF_6^-, ClO_4^-, and so on. The ion pairs are dissolved in the amorphous polymer, so that in a sense a molten salt has been diluted by a polyether.

There are some interesting advantages to polymer electrolytes for battery use. Although some polymer electrolytes require elevated temperatures of operation, those temperatures are still usually below 100°C, and in fact some polymer-electrolyte cells have been successfully demonstrated at room temperature. The electrolyte is a chemically inert solid, which offers few corrosion problems and withstands recharge cycles well. Perhaps their greatest difficulty is that the "molten salt" must be kept to a low concentration in the polymer to prevent strong ion pairing, which keeps the cell internal resistance high enough that high currents cannot be drawn from the cell.

8.8 REDUCING SOLVENT SYSTEMS: THE SOLVATED ELECTRON

We noted earlier that strongly basic solvents tend to be more hospitable to strong reducing agents and briefly surveyed some reductions carried out in Lewis-base organic solvents. We shall now explore a remarkable observation: The ultimate reducing agent, the free electron, exists as a solvated species in some simple solvent liquids. Although it will reverse the chronological order of discovery, we shall begin with the simplest solvent liquids, the noble gases.

An alpha-particle source immersed in liquid argon (where the alpha particles have a range of less than a millimeter) can be used to generate free electrons dissolved in the liquid up to a concentration of perhaps $10^8 e^-$/mL. In ordinary concentration terms, this is still extremely dilute: 10^{-14} M or less. However, if the noble-gas solvents have been rigorously purified, their intrinsic electrical conductivity will be low enough to allow the conductance properties of the dissolved electrons to be measured. The apparent mobility of the electrons at liquid-argon temperatures (around 100 K) is about 400 cm/sec per V/cm applied electric field, or 400 cm^2/V sec. In liquid krypton, the apparent electron mobilities are even higher—up to 2200 cm^2/V sec at 180 K. These values can reasonably be considered the behavior of free electrons, since the solution is so dilute that the average electrostatic interaction between dissolved electrons is smaller than kT thermal energy at the temperature of the solution. By way of comparison, the mobilities of ordinary ions in liquid water are roughly 5×10^{-4} cm^2/V sec. Even the proton in water, which benefits from the hydrogen-bond transfer shown in Fig. 6.3, only has a mobility of about 3×10^{-3} cm^2/V sec. From the high mobility values and from the dependence of the electron velocity on applied field strength, it can be deduced that the dissolved electrons are essentially free electrons analogous to the quantum-mechanical particle in a box, with little or no structural rearrangement of the liquid argon to accommodate the electrons in any geometrically solvated manner. This, of course, is just what we would expect, given the spherical nonpolar nature of argon atoms.

Free electrons can also be prepared in a number of molecular-solvent liquids, where solvation occurs and the solution structure is therefore more complex. The classic solvated-electron solutions are those involving group Ia or IIa metals (except Be) dissolved in liquid ammonia. These were first prepared in 1864 by Weyl. A great deal of experimental work has been done on metal–ammonia solutions since that time. Because dilute, clean, cold metal–ammonia solutions are quite stable, they have been much more thoroughly investigated than solutions in other solvents. It is worth noting, however, that solvated-electron solutions have been prepared in water, organic amines, organic ethers and polyethers, and even hexamethylphosphoramide. It seems likely that any polar Lewis-base solvent can serve as a host medium for solvated free electrons.

Metal-ammonia solutions are prepared by simply dissolving a freshly cleaned portion of alkali or alkaline-earth metal in liquid ammonia. If the ammonia is pure and anhydrous, no gas is evolved (as might be expected if the metal were reducing

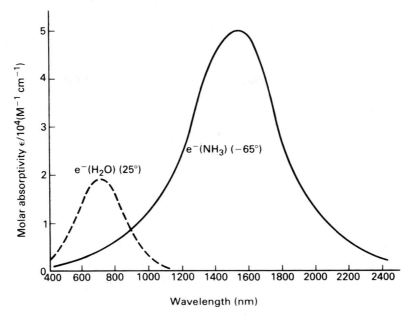

Figure 8.3   Electronic absorption spectra of the solvated electron in water and ammonia.

the solvent: $Na^0 + NH_3 \rightarrow Na^+ + NH_2^- + \frac{1}{2}H_2$). A dilute solution has a strong blue color resulting from an intense absorption at a wavelength of about 1500 nm and tailing into the visible range. The absorption peak is influenced only slightly by temperature, and essentially not at all by the identity of the metal dissolved. Figure 8.3 compares the spectra of the solvated electrons in ammonia and water. The dilute solution is an excellent electrical conductor, as Fig. 8.4 indicates. It is worth noting, however, that the mobility of the dissolved electrons is only about 1.2×10^{-2} cm^2/V sec, about 10^5 times smaller than the value in liquid krypton. The electron's mobility is only seven or eight times greater than that of the solvated alkali-metal ion, which strongly suggests substantial geometric solvation by ammonia molecules. The equivalent conductance, however, is greater than that of any other solute in liquid ammonia.

As the concentration of the metal–ammonia solution increases, the equivalent conductance decreases gradually (presumably because of increasing cation–electron association) up to a concentration of about 0.05 M. At higher concentrations, the conductance increases sharply as shown in Fig. 8.4. Near 1 M, the electrical conductance is comparable to that of the pure metal itself. As the conductance increases, the color of the solution changes from blue to a metallic bronze color. Still no net chemical reaction has occurred; if the solution is evaporated to dryness, the metal is recovered unchanged.

These properties can be explained by assuming that in liquid ammonia, the electron occupies a cavity in the liquid lattice with a radius of about 3.3 Å. Of

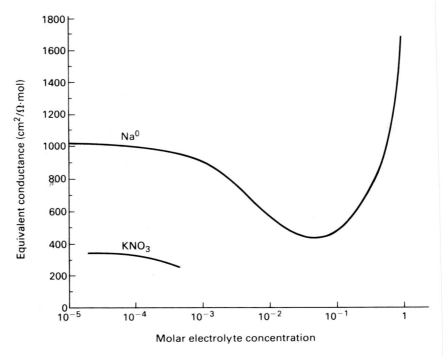

Figure 8.4   Equivalent conductance of Na metal and KNO_3 dissolved in liquid NH_3.

course, it is repelled by the nearest bonding electrons of the surrounding ammonia molecules, but it is stabilized overall by the longer-range effect of the oriented dipoles. If the electron is held in the cavity by the electrostatic potential energy of the dipoles, wavefunctions can be written that describe the electron's distribution by spatial functions roughly resembling the orbitals of a hydrogen atom. The blue color of dilute solutions corresponds to a transition from a "$1s$" to a "$2p$" wave function. In more concentrated solutions, the electrons are so near that their wave functions overlap. The solution must now be considered a liquid "expanded metal," which is consistent with the color change and with the metallic reflectivity of concentrated solutions.

Liquid ammonia, as a basic solvent, is generally a good host medium for reductions. The presence of free electrons in metal–ammonia solutions makes them very strong reducing media. In some reductions, electrons are simply added to a given species. Where the species is an element, the product is usually a polymeric anion such as one of the Zintl anions we have already considered:

$$4e^- + 9Sn^0 \rightarrow Sn_9^{4-} \quad \text{or} \quad 4Na + 9Sn \rightarrow Na_4Sn_9$$

$$K + 5P + 3NH_3 \rightarrow KP_5 \cdot 3NH_3$$

Some molecular species reduce simply,

$$e^- + MnO_4^- \rightarrow MnO_4^{2-}$$

$$e^- + Ni(CN)_4^{2-} \rightarrow Ni(CN)_4^{3-} \xrightarrow{e^-} Ni(CN)_4^{4-}$$

whereas others polymerize to some extent:

$$2Na + 2NO \rightarrow Na_2N_2O_2 \qquad (cis \text{ configuration of hyponitrite})$$

$$6K + 6CO \rightarrow K_6C_6O_6 \qquad (\text{salt of hexahydroxybenzene})$$

$$9Na + 4Zn(CN)_2 \rightarrow 8NaCN + NaZn_4$$

With hydrides and organometallic compounds, bond cleavage usually occurs:

$$K + PH_3 \rightarrow KPH_2 + \tfrac{1}{2}H_2$$

$$Na + SnH_4 \rightarrow NaSnH_3 + \tfrac{1}{2}H_2 \xrightarrow{Na} Na_2SnH_2$$

$$2Na + Me_3SnBr \rightarrow NaBr + Me_3SnNa$$

$$2K + Si_2Ar_6 \rightarrow 2KSiAr_3$$

The latter two compounds are particularly useful in organometallic synthesis because of the ease with which the alkali metals can be replaced by other organic groups:

$$Me_3SnNa + RX \rightarrow NaX + Me_3SnR$$

Although metal–ammonia solutions are remarkably stable, they decompose slowly when pure—and rapidly in the presence of dissolved transition-metal ion catalysts—to form hydrogen gas and the amide ion. If we represent the ammoniated electron as $(NH_3)^-$, the reaction becomes

$$(NH_3)^- + (NH_3)^- \rightarrow H_2 + 2NH_2^-$$

Perhaps the most remarkable feature of metal–ammonia solutions is that this reaction is so slow. By contrast, the rate constant for the comparable reaction in water,

$$(H_2O)^- + (H_2O)^- \rightarrow H_2 + 2OH^-$$

is about 6×10^9 M^{-1} sec^{-1}, fast enough that a comparable rate constant in ammonia would cause the characteristic blue color to disappear in a microsecond.

The rapid disappearance of the hydrated electron, both by the mechanism above and by others such as

$$(H_2O)^- + H_2O \rightarrow H\cdot + OH^- + H_2O$$

$$H\cdot + (H_2O)^- \rightarrow H_2 + OH^-$$

means that the hydrated electron must be studied on a microsecond time scale. Although alkali metals dissolving in water probably produce the hydrated electron as a transient product, such electrons cannot diffuse into the bulk solution rapidly enough for experimental study. Accordingly, hydrated electrons are usually produced either by irradiating water or aqueous solutions with gamma rays produced by ^{60}Co, or by flash photolysis using a xenon flash lamp. The latter is particularly convenient in basic solution:

$$OH^- + h\nu \rightarrow \cdot OH + e^-$$

Concentrations on the order of 10^{-7} M can be achieved. The half-life of the resulting hydrated electrons in neutral water (pH 7) is about 300 μsec. This is long enough that the absorption spectrum can be measured (see Fig. 8.3). From the absorption spectrum, structural inferences can be drawn as to the nature of the hydration sphere. A reasonable fit can be achieved for the kinetic data and for the spectrum by assuming the same model as for ammonia—a cavity in the liquid-water lattice surrounded by water dipoles oriented toward the center of the electron distribution. However, the shorter wavelength (higher frequency) of the absorption peak in water implies a smaller cavity. The best fit to λ_{max} is achieved, in fact, by assuming zero radius for the cavity—that is, touching dipoles. The wave function for the hydrated electron does not have zero radius, however, but is roughly the same size as a water molecule. Therefore, appreciable electron density extends out past the primary solvation sphere. This difference between a hydrated electron and an ammoniated electron is perhaps the major reason for the great instability of the hydrated electron.

In spite of the vanishingly small cavity size in liquid water, the rather low mobility of the hydrated electron, 1.8×10^{-3} cm^2/V sec, leaves little doubt that geometric solvation is occurring. This mobility value is only about a tenth that of the ammoniated electron. This presumably represents the increased difficulty of breaking hydrogen bonds to the solvated water molecules as the electron moves. There is a substantial solvation energy, which can be calculated from the following cycle:

$$
\begin{array}{ccccccc}
e^-(g) & + & H^+(g) & \xrightarrow{-IP} & H(g) & \xrightarrow{\frac{1}{2}BE} & \frac{1}{2}H_2(g) \\
-\Delta G_{hyd} \uparrow & & -\Delta G_{hyd} \uparrow & & & & \downarrow \Delta G_{soln} \\
e^-(aq) & + & H^+(aq) & \xrightarrow{\Delta G_{red}} & & & \frac{1}{2}H_2(aq)
\end{array}
$$

This yields

$$\Delta G_{red} = -\Delta G_{hyd}(e^-) - \Delta G_{hyd}(H^+) - IP + \tfrac{1}{2}BE(H_2) + \Delta G_{soln}(H_2)$$

or

$$\Delta G_{hyd}(e^-) = -\Delta G_{red} - \Delta G_{hyd}(H^+) - IP + \tfrac{1}{2}BE(H_2) + \Delta G_{soln}(H_2)$$

$$= -\Delta G_{red} + 1103 - 1311 - 223 + 18 \text{ kJ/mol rn}$$

To make further progress, we need a value for ΔG_{red}, the absolute free-energy change for the reduction of the aqueous proton. This is an interesting quantity in itself, because it represents the standard electrode potential of the hydrated electron. We can use the reaction

$$e^-(aq) + H_2O(l) \underset{k_{-1}}{\overset{k_1}{\rightleftharpoons}} H(aq) + OH^-(aq)$$

for which both rate constants, k_1 and k_{-1}, are known. This allows us to calculate K_{eq}, and thus ΔG^0, for the reaction

$$K_{eq} = \frac{k_1}{k_{-1}} = \frac{16 \text{ M}^{-1} \text{ sec}^{-1}}{2.2 \times 10^7 \text{ M}^{-1} \text{ sec}^{-1}} = 7.3 \times 10^{-7}$$

$$\Delta G^0 = -RT \ln K_{eq} = -8.314 \times 10^{-3} \times 298.2 \times \ln(7.3 \times 10^{-7})$$

$$= +35 \text{ kJ/mol reaction}$$

We can now use Hess's law to sum the free energies for the following reactions:

$$e^-(aq) + H_2O \rightarrow H + OH^- \qquad \Delta G^0 = +35 \text{ kJ}$$

$$H \rightarrow \tfrac{1}{2}H_2 \qquad\qquad -223$$

$$H^+(aq) + OH^-(aq) \rightarrow H_2O(l) \qquad\qquad -80$$

$$\overline{ e^-(aq) + H^+(aq) \rightarrow \tfrac{1}{2}H_2 \qquad \Delta G^0 = -268 \text{ kJ}}$$

Using the equivalence between free energy and reduction potential, we can calculate that this corresponds to a standard potential of 2.77 V for the hydrated electron—a strong reducing agent indeed!

Returning to our solvation energy calculation, we note that the ΔG^0 we have just obtained is the ΔG_{red} needed for the cycle calculation:

$$\Delta G_{hyd}(e^-) = 267 + 1103 - 1311 - 223 + 18 = -146 \text{ kJ/mol}$$

This is a smaller value than the hydration energy of any of the halide ions, but it is substantial nonetheless. It is clear that geometric solvation of the aqueous "free" electron is quite extensive.

It is interesting that the geometric solvation of the "free" electron has its counterpart in the solid state. In 1974 Dye prepared the first solid natride using sodium metal and a cryptand ligand, as we have already seen in Chapter 7. In this and other such alkalide compounds, the electron stripped from the cationic Na^+ is specifically associated with another sodium atom as the Na^- anion. However, it is apparently unnecessary to have an atom as a home for the electron. Just as in liquid ammonia the large stoichiometric excess of solvating ammonia molecules can coordinate all the M^+ ions that M metal can form, if a stoichiometric excess of cryptand or crown ether is added to the alkalide reaction mixture, crystalline salts of ML^+e^-, a metal *electride*, are formed:

$$K_{(s)} + N(CH_2CH_2OCH_2CH_2OCH_2CH_2)_3N \xrightarrow{\text{Me}_2\text{O}, -50°} K(2,2,2\text{-crypt})^+ e^-$$

Here M^+ is the alkali metal cation, L is the cryptand or crown ether, and e^- is a "free"—though often localized in a geometric anion site—electron within the regular crystal structure. Figure 8.5 shows two coordinated cations that have been used, along with the reported crystal structure of one. It can be seen in the crystal structure that the large open spaces between $Cs(15\text{-crown-}5)_2^+$ cations are electrostatically ideal locations for an electron serving the lattice-forming role of an

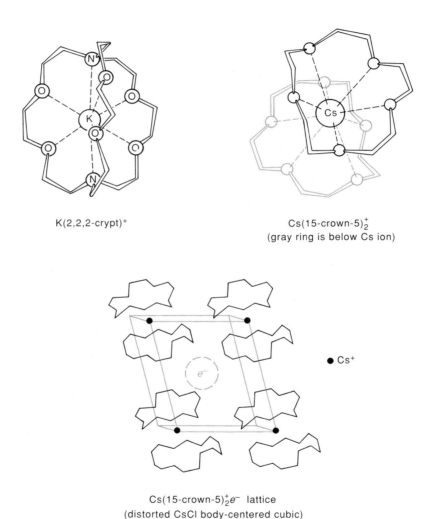

K(2,2,2-crypt)$^+$

Cs(15-crown-5)$_2^+$
(gray ring is below Cs ion)

Cs(15-crown-5)$_2^+ e^-$ lattice
(distorted CsCl body-centered cubic)

Figure 8.5   Structures of cations in alkalide and electride lattices. Note that the small 15-crown-5 ring cannot encompass the large Cs$^+$ ion and coordinates it as a sandwich with 2:1 stoichiometry.

anion. In some of these crystalline electrides, the electron is essentially an isolated localized "anion," as for $Cs(18\text{-crown-}6)_2^+e^-$. For others with less bulky encrypted cations such as $K(2,2,2\text{-crypt})^+e^-$, the electrons interact fairly strongly with each other so that the predominant binding is weak trapping of electron pairs. The $Cs(15\text{-crown-}5)_2^+e^-$ compound seen in Figure 8.5 shows an intermediate stage of electron–electron interaction in which the anion–electron spins are held antiparallel, making the crystal antiferromagnetic. It seems clear that the "free" electron can exist in crystals, geometrically bound by lattice forces just as it is solvated in ammonia or water, but that variation of the cation structure can tailor the degree of interaction of the lattice electrons.

8.9   REDUCTION PROCESSES FOR METAL ORES

By far the largest-scale redox-reaction processes are the commercial processes by which metals and metalloids are extracted from their ores or raw materials. Our whole civilization is based on the use of abundant metals; indeed, the transition from prehistory coincides with the introduction of the Bronze Age, followed by the Iron Age. Worldwide, nearly a billion tons of metals and metalloids are extracted each year. Iron, copper, and aluminum are produced in the largest amounts, by high-temperature carbon reduction, pyrolysis and oxidation, and molten-salt electrochemical reduction, respectively. These processes, described in most introductory texts, are not repeated here. One might note, however, that there are only a few fundamentally different processes for reducing metals and metalloids, even though some three-quarters of the elements in the periodic table fall into those categories. Iron, copper, and aluminum represent three of these fundamental processes.

Table 8.5 summarizes the commercial reduction processes in a rather generalized fashion, and indicates the metals produced by each. The categories are arranged in the table in an order that very roughly corresponds to increasing ease of reduction for the metals involved. The first category, electrochemical reduction, is used for the most electropositive metals (though not for all of them). Generally speaking, electrochemical reduction is the most strenuous reducing condition that can be applied. It is therefore necessary to choose a solvent medium that is both electrically conductive and inert toward strongly reducing conditions. The usual choice is a molten salt, either the fluoride or chloride of the metal involved. Aluminum is reduced from the AlF_6^{3-} species dissolved in molten Na_3AlF_6 (though Al_2O_3 is the formal starting material). Similarly, Li, Na, and Mg are reduced from their molten chlorides, sometimes with KCl added to lower the melting point of the electrolyte. A chloride-melt process has also been demonstrated for Al.

The slightly less active metal Ra can be reduced from an aqueous solution at a mercury cathode, forming radium amalgam from which the mercury is subsequently distilled. Other less-active metals are conveniently reduced from aqueous solution, often from solutions of the metal sulfate. Thallium is produced in this

Table 8.5   Reduction Processes Yielding Metals and Metalloids from Their Ores

Electrochemical reduction

Al, Li, Mg, Na, Ra, Tl, Zn (and purification of Cu, Ni, V, Mn, Pb).

$$M^{n+} + n\, e^- \rightarrow M^0 \text{ (usually in molten salt)}$$

Active-metal reduction (usually from halide; active metal in parentheses)

Ac(Li), Sb(Fe), Ba(Al), Be(Mg), Ca(Al), Ce(Ca), Cs(Na), La–Lu(Ca), Au(Zn), Ti–Zr–Hf(Mg), K(Na), Rb(Ca), Sc(Ca), Ag(Zn), Th(Ca), U(Ca), V(Ca or Al–Fe).

$$MCl_n + n\,Li-Na \xrightarrow{\Delta} M^0 + n\,LiCl-NaCl$$

$$MF_n + \tfrac{n}{2}Mg-Ca \xrightarrow{\Delta\Delta\Delta} M^0 + \tfrac{n}{2}MgF_2-CaF_2$$

$$2M(CN)_2^- + Zn \longrightarrow 2M^0 + Zn^{2+} + 4CN^-$$

Carbon reduction (usually from oxide)

As, Cr, Co, Fe, Pb, Mn, Nb–Ta, P, Si, Sn, W, Zn.

$$MO_n + n\,C \xrightarrow{\Delta\Delta\Delta} M^0 + n\,CO$$

Hydrogen reduction (usually from oxide)

Ge, Ir, Mo, Os, Re, W.

$$MO_n + n\,H_2 \xrightarrow{\Delta\Delta\Delta} M^0 + n\,H_2O$$

Internal redox pyrolysis (usually with gas formation)

As, Cs, Cu, Pb, Hg, Ni, Pd, Pt.

$$MS + O_2 \xrightarrow{\Delta\Delta\Delta} M^0 + SO_2$$

$$(2CsCl(g) + CaC_2 \xrightarrow{\Delta\Delta\Delta} CaCl_2 + 2C + 2Cs^0(g))$$

$$(3Pd(NH_3)_2Cl_2 \xrightarrow{\Delta\Delta\Delta} 3Pd^0 + 6HCl + N_2 + 4NH_3)$$

way from Tl_2SO_4 solution, and copper, lead, and manganese are purified from the crude metals by electrolytic dissolution and redeposition.

Once they are made available through electrochemical reduction, active metals such as Al and Na can be used to produce a wide variety of other metals. The category "Active-metal reduction" in Table 8.5 relies heavily on Na, Mg, and Ca to reduce other less electropositive metals. The first two of these reductants are produced electrochemically, but Ca is itself produced primarily by reducing the oxide with the active metal Al. The reaction conditions are fairly typical for this type of reduction: Calcium oxide is heated to about 1200°C under vacuum with finely divided aluminum. The elemental calcium distills out at that temperature, leaving part of the Ca behind as calcium aluminate:

$$5CaO + 2Al^0 \rightarrow 3\,Ca^0 + Ca_2Al_2O_5$$

An interesting feature of active-metal reductions is that they can frequently be caused to proceed more or less quantitatively even when they are not thermodynamically favorable. The commercial production of potassium metal using sodium as reductant is an example. The reaction vessel is held at about 850°C or 1120 K, a temperature high enough to allow reaction between molten salts and metal vapor:

$$Na^0(g) + KCl(l) \rightarrow K^0(g) + NaCl(l)$$

Approximating the free energies of formation of these species at 1120 K by those at 298 K for gaseous metals and solid salts, we have

$$\Delta G^0 = 61.2 - 384.0 + 408.3 - 78.1 = +7.4 \text{ kJ/mol rn}$$

The reaction is thus unfavorable, though not as much as one might expect. However, the potassium metal product can be distilled out preferentially because of its higher volatility. Since ΔG for the reaction depends on ΔG^0 but also on the vapor-pressure ratio of the metals, we have at equilibrium

$$\Delta G = \Delta G^0 + RT \ln\left(\frac{P_K}{P_{Na}}\right)$$

$$0 = 7400 + 9310 \ln\left(\frac{P_K}{P_{Na}}\right)$$

$$\frac{P_K}{P_{Na}} = e^{-7400/9310} = 0.452$$

An efficient fractionating column for the two metallic vapors can easily reduce P_K well below this fraction of P_{Na} and thus provide a negative free-energy change for the reaction as a whole.

Carbon reduction and hydrogen reduction are very similar processes. Both rely on the high-temperature formation of the gaseous oxides, CO in the one case and H_2O in the other. Energetically, most of the enthalpy required to break the very stable metal-oxide lattice is provided by the formation of the strong O—H and C≡O bonds. In many cases, however, the reduction is quite endothermic even with these strong bonds. Consider the hydrogen reduction of tungsten:

$$WO_3(s) + 3H_2(g) \xrightarrow{1100°C} W(s) + 3H_2O(g)$$

ΔH_{298} for this reaction is +115 kJ/mol rn, which is clearly unfavorable. On the other hand, ΔS^0_{298} for the process is also positive (+125 J/K), so that the reaction is favored by high temperatures. We can estimate the breakeven point—the temperature at which ΔG would equal zero—by assuming that neither ΔH nor ΔS is affected significantly by temperature as the reaction mixture is heated:

$$\Delta G = \Delta H - T\Delta S$$

$$0 = 115,000 - T(125)$$

$$T = 920 \text{ K or } 650°C$$

The usual reaction temperature of 1100°C is obviously more than high enough to compensate for the unfavorable energy relationships. In Chapter 9 we shall look more closely at entropy-driven reactions, but it is worth adding here that carbon and hydrogen reductions are always run at high temperatures, both to maximize the thermodynamic benefit of the positive entropy change, and to provide thermal activation energy to gain kinetic advantage of rapid reaction.

Most of the metals in the final category of Table 8.5 are soft Lewis acids occurring naturally as the sulfide ores. The process of roasting them to the molten metal and SO_2 relies on the fact that they are relatively electronegative and that the soft base S^{2-} is readily oxidized to the fairly hard acid S^{4+} in SO_2. This oxidation takes all the oxygen from a limited supply, leaving the metal rather than the metal oxide. These reactions are usually energetically favorable because of the strong $S{=}O$ bonds (531 kJ/mol bond) in SO_2. An interesting variant is the Mathieu process for producing cesium (shown in Table 8.5), in which CsCl is vaporized in a vacuum chamber and passed over very hot CaC_2. The acetylide ion reduces the cesium, forming molten calcium chloride, and cesium metal distills away.

8.10 REDOX REACTIONS IN THE GAS PHASE AND IN THE ATMOSPHERE

Thus far we have confined our attention almost entirely to redox reactions occurring in liquid media. In the laboratory, redox reactions rarely occur in any other way. However, gas-phase redox reactions are extremely important, both industrially and environmentally. We can begin with direct gas-phase redox reactions and move on to fuel cells as electrochemical current sources.

Perhaps the most famous gas equilibrium is that of the Haber ammonia synthesis, $N_2 + 3H_2 = 2NH_3$, which is clearly a redox process. This is essentially the only commercial nitrogen fixation process in use in the world, but it is curiously inefficient thermodynamically as a source of nitrogen for plants. The reaction as it is written is exothermic, but the H_2 must be produced somehow. Furthermore, although plants use nitrogen in the amine form in plant protein, they must absorb it from the soil as NO_3^- and photoreduce it in the cells to the amine form. Consequently, soil bacteria must oxidize NH_3 fertilizer to NO_3^- before it can be absorbed. It would obviously be advantageous to oxidize the N_2 in the air rather than reduce it. Thermodynamically, this is quite feasible, as Table 8.6 indicates.

Unfortunately, there are no kinetically feasible nitrogen-oxidation reactions. Direct oxidation to HNO_3, which would yield energy, has never been demonstrated, and the reactions producing NO and NO_2 are feasible only at such high temperatures that they are uneconomic. The Haber equilibrium (reaction 4 in Table 8.6) is slightly exothermic, but if the H_2 must be obtained from water (reaction 5), the ammonia is obtained only at a very large energy cost. Early Haber plants produced H_2 by passing steam over hot coke (reaction 6), but the reaction is so endothermic that much of the coke must be burned to keep the temperature high enough for the reaction with water. Additional H_2 is obtained from reaction 7, the *water–gas shift* reaction, which can be catalyzed at moderate temperatures. Reactions 6 and 7 have become more important since high gas and oil prices have made the conversion of coal to syngas or syncrude attractive, but few ammonia plants use coke as an energy source for H_2 generation any more. Instead, methane

Table 8.6 Thermodynamics of Nitrogen-Fixation Reactions

	ΔH^0_{298} (kJ/mol rn)	ΔS^0_{298} (J/deg mol rn)
Oxidation		
1. $\frac{1}{2}N_2 + \frac{1}{2}H_2O + \frac{5}{4}O_2 \rightarrow HNO_3$	−30	−291
2. $\frac{1}{2}N_2 + O_2 \rightleftharpoons NO_2$	+33	−60
3. $\frac{1}{2}N_2 + \frac{1}{2}O_2 \rightleftharpoons NO$	+90	+12
Reduction		
4. $\frac{1}{2}N_2 + \frac{3}{2}H_2 \rightleftharpoons NH_3$	−46	−99
5. $\frac{1}{2}N_2 + \frac{3}{2}H_2O \rightarrow NH_3 + \frac{3}{4}O_2$	+383	−33
6. $C + H_2O \rightarrow CO + H_2$	+175	+134
7. $CO + H_2O \rightleftharpoons CO_2 + H_2$	−41	−42
8. $CH_4 + H_2O \rightleftharpoons CO + 3H_2$	+206	+215
Overall		
9. $(\frac{1}{2}N_2 + \frac{1}{8}O_2) + \frac{3}{2}H_2O + \frac{7}{8}C \rightarrow NH_3 + \frac{7}{8}CO_2$	+38	
10. $(\frac{1}{2}N_2 + \frac{1}{8}O_2) + \frac{5}{8}H_2O + \frac{7}{16}CH_4 \rightarrow NH_3 + \frac{7}{16}CO_2$	−7	

from natural gas (reaction 8) or naphtha (liquid hydrocarbons around C_7) are used, even though about as much hydrocarbon must be burned to provide energy as is converted to H_2. Reactions 9 and 10 of Table 8.6 represent somewhat idealized stoichiometries for the overall ammonia synthesis, using air as the N_2 source and coke or methane as the energy source. Note that the coke reaction is still endothermic and requires the burning of additional coke. In fact, both processes require substantial additional fuel combustion because of the inevitable heat losses and because the gases must be compressed to quite high pressures to give favorable conversion in the final N_2—H_2—NH_3 equilibrium (reaction 4).

The equilibrium of reaction 4 is favored thermodynamically by low temperatures because of its negative entropy change, but even the effective catalysts that have been developed (iron metal on the surface of potassium aluminosilicates) do not function well below about 400°C. As Table 8.7 suggests, this drastically limits the yield of NH_3 possible in a single pass of the reactants over the catalyst. High pressures favor the NH_3 product, but there are mechanical limitations on attainable pressures. Newer ammonia plants use relatively low-cost rotary compressors that yield only about 200 atm pressure (still 3000 psi!). The yield is thus usually only about 15–20% ammonia, but the product can be separated as liquid ammonia by chilling and the reactants can then be recirculated.

Table 8.7 Equilibrium NH_3 Percentages in the Haber Process

T (°C)	Pressure (atm)				
	25	50	100	200	400
100	91.7	94.5	96.7	98.4	99.4
200	63.6	73.5	82.0	89.0	94.6
300	27.4	39.6	53.1	66.7	79.7
400	8.7	15.4	25.4	38.8	55.4
500	2.9	5.6	10.5	18.3	31.9

The perception of long-range exhaustion of reserves of crude oil and natural gas has revived interest in coal gasification and liquefaction, since the U.S. has enormous coal reserves. Most of the available gasification processes rely on reactions 6–8 of Table 8.6. If the CO—H_2 product from reaction 6 is passed over Raney nickel catalyst, reaction 8 is driven strongly to the left in what is called a *methanation* reaction. If the water–gas shift reaction is then applied to the CH_4—CO—H_2O mixture and the CO_2 product is stripped out, the result is a high-quality synthetic pipeline gas. This is essentially the Bureau of Mines *Synthane* process, but other processes also rely on the water–gas shift and methanation.

Much of the ammonia produced in commercial Haber-process plants is oxidized to nitric acid because of the advantages of the nitrate ion as a plant nutrient. The reaction sequence is

$$NH_3 + \tfrac{5}{4}O_2 \xrightarrow{Pt} NO + \tfrac{3}{2}H_2O \qquad \Delta H = -292 \text{ kJ/mol rn}$$

$$NO + \tfrac{1}{2}O_2 \rightleftharpoons NO_2 \qquad\qquad\qquad -57 \text{ kJ/mol rn}$$

$$3NO_2 + H_2O \rightleftharpoons 2HNO_3 + NO \qquad -70 \text{ kJ/mol rn}$$

or, overall

$$NH_3 + 2O_2 \rightarrow HNO_3 + H_2O \qquad \Delta H = -413 \text{ kJ/mol rn}$$

Unfortunately, this very large overall energy release cannot be used as the energy input for the ammonia synthesis, because the platinum-gauze catalysts cannot withstand extremely high temperatures. The heat must therefore be released at an uneconomically low temperature. There are interesting kinetic constraints on the process at each stage. In the ammonia-oxidation step, although platinum gauze is a highly specific catalyst for the NO product, almost any solid surface will catalyze the oxidation of NH_3 to N_2 and H_2O. Consequently, the container walls must be kept cool to prevent loss of ammonia. Furthermore, there is a facile reaction between NH_3 and NO:

$$4NH_3 + 6NO \rightarrow 5N_2 + 6H_2O$$

so that the gas must remain in contact with the catalyst surface long enough to remove all NH_3 but not so long that the NO can diffuse back into the NH_3 stream.

In the second step, the oxidation of NO to NO_2, both thermodynamic and kinetic limitations apply to the reaction. Although the reaction is exothermic, it has quite an unfavorable entropy change (−73 J/deg mol rn). It thus has only a small favorable free-energy change and a correspondingly modest equilibrium constant, which because of the negative ΔS becomes even less favorable at elevated temperatures. The reaction must be run near room temperature, which imposes cooling problems. Kinetically, the reaction is slow at these temperatures. It is also first-order in P_{O_2}, so that high pressures and pure oxygen are desirable. However, the NH_3 oxidation uses air so that the associated N_2 will limit NH_3 oxidation to N_2, and this N_2 diluent slows the NO oxidation rate.

Finally, in the third step, the absorption of NO_2 into water and the disproportionation are both favored by low temperatures, so that the system must actually be refrigerated. The result of all these conditions is that the overall oxidation of ammonia to nitric acid requires energy expenditure even though it is strongly exothermic, and the heat produced is essentially used only to warm the surroundings.

Nitrogen-based redox reactions are also extremely important in the chemistry of air pollution. Whenever carbon-based fuels are burned in air, the resulting high flame temperatures make it both thermodynamically and kinetically possible to partially oxidize the air's N_2 to NO if excess oxygen is present, which it normally is. Reaction 3 of Table 8.6 begins to occur appreciably at temperatures above about 1200°C. Even in mixtures containing only a few per cent O_2, temperatures near 2000°C yield as much as 0.5% NO in only a few tenths of a second. As the mixtures cools (in automobile exhaust or power plant stack gases), further oxidation of NO to NO_2 occurs. The resulting mixture is the NO_x pollutant category described in Chapter 5. However, the reaction $2NO + O_2 = 2NO_2$ proceeds via a preliminary dimerization of NO to N_2O_2. This causes the rate constant of the NO → NO_2 oxidation to decrease at higher temperatures, since the dimerization is less favored. Cooling speeds the reaction up, but the dimerization also makes the reaction second-order in NO, which lowers the overall rate as the gas mixture is diluted. Since even dangerous concentrations of NO_x are quite dilute in absolute terms (on the order of ppm), little net oxidation occurs by thermal reaction. However, Chapter 5 described the photochemical processes that lead to NO oxidation in sunlight in the presence of hydrocarbons. The overall result is that the concentrations of NO_2 and O_3 in the atmosphere are increased.

Since the redox chemistry of gaseous nitrogen species is responsible for NO_x pollution, it is interesting that similar processes can be used to reduce it. In the Exxon DeNO$_x$ process, for example, ammonia is injected into stack gases in slight excess over the NO present. Since some 2–3% O_2 is also present, some of the ammonia is oxidized to NO by the same reaction as that used in the HNO_3 synthesis; but a competing reaction eliminates NO:

$$NO + NH_3 + \tfrac{1}{4}O_2 \rightarrow N_2 + \tfrac{3}{2}H_2O$$

The fractional coefficients should suggest that a complex mechanism governs this reaction. Although the thermal DeNO$_x$ process has been studied extensively, its mechanism is still controversial. It appears likely that the fundamental elementary process is the reaction of NH_2 with NO, which can have several products: N_2 + H_2O, H_2 + NNO, or NNH + OH. Very roughly half the reacting materials yield the desired N_2 + H_2O, possibly by the rearrangement below, which is unusually complex for a single elementary process:

$$NH_2 + NO \rightarrow H_2NNO \rightarrow \textit{trans } HNN-OH \rightarrow \textit{cis } HNN-OH \rightarrow N_2 + H_2O$$

Because of the competing reactions and mechanisms, the thermal behavior of the process is critical; at temperatures above about 950°C, the oxidation of ammonia is too rapid, and NO removal is inefficient. Below about 850°C the reaction kinetics prevent the ammonia from reacting at all. But at about 920°C, if the relative concentrations of NO and NH_3 are carefully controlled, about 98% of the NO can be eliminated without leaving significant quantities of ammonia in the stack gas. To essentially eliminate NO_x emissions from power-plant stacks, it is thus only necessary to inject NH_3 at the point where the gases have cooled to the correct temperature.

An interesting alternative NO removal process is known as the RapReNO$_x$ process, which produces a mixture of N_2O (a less objectionable product) and N_2. The key reactant is HNCO, which is produced by heating cyanuric acid (1, 3, 5-triazine-2, 4, 6-triol) above 300°C; at slightly higher temperatures, HNCO yields CO and the NH radical:

$$HNCO \rightarrow NH + CO$$
$$NH + NO \rightarrow H + NNO$$
$$H + HNCO \rightarrow NH_2 + CO$$
$$NH_2 + NO \rightarrow N_2 + H_2O$$
$$NH_2 + NO \rightarrow NNH + OH$$
$$NNH \rightarrow N_2 + H$$
$$OH + CO \rightarrow H + CO_2$$

A test of the process on a small diesel engine yielded 99% NO removal, predominantly to N_2. Efforts are under way to commercialize the process for varied combustion applications.

Another gas-phase redox equilibrium of great industrial importance is the catalytic oxidation of SO_2 to SO_3:

$$SO_2 + \tfrac{1}{2}O_2 = SO_3 \qquad \Delta H^0 = -98.3 \text{ kJ/mol rn}, \qquad \Delta S^0 = -94.8 \text{ J/deg mol rn}$$

The SO_3, of course, is subsequently converted to H_2SO_4 by dissolving it in H_2SO_4 solvent and reacting the $H_2S_2O_7$ product with water—at a rate of roughly 40 million tons a year in the United States alone. Most of the SO_2 is obtained by burning molten sulfur. Initially it seems curious that this combustion does not proceed all the way to SO_3. However, the formation of SO_2 from liquid sulfur is overwhelmingly exothermic and involves essentially zero entropy change, so that the equilibrium constant is very large and is not strongly affected by temperature changes. On the other hand, the large negative ΔS^0 for the further oxidation to SO_3 and the relatively modest ΔH^0 make the equilibrium favorable at room temperature but much less so at elevated temperatures. The K_p value shrinks to 1 at about 790°C, and is much smaller at typical flame temperatures. The kinetics of the reaction, unfortunately, are such that it does not proceed at any appreciable rate at temperatures low enough to have a favorable K_p. The SO_2—SO_3 oxidation is catalyzed, however, by V_2O_5 at temperatures above about 450°C, where the equilibrium constant is still about 10^2. However, the heat release raises the gas temperature and represses the oxidation, so it is necessary to catalyze the oxidation in several stages, cooling the gas mixture between passes to about 430°C to balance thermodynamic yield against catalytic efficiency. After three passes the yield is up to about 96%, but in view of the high tonnage and the severe air-pollution qualities of SO_2, the remaining 4% cannot be emitted to the atmosphere. In current practice, the SO_3 is absorbed in H_2SO_4 after the third pass, the remaining SO_2 is reheated, and the SO_2 is oxidized by one more pass over the catalyst. This gives an overall yield of about 99.7%, which satisfies air-pollution standards.

Still another important gas-phase redox process is that occurring in commercial fuel cells. Fuel cells are attractive, of course, because the absence of a mechanical working fluid means that the ordinary Carnot thermodynamic limitations on energy-conversion efficiency do not apply to a fuel cell. Although extensive efforts have been made to run fuel cells on methane/air, existing systems use steam reforming to convert the methane to hydrogen gas, so that the basic redox reaction is the oxidation of H_2 gas: $2H_2 + O_2 \rightarrow 2H_2O$. Although a number of experimental systems are under study, the two designs that have received greatest use (for example, in space probes) are the phosphoric acid fuel cell and the molten carbonate fuel cell.

The phosphoric acid fuel cell runs at approximately 200°C, with electrode half-reactions

cathode: $\frac{1}{2}O_2 + 2H^+ + 2e^- \rightarrow H_2O$
anode: $H_2 \rightarrow 2H^+ + 2e^-$

The role of the phosphoric acid electrolyte is to transfer protons from the anode to the cathode; the phosphoric acid is held in a silicon carbide matrix. Carbon electrodes are used, but they must be platinized to prevent kinetic problems due to overvoltage. Because fuel cells are expected to provide power for extended periods of time, the slow poisoning of the Pt catalyst sites by traces of either CO or S from the fuel gas constitutes a problem for cell design. However, cells up to 26 MW have been built and successfully operated for up to four years. Although the thermodynamic efficiency is high and maintenance is low, the cells are still expensive to construct and lose some efficiency in the dc-to-ac conversion that is necessary in large applications.

The molten-carbonate fuel cell relies on a different set of electrode half-reactions:

cathode: $CO_2 + \frac{1}{2}O_2 + 2e^- \rightarrow CO_3^{2-}$
anode: $H_2 + CO_3^{2-} \rightarrow CO_2 + H_2O + 2e^-$

Here the role of the electrolyte is to transport carbonate ions from the cathode to the anode. The electrolyte melt is suspended in a porous tile of sintered lithium–aluminate/aluminum–oxide fibers resembling the Nextel space shuttle tiles described in Chapter 3. Engineering design problems are in some ways more severe for the molten-carbonate fuel cell, but there are advantages. The high temperature (about 600°C) needed to keep the Li_2CO_3/K_2CO_3 electrolyte molten means that catalytic electrode surfaces are not necessary; nickel metal electrodes are used. Remembering the casual version of the Second Law, "Heat's no good unless it's hot," we can also see the advantage that waste heat can be more readily used in co-generation than the heat from a phosphoric acid cell. However, the cell design must recycle CO_2 gas, which complicates the design and leads to loss of electrolyte in even the best designs, and electrode surface poisoning by H_2S still leads to loss of efficiency and cell lifetime. The nickel cathode also suffers corrosion through formation of nickel oxide, which is soluble in the molten carbonate electrolyte. A possible remedy is the substitution of lanthanum perovskites as cathode materials, although the cell's internal resistance is raised.

Other fuel cell designs continue to be proposed and developed, including an alkaline cell with KOH/water electrolyte, a superacid fuel cell with $HOSO_2CF_3$ electrolyte, a solid polymer cell with perfluorosulfonic acid copolymer (Nafion), and the high-temperature solid oxide cell with stabilized zirconia electrolyte that we saw in Chapter 3. Fuel cell research and development appears to have been characterized by careful improvement rather than dramatic breakthroughs, but we may be near economically competitive, commercially successful systems.

PROBLEMS

A. DESCRIPTIVE

A1. The Latimer diagram for acidic plutonium solutions is given below. How many species are unstable toward disproportionation? Which species should react with water?

$$PuO_2^{2+} \xrightarrow{0.91} PuO_2^{+} \xrightarrow{1.17} Pu^{4+} \xrightarrow{0.98} Pu^{3+} \xrightarrow{-2.03} Pu^{0}$$
$$\underset{1.04}{\rule{3cm}{0.4pt}}$$

A2. What reaction do you expect in an acidic aqueous medium when the following reactions are carried out?

(a) BrO_3^- is added dropwise to a hydrazine solution.

(b) I_2 is added dropwise to a thiosulfate solution.

(c) Br_2 is added dropwise to a thiosulfate solution.

(d) Solutions of Fe^{2+} and H_2O_2 are mixed.

(e) Phosphorous acid is slowly added to a solution of $CuSO_4$.

(f) VO_2^+ is added to a Sn^{2+} solution.

(g) Elemental selenium is stirred into a Mn^{2+} solution.

(h) Chlorine is bubbled through a solution of $SbCl_3$.

(i) SO_2 is bubbled through Mn^{3+}.

(j) NO_2 is bubbled through Ti^{3+}.

A3. What reaction do you expect in a basic aqueous solution when the following reactions are carried out?

(a) ReO_2 is added to a chromate solution.

(b) Potassium superoxide is added to a bromate solution.

(c) Solutions of nitrite and hypochlorite are mixed.

(d) A stannite solution $(Sn(OH)_3)^-)$ is added slowly to a hypophosphite solution.

(e) SO_2 is bubbled through an arsenite solution.

(f) O_2 is bubbled through a suspension of finely divided selenium.

(g) Nickel hydroxide is treated with hypochlorite.

(h) An excess of a chlorine solution is added to an iodine solution.

(i) NO is bubbled through a phosphite solution.

(j) $Cr(OH)_2$ is stirred into an arsenite solution.

A4. If mercury metal is added to a solution 1 M in both HgI_4^{2-} and I^-, what should the equilibrium composition of the solution be?

A5. Why is $AuCl_4^-$ a stronger oxidant than $AuBr_4^-$?

A6. Why is Cu^+ a much stronger oxidant than Pb^{2+} in water, but substantially weaker in acetonitrile?

A7. Under the appropriate circumstances, the following reactions can both be made to occur *quantitatively* in wet analytical chemistry:

$$As_2O_3 + I_2 \rightarrow H_3AsO_4 + I^-$$

$$H_3AsO_4 + I^- \rightarrow As_2O_3 + I_2$$

Balance each skeleton equation, explain how it is possible to obtain a quantitative reaction in each case, and indicate what solution conditions are necessary.

A8. Suggest an appropriate solvent for each of the following reactions:

(a) $2S_4N_4 + S_8 \rightarrow 4S_4N_2$

(b) $IO_2F + AsF_5 \rightarrow IO_2^+AsF_6^-$

(c) $CuCl_2 + ClONO_2 \rightarrow Cu(NO_3)_2 + 2Cl_2$

(d) $K_2PdCl_4 + 2BrF_3 \rightarrow K_2PdF_6 + Br_2 + 2Cl_2$

(e) $2Br_2 + 3AgNO_3 \rightarrow Br(NO_3)_3 + 3AgBr$

(f) $H_3NBF_3 + K \rightarrow H_2NBF_2 + KF + \frac{1}{2}H_2$

(g) $24Cl^- + 10CrO_3 + 3I_2 \rightarrow 6IO_3^- + 4CrCl_6^{3-} + 6CrO_2^-$

(h) $LiAlH_4 + V(bipyridyl)_3 \rightarrow V(bipyridyl)_3^- + \frac{1}{2}H_2 + AlH_3 + Li^+$

A9. What product(s) would you expect from the reaction of K and Bi in liquid ammonia?

A10. Suggest an approximate temperature limit at which the direct oxidation of N_2 to HNO_3 would be feasible thermodynamically if a suitable catalyst were available.

B. NUMERICAL

B1. Calculate the standard potential for the reduction of sulfate ion to H_2S in acid solution.

B2. Construct a Born–Haber cycle to estimate ΔH^0 at 298 K for the high-temperature commercial reaction that produces rubidium metal:

$$2RbF + Ca \rightarrow 2Rb + CaF_2$$

At 298 K, all species are solids. How does the result compare with a prediction based on the relative electronegativities of the two metals? What is the largest single driving force in your cycle? What effect should very high temperatures have on ΔH? On ΔG?

B3. Calculate ΔG^0 and K_{eq} for the reaction of copper(II) and copper metal in the presence of bromide ion to make CuBr.

B4. The Latimer diagram for acidic phosphorous species indicates that H_3PO_3 should *not* disproportionate to H_3PO_4 and PH_3—but the potentials are very close. Calculate the equilibrium concentrations of all three species in a solution initially 1 M in H_3PO_3. (A good bit of disproportionation *does* occur.)

B5. The molten-salt electrolyte LiCl–KCl is sometimes used as an inert solvent for high-energy batteries. One such combination involves the electroactive species Mg^0 and NiO:

$$Mg^{2+} + 2e^- \rightarrow Mg^0 \qquad \mathscr{E} = -3.58 \text{ V}$$
$$NiO + 2e^- \rightarrow Ni^0 + O^{2-} \qquad \mathscr{E} = -1.23 \text{ V}$$

Calculate the energy density of such a cell (*a*) considering only the weight of the electroactive species; (*b*) including 50 g electrolyte per mole of electroactive material; and (*c*) including the electrolyte and 3 kg of container and heater elements for 50 moles of electroactive material.

B6. Manganate(VI) disproportionates in both acidic and basic solution, but much less strongly in base. Calculate the pH at which a solution 1 M in both MnO_4^- and MnO_4^{2-} (over MnO_2) would be stable.

B7. Use a Born–Haber cycle and the Kapustinskii lattice-energy treatment to calculate an approximate ΔH^0 for the reaction

$$MCl_2(s) + \tfrac{1}{2}Cl_2(g) \rightarrow MCl_3(s)$$

where M is (*a*) Mg, (*b*) Al, and (*c*) Fe. What do the calculated ΔH values indicate about the relative stabilities of MCl_2 and MCl_3? Repeat the calculation for FeI_2 being oxidized by gaseous I_2 to FeI_3. What is the effect of anion radius on oxidation-state stability?

C. EXTENDED REFERENCE

C1. Construct the Latimer diagram for sodium in its 2,2,2-crypt:

$$\text{Na}^+(\text{crypt}) \xrightarrow{\;\mathscr{E}_1\;} \text{Na}^0(\text{s}) \xrightarrow{\;\mathscr{E}_2\;} \text{Na}^-(\text{s})$$

For details on the structure of $\text{Na}(\text{crypt})^+\text{Na}^-$, see Chapter 7 and *J. Am. Chem. Soc.* (1974), *96*, 7203. Use a Born–Haber cycle for the disproportionation of Na^0 to $\text{Na}(\text{crypt})^+\text{Na}^-$ with a reasonable approximation for the energy of encrypting Na^+. Get ΔG^0 by approximation from ΔH^0 for the disproportionation. Choose a reasonable medium for the reference $\mathscr{E}_1$, and solve for $\mathscr{E}_2$.

C2. If the ammoniated electron is considered a quantum mechanical particle in a box, its energy levels are given by

$$E = (n_x^2 + n_y^2 + n_z^2)h^2/8mL^2$$

where m is the electron mass and L is the length of one side of a cubic box (roughly twice the radius of the cavity). Estimate the cavity radius from the observed spectrum of $e^-(\text{NH}_3)$. Use an introductory quantum mechanics text to decide what the effect on energy-level spacing will be if the "box" (infinite potential) is replaced by a finite potential well. Does this change of the model move your predicted cavity radius in the right direction?

C3. In the Latimer diagram for nitrogen, the great stability of N_2 causes it to thermodynamically dominate the redox products of nitrogen species. However, kinetic mechanisms frequently bypass N_2, as in the reaction between nitrous acid and the bisulfite ion in acid medium:

$$\text{HNO}_2 + 2\text{HSO}_3^- + \text{H}_2\text{O} \rightarrow \text{HONH}_2 + 2\text{HSO}_4^-$$

Propose a mechanism for this reaction consistent with the fact that the rate of the reaction is first-order in $[\text{H}_3\text{O}^+]$, and that two intermediates can be isolated in the order given:

$$\text{O}-\text{N} \Big\langle {}^{\displaystyle \text{SO}_2\text{OH}}_{\displaystyle \text{SO}_2\text{OH}} \qquad\qquad {}^{\displaystyle \text{H}}_{\displaystyle \text{HO}}\Big\rangle \overset{+}{\text{N}} \Big\langle {}^{\displaystyle \text{H}}_{\displaystyle \text{SO}_3^-}$$

[See G. Stedman, *Adv. Inorg. Chem. Radiochem.* (1979), *22*, 113.]

C4. The oxidation of N_2H_4 by two-electron oxidants usually produces only N_2, whereas one-electron oxidants produce a mixture of N_2 and NH_3. Suggest mechanisms to account for this behavior using the two-electron oxidant T and the one-electron oxidant $O\cdot$. [See G. Stedman, *Adv. Inorg. Chem. Radiochem.* (1979), *22*, 113.]

9

Entropy-Driven Reactions ────────

All of the chemical reactions observed to occur are, by definition, spontaneous. One cannot force a reaction to occur; one can only arrange conditions so that the desired reaction is the most favorable, most probable result. Even when reaction conditions are thus optimized, the reaction may not occur for either thermodynamic or mechanistic reasons. The thermodynamic index of spontaneity is the entropy of the universe, of course. If $\Delta S_{univ} > 0$ for a given chemical process, it is thermodynamically possible, but not otherwise. For processes occurring at constant temperature and constant pressure, the entropy change of the universe is more conveniently expressed using the Gibbs free-energy function,

$$0 > \Delta G = \Delta H - T\,\Delta S$$

where all of the properties now refer to the system alone. The acid–base and redox reactions that we have considered in Chapters 7 and 8 are, with few exceptions, enthalpy-driven—that is, their associated free-energy changes are negative primarily because their enthalpy changes are negative, regardless of the contribution made by the entropy change. However, there are several kinds of reactions that are essentially entropy driven. Their free-energy changes are negative primarily because their entropy changes are positive and large enough either to make a major contribution along with a favorable enthalpy change, or to overcome an unfavorable enthalpy change. Such reactions are the subject of this chapter.

──────── ## 9.1 THE ROLE OF ENTROPY CHANGES IN FREE ENERGY

For a very large number of reactions, the enthalpy change and the entropy change oppose each other in the ΔG expression. For example, a spontaneous reaction might evolve heat (ΔH negative) because an extremely stable ionic lattice is formed from, say, a gas and a liquid that are less strongly bonded. Such a reaction, however, has a negative ΔS because the product's orderly lattice has a much smaller degree of randomness than the disordered reactants. The $-T\Delta S$ quantity in ΔG, then, is positive, and the entropy change is working *against* the spontaneity of the reaction. On the other hand, a gas might be driven off by heating a solid reactant: $2CsN_3 \rightarrow 2Cs + 3N_2$. Such a reaction has a positive ΔH (heat is flowing into the endothermic process) but also a positive ΔS, because the highly random gas is being formed from an ordered solid. This reaction is entropy-driven, because the enthalpy change is working against the spontaneity of the reaction; an input of energy is required to break the highly stable lattice. For both kinds of reaction, the essential consideration is that orderly systems maximize atom–atom contacts and bonding possibilities. They thus have low entropy, but tend to release energy when formed. When a disorderly system becomes orderly (ΔS negative), it tends to form additional bonds and release the bond energy (ΔH negative), and vice versa. The exceptions to the rule that ΔS and ΔH have the same sign form an interesting class; we shall return to these later in the chapter.

It is intrinsic to the strengths of chemical bonds in general and to the patterns of order–disorder for molecules and lattices that most chemical reactions we observe have ΔH larger than $T\Delta S$ at ordinary laboratory temperatures. They are thus enthalpy-driven and exothermic. Very, very roughly, ΔH values often are on the order of 50 kJ/mol reaction, and ΔS values are on the order of 50 J/deg mol reaction. At 300 K, near room temperature, $T\Delta S$ is thus on the order of 15 kJ/mol reaction, significantly smaller than ΔH. Therefore, most spontaneous reactions near room temperature are exothermic. Clearly, the situation can change dramatically as the temperature of the reaction changes.

Near room temperature we are justified in using standard-state enthalpies and entropies, because standard state is quite near that temperature. However, what happens to ΔH and ΔS values when the temperature is raised to say, 1000 K? There are two important effects. First, phase changes often occur—solids melt or sublime, liquids vaporize. In that case, ΔH for the reaction must be corrected for the enthalpies of each phase change:

$$\Delta H_T^0 = \Delta H_{298}^0 + \sum_{\text{prod}} \Delta H_{\text{phase}} - \sum_{\text{react}} \Delta H_{\text{phase}}$$

Similarly, when phase changes occur, ΔS changes for each reaction species. Since ΔS for a phase transition on heating is given by $\Delta H_{\text{phase}}/T_{\text{transition}}$, we can write

$$\Delta S_T^0 = \Delta S_{298}^0 + \sum_{\text{prod}} \frac{\Delta H_{\text{phase}}}{T} - \sum_{\text{react}} \frac{\Delta H_{\text{phase}}}{T}$$

where it is understood that T refers to the appropriate temperature for each individual phase transition. These changes can be quite large. In estimating free-energy changes at high temperatures, it is essential to use the correct phases. Consider the production of cesium metal by the Mathieu process, for which the reaction equation is

$$2\,CsCl + CaC_2 \rightarrow CaCl_2 + 2C + 2Cs$$

At 298 K, all species in the equation are solids, and $\Delta G^0_{298} = +124$ kJ/mol reaction (substantially unfavorable). This value has components $\Delta H^0 = +134$ kJ/mol reaction and $\Delta S^0 = +33$ kJ/mol reaction. However, the reaction is actually run at about 1600 K, at which temperature CsCl and Cs are gases and $CaCl_2$ is a liquid. Under these circumstances, using experimental heats of fusion and vaporization we have

$$\Delta H^0_{1600} = +134 + (\Delta H_{fus})_{CaCl_2} + 2(\Delta H_{fus} + \Delta H_{vap})_{Cs} - 2(\Delta H_{fus} + \Delta H_{vap})_{CsCl}$$

$$= +134 + 26 + 267 - 329 = -2 \text{ kJ/mol reaction}$$

$$\Delta S^0_{1600} = +8.0 + \left(\frac{\Delta H_{fus}}{T_{mp}}\right)_{CaCl_2} + 2\left(\frac{\Delta H_{fus}}{T_{mp}} + \frac{\Delta H_{vap}}{T_{bp}}\right)_{Cs} - 2\left(\frac{\Delta H_{fus}}{T_{mp}} + \frac{\Delta H_{vap}}{T_{bp}}\right)_{CsCl}$$

$$= +33 + 24 + 180 - 223 = +14 \text{ J/K mol reaction}$$

The high-temperature ΔH value is markedly different from the value at 298 K. Although the ΔS value has changed little, it is clear that the phase changes involve larger entropy changes than the original reaction did. From the high-temperature ΔH and ΔS values we calculate:

$$\Delta G^0_{1600} = \Delta H^0_{1600} - 1600 \cdot \Delta S^0_{1600}$$

$$= -2000 - 22,400 = -24,400 \text{ J} \qquad \text{or} \quad -24 \text{ kJ/mol reaction}$$

This, although modest, is clearly a favorable value for the reaction.

The second consideration in adjusting thermodynamic quantities for temperature is the change due to heat absorption by a given phase as the temperature is increased. Kirchhoff's law governs ΔH, and a comparable expression applies to ΔS:

$$\Delta H^0_T = \Delta H^0_{298} + \int_{298}^T C_P(T)\, dT$$

$$\Delta S^0_T = \Delta S^0_{298} + \int_{298}^T \frac{C_P(T)}{T}\, dT$$

If no phase changes occur, these effects are usually fairly small. For example, if we oxidize lead metal at room temperature,

$$2Pb + O_2 \rightarrow 2PbO$$

$$\Delta H^0 = -435 \text{ kJ/mol reaction}$$

$$\Delta S^0 = -197 \text{ J/K mol reaction}$$

The *change* in ΔH when the reaction is run at 600 K (just below the melting point of lead) amounts to only 2.8 kJ. Thus, $\Delta H^0_{600} = -432$ kJ/mol reaction. For ΔS, the change is +6.4 J/K so that $\Delta S^0_{600} = -191$ J/K mol reaction. Because the phase-change corrections to ΔH and ΔS are fairly large and the heat-capacity corrections are small, we shall in general allow for the first but ignore the second in approximate thermodynamic calculations. That is, we shall do calculations for the correct phases, but otherwise assume that ΔH and ΔS are constant with temperature.

With all this in mind, the question is still before us: What kinds of reactions are entropy-driven? Since in the most general case ΔH outweighs $T \Delta S$ in the free-energy function, we must seek entropy-driven functions in two areas—reactions having little or no energy change (so that ΔH is small), and reactions occurring at high temperatures with an entropy increase (so that $T \Delta S$ is large).

Included in the category of reactions with small ΔH are many solubility relationships, the related area of metal separation and purification through liquid–liquid extraction, and the formation and dissociation of nonstoichiometric *clathrate* compounds. A clathrate compound is a crystalline solid with lattice cavities in which guest molecules are held by polarizability forces. We shall briefly consider all three of these areas. Included in the category of high-temperature reactions are a number of industrially important processes that usually involve the endothermic formation of a gas, such as the pyrolysis of limestone ($CaCO_3$) to quicklime (CaO), and several forms of metal smelting. In an academic context, some very interesting chemistry has been done with unusual atomic or molecular-fragment species produced at high temperatures and condensed by rapid chilling. The resulting chemistry follows from the initial entropy-driven production of the fragments.

9.2 LOW-ENTHALPY PROCESSES

Perhaps the most obvious low-enthalpy chemical process is the mixing of two mutually soluble liquids to form a solution. If the solution is ideal (which is never absolutely true), $\Delta H_{mix} = 0$ by definition, and the free energy of mixing is given simply by the entropy terms for the two components,

$$\Delta G_{mix} = X_1 RT \ln X_1 + X_2 RT \ln X_2$$

where ΔG_{mix} refers to the free-energy change on mixing a quantity of material totalling one mole. Because the mole-fraction quantities are less than 1, each term will always be negative, and ΔG_{mix} must always be negative. Figure 9.1 gives the concentration dependence of ΔG_{mix}.

It can be seen that the most stable composition corresponds to $X_1 = X_2 = 0.5$. However, this entropy-driven process is not accompanied by a very large free-energy change in absolute terms—roughly 500–1500 J/mol solution for ordinary concentrations. It does not require much ΔH to alter the situation substantially. Hildebrand has derived an expression for the mixing of two components in a nonideal solution,

$$\Delta G_{mix} = X_1\, RT \ln X_1 + X_2\, RT \ln X_2 + V(\delta_1 - \delta_2)^2 \phi_1 \phi_2$$

where the first two terms represent the entropy contribution and the third is the enthalpy contribution. In the enthalpy term, V is the volume of the solution (on the same molar basis as ΔG_{mix}), ϕ_1 and ϕ_2 are volume fractions of the individual

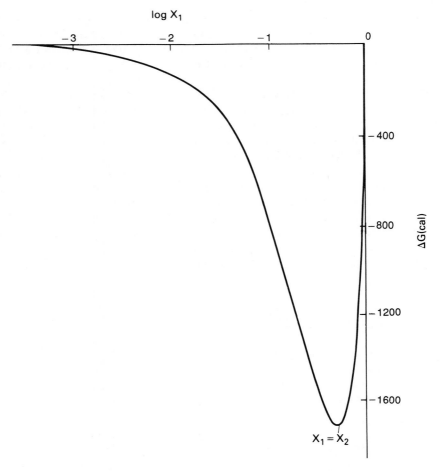

Figure 9.1   Free energy of mixing ideal solution of components 1 and 2 as a function of mole-fraction composition at 298 K.

components, and δ_1 and δ_2 are solubility parameters defined in terms of the *cohesive energy density* of each individual pure liquid, which is its internal energy of vaporization per milliliter of liquid:

$$\delta_1 \equiv (CED)^{\frac{1}{2}} = \left(\frac{\Delta E_{vap}(1)}{\text{molar volume}} \right)^{\frac{1}{2}} = \left(\frac{\Delta H_{vap}(1) - RT}{MW/\rho} \right)^{\frac{1}{2}}$$

The cohesive energy density (CED) is a measure of the energy required to separate the molecules in 1 mL of liquid. It is thus similar to the intermolecular-force parameters of Table 6.1. Table 9.1 gives CED values, and the corresponding solubility parameters δ, for a number of solvents. Most δ values fall in the range

$15-30$ $(J/mL)^{1/2}$ for common solvents. If we calculate the enthalpy of mixing for two solvents that have equal volume fractions, a total volume of 50 mL, and solubility parameters that differ by 12 units, the result is $\Delta H_{mix} = +1900$ J/mol solution, or more than enough to prevent such a solution from forming even with the most favorable entropy change in Fig. 9.1. This represents the net energy cost of making holes in one liquid to accommodate molecules of the other, but does not include specific attractions or chemical interaction between the two components, such as Lewis acid–base interaction.

For two liquids that have no such interaction, the balance between the entropy and enthalpy of mixing is often delicate, as the example above has suggested. Because absolute temperature appears in the entropy term but not in the enthalpy term, heating two partially miscible liquids always increases their mutual solubility. Similar arguments apply to the dissolution of nonpolar solids in solvents of low polarity, as for instance sulfur in CS_2.

An interesting parallel can be drawn for many ionic solids dissolving in water or in some other highly polar solvent. The arguments of Chapter 6 showed that for alkali halides lattice energies and solvation energies are nearly equal. NaCl has

Table 9.1   Cohesive Energy Density and Solubility Parameters for Common Solvents

Solvent	CED (J/mL)	δ $(J/mL)^{1/2}$
NaCl (molten)	4341	65.9
H_2O	2302	48.0
CH_3OH	874	29.6
NH_3	831	28.8
BrF_3	826	28.7
Propylene carbonate	763	27.6
Dimethylsulfoxide	705	26.6
C_2H_5OH	675	26.0
HCN	588	24.2
CH_3CN	582	24.1
$AlCl_3$	581	24.1
SO_2	511	22.6
CS_2	418	20.5
Acetone	395	19.9
$POCl_3$	341	18.5
CCl_4	308	17.5
Ethyl ether	251	15.8
HF	249	15.8
Hexane	219	14.8

an experimental heat of solution $\Delta H = +4.9$ kJ/mol, CsI $+34.5$ kJ/mol, and LiBr -47.1 kJ/mol; the others are generally smaller. The result is that the entropy change of solution can have a significant effect on solubility even though individual lattice energies and solvation energies are quite large. Chapter 6 developed these arguments; they need not be expanded here.

Less ionic systems, particularly metal salts that can exist as neutral molecules or ion pairs in aqueous solution, can in some cases be purified or separated from salts of close periodic-table neighbors by liquid–liquid extraction into an organic solvent that is immiscible with water, but is a good enough Lewis base to form metal-atom complexes more or less equal in strength to the hydrate complex in water. The distribution between the two solvents, then, will have only a small enthalpy change and can to a considerable extent be entropy-driven. An example of the industrial application of the solvent-extraction technique is the purification of uranium for nuclear fuel by extracting uranyl nitrate from nitric-acid solution into tributyl phosphate as the complex $UO_2(NO_3)_2(OP(OBu)_3)_2$. In this extraction, the oxygen atom in the O=P bond serves as the electron donor.

Suppose a metal M is to be extracted in a single pass from a heavy solvent H into a light solvent L (the algebra is the same for extraction in either direction). Let the volume of the heavy solvent be V, and the volume of the light solvent be $R \cdot V$, where R is the ratio of the two volumes used (see Fig. 9.2). If the original

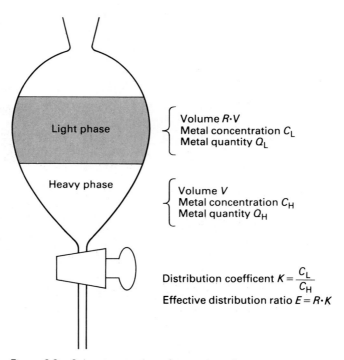

Figure 9.2 Solvent extraction of a metal species.

quantity of M is Q_O, then the quantities in each phase after equilibrium has been reached are Q_L and Q_H. Obviously, $Q_O = Q_L + Q_H$. If C represents a general concentration, $C_H = Q_H/V$ and $C_L = Q_L/RV$, or $Q_H = VC_H$ and $Q_L = RVC_L$.

Now consider the fraction of the metal that remains in the heavy phase after the extraction, F_H, defined as $F_H = Q_H/Q_O$. Substituting the concentration definitions and dividing out V, we have

$$F_H = \frac{Q_H}{Q_O} = \frac{Q_H}{Q_H + Q_L} = \frac{VC_H}{VC_H + RVC_L} = \frac{C_H}{C_H + RC_L}$$

The distribution coefficient K is the equilibrium constant for the transfer from the heavy solvent to the light solvent, $K = C_L/C_H$. Dividing the fraction for F_H by C_H and using the definition of K, we have

$$F_H = \frac{1}{1 + RK} = \frac{1}{1 + E}$$

where E is an effective distribution ratio proportional to both K and the volume ratio R. A similar manipulation of the fraction extracted, F_L, yields $F_L = E/(E + 1)$. Note that even where K is inconveniently small, effective extraction can sometimes be achieved by using an increased volume of the extracting solvent, since the fraction extracted depends on R as well as on K.

When the extraction is intended to separate two metals M and N, both will generally be extracted and the separation depends on the *ratio* of the distribution coefficients. If they are equal, no separation can be achieved. The ratio is called the *separation factor* β ($\beta = K_M/K_N$). Substituting β into the F_L expression for the extracted fraction reveals that efficient extraction requires not only a given value of β, but also that K_M and K_N be fairly near 1 (see study problem B1). This is equivalent to requiring a small value of ΔG for the extractions and thus requires a low-enthalpy process.

When separating intrinsically similar metals, such as uranium from the other actinide rare earths, niobium from tantalum, or any of the lanthanide rare earths from the naturally occurring mixture, it may be necessary to work with β values as low as 2. Under these circumstances, even ideal K values do not yield satisfactory separation. Multiple extractions (original H phase with new L solvent) will improve the yield of such a separation, and multiple scrubs of the extract (extracted L phase with new H solvent) will improve the purity of the metal with the larger K value. Countercurrent extractors have been devised to perform the multiple extraction and scrubbing operations continuously. Even at $\beta = 2$, better than 99% yield and 99% purity can be achieved with a 27-stage extractor if K_M and K_N have the optimum relation to each other ($K_M = 1/K_N$).

Another low-enthalpy process for which the entropy change is an important factor is the formation (and decomposition) of clathrates. In a clathrate, the crystal lattice of a host molecule accommodates isolated guest molecules in cavities produced symmetrically by the lattice geometry. The lattice symmetry guarantees a "stoichiometry" of cavities to host molecules, but the cavities are frequently only partially filled. Perhaps the best-known examples are the noble gas hydrates, which

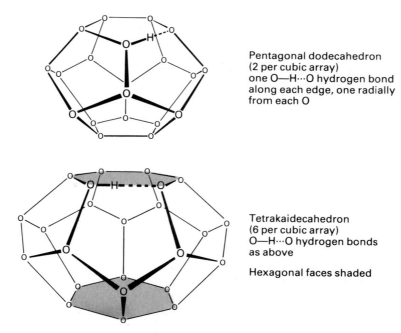

Pentagonal dodecahedron
(2 per cubic array)
one O—H⋯O hydrogen bond
along each edge, one radially
from each O

Tetrakaidecahedron
(6 per cubic array)
O—H⋯O hydrogen bonds
as above

Hexagonal faces shaded

Figure 9.3   Gas-hydrate crystal polyhedra.

have an idealized formula of $8G \cdot 46H_2O$ (G = Ar, Kr, Xe, Rn). In these, the water host-lattice is not that of ordinary ice but is a cubic assembly of dodecahedra and tetrakaidecahedra (see Fig. 9.3) with O atoms at the corners and edges formed by O—H⋯O hydrogen bonds. The dodecahedral geometry is not unique to a solid lattice; water vapor at high pressure forms clusters in which a dodecahedral $(H_2O)_{20}$ cluster hosts either a proton or an oxonium ion, yielding formulas $H^+(H_2O)_{20}$ or $H_3O^+(H_2O)_{20}$. The noble-gas atoms are held in these large polyhedra in clathrates by London dispersion forces, which is to say the mutual polarizability of the gas atom and the water molecules.

Small molecules such as Cl_2, SO_2, and CH_3Cl can also serve as guests in this water lattice, usually with a preferred orientation. In particular, small hydrocarbons form solid water clathrates at room temperature, which is to say well above the temperature at which water would be a liquid. Methane forms clathrate hydrates of this sort so well that if natural gas pipelines are not kept rigorously dry they can be blocked by the formation of solid methane hydrates. An interesting parallel to this observation is the realization that this stability, coupled with the enormous amount of natural gas and of water and ice under high pressure, means that much of the fossil fuel content of the earth's crust is in the form of natural gas clathrate hydrates. It has been estimated that the energy content of naturally occurring methane hydrates might be as much as twice that of all other fossil fuel reserves.

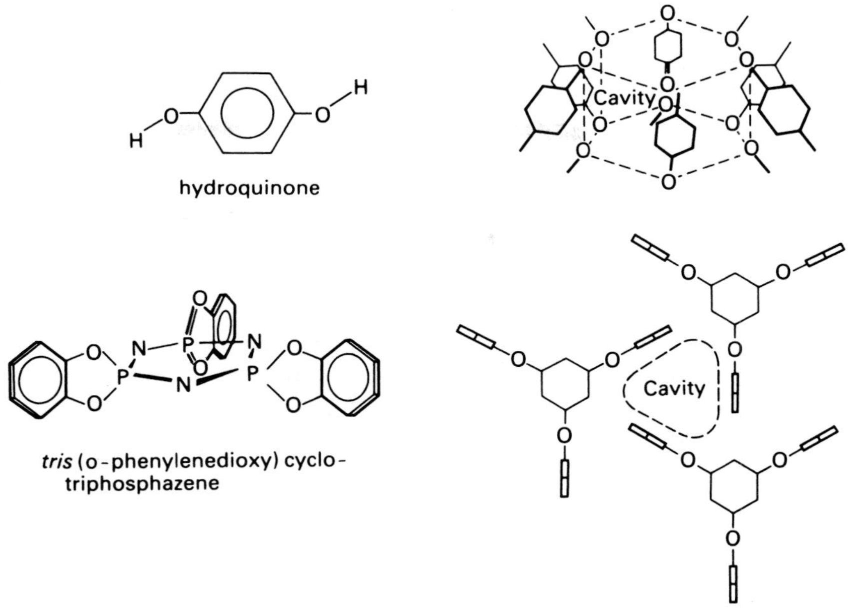

Figure 9.4 Hydroquinone and cyclotriphosphazene clathrates.

Two other host lattices that form a variety of clathrates are those of the molecules hydroquinone and trisubstituted cyclotriphosphazenes (Fig.9.4). Hydroquinone forms hexagonal prisms of O atoms surrounded by benzene rings, whereas the phosphazene species form open triangular shafts surrounded by benzene rings. Hydroquinone accepts Ar, Kr, Xe, N_2, H_2S, CO, CH_3OH and some other small molecules; the larger phosphazenes accept benzene and xylenes.

The host molecule is normally in a lattice different from the one it adopts when pure. This entails a small enthalpy cost, on the order of 1–2 kJ/mol host. However, the mutual-polarizability attractions between guest and host can be significantly larger than this, though the enthalpies are usually still small:

$$H^\beta \cdot G(s) \rightarrow H^\alpha(s) + G(g)$$

Host	Guest	ΔH (kJ)
H_2O	Kr	27
H_2O	Xe	39
H_2O	CH_4	19
$Q(OH)_2$	Ar	19
$Q(OH)_2$	O_2	17
$Q(OH)_2$	HBr	37

Here α refers to the normal lattice for the host and β to that in the clathrate. $Q(OH)_2$ is hydroquinone, H is the host, and G is the guest. Since these interaction energies are modest, the nonstoichiometric nature of the clathrates (in which 20–80% of the cavities are unfilled) reflects the additional stability provided by the configurational entropy of random cavity filling. Presumably the final composition of the clathrate represents a compromise between the London binding energy, which increases as cavities are filled, and configurational entropy, which increases as the number of vacancies increases. The resulting stability is considerable: At the normal boiling point of argon, the dissociation pressure of the argon–hydroquinone clathrate is only 2×10^{-8} atm. On the other hand, at 298 K, crystals of this clathrate lose roughly 10% of their bound argon after a week of exposure to the open air, which clearly represents an entropy-driven process.

A different and quite unusual low-enthalpy process in which the observed reaction is entropy-driven is the spontaneous formation of a distannene ($Sn{=}Sn$ bond) from a polystannane,

$$2 \; \underset{Ar_2Sn\overline{}SnAr_2}{\overset{SnAr_2}{\triangle}} \quad \xrightarrow{\;hv,\,-78°C\;} \quad 3Ar_2Sn{=}SnAr_2$$

where *Ar* represents a 2,4,6-triisopropylphenyl group. We have seen in Chapter 5 that sterically protected $Sn{=}Sn$ bonds are possible, but that in general pi overlap is weak enough for heavy atoms that rearrangement to sigma-bonded systems will be energetically preferable. Here we may assume that bond strain in the Sn_3 rings destabilizes the reactant's set of six Sn–Sn bonds enough to make the product's set of three $Sn{=}Sn$ bonds nearly equivalent. ΔH for the reaction should thus be nearly zero, and photochemical initiation (which presumably breaks a Sn–Sn bond) allows the favorable entropy change for the reaction to the right to dominate the process even at a temperature as low as 195 K. On warming to 0°C in the dark, the reaction is cleanly reversed, which suggests that the Sn–Sn ring system is slightly favored by ΔH, absent the bond-breaking effect of the radiation. But, remarkably, further warming above 50°C shifts the reaction to the right in a reversible equilibrium; at 90°C roughly two-thirds of the mixture is in the distannene form. It is unusual to see as clear an example of the effect of temperature in magnifying the thermodynamic significance of the $T\Delta S$ term in a reaction's driving force.

9.3 HIGH-TEMPERATURE PROCESSES

For the observed magnitudes of ΔH and ΔS, we can see entropy-driven reactions not only for those cases in which ΔH is unusually low, but also (for most reacting systems) at high temperatures. In fact, at sufficiently high temperatures (which, for the most recalcitrant systems, means 2500–3000°C) *all* observed reactions have a positive ΔS. This is achieved by forcing condensed phases into gaseous form, usually by breaking bonds to form smaller molecules, molecular fragments, or isolated atoms. There is, then, a bond-energy cost or positive ΔH that is outweighed by the large $T\Delta S$.

One of the most common—and industrially quite important—high-temperature reactions is the pyrolysis of $CaCO_3$ to CaO and CO_2. Ionic salts of most oxoanions can be pyrolyzed in a parallel fashion; we shall look at other examples later. Here we shall examine the energy relationships for the carbonate decomposition. Consider the general pyrolysis of alkaline-earth carbonates:

$$MCO_3(s) \rightarrow MO(s) + CO_2(s)$$

For the two metals Mg and Ba, the relevant thermodynamic data are as follows:

	Theoretical			Experimental		
	U_{MCO_3}	U_{MO}	ΔU	ΔH^0	ΔS^0 (J/mol K)	ΔG^0 (kJ/mol)
Mg	3238	3795	−557	+118	+175	+ 65
Ba	2678	3054	−376	+267	+172	+216

The substantial positive ΔH values make it clear that the reaction must be entropy-driven in either case. In both cases the metal oxide is more stable than the carbonate, essentially because the O^{2-} ion is smaller than CO_3^{2-} but carries the same charge. However, this lattice-energy advantage is not enough to drive the reaction, because the carbonate C=O bond must be broken. Furthermore, because the radii of the carbonate and oxide ions are presumably constant, there is an interesting effect of the metal radii on the lattice energies: the large Ba^{2+} ion experiences a smaller proportional change in going from the carbonate lattice to the oxide lattice than the small Mg^{2+} does. The lattice-energy change thus helps the reaction less for barium, and its free-energy change is much more unfavorable than that of magnesium. The result is that, since ΔS for the pyrolysis reaction is very nearly the same for both reactions, a much higher temperature is required before $T\Delta S$ can overcome the unfavorable ΔH of the barium reaction. Table 9.2 gives room-temperature equilibrium constants and approximate decomposition temperatures for the alkaline-earth carbonates.

The decomposition temperature can be estimated by solving the free-energy definition for the temperature at which $\Delta G = 0$, assuming that ΔH and ΔS are constant:

$$\Delta G = \Delta H - T\Delta S = 0$$

$$T \simeq \frac{\Delta H}{\Delta S}$$

For $MgCO_3$, this procedure yields an estimate of 400°C; for $BaCO_3$, it yields 1280°C. Both of these are good approximations to experiment.

Table 9.2 Pyrolysis of Carbonates

	$MCO_3(s) \rightarrow MO(s) + CO_2(g)$	
Metal	K (25°C)	T_{dec} (°C)
(Be)	—	(forms basic carbonate at 25°)
Mg	3×10^{-12}	400
Ca	2×10^{-23}	900
Sr	6×10^{-33}	1280
Ba	1×10^{-38}	1360
Compare:		
K_2CO_3	4×10^{-63}	1400
$NiCO_3$	2.5×10^{-1}	140 (to basic carbonate)

The industrial pyrolysis of limestone ($CaCO_3$) to lime (CaO) is an extremely energy-intensive operation, since the reaction must be entropy-driven. The energy (enthalpy) input is 178 kJ/mol reaction, even disregarding the heat required to take the cold rock up to the temperature of reaction. From Table 9.2, the temperature of reaction is 900°C but decomposing the center of large lumps can require temperatures of 1200°C or even higher. In units common to the chemical industry, the pyrolysis usually requires at least 4.25 million Btu per ton of lime produced. This is the heat produced in burning about 28 gallons of fuel oil. Since the United States produces about 20 million tons of lime a year, the lime-burning process uses over half a billion gallons of fuel oil or the gas equivalent.

Other oxoanions besides the carbonate can be pyrolyzed to a gas and a more stable lattice, but the lattice is not always the oxide. Table 9.3 gives the principal species that have been investigated and their reactions on pyrolysis. Besides carbonates, some nitrates and sulfates decompose to oxides, as do peroxides. However, most oxoanion salts decompose to other oxoanions in which there is a better size match within the lattice for cation and anion. For example, lithium nitrate decomposes to the oxide, in which two small Li^+ ions can reasonably be accommodated in the lattice for each O^{2-}. The nitrates of larger alkali metals, however, decompose to the nitrites instead because the two M^+ ions form a relatively unstable lattice with a single oxide ion.

Exactly the same pattern of ionic-lattice rearrangement and gas generation can be seen for polyatomic anions that contain no oxygen. For example, Rb and Cs form crystalline tri-iodides $M^+I_3^-$, but no other alkali metals do. If we set up the same sort of Born–Haber cycle for the thermal decomposition of MI_3 to MI and I_2 vapor as for the carbonate decomposition, we find again that the smaller ions experience a much more favorable lattice-energy change on going to the lattice with the smaller I^- ion than do the larger ions. Presumably the nonexistence of tri-iodides for Li^+, Na^+, and K^+ means that thermal decomposition of those compounds occurs at room temperature. For the decomposition reaction $MI_3 \rightarrow MI + I_2$,

Table 9.3 Pyrolysis of Oxyanion Salts

Group IV oxyanions: CO_3^{2-}, HCO_3^-, $C_2O_4^{2-}$

$$M^{II}CO_3 \rightarrow M^{II}O + CO_2$$
$$2M^IHCO_3 \rightarrow M_2^ICO_3 + H_2O + CO_2$$
$$M^{II}C_2O_4 \rightarrow M^{II}CO_3 + CO$$

Group V oxyanions: NO_3^-, HPO_4^{2-}

$$M^INO_3 \rightarrow M^INO_2 + \tfrac{1}{2}O_2$$

but

$$2LiNO_3 \rightarrow Li_2O + 2NO + \tfrac{3}{2}O_2$$
$$M^{II}(NO_3)_2 \rightarrow M^{II}(NO_2)_2 + O_2$$

but also

$$Pb(NO_3)_2 \rightarrow PbO + 2NO_2 + \tfrac{1}{2}O_2$$
$$Mn(NO_3)_2 \rightarrow MnO_2 + 2NO_2$$
$$2M_2^IHPO_4 \rightarrow M_4^IP_2O_7 + H_2O$$

Group VI oxyanions: O_2^-, O_2^{2-}, SO_4^{2-}, HSO_4^-, HSO_3^-

$$M^IO_2 \rightarrow \tfrac{1}{2}M_2^IO_2 + \tfrac{1}{2}O_2$$

but

$$LiO_2 \text{ unknown}$$
$$M^{II}O_2 \rightarrow M^{II}O + \tfrac{1}{2}O_2$$
$$M^{II}SO_4 \rightarrow M^{II}O + SO_3$$
$$2M^IHSO_4 \rightarrow M_2S_2O_7 + H_2O$$
$$2M^IHSO_3 \rightarrow M_2S_2O_5 + H_2O$$

Group VII oxyanions: XO_2^-, XO_3^-, XO_4^-

$$3M^IClO_2 \rightarrow 2M^IClO_3 + M^ICl$$

but

$$M^IBrO_2 \rightarrow M^IBr + O_2$$
$$4M^IClO_3 \rightarrow 3M^IClO_4 + M^ICl$$
$$M^{II}(ClO_3)_2 \rightarrow M^{II}Cl_2 + 3O_2$$

but

$$Mg(ClO_3)_2 \rightarrow MgO + Cl_2 + \tfrac{5}{2}O_2$$
$$Ba(BrO_3)_2 \rightarrow Ba(BrO_2)_2 + O_2$$
$$5M^{II}(IO_3)_2 \rightarrow M_5^{II}(IO_6)_2 + 4I_2 + 9O_2$$
$$M^IClO_4 \rightarrow M^ICl + 2O_2$$

but

$$2Al(ClO_4)_3 \rightarrow Al_2O_3 + Cl_2 + \tfrac{21}{2}O_2$$

Note: M^I is alkali metal; M^{II} is alkaline-earth metal.

ΔH^0 is +13 kJ/mol reaction for M = Rb and +15 kJ/mol reaction for M = Cs. For smaller alkali metal cations it will be quite small, perhaps even negative. At the same time, ΔS^0 is +10 J/K mol reaction for M = Rb and +4 J/K mol reaction for M = Cs even if I_2 is taken to be a solid, but much higher for I_2 vapor. Other polyhalides follow much the same pattern: $M^+ClF_2^-$ and $M^+ClF_4^-$ salts are known only for K, Rb, and Cs, not for Li or Na even though the smaller cations are even harder acids. Similarly, alkali fluorides dissolve in XeF_6 to give $MXeF_7$ salts for the large Cs and Rb, M_2XeF_8 for the smaller K and Na, but no isolatable salt for Li.

Thermal decomposition ($M_2XeF_8 \rightarrow 2MF + XeF_6$) requires about 400°C for Rb and Cs, roughly 250°C for Na and K, and—following our earlier argument—perhaps only room temperature for Li because of the great lattice-energy advantage for the small Li^+ with the small F^- instead of the large XeF_7^- or XeF_8^{2-}.

A familiar related reaction is the decomposition of azides to yield N_2 gas. Normally this reaction also produces the elemental metal from the azide lattice, but lithium and magnesium, which form stable nitrides with the N_3^- ion, yield the nitride instead. The most electropositive elements form clearly ionic azides that are not normally explosive, but the more electronegative metals form azides of substantial covalent character that are very sensitive explosives. The ΔH^0 values for the composition reactions indicate the distinction:

$$M(N_3)_n(s) \rightarrow M^0(s) + \tfrac{3n}{2}N_2(g)$$

Metal	Li	Na	K	Rb	Cs
ΔH^0 kJ/mol rn	−77	−21	+1.2	+0.4	+10
	(to Li_3N)				

Metal	$Cu^{(I)}$	Ag	$Pb^{(II)}$	Ba
ΔH^0 kJ/mol rn	−281	−310	−483	+22

Very pure N_2 can be conveniently prepared by gently heating $Ba(N_3)_2$. The reaction is endothermic because of the large amount of energy to destroy the ionic lattice, even though the formation of the N≡N triple bond yields a great deal of energy.

Still another entropy-driven process common in the chemical laboratory is the regeneration of desiccants by heating. Most of the common desiccant materials are salts of 2+ or 3+ cations that serve as strong Lewis acids toward the Lewis base water. Magnesium sulfate, magnesium perchlorate, and calcium chloride are all used in various applications. Although it is not always economically worthwhile to do so, they can be regenerated in an entropy-driven reverse acid–base process:

$$CaSO_4 \cdot 2H_2O(s) \rightarrow CaSO_4(s) + 2H_2O(g)$$
$$\Delta H^0 = +105 \text{ kJ/mol rn}$$
$$\Delta S^0 = +290 \text{ J/K mol rn}$$

Calcium sulfate is readily regenerated by heating to about $250°C$, at which temperature ΔG for the thermal decomposition above is -46 kJ/mol rn in spite of the highly unfavorable enthalpy change.

Some unusual and interesting synthetic reactions can be carried out following an initial high-temperature entropy-driven process. At very high temperatures all condensed phases break most or all of the chemical bonds present in their molecules or lattices. The resulting individual atoms or molecular fragments (usually only two or three atoms) pass into the gas phase. Since all of these species have coordination numbers far short of their usual maximum bonding capability, most are quite chemically reactive and must be kept at low pressure to minimize gaseous collisions. If this reactive vapor is immediately condensed on a cold surface with another reactant, some reactions can occur that have no counterpart in the ordinary chemistry of the same starting materials. For example, boron atoms produced at very high temperatures by electron bombardment of a boron rod were condensed with CO_2 at $-196°C$, then allowed to warm. At $-150°C$, an explosive redox reaction and polymerization occurred:

$$B(atom) + CO_2(s) \rightarrow \tfrac{1}{n}(BO)_n(s) + CO(g)$$

Many studies have been made of the reactivity of high-temperature species on condensed phases, using both main-group and transition-metal vapor species. Here we shall consider only the main-group species, reserving transition-metal reactions until Chapter 13.

Table 9.4 gives some high-temperature species for main-group elements that have been identified and considered for condensed-phase reactions. Some of these, such as CCl_2, are available via more conventional reactions and offer no particular advantage through the high-temperature process. Most, however, are unique to this technique and offer high reactivity and unusual mechanisms. One problem is

Table 9.4   High-Temperature Species of Interest in Condensation Reactions

Be		**B**		**C**					
Be	B		HBS	C	CS				
	BF		BC_2	C_2	CF_2				
	BCl		B_2O_2	C_3	CCl_2				
				C_4	CBr_2				
Mg		**Al**		**Si**		**P**		**S**	
Mg	AlF			Si	SiO	P_2	PN	S_2	
				SiF_2	SiS	PF	PF_2		
				$SiCl_2$	SiC				
Ca				**Ge**		**As**		**Se**	
Ca				Ge	GeO	As_2		Se_2	
				Sn					
				Sn	SnO				

that the high reactivity produces a strong tendency to polymerize, since the vapor species are mobile on the surface of the condensed phase. The polymerization reaction thus competes with the reaction of interest.

Mg, Ca, and Zn atoms react with alkyl halides by insertion in the C—X bond. For Mg, the product is a crystalline, unsolvated Grignard reagant R—Mg—X that has somewhat different chemical properties from the traditional ether-solvated species. Ca atoms react with unsaturated perfluoroolefins to yield R_F—Ca—F systems, and Zn atoms react with perfluoroalkyl iodides to yield R_F—Zn—I systems.

B atoms react with BF_3, BCl_3, and PCl_3 to give the catenated species B_2F_4, B_2Cl_4, and P_2Cl_4. B_2F_4, however, is more cleanly prepared using the BF radical, which is formed when BF_3 is passed over hot boron (2000°C):

$$BF_3(g) + 2B(s) \rightarrow 3BF(g)$$
$$BF + BF_3 \rightarrow B_2F_4$$

Larger molecules form in a continuing sequence:

$$BF + B_2F_4 \rightarrow B_3F_5$$
$$4B_3F_5 \rightarrow B_8F_{12} + 2B_2F_4$$

B_8F_{12} is a liquid stable up to about −10°C, probably with a structure analogous to that of diborane with terminal and bridging BF_2 groups. It readily reacts with soft bases such as CO and PF_3 to yield compounds such as OC→B$(BF_2)_3$. The high-temperature BF species also reacts readily with unsaturated hydrocarbons:

$$2BF + 2CH_3C{\equiv}CCH_3 \rightarrow$$

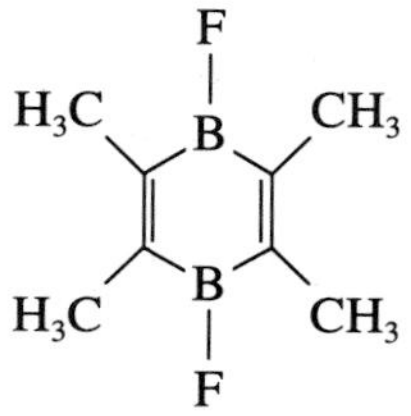

$$2BF + 2HC{\equiv}CH \xrightarrow{30°C} FB \underset{CH=CH-BF_2}{\overset{CH=CH-BF_2}{<}}$$

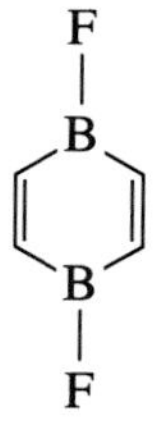

$$BF + BF_3 + H_2C{=}CH-CH_2-CH_3 \rightarrow H_2C\underset{|}{\overset{}{-}}CH\underset{|}{\overset{}{-}}CH_2-CH_3$$
$$\qquad\qquad\qquad\qquad\qquad\qquad\qquad BF_2 \quad BF_2$$

Carbon vaporizes to a mixture of C, C_2, C_3, and C_4. These species react readily with a variety of molecular halides, usually by insertion:

$$C + BCl_3 \rightarrow Cl_2C(BCl_2)_2 + ClC(BCl_2)_3$$

$$C + B_2Cl_4 \rightarrow C(BCl_2)_4$$

$$C + PCl_3 \rightarrow Cl_3CPCl_2 + Cl_2C(PCl_2)_2$$

$$C_2 + BCl_3 \rightarrow Cl_2BC\!\!=\!\!CBCl_2$$
$$\underset{Cl \quad Cl}{| \quad |}$$

$$C_2 + B_2F_4 \rightarrow (F_2B)_2C\!\!=\!\!C(BF_2)_2$$

$$C_3 + B_2F_4 \rightarrow (F_2B)_2C\!\!=\!\!C\!\!=\!\!C(BF_2)_2$$

Silicon vaporizes predominantly to single Si atoms, which form an interesting product with trimethylsilane:

$$Si + (CH_3)_3SiH \rightarrow (CH_3)_3Si\!-\!SiH_2\!-\!Si(CH_3)_3$$

However, the SiF_2 vapor species (prepared by passing SiF_4 over hot Si) has a more varied chemistry:

$$SiF_2 + H_2O \rightarrow HSiF_2OSiF_2H$$

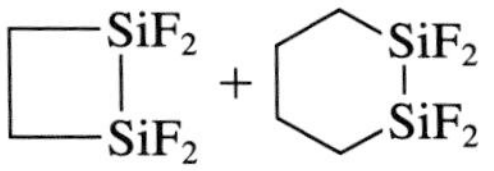

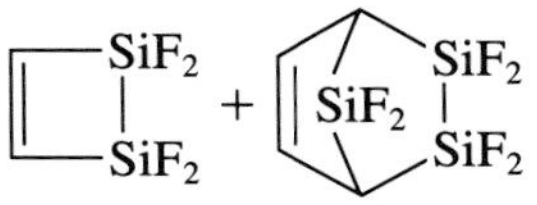

$$SiF_2 + BF_3 \rightarrow SiF_3SiF_2BF_2 + SiF_3(SiF_2)_2BF_2$$

The preceding discussion suggests that high-temperature, high-entropy species tend to form catenated reaction products. Several important reactions carried out at more modest temperatures show the same property. Although thermochemical data are scarce, the two reactions below are strongly endothermic and thus entropy-driven:

$$6S_2Cl_2 + 4NH_4Cl \xrightarrow[\text{Tetrasulfur tetranitride}]{160^\circ C} S_4N_4 + S_8 + 16HCl$$

$$\Delta H^0 = +428 \text{ kJ/mol rn}$$

$$\Delta S^0 = \text{About} +1080 \text{ J/K mol rn}$$

$$3BCl_3 + 3NH_4Cl \xrightarrow{175^\circ C} \text{[B-trichloroborazine]} + 9HCl$$

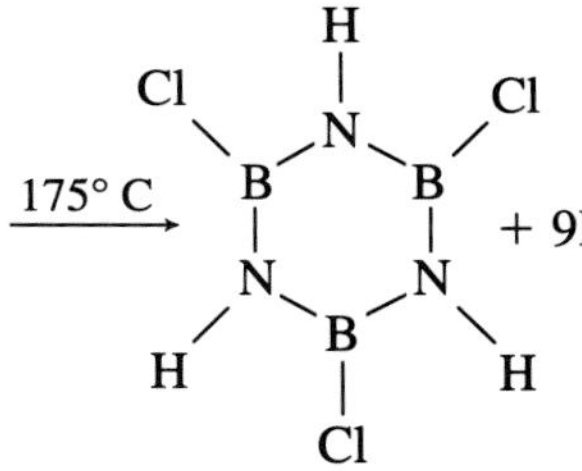

B-trichloroborazine

$$\Delta H^0 = +322 \text{ kJ/mol rn}$$

$$\Delta S^0 = \text{about } +1000 \text{ J/K mol rn}$$

The S_4N_4 molecule, whose structure is shown in Fig. 5.24, is the starting compound for most sulfur–nitrogen chemistry (Table 5.16) and has received a great deal of study. In the same sense, the borazines represent some of the most interesting boron–nitrogen compounds. B-trichloroborazine is the usual starting compound for these studies, as Chapter 5 suggested. A great deal of recent nonmetal chemistry thus depends on two very endothermic, entropy-driven reactions.

9.4 ENERGY AND ENTROPY IN NOBLE-GAS CHEMISTRY

Because entropy-driven reactions so often involve gas formation, the reactions of noble gases and their compounds are of obvious interest. The decomposition of noble-gas compounds always produces gaseous noble-gas atoms with the formation of no new bonds. Are these "pure" entropy-driven reactions? The answer is "no," because the bonds formed by other atoms have a dominant effect on the reactions of noble gases, both in the formation of their compounds and in their decomposition. In Chapter 5 we reviewed the discovery of noble-gas compounds and related their bond energies to those of nearby groups of the periodic table. Here we can look more closely at the balance between entropy and enthalpy in these reactions.

The xenon fluorides all form exothermically in reactions that thermodynamically resemble that between NH_3 and HCl to give NH_4Cl: Two gases disappear into a solid product whose enhanced bonding more than makes up for the unfavorable loss of entropy. For xenon fluorides, the basic energy condition is that two moles of Xe—F bonds (134 kJ/mol each) are more stable than one mole of F—F bonds (159 kJ/mol). In addition, the reactant gases have no van der Waals forces to bind molecules together as in the solid fluorides. The following result is seen:

$$Xe(g) + F_2(g) \rightarrow XeF_2(s)$$

$$\Delta H^0 = -164 \text{ kJ/mol rn}$$

$$\Delta S^0 = -249 \text{ J/K mol rn}$$

$$Xe(g) + 2F_2(g) \rightarrow XeF_4(s)$$

$$\Delta H^0 = -277 \text{ kJ/mol rn}$$

$$\Delta S^0 = -428 \text{ J/K mol rn}$$

$$Xe(g) + 3F_2(g) \rightarrow - XeF_6(s)$$

$$\Delta H^0 = -412 \text{ kJ/mol rn}$$

$$\Delta S^0 = -567 \text{ J/K mol rn}$$

These substantial negative enthalpy changes explain why none of the xenon fluorides is explosive and why very high temperatures are required to pyrolyze the compounds.

The only known fluoride of krypton, KrF_2, is a different matter. The smaller krypton atom is distinctly less polarizable and the Kr—F bond energy is only about 50 kJ/mol. The formation reaction thus has unfavorable thermodynamic qualities:

$$Kr(g) + F_2(g) \rightarrow KrF_2(s)$$

$$\Delta H^0 = +19 \text{ kJ/mol rn}$$

$$\Delta S^0 = -254 \text{ J/K mol rn}$$

The reaction is neither enthalpy-driven nor entropy-driven; indeed it should be thermodynamically impossible. It does not in fact occur as written, but rather relies on a variation of the high-temperature condensation technique described in the previous section. Fluorine atoms are produced at low temperatures by electric discharge, ultraviolet light, or other forms of irradiation:

$$Kr(g) + 2 F(g) \rightarrow KrF_3(s)$$

$$\Delta H^0 = -139 \text{ kJ/mol rn}$$

$$\Delta S^0 = -340 \text{ J/K mol rn}$$

If the KrF_2 product were in the gas phase, the sublimation energy would reduce ΔH to −98 kJ. This value is so small that we should see substantial thermal decomposition ($\Delta H/\Delta S \simeq 285$ K or ca. 10°C) well below room temperature. Experimental results show that KrF_2 vapor does decompose (though not as an equilibrium—the F atoms recombine to F_2). The solid is, as we might expect, much more stable.

In a similar sense, it is thermodynamically impossible to form xenon trioxide directly from Xe and O_2:

$$Xe(g) + \tfrac{3}{2}O_2(g) \rightarrow XeO_3(s)$$

$$\Delta H^0 = +402 \text{ kJ/mol rn}$$

$$\Delta S^0 = -393 \text{ J/K mol rn}$$

These values are so spectacular that it is not surprising that XeO_3 is a violent explosive, reversing the above reaction under the least provocation. How is it possible to prepare it at all?

The synthesis relies on the high H—F bond energy and the strong hydrogen bonding of HF to water:

$$XeF_6(aq) + 3H_2O(l) \rightarrow XeO_3(aq) + 6HF(aq)$$

$$\Delta H^0 = -298 \text{ kJ/mol rn}$$

This hydrolysis of XeF_6 is so facile that water vapor must be rigorously excluded from XeF_6 storage or reaction vessels to prevent accidental detonation of traces of unwanted XeO_3.

In an attempt to prepare a xenon chloride, a reaction analogous to the hydrolysis of XeF_6 was attempted with HCl instead of water. However, XeF_6 is a strong
enough oxidizing agent to oxidize the chlorine instead:

$$XeF_6(s) + 8HCl(g) \rightarrow Xe(g) + 3Cl_2(g) + 6HF(g)$$

$$\Delta H^0 = -726 \text{ kJ/mol rn}$$

$$\Delta S^0 = +548 \text{ J/K mol rn}$$

Since all the products are gases, the entropy change obviously assists this reaction.
However, in a comparable reaction between XeF_6 and ammonia, the acid–base reaction between NH_3 and the HF product adds to the enthalpy yield but makes the
whole system much more ordered:

$$XeF_6(s) + 8NH_3(g) \rightarrow Xe(g) + 6NH_4F(s) + N_2(g)$$

$$\Delta H^0 = -2069 \text{ kJ/mol rn}$$

$$\Delta S^0 = -957 \text{ J/K mol rn}$$

Another interesting example of an entropy-aided xenon reaction is the disproportionation of XeF_6 to XeO_6^{-4} and Xe in basic aqueous solution:

$$2XeF_6(s) + 4Na^+(aq) + 16OH^-(aq) \rightarrow$$
$$Na_4XeO_6(s) + Xe(g) + O_2(g) + 12F^-(aq) + 8H_2O(l)$$

For this reaction, ΔS^0 can be estimated as about +1500 J/K mol rn, due predominantly to the formation of the two moles of gas and the decreasing number of ions
to order solvent water molecules. (An accurate ΔH^0 cannot be determined; not
enough thermochemical data exist.) Note that the reduction potential for the
Xe(VIII)–Xe(VI) couple is +3.0 V in acid but is only +0.9 V in base. Figure 8.1
reveals that it is only very near pH 14 that this couple becomes a sufficiently weak
oxidant to be at least metastable in water solution. This explains why the disproportionation proceeds only in very strongly basic solution.

9.5 REACTIONS DRIVEN BY LARGE ΔH AND ΔS: EXPLOSIVES

Although they are relatively rare, reactions do exist that are characterized by large
negative ΔH values and at the same time by large positive ΔS values—that is, they
are becoming more stable and generating gas at the same time. Such reactions, of
course, are thermodynamically favored at any temperature, though a convenient
reaction mechanism may not exist. When the reaction is kinetically possible, the
reactants obviously cannot be allowed to contact each other. An example is the
hypergolic (self-igniting) reaction between the rocket propellant hydrazine and
the oxidizer dinitrogen tetroxide:

$$2N_2H_4(l) + N_2O_4(l) \rightarrow 4H_2O(g) + 3N_2(g)$$

$$\Delta H^0 = -1040 \text{ kJ/mol rn}$$

$$\Delta S^0 = +912 \text{ J/K mol rn}$$

Burning at 500 psi pressure inside a rocket motor, this reaction produces a flame temperature of over 2700°C. About a third of a pound of liquid yields seven moles of gas. A number of rocket propellant-oxidizer combinations are self-igniting, but most have enough kinetic metastability to require thermal ignition at the start of the motor firing. All, however, have more or less comparable ΔH and ΔS values and the corresponding overwhelming thermodynamic spontaneity.

Explosives are very similar to rocket propellants in that both must generate large amounts of gas at high temperatures. One difference is that high explosives must detonate—that is, they must be able kinetically to react as rapidly as the shock wave from the initial explosion can pass through the solid or liquid explosive. If this is possible, the shock wave will accelerate until its passage through the explosive is limited by the reaction rate at what is called the *detonation velocity*. If the characteristic reaction rate is slower, the shock wave will dissipate and the only reaction will be the surface burning of the explosive. Since the reaction rate generally increases with pressure, many high-explosive compositions burn in the open air but detonate in a sealed chamber, particularly if initiated by a small explosion such as that of a blasting cap.

The first chemical explosive, black powder, does not detonate at any pressure and thus is not a high explosive. Black powder is a mixture of charcoal, sulfur, and potassium nitrate in proportions to burn approximately as

$$14KNO_3 + 18C + 5S \rightarrow 5K_2CO_3 + K_2SO_4 + K_2S + 3S + 10CO_2 + 3CO + 7N_2$$

Such a reaction has $\Delta H^0 = -4948$ kJ/mol reaction and $\Delta S^0 \simeq +3260$ J/K mol reaction. The large numbers, however, are somewhat misleading, because a mole reaction corresponds to burning almost four pounds of black powder. Weight-for-weight, the reaction is only about one-third as energetic or as entropic as the N_2H_4–N_2O_4 reaction.

The black-powder reaction is suggestive in that much of the energetic stability and gas generation come from the formation of the strongly bonded gases CO_2, CO, and N_2, featuring the very strong C=O, C≡O, and N≡N bonds at nearly 1000 kJ/mol bond. Essentially no other gases are bonded strongly enough to yield a large negative enthalpy change as the gas forms. Accordingly, all commercial high explosives are organic molecules containing —NO_2, —ONO_2, or —$NHNO_2$ groups, as the examples in Fig. 9.5 indicate.

The detonation of nitroglycerin follows the approximate reaction

$$C_3H_5N_3O_9 \rightarrow \tfrac{3}{2}N_2 + \tfrac{5}{2}H_2O + 3CO_2 + \tfrac{1}{4}O_2$$

$$\Delta H^0 = -1809 \text{ kJ/mol rn}$$

$$\Delta S^0 \simeq +920 \text{ J/K mol rn}$$

The stoichiometry of the above reaction reveals that nitroglycerin is *overoxidized*—that is, it contains more oxygen than is needed to burn all its carbon and hydrogen to CO_2 and H_2O. Most explosive materials, by contrast, are underoxidized because they have a relatively large carbon framework. In particular, nirocellulose

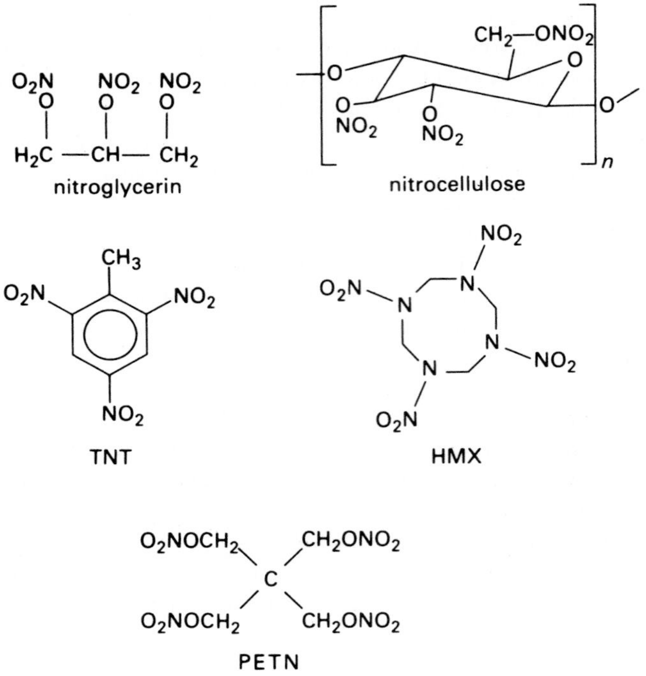

Figure 9.5   Commercial high explosives.

(gun-cotton) is severely underoxidized. The oxygen balance of nitrocellulose can be improved by plasticizing it into a rubbery mass with liquid nitroglycerin. The soft high-nitroglycerin mixture is called *blasting gelatin*, whereas the harder material with a higher proportion of nitrocellulose is a *double-base* solid propellant in rockets.

Another material commonly used to improve the oxygen balance of underoxidized explosives is ammonium nitrate, which yields N_2, H_2O, and excess O on thermal composition. NH_4NO_3 is itself a high explosive (though difficult to detonate) and is quite inexpensive to produce. In the years since World War II mining and construction explosive use has turned from dynamite (nitroglycerin plus wood pulp and $NaNO_3$ or NH_4NO_3) to mixtures of NH_4NO_3 and fuel oil known as ANFO, which can be detonated by a powerful initiator. The fuel oil, of course, is not an explosive itself, but it is burned to CO_2 and H_2O by the excess oxygen in the ammonium nitrate.

Ordinary hydrocarbons are also used as fuel in rubber-base solid propellants. The rubber binder in these must contain a large proportion of oxidizer (NH_4NO_3 or NH_4ClO_4) and, to increase the energy yield, a metal forming an extremely stable oxide lattice (normally aluminum):

$$6\ NH_4ClO_4 + 10\ Al \rightarrow 3N_2 + 9H_2O + 5Al_2O_3 + 6HCl$$

$$\Delta H^0 = -9337 \text{ kJ/mol rn}$$

$$\Delta S^0 = +2246 \text{ J/K mol rn}$$

The enormous enthalpy yield of this reaction (higher, even weight for weight, than the N_2H_4–N_2O_4 reaction) comes from the extremely high lattice energy of the Al_2O_3 system. Experimentally, the temperatures attained in burning an aluminized propellant are so high that the product molecules are not stable toward dissociation, and the pyrolysis fragments such as OH and AlO are much less stable than the species indicated in the reaction equation. The energy yield is thus lower than that suggested by the high ΔH^0, and the combustion chamber temperatures are correspondingly lower. On the other hand, the fragments actually increase the entropy yield of the reaction, so ΔS takes on increased importance in driving the reaction.

Propellants are important not only in rocket-motor applications, but in their original application as gunpowder. High explosives cannot be used in guns because the detonation pressure would shatter the breech or barrel. Instead, a relatively slow-burning propellant is used, and the size of the particles is chosen so that they will burn for approximately the length of time the bullet or shell is in the barrel. Black powder was used from the thirteenth to late nineteenth century, but as its reaction equation reveals, over half of the weight of powder burned results in solid products. The resulting smoke obscures the field of view and reveals the position of the shooter. This defect was remedied and the strength of the propellant was increased by the invention of smokeless powder, which is predominantly nitrocellulose plasticized with a small amount of nitroglycerin and molded into small pellets.

Although propellant explosives, which only burn, can be ignited by a flame or even a spark (as in a flintlock musket), high explosives usually require a small initiating explosion in order to detonate. Several compounds with high sensitivity to flame or mechanical shock are used as initiators or *primary detonators*: mercuric fulminate, $Hg(ONC)_2$, lead azide, $Pb(N_3)_2$, and lead styphnate, $PbO_2C_6H(NO_2)_3$ (the lead salt of 2,4,6-trinitroresorcinol). Of these, lead azide is by far the most commonly used, usually with some added lead styphnate to increase flame sensitivity. Lead azide is predominantly covalent, with a substantial positive enthalpy of formation (see the discussion of azides in Section 9.3 on high-temperature reactions). Since the nitrogen atoms are already bonded to each other, there is little mechanistic hindrance to the decomposition of $Pb(N_3)_2$ into Pb and N_2. It is accordingly quite sensitive to shock and to electrical ignition. Actually, both of these are special cases of thermal decomposition: The electric fuse wire or a sharp blow produces a momentary hot spot about 10^{-3} mm across in the $Pb(N_3)_2$ crystal, and the exothermic decomposition reaction that occurs within this hot spot carries both heat and a mechanical shock wave beyond its borders through the crystal. Detonation occurs in less than a microsecond after the initial shock.

Several inorganic species are sensitive to mechanical shock in much the same manner as lead azide. Metal azides that are not stabilized by a high ionic-lattice energy and thus have positive enthalpies of formation usually fall into this category, along with molecular azides such as dicyandiazide, $N\equiv C-N\equiv C(N_3)_2$. (The latter can be detonated by the force of a cotton ball wiping against it.) The most familiar inorganic explosives, however, are the nitrogen halides NCl_3, $NBr_3 \cdot 6NH_3$, and $NI_3 \cdot NH_3$, which in their instability contrast dramatically with

Table 9.5   Born–Haber Cycle Energies for Nitrogen Halides (kJ/mol)

X	$\Delta H_{at}(N)$	$\Delta H_{at}(X)$	BE_{N-X}	$\Delta H_{vap}(NX_3)$	ΔH_f^0(theor.)	ΔH_f^0(expt.)
F	473	79	−272	0	−105	−126
Cl	473	121	−192	29	+230	+230
Br	473	113	−155	42	+305	
I	473	109	−138	42	+343	

Note: $\Delta H_f^0 = \Delta H_{at}(N) + 3\Delta H_{at}(X) + 3BE_{N-X} - \Delta H_{vap}(NX_3$ except F).

the very inert NF_3 (though many explosive —NF_2 compounds are known). Liquid NCl_3 and the solid ammoniated bromide and iodide are all sensitive to even the slightest shock, whereas gaseous NF_3 is not shock sensitive and even resists thermal decomposition fairly well. NF_3 also resists hydrolysis; it reacts slowly with aqueous base even at 100°C. Compare the ready hydrolysis of $NI_3 \cdot NH_3$:

$$2NF_3 + 6OH^- \xrightarrow{100°C} 6F^- + NO + NO_2 + 3H_2O$$

$$NI_3 \cdot NH_3 + 5H_3O^+ + 3Cl^- \longrightarrow 3ICl + 2NH_4^+ + 5H_2O$$

NF_3 will explode, however, if mixed with a gaseous reducing agent and sparked:

$$NF_3 + NH_3 \xrightarrow{Spark} N_2 + 3HF$$

The energetic basis for the striking differences between NF_3 and the other nitrogen halides can be seen from a Born–Haber cycle for their formation:

$$2N(g) \quad + \quad 6X(g) \quad \xrightarrow{6BE_{N-X}} \quad 2NX_3(g)$$
$$2\Delta H_{at}\uparrow \qquad 6\Delta H_{at}\uparrow \qquad\qquad \downarrow -2\Delta H_{vap}(Cl,Br,I)$$
$$N_2(g) \quad + \quad 3X_2 \quad \xrightarrow{2\Delta H_f^0} \quad 2NX_3$$
$$[X = F(g),Cl(g),Br(l),I(s)] \qquad [X = F(g),Cl(l),Br(s),I(s)]$$

Table 9.5 gives the relevant energy data for this cycle. The difference between NF_3 and the other species is the weak F—F bond and the strong N—F bond. The latter is caused by the good sigma overlap between the N and F valence orbitals, which are of comparable sizes and extend well past the small inner cores. From these data, it is not surprising that NCl_3 and NBr_3 are treacherously explosive, and that NI_3 is unknown as a pure compound.

PROBLEMS

A. DESCRIPTIVE

A1. Which of the following spontaneous halogen reactions are enthalpy-driven and which are entropy-driven? Why? (Do not look up thermodynamic data.)

(a) $2ClO_3^-(aq) + C_2O_4^{2-}(aq) + 4H_3O^+(aq) \rightarrow 2ClO_2(g) + 2CO_3(g) + 6H_2O(l)$

(b) $2IOF_3(s) \xrightarrow{110°C} IO_2F(s) + IF_5(l)$

(c) $Cl_2O(g) + H_2O(l) \rightarrow 2HOCl(aq)$

(d) $I_2(s) + (OSO_2F)_2(l) \rightarrow 2IOSO_2F(s)$

(e)
(f)
$$ClOClO_3 \begin{cases} \xrightarrow{\Delta} Cl_2O_6(80\%) + ClO_2 + Cl_2 + O_2 \\ \xrightarrow{Br_2} BrOClO_3 + Cl_2 \end{cases}$$

(g) $5IPO_4(s) + 9H_2O(l) \rightarrow I_2(s) + 3HIO_3(aq) + 5H_3PO_4(aq)$

(h) $2BrF_3(l) \rightleftharpoons BrF_2^+(solv) + BrF_4^-(solv)$

(i) $4BrF_3(l) + 2B_2O_3(s) \rightarrow 4BF_3(g) + 2Br_2(solv) + 3O_2(g)$

(j) $2NH_4ClO_4(s) \xrightarrow{400°C} Cl_2(g) + 4H_2O(g) + 2NO(g) + O_2(g)$

A2. The two molten-salt solvents NaCl and AlCl$_3$ are completely miscible, even though their solubility parameters differ by over 40 units. Why do they mix?

A3. By controlling the acidity of fluoride-containing solutions of niobium and tantalum, it is possible to control the degree to which each metal is complexed by the fluoride ion. For example, in a solution 4 M in NHO$_3$ and 0.4 M in HF, the species are thought to be NbF$_4$(H$_2$O)$^+$ and TaF$_5$(H$_2$O). From such a solution, tantalum is extracted into diisopropyl ketone with a distribution coefficient of 3.8; niobium is extracted with a coefficient of only 4.3×10^{-3}. This corresponds to a β of 880 for the system, and allows a good separation of the two metals. Why is tantalum so much more strongly extracted than niobium under these circumstances?

A4. For many clathrates, the enthalpy of decomposition is roughly proportional to the absolute entropy of the departing guest in the gas phase, without regard to the nature of the host lattice. Why is this true?

A5. The HSAB principle suggests that the hard acid Na$^+$ will prefer to surround itself with hard bases. Yet when NaClO$_4$ is pyrolyzed, the product is NaCl, not Na$_2$O, even though oxide is a harder base than chloride. What other factors must be considered?

A6. What gaseous products do you expect when SiF$_4$ gas is passed over solid boron at 1800°C? What product should result when the gaseous mixture is condensed on a cold surface?

A7. Do you expect the formation of the phosphonitrile trimer (NPCl$_2$)$_3$ to be enthalpy-driven or entropy-driven? (See Chapter 5, Section 5.11.)

A8. What reaction should occur between XeF$_6$ and PH$_3$?

B. NUMERICAL

B1. Consider the separation of two metals M and N by solvent extraction. For a given pair of solvents, $K_M = 10^6$ and $K_N = 10^2$, so that $\beta = 10^4$. Assuming that the initial concentrations of the two metals are equal, calculate the fraction of each extracted. What do the results indicate about the quality of separation achieved? For another pair of solvents, β also equals 10^4, but now $K_M = 10^2$ and $K_N = 10^{-2}$; calculate the fraction extracted for each metal and discuss the separation. Finally, a third pair of solvents has $\beta = 10^4$ with $K_M = 10^{-2}$ and $K_N = 10^{-6}$; again calculate the fraction extracted for each metal and comment. What would an ideal relation be between K_M and K_N?

B2. Write a Born–Haber cycle for each of the following reactions:

(a) $KClO_2 \rightarrow KCl + O_2$ (include the two lattice energies and use bond energies from Table 4.5)

(b) $3KClO_2 \rightarrow 2KClO_3 + KCl$ (use lattice energies only)

Reaction (b) predominates over reaction (a) in the pyrolysis of $KClO_2$. Calculate lattice energies for the three ionic species using the Kapustinskii approximation, making a reasonable approximation for the thermochemical radius of ClO_2^-. Show that the thermochemical radius of ClO_2^- must be at least as large as that of ClO_3^- if reaction (b) is to be thermodynamically favored over reaction (a).

B3. From mass spectrometric data, the energy change for $XeF_2(g) \rightarrow XeF^+(g) + F(g) + e^-$ is 12.8 eV. From vapor-pressure data, $\Delta H_{subl}(XeF_2) = 54$ kJ/mol. Using these data and bond energies, ionization energies, and enthalpies of formation from this and previous chapters, construct a Born–Haber cycle from which it is possible to calculate a bond energy for XeF^+. Use a qualitative MO energy-level diagram to compare this value with the bond energy for neutral XeF (XeF $\rightarrow$ Xe + F). How does the bond energy of XeF compare with the bond energy broken in the first dissociation of XeF_2 ($XeF_2 \rightarrow$ XeF + F)? The total atomization energy of XeF_2 can be obtained from your cycle data.

B4. Calculate the weight percent of black powder that remains as solids after firing.

B5. The alkali-metal superoxides are pyrolyzed to the peroxides as follows:

$$MO_2(s) \rightarrow \tfrac{1}{2}M_2O_2(s) + \tfrac{1}{2}O_2(g)$$

Assume that the thermochemical radius of O_2^- is 1.50 Å. Given the following electron affinity data,

$$O_2 + e^- \rightarrow O_2^- \qquad EA = +0.43 \text{ eV}$$
$$O_2 + 2e^- \rightarrow O_2^{2-} \qquad EA(23) = -7.0 \text{ eV}$$

calculate ΔH for the pyrolysis reaction for (a) NaO_2 and (b) CsO_2. If the entropy change for the reaction as written is +105 J/K mol rn, estimate the pyrolysis temperature for NaO_2 and CsO_2. How do your numbers compare with the experimental values of 100°C and 900°C?

B6. Propylene carbonate solvent has a molecular weight of 102 g/mol and a density of 1.20 g/mL. At 25°C, water is soluble in PC to the extent of 8.3 g H_2O/100 g PC. Using these data and Hildebrand's expression for the free energy of mixing of liquids, calculate the free energy of formation of such a solution. Does your result suggest that there is an acid–base interaction between H_2O and PC?

C. EXTENDED REFERENCE

C1. Aluminum metal is an extremely powerful reducing agent. In the Goldschmidt reaction, it reduces iron oxides to iron metal so vigorously that all reaction species become molten:

$$2Al(l) + 3FeO(l) \rightarrow 3Fe(l) + Al_2O_3(l)$$

Find the enthalpy of transition and heat-capacity function of each reactant in a recent edition of the *CRC Handbook of Chemistry and Physics*. Using these values, calculate ΔH and ΔS at 2400 K for the Goldschmidt reaction. Compare these with ΔH^0 and ΔS^0 at 298 K. Is the reaction entropy-driven at 2400 K?

C2. Assuming that all the energy released in the N_2H_4–N_2O_4 reaction is used to heat the gaseous products, calculate the flame temperature of the reaction for ideal stoichiometry. Calculate the flame temperature again under the assumption that 10% of the H_2O dissociates to the ideal-gas species OH and H, breaking an O—H bond in the process. (Heat capacities can be found in the *CRC Handbook of Chemistry and Physics*.)

C3. Hexamethylenetetramine is more soluble in cool water than in warm water. Below about 10°C, a hexahydrate HMT · $6H_2O$ can be crystallized. What kind of bonding is going on in the compound? What relation does this have to the unusual temperature dependence of the solubility? [See T. C. W. Mak, *J. Chem. Phys.* (1965), *43*, 2799.]

HO   NOH
  | ||

C4. Alpha-hydroxyoximes R_2C—CR are used to extract copper from ore containing a low percentage of the metal. The ore is leached by a dilute sulfuric acid solution, then extracted by the oxime dissolved in kerosene. What chemical reaction is occurring? How is the copper recovered? How is the oxime recovered for reuse? [See *Kirk–Othmer Encyclopedia of Chemical Technology,* 3rd ed., Wiley-Interscience, New York, 1978–1983.]

Part **IV**

Transition-Metal Compounds

10

The Properties of Transition Metals and Their Compounds

Most of our discussion thus far has dealt with the structure and reactions of main-group elements—elements having s and p valence electrons only. The remainder of the book will be devoted to the chemistry of transition elements, those having valence d and f electrons. This perhaps seems a rather trivial distinction to make, but in fact this single change opens up entirely new types of orbital overlap and bonding, allowing the formation of compounds with quite different chemical properties. The inorganic chemist's attraction has focused particularly on d-electron transition elements, usually called simply the transition metals. The $4f$-electron transition elements are usually known as lanthanides or rare earths; those with $5f$ valence electrons are known as actinides. This chapter deals primarily with the d-electron transition metals, and makes a few comparisons with lanthanides and actinides. We shall look at the nature of d orbitals and electrons, the bonding of these electrons in metals, the nature of the metals themselves, typical ionic lattices for the d block metals, and the basis for superconductor behavior in these lattices.

Since the distinction between main-group metals and transition metals depends on the orbitals occupied, there is little difference in the properties of the predominantly ionic compounds of the two groups. Ionic bonding, of course, can be described quite well on a "billiard-ball" basis without explicitly considering orbitals or overlap. Table 10.1 suggests that the predominantly ionic transition-metal fluorides have much the same physical properties as the main-group ionic fluorides with the same metal charge and thus comparable lattice and solvation energies. For both groups, the chlorides are significantly less ionic, but the parallels are still quite strong. There are some differences in transition-metal ionic lattice formation that can be explained on purely electrostatic grounds, because of the directional character of d orbitals and the fact that most transition metal ions have remaining valence

489

Table 10.1   Properties of Main-Group and Transition-Metal Ionic Halides

	Main group		Transition metal		Main group	Transition metal
	MgF$_2$	**CaF$_2$**	**MnF$_2$**	**CuF$_2$**	**AlF$_3$**	**FeF$_3$**
mp(°C)	1263	1418	920	785	1040	>1000
solubility (g/100 g H$_2$O)	0.013	0.002	0.66	4.7	0.56	slightly soluble
	MgCl$_2$	**CaCl$_2$**	**MnCl$_2$**	**CuCl$_2$**	**AlCl$_3$**	**FeCl$_3$**
mp(°C)	714	782	652	620	194	306
solubility (g/100 g H$_2$O)	35	74	72	70	70	74

d electrons; this is the basis of crystal field theory, which we shall delay until Chapter 11.

Transition-metal compounds in which the bonding is predominantly covalent are sometimes quite like main-group compounds with similar stoichiometry, but they can show striking differences. For example, the compounds CH_3TiCl_3 and CH_3SnCl_3 are very similar: The first melts at 29°C, the second at 46°C, and both are readily soluble in hexane. On the other hand, main-group metals form no compounds that resemble the transition-metal carbonyls such as $Fe(CO)_5$, $Ni(CO)_4$, and others. The carbonyls represent a group of compounds in which the bonding is uniquely dependent on the presence of valence d electrons and the overlap of d orbitals. We shall explore this distinction in later chapters; it is in fact the basis for the organization of the book.

The essential differences in d-orbital overlap arise from the angular dependence of the orbitals, and also from their radial dependence relative to the other valence electrons and the inner core of the atom. As Fig. 1.4 indicated, angular nodes are more numerous for d orbitals, and their overlap is correspondingly more directional in its character. The numerous nodes make pi overlap convenient, but this pi overlap is in itself somewhat different from pi overlap of p orbitals, as Fig. 5.28 and the associated discussion of phosphonitrilic compounds has indicated. The radial dependence (shown in Fig. 1.5 for hydrogen-like orbitals) is different from that of most s and p orbitals in that the electron does not penetrate the inner core much but is still in general closer to the nucleus than the $(n + 1)$ s electrons also present in the neutral atom. This affects the ionization energy of the d electrons and thus their availability for chemical combination. Ultimately this produces more-or-less nonbonding d electrons, often unpaired, in many transition metal compounds. The energies of these electrons and the possibility that they will give the atom or molecule a net electron spin are the foundations of several important experimental techniques, particularly ultraviolet–visible spectrophotometry and magnetochemistry.

10.1 THE ELEMENTS WITH d AND f VALENCE ELECTRONS

The Schrödinger equation for hydrogenlike atoms leads to series solutions in r^{n-1} and $\cos^l \theta$ and to a group of solutions $e^{im\phi}$ for the quantum numbers n, l, and m, where the relations between the separated one-dimensional differential equations require $n > l \geq |m|$ for a given three-dimensional solution or wave function. The m quantum number counts angular nodes containing the z axis in the atom's coordinate system for a given wave function or orbital. The l quantum number counts the total number of angular nodes for that orbital. The n quantum number counts the total number of nodes (angular + radial), including one that exists for every orbital at $r = \infty$. All d orbitals thus have two angular nodes, and f orbitals have three. There are five nd orbitals (where n must be at least 3) corresponding to m values of +2, +1, 0, −1, and −2, and seven nf orbitals (where n must be at least 4) corresponding to m values of +3, +2, +1, 0, −1, −2, and −3. These are shown in schematic form in Fig. 10.1. To understand our later discussion, the reader should be able to sketch the d orbitals in the proper relation to their coordinate axes. The f orbitals are more esoteric chemically as well as being harder to draw; they can be looked up when required.

The radial dependence of the d orbitals is responsible for the position of the transition metals in the periodic table. For potassium at $Z = 19$, with an argon inner core of $1s^2 2s^2 2p^6 3s^2 3p^6$, the nineteenth electron is most stable in the $4s$ distribution because that orbital has small inner lobes that place part of the overall $4s$ electron density near the nucleus with a charge of 19+, even though most of the $4s$ density is well outside the $3s$–$3p$ edge of the inner core. By contrast, the $3d$ distribution for K is on the average closer to the nucleus than the $4s$—only slightly farther away than the $3s$ and $3p$, according to SCF calculations. However, because the $3d$ has no radial nodes, the electron cannot penetrate near the highly charged nucleus. Although the $4s$ orbital is more diffuse than the $3d$, it is more stable—for potassium—because not all of its distribution is effectively shielded from the nuclear charge by the inner core.

The relative energies of the $4s$ and $3d$ orbitals change, however, as the nuclear charge and the number of valence electrons increase. Figure 10.2 shows the inner-core distribution and the relative positions of the $3d$ and $4s$ orbitals for Ti at $Z = 22$, where there are now four electrons beyond the Ar inner core. For a $4s$ electron, the small inner lobe that provides a major portion of the electron's stability has experienced only a small increase in the effective nuclear charge (from potassium's 19+ to 22+). For the $3d$ electrons, which lie almost entirely inside the $4s$, the effective nuclear charge has increased from about 3+ to about 8+, as shown in Fig. 1.8.

This sharp increase occurs for two reasons: First, the nuclear charge balanced by the two $4s$ electrons is almost entirely transmitted to the $3d$. Second, the $3d$ probability maximum progressively moves from the outer edges of the $3s$–$3p$ core density toward the inside of it as Z increases (compare Ti and Zn in Fig. 10.2). Thus after the $4s$ orbital is filled the $3d$ is greatly preferable to the $4p$, so the transition

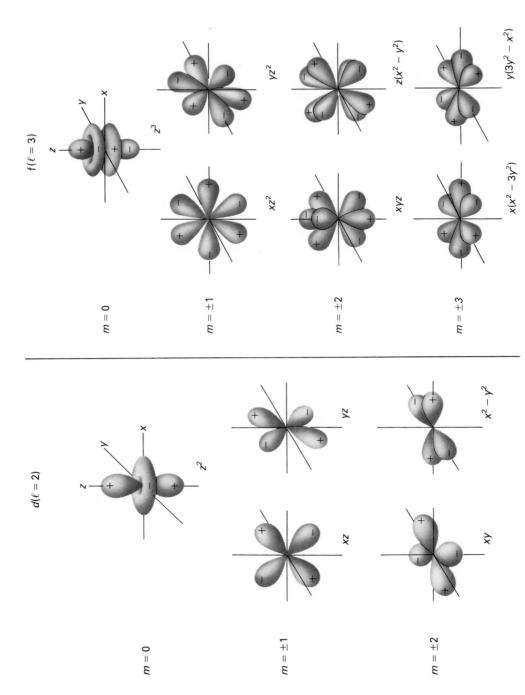

Figure 10.1 *d* and *f* orbitals.

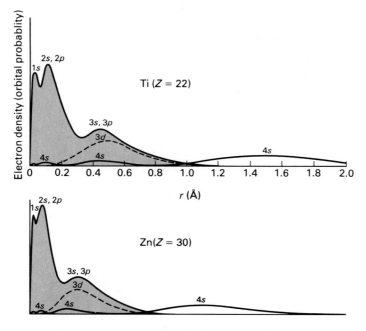

Figure 10.2   Radial dependence of orbitals and electron densities for transition metals. (3d and 4s orbitals are vertically exaggerated for clarity relative to the inner core.)

metals occupy the next 10 spaces in the periodic table. With minor exceptions, all of the transition metals have the electron configuration s^2d^n as neutral atoms (see Table 1.6).

However, consider the relative orbital energies in a transition-metal atom ionized down to its inner core, such as Ti^{4+}. If this ion gains an electron (is reduced to Tl^{3+}), the electron will adopt a 3d distribution, not the 4s. The net charge on the ion dominates the effective nuclear charge, and the close 3d distribution is preferable to the distant 4s distribution. If another electron is added, the net charge decreases to 3+, but this is still large enough to favor the 3d distribution for the next electron. At 2+, however, the net charge is no longer large enough to favor the 3d distribution, so the last two electrons occupy the 4s orbital. Thus, for titanium we find the following ground states for the neutral atom and ions:

Ti^0	Ti^+	Ti^{2+}	Ti^{3+}	Ti^{4+}
$Ar4s^23d^2$	$Ar4s^13d^2$	$Ar3d^2$	$Ar3d^1$	Ar

Although the 4s electrons go in first when the periodic table is built up, they also come off first when any individual transition metal is ionized.

As Z increases going across the row of transition metals for Ti to Zn, Fig. 1.8 gives the changes in effective nuclear charge experienced by the valence electrons. The $4s$ electrons become only modestly more stable, because each added proton is almost completely shielded from the $4s$ electrons by the added $3d$ electron. Only the small fraction of $4s$ electron that penetrates near the nucleus is affected by the slow increase in true nuclear charge (36% increase from Ti to Zn). The $3d$ electrons, however, become much more stable. As the number of protons increases, the added electrons are all in $3d$ distributions at about the same distance from the nucleus. They do not shield each other effectively, so the effective nuclear charge increases sharply: from about 8+ for Ti to about 14+ for Zn. The result is that $3d$ electrons are readily available for chemical combination on the left side of the transition-metal d-block, but much less so on the right side. The common oxidation states reflect this pattern. Titanium is usually in the 4+ oxidation state. The 3+ and 2+ oxidation states are known, but they are good reducing agents (electrons are readily removed). The common oxidation states of manganese are 2+, 4+, and 7+, and all the intermediates are known as well. Mn^{4+} is a good oxidizing agent, and Mn^{7+} is a very strong one. Nickel is known almost exclusively as Ni^{2+}. Ni^{3+} and Ni^{4+} are rare, and are powerful oxidizing agents. Zinc is known only in the 2+ oxidation state. The $4s$ electrons are lost fairly easily, but then ten $3d$ electrons never are.

This discussion suggests accurately that multiple oxidation states will be possible for most transition metals, save only those at the far left and far right of the d block. For simple compounds such as chlorides and sulfates, Table 10.2 indicates the common oxidation states that are seen, organized as the d-electron configurations seen in more-or-less ionic compounds. It must be emphasized that these are by no means all the known formal oxidation states of the transition metals, which are shown in Table 10.3 by element. In general, very low oxidation states and very high oxidation states are known only in predominantly covalent compounds, whose bonding and properties we will delay for later chapters. Most elements appear in Table 10.2 more than once—that is, multiple oxidation states are stable in ionic surroundings for most transition metals.

Why do we see this variety of oxidation states, when most main-group elements have only a single oxidation states for ionic compounds and often only a

Table 10.2 *d*-Electron Configurations in Ionic Surroundings

d^1	d^2	d^3	d^4	d^5	d^6	d^7	d^8	d^9
Ti^{3+}	V^{3+}	Cr^{3+}	Mn^{3+}	Mn^{2+}	Fe^{2+}	Co^{2+}	Ni^{2+}	Cu^{2+}
	Ti^{2+}	V^{2+}	Cr^{2+}	Fe^{3+}	Co^{3+}		Cu^{3+}	
		Mn^{4+}						
Zr^{3+}	Mo^{4+}	Mo^{3+}		Ru^{3+}	Rh^{3+}		Pd^{2+}	Ag^{2+}
		Te^{4+}						
Hf^{3+}	W^{4+}	Re^{4+}	Os^{4+}	Ir^{4+}	Pt^{4+}		Pt^{2+}	
					Ir^{3+}		Au^{3+}	

Table 10.3 Formal Oxidation States of Transition Metals

Predominant species	Oxidation state	First row								Second row								Third row							
		Ti	V	Cr	Mn	Fe	Co	Ni	Cu	Zr	Nb	Mo	Tc	Ru	Rh	Pd	Ag	Hf	Ta	W	Re	Os	Ir	Pt	Au
Oxides, oxyanions, fluorides ⎰	8+													×								×			
	7+				×								◯								◯	×			
⎱	6+			×	×	×						◯	×	×						◯	×	×	×	×	
	5+		×	×	×						◯	×	×	×					◯	×	×	×	×	×	×
	4+	◯	◯	×	×	×	×	×		◯	×	×	×	◯	×	×		◯	×	×	×	◯	×	◯	
	3+	×	×	◯	×	◯	×	×	×	×	×	×	×	×	◯	×	×	×	×	×	×	×	◯		◯
	2+	×	×	×	◯	×	◯	◯	◯	×	×	×	×	×	×	◯	×	×	×	×	×	×	×	×	
Carbonyls, clusters, organometallics ⎰	1+								×						×	×	◯				×		×		×
	0	×	×	×	×	×	×	×		×	×	×	×	×	×	×		×	×	×	×	×	×	×	
	1−	×	×	×	×	×	×	×		×	×	×	×	×	×	×		×	×	×	×	×	×	×	
	2−		×	×	×	×				×	×	×	×	×				×	×	×	×	×	×		
	3−		×	×	×	×	×			×	×	×			×				×	×	×	×	×		
⎱	4−			×							×									×					

Note: Circles show most stable oxidation state under common conditions.

single oxidation state in any compound? The enthalpy data in Table 10.4 give the thermodynamic basis of this stability and the distinction between the oxidation states of transition metals and main-group metals. The three transition metals and aluminum all form the 2+ and 3+ oxides exothermically (presumably spontaneously, since the enthalpy changes are large). Ti and Mn form the 4+ oxide spontaneously, but Ni does not (presumably because its fourth ionization potential is so large that the NiO_2 lattice energy cannot overcome it). Neither, of course, does Al, because its fourth ionization potential is enormous. Atmospheric oxidation of MO to M_2O_3 should also occur spontaneously, though it is a close question for NiO. The oxidation of M_2O_3 to MO_2 should be spontaneous for Ti and Mn, but certainly not for either of the other two metals.

The proof that the multiple oxidation states are stable for the transition metals, however, lies in the fact that all of the disproportionation reactions are endothermic for the transition metals (TiO, in fact, is prepared by heating Ti and TiO_2), whereas the lower oxidation state of Al (in the hypothetical AlO) should disproportionate to Al_2O_3 and Al almost explosively. The energy costs for disproportionation to AlO_2 and Al, of course, are so high that they are totally beyond the range of chemical binding energies. Because d-electron ionization energies are intermediate between the very low energies required to ionize the alkali and alkaline-earth metals and the very high energies required to ionize closed (noble-gas) shells, the increased lattice energy for a higher charge just about balances the ionization-energy cost. The intermediate character of the ionization energy reflects the radial distribution of most d electrons: closer to the nucleus and more tightly bound than most of the s electron density, but with no inner lobe of density near the highly charged nucleus and thus

Table 10.4 ΔH^0 Values for Ionic Transition-Metal Oxides (kJ/mol reaction)

	Ti	Mn	Ni	Al
Direct formation				
$M + \frac{1}{2}O_2 \rightarrow MO$	−519	−385	−238	(−184)
$2M + \frac{3}{2}O_2 \rightarrow M_2O_3$	−1519	−958	−490	−1674
$M + O_2 \rightarrow MO_2$	−941	−519	(+653)	(+4745)
Oxidation of lower oxide				
$2MO + \frac{1}{2}O_2 \rightarrow M_2O_3$	−481	−188	−12	(−1305)
$M_2O_3 + \frac{1}{2}O_2 \rightarrow 2MO_2$	−368	−84	(+1799)	(+11160)
Disproportionation				
$3MO \rightarrow M + M_2O_3$	+38	+197	+226	(−1121)
$2MO \rightarrow M + MO_2$	+92	+251	(+1130)	(+5113)
$2M_2O_3 \rightarrow M + 3MO_2$	+205	+351	(+2937)	(+17590)
$M_2O_3 \rightarrow MO + MO_2$	+59	+50	(+904)	(+6234)

Note: The values in parentheses are calculated by Kapustinskii lattice-energy approximation for non-existent NiO_2, AlO, and AlO_2.

less tightly bound than closed-shell *p* electrons. Accordingly, several oxidation states are usually stable for any given transition metal in ionic compounds such as oxides and fluorides.

Digressing briefly from ionic compounds, we find that the multiplicity of oxidation states for transition metals is quite striking if all the formal states found in complexes and organometallic compounds are included. Table 10.3 has indicated these schematically and noted the oxidation state most stable under common conditions. Note that the heavier transition metals tend to have higher oxidation states than the lighter ones, particularly toward the left side of each row. Thus chromium is usually more stable as Cr(III), but molybdenum and tungsten are most commonly found as Mo(VI) and W(VI). In each row, the increasing effective nuclear charge as the *d* orbitals are filled makes the later elements stable in lower oxidation states. Rhenium is found in many Re(VII) species, but platinum is most stable as Pt(IV), and PtF_6 is one of the most powerful oxidizing agents known.

Similar arguments on relative valence electron stability apply to the lanthanide metals as the set of 4*f* orbitals is filled, but in terms of available oxidation states, the outcome is quite different. Going beyond the Xe core, cesium favors the 6*s* orbital for its valence electron (again because of penetration) over the 5*d* or 4*f*. As Fig. 10.3 suggests, the 5*d* orbital penetrates the inner core to some extent, but not

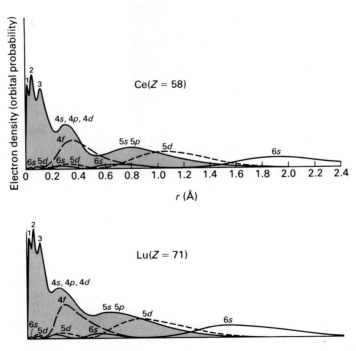

Figure 10.3   Radial dependence of orbitals and electron densities for lanthanides. (4*f*, 5*d*, and 6*s* orbitals are vertically exaggerated for clarity relative to the inner core.)

significantly inside the $n = 3$ shell where the highest nuclear charge is experienced. On the other hand, the $4f$ orbital is completely buried inside the inner core for Cs or any of the lanthanides; it does not fill earlier because its r^6 electron-probability dependence so completely prevents it from penetrating inside the $n = 4$ shell that it does not experience as great an effective nuclear charge as the $5p$, $6s$, or $5d$. Even though the $4f$ orbitals are well shielded, however, they become more attractive as Z increases. Cs and Ba fill the $6s$ orbital and La narrowly prefers the $5d$ to the $4f$, but at $Z = 58$ (Ce), the $4f$ orbitals become more stable than the $5d$ orbitals and the lanthanide series begins. As with the transition metals, low net atomic charges favor the s orbital while high net charges stabilize the f orbitals, so all of the lanthanide neutral atoms have the configuration $6s^2 4f^n$ except where the minimization of electron–electron repulsion associated with a set of orbitals completely filled by parallel spins makes the f^7 configuration preferable (gadolinium, Gd). The ions, however, lose the $6s$ electrons first—but in general lanthanides do not have a stable 2+ oxidation state in compounds. Instead, they all show a stable 3+ oxidation state, which is the net charge that strikes the best balance between the ionization-energy cost and the lattice-energy or solvation-energy stabilization of the ion.

Among the lanthanides, departures from the 3+ oxidation state occur for elements that can approach f^0, f^7, or f^{14} configurations by adopting a 2+ or 4+ oxidation state. Ce, Pr, and Nd (just past f^0) and Tb and Dy (just past f^7) show 4+ oxidation states; Sm and Eu (just short of f^7) and Yb (just short of f^{14}) show 2+ oxidation states. However, all 4+ species are strong oxidizing agents, and all 2+ species are strong reducing agents. Ce^{4+} is a widely used analytical oxidant; YbI_2 dissolves in liquid ammonia to yield Yb^{3+} and ammoniated electrons, as the alkali metals do.

The trends in radial electron distribution visible in Figs. 10.2 and 10.3 for the transition metals and lanthanides are reflected most clearly in the ionic radii of these cations. Figure 10.4 shows these trends in data taken from Table 2.3. For the transition metals there are obviously two factors at work; initially the increase in nuclear charge and the partially buried nature of the $3d$ orbitals leads to a decrease in ionic radius, as Fig. 10.2 will have suggested. However, this trend reverses sharply at Fe because of the symmetry properties of the $3d$ orbitals, which are not entirely buried in the inner core. The symmetry factor is intimately connected to the potential-energy field of the crystal in which the radii are measured, a topic for the next chapter.

For the lanthanides, whose $4f$ electrons are entirely buried in the inner core, the increasing nuclear charge leads to a smooth contraction from $Z = 57$ to 71. This trend is known as the *lanthanide contraction*. It has some chemical effects of interest. For example, the proton-acid pK_a of hydrated Lu^{3+} is significantly smaller than that for La^{3+} (7.6 versus 8.5) because the smaller Lu^{3+} ion can polarize surrounding water molecules more strongly. Just beyond the lanthanides, the lanthanide contraction leads to essentially identical radii for Zr and Hf (0.86 versus 0.85 Å), which makes the chemical separation of Zr from Hf one of the most difficult in the whole periodic table.

In general, the chemical properties of the transition metals in their most stable oxidation states (usually 2+ or 3+) resemble the properties of the main-group met-

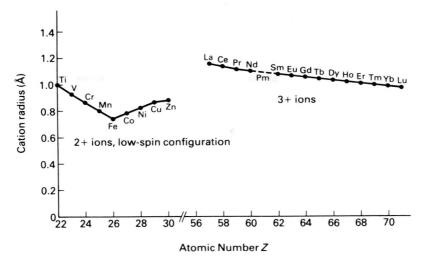

Figure 10.4 Six-coordinate cation crystal radii for transition metals and lanthanides.

als that have the same charge and roughly the same electronegativity (see Fig. 1.17), particularly aluminum. Table 10.1 has already noted some of these similarities. Because transition metals have low-energy unfilled *d* orbitals, however, they are particularly good Lewis acids toward an incredible variety of electron-pair donors; Chapter 12 will explore this behavior in much more detail.

The directional nature of partially filled *d* orbitals has another effect that distinguishes the transition metals from the main-group metals. Within elemental crystals—the bulk metals—the overlap of directional *d* orbitals provides a three-dimensional network of covalent bonds within the metal that is quite different from the delocalized "electron gas" that characterizes an idealized metal. The result is that transition metals are physically stronger and have much higher melting points than main-group metals. We need to consider the origin of this stability in the context of band theory, and to examine some of the consequences in the nature of metals.

10.2 BONDING AND MAGNETISM IN *d*-ELECTRON TRANSITION METALS

Section 4.10 of Chapter 4 has set up the basic structure of band theory for solid materials and considered some of its consequences for the physical properties of main-group metals. The principles are the same for *d*-block metals, but the more complex geometry of *d* orbitals yields some differences that have profound physical consequences.

In our approach to basic band theory we saw that Avogadro's number of atomic orbitals for atoms in a lattice will combine to yield Avogadro's number of molecular orbitals for the lattice. This very large number of MOs is spread out over a limited energy range, however, because the coordination number of each atom remains a small number no matter what lattice symmetry is adopted. The MOs thus have energies so closely spaced that they effectively form a continuous band from lowest energy (most strongly bonding) to highest (most strongly antibonding). The width of the band will depend on the coordination number, but also on the degree of overlap of the AOs. This overlap is influenced by distance, as in Fig. 4.31, but also by orbital symmetry (sigma versus pi, etc.) and by lattice symmetry. Still, in all crystals of metals, the orbital energy spacing is so close that thermal energy kT vastly exceeds energy level spacing. As a result, we saw that for sodium metal the band of N_A energy levels (with capacity for $2N_A$ electrons) is half filled by the N_A electrons from the sodium atoms. The electrons are not, however, all paired; thermal excitation fuzzes out the occupation within about kT of the highest formally occupied level, known as the Fermi level, as Fig. 4.32 has indicated. As a result of the vanishingly small energy gap from HOMO to LUMO in sodium, it is extremely easy to insert an electron into one end of the lattice and withdraw one from the other end, which is the process of electrical conduction. Metals are conductors because the Fermi energy occurs not at the boundary of a band but in the middle of it, or in the overlapping envelope of two bands.

With this general reminder in place, we need to consider the nature of these bands in more detail. The first consideration is that the width of a band will not be the same in all directions unless the lattice symmetry is also the same—that is, cubic. As we have seen in Chapter 4, the running index that counts MOs in the band, starting at the bottom, is the number of nodes in the MO: in effect, the wavenumber for the electron wave. In a noncubic crystal this wavenumber k will increase at a different rate in each direction. As a result, the Fermi level will be reached at a different value of k_x from k_y or k_z. We can imagine a three-dimensional graph of energy levels in "k space," in which the Fermi level would be a three-dimensional surface—spherical for a cubic crystal but some other shape for different lattice symmetry. Such a shape is the *Fermi surface* for the lattice. We will need to think in terms of this Fermi surface rather than a one-dimensional Fermi level to understand the electrical and magnetic properties of the transition metals.

The second consideration is that the varying symmetries and overlap of AOs combining to form a band mean that the spacing of energy levels within a band will not be uniform from top to bottom of a band. A glance at Fig. 4.30 will show that, for 10 orbitals in one dimension, the resulting MO energies are more closely spaced at the outer edges of the "band" than in the center. For a one-dimensional system this will be the case generally, but for three-dimensional crystals calculations show that spacing is smallest roughly in the middle of a band because that is where the most variations of the n_x, n_y, and n_z quantum numbers independent of each other yield the same overall bonding effect and thus the same energy. Figure 10.5 shows this behavior by introducing a new term, the *density of states,* meaning the relative number of energy levels in an arbitrarily small energy increment. Figure 10.5 shows several density-of-states diagrams—the one-dimensional case but also the diagrams for Na, Ca, and an electrical insulator. In each case, the Fermi

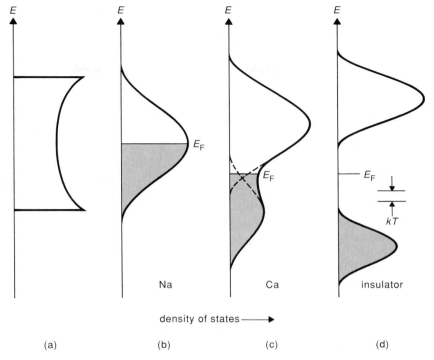

Figure 10.5 Density-of-states diagrams: (a) one-dimensional array of atoms; (b) *s* band for Na metal; (c) overlapping *s* and *p* bands for Ca metal; (d) well-separated bands for an electrical insulator.

level is shown; the diagrams have an obvious relationship to those of Fig. 4.32, which uses fractional orbital occupancy rather than the density of states. Density-of-states diagrams are sometimes drawn with the energy axis horizontal, which gives them the appearance of an absorption or emission spectrum that corresponds well with observations from photoelectron spectroscopy. However, our density-of-states diagrams will keep a vertical energy axis to emphasize their origin in MO energy-level diagrams.

For transition metals at least some of the overlapping AOs that form bands will be *d* orbitals, which have more angular nodes and are thus more directional than *s* or *p* orbitals. This property means that there will be a relatively large number of AO combinations that have essentially nonbonding overlap and nearly the same energy. Figure 10.2 also indicates that *d* orbitals are to some degree buried in the 3*s* and 3*p* core, reducing overlap somewhat even for ideal overlap orientations. To the extent this is true, the binding energy integral β is reduced. Both of these effects tend to narrow the *d* band in bulk transition metals. Coupled with the fact that each atom has five *d* orbitals, the overall result is that the density of states in the *d* band will be quite high for these metals.

In narrow bands containing a large number of orbitals with nearly the same energy, and for which the Fermi surface falls in the high density-of-states region, a correspondingly large number of electrons near the Fermi surface will have the same energy—that is, will be accidentally degenerate. This is the situation commonly found for transition metals with their d orbitals mostly but not completely filled, such as Fe–Ni. When this extensive accidental degeneracy occurs, exchange energies become important in determining the spatial and spin properties of each electron. As a result, there is an energy benefit available to the transition metal analogous to that shown in Fig. 1.7; the exchange energies favor parallel spins. We can represent this situation in a density-of-states diagram if we use a two-sided diagram in which one side shows the density of states available to electrons with spin up (α) and the other side shows the density of states available to electrons with spin down (β); one arbitrary spin orientation will be preferable to the other though the two density-of-states diagrams will still have the same general shape. Figure 10.6 shows such a diagram for Ni, which has its bands and Fermi surface appropriately located for this argument. Note that the Ni crystal has a substantial surplus of spin α electrons over spin β electrons. Such a crystal is said to be *ferromagnetic;* a large proportion of its valence electrons are locked into a parallel orientation, and as a result the crystal has a large net magnetic mo-

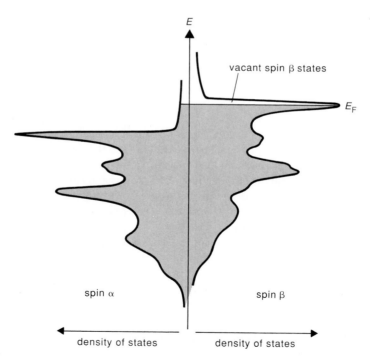

Figure 10.6 Calculated density-of-states diagram for Ni metal, showing spin-up and spin-down states separately.

ment. It will appear from the calculated energy bands that the Fermi energy is somewhat too high in that an even greater disparity between total spin α and spin β electrons could exist if there were fewer electrons, lowering E_F with respect to the energy bands. In fact, although nickel is ferromagnetic it has a relatively low density of aligned spins: about 0.6 net aligned spin per atom. Cobalt, with one fewer electron, has more aligned spins (about 1.6 per atom) and iron still more (about 2.1 spins per atom). Alloys can provide fractional-electron adjustments to band occupancy, and Fe–Co alloys yield about 2.4 aligned spins per atom at about 40% Fe, 60% Co.

The ferromagnetism observed for the iron group of metals does not usually lead to observable net magnetic moments for metal samples, even though the spin alignment is real. If the metal object is cast from molten metal, the aligned spins will exist in magnetic *domains*—regions of the metal crystal that have spins aligned in a random direction relative to a neighboring domain. Most domains have diameters of only 1–100 microns, so that only metal samples substantially smaller than that, 0.1 micron or less, can be expected to show spontaneous magnetization. However, when a ferromagnetic metal object is placed in an external magnetic field its domains with net magnetic moment parallel to the applied field will grow at their edges with respect to other domains with other spin orientations; if the applied field is strong enough whole domains will rotate their magnetization axes. The aligned-spin values mentioned above represent the *saturation magnetic moment* when maximum alignment with the external field has been achieved.

As a result of the spontaneous ferromagnetism of the later transition metals, permanent magnets are possible, and a number of magnetic alloys have been developed in addition to the ferrites described in Chapter 3. These fall into two categories, soft and hard. Soft magnetic materials are used for applications in which we want to induce a high magnetic field as a result of an electrical signal and have it go away just as quickly. Hard magnetic materials are used when we wish to create a net magnetic moment in an object and have that net magnetization remain permanently. To understand the physical properties that allow these varied behaviors, we need a graphical representation of the behavior of materials in an applied magnetic field. The defining equation is

$$B = \mu_0(H + M)$$

where B is the magnetic induction or flux density, H is the applied field, and M is the magnetization. If we start with a virgin unmagnetized magnetic material and slowly apply a field, a graph of B versus H follows the path shown in Figure 10.7 from the origin up to point S. At point S the high applied field H has achieved saturation of the magnet. A soft magnetic material will rise steeply along its path; B/H, its magnetic *permeability*, will be high. If we now decrease the applied field H, we do not see the magnetization M fall significantly until we have actually reversed the direction of the applied field. Point C represents the *coercivity*, the opposing applied field H_c required to neutralize the net magnetization of the material. Soft magnetic materials will have very low values of coercivity, so that their temporary magnetization is readily removed; hard magnetic materials will have high

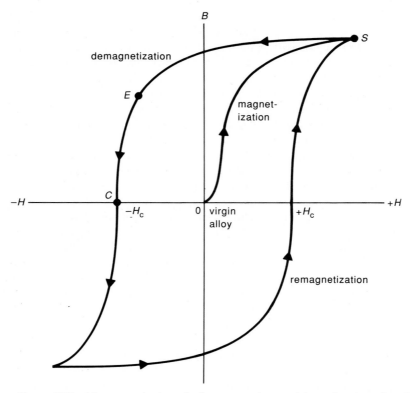

Figure 10.7   Net magnetization of a ferromagnetic material as a function of applied external magnetic field. Points S, E, and C defined in text.

coercivity. An important characteristic of permanent magnets is their *maximum energy product B · H,* represented by point E on the graph; hard magnets will in general have a high energy product, which means that a given magnetic flux density can be achieved with the smallest possible magnet.

In commercial applications, soft magnetic alloys include soft iron (low carbon content) and low-carbon steel, Fe–Si, Fe–Al, Fe–Al–Si, Fe–Co, and Ni–Fe compositions. The magnetic recording head material Sendust is 85% Fe, 5.6% Al, 9.5% Si, for example. Other soft magnetic alloys include the Permalloy family (Fe with 35–90% Ni), which is widely used for transformer cores and magnetic shielding; Hiperco (Fe with about 35% Co), which has the highest saturation magnetization of some 24,500 gauss; and Permendur (equal Fe and Co with 2% V). All of these magnetic alloys are reasonably good electrical conductors, which means that at high ac frequencies they will lose energy because of eddy currents. As a result, for applications in which frequencies are above about 10 kHz soft ferrites are used, which have high electrical resistivity and thus lose little energy in eddy currents.

Hard magnetic alloys include the Alnico family, which have compositions in the following general ranges: 7–12% Al, 14–25% Ni, 5–40% Co, ~3% Cu, 0–8%

Ti, and the rest iron. A typical Alnico magnet might be 8% Al, 15% Ni, 25% Co, 3% Cu, and 49% Fe. Ferromagnetism is also seen in the *f* block metals, and there are some commercial rare-earth magnets with compositions such as $SmCo_5$ and $(RE)_2Co_{17}$. The rare-earth magnetic materials have very high energy products and are used extensively in loudspeaker magnets. Other hard magnetic alloys include Cr–Co–Fe, Cu–Ni–Fe, Pt–Co, Mo–Co–Fe, and V–Co–Fe compositions. Vicalloy (10% V, 52% Co, 38% Fe) has the unusual property for magnetically hard alloys of being ductile, so that it can be rolled into thin sheets; it is widely used for antitheft tags by department stores and for book security by libraries.

We should not leave the concept of density-of-states diagrams without pointing out that the diagrams' usefulness is not limited to elements or alloys. Band theory is used to describe bonding in solids of all compositions, and takes a simple intuitive form for solids with primarily ionic bonding such as TiO_2. In the TiO_2 rutile lattice (Fig. 2.15) the atom–atom contacts are between Ti and O, whose electronegativities and valence orbital ionization potentials differ so much that the resulting molecular orbitals will be very lopsided in electron ownership: The low-energy orbitals will be almost entirely O 2*s* and 2*p*, some medium-energy orbitals will be Ti 3*d* orbitals that are geometrically nonbonding, and the high-energy antibonding orbitals will be mostly Ti 3*d*. Figure 10.8 shows the density-of-states diagram for TiO_2 in three forms. Part (a) shows the basic diagram, part

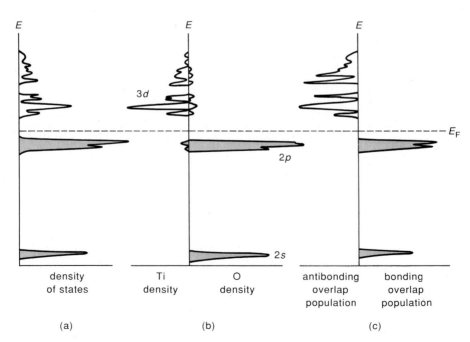

Figure 10.8 Density-of-states diagram for TiO_2: (a) basic diagram; (b) diagram showing relative atomic electron ownership for each band; (c) diagram showing net bonding or antibonding contribution to overall lattice within each band.

(b) uses a two-sided diagram to assign electron ownership to Ti and O in the lattice, and part (c) uses a two-sided diagram to indicate bonding/antibonding character for each band. Because the VOIP difference is so great, electron transfer is extensive as part (b) shows, and the magnitude of the bonding integral is small so that the bands are narrow. The Fermi energy is just above the band that corresponds primarily to O $2p$ electrons (in the $Ti^{4+}O^{2-}$ oxidation state formalism, O^{2-} will have its $2p$ band filled). The energy gap to the lowest vacant band means that TiO_2 should be an insulator. In fact, stoichiometric TiO_2 has a resistivity of 10^{13} ohms—but if the lattice has some oxygen vacancies, corresponding to partial reduction to Ti^{3+} or even Ti^{2+}, the missing O valence electrons constitute holes in the highest filled band and the solid becomes a semiconductor with resistivity as low as 10^{-1} ohm.

In less ionic transition metal compounds such as the sulfides, the increased covalent bonding would broaden the bands, cause some overlapping of bands, and more evenly divide the bonding electrons, which would complicate the simple assignments we have made for TiO_2. Still, in principle the same approach can be used to describe the electrons within these crystals. We shall shortly see that the electron distribution in such lattices has important consequences for contemporary superconductor research.

10.3 TRANSITION METALS AS METALS

First copper metal, then iron metal characterized the emergence of human societies from a primitive state; massive industrialized iron and steel production was the central event of the Industrial Revolution. How do we obtain the transition metals, and what uses do they find today? Given the properties of physical strength, high melting point, electrical conductivity, and possible ferromagnetism that we have considered, it is obvious that the uses will be diverse in a technological society.

In nature, most transition metals are found in a wide variety of surroundings, primarily depending on whether the mineral was deposited under reducing or oxidizing conditions. Transition metals with few d electrons are most stable in fairly high oxidation states and, for both of these reasons, are hard Lewis acids. In minerals, they are almost always found in an environment of hard-base oxygen atoms either as an oxide, a silicate, or a mixed oxide such as ilmenite, $FeTiO_3$. Conversely, the transition metals toward the right side of the d block have many d electrons and are found in sulfide or arsenide minerals—that is, combined with a soft base. The very softest Lewis-acid elements—copper, silver, gold, and platinum—are found to some extent as the uncombined native metals. Table 10.5 gives the predominant distribution of the transition metals in economically significant minerals, reflecting the pattern just described. However, not all of them are mined from a mineral source of that specific element, which is the way we usually think of ores. At least seven, principally the platinum-group metals, are normally obtained by separation and purification of the insoluble sludge residue from electrolytic

Table 10.5   Economically Important Minerals of the Transition Metals

Ti	V	Cr	Mn	Fe	Co	Ni	Cu
$FeTiO_3$ ilmenite	VS_4 patronite	$FeCr_2O_4$ chromite	MnO_2 pyrolusite	Fe_2O_3 hematite	$CoAs_2$ smaltite	$(Ni,Fe)_9S_8$ pentlandite	$CuFeS_2$ chalcopyrite
TiO_2 rutile	$Pb_9(VO_4)_6Cl_2$ vanadinite				$CoAsS$ cobaltite		$Cu_{12}Sb_4S_{13}$ tetrahedrite
	$KUO_2VO_4 \cdot \frac{3}{2}H_2O$ carnotite						native Cu

Zr	Nb	Mo	Tc	Ru	Rh	Pd	Ag
$ZrSiO_4$ zircon	$(Fe,Mn)(Ta,Nb)_2O_6$ columbite	MoS_2 molybdenite	synthetic element		All platinum metals occur mixed in minerals with formulas $(Pt)As_2$ sperrylite, $(Pt)S$ cooperite, braggite		Ag_2S argentite
ZrO_2 baddeleyite							$Ag(Cl,Br)$ embolite
							native Ag

Hf	Ta	W	Re	Os	Ir	Pt	Au
mixed in Zr minerals	$(Fe,Mn)(Ta,Nb)_2O_6$ tantalite	$CaWO_4$ scheelite	traces from copper-bearing molybdenites			native Pt	Native Au
		$(Fe,Mn)WO_4$ wolframite					

⟵——— predominantly oxides ———   ⟵——— predominantly sulfides, arsenides ———⟶

refining of copper or zinc, where they appear because they occur mixed in low concentration in the sulfide copper and zinc ores. This pattern is consistent with these metals' preference for a soft-base environment.

The extraction of metals from their ores has been described in general terms in Chapter 8. Table 10.6 presents specific data on abundance, commercial source, and reduction reactions for the more important transition metals. Because the principal use of several transition metals is in steel alloying, the preparative reactions include in several cases industrial reactions yielding products such as "ferrovanadium," which are up to half iron and are usually carbon-rich as well. Several of the metals can be purified by the *van Arkel–de Boer process,* which cycles a reversible reaction between the metal and iodine in the energy-favored direction at low temperatures and in the entropy-favored direction at high temperatures:

$$Ti_{(s,\ crude)} + 2I_{2(g)} \xrightarrow{\ 200°C\ } TiI_{4(g)} \xrightarrow[\text{W filament}]{\ 1300°C\ } Ti_{(s)} + 2I_{2(g)}$$

In a sealed container under vacuum or argon, a tungsten filament decomposes the TiI_4 vapor formed at a lower temperature by the crude titanium sponge and iodine vapor. This is in effect the same process by which a quartz–halogen lamp repairs its filament: a thin spot in the filament will be hotter than a neighboring region, and the halogen vapor present transports metal atoms from the cooler region of the filament to the thin spot, filling it in so that it cools.

The transition metals are in general hard and have relatively high melting points as a result of the network of covalent bonds within the metal lattice, as Fig. 4.33 and the associated discussion have suggested. In addition, they are reasonably good to very good electrical conductors, and as the previous bonding discussion has indicated several are ferromagnetic at room temperature. Table 10.7 provides a brief collection of data on these physical properties. Note that melting points are highest by far for third-row metals, that densities increase to the right but are markedly lower for the last one or two metals in each row, and that electrical conductivity is a rather erratic property whose only reliable feature is that it is best for the group Ib coinage metals with Mn and Hg being particularly poor conductors by metallic standards.

With respect to the chemical properties of the transition metals, most are reasonably electropositive so that they will dissolve in acids with evolution of hydrogen, although aqua regia ($HCl + HNO_3$) is required for the noble metals in the platinum group (Ru–Pd, Os–Pt) and Ru, Os, and Ir will not even dissolve in aqua regia. The metals all react with elemental F_2 or Cl_2, yielding the fluoride in the highest recognized oxidation state and a chloride in a lower oxidation state or a mixture of lower oxidation states. Most react with oxygen gas, and because we live in an oxygen atmosphere the nature of the oxides that form is critically important to the deterioration of metal objects by corrosion, and is responsible for a number of the uses of the metals. The electropositive metals at the left of the *d* block react very rapidly with atmospheric O_2 to form an oxide, but the oxide lattice is a good fit for the metal lattice at the surface and as a result the oxide film is bound to the metal by the strong forces of an ionic lattice. Once a complete film is

Table 10.6 Abundance, Sources, and Reduction Reactions for d-Block Metals

Element	Abundance in crust (ppm)	Commercial sources	Reduction reactions
Ti	6320	$FeTiO_3$ ilmenite TiO_2 rutile	$TiO_2 + C + Cl_2 \rightarrow TiCl_4 + CO$ $TiCl_4 + Mg \rightarrow Ti + MgCl_2$ (Kroll process)
V	136	VS_4 patronite Venezuelan crude oil	$VS_4 + O_2 + Na_2CO_3 \rightarrow NaVO_3 + SO_2 + CO_2$ recrystallize, fuse in air $\rightarrow$ crude V_2O_5 $V_2O_5 + Fe + C \rightarrow$ ferrovanadium $+ Co_2$
Cr	122	$FeCr_2O_4$ chromite	$FeCr_2O_4 + C \rightarrow$ ferrochrome $+ CO_2$ $Fe \cdot Cr + H_2SO_4 \rightarrow Cr(SO_4)_2^{2-}$ electrolysis $\rightarrow$ Cr
Mn	1060	MnO_2 pyrolusite	$MnO_2 + Fe + C \rightarrow$ ferromanganese $MnO_2 \rightarrow MnO + O_2 \rightarrow Mn(SO_4)_3^{2-}$ (H_2SO_4) electrolysis $\rightarrow$ Mn
Fe	62,000	Fe_2O_3 hematite Fe_3O_4 magnetite $FeCO_3$ siderite	$Fe_2O_3 + C \rightarrow Fe + CO_2$ pure Fe by electrodeposition from Fe^{3+}
Co	29	Cu, Pb ores	$CuCo_2S_4 + O_2 \rightarrow Cu \cdot Co + SO_2$ $Cu \cdot Co \rightarrow Cu^{2+} + Co^{2+} \rightarrow Co(OH)_3(OCl^-)$ electrolysis $\rightarrow$ Co
Ni	99	$(Fe,Ni)_9S_8$ pentlandite Cu sulfide ores	$(Cu,Ni)S + O_2 \rightarrow Cu \cdot Ni + Ni_3S_2 + SO_2$ $Ni_3S_2 + O_2 \rightarrow NiO + SO_2 \rightarrow Ni^{2+}$ (acid) electrolysis $\rightarrow$ Ni
Cu	68	Cu_2S chalcocite $CuFeS_2$ chalcopyrite	$Cu_2S + O_2 \rightarrow Cu + SO_2$ (roasting) Cu (crude, electrolysis) $\rightarrow$ Cu other metals in electrolysis sludge
Zn	76	ZnS sphalerite	$ZnS + O_2 \rightarrow ZnO + SO_2$ $ZnO + C \rightarrow Zn + CO_2$
Zr	220	$ZrSiO_4$ zircon ZrO_2 baddeleyite	$ZrSiO_4 + C \rightarrow ZrC + SiO + CO$ (arc furnace) $ZrC + Cl_2 \rightarrow ZrCl_4 + CCl_4 \rightarrow Zr$ (Kroll)
Mo	1.2	MoS_2 molybdenite Cu sulfide ores	$MoS_2 + O_2 \rightarrow MoO_3 + SO_2$ $MoO_3 + Fe + C \rightarrow$ ferromolybdenum $MoO_3 + H_2 \rightarrow Mo + H_2O$
Pd	0.015	Cu,Ni sulfide ores	Carried into Cu·Ni metal on roasting; separated from sludge in electrolytic refining
Ag	0.08	Ag_2S in Cu,Pb sulfide ores	Carried into Cu metal on roasting; separated from sludge in electrolytic refining
Cd	0.16	Zn,Cu sulfide ores	Separated from crude Zn metal by distillation (bp Zn = 905°C, Cd = 767°C)
W	1.2	$CaWO_4$ scheelite $(Fe,Mn)WO_4$ wolframite	$CaWO_4 \rightarrow H_2WO_4$ (from HCl sol'n) $H_2WO_4 + H_2 \rightarrow W + H_2O$ $FeWO_4 + Fe + C \rightarrow$ ferrotungsten
Ir	0.001	Cu,Ni sulfide ores	Carried into Cu·Ni metal on roasting; separated from sludge in electrolytic refining
Pt	0.01	Cu,Ni sulfide ores	Carried into Cu·Ni metal on roasting; separated from sludge in electrolytic refining
Au	0.004	native Au $AuTe_2$ calaverite	$Au + CN^- + O_2 + H_2O \rightarrow Au(CN)_2^-$ $Au(CN)_2^- + Zn \rightarrow Au + Zn(CN)_2$ (concentration)
Hg	0.08	HgS cinnabar	$HgS + O_2 \rightarrow Hg + SO_2$

Table 10.7 Physical Properties of the d-Block Metals

	Ti	V	Cr	Mn	Fe	Co	Ni	Cu	Zn
MP (°C)	1660	1887	1860	1244	1535	1495	1453	1084	420
density (g/cm^3)	4.54	6.11	7.19	7.44	7.87	8.90	8.90	8.96	7.13
resistivity (microohm-cm)	42	25	13	185	10	6.2	6.8	1.7	5.9
magnetism	—	—	—	—	ferromagnetic	ferromagnetic	ferromagnetic	—	—

	Zr	Nb	Mo	Tc	Ru	Rh	Pd	Ag	Cd
	1852	2468	2617	2172	2310	1966	1552	962	321
	6.51	8.57	10.22	11.5	12.41	12.41	12.02	10.50	8.65
	40	12	5.2	23	7.6	4.5	11	1.6	6.8
	—	—	—	—	—	—	—	—	—

	Hf	Ta	W	Re	Os	Ir	Pt	Au	Hg
	2230	2996	3400	3180	3054	2410	1772	1064	−39
	13.31	16.65	19.30	21.02	22.57	22.42	21.45	19.32	13.55
	35	12	5.7	19	8.1	5.3	11	2.4	94
	—	—	—	—	—	—	—	—	—

in place, further oxidation requires O_2 diffusion through the oxide, which is very slow. Accordingly, even though Ti is a very electropositive metal, its adherent film of TiO_2 protects it so well that Ti is, practically speaking, noncorroding. The same is true for most of the other transition metals, except iron. Even the brown film of Cu_2O that slowly forms on copper protects the metal fairly well, though it is basic enough to slowly absorb CO_2 to form the blue-green $Cu_2CO_3(OH)_2$ (or the equivalent sulfate) seen on copper roofs and bronze statues.

Unfortunately, iron is the great exception to the protective-oxide-film pattern. Its oxides form lattices that are a poor fit for the underlying iron metal lattice, so that they are not adherent and split away leaving a site for further oxidation. Dry iron or steel objects will scale very slowly, presumably by direct oxidative attack; the Delhi Pillar in India is almost pure iron and has been in place for 1500 years with only a thin film of oxide in place. However, in the presence of water electrochemical attack produces a much more rapid oxidation; dissolved O_2 is reduced to OH^-, and the four electrons required per O_2 molecule are withdrawn from an anodic site elsewhere on the polycrystalline metal, yielding two Fe^{2+} ions, which in the presence of more dissolved O_2 oxidize to Fe^{3+}, precipitating hydrous iron(III) oxide at any pH above about 1.5. The presence of SO_2 in the atmosphere greatly accelerates the rusting process, possibly by forming nuclei of $FeSO_4 \cdot 4H_2O$ that may catalyze entry of Fe^{2+} ions into the Fe^{III} oxide. If the sulfate is present, painting rust does little to delay further rusting if the paint film is permeable to O_2 because the oxide lattice can grow from below, eventually breaking the paint film.

Given the overwhelming role of iron as a structural metal in our society, it is not surprising that a number of the applications of other transition metals are oriented toward modifying the properties of iron for specific purposes, both to change its physical properties and to change its chemical capacity for corrosion. Table 10.8 indicates the primary uses of a number of the transition metals, and it is clear from the entries in the table that iron- and steel-related applications are dominant. A few of the more unusual applications deserve comment. Shape-memory alloys such as Nitinol can be formed hot into a desired shape, cooled and deformed by up to about 8%, and when reheated above a "shape-memory temperature" will regain their original shape. These alloys depend on a composition that has two separate high-temperature and low-temperature phases such that formation of the low-temperature phase requires diffusion of one element in the alloy, which will necessarily be slow in a solid. When a polycrystalline object in the high-temperature phase is cooled, new microcrystals of the low-temperature *martensitic* phase form in plates of symmetry-related orientations, with their boundaries in locations defined by the original high-temperature crystal layers because diffusion cannot allow random growth. Cold deformation causes the martensitic plates with favorable orientations (to the deformation) to grow at the expense of plates with less favorable orientations, but still with their original orientation relative to the high-temperature parent microcrystal. Reheating causes the high-temperature phase to form again with its crystal boundaries defined by those of the martensitic plates—but those are the boundaries of the original form of the

Table 10.8 *Principal Applications of the Transition Metals as Metals*

Ti: lightweight alloys with high-temperature strength (Ti–Al–V), shape-memory alloys (Ni–Ti, Nitinol), hydrogen-storage alloys (Fe–Ti)

V: steel hardening ingredient, nuclear reactor fuel cladding, magnetic alloys (V–Co–Fe Vicalloy)

Cr: steel electroplating, stainless steel (Fe–Cr–Ni), hard deposited alloys (Cr–Co Stellite), high-temperature alloys (Ni–Cr Nichrome)

Mn: steel toughness ingredient and sulfur remover, aluminum strengthening alloy ingredient

Fe: wide variety of steel alloys, magnetic alloys

Co: magnetic alloys, hard alloys (Stellite and Mo–Co or W–Co) for cutting tools

Ni: stainless steel, corrosion-resistant alloys (Ni–Cu Monel, Ni–Mo Hastelloy), high-temperature superalloys (Ni–Cr–Mo–Nb and variations)

Cu: electrical wiring, brass (Cu–Zn), tubing and piping, bronze (Cu–Sn)

Zn: steel corrosion-resistant coating, die-casting alloys (Zn–Al), brass, dry cells

Zr: nuclear reactor fuel cladding, corrosion-resistant chemical reactor material

Mo: steel strength ingredient

Pd: catalyst (primarily for hydrogenation and C_1 reactions)

Ag: brazing alloys (silver solder), oxidation catalysts, electrical contacts

Cd: steel electroplating, dry cells (Ni–Cd), brazing alloys, fusible alloys for sprinkler systems

Hf: nuclear reactor control rods

Ta: surgical implant prostheses, superalloys

W: high-temperature lamp filaments, superalloys, steel hardness ingredient

Pt: catalyst (primarily for hydrogenation, oxidation, and reforming reactions), resistance thermometers and thermocouples

Au: decorative and coinage alloys, electrical contacts, infrared reflector films, glass colorant (colloidal red)

Hg: vapor lamps, thermometer and barometer fluid, amalgam host metal for dental fillings and chloralkali process for NaOH production

object, so that a kind of thermally controlled shape memory has been demonstrated. Nitinol alloys are used in "shrink-fit" (actually expanding from inside) hydraulic line couplings, in orthodontic braces (where body temperature is above the shape-recovery temperature so that a force is continually exerted on displaced teeth) and bone-fracture repair plates. Comparable shape-memory alloys with Cu–Zn–Al composition are used in thermal-overload circuit breakers, thermostatic valves for hydronic home heating, and other thermal controls such as those for greenhouse windows.

Zirconium and hafnium, close neighbors in the periodic table, share primary uses in nuclear power reactors. However, zirconium is used as a corrosion-resistant container or cladding material for the fuel pellets because it is nearly transparent to neutrons, while hafnium is used in control rods because it absorbs neutrons so effectively. Zinc is extensively used in die-casting alloys for objects such as automobile door handles and trim because the Zn−Al alloy melts at a convenient temperature and readily fills every crevice of the mold to give an accurate finished shape requiring little further surface treatment except painting or chrome plating. The superalloys (usually Ni, Cr, and Nb, Ta, Mo, or W) show exceptional strength at high temperatures with good oxidation resistance, so that they are essential to applications such as rotors for turbojet engines. Finally, as we shall see in later chapters, many transition metal compounds show important catalytic properties; palladium and platinum are unusual in their remarkable ability to catalyze gas-phase reactions as the metals. We have already seen Pt gauze as the catalyst for the key reaction in the oxidation of ammonia to NO, but it is also used extensively in crude oil refining reactions and both metals are used in large amounts in automobile catalytic converters.

10.4 IONIC SURROUNDINGS: TRANSITION-METAL LATTICES AND MINERALS

Transition-metal cations form nearly ionic lattices with the oxide and fluoride ions in a variety of symmetries, mostly depending on the stoichiometry of the electrically neutral crystal. For simple stoichiometries, most of the lattices adopted have already been described in Chapter 2. The NaCl lattice is adopted by most of the MO monoxides, by MnS, and even in a modified form by FeS_2 (pyrites). The rutile lattice (Fig. 2.15) is adopted by most MO_2 dioxides, by first-row MF_2 difluorides, and (in a slightly distorted form) by such ionic oxyfluorides as TiOF and FeOF. In the rutile structure, the cations are in a six-coordinate octahedral environment. If the cations can accommodate more neighbors, an alternative is the fluorite structure (also Fig. 2.15), which is adopted by the lanthanides and actinides that can form dioxides. Note that second- and third-row transition metals do not in general form MF_2 difluorides. Their other "dihalides" are often cluster compounds with formulas that approximate MX_2 (we shall examine these in Chapter 13). Even when they adopt a conventional lattice, the increased covalent character of the bonding requires a geometry other than that of the rutile structure.

When the "anion" is less electronegative than F or O (as with the other halides and sulfide), lattices with less three-dimensional symmetry of coordination begin to appear. For 1:1 stoichiometry, transition-metal compounds such as sulfides, tellurides, arsenides, and so on most often adopt the NiAs lattice (see Fig. 2.14). In the NiAs lattice, the octahedra are often distorted enough to allow chains of metal atoms to form, producing electrical conductivity and metallic lustre. Some transition-metal ions, particularly d^8 configurations, strongly prefer square-planar coordination. The PtS lattice shown in Fig. 10.9 provides square-planar coordination around the Pt and tetrahedral coordination around the S. In PdS_2 (also shown in Fig. 10.9), the metal has essentially square-planar coordination, but the S atoms above and below a given Pd form a very elongated octahedron. We saw examples

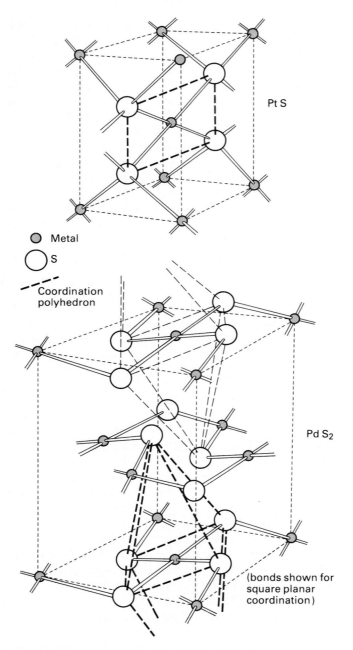

Figure 10.9   Square-planar coordination in transition-metal sulfides.

of layer lattices in Chapters 2 and 3. For MX_2 stoichiometry, Fig. 2.14 shows the CdI_2 lattice (metals octahedrally coordinated by hcp halides), which is related to the NiAs symmetry. There is also a $CdCl_2$ lattice, which is similar but has ccp halides. For MX_3 stoichiometry, Fig. 3.2 shows (for $InCl_3$) the layer structure usually known as the BiI_3 lattice, which is hcp; the equivalent ccp structure is known as the YCl_3 lattice. For a 3:1 stoichiometry with strongly electronegative anions, the ReO_3 lattice often appears (also in Fig. 3.2). In some cases, variants of the ReO_3 lattice appear in which the octahedra are somewhat distorted.

Transition metals also form a very wide variety of mixed oxides and fluorides (and, to some extent, other halides). Figures 3.11 and 3.12 show two of the most common of these: spinel (AB_2O_4), with octahedrally coordinated B and tetrahedrally coordinated A, and perovskite, ABO_3, with octahedrally coordinated A and a 12-coordinate B. As we have seen in Chapter 3, without the B atom the perovskite lattice would simply be an ReO_3 network of octahedra linked at every corner. The ReO_3 and perovskite lattices are fairly open and allow stacking variations or superstructures. Figure 10.10 shows a superstructure related to the perovskite

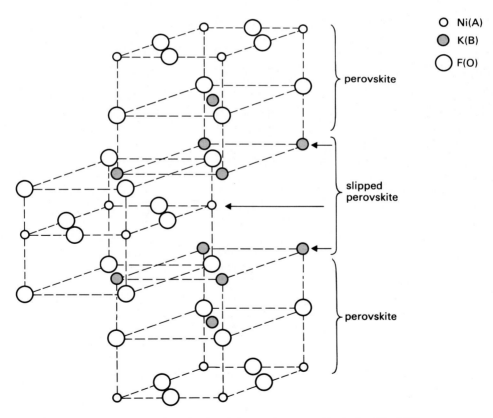

Figure 10.10 The K_2NiF_4 lattice in relation to the perovskite lattice (Fig. 3.12).

ABO_3 lattice, the K_2NiF_4 (AB_2O_4) lattice. The layers and their sequence are the same as those in the perovskite lattice, but the middle layer of the perovskite cell has been repeated and slipped sideways (45°) half a cell spacing. Since the repeated layer has the composition KF (BO) the lattice composition changes from $KNiF_3$ in the perovskite lattice, to K_2NiF_4 in the new lattice. Both complex fluorides and oxides are found in the K_2NiF_4 lattice. Ni can be replaced by Mg, Co, or Zn, and oxides with the same structure include K_2UO_4 and Sr_2MO_4 (M = Ti, Mn, Mo, Ru, Rh, Ir, Sn).

In the first edition of this text the K_2NiF_4 structure was presented as an interesting example of a slipped-layer superstructure. In 1986 that structure—though not that compound—suddenly took on enormous importance as the basis of the first high-temperature layered cuprate superconductors, the background for which we will explore in the next section. Accordingly, we need to look at exactly how the structure of the layered cuprate superconductors relates to the idealized K_2NiF_4 lattice and to the fundamental perovskite lattice. Figure 10.11 redraws the K_2NiF_4 lattice as a vertical stack of cubic units and compares it first to three units of the perovskite lattice of which it is a superstructure, then to the idealized lattice of one high-temperature superconductor derived from it, $YBa_2Cu_3O_7$; finally, to a more complex superconductor lattice, $Bi_2CaSr_2Cu_2O_8$. The similarities are obvious, but both they and the differences deserve comment. Although the perovskite lattice (a) has cubic symmetry, it is convenient to describe the composition of each of its layers for comparison with the other structures; accordingly, the stoichiometric composition of each layer is given in the figure. (This amounts only to counting the fractional spheres inside each cube, but we have doubled the top and bottom layer formulas to avoid fractions.) The layer formulas add up to $A_3B_3O_9$ (allowing for the factor of $\frac{1}{2}$ on the top and bottom), which correlates to the perovskite formula ABO_3.

Moving to the K_2NiF_4 lattice (b), the layer formulas add up to $A_2B_4O_8$ (allowing for the factor of $\frac{1}{2}$ on the top and bottom layers), which yields AB_2O_4 as the empirical formula, matching the stoichiometric formula K_2NiF_4. Shifting the center block of the perovskite lattice (as in Figure 10.10) has yielded very different coordination geometries: the top and bottom B atoms (K in the formula) are now 9-coordinate in a capped square antiprism, and the two central layers of B atoms are 5-coordinate in a square pyramid. The AO_2 and BO designations make the layered superstructure nature of the lattice clear. Note that the center cube now has NaCl (rock salt) symmetry.

In the $YBa_2Cu_3O_7$ superconductor lattice (c), the layer designations are slightly misleading in that the top and bottom AO layers are really CuO layers; copper atoms occupy all the original A sites in the perovskite lattice, though their coordination geometries have changed markedly. The top and bottom copper atoms are now 4-coordinate (square planar), and the central coppers are again 5-coordinate in a square pyramid, though now in a flatter pyramid than in K_2NiF_4. The central M atom (Y for this formula) is 8-coordinate cubic, as in CsCl. Allowing for the factor of $\frac{1}{2}$ at top and bottom, the layer formulas add to $MB_2Cu_2AO_7$, consistent with $YBa_2Cu_3O_7$.

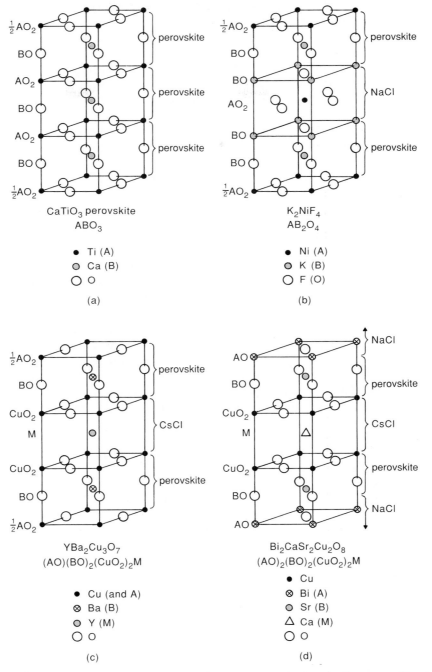

Figure 10.11 Derivation of layered cuprate superconductor lattices from perovskite through K_2NiF_4 superstructure and progressive layer modification: (a) three perovskite cubes; (b) K_2NiF_4 modification of perovskite; (c) $YBa_2Cu_3O_7$ modification of perovskite; (d) $Bi_2CaSr_2Cu_2O_8$ modification of perovskite and K_2NiF_4 lattices. Parts (c) and (d) superconductor structures require sheets of 5-coordinate square pyramidal CuO_5 units, though intermediate sheets of 4-coordinate square planar CuO_4 units may also be present.

Finally, in the $Bi_2CaSr_2Cu_2O_8$ superconductor lattice (d), we see that the layer compositions and sequence are the same as for (c). This sequence seems to be common to layered cuprate high-temperature superconductors, although extra $CuO_2-M-CuO_2$ sandwiches can be inserted in the middle and extra AO layers can be added at top and bottom. In fact, this superconductor lattice has one extra AO layer, which requires the whole three-cube stack to be diagonally offset with respect to the three-cube stacks above and below it just as the central cube in the K_2NiF_4 stack was offset relative to its neighbor perovskite cubes. The copper atoms occur again in layers of 5-coordinate square pyramids, the central M (Ca) atom is still 8-coordinate in CsCl symmetry, the B (Sr) atoms are 9-coordinate as in K_2NiF_4 though the orientation of the capped square antiprism has been reversed, and the A (Bi) atoms now occur in a thin slice of NaCl lattice. All of these coordination-geometry features seems to be common to the numerous layered cuprate high-temperature superconductors that have been described, yielding the general formula $(AO)_m(BO)_2(M)_{n-1}(CuO_2)_n$.

Given that in the most complex of these layered cuprates there are four different cations that must be accommodated within a lattice of oxide anions, it would be a miraculous coincidence if the cationic crystal radii were all perfect fits for their sites. They are not, of course, and the lattices shown are idealized from the true geometry. The oxygen atoms must pucker around the A atoms, which has the effect of causing the CuO_5 square pyramids to tilt alternately toward and away from each other in their layers. Although the theory of superconduction in these structures is still under intense investigation, it seems likely that this buckling of the CuO_2 layers is important to the phenomenon. Also important are departures from ideal stoichiometry in the form of occasional oxygen vacancies in $YBa_2Cu_3O_7$ and excess oxygen in the NaCl layers of $Bi_2CaSr_2Cu_2O_8$ (or cation vacancies). These departures from ideal stiochiometry have the effect of doping holes into the valence band of the solids, which bears on the electronic properties of the lattices.

In addition to the layered oxide lattices that are important for their superconducting properties, several transition-metal sulfide structures have the same property. As we have seen before, basic lattice symmetries are likely to be different for sulfides and oxides, both because the sulfide ion is larger than oxide and because S is much less electronegative than O. Our earlier principle applies here: anisotropic lattices—layer, chain, and zero-dimensional molecular—are more likely for a smaller electronegativity difference between metal and chalcogenide. For example, TiO_2 has the rutile lattice of Fig. 2.15 but TiS_2 has the layered CdI_2 lattice of Fig. 2.14. Between the TiS_2 layers weak van der Waals attractions are the primary binding force, and this attraction would be overcome if there were a large net charge on the facing "anions." The low net charge on the sulfur atoms can be stabilized by the vdW attraction between layers. By contrast, stabilizing an oxide layer lattice requires intercalated cations, as we have seen in several contexts. Although polymeric lattices are favored by less electronegative chalcogens, the anisotropy of physical properties that results from a polymeric lattice becomes less and less as the electronegativity difference decreases; metal tellurides tend to have essentially isotropic metallic properties because Te is a metalloid—the metal telluride compound is almost an alloy.

Chapter 3 has already noted the prevalence of layered transition-metal sulfides: MoS_2 with trigonal prismatic coordination of the metal atom, NbS_2 with the same coordination but with a different stacking sequence (a *polytype*) of the double layers so that they are ultimately ccp instead of hcp, and TaS_2 with the layered CdI_2 lattice. These compounds and their intercalated derivatives are superconductors at low temperature, but at room temperature their electronic properties are not the same. This is easiest to see in the density-of-states diagrams for these lattices, shown in Figure 10.12 on a qualitative basis. Part (a) of the figure shows the diagram for the octahedral coordination of the metal atom found in TiS_2, ZrS_2, and TaS_2. In ZrS_2 the band of sigma bonding states is filled, and there are no d electrons on Zr^{4+}, so the Fermi energy lies in the narrow band gap below the nearest portion of the d band; ZrS_2 is a semiconductor. By contrast, TaS_2 has Ta^{4+} with one remaining d electron, so that the bottom of the d band is filled. This should make TaS_2 a metallic conductor, and above 270°C it is. At room temperature, however, the lattice distorts slightly in a periodic way so as to split the d band still further and yield a Fermi surface that corresponds to a filled band. Such a periodic distortion of the lattice symmetry is called a *charge density wave*.

Part (b) of the figure is a comparable diagram for NbS_2 and MoS_2, which have layered lattices like those of the other sulfides but with trigonal prismatic instead

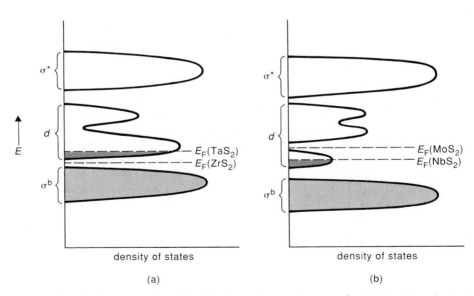

density of states

(a)

density of states

(b)

Figure 10.12 Density-of-states diagrams for early transition metal sulfides: (a) octahedral MS_6 coordination of metal found in TiS_2, ZrS_2, and TaS_2; (b) trigonal prismatic MS_6 coordination of metal found in NbS_2 and MoS_2.

of octahedral coordination of the metal atom. Because of the highly directional nature of the d AOs, this geometry change yields a different splitting pattern for the band of d orbitals in the crystal; Chapter 11 will look into the patterns of orbital splitting by crystal fields. Here we may note that trigonal prismatic coordination yields, for N atoms, a band with N orbitals just above the sigma-bonding band. Such a band can accommodate two electrons from each atom, and in MoS_2 (where Mo^{4+} has a d^2 configuration) the Fermi level is indeed just above this band, making MoS_2 a semiconductor. However, NbS_2 has only one remaining d electron per Nb, so that its band is half full. As a result, NbS_2 is a metallic conductor.

There are some very interesting one-dimensional chain structures found among transition metal sulfides. Patronite, the mineral source of vanadium, has V^{4+} and S_2^{2-} formal oxidation states; each V has two S_2 units around it in a planar rectangle with S–S bonds along the short sides. These VS_4 rectangles are stacked along a central V–V–V–V axis, with each rectangle being rotated by 45° relative to the previous one, as in Fig. 10.13. In the figure, $r_2 > r_1$, which represents a spontaneous distortion of the V–V–V–V chain called a *Peierls distortion,* analogous to the charge density waves in the TaS_2 lattice. As a result, pairs of rectangles form rectangular antiprisms along the chain. In general, chain lattices will require a higher stoichiometric ratio of S to M than layer structures, and this is borne out for VS_4.

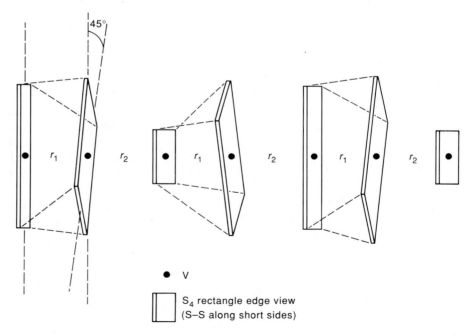

Figure 10.13 VS_4 (patronite) one-dimensional polymeric lattice as stack of alternating V atoms and S_4 rectangles. Each S_4 rectangle is rotated relative to the previous one by 45°. Peierls distortion stabilizes the lattice by alternating V–V distances along chain.

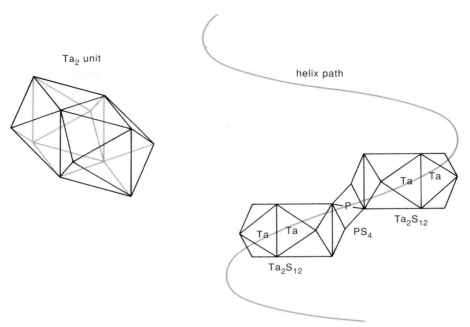

Figure 10.14   One-dimensional helix formed in $Ta_4P_4S_{29}$ lattice. Four Ta_2S_{12} units, each with 9-coordinate Ta, are linked into a chain by PS_4 tetrahedra in which two S from each Ta_2S_{12} unit are used to form the tetrahedron; asymmetry of Ta_2 attachment to PS_4 tetrahedron coils chain into helix with four Ta_2S_{12} units per turn.

Another remarkable one-dimensional structure is shown by the mixed sulfide $Ta_4P_4S_{29}$ (which also has a high S:M ratio.) In this lattice the tantalum atoms are 9-coordinate in a tricapped trigonal prism like that seen in the $PbCl_2$ lattice (Fig. 3.4), except that one of the capping atoms is another tantalum, so that the tantalum atoms occur in pairs: each a bicapped trigonal prism, sharing a square face, as seen in Fig. 10.14. In the lattice, one of these Ta_2S_{12} units is linked to the next by a tetra-hedral PS_4 unit, but the P links the top of one prism (or biprism) to the bottom of the next. When this is repeated, a helix is generated with four Ta_2 units per turn, as the figure indicates. This helix has quite a large diameter, but its stoichiometry does not add up to 29 S per 4 Ta. In fact, the large helix contains another smaller one at its center, consisting of a tightly wound S_n helical chain. The unit cell also contains a smaller Ta–S–P helix that is empty at its core, and there is an interesting possi-bility that such sulfur-lined cavities could provide a soft-base catalytic structure equivalent to the hard-base zeolites.

If chain sulfides have more sulfurs per metal than layer sulfides, it is reason-able to expect that molecular (zero-dimensional) sulfides will have still more. This is partly true: $Mo(S_2)_6^{2-}$, with Mo:S = 1:12, is an isolated species in crystals with a structure roughly equivalent to a single V_2S_{12} pair in patronite (Fig. 10.13). How-ever, it is also possible to see isolated species at low S:M if metal clustering

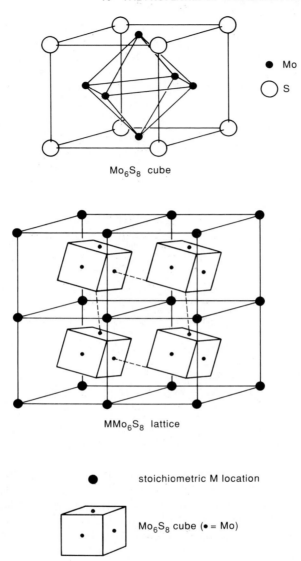

Figure 10.15   Chevrel phase ($M_nMo_6Z_8$) lattice. Each unit cell of the set of four shown contains one S_8 cube which is concentric with a Mo_6 octahedral cluster; each Mo bonds to a Z atom in a neighbor cube as well as to the surrounding Z atoms in its own cube.

occurs. The *Chevrel phases,* with a general formula $M_nMo_6Z_8$ where Z is a soft chalcogen, appear at first sight to contain isolated Mo_6S_8 units, each as a cube of S atoms containing an octahedron of Mo atoms as in Figure 10.15. But as we have noted before it is often difficult to distinguish a lattice of isolated molecular units from a three-dimensional network. Closer inspection of the figure reveals that the

Mo_6S_8 cubes are uniformly skewed in the lattice in such a way that each Mo, besides being coordinated to its Mo neighbors and immediate S_4 environment, is also oriented almost directly toward a sulfur from a neighboring Mo_6S_8 unit in the lattice. Furthermore, the Mo–S distance is little greater between cubes than within a cube—about 2.6 Å versus 2.5 Å. This geometry suggests that a three-dimensional network of cubic clusters might be a better description.

In the Chevrel phases the M atom is intercalated into the cluster lattice, so that the stoichiometric subscript n in the general formula need not be an integer; indeed, the Mo_6S_8 stoichiometry can be prepared with no intercalated metal at all. Valence electrons on the M atom enter a band that is more or less nonbonding, so that their number can vary without substantially affecting the overall lattice bonding. The three-dimensional character of the bonding is emphasized by the observation that when the stoichiometric MMo_6S_8 lattice is changed from $M = Eu^{2+}$ with crystal radius 1.39 Å (C.N. 8) to Sr^{2+} (1.40 Å) to Ba^{2+} (1.56 Å) the diameter of the site for M^{2+} must expand by 0.34 Å, but the Mo–S distance between clusters increases by only 0.04 Å—the extra space is created predominantly by twisting the Mo_6S_8 cubes away from the M atom, not by stretching the lattice.

A number of the Chevrel phases display superconductivity, though not at unusually high temperatures. Interest in them continues in spite of the layered cuprates with high critical temperatures for superconduction because the Chevrel phases can maintain superconduction in the presence of exceptionally high magnetic fields, which would destroy superconduction in other materials. Inasmuch as the principal commercial use of superconductors at the present time is as solenoid windings in high-field magnets for chemical NMR instrumentation and medical magnetic resonance imaging, the limiting magnetic field for the windings is of critical importance.

10.5 ELECTRONS IN LATTICES: SUPERCONDUCTORS

We have indicated at several points in this chapter that certain materials or lattices were important because of their ability to behave as superconductors under certain circumstances. Superconduction is a property of some solids that has been observed for more than 80 years, since its discovery by Kamerlingh Onnes for metallic mercury in 1911; it refers to the fact that as the temperature is reduced the resistance of an ordinary metal decreases steadily, but remains in the general vicinity of the values in Table 10.7. For mercury and other superconductors, however, there is a temperature called the *critical temperature* T_c at which the resistivity of the material drops suddenly to an unmeasurably low value, effectively zero, as in Figure 10.16. This is potentially enormously useful: About a third of the electric power generated in the United States is lost in transmission because of the resistive loss of the cables, so that superconducting cables would constitute an enormous conservation measure. Very high currents in a superconducting coil or solenoid could yield very high magnetic fields for chemical and medical purposes, as has been suggested. In addition, high magnetic fields could levitate passenger vehicles or freight containers over a rail system, so that "maglev" transportation could operate with no

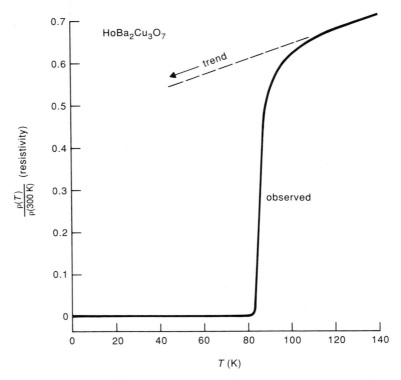

Figure 10.16 Resistivity (expressed as a ratio to the 300 K value) for superconductor $HoBa_2Cu_3O_7$ as a function of temperature. T_c is just over 80 K but will vary depending on whether it is quoted for loss of half the trend-line resistivity or 90% of the trend-line resistivity.

rolling friction and no vibration. Of all these, only the instrumental and medical uses have been commercially developed, because the technical problems of using most superconducting materials are severe.

Until 1986 all materials known to be superconductors had T_c values below about 20 K, requiring them to be cooled by liquid helium to function. Most were metals; the primary material used in superconducting magnet windings was (and is) a Nb–Sn alloy. The discovery in 1986 of the layered cuprate superconductors brought T_c values to roughly 100 K, which seems only modestly helpful even though it represents a fivefold advance. However, these T_c values are above the boiling point of liquid nitrogen (77 K), and liquid nitrogen is much cheaper and easier to handle than liquid helium. The layered cuprates obviously represent a completely different category of superconductor, and it is not entirely clear that their behavior can be adequately modeled by the theory that had been shown to govern low-temperature superconductors, the *Bardeen–Cooper–Schrieffer (BCS) theory*. It is beyond the purposes of this treatment to outline BCS theory, but its physical basis is not too hard to understand, and it may still be the governing theory of superconduction for layered cuprate structures.

Suppose we have a crystal with electrons in delocalized and therefore fairly diffuse orbitals, as in a metal or other material with energy bands such that there is only a small gap above the Fermi surface. Then near the Fermi surface there will be unpaired electrons and holes in the valence band. If the crystal has its full normal electrical symmetry, an electron will be bound within the crystal by the attraction of the atomic cores and repelled by the surrounding valence electrons, but its energy will be that of the delocalized orbital it occupies; no one location in the crystal will be better than another. However, there may well be a location within the crystal in which the electron could induce dipoles around itself, deforming the lattice in the process so as to stabilize itself in that location. It has created a temporary dimple in the potential energy surface for electrons within that crystal. That low-energy dimple will attract another diffuse valence electron with opposed spin, creating a *Cooper pair* partially localized near the distortion in the lattice.

Now suppose that the normal, full-symmetry lattice has a normal mode of lattice vibration that creates, at one instant in the vibration, the same distortion at one point in the lattice that stabilizes the Cooper pair. Further, suppose that the lattice mode is such that the distortion—the dimple in the local potential energy surface—moves through the lattice rapidly. Then the Cooper pair will be transported through the lattice without scattering and without resistance as long as thermal excitation kT is less than the depth of the dimple, the stabilization due to the lattice deformation. This is the process of superconduction, but if only two electrons are transported the current will be immeasurably small. The importance of the Cooper pair is that if two electrons with spin $\frac{1}{2}$ are locked in a pair the pair can be treated as a single particle with spin 1, which is to say a boson instead of a fermion. A boson is not subject to the exclusion principle, because its wave function does not change sign when particle labels are interchanged. As a result, large numbers of bosons can occupy a single quantum state. And that means that large numbers of Cooper-pair electrons can be transported through a superconducting crystal by a normal mode with the right symmetry, without resistance.

For the layered cuprate superconductors, the possibility of unusually high T_c values has led to intense study of the detailed structural parameters of those systems and their influence on T_c. Because Cooper pairs must be created from formally unpaired electrons, the electronic nature of the states within kT of the Fermi surface is critically important. Photoemission spectroscopy studies of series of cuprates have shown that the states near the Fermi surface involve extensive participation by Cu $3d$ orbitals and in-plane O $2p$ orbitals, implying that superconduction occurs primarily within those planes (and presumably by normal vibrational modes involving distortion of those planes). In particular, taking a convenient coordinate axis definition, the d_{x2-y2} has a strong sigma antibonding interaction with p_x and p_y near the Fermi surface. We might well expect that T_c would be strongly affected by the Cu–O distance in that plane, because shrinking r_{Cu-O} would enhance the antibonding effect and change the character of vibrational modes involving those bonds. This is indeed observed, as Whang-bo has shown in Figure 10.17. The figure indicates a good functional relationship between T_c and r_{Cu-O}, but the existence of widely differing T_c values for the same r_{Cu-O} value in the La

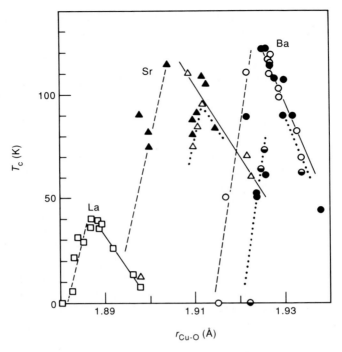

Figure 10.17 Observed T_c for layered cuprate superconductor $[(AO)_m(BO)_2(M)_{n-1}$ $(CuO_2)_n]$ families in which B = La, Sr, or Ba and Cu−O spacing changes as B radius increases. [M. H. Whangbo and C. C. Torardi, *Acc. Chem. Res.* (*1991*), *24*, 127.]

and Sr families, or in the Sr and Ba families, clearly means that other factors are at work—probably many other factors. Still, the progress that has been made in understanding the properties of these materials offers great promise for practical superconduction applications in the future.

PROBLEMS

A. DESCRIPTIVE

A1. The melting points of CrX_2 and CoX_2 halides show different trends. Why does CoF_2 melt at a significantly higher temperature than CrF_2, even though CrI_2 melts at a higher temperature than CoI_2?

	X = F	Cl	Br	I
MP(CrX$_2$)(°C)	894	820	842	868
MP(CoX$_2$)(°C)	1,200	724	678	515

A2. Why is Mn(III) a moderately good oxidizing agent, even though its neighbors Cr(III) and Fe(III) have little or no oxidizing ability under ordinary circumstances?

A3. What electronic effect results when Al is added to a Fe-Co-Ni alloy (as in Alnico magnet compositions), in terms of valence electron density per atom? Is there a sense in which Al can help maximize the ferromagnetic saturation magnetic moment?

A4. What is the reaction when solid $FeCl_3$ is dissolved in water? in liquid ammonia?

A5. Nb^{5+} is significantly larger than V^{5+} (0.78 Å versus 0.68 Å), but Ta^{5+} is no longer than Nb^{5+} (0.78 Å for both ions). Why?

A6. From the Latimer diagrams of Table 8.2, which transition-metal ions should be able to reduce water to H_2 in acid solution? Which should be able to oxidize water to O_2?

A7. Cr(VI) is a powerful oxidizing agent in water, but W(VI) is essentially redox-stable in water. What does this experimental evidence suggest about the relative trends in shielding and Z_{eff} for $5d$ electrons compared to $3d$ electrons?

A8. To achieve ferromagnetism in a bulk metal, why is it important that the Fermi level occur at an energy at which the metal has a very high density of states?

A9. Why is it important to have high permeability in a magnetic recording tape? Could the permeability (in principle) be *too* high? Why or why not?

A10. Sketch a qualitative density-of-states diagram for NaCl. Use the VOIPs in Table 1.8 for general guidance on relative energies. Where should the Fermi energy be?

A11. What combination of structural and thermodynamic properties is responsible for the fact that zirconium is used as a corrosion-resistant chemical reactor material?

A12. Why can't a pure metallic element form a shape-memory metal?

A13. If a layered cuprate superconductor has a lattice in which there are three CuO_2 central layers and three layers in the rock-salt slab, what is its general formula?

A14. Suggest a preparative reaction for $Ta_4P_4S_{29}$.

B. NUMERICAL

B1. Set up a Born–Haber cycle for a typical lanthanide halide LX_3 that is disproportionating into LX_2 and LX_4. (Assume that all are ionic crystals.) Estimate the cationic radii for L^{2+}, L^{3+}, and L^{4+} and the ionization energies involved for a typical lanthanide. Solve for $\Delta H_{disprop}$ for LF_3 and for LI_3. Estimate the radius of a hypothetical X^- ion that would yield $\Delta H_{disprop} = 0$. How likely is it that the lanthanide halides will show a nonstoichiometric composition?

B2. Entropy changes for chemical reactions can be estimated with fair accuracy simply by assuming that all solids have an absolute entropy of 60 J/mol K, liquids 100 J/mol K, and gases 200 J/mol K. Use this approximation to estimate the lowest value of the Ti–I bond energy that would allow the van Arkel–de Boer process to be spontaneous toward TiI_4 at 200°C. Use it again to estimate the highest value of the Ti–I bond energy that would still allow decomposition of TiI_4 at 1300°C. The tabulated value of $D_{Ti–I}$ is 296 kJ/mol bond; how does this value compare with your calculated outer limits?

B3. Zinc is often considered not to be a transition metal because it does not use d electrons in bonding; for example, it has no observed oxidation state higher than 2+. How realistic thermodynamically is the 3+ oxidation state for Zn? Set up a Born–Haber cycle for the reaction $ZnF_2(s) + \frac{1}{2}F_2(g) \rightarrow ZnF_3(s)$, assuming the zinc fluorides to be ionic, and calculate ΔH for the reaction. Make reasonable assumptions about radii and coordination numbers in lattice energy calculations. Does your result suggest that Zn^{3+} might be stable?

B4. Use data in the chapter to construct a graph of net ferromagnetic spins per atom versus atomic number for the late transition metals. What does the graph suggest should be the magnetic effect of adding 10% Ni to pure Fe? of adding 10% Mn to pure Fe?

B5. For the idealized structure of the superconductor $YBa_2Cu_3O_7$ shown in Fig. 10.11b, use simple geometry to calculate the ideal ratio of ionic radii $r_A:r_B:r_{Cu}:r_O$, assuming that the lattice has touching oxides. How well are these radii matched by the crystal radii of Table 2.3?

C. EXTENDED REFERENCE

C1. What crystal lattice symmetry do you expect for the superconductor $Tl_2Ca_2Ba_2Cu_3O_{10}$? [See C. N. R. Rao and B. Raveau, *Acc. Chem. Res.* (1989), *22*, 106.]

C2. Kaner has reported a convenient synthesis of NiS_2 [P. R. Bonneau, R. K. Shibao, and R. B. Kaner, *Inorg. Chem.* (1990), *29*, 2511]. What are the oxidation states of Ni and S in this compound? What evidence does the article give?

C3. Gold has common oxidation states +1 and +3, and as a soft acid usually combines with soft bases in predominantly covalent systems with low coordination numbers for Au, typically 2 or 4. Selenium, like sulfur, readily forms polyselenide ions Se_n^{2-} with $n = 3$ and 5 particularly stable. Recently two interesting Au–Se compounds have been reported: K_nAuSe_5 and K_mAuSe_{13} [Y. Park and M. G. Kanatzidis, *Angew. Chem. Int. Ed.* (1990), *29*, 914]. Should these compounds have three-dimensional, layer, or chain structures? Suggest structures and values of m and n for the two compounds.

C4. How are the six platinum group metals separated from the naturally occurring mixture obtained from refining nickel ore into individual pure metals? [See R. D. Lanam and E. D. Zysk, "Platinum-Group Metals," *Kirk–Othmer Encyclopedia of Chemical Technology,* Third Edition, vol. 18, Wiley, 1982.]

11

Ligand-Field Theory, Spectroscopy, and Magnetism

In Chapters 2 and 6 we dealt successfully with the energies of main-group ionic compounds without considering the quantum-mechanical nature of the ions themselves—that is, we considered them simply as point charges. This was possible because species such as the alkali metal ions are symmetric spheres of charge, which is the mathematical equivalent of a point charge. The situation is quite different when we consider ionic compounds of the transition metals, because in general their cations contain varying numbers of d electrons in orbitals that are *not* spherically symmetrical. The shape or symmetry of those orbitals thus becomes important to an accurate description of the energy of a transition-metal ion in a crystal, even if the bonding is considered completely ionic. In the last chapter we considered briefly the origin of the shapes of the d orbitals, but essentially all the succeeding discussion dealt only with the radial distribution of d electrons relative to the atomic core, not with their angular distribution relative to neighbor atoms or ions. In this chapter we will develop the theoretical and experimental background by which we understand the behavior of d electrons with specific reference to their angular distribution: We will consider their orbital shapes as they affect the properties of transition metal ions in ionic lattices or solvates (known as *crystal-field theory*), move on to the equivalent molecular orbital theory for more covalent species (*ligand-field theory*), and then look at the ways in which these theoretical structures allow us to understand two of the most important experimental properties of transition metals—their electronic spectra and their paramagnetism.

11.1 IONIC SURROUNDINGS: CRYSTAL-FIELD THEORY

We begin with the behavior of transition-metal ions in ionic crystals: The presence of cation d electrons in nonspherical distributions raises the question of the effect of their orbital symmetry on their repulsion of the surrounding anions. For example, many transition-metal monoxides MO adopt the NaCl crystal lattice. What can we say about the repulsion of the oxide ions by the two $3d$ electrons on Ti^{2+} in TiO? How will this repulsion affect the lattice energy of the crystal, for instance?

Each Ti^{2+} is surrounded by six O^{2-} at the corners of an octahedron. If the Cartesian coordinate system for the Ti orbitals is oriented so that the axes pass through the O nuclei, an electron in each of the five d orbitals is not repelled equally by the six anions (see Fig. 11.1). Three of the d orbitals tuck the electron conveniently into the spaces between anions, while the other two force the electron to orient itself directly toward the repelling anion. In fact, the three favorable (low-repulsion) orbital distributions are even better than a spherical distribution with the same average radius would be, while the other two are worse. The d_{xy}, d_{xz}, and d_{yz}

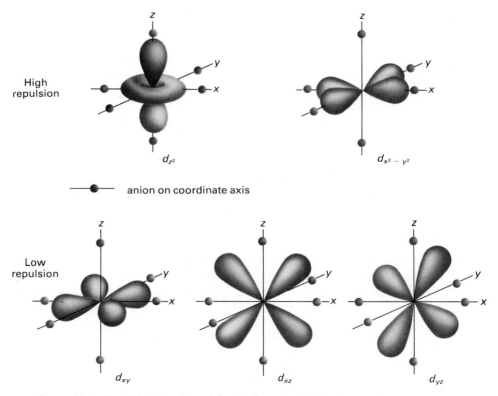

Figure 11.1 Spatial relationship of d orbital angular distributions to the surrounding anions in NaCl lattice (octahedral crystal field).

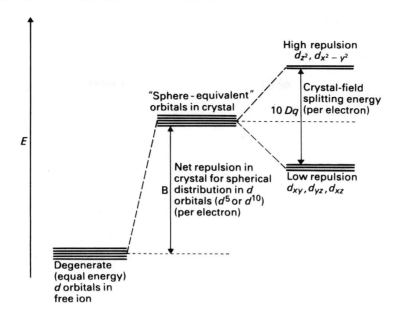

Figure 11.2 d-Electron energies in an octahedral crystal field.

are obviously equivalent in their repulsion and represent relatively low energy, whereas the d_{z^2} and $d_{x^2-y^2}$ are also equivalent (somewhat less obviously) but represent much greater repulsion and therefore higher energy. Figure 11.2 qualitatively indicates the effect of crystal repulsion on d-orbital energies. The magnitudes of the energies involved are suggested by the Born repulsion energy B for *all* the electrons on a cation, which is about 10% of the lattice energy, or about 400 kJ/mol for a divalent oxide such as TiO. As we shall see later, the crystal-field splitting energy symbolized in Fig. 11.2 as $10\ Dq$ is usually on the order of 200 kJ/mol. So these are substantial energy effects.

It is obvious that the symmetry relationships between anions and orbital angular distributions are of crucial importance. Less obvious is the fact that for high symmetry (such as that shown by the octahedral ions in the NaCl lattice), only the d and f sets of orbitals show crystal-field splitting. Figure 11.3 shows the orientation of a set of p orbitals along the same axes used in Fig. 11.1. Note that each of the three p orbitals has exactly the same relationship to the repelling anions, so that, while repulsion effects will be quite large, there will be no splitting in the high symmetry of an octahedral environment. Of course, a single s orbital cannot be split by a crystal field of any symmetry.

Our qualitative appreciation of the crystal-field energy splitting shown in Fig. 11.2 can be amplified by a brief consideration of the algebraic treatment within the quantum-mechanical model that leads quantitatively to the same result. What we seek is the energy of the coulomb law interaction between an electron in a given

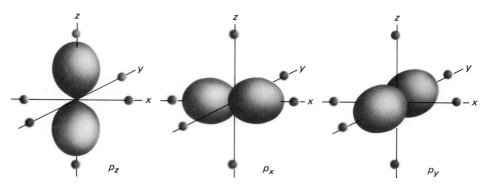

Figure 11.3 Spatial relationship of p-orbital angular distributions to the surrounding anions in NaCl lattice (octahedral crystal field).

d-orbital distribution and a set of point-charge anions arranged octahedrally about the nucleus of the transition-metal atom or ion. Suppose the orbital we are interested in is the d_{z^2}, one of the two high-energy orbitals. Its interaction energy is represented quantum mechanically by the integral

$$E(d_{z^2}) = \int_{\text{all space}} (d_{z^2})^* V_{\text{oct}} (d_{z^2}) \, d\tau$$

where (d_{z^2}) represents the algebraic wave function for that orbital, $(d_{z^2})^*$ represents its complex conjugate, V_{oct} is an algebraic and trigonometric representation of the total coulomb-law potential energy of repulsion between six octahedrally arranged point charges and a point-charge electron in the metal coordinate system, and $d\tau$ is the volume element in the coordinates used. Obviously, V_{oct} is a messy algebraic quantity, but it can be written in a reasonably convenient form by using the algebraic form of the d wave function that is also in the integral. When the integral is evaluated, it contains three factors: (1) a collection of constants from the geometric factors in V_{oct}, $35ze^2/4a^5$, symbolized as D; (2) the result of the integral over r (in polar coordinates), which is the same for all $3d$ orbitals in a given transition-metal ion and is symbolized as $q = \frac{2}{105} <r^4>$; and (3) the result of the integral over θ and ϕ, which is unique to each d orbital with its particular angular distribution. If we take the product of D and q as a sort of natural theoretical unit for crystal-field splitting energies, the result of evaluating the integral for the d_{z^2} orbital is

$$E(d_{z^2}) = +6Dq$$

Repeating the calculation for each of the other d orbitals gives

$$E(d_{x^2-y^2}) = +6Dq$$
$$E(d_{xy}) = -4Dq$$
$$E(d_{yz}) = -4Dq$$
$$E(d_{xz}) = -4Dq$$

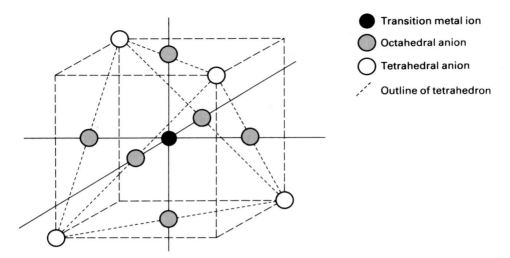

Figure 11.4   Cubic symmetry of octahedral and tetrahedral crystal fields.

This, of course, is equivalent to the result we inferred from the sketches in Fig. 11.1, and explains why the crystal-field energy splitting was defined as $10Dq$ in Fig. 11.2. The combined Dq quantity is defined as above,

$$Dq = \frac{1}{6}\left\{\frac{ze^2 <r^4>}{a^5}\right\}$$

where ze is the charge on the anion, a is the bond length or internuclear distance between the metal and the anion, and $<r^4>$ is the average value of the fourth power of the distance of the electron from its nucleus. The $<r^4>$ quantity is not accurately known from theory for even $3d$ electrons in first-row transition-metal ions, let alone the heavier transition metals. However, if it is taken as the fourth power of 1.1 Å (half the approximate $M^{2+}-O^{2-}$ distance in the monoxides), Dq can be calculated to a reasonable approximation of the experimentally observed $10Dq$ splitting in crystals.

If we pursue a similar derivation for the d orbitals in a tetrahedral crystal field, the arguments are exactly the same except that there are now only two-thirds as many repelling charges, and they are located at different θ and ϕ angles. A tetrahedral crystal field still has what is called *cubic symmetry*—that is, a tetrahedron and an octahedron can both be defined in terms of a parent cube, as in Fig. 11.4. The change in the angle from the octahedral positions to the tetrahedral positions applies a geometric factor of $-\frac{2}{3}$ to the spherical harmonic function for V_{oct}. So we have

$$V_{\text{tet}} = \left(-\frac{2}{3}\right)\left(\frac{2}{3}\right)V_{\text{oct}} = -\frac{4}{9}V_{\text{oct}}$$

The final energy result for the tetrahedral field, then, is

$$E(d_{z^2}) = -\tfrac{4}{9}(6Dq) = -2.67Dq$$

$$E(d_{z^2-y^2}) = -\tfrac{4}{9}(6Dq) = -2.67Dq$$

$$E(d_{xy}) = -\tfrac{4}{9}(-4Dq) = +1.78Dq$$

$$E(d_{yz}) = -\tfrac{4}{9}(-4Dq) = +1.78Dq$$

$$E(d_{xz}) = -\tfrac{4}{9}(-4Dq) = +1.78Dq$$

There are two important differences between the tetrahedral and octahedral cases. First, the total crystal-field energy splitting is less than half as large for the tetrahedral case, for the reasons already mentioned. Second, the relative energies of the d orbitals are inverted in the tetrahedral field. The d_{z^2} and $d_{x^2-y^2}$ orbitals now keep the electron between the point-charge anions, while the d_{xy}, d_{yx}, and d_{xz} direct an electron somewhat obliquely at the anions.

One more important crystal-field symmetry should be mentioned; the square-planar case. This is the extreme of a distorted octahedron in which two opposing anions (ligands) have been pulled farther and farther away from the transition metal. Since the six faces of the parent cube are no longer equivalent, the symmetry is no longer cubic. Therefore, there should be a greater variety of energy splitting because the d orbitals now see different electrostatic environments. The algebraic derivation is more complicated, since the anion symmetry no longer allows as much simplification in V_{sq} as in V_{oct}, and the result must be expressed in terms of two parameters Ds and Dt instead of the single quantity Dq. However, if simplifying assumptions are made about the radial dependence of the d electrons, the orbital energies can be expressed in terms of Dq for comparison:

$$E(d_{z^2}) = -4.28Dq \text{ (anions along } x, y \text{ axes)}$$
$$E(d_{x^2-y^2}) = +12.28Dq$$
$$E(d_{xy}) = +2.28Dq$$
$$E(d_{yz}) = -5.14Dq$$
$$E(d_{xz}) = -5.14Dq$$

The energy levels for the d orbitals in the three crystal-field symmetries discussed thus far are shown in Fig. 11.5. Table 11.1 repeats these energies, and gives others for geometries seen in some transition-metal compounds. Note from Fig. 11.5 that one can infer qualitative energy splittings for distorted geometries such as a tetragonally distorted octahedron (anions on z axis pulled away) or a flattened tetrahedron.

In general, d electrons in a crystal field will arrange themselves to achieve the lowest energy possible. For a d^1 configuration, this means placing the electron in the lowest-energy orbital in the appropriate energy-level diagram. When more than one electron is present in the set of d orbitals, however, repulsion energies between the d electrons must be taken into account as well as the crystal-field repulsions. In an oc-

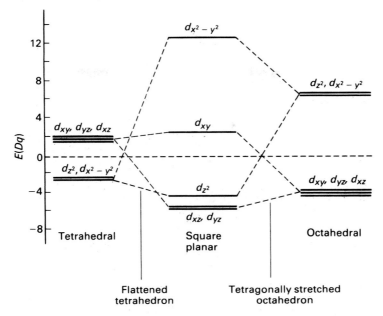

Figure 11.5 Relationship of crystal-field energies for d orbitals, including intermediate geometries.

tahedral field, the first three d electrons can be placed in the three low-energy orbitals with parallel spins to minimize repulsion, but two results are possible for a fourth electron. If $10Dq$ is quite large, it will be energetically preferable for the electron to enter one of the three half-filled low-energy orbitals by pairing spins with the existing electron, even though a considerable repulsion between d electrons results.

Table 11.1 d-Orbital Energy Levels in Crystal Fields of Different Symmetries (Dq)

Coord. No.	Geometry	d_{z^2}	$d_{x^2-y^2}$	d_{xy}	d_{yz}	d_{xz}
2	Linear	+10.28	−6.28	−6.28	+1.14	+1.14
3	Trigonal planar	−3.21	+5.46	+5.46	−3.86	−3.86
4	Tetrahedral	−2.67	−2.67	+1.78	+1.78	+1.78
4	Square planar	−4.28	+12.28	+2.28	−5.14	−5.14
5	Trigonal bipyramid	+7.07	−0.82	−0.82	−2.72	−2.72
5	Square pyramid	+0.86	+9.14	−0.86	−4.57	−4.57
6	Octahedral	+6.00	+6.00	−4.00	−4.00	−4.00

Note: Bonds lie along coordinate axes where possible. The unique axis is the z axis for all geometries except tetrahedral and octahedral.

Table 11.2 Crystal-Field Stabilization Energies for d-Electron Configurations (Dq)

	d^1	d^2	d^3	d^4	d^5	d^6	d^7	d^8	d^9
Examples	Ti^{3+}	V^{3+}, Ti^{2+}	Cr^{3+}	Cr^{2+}, Mn^{3+}	Mn^{2+}, Fe^{3+}	Fe^{2+}, Co^{3+}	Co^{2+}	Ni^{2+}	Cu^{2+}
CFSE (octahedral high-spin)	4.00	8.00	12.00	6.00	0	4.00	8.00	12.00	6.00
CFSE (octahedral low-spin)	4.00	8.00	12.00	16.00	20.00	24.00	18.00	12.00	6.00
CFSE (tetrahedral high-spin)	2.67	5.33	3.56	1.78	0	2.67	5.33	3.56	1.78

This is the *high-field* or *low-spin* case. Alternatively, if the energy input necessary to pair electrons is greater than $10Dq$, the fourth d electron will enter one of the vacant high-energy orbitals with its spin parallel to the first three electrons to reduce repulsion (as in Fig. 1.7). This is the *low-field* or *high-spin* case. Similar choices exist for configurations from d^4 through d^7 in octahedral fields. Of course, this filling process, whether low-spin or high-spin, leads only to the ground state of the ion in its crystal field. There will be excited states corresponding to different electron configurations in the crystal-field energy levels, with higher total energy.

For the ground state, we can establish a total *crystal-field stabilization energy* (CFSE) for the ion simply by adding up the Dq energies for each electron after the low-spin or high-spin configuration has been established. For example, Fig. 11.6(a) shows the ground-state configuration of Cr^{3+} (d^3) in an octahedral field. Each of the three electrons is in an orbital with energy $-4.00Dq$, so the total CFSE is $(-4.00) + (-4.00) + (-4.00) = -12.00Dq$. That is, the ion is $12.00Dq$ units more stable than it would be if the electrons were in hypothetical spherical orbitals. Figure 11.6(b) compares the ground-state configurations of Fe^{3+} (d^5) in strong and weak crystal fields (the low-spin and high-spin cases, respectively). Adding up the individual electron energies, CFSE $= 0$ for the high-spin case, which is reasonable since the equally occupied orbital set constitutes a spherically symmetrical distribution. For the low-spin case, CFSE $= -20.00Dq$—a substantial stabilization that must be balanced against the pairing energy required to force d electrons into the same orbital. Table 11.2 gives CFSE values for d-electron configurations in several cases. (No low-spin tetrahedral data are presented here because the splitting is so small for tetrahedral geometry that the low-spin case is never preferred.)

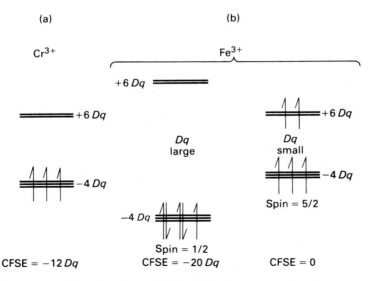

Figure 11.6   Examples of crystal-field stabilization energy calculation.

These points suggest that Dq is variable for different environments, even within the same geometry. This is experimentally true, even though it cannot be accounted for simply by the known changes in ionic charge, ion spacing, and d-electron radius. For first-row 2+ transition metal ions, Dq (as determined from spectrophotometric data) varies from about 700 to about 3000 cm^{-1}, depending on the nature of the ligands around the metal ion. For first-row 3+ ions, Dq varies from about 1200 to about 3500 cm^{-1}; and for second- and third-row 3+ ions, Dq varies from about 2000 to about 4000 cm^{-1}. Although the simple electrostatic theory can be extended to include the effects of dipole ligands with no net charge, no strictly ionic model will explain this variation. We shall develop a more comprehensive theory in a subsequent section, after examining the properties of some predominantly ionic transition-metal systems.

11.2 IONIC SURROUNDINGS: LATTICE ENERGIES AND REDOX STABILITY

The stability of ionic transition-metal compounds in various oxidation states is substantially influenced by the crystal-field effects we have just been considering. The energy required to create cations in the gas phase—the ionization energy—increases fairly smoothly across the first row, as Fig. 11.7 indicates. A discontinuity

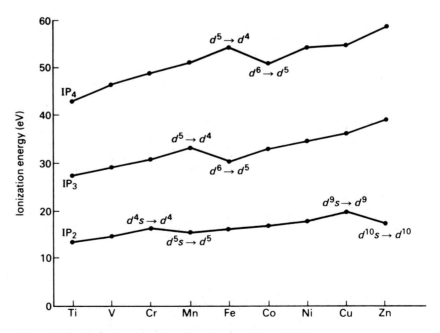

Figure 11.7   Ionization energies of gaseous transition-metal ions.

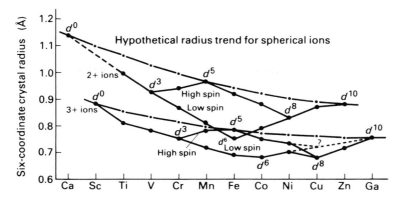

Figure 11.8 Crystal radii for transition-metal ions in octahedral surroundings.

occurs at the d^5 configuration for each ionization energy and at d^{10} for the second-ionization energy. These discontinuities are caused by the spherical symmetry of shielding by the d^5 and d^{10} configurations, which is lost for fewer than five or fewer than 10 electrons. However, the stabilization of these ions does not follow as uniform a trend in crystals, because the cations are nonspherical and the crystal-field stabilization energy is an additional factor. Figure 11.8 shows the trends in crystal radii for the 2+ and 3+ ions (see also Fig. 10.4). The d^0, d^5 high-spin, and d^{10} ions are all spherically symmetrical, and we can draw a smooth curve to represent the hypothetical radii for "spherical" transition-metal ions. In fact, however, when we create ions with one, two, or three d electrons, they are smaller in crystals than a spherical ion would be. This is because the d electrons have been placed in the d_{xy}, d_{xz}, and d_{yz} orbitals, which do not repel surrounding octahedral anions as strongly as a spherical electron distribution would. For a d^4 configuration, there are two possible radii. If the ion is high spin, one electron is in the strongly repelling d_{z^2} or $d_{x^2-y^2}$ orbital and the cation appears relatively large. If the ion is low spin, the fourth electron is still distributed so as to avoid the anions and allows them to approach closer to the transition-metal nucleus, yielding a smaller crystal radius. The d^6 low-spin configuration yields the smallest ion, whether 2+ or 3+, because it gives the maximum nuclear charge that can accommodate all d electrons in the low-repulsion orbitals. Similar arguments apply to the other radii and their trends. Note that a high-spin Fe^{3+} actually appears larger in a six-coordinate crystal than a low-spin Fe^{2+} ion, even though it has one fewer electron!

The lattice energies of ionic transition-metal compounds reflect both the trends in ionic radius and the changes in CFSE that occur as d electrons are added to the ion. Figure 11.9 shows both these influences at work. In the figure, solid points represent "experimental" lattice energies from a Born–Haber cycle for the enthalpy of formation of the difluoride; open circles represent lattice energies calculated from the crystal radii and scaled to fit the Born–Haber values at d^0, d^5,

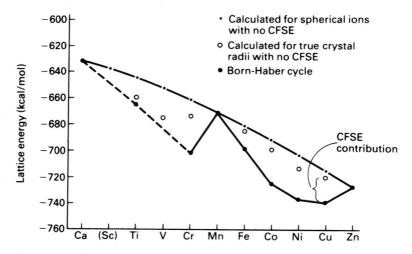

Figure 11.9   Lattice energies for transition-metal difluorides MF_2 (all high spin).

and d^{10}; and small circles represent the lattice energy that spherical ions with no crystal-field stabilization energy would presumably have. The nonspherical nature of the ions is clearly responsible for about half the additional stability shown in the Born–Haber energies. The rest represents the crystal-field stabilization energy for each lattice. Comparison with Table 11.2 suggests that the fit is fairly close if $Dq = 12$ kJ/mol, which is quite reasonable when compared to spectroscopic values.

In Chapter 10 we considered the relative stability of successive oxidation states for transition metal atoms in ionic lattices (see Table 10.4), and saw that multiple oxidation states are usually almost equally stable in these lattices. If the lattice energy increase for higher charge *exactly* compensated for the ionization energy, the energy equivalence in the crystal for differently charged ions should lead to a nonstoichiometric crystal: The total anionic charge in the crystal must equal the total cationic charge, so a few Ti^{3+} ions among the Ti^{2+} in "TiO" will require a few more O^{2-} to compensate, and the stoichiometry will no longer be 1:1. We have noted in Chapter 3 the possibility of such nonstoichiometric crystals for transition metal ions; here we may note that a Born–Haber cycle treatment of the disproportionation reaction

$$4TiO(s) \rightarrow Ti_2O(s) + Ti_2O_3(s)$$

yields an estimated ΔH of only about 20 kJ—almost a flawless balance considering that aggregate lattice energies are well over 10 megajoules. A single Ti–O phase exists all the way from $TiO_{0.69}$ to $TiO_{1.33}$, and the stoichiometric composition $TiO_{1.000}$ actually has about 15% vacancies in *each* site.

In aqueous solution, the relative stability of transition-metal oxidation states can be related fairly nicely to gas-phase ionization energies as modified by hydration energies and, in particular, changes in CFSE for the different oxidation states. For example, stable ionic compounds in both the 2+ and 3+ oxidation states are known for every first-row transition metal from Ti through Cu. Both states are stable in water for the metals from V through Co [though the hexaaquacobalt(III) ion slowly oxidizes water to O_2]. The reduction potentials $\mathscr{E}_M^0$ are well known:

$$M(OH_2)_6^{3+}(aq) + e^- \rightleftharpoons M(OH_2)_6^{2+}(aq) \qquad \mathscr{E}_M$$

In thermodynamic terms, the $\mathscr{E}^0$ of this half-reaction is equivalent to the free-energy change for the reaction

$$M^{3+}(aq) + \tfrac{1}{2}H_2(g) \rightarrow M^{2+}(aq) + H^+(aq)$$

For this reaction, we can construct the following Born–Haber cycle:

$$
\begin{array}{ccccccc}
M^{3+}(g) & + & H(g) & \xrightarrow{\text{IP}_H - \text{IP}_3} & M^{2+}(g) & + & H^+(g) \\
{\scriptstyle -U_{3hyd}}\Big\uparrow & & {\scriptstyle \Delta H_{at}}\Big\uparrow & & {\scriptstyle U_{2hyd}}\Big\downarrow & & {\scriptstyle U_{Hhyd}}\Big\downarrow \\
M^{3+}(aq) & + & \tfrac{1}{2}H_2(g) & \xrightarrow{\Delta H^0} & M^{2+}(aq) & + & H^+(aq)
\end{array}
$$

We thus have

$$\Delta H^0 = U_{2hyd} - U_{3hyd} + U_{Hhyd} + \Delta H_{at}(H) + \text{IP}_H - \text{IP}_3(M)$$

In Chapter 6, we calculated hydration energies with fair success for spherical ions on a purely classical electrostatic basis, assuming that the enthalpy of hydration of a proton is −1131 kJ/mol. However, gaseous transition-metal ions will gain more stability than spherical ions as octahedral hydrates in solution because of the CFSE arising from their nonspherical electron distributions. We can, therefore, break down U_{2hyd} and U_{3hyd} into spherical and nonspherical components:

$$U_{2hyd} = U_{2sph} + \text{CFSE}_2 \qquad \text{and} \qquad U_{3hyd} = U_{3sph} + \text{CFSE}_3$$

Inserting these into the Born–Haber cycle enthalpy summation and grouping terms, we have

$$\Delta H^0 = [U_{2sph} - U_{3sph} + U_{Hhyd} + \Delta H_{at}(H) + \text{IP}_H] - \text{IP}_3 + [\text{CFSE}_2 - \text{CFSE}_3]$$

where all the terms are known experimentally or can be calculated to good accuracy. Note that a given transition metal will not only have a different charge and radius in the 2+ and 3+ oxidation states, but also will have a different number of d electrons (and thus a different CFSE), and a larger value of Dq for the 3+ ion than for the 2+ ion. The results of a series of such calculations are shown in Fig. 11.10. In the figure, the open circles represent primarily the smooth trend in IP_3 with sharp break at d^5 shown in Fig. 11.7. The points labeled E, however, are experimental aqueous potentials (free energies) that show a much smaller change and less regular pattern. As the arrows suggest, the change in CFSE from the 3+ to the 2+ oxidation state is almost adequate to account for the observed difference; in every case, the CFSE change is in the correct direction and is nearly of the right magnitude. It is clear that although the crystal-field picture of transition-metal ion stabilities

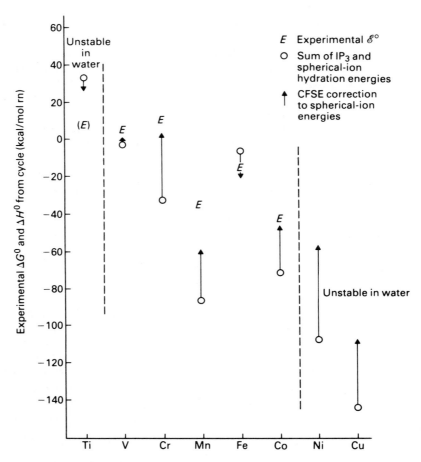

Figure 11.10 Experimental free energies and calculated enthalpies for the reaction $M^{3+}(aq) + \frac{1}{2}H_2 \rightarrow M^{2+}(aq) + H^+(aq)$.

is inadequate since it makes no provision for covalent bonding, it is an important first step toward understanding the differences between transition metals and main-group elements.

11.3 COVALENT SURROUNDINGS: TRANSITION-METAL MOs AND LIGAND-FIELD THEORY

In spite of the successes of crystal-field theory, which relies on a model of point-charge electrostatic repulsions of d electrons in atomic orbitals by ligands, the theory has some severe shortcomings and limitations. Some of these are obvious, and some appear only in the interpretation of specific experiments. Ligands—atoms or molecules surrounding a transition-metal atom—are not point charges. This is obvious for molecules, but even monatomic ions such as Cl^- will be polarized into a nonspherical electron distribution when placed next to a cation, so their overall electric nature can no longer be represented by a monopole point charge. In addition to the point-charge fallacy, no purely electrostatic (ionic) model can possibly represent real compounds of metals with good accuracy if they have intermediate electronegativities. Some provision must be made for describing covalent bonding in transition-metal compounds. Finally (perhaps as a consequence of these two difficulties) the value of Dq, as measured by either calorimetric or spectroscopic means, is *not* proportional to the net charge on the ligand as the derived expression for D (in Dq) requires. For example, the electrically neutral ligand CO usually yields a larger energy splitting for a given transition metal than the singly charged anion ligand NO_2^-, which in turn usually yields a larger splitting than the doubly charged ligand $C_2O_4^{2-}$. This is precisely the reverse of the trend predicted by crystal-field theory. We must develop a more comprehensive theory to describe the bonding in transition-metal compounds.

As in Chapter 4, we will develop qualitative molecular orbitals and their associated energy-level diagrams. The reader should review the basic patterns of sigma and pi overlap described in Chapter 4, along with the rules for forming qualitative MO energy-level diagrams. Let us begin by considering the formation of an octahedral species, such as TiF_6^{3-}. Initially we shall consider only sigma overlap between a single orbital on each ligand atom—either a p orbital or a hybrid—and the set of transition-metal atomic orbitals to be considered (the *basis set*). For first-row transition metals, the basis set usually includes not only the $3d$ and $4s$ orbitals, but also the $4p$. The latter are clearly close in energy—there is no great discontinuity in properties between zinc at the end of the d block and gallium at the beginning of the p block. We need, then, to consider the overlaps between the nine atomic orbitals of the transition-metal basis set with different combinations of the six ligand sigma orbitals, L_1 through L_6.

For octahedral geometry, the most convenient axes pass directly through each ligand orbital. Qualitative overlaps in this geometry are sketched in Fig. 11.11. The ligands have been numbered as above, and the individual wave function signs have been chosen to give maximum favorable overlap (optimum bonding). Note that no signs are indicated for ligand orbitals that, because of their symmetry, can have no net overlap. A single (normalized) combination of ligand-orbital signs can be considered a single symmetry-equivalent ligand, so that if that combination overlaps only one metal orbital, one bonding and one antibonding MO result. Since there are six initial ligand sigma orbitals, there must be six different sign combinations that form symmetry-equivalent ligand orbitals. However, they won't all have the right signs to yield any net overlap with metal orbitals, and if they don't have any net overlap they will simply constitute a nonbonding orbital with the energy of the ligand combination (at least as far as sigma overlap is concerned).

The metal $4s$ orbital overlaps the ligand combination $L_1 + L_2 + L_3 + L_4 + L_5 + L_6$, and no other metal orbital matches all those signs on L_1 through L_6. So a bonding MO and an antibonding MO will originate from the metal $4s$, as

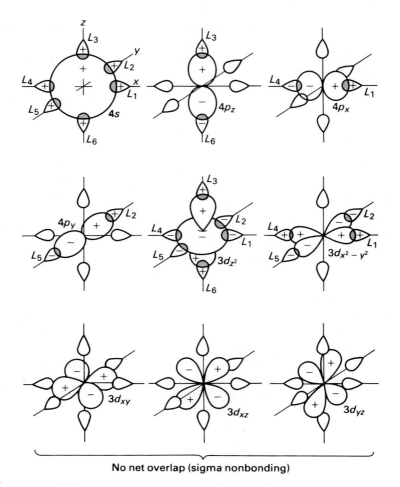

No net overlap (sigma nonbonding)

Figure 11.11   Octahedral sigma overlap for first-row transition-metal orbitals.

Fig. 11.12(a) indicates. Because of the symmetry of the signs chosen, the two MOs are said to have a_{1g} symmetry in the language of group theory, which we introduce here only as a label.

In the same way, the metal $4p$ orbitals (which of course have the same energy as atomic orbitals) each overlap a single combination of ligand orbitals. The $4p$ overlaps (L_1–L_4), the $4p_y$ overlaps (L_2–L_5), and the $4p_z$ overlaps (L_3–L_6). Each of these overlaps will produce a bonding and an antibonding MO. Since the overlaps are equivalent, there will be a set of three *degenerate* bonding MOs (MOs having the same energy) and a set of three degenerate antibonding MOs. Each of these sets of three is called t_{1u} in symmetry notation. These MOs are shown in Fig. 11.12(a).

The $3d$ orbitals on the metal atom are not all equivalent in their sigma-overlap possibilities, as Fig. 11.11 indicates. The d_{xy}, d_{xz}, and d_{yz} atomic orbitals stick out in all the wrong places; they have no possible overlap with ligand sigma orbitals, no matter what signs are chosen. Those three metal orbitals constitute a t_{2g} set of nonbonding orbitals with the same energy as the three metal atomic orbitals. On the other hand, the $d_{x^2-y^2}$ and d_{z^2} have good sigma overlap with different ligand-sign combinations and will each produce a bonding and an antibonding MO. Although it is not obvious from a qualitative sketch, the net overlap is the same in each case, so there will be a set of two degenerate bonding MOs and a set of two degenerate antibonding MOs. These are given the symmetry designation e_g. Figure 11.12(b) gives the full MO energy-level diagram (on a sigma-only basis) for the octahedral complex ML_6. The basis set of atomic orbitals has 15 AOs (9 metal + 6 ligand); likewise, there are 15 MOs in the diagram. The sequence of energies might change slightly in a detailed numerical calculation, but the general order and the number of bonding MOs (six) would not be affected.

With the energy-level diagram established, we must next consider how they are populated by electrons in the octahedral molecule. After all, orbitals are meaningless unless they actually describe the distribution of electrons in space. Molecular-orbital approaches such as this are commonly used to describe electron distribution in coordination compounds or complexes in which a transition-metal atom or ion serves as a Lewis acid and several (here, six) Lewis bases each donate a pair of electrons to the transition-metal atom. In TiF_6^{3-}, for example, six fluoride ions each contribute a pair of electrons to a Ti^{3+} ion. Of course, the origin of the electrons in the complex ion cannot be identified, but for the sake of bookkeeping we usually assume that the six pairs of electrons contributed by the ligands occupy the six sigma bonding orbitals (a_{1g}, e_g, and t_{1u}). The remaining transition-metal valence electrons, anywhere from d^0 to d^{10}, are then distributed among the t_{2g} nonbonding and e_g antibonding orbitals. In TiF_6^{3-}, the one remaining Ti^{3+} valence

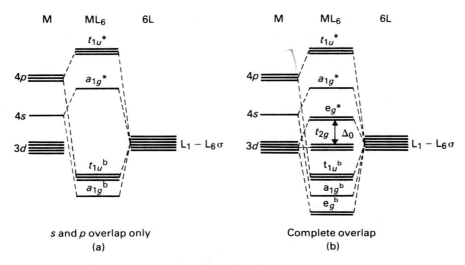

s and p overlap only
(a)

Complete overlap
(b)

Figure 11.12 Sigma-only MOs for octahedral overlap.

electron would logically be placed in one of the three t_{2g} MOs to give the overall lowest electronic energy for the complex. In an excited state of the complex, that last electron would be in an e_g^* MO. Note that the t_{2g} and e_g^* MOs have exactly the same atomic-orbital connections and energy relationship as the d orbitals considered in the crystal-field theory model. The energy separation that was called $10Dq$ in crystal-field theory is usually called Δ_o (subscript "o" for octahedral) in the molecular-orbital model. For a strongly ionic complex—which TiF_6^{3-} probably is—the MO model yields results indistinguishable from crystal-field theory. For this reason, the d-electron MO approach is often called *ligand-field theory* (LFT) to indicate the similarities.

The MO approach is more powerful than crystal-field theory, however. Partially covalent or even predominantly covalent bonding is handled naturally through the varying atomic-orbital coefficients in each MO. In a numerical calculation, they would be varied systematically to yield the lowest overall electronic energy for the complex. The resulting coefficient values would indicate how the electrons were shared or delocalized through the molecule. If we consider only sigma overlap, the energy levels of Fig. 11.12(b) are as applicable to $Co(CN)_6^{3-}$ as they are to TiF_6^{3-}—even though it is certainly a very bad approximation to say that $Co(CN)_6^{3-}$ is ionic. However, the different relative VOIP values and different degrees of overlap for the various atomic orbitals might make Δ_o quite different for $Co(CN)_6^{3-}$ and TiF_6^{3-}. However, the metal oxidation states and ligand ion charges are the same, so crystal-field theory might well predict nearly the same value of $10Dq$. Pi bonding, which we take up in the next section, alters Δ_o substantially. This is readily handled via MO theory, but next to impossible to describe via crystal-field theory.

Another important molecular geometry for transition-metal complexes is that of the tetrahedron, such as $CoCl_4^{2-}$. Figure 11.13 shows the atomic-orbital sigma overlaps for such a system. Figure 4.15 showed s and p sigma overlaps for tetrahedral geometry, and these are the same here, of course. In addition, Fig. 11.13 shows the d-orbital sigma overlaps, and these are quite different from the octahedral case. The three d orbitals that were nonbonding for octahedral geometry now show sigma overlap, while the two d orbitals that showed sigma overlap for the octahedral case are now nonbonding. If we choose ligand-sign combinations that provide optimum overlap for the three p orbitals on the metal, we find that each combination matches not only a p orbital, but also one of the sigma-bonding d orbitals—that is, the same ligand sign combinations are required for $4p_x$ and $3d_{yz}$, $4p_y$ and $3p_{xz}$, and $4p_z$ and $3d_{xy}$. Figure 11.14a shows the MO energies resulting from sigma overlap of the $4s$ orbital (which overlaps a unique combination of ligand orbitals) and the two nonbonding $3d$ orbitals. In symmetry notation, these are a_1 and e respectively. Figure 11.14b adds the p- and d-orbital overlaps. For each p orbital, there is a unique ligand-symmetry combination and a d orbital having the same overlap (that is, a set of three net atomic orbitals overlapping), so a strongly bonding MO, an approximately nonbonding MO, and a strongly antibonding MO will result. Since these arise from sets of atomic orbitals with the

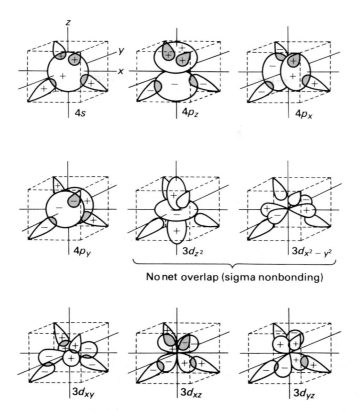

Figure 11.13 Tetrahedral sigma overlap for first-row transition-metal orbitals.

same energies, the resulting MOs will come in sets of three, each set with the t_2 symmetry designation: t_2^b, t_2, and t_2^*. Now there is an energy separation Δ_t ("t" for tetrahedral) corresponding to the crystal-field splitting of $4.44Dq$ implied in Fig. 11.5 for the tetrahedral geometry. Δ_t is the energy separation between two more-or-less nonbonding sets of orbitals, so it is probably not large. However, its magnitude is a sensitive function of atomic-orbital sizes and overlaps, so we should not expect the rigid prediction of four-ninths the octahedral splitting that comes from crystal-field theory.

One final geometry of general interest for coordination compounds of transition metals is the square-planar conformation. The sigma overlaps of atomic orbitals for this conformation are shown in Fig. 11.15 (ligands on the x and y axes). This geometry does not have the high symmetry of the two previous cubic cases, and there are fewer equivalent overlaps. Note that the s orbital and the d_{z^2} orbital on the metal overlap the same symmetry combination of ligand sigma orbitals; three MOs must result. The p_x and p_y are equivalent in their overlap, but the p_z is strictly nonbonding. The $d_{x^2-y^2}$ has strong sigma overlap with a unique ligand combina-

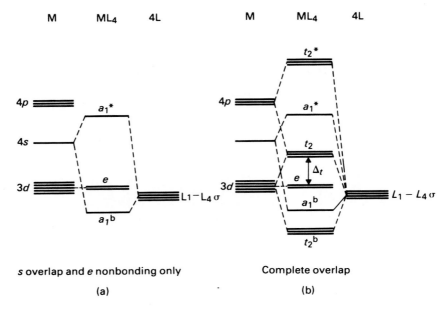

Figure II.14   Sigma-only MOs for tetrahedral overlap.

tion, but the other three d orbitals are nonbonding. The d_{xy} lies in the plane of the ligands, but their sigma orbitals lie in its nodes; on the other hand, the d_{xz} and d_{yz} are equivalent to each other but lie out of the plane of the ligands.

We briefly summarize the symmetry designations for these orbitals with reference to Fig. 11.16. In part (a) of the figure, the nonbonding metal orbitals are shown: the p_z is a_{2u}, the d_{xy} is b_{2g}, and the d_{xz} and d_{yz} together are e_g. Also in part (a) are the bonding and antibonding MOs resulting from $d_{x^2-y^2}$ overlap (with b_{1g} symmetry) and the two degenerate bonding and antibonding MOs from p_x and p_y overlap (with e_u symmetry). Finally, in part (b), the mixed overlap of the s and d_{z^2} to form a strongly bonding, a nonbonding, and a strongly antibonding MO is added, each with a_{1g} symmetry.

If a square-planar complex is formed by four ligands that each donate a pair of electrons to the metal, these ligand electrons will fill the sigma bonding MOs (a_{1g}^b, b_{1g}^b, and e_u^b). The d electrons on the metal atom or ion must then be apportioned among the e_g, b_{1g}, and b_{1g}^* MOs. The reduced symmetry means that there is no unique energy splitting analogous to Δ_o or Δ_t, but comparison with Table 11.1 shows that these vacant MOs lie in much the same energy sequence as the crystal-field energies. From this portion of the energy-level diagram, it should be apparent why square-planar complexes are much more common for d^8 electron configurations than for any other. The four approximately nonbonding MOs lie much lower in energy than the b_{1g}^*, and the d^8 system is more stable than it would be in a perfect octahedral complex with two electrons in the e_g^* MOs. Most square-

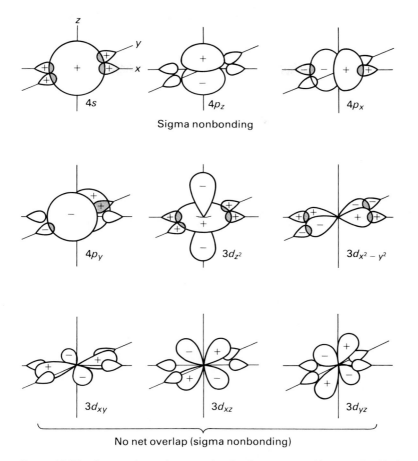

Sigma nonbonding

No net overlap (sigma nonbonding)

Figure 11.15   Square-planar sigma overlap for first-row transition-metal orbitals.

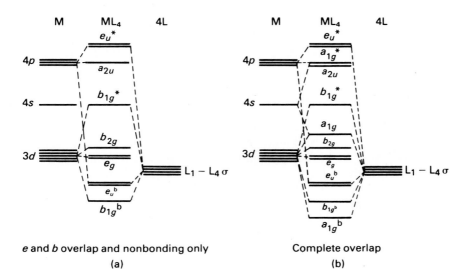

e and *b* overlap and nonbonding only

(a)

Complete overlap

(b)

Figure 11.16   Sigma-only MOs for square-planar overlap.

planar complexes involve Ni^{2+}, Pd^{2+}, Pt^{2+}, and Au^{3+}, all of which are d^8 configurations.

As the next chapter will suggest, transition-metal complexes have many possible geometries. The most common, however, are octahedral, tetrahedral, and square planar. We are now equipped to describe (on a sigma-only basis) their bonding through the MOs in Figs. 11.12, 11.14, and 11.16. However, for complexes or molecules in which covalent bonding is predominant, pi overlap of atomic orbitals may be quite important. We must consider the effect of pi bonding on the sigma MOs we have just developed.

11.4 COVALENT SURROUNDINGS: TRANSITION-METAL PI BONDING

Our MO model qualitatively reproduces the energy-level patterns of crystal-field theory, but within a framework that allows partial or extensive covalent bonding. However, the absence of any description of pi overlap is a serious omission, particularly because (as we noted earlier) the frequent angular nodes in d orbitals make pi overlap convenient, and the d orbitals are not so deeply buried in the core that their overlap with ligand orbitals is poor.

A related problem we have not mentioned is that if a transition-metal atom or ion in a low oxidation state is to accept electrons from, say, six ligands with any substantial degree of covalent bonding, its share of the total electron distribution will build up to an entirely unrealistic level. Suppose a metal 2+ ion has a 40% share of each electron pair from six ligands. Its net charge will be 2.8−, and the ligands might well have a net positive charge even though they are significantly more electronegative than the transition-metal atom. This makes no sense; such complexes are quite common and extremely stable.

Consider the sequence of orbital overlaps shown in Fig. 11.17. Suppose the metal atom has two electrons in the d_{xy} orbital, but none in the $d_{x^2-y^2}$. Suppose further that the ligand has a nonbonding pair with sigma symmetry, but also has a vacant pi orbital that can overlap the d_{xy}. The sigma overlap is shown in part (a) of the figure; shading indicates initial electron density. The pi overlap is shown in part (b) of the figure, where the potential electron donor–acceptor relationship is reversed. Finally, part (c) shows the simultaneous sigma and pi overlap and the mutual donor–acceptor pattern of electron flow and bonding that results. The metal–ligand bond is strengthened, since it now approximates a conventional double bond. Also, it should be clear that the overall degree of electron transfer is reduced and there is no longer a large buildup of negative charge on the transition-metal atom. In the molecular-orbital model, pi overlap of this sort is critically important to the stability of these complexes. The availability of d electrons for this charge compensation makes transition metals uniquely strong Lewis acids toward ligands that have pi-acceptor capabilities.

What effect does such pi bonding have on sigma-only MO energies, such as those for the octahedral geometry in Fig. 11.12? As part (b) of Fig. 11.17 perhaps

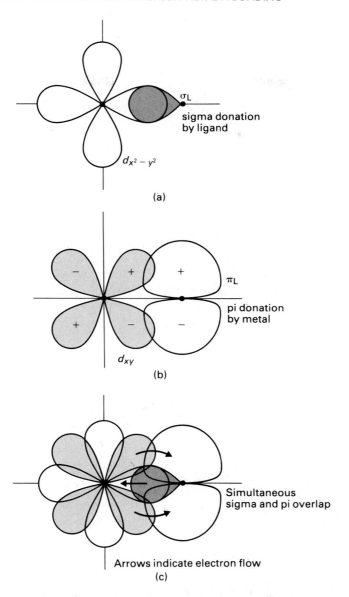

Figure 11.17 Charge delocalization through sigma and pi overlap in transition-metal orbitals.

suggested, in an octahedral complex some of the pi-type ligand orbitals will have the same overall symmetry as the sigma-nonbonding d_{xy}, d_{xz}, and d_{yz} orbitals on the metal—that is, t_{2g}. A t_{2g} symmetry-equivalent ligand pi-orbital combination will combine with each of these previously nonbonding d orbitals to yield a bonding and an antibonding MO. (If there are two p orbitals with pi symmetry on each of the six ligand atoms, the set of twelve pi-symmetry orbitals will have other symmetry-equivalent combinations. However, none of these can interact with the metal sigma nonbonding MOs, so we need not consider them.)

If the ligand atoms have vacant p or pi-symmetry orbitals, they will in general lie at higher energies than the metal d orbitals and thus at higher energy than the t_{2g} nonbonding set in the sigma-only MO diagram. Figure 11.18 shows the result: The e_g^* energy is unchanged, since no ligand pi combinations have E_g symmetry, but bonding and antibonding combinations now exist for the t_{2g} orbitals. Δ_o is significantly increased, and the octahedral molecule or complex is stabilized to the extent that it has electrons in the t_{2g} orbitals. Few if any ligand atoms have vacant p orbitals, but any vacant orbital with pi symmetry will do as well. Carbon monoxide, for example, is a good sigma donor because of a nonbonding pair on the C atom, but it is also a good pi acceptor because of the two vacant π^* orbitals accompanying its triple bond. The same is true for the isoelectronic CN^- ion and (to a lesser extent) for the N_2 molecule as a ligand. In the trialkylphosphines, PR_3, the P atom has valence $3s$ and $3p$ electrons and thus must also have vacant $3d$ orbitals that can serve as pi acceptors. Figure 11.19 gives the geometry of these pi overlaps, which is obviously equivalent to that shown in Fig. 11.17.

Under certain circumstances, ligands can also be pi *donors* to a transition metal in a complex. Halide ions, for example, have filled p orbitals with pi symmetry and F^-, at least, has no low-energy d orbitals that could serve as pi acceptors. The valence p orbitals for halides lie at quite low energies, normally lower than the

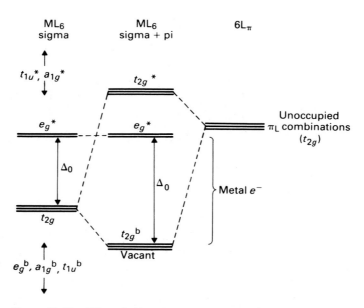

Figure 11.18 Effect of pi bonding to pi-acceptor ligands on sigma-only O_h energy levels.

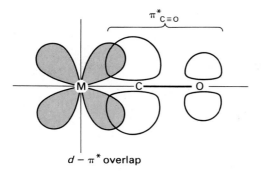

$\pi^{*}_{C\equiv O}$

$d - \pi^{*}$ overlap

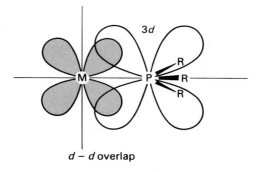

$3d$

$d - d$ overlap

Figure 11.19 Pi-acceptor ligand orbitals and overlaps.

metal d orbitals and the sigma-nonbonding t_{2g} orbitals. Pi overlap still leads to t_{2g} bonding and antibonding orbitals, but now the t_{2g} bonding orbitals are filled by ligand electrons, and the metal d electrons must occupy the t_{2g}^{*} antibonding orbitals. This decreases Δ_{o}, as Fig. 11.20 shows.

We now have a reasonably complete qualitative picture of the bonding in transition-metal compounds by the molecular orbital model. We have successfully reproduced the d-orbital energy-level patterns of crystal-field theory in a flexible formalism that allows for covalence, and have seen in addition that the Dq or Δ quantity is sensitive to the quantum-mechanical nature of the ligand atom. From this we should be able to explain the different energy splittings observed experimentally in transition-metal complexes. However, before we can deal directly with those experiments, we must develop our theory one more step. In general, techniques for observing electronic energies in molecules yield the energy of the whole polyelectronic molecule, not the energy of a single electron. We must move from a "one-electron" approximation to the polyelectronic states of an atom or molecule.

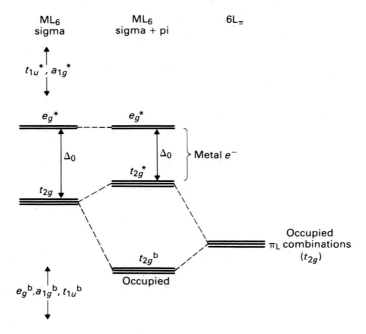

Figure 11.20   Effect of pi bonding to pi-donor ligands on sigma-only O_h energy levels.

11.5 ELECTRONIC STATES AND TERMS FOR TRANSITION-METAL ATOMS

We shall develop polyelectronic states and terms for transition-metal atoms in two stages. First, we need to understand their origin for a free atom—one in which electrons are interacting, but without bonds or surrounding atoms to influence them. Second, with the free-atom or free-ion states and terms established, we need to see how they are changed or split by the added influence of chemical bonds or a ligand field with the common symmetries.

First, what is the overall arrangement of electrons in a free atom, and what are the possible aggregate electronic energies? For a one-electron atom, two pieces of information follow simply from the orbital designation: $3d_{z^2}$ tells us that $n = 3$, $l = 2$, $m_l = 0$. The spin is $\frac{1}{2}$ for a single electron, but m_s can be $\pm\frac{1}{2}$, so there are two spin possibilities. We describe this by saying that the one-electron atom is in a *doublet state*. The one-electron atom's energy is completely determined by the quantum number n, while l and m_l tell us about the orbital angular momentum of the electron (the entire atom, in this case). Specifically, L^2, the square of the total orbital angular momentum, is given by $l(l + 1)h^2/4\pi^2$, and L_z, the z component of the orbital angular momentum, is given by $mh/2\pi$.

For a polyelectronic atom, we describe the overall electronic distribution in a parallel fashion. That is, we designate the total orbital angular momentum of all the electrons and the total spin angular momentum of all the electrons. Particularly for first-row transition metals, the most convenient way to arrive at this designation is by taking the vector sum of the individual electron l values to yield a total orbital angular momentum for the polyelectronic atom of L, with corresponding M_L values, and taking the vector sum of the individual electron-spin angular momenta to yield a total spin S. The spin and orbital angular momenta can then interact through what is called $L \cdot S$ coupling, or *Russell–Saunders coupling*. A given state of a polyelectronic atom will be characterized by a capital letter S, P, D, F, G, ... indicating the state's orbital angular momentum L just as l does for a single electron ($L = 0$ for an S state, 1 for a P state, and so on). The spin of the polyelectronic state is indicated by a superscript prefix giving the numerical value of $2S = 1$. Thus, one electron yields a spin doublet, and $2(\frac{1}{2}) + 1 = 2$. An atom might have a ^3P state (read "triplet P"), indicating that there is a total spin S of 1 and a total spin orbital angular momentum L of 1. Actually, there will be several ^3P states for a given atom. This reflects the fact that there are different z components of the spin and orbital angular momentum for the polyelectronic system, just as there are different m_l and m_s values for a single electron. In the absence of a magnetic field, all these ^3P states will have the same energy, and we group them as a ^3P *term*. There are thus nine states in a ^3P term: a spin multiplicity of three ($M_S = 0, \pm 1$) times an orbital multiplicity of three ($M_L = 0, \pm 1$).

For transition metals (with d electrons), what states and terms are possible? First, note that for any atom with all the orbitals in a given l subshell filled by paired electrons (all three $2p$ orbitals, for instance), the only possible value of L is zero, since all the possible m_l values are equally represented. For every electron with $m_l = +1$ there is an electron with $m_l = -1$, and so on, so that when the m_l values for all the electrons are summed, they must yield zero. Similarly, the paired electrons have an $m_s = -\frac{1}{2}$ for every $m_s = +\frac{1}{2}$, and the only possible value of S is zero. Such an atom, then, necessarily has a ground state or ground term of ^1S (singlet S). Any terms other than ^1S can only arise from the presence of unequally filled orbitals in an l subshell. For transition metals with varying numbers of d electrons, the possible electronic terms will arise from combinations of the appropriate number of electrons, each of which has $m_l = +2, +1, 0, -1, -2$, and $m_s = +\frac{1}{2}$, $-\frac{1}{2}$. If the number of d electrons is large (say four or five) the number of permutations of these m_l and m_s possibilities becomes enormous. Every permutation is a possible state of the atom, and these states are linked as described earlier into terms.

A free atom or ion with one d electron, such as gaseous Ti^{3+}, yields a ^2D term, since L must equal l and there are two spin possibilities. For an atom with two d electrons, the l values can add to give L values ranging all the way from $l_1 - l_2$ to $l_1 + l_2$—that is, L can be anywhere from 0 (2 − 2) to 4 (2 + 2). So the possible terms are S, P, D, F, and G. We might at first imagine that the two electron spins could either be paired, yielding a singlet, or unpaired, yielding a triplet, for each of these L values. In fact, however, the exclusion principle rules out some of the permutations, so the possible terms are ^1S, ^3P, ^1D, ^3F, and ^1G. The process of establishing the permutations (states) and grouping them into terms is tedious, but

Table 11.3   Electronic Terms Arising from Transition-Metal Atoms

Configuration	Terms
d^0, d^{10}	^1S
d^1, d^9	^2D
d^2, d^8	$^3F, {}^3P$
	$^1G, {}^1D, {}^1S$
d^3, d^7	$^4F, {}^4P$
	$^2H, {}^2G, {}^2F, {}^2D, {}^2D', {}^2P$
d^4, d^6	^5D
	$^3H, {}^3G, {}^3F, {}^3F', {}^3D, {}^3P, {}^3P'$
	$^1I, {}^1G, {}^1G', {}^1F, {}^1D, {}^1D', {}^1S, {}^1S'$
d^5	^6S
	$^4G, {}^4F, {}^4D, {}^4P$
	$^2I, {}^2H, {}^2G, {}^2G', {}^2F, {}^2F', {}^2D, {}^2D', {}^2D'', {}^2P, {}^2P, {}^2S$

Note: Primes indicate terms occurring more than once for a given configuration.

the process is fortunately shortened by the fact that the permutations of two holes in 10 boxes (spin-orbital combinations) are the same as the permutations of two electrons in 10 boxes, so that the terms arising from eight d electrons are the same as the terms arising from two d electrons. The terms arising from a d^n configuration, then, are the same as the terms arising from a d^{10-n} configuration. Table 11.3 shows the terms for each configuration.

In a one-electron atom, all orbitals with the same n value have the same energy. If $n = 3$, for instance, the energies of the s, p, and d orbitals are equal. But in a polyelectronic atom, the energies of S, P, D, and other terms will in general be different, and even two terms with the same L—two P terms, say—will have different energies if their spin multiplicity is different. The obvious reason is that repulsions between the electrons are different for different overall orbital or spin angular momenta. This is true even though the atom's orbital *configuration* of electrons (for example p^2) may be the same for the different terms. The differing repulsions lead to changes in the orbital shapes that reduce electron–electron repulsion but also reduce electron–nucleus attraction; if the repulsions differ, the orbital shape changes and attraction changes will differ too.

A quantitative theoretical approach to this problem begins by calculating the magnitude of the repulsion between two electrons in a single atom. For electrons numbered 1 and 2, this is given by

$$E_{\text{repulsion}} = \int (\psi_1^2)\left(\frac{1}{r_{1-2}}\right)(\psi_2^2)$$

where ψ_1 is the wave function of electron 1, ψ_2 is the wave function of electron 2, and r_{1-2} is the distance between the two electrons. The integral is taken over all the coordinates of both electrons. A series of these integrals must be evaluated for all possible pairs of electrons in the atom, a task that requires some mathematical elegance. For pairs of d electrons, however, all repulsions are represented by one of three integral values, F_0, F_2, or F_4, regardless of m_l and m_s. These three values are not universal, but are specific for each particular transition-metal atom or ion. Furthermore, the exclusion principle for polyelectronic systems requires that the total polyelectronic wave function change sign if two electron numbers are interchanged. This sign change means that some of the repulsion integrals will have a negative sign, representing a *reduction* of repulsion. For example, the terms arising from a d^2 configuration have the following repulsion-integral totals:

$$^1S \text{ states' repulsion total} = F_0 + 14F_2 + 126F_4$$
$$^1G \text{ states' repulsion total} = F_0 + 4F_2 + F_4$$
$$^3P \text{ states' repulsion total} = F_0 + 7F_2 - 84F_4$$
$$^1D \text{ states' repulsion total} = F_0 - 3F_2 + 36F_4$$
$$^3F \text{ states' repulsion total} = F_0 - 8F_2 - 9F_4$$

These are listed in order of increasing reduction of repulsion. That is, the ^3F term has the lowest energy, and the ^1S term the highest. Hund's rule for predicting the ground state of a polyelectronic atom or ion is a generalization of this result:

The ground term of a polyelectronic atom or ion is the term having the greatest spin multiplicity; if there are several with this spin multiplicity, the ground term is the one with the highest orbital angular momentum L.

As we shall see, in interpreting electronic spectra we are rarely interested in the absolute energies of terms; rather, we are interested in the energy differences between them. The F_0, F_2, and F_4 integrals are usually lumped together as the *Racah parameters A, B,* and *C*:

$$A = F_0 - 49F_4$$
$$B = F_2 - 5F_4$$
$$C = 35F_4$$

The coefficients here have been chosen to simplify the results for each atom as much as possible. Since F_0 occurs equally in the total repulsion of all states, it will disappear when the energy difference between terms is taken. Accordingly, the relative energies of terms can be expressed in terms of the Racah parameters B and C alone. These are usually not evaluated theoretically from repulsion integrals, but are obtained empirically by fitting spectra. In common spectroscopic energy units, B is about 1000 cm^{-1} and C is about 4000 cm^{-1}. Specific values will be different for different ions, of course. Table 11.4 gives free-ion B values from atomic spectroscopy.

We usually need not evaluate the relative energies of all the terms of a given ion. There is a spectroscopic selection rule that electronic transitions can only occur between two states that have the same spin. If we examine Table 11.3 in light of Hund's rule, we see that in each case the ground term is the first one listed for

Table 11.4 Racah Parameters of Electron Repulsion for Transition-Metal Free Ions

	Ti	V	Cr	Mn	Fe	Co	Ni	Cu
neutral atom:	560, 3.3	580, 3.9	790, 3.2	720, 4.3	805, 4.4	780, 5.3	1025, 4.1	
charge 1+:	680, 3.7	660, 4.2	710, 3.9	870, 3.8	870, 4.2	880, 4.4	1040, 4.2	1220, 4.0
charge 2+:	720, 3.7	765, 3.9	830, 4.1	960, 3.5	1060, 4.1	1120, 3.9	1080, 4.5	1240, 3.8
charge 3+:		860, 4.8	1030, 3.7	1140, 3.2				

	Zr	Nb	Mo	Tc	Ru	Rh	Pd	Ag
	250, 7.9	300, 8.0	460, 3.9		600, 5.4			
	450, 3.9	260, 7.7	440, 4.5		670, 3.5			
	540, 3.0	530, 3.8			620, 6.5		830, 3.2	
		600, 2.3						

	Hf	Ta	W	Re	Os	Ir	Pt	Au
	280	350, 3.7	370, 5.1	850, 1.4				
	440, 3.4	480, 3.8		470, 4.0				

Note: For each element and oxidation state, the first entry is B in cm^{-1} and the second entry is the C/B ratio.

Value in J/mol = tabulated value (in cm^{-1}) $\times$ 11.96.

that configuration. Only for d^2/d^8 and d^3/d^7 is there another term with the same spin, in each case a P term. For both of these cases, theoretical analysis shows that the energy of the P term should be $15B$ higher than the energy of the ground term. This relatively simple result will be put to good use in interpreting the electronic spectra of high-spin transition-metal complexes for the simple case in which the spin selection rule is obeyed. At the end of this section, we shall look at a more complete set of energy-level diagrams for symmetry terms arising from all the different free-ion terms, which allow us to interpret more complex spectra.

With the free-ion terms established for different d^n configurations, we can move on to the second stage of the theoretical development. How are the free-ion terms affected or split by the presence of a ligand field? The answer will, of course, depend on the symmetry of the ligand field, so there will be a different result for each molecular geometry. Here we shall focus on the two cubic fields, octahedral ML_6 and tetrahedral ML_4.

Our primary interest is the ground terms of the free ions and the few excited states with the same spin. Note that in Table 11.3, there is a good deal of symmetry in the tabulated ground states. For d^0 (filled shell), d^5 (uniformly half-filled shell), and d^{10} (filled shell), the ground states are all S terms, indicating spherical electron distributions. For d^1, d^4, d^6, and d^9, the ground states are D terms. We should not be surprised that a single d electron gives a D term, but also a spherical distribution plus one d electron (d^6) gives a D term, and so do spherical distributions *minus* one electron (d^4 and d^9). This introduces the idea that a "hole"—a specific vacancy in a set of d orbitals—has the same effect as an electron in determining the symmetry and multiplicity of the ground state. A d^9 atom is most easily thought of as a hole roaming around in a set of d orbitals. In exactly the same sense, the ground states for d^2, d^3, d^7, and d^8 are F terms; d^2 and d^7 are spherical distributions plus two electrons, whereas d^3 and d^8 are spherical distributions plus two holes. One d electron or hole gives a D term as ground state; two d electrons or holes give an F term as ground state.

Now, in an octahedral ligand field that maintains the high spin of the free-ion ground state, what sort of energy splitting occurs for the various possible ground terms S, D, and F, and for the only type of important excited term, $P?$ S terms are easy. Because they represent spherically symmetrical electron distributions, they are not affected by ligands in *any* geometry, just as an s orbital cannot have its energy split by a ligand field. Similarly, an octahedral ligand field will not split P terms for essentially the same reason it does not split p orbitals (see Fig. 11.3 and the associated discussion).

D terms behave like d orbitals. Of the five d orbitals that a single electron could adopt, three are sigma-nonbonding in an octahedral molecule and lie at low energy, and two are sigma-antibonding and lie at high energy. The D term, composed of five states, produced by one electron (or by six) is thus split into a low-energy T_{2g} term (three states) and a high-energy E_g term (two states). (Here we use capital letters in the symmetry designations to indicate that the symmetry refers to the state of the polyelectronic atom in a molecule, not just to the symmetry of a single orbital.) Figure 11.21(a) shows the resulting states for a d^1 system,

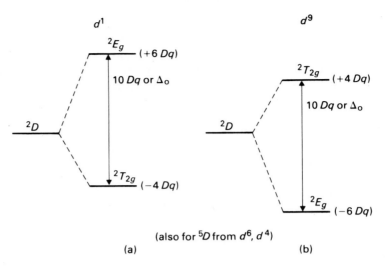

Figure II.21   Splitting of *D* terms by an octahedral field.

such as $TiCl_6^{3-}$, or for a d^6 system that has a ^5D ground term and is not split enough to pair the *d* electrons, such as CoF_6^{3-}. When we turn to *D* terms arising from d^4 or d^9 configurations, however, there is an important difference. Such terms arise from the behavior of a hole distributed among the *d* orbitals. Since electrons spontaneously distribute themselves to achieve low energy, they force holes into high-energy orbitals. That is, whereas a single *d* electron will prefer to be in the t_{2g} orbitals, creating a T_{2g} state, a single hole will prefer to be in the e_g orbitals, creating an E_g state. For the d^4 or d^9 *D* terms, then, the state splitting in an octahedral ligand field is reversed so that the E_g term lies at low energy and the T_{2g} term at high energy, as in Fig. 11.21(b).

F terms behave like *f* orbitals, but with a difference. We did not specifically consider the LFT splitting of *f* orbitals, but from Fig. 10.1 it should be clear that f_{xyz} is uniquely low-energy because it avoids all six coordinate axes and ligands. Less obviously, the other six *f* orbitals can be combined to a set of three that has pi symmetry along the bond axes, like $f_{z(x^2-y^2)}$, and another set of three that has sigma symmetry along one axis, like f_{z^3}. If we consider the *d* wave functions that combined yield an *F* term, they are also split into a single state (A_{2g}) and two sets of three states (T_{1g} and T_{2g}). In this case, however, energies of the term components or symmetry terms are in inverse order from the energies of the *f* orbitals. As Fig. 11.22a indicates, the *F* terms arising from d^2 or d^7 configurations have the A_{2g} term at highest energy ($+12Dq$), the T_{2g} term at $+2Dq$, and the T_{1g} term at lowest energy ($-6Dq$). Note that these symmetry terms will be spin triplets for d^2, but spin quartets for d^7. As for the *D* terms, an *F* term arising from two holes in an otherwise spherical electron distribution is split into octahedral-field terms with the same symmetry but with inverted energies from those arising from

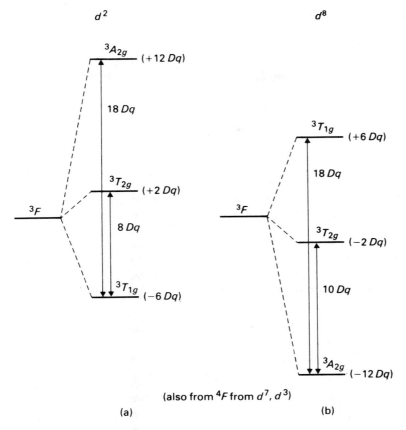

Figure 11.22   Splitting of F terms by an octahedral field.

two electrons. Figure 11.22b shows this behavior for the F terms from d^3 and d^8 configurations.

This discussion has referred specifically to an octahedral ligand field, but it can be readily transferred to tetrahedral complexes. The d orbitals have the same symmetry properties in a tetrahedral field as they do in an octahedral field, but the sign of Dq is reversed so that the e levels lie at low energy and the t_2 levels at high energy. Therefore, all of the splittings we have just worked out for octahedral complexes can be used directly. To construct symmetry-term energy-level diagrams for d^n configurations in tetrahedral complexes, we only need to invert the splitting for each atomic term. Of course, the splittings will in general be smaller because Δ_t is smaller than Δ_o. Table 11.5 gives the general symmetry terms for the important atomic terms in a cubic field, whether octahedral or tetrahedral.

We can usefully summarize our term-splitting discussion in a tabular format, with references to Fig. 11.21 and 11.22. Table 11.6 gives the energy-level sequence

Table 11.5 Symmetry Terms Arising from
Atomic Terms in a Cubic Field

L value	Term	Symmetry terms (splitting)
0	S	A_1
1	P	T_1
2	D	$E + T_2$
3	F	$A_2 + T_1 + T_2$

for the symmetry terms appropriate to d^n metal ions in weak-field complexes—that is, in complexes for which Dq or Δ is not large enough to force electron pairing. The table is expressed in fractions of Δ_o or Δ_t. To convert to Dq equivalents, multiply by 10 for an octahedral complex or by 4.44 for a tetrahedral complex.

There is one last energy effect to consider before we attempt to draw overall energy-level diagrams for d^n configurations. All four of the configurations (d^2, d^3, d^7, d^8) that yield F terms in their ground state have nearby excited-state P terms, as Table 11.3 indicates. Since the F terms yield, among others, T_{1g} terms in an octahedral environment, and since the excited-state P terms also yield T_{1g} terms, two terms with the same symmetry are fairly close in energy. We have seen an analogous situation before: When two overlapping atomic orbitals have the same symmetry, they combine to yield two new molecular orbitals with the same symmetry, one higher in energy than either of the original orbitals, and the other lower than either. Exactly the same sort of interaction occurs here. Two nearby terms (close in energy) with the same symmetry interact to yield two new terms with the same symmetry, but with energies higher and lower than the original terms. So whenever we have a T_{1g} term from a P excited state, it will lie at somewhat higher energy than it had in the free ion because it interacts with the T_{1g} term from the F ground state below it. The $T_{1g}(F)$ term will, in turn, lie at somewhat lower energy than it

Table 11.6 Symmetry-Term Energies for Transition-Metal Complexes with Cubic Ligands Yielding Weak Ligand Fields

Geometry and configuration	Symmetry terms and energies	Figure reference
Octahedral d^1, d^6 Tetrahedral d^4, d^9	$T_2(-0.4\Delta)$, $E(+0.6\Delta)$	Fig. 11.21(a)
Octahedral d^2, d^7 Tetrahedral d^3, d^8	$T_1(-0.6\Delta)$, $T_2(+0.2\Delta)$ $A_2(+1.2\Delta)$, $T_1(P)(15B)$	Fig. 11.22(a)
Octahedral d^3, d^8 Tetrahedral d^2, d^7	$A_2(-1.2\Delta)$, $T_2(-0.2\Delta)$ $T_1(+0.6\Delta)$, $T_1(P)(15B)$	Fig. 11.22(b)
Octahedral d^4, d^9 Tetrahedral d^1, d^6	$E(-0.6\Delta)$, $T_2(+0.4\Delta)$	Fig. 11.21(b)

otherwise would. [These effects can be inverted if the $T_{1g}(F)$ term is split to a higher energy than the $T_{1g}(P)$ term before interaction.] The size of this energy interaction will depend on how close the two original T_{1g} terms are together. As for combining atomic orbitals, the energy effect is greater the closer together the two original orbitals/terms. It is not possible to predict the exact energy quantitatively in any simple model, but it can be estimated from UV-visible spectra. It is roughly the size of the Dq quantity, so it is not negligible.

All of the term discussion so far has dealt with electronic states of complexes with high spin, meaning that the magnitude of $10Dq$ or Δ_o is not great enough to make spin pairing preferable. How do we deal with low-spin (high-field) complexes? As Dq increases, the overall CFSE and thus the e_g/t_{2g} electron configuration come to dominate the energy diagram. For d^4 complexes, for instance, in low field the electron–electron repulsion reduction arising from parallel spins makes the ^6S term and the resulting symmetry term $^6A_{1g}$ the ground state; five parallel spins can only arise from the configuration $t_{2g}^3 e_g^2$. But in a really strong octahedral crystal field (or ligand field) pairing in the low-energy t_{2g} orbitals will be preferable: t_{2g}^5. This will yield a high-field ground term $^2T_{2g}$. We can expect some intermediate-energy terms arising from the intermediate-energy configuration $t_{2g}^4 e_g^1$, and because numerous permutations of electron/orbital occupancy (states) are possible there will be several such terms. Now, each term possible for the complex arises from a particular set of states that depend only on orbital occupancy and the geometric symmetry of the complex—not from the size of Dq. So as Dq increases the number and identity of terms doesn't change, only their relative energies. We can create a term correlation diagram, as in Fig. 11.23 for the d^5 case, to show the general way in which the relative term energies change with crystal field or ligand field. At the left side of the diagram the terms are in essentially Hund's rule order, grouped in order of increasing electron–electron repulsions. As Dq increases CFSE becomes more important, and eventually terms arising from the same orbital occupancies cluster together.

Useful approximate calculations can be done to yield a quantitative diagram analogous to Fig. 11.23, from which term-to-term energy differences could be measured and electronic-excitation spectra interpreted. Because the Racah B parameter and the size of Dq for a given ligand vary from one transition-metal ion to another, it might be expected that an enormous number of such calculations and diagrams would be necessary. However, if relative energies are expressed in multiples of B—that is, if total energy E is plotted as E/B and Dq is plotted as Dq/B—the quality of the approximation deteriorates a bit, but the correlation diagram for each d^n configuration becomes applicable to different metals and different ligands in the same sense that Figs 11.21 and 11.22 were. These diagrams are known as *Tanabe–Sugano diagrams,* and a set of them appears as Appendix C at the back of the book. Here we can consider the construction and use of one Tanabe–Sugano diagram, that of the d^5 configuration shown in Fig. 11.24.

Note first that the axes are in E/B and Dq/B units so that if, for a given metal in a given complex, B is 1000 cm^{-1} a Dq value of 1250 cm^{-1} would correspond to 1.25 on the horizontal axis. $Dq/B = 0$ at the left of the diagram simply refers to the

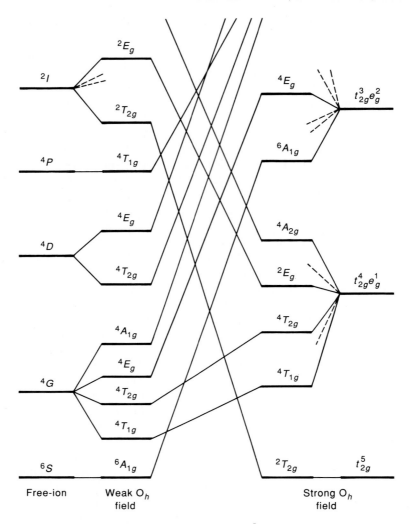

Figure 11.23 Correlation of term energies for d^5 configuration in weak octahedral field with energy groupings in strong octahedral field. [From M. Gerloch and R. C. Slade, *Ligand-Field Parameters*, Cambridge Univ. Press (1973).]

free-ion atomic term energies. For measuring convenience on the diagram, the horizontal axis is always the energy of the ground term, and other term energies are all placed on the diagram relative to the ground term. The calculated energies in each case assume that the Racah parameter C is a constant multiple of B, about $4B$, to keep the diagram two-dimensional; this is not a bad approximation as Table 11.4 will indicate, but it does limit the detailed accuracy of the diagram for a particular complex. For values of Dq/B lower than about 2.8 the ^6S (or $^6A_{1g}$)

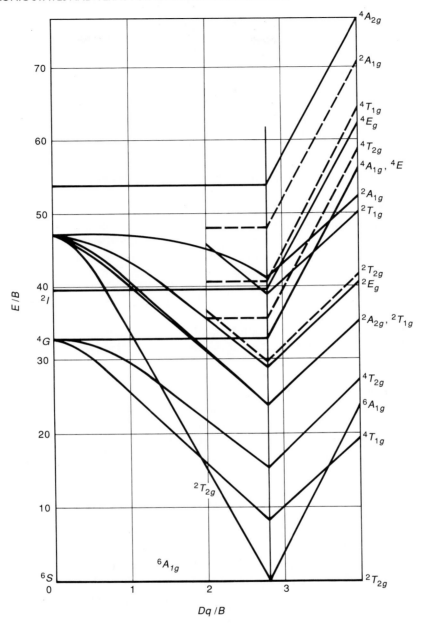

Figure 11.24 Tanabe–Sugano term energy diagram for d^5 configuration, assuming $C/B = 4.48$. [From Y. Tanabe and S. Sugano, *J. Phys. Soc. (Japan)* (1954), *9*, 766.]

term is the ground term, but when Dq increases above that value the low-spin $^2T_{2g}$ term becomes the ground term, so there is a vertical discontinuity in the diagram at that point. (You can see the $^2T_{2g}$ term nose-diving to the left of the discontinuity.) The diagram has been simplified somewhat (believe it or not), and the dashed lines represent symmetry terms arising from free-ion terms not on the diagram. These features appear on the other Tanabe–Sugano diagrams as well, though not all show different ground terms at high field. We will return to the use of these diagrams in the next section.

11.6 ELECTRONIC SPECTRA AND STRUCTURES

Two major techniques are available for the experimental study of electron energies in transition-metal compounds. In the older method, ultraviolet–visible–near-infrared absorption spectroscopy, energy is supplied to the molecule as electromagnetic radiation near the 10^{14}–10^{15} Hz frequency region characteristic of visible light. The molecule absorbs this energy, uses it to rearrange its bonding electrons, and in so doing changes from its ground state to an electronic excited state—from one of the terms developed in the last section to another term at higher energy. All of the colors of ordinary nonmetallic substances arise from this sort of transition, and in fact transition-metal compounds are often used as commercial pigments because of the atmospheric stability of many of their intensely colored compounds.

A newer experimental technique is that of *photoelectron spectroscopy*. In this technique, highly monochromatic high-energy electromagnetic radiation (usually from a DC discharge through pure gaseous helium, which radiates almost exclusively at 21.21 eV or 171,070 cm^{-1} compared to the 15,000–25,000 cm^{-1} of visible light) supplies enough energy to the molecule to ionize electrons away from it. The velocity of the photoionized electrons is measured, which yields their kinetic energy. The ionization energy of the electron from its original distribution is the difference between the energy of the original ionizing radiation and the kinetic energy of the ejected electrons. Because the experimental measurement is being made on the ejected electron, not on the polyelectronic atom or molecule, the measured ionization energy does not include changes in repulsion for the other electrons, and what is produced is not a term energy difference but a one-electron orbital energy (although this is, of course, influenced by repulsions). Information from photoelectron spectroscopy is thus complementary to that from absorption spectra, and it is beginning to be used extensively as a method for characterizing molecular structure (particularly for organometallic species, which are usually volatile enough to make the measurement possible). Although both techniques yield useful information, we shall be primarily concerned with interpreting absorption spectra.

Most transition-metal compounds are colored if the d-electron configuration is anything other than d^0 or d^{10}. The absorptions are transitions between terms that represent rearrangements of the d electrons, and most peaks appearing in the near-infrared and visible ranges of the spectrum—and even out into the ultraviolet—represent *d–d transitions*. In the UV (high-energy) range and sometimes down into the visible are peaks that represent the transfer of an electron from a ligand-based orbital to a metal-based orbital or vice versa (*charge-transfer transitions*). Charge-transfer peaks are usually quite intense, but they are difficult to predict quantitatively, so we shall restrict ourselves to interpreting the $d–d$ transitions. These are normally found between about 6000 cm^{-1} and 40,000 cm^{-1}, depending on the identity of the metal, the oxidation state of the metal, and the nature of the ligands. In wavelength terms, this corresponds to 250–1600 nm, a convenient range for available instrumentation.

Even though the energy differences between several terms may lie in the accessible spectral range, experimental spectra do not consist of several sharply defined lines as atomic spectra do. Instead, each "line" is a broad envelope covering different vibrational states of the molecule. Sometimes individual vibrational peaks within such an envelope can be partially resolved (particularly at low temperatures), but in most cases only the smooth envelope is seen. The shape of an isolated peak envelope usually closely approximates a Gaussian (e^{-x^2}) function. Often, however, several electronic transitions overlap, and *deconvoluting* the spectrum, or establishing the wavelengths and intensities of the individual Gaussian components, can be quite difficult.

Let us begin by considering a couple of spectra of transition-metal complexes in the context of the term discussion given in the last section and the theoretical splittings summarized in Table 11.6. Figure 11.25 shows the experimental electronic spectrum (UV-visible) of a fairly easy case, Ti(H$_2$O)$_6^{3+}$, and that of a slightly more sophisticated case, Cr(H$_2$O)$_6^{3+}$.

For the hexaaquatitanium(III) case, Fig. 11.25a, we begin by guessing octahedral coordination geometry from the stoichiometry of the complex. Since Ti^{3+} is a d^1 ion, Table 11.6 tells us that the appropriate term energies are those of Fig. 11.21a, consisting of the two terms $^2T_{2g}$ and ^2E_g. This suggests that there is only one transition, the notation for which is $^2E_g \leftarrow {}^2T_{2g}$. (The excited state is written first.) In theoretical terms, the energy separation between these two states is $10Dq$, so if we assume that the peak represents this transition, $10Dq = 21,000$ cm^{-1}, or $Dq = 2100$ cm^{-1} for this complex. There are, however, two additional features of interest in this spectrum. First, at high wavenumber (beyond 30,000 cm^{-1} in the ultraviolet) or high energy there is the edge of a strong peak. We should assign this absorption to a charge-transfer transition, probably one in which an electron is being transferred from Ti^{3+} to O since Ti^{3+} is a reducing agent anyway. Second, the single peak isn't really a single peak—there is a maximum at about 21,000 cm^{-1}, but also a prominent shoulder that presumably corresponds to an overlapping peak. Using a little artistic judgment to deconvolute these, the shoulder probably represents a peak at about 19,500 cm^{-1}. This probably represents a splitting of either the ground term or the excited term because of a deviation from perfect octahedral symmetry. We will say more shortly about splitting through symmetry

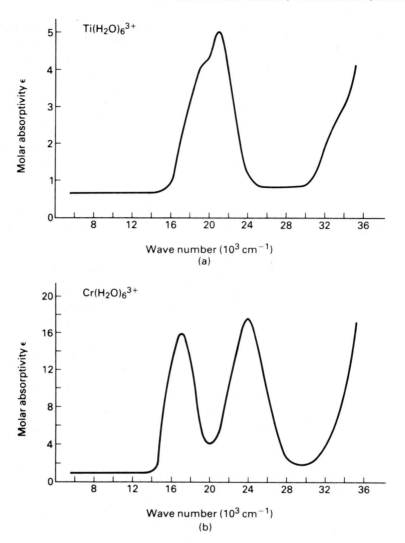

Figure 11.25   Electronic spectra of $Ti(H_2O)_6^{3+}$ and $Cr(H_2O)_6^{3+}$.

reduction; for now, note that the appropriate approximation of a single "ideal oc-tahedral" peak is the average of the two, at about 20,300 cm^{-1}. So Dq is given the value 2030 cm^{-1} when the splitting is allowed for.

For the hexaaquachromium(III) case, Fig. 11.25b, we again guess octahedral geometry for the d^3 complex. From Table 11.6 and Fig. 11.22b, we see that there are three symmetry terms arising from the ground term (^4F) and another symmetry term from the atomic excited term ^4P. Three transitions should thus be possible: $^4T_{2g} \leftarrow {}^4A_{2g}$, $^4T_{1g} \leftarrow {}^4A_{2g}(P) \leftarrow {}^4A_{1g}$. At least two of these are easy to find in the spectrum, one at 17,000 cm^{-1} and another at 24,000 cm^{-1}. A rise that suggests a peak beyond 35,000 cm^{-1} could be either the third peak or a charge-transfer peak. In fact, a slight shoulder at about 37,000 cm^{-1} probably represents the third $d-d$ peak. These peaks are often numbered from low energy to high energy as ν_1, ν_2, and ν_3. If our assignment of ν_1 is correct, Fig. 11.22b tells us that ν_1 should represent $10Dq$, so $Dq = 1700$ cm^{-1}. On the other hand, ν_2 (24,000 cm^{-1}) should represent $18Dq$, which means that $Dq = 1333$ cm^{-1}. This is pretty unsatisfactory agreement of Dq values; what has gone wrong? Figure 11.26 provides the answer. In the previous section we mentioned that two terms close in energy with the same symmetry will interact to yield two new terms with energies lower and higher than the two original terms. In Fig. 11.26, the magnitude of the interaction is designated x. For this case, if Dq is truly 1700 cm^{-1} (from ν_1), $\nu_2 = 24,000 = 18Dq - x = 30,600 - x$. Solving for x, $x = 6600$ cm^{-1}. We can now go on to estimate the Racah parameter B in the complex from ν_3. The ν_3 transition should have energy $12Dq$ (the stabilization of the $^4A_{2g}$ ground term), plus $15B$, plus x [the destabilization of the $^4T_{1g}(P)$ excited term]: $\nu_3 = 37,000 = 12Dq + 15B + x = 12(1700) + 15B + 6600$. Solving for B, $B = 667$ cm^{-1}.

This last result is particularly interesting. We interpreted the experimental spectrum of the $Cr(H_2O)_6^{3+}$ complex by assigning transitions to the observed peaks and using the theoretical quantities Dq and B as parameters to be established by

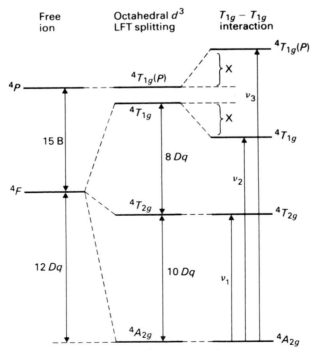

Figure 11.26 Electronic energy levels and transitions for an octahedral d^3 complex.

fitting them to the spectrum. The significance of the Dq quantity can only be established by comparing various ligands and various metals, which we shall do in the next section. The B parameter, however, we can immediately compare with the value for the free ion (Table 11.4), which is 1030 cm^{-1}. In the complex, the value of B has been reduced to only about 65% of the free-ion value. Remembering that B represents the magnitude of d-electron/d-electron repulsion, this result implies that the repulsion was significantly reduced when the complex formed. This has been interpreted as an expansion of the d-electron cloud out onto the ligand atoms (oxygens in this case). If the cloud of electrons is larger, more diffuse, the repulsion within it is less. The d electrons are thus delocalized into at least partially covalent bonds with the ligand atoms—a strong argument for the MO approach. The cloud-expanding effect is called the *nephelauxetic effect* (Greek: "cloud-expanding"). We shall compare the nephelauxetic ability of different ligands in our later discussion.

So far, our interpretation of spectra has used only the basic symmetry terms from the discussion of the last section. It is important to understand the origin of the symmetry terms, but frequently Tanabe–Sugano diagrams represent a convenient shortcut to spectral assignment (values of Dq and B). How can they be applied to the two spectra of Fig. 11.25? The Ti(H$_2$O)$_6^{3+}$ spectrum of Fig. 11.25a is a d^1 spectrum, for which no Tanabe–Sugano diagram exists; the d^1 and d^9 cases are so simple that Fig. 11.21 is adequate by itself. For the Cr(H$_2$O)$_6^{3+}$ spectrum of Fig. 11.25b, we use the appropriate d^3 Tanabe–Sugano diagram from Appendix C. Because that diagram has a quartet ground term ($^4A_{2g}$) at all Dq values we can focus our attention on the quartet excited terms, of which there are three. The interaction we represented by x in Fig. 11.26 causes the two highest terms to curve away from each other as Dq increases. As a result, the best-fit value of Dq/B will be fairly sensitive to the *ratio* of the wavenumbers of the transitions to those two excited terms. To get this we need the value of 37,000 cm^{-1} for ν_3 that we mentioned earlier. The ratio $\nu_3/\nu_2 = 1.54$, which can only be matched in vertical distances on the Tanabe–Sugano diagram by taking $Dq/B = 2.45$. At that Dq/B value ν_1 would intercept the $^4T_{2g}$ line at an E/B value of 24.5:

$$\frac{E}{B} = \frac{\nu_1}{B} = \frac{17,000}{B} = 24.5$$

$$B = 695 \text{ cm}^{-1}$$

Inserting this value in $Dq/B = 2.45$, we find that $Dq = 1700$ cm^{-1}. This is the same value of Dq that we obtained directly from the symmetry state diagrams in Fig. 11.26, but a slightly different value of B. In either interpretation, however, we see a great reduction of B from the free-ion value of 1030 cm^{-1}.

As we have seen, some structural inferences are possible from spectroscopic information on transition-metal complexes. So far, however, we have only considered the wavelengths or frequencies of the electronic absorption peaks. Another feature of UV-visible spectra that yields valuable information about complexes in solution is the intrinsic intensity of each peak as measured by its *molar absorptivity* ϵ. The molar absorptivity should be familiar from Beer's law for optical absorbance: $A = \epsilon \cdot l \cdot c$, where A is absorbance, l is the optical path length through the dissolved sample, and c is its molar concentration. ϵ can vary over a wide range for peaks arising from electronic transitions of different types: Very intense absorptions, such as charge-transfer peaks, can have molar absorptivities greater than 10,000, while peaks that violate several selection rules may have molar absorptivities less than 0.01 and be just barely visible. Where do the selection rules come from, and what does their influence on peak intensities tell us about the structure of transition-metal complexes?

In theoretical terms, the absorption of electric-dipole radiation is governed by the quantum-mechanical *transition-moment integral Q:*

$$Q = \int \psi_1 \cdot r \cdot \psi_2$$

where ψ_1 and ψ_2 are the ground-state and excited-state wave functions, and r is the electric-dipole vector for the radiation being absorbed. A large Q indicates that there is a strong interaction between the electron that is redistributing itself and the radiation that is providing the energy required for the redistribution. The square of Q is proportional to the absolute intensity of an absorption, which is really given by the area under a peak rather than simply by its height ϵ. However, most electronic absorption peaks for transition metals have roughly the same width, so it is usually not a bad approximation to say that there is a direct relationship between Q^2 and ϵ. We need, then, to see what properties of ψ_1, ψ_2, and r in the transition-moment integral lead to selection rules.

One of the most important of these is the *spin selection rule* mentioned earlier: A strong electronic absorption arises from a transition in which the spin of the excited state is the same as the spin of the ground state. This comes very simply from the transition-moment integral, in which the r operator is only an electric dipole that cannot alter the spin orientation of the electron undergoing the spatial redistribution. Thus, although excitations from, say, a quartet term to a doublet term can occur, they are very weak, with a molar absorptivity perhaps only 0.1 to 1. The only reason they are seen at all is that the orbital angular momentum of a given state can couple with the spin angular momentum of the electron to some extent, thus partially destroying the "pure quartet" or "pure doublet" spin multiplicity. We will consider spin-orbit coupling further in the later section on the magnetic properties of transition-metal ions.

Another selection rule arises from the directional, vectorial quality of the r operator for the electric-dipole radiation. In infrared spectra, this property requires that the dipole moment of the molecule must change in the molecular vibration being excited. In electronic spectra, the equivalent is that the algebraic signs of the component orbitals must change in a dipole-like fashion on going from the ground state to the excited state. For example, a +/− arrangement of orbital lobes must change to −/+. Now d orbitals are symmetric with respect to the nucleus; a + lobe on one side of the nucleus is balanced by a + lobe on the opposite side (see Fig. 10.1). Therefore, a d–d transition does not have the right sign arrangement and is in general forbidden by

Table 11.7 Approximate Intensities for Bands Governed
by Major Selection Rules

Type of transition	Molar absorptivity, ϵ
Spin-forbidden, Laporte-forbidden	0.1
Spin-allowed, Laporte-forbidden	10
Spin-allowed, tetrahedral species	100
Spin-allowed, orbitally allowed (charge transfer)	10,000

the *Laporte selection rule:* A strong electronic absorption arises from a transition in which orbital signs change in a dipole-like fashion in passing from the ground state to the excited state. Octahedral complexes have MOs in which Δ_o corresponds to a change from a pure d orbital to a d-based MO, so $d-d$ transitions for octahedral complexes are Laporte-forbidden. On the other hand, tetrahedral complexes have MOs (see Fig. 11.14) in which Δ_t corresponds to a change from a pure d orbital to a mixed d- and p-based MO. Since p orbitals have the right "dipole" character, $d-d$ transitions are Laporte-allowed for tetrahedral complexes—(precisely because they are not strictly $d-d$). Again, there is a mechanism for getting around the Laporte selection rule for octahedral complexes; when the molecule vibrates it is no longer octahedral, and the basis of its MOs changes. Nonetheless it is true in general that tetrahedral complexes have much stronger colors than octahedral complexes.

The opposite of a selection rule, in a sense, is shown by charge-transfer transitions, which are very intense. In such a transition, the "dipolar" character of the change in electron distribution between the ground state and the excited state is very great, and the interaction with the radiation electric dipole is correspondingly intense. Table 11.7 gives some useful rough approximations for ϵ values under different selection rules. The intensity of a peak is obviously helpful in identifying its electronic origins. For example, the visible–UV spectrum of Mn^{2+} in water yields a series of seven peaks between 18,000 and 32,000 cm^{-1}, all with ϵ about 0.02. This intensity pattern strongly suggests spin-forbidden, Laporte-forbidden peaks; the transitions would be Laporte-forbidden if the manganese hydrate were octahedral, and spin-forbidden if the d^5 configuration of Mn^{2+} were in its high-spin ^6S ground term so that any excitation would necessarily take it to a quartet or doublet state. In fact, the peaks fit quite well to the Tanabe–Sugano diagram for octahedral d^5 species at $Dq/B = 1.1$, well down in the high-spin region of the diagram.

11.7 EMPIRICAL SPECTRAL CORRELATIONS AND DISTORTED SYMMETRY

If we examine and interpret the spectra of a variety of metal ions complexed by a variety of ligands, we see that Dq varies over a fairly wide range. For example, we just saw that for Cr^{3+} with six H_2O ligands, $Dq = 1700$ cm^{-1}. However, if the waters are replaced by six Cl^- ligands, Dq shrinks to 1300 cm^{-1}, and if the ligands are six CN^- ions instead, Dq is 2600 cm^{-1}. Similarly, for the water ligand, Cr^{3+} has $Dq = 1700$ cm^{-1}, but Ni^{2+} has $Dq = 900$ cm^{-1} and Rh^{3+} has $Dq = 2700$ cm^{-1}. Furthermore, these effects are fairly consistent—Cl^- produces a smaller Dq than H_2O does for almost every metal ion, and Rh^{3+} has a larger Dq than Cr^{3+} for almost every ligand. By studying the spectra of a very large number of complexes, a consistent ordering, the *spectrochemical series,* has been produced. The spectrochemical series for ligands arranges them in order of their ability to cause increasing Dq values in complexes with any metal ion:

$$I^- < Br^- < OCrO_3^{2-} < Cl^- \simeq SCN^- < N_3^- < dtp^- < F^- \simeq SSO_3^{2-}$$

$$\simeq urea < OCO_2^{2-} < OCO_2R^- < ONO^- \simeq OH^- < OSO_3^{2-} < ONO_2^-$$

$$< O_2CCO_2^{2-} < H_2O < NCS^- < gly^- \simeq EDTA^{4-} < py \simeq NH_3 < en$$

$$< SO_3^{2-} < dip < phen < NO_2^- < cp < CN^-$$

Some translation is obviously necessary here. The abbreviations represent the following molecules or ions. The electron-donor atoms are given in parentheses:

dtp$^-$ = diethyldithiophosphate $(EtO)_2PS_2^-$ (2 S)

urea = H_2NCONH_2 (1 O)

gly$^-$ = glycinate $H_2NCH_2COO^-$ (1 O, 1 N)

EDTA^{4-} = ethylenediaminetetraacetate $(O_2CCH_2)_2NCH_2CH_2N(CH_2COO_2)_2^{4-}$
(4 O, 2 N)

py = pyridine C_5H_5N (1 N)

en = ethylenediamine $H_2NCH_2CH_2NH_2$ (2 N)

dip = 2,2'-dipyridyl 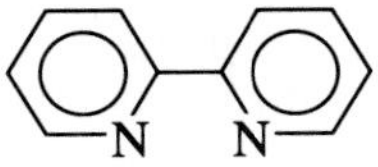 (2 N)

phen = o-phenanthroline 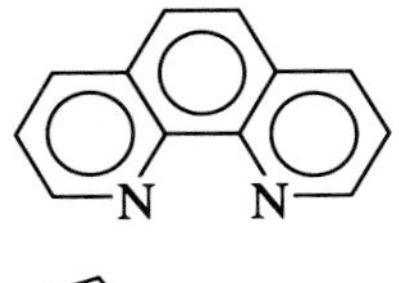 (2 N)

cp = cyclopentadienyl  (usually 5 C)

The spectrochemical series is not an absolute ordering under all circumstances, but it is a good approximation and flexible in its application. For example, it is valid for tetrahedral and noncubic geometries as well as for octahedral geometry—and even for complexes that strictly speaking, have destroyed symmetry by substitution. For pseudooctahedral complexes MX_3Y_3 or, in general, MX_nX_{6-n}, there is a *rule of average environment* for the frequency or wavenumber of absorption peaks:

$$\nu(MX_n Y_{6-n}) = \frac{n}{6}[\nu(MX_6)] + \frac{6-n}{6}[\nu(MY_6)]$$

It is possible to simplify the spectrochemical series by noting a rough, imperfect ordering in terms of the identity of the donor atoms. That is, ligands tend to fall in the order of their donor atoms: $I < Br < Cl < S < F < O < N < C$. This is an interesting series, since it is not in order by electronegativity (the three least electronegative elements fall at both ends and in the center) and also not by "hardness" or "softness" in Lewis-base terms. It is, in fact, in order of pi-acceptor capability. An iodine atom is an excellent pi *donor;* a carbon atom in CO, CN⁻, or an olefin is an excellent pi *acceptor.* The spectrochemical series thus superimposes fairly nicely on the pi-donor and pi-acceptor MOs we generated in Figs. 11.18 and 11.20, since pi-donor ligands and pi-acceptor ligands shift the sigma-only Δ_o in opposite directions. The donor-atom dependence within the spectrochemical series is also seen in two ligands, nitrite and thiocyanate, whose positions in the series depend on which of two atoms they are donating through. These ligands are said to be *ambidentate:* SCN⁻ can donate through either the S or the N, and NO_2 through either the O or the N. Presumably the sulfur end of the thiocyanate ion is a better pi donor than the nitrogen end, and similarly for the nitrite ion.

Jorgensen has also produced a *spectrochemical series of central ions.* Here, transition metals are ordered by their $10Dq$ values with a given ligand:

$$Mn^{2+} < Ni^{2+} < Co^{2+} < Fe^{2+} < V^{2+} < Fe^{3+} < Cr^{3+} \simeq V^{3+} < Co^{3+} < Mn^{4+}$$
$$< Mo^{3+} < Rh^{3+} \simeq Ru^{3+} < Pd^{4+} < Ir^{3+} < Re^{4+} < Pt^{4+}$$

These are in general order of increasing charge, but in strict order of principal quantum number. All of the $5d$ species are higher than all the $4d$ species, which are in turn higher than the $3d$ species. These two series, for the ligands and the central ions, can be combined into a useful empirical rule (also due to Jorgensen):

$$10Dq = \Delta_o = f(\text{ligand}) \times g(\text{metal})$$

where f and g are given in Table 11.8 on a numerical basis that yields Δ_o in thousands of reciprocal centimeters. The $f \cdot g$ rule is only approximate, but it usually gives Δ_o values within perhaps 5%.

The other series of constants in Table 11.8, h and k, allow a rough estimation of B for an octahedral complex from the value of B for the free metal ion (Table 11.4). They place metal ions and ligands, separately, in *nephelauxetic series* in which electron delocalization increases and B becomes smaller:

(large B) $Mn^{2+} \simeq V^{2+} > Ni^{2+} \simeq Co^{2+} > Mo^{3+} > Re^{4+} \simeq Cr^{3+} > Fe^{3+}$

$\simeq Os^{4+} > Ir^{3+} \simeq Rh^{3+} > Co^{3+} > Pt^{4+} \simeq Mn^{4+} > Pt^{6+}$ (small B)

Table 11.8   Empirical Constants for Estimating Δ_0 (10Dq) and B for Octahedral Complexes (10^3 cm^{-1})

10Dq = $\Delta = f \cdot g$			$B = B_{\text{free ion}}(1 - h \cdot k)$		
Metal ion	g	k	Ligand group	f	h
Co^{2+}	9.3	0.24	6Br	0.76	2.3
Co^{3+}	19.0	0.35	6CH$_3$·COO	0.96	—
Cr^{2+}	14.1	—	6Cl	0.80	2.0
Cr^{3+}	17.0	0.21	6CN	1.7	2.0
Cu^{2+}	12.0	—	6NCS	1.03	—
Fe^{2+}	10.0	—	edta	1.20	—
Fe^{3+}	14.0	0.24	3 dtp	0.86	2.8
Ir^{3+}	32	0.3	3 dip	1.43	—
Mn^{2+}	8.5	0.07	3 en	1.28	1.5
Mn^{3+}	21	—	6F	0.90	0.8
Mn^{4+}	23	0.5	3 glycine	1.21	—
Mo^{3+}	24	0.15	6H$_2$O	1.00	1.0
Ni^{2+}	8.9	0.12	6NH$_3$	1.25	1.4
Pt^{4+}	36	0.5	6NO$_2$	1.5	—
Re^{4+}	35	0.2	6OH	0.94	—
Rh^{3+}	27.0	0.3	3 ox	0.98	1.5
Ti^{3+}	20.3	—	3 phen	1.43	—
V^{2+}	12.3	0.08	6 py	1.25	—
V^{3+}	18.6	—	6 urea	0.91	1.2

Source: From B. N. Figgis, *Introduction to Ligand Fields,* Interscience Publishers, New York, 1966. By permission from John Wiley & Sons, Inc.

Estimated values in kJ/mol = *product* of tabulated values × 11.96.

and

$$\text{(large } B) \quad F^- > H_2O > \text{urea} > NH_3 > \text{en} \simeq C_2O_4^{2-} > {-}NCS > Cl$$
$$\simeq CN > Br > I > \text{dtp (small } B)$$

In the ligand series, the ligands lie to a reasonable approximation in order of *decreasing* electronegativity of the donor atom: F, O, N, Cl, C, Br, S. That is, the less electronegative a donor atom, the more it reduces the repulsions between d electrons already on the metal. To reduce d-electron repulsion (the B parameter), the d electrons must expand into a larger volume in the molecule than in the atom so they can stay farther apart. In other words, B is reduced when electrons delocalize from the original compact atomic orbitals into diffuse molecular orbitals. Now consider Fig. 11.12b. In these octahedral MOs, the d electrons will be distributed among the t_{2g} and e_g^* MOs. The less electronegative L is, the higher the energy of its sigma-donor orbitals and the larger their coefficients will be in the e_g^* MOs. A larger coefficient for the ligand-donor orbitals means that the electrons are delocalized more onto the ligand atoms, which is exactly the effect we see. The effect on coefficients and delocalization is the same for pi-donor ligands (Fig. 11.20), but it is reversed for pi-acceptor ligands (the MOs that acquire more ligand character are in general unoccupied). However, the effect is small anyway in the latter case because of the high energy of the pi-acceptor orbitals; it is thus likely to be swamped by the larger sigma effect.

To sum up, in complexes we normally expect to see the free-ion B value reduced to some new smaller value B'. By analogy with the $f \cdot g$ product for Δ_o we can calculate B' as follows.

$$B' = B[1 - \{h(\text{ligand}) \times k(\text{metal})\}]$$

where the constants are those given in Table 11.8. (Sometimes the ratio B'/B is designated β in discussing the reduction of repulsion.) The $f \cdot g$ product and the $h \cdot k$ product give us an empirical, but effective, way to approximate the Dq and B (B') parameters that must be combined with the term energies from the last section to predict transition-metal complex-ion spectra.

To see how these rule-of-thumb calculations are done, we can predict the Dq and B values for the $Cr(H_2O)_6^{3+}$ complex we have just analyzed. For Dq we have

$$10Dq = f \cdot g = 1.00(H_2O) \times 17.0(Cr^{3+})$$
$$= 17{,}000 \text{ cm}^{-1}$$

For B we have

$$B' = B[1 - h \cdot k] = 1030(1 - 1.0(H_2O) \times 0.21(Cr^{3+}))$$
$$= 814 \text{ cm}^{-1}$$

The Dq match is perfect, but the B match is rather crude, which is not too unusual for this approximation.

One final influence on the electronic structure of transition-metal complexes and their spectra should be described. We have been considering highly symmetrical structures and thus have seen a good deal of MO degeneracy, such as the d_{xy}, d_{yz}, and d_{xz} in cubic fields. If some of this symmetry is lost by distortion, the orbitals' energies will diverge. Sometimes symmetry is lost because of the crystal environments of the complexes, but in some d^n electron configurations, the complex distorts spontaneously. Consider a d^7 low-spin octahedral complex such as NiF_6^{3-}, whose spectrum is shown in Fig. 11.27. The electron configuration of such an ion in strictly octahedral geometry is shown in Fig. 11.28a, where the t_{2g} orbitals are uniformly filled, but only one of the e_g orbitals is populated by an electron. If that electron is in the d_{z^2} orbital, it is oriented strongly toward the two ligands on the z axis and much less so toward the four ligands on the x and y axes. Such an arrangement will push the two z ligands away from the transition-metal atom, allowing the four x and y ligands to sag a little closer to the transition metal.

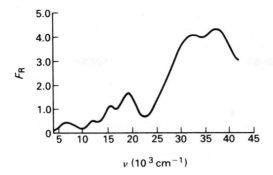

Figure 11.27 Electronic absorption spectrum of NiF_6^{3-} [Reprinted with permission from G. C. Allen and K. D. Warren, *Inorg. Chem.* (1969), *8*, 1895. Copyright © 1969 American Chemical Society.]

Since the z ligands are now farther away, the d_{z^2} orbital is somewhat more stable than it would have been in a strictly octahedral geometry, and since the x and y ligands are now closer, the $d_{x^2-y^2}$ orbital is somewhat less stable than it would have been in octahedral geometry. The energy levels are now those shown in Fig. 11.28b: The complex has stabilized itself electronically by distorting away from perfect octahedral geometry. This will occur whenever orbitals that CFT/LFT describes as strongly repelling or antibonding are unequally filled. The formal statement of this tendency is the *Jahn–Teller theorem:*

For an orbitally degenerate electronic state, the totally symmetric molecule geometry is unstable with respect to distortion that will remove the orbital degeneracy (except for linear molecules).

The extreme version of distortion in this sense is the complete removal of the z-axis ligands to leave a square-planar complex. Figure 11.5 has shown the energy-level

Figure 11.28 Energy levels for orbitals of low-spin d^7 complex in octahedral and tetragonally distorted geometry.

patterns for this sort of distortion, and it is clear from that diagram that extensive splitting of the octahedral e_g energy levels has occurred.

The Jahn–Teller distortion we have just described is a tetragonal distortion, so called because it occurs along a fourfold rotation axis of the molecule. A molecule could also be distorted in the opposite direction—that is, the z ligands could be closer to the transition-metal atom and the x and y ligands farther away. This would still be a tetragonal distortion, but the resulting molecule would be oblate (tomato shaped) rather than prolate (football shaped). This distortion is less common than tetragonal elongation. Also less common is trigonal distortion, in which an octahedral molecule is stretched or compressed along a threefold axis.

In the case of the low-spin complex NiF_6^{3-} (Fig. 11.27), the spectrum cannot be fitted to the d^7 term energies of Fig. 11.22a if complete octahedral symmetry is assumed (nor to the octahedral Tanabe–Sugano diagram, of course). The four peaks between 4,000 and 20,000 cm^{-1} simply do not mesh with the energy levels given in the diagram. If a tetragonal distortion (elongation) is assumed, however, the ^2E_g ground state is split into two terms and so is the lowest excited state of the same spin, the $^2T_{1g}$. With this splitting, the peaks can be fitted fairly well.

There should be a substantial Jahn–Teller distortion whenever the "octahedral" e_g orbitals are unequally filled. This will occur for low-spin d^7, as we have seen, and also for high-spin d^4 (Cr^{2+} and Mn^{3+} in some complexes). However, d^9 complexes (primarily Cu^{2+}) should show Jahn–Teller distortion regardless of the size of Δ_o, and Cu(II) complexes provide most of the examples of this spontaneous distortion. They are almost always prolate, with four ligands on the x and y axes about 0.2 Å to 0.5 Å closer to the Cu nucleus than two equal ligands on the z axis. In fact, the very small number of Cu(II) complexes that *do* show ideal octahedral geometry, such as some chelates and the $Cu(NO_2)_6^{4-}$ ion in $K_2Pb[Cu(NO_2)_6]$, are thought to show a *dynamic Jahn–Teller distortion* in which the molecule oscillates between the three possible tetragonally distorted shapes so rapidly that experimental techniques show only the time-average geometry, which is octahedral. It is also possible, however, that other heavy atoms or bonds within the molecule or lattice force the Cu to remain in an octahedral environment, even though it could otherwise lower its energy by distorting. This would be the opposite of distortion induced by crystal packing influences—it would be symmetry induced by crystal packing (or chelate bonding).

The extreme of a tetragonal elongation distortion is the complete removal of the z ligands leaving a square-planar complex, as we have noted. Cu^{2+} often forms square-planar complexes or crystal environments, as in the mineral CuO and the complex $[Cu(Opy)_4](BF_4)_2$, where Opy = pyridine 1-oxide. We have already pointed out that most square-planar complexes are formed by d^8 species. Ni^{2+} forms square-planar complexes only with the ligands highest in the spectrochemical series, since a large Dq value (a large f value in Table 11.8) is needed to overcome the pairing energy and keep electrons out of the high-energy b_{1g}^* orbital of Fig. 11.16b. Thus $NiCl_4^{2-}$ is a slightly distorted tetrahedron, but $Ni(CN)_4^{2-}$ is square planar. For Pd^{2+} and Pt^{2+}, however, the $4d$ and $5d$ electrons have much larger Dq values (large g values in Table 11.8), and the square-planar geometry is almost the only one ever seen.

11.8 TRANSITION-METAL MAGNETOCHEMISTRY

As we have seen, transition-metal atoms in molecules have varying numbers of d electrons. This gives us a powerful tool for probing the electronic structure of these molecules through electronic absorption spectra, as the previous section has outlined. The magnetic properties of these electrons are also accessible to experiment, however, and provide a great deal of information complementary to that available from spectra. We outline the most rudimentary theory of magnetochemistry here, although most of the subtleties of magnetic interpretation are beyond the scope of this chapter. One special case, the ferromagnetism of bulk transition metals, has been discussed in the context of band theory and density-of-states diagrams in the previous chapter; here we will consider only isolated ions or molecules.

The presence of mobile electrons in atoms guarantees that all matter will interact with an applied magnetic field. Nearly all main-group compounds have filled sets of orbitals with completely paired electron spins, so there is no net spin or orbital angular momentum for such a molecule. Even so, the electron pairs are mobile within the molecule, and an applied magnetic field will cause them to circulate so that they produce a small induced magnetic field opposing the applied field. This is the effect of Lenz's law in basic physical theory. Within a sample of this paired-spin matter, the field is called the *magnetic induction B*. Because of the Lenz's law circulation, B will be less than the applied field H in a vacuum. The difference—the induced field—is proportional to the *intensity of magnetization I* within the substance:

$$B = H + 4\pi I$$

In the case we are describing, I is negative and the substance is said to be *diamagnetic*. All matter that contains paired electrons shows diamagnetism, but studies of diamagnetism offer little insight into the nature of bonding or the structure of matter.

However, if the molecules present in our sample have unpaired electrons, the physical situation is quite different. An unpaired electron has spin angular momentum and may also have orbital angular momentum. Both of these reinforce the applied field by orienting the electron's magnetic moment parallel to the applied field:

$$\mu_1 = \frac{eh}{4\pi mc}\sqrt{l(l + 1)} = \beta\sqrt{l(l + 1)}$$

$$\mu_s = 2\beta\sqrt{s(s + 1)}$$

where μ_1 is the orbital magnetic moment and μ_s is the spin magnetic moment (which is twice as great for a given value of the angular momentum). The quantity

β ($eh/4\pi mc$) is the *Bohr magneton,* a natural unit for magnetic moment. A sample with these properties is *paramagnetic,* with positive *I*. *I* is positive even though the paired electrons in the molecule have an underlying diamagnetism, because the paramagnetic effect is much larger than the diamagnetic effect. For transition metals, studies of the paramagnetism due to the unpaired *d* electrons are of particular interest.

Returning to the intensity of magnetization *I*, it is normally proportional to the strength of the applied field *H*:

$$I = \kappa \cdot H$$

where κ is the volume magnetic susceptibility of the substance. κ is the basis for the two common experimental methods for determining the magnetic moments of transition-metal compounds. If a differentially small volume *dv* of matter is placed in an applied field *H* that has a gradient in some direction *x* of $\partial H/\partial x$, it will experience a force *df* in the *x* direction:

$$df = \kappa H(dv)\left(\frac{\partial H}{\partial x}\right)$$

Both experimental methods measure this force as a change in the weight of the sample by suspending the sample from a balance pan, but with different arrangements.

A *Gouy balance* uses a long sample reaching vertically from the center of the magnet pole faces up to a height above the magnet at which the field is essentially zero. This corresponds to integrating the equation over *x* from $H = H_0$ to $H = 0$:

$$\int df = F = \tfrac{1}{2}H_0^2 \cdot \kappa V$$

Here *V* is the total volume of the sample and H_0 is the applied field strength. A Gouy balance can be calibrated by using a standard substance with a known κ to solve for the instrument constant $\tfrac{1}{2}VH_0^2$. The stable solid $HgCo(SCN)_4$, which has a susceptibility at 20°C of 16.44×10^{-6} cgs emu, is often used. The sample is weighted with the magnet turned off and again with it turned on. The weight difference is the force *F* from which κ can be obtained, given a calibrated balance and sample tube.

The other experimental method uses a *Faraday balance,* in which the magnet pole faces have been shaped to keep $H \, \partial H/\partial x$ constant over the volume occupied by the sample, which can be quite small. In these circumstances the differential equation for magnetic force integrates as

$$F = \kappa V\left(\frac{H \, \partial H}{\partial x}\right)$$

and κ can be obtained from *F* if the Faraday balance has been calibrated using a standard substance, as above. A Faraday balance is usually more useful than a Gouy balance because the small sample makes it easy to thermostat the sample over a wide range of temperature—and the temperature dependence of paramagnetism is extremely important.

Although the force relates directly to the volume susceptibility, it is usually more convenient to measure *mass susceptibility* χ_g, which is related to the volume susceptibility through the density of the sample:

$$\chi_g = \frac{\kappa}{\rho}$$

Finally, to get our measured quantity on a basis that can be related to atomic properties, we convert to a *molar susceptibility* χ_M,

$$\chi_M = \chi_g \cdot MW$$

by using the molecular weight MW. The χ_M value includes the underlying diamagnetism of the paired electrons, however, so it is necessary to subtract the diamagnetic portion of χ_M to get the pure paramagnetic susceptibility χ_A:

$$\chi_A = \chi_M - \chi_{dia}$$

To a good approximation, χ_{dia} is simply the sum of a contribution from each atom in the diamagnetic system, plus a contribution from each of certain kinds of bonds present. These contributions are tabulated as *Pascal's constants* in Table 11.9. By convention, the paramagnetic transition-metal atom itself is not included in the summation.

With a measured value for the paramagnetic susceptibility, we only have to connect it with the magnetic moment μ to have a direct experimental probe for the spin angular momentum and the orbital angular momentum of a transition-metal atom in a molecule. We start by recognizing that the intensity of magnetization I is the rate of change of the energy of an atom with magnetic field:

$$I = -\frac{\partial E}{\partial H}$$

We now consider the statistics of a mole of such atoms in thermal equilibrium with the applied field:

$$\chi_A = \frac{N\beta^2}{3kT}\{L(L + 1) + 4S(S + 1)\}$$

where N is Avogadro's number and L and S are the total orbital angular momentum and spin angular momentum quantum numbers of the ground state of the transition-metal ion. Since we have immediately that

$$\mu = \beta\sqrt{L(L + 1) + 4S(S + 1)}$$

$$\chi_A = \frac{N\beta^2}{3kT} \cdot \mu^2 \qquad \text{or} \qquad \mu = \left(\frac{3kT}{N\beta^2} \cdot \chi_A\right)^{1/2}$$

Table 11.9   Pascal's Constants (10^{-6} cgs emu)

Cations		Anions	
Li^+	−1.0	F^-	−9.1
Na^+	−6.8	Cl^-	−23.4
K^+	−14.9	Br^-	−34.6
Rb^+	−22.5	I^-	−50.6
Cs^+	−35.0	NO_3^-	−18.9
Tl^+	−35.7	ClO_3^-	−30.2
NH_4^+	−13.3	ClO_4^-	−32.0
Hg^{2+}	−40.0	CN^-	−13.0
Mg^{2+}	−5.0	NCS^-	−31.0
Zn^{2+}	−15.0	OH^-	−12.0
Pb^{2+}	−32.0	SO_4^{2-}	−40.1
Ca^{2+}	−10.4	O^{2-}	−12.0

Neutral atoms			
H	−2.93	As(III)	−20.9
C	−6.00	Sb(III)	−74.0
N (ring)	−4.61	F	−6.3
N (open chain)	−5.57	Cl	−20.1
N (imide)	−2.11	Br	−30.6
O (ether or alcohol)	−4.61	I	−44.6
O (aldehyde or ketone)	1.73	S	−15.0
P	−26.3	Se	−23.0
As(V)	−43.0		

Some common molecules			
H_2O	−13	$C_2O_4^{2-}$	−25
NH_3	−18	Acetylacetone	−52
C_2H_4	−15	Pyridine	−49
CH_3COO^-	−30	Bipyridyl	−105
$H_2NCH_2CH_2NH_2$	−46	o-Phenanthroline	−128

Constitutive corrections			
C=C	5.5	N=N	1.8
C=C−C=C	10.6	C=N−R	8.2
C≡C	0.8	C−Cl	3.1
C in benzene ring	0.24	C−Br	4.1

Note: Pascal's constants are defined as the diamagnetic susceptibilities per mole of atoms, ions, or bond groupings.

Combining the constants from this last expression, we have

$$\mu = 2.828(\chi_A T)^{1/2}$$

which is a convenient relationship between a measured magnetic susceptibility and a theoretical set of total angular-momentum quantum numbers.

It is easy to make an experimental comparison, since we already know L and S for the ground-state terms of the various possible d-electron configurations. What we find is somewhat surprising: The orbital angular momentum does not contribute very much to the magnetic moment, since the observed magnetic moments are usually fairly close to those calculated on a spin-only basis (see Table 11.10). This series of observations is described as the *quenching of orbital angular momentum.* Why does this quenching occur?

In the wave picture of atomic electron distribution, the existence of orbital angular momentum for an electron around a certain axis requires that another identical orbital exist at the same energy into which the electron, with its given spin, can be transformed by rotation around that axis. For a free d^1 ion in the gas phase, if the electron is in the d_{xz} orbital with spin α, a 90° rotation about the z axis places it in the previously vacant d_{yz} orbital. The rotation of charge has occurred, and orbital angular momentum exists. The same argument applies to the d_{xy} and the $d_{x^2-y^2}$ orbitals, with a 45° rotation. Now suppose that the two orbitals interchanged by this rotation no longer have the same energy (for instance, the d_{xy} and $d_{x^2-y^2}$ are split by an octahedral crystal field), or that the new orbital already contains an electron of the same spin (so that the exclusion principle prevents the transformation). In such cases, the rotation cannot occur and the portion of the orbital angular momentum represented by that particular rotation will have been quenched. For a complex with a given symmetry, quenching does not have to be complete—but for the octahedral cases described in Table 11.10, quenching is extensive. If we ignore the orbital contribution to the magnetic moment, that expression becomes particularly simple:

$$\mu = \sqrt{4S(S + 1)} \quad \text{(in } \beta \text{ units)}$$

or, since $S = 2n + 1$, where n is the number of unpaired electrons,

$$\mu = \sqrt{n(n + 2)} \quad \text{(still in Bohr magneton units)}$$

Thus, the paramagnetic moment of a complex will in most cases reveal the number of unpaired electrons, telling immediately whether the complex is high spin or low spin.

When the deviations from the spin-only magnetic moments are not too large, it is usually because there is a small contribution arising from *spin-orbit coupling.* If orbital angular momentum exists for an electron, the magnetic field arising from the circulation of the electric charge will interact somewhat with the spin magnetic moment. For a single electron with given n and l quantum numbers, the size of the interaction in energy terms is designated $\zeta_{n,l}$ (zeta), the single-electron spin-orbit coupling constant. When we measure the magnetic moment of a specific complex,

Table 11.10 Magnetic Moments of d-Electron Configurations in High-Spin Octahedral Complexes

Configuration	d^1	d^2	d^3	d^4	d^5	d^6	d^7	d^8	d^9
Free-ion ground term	^2D	^3F	^4F	^5D	^6S	^5D	^4F	^3F	^2D
L	2	3	3	2	0	2	3	3	2
S	$\frac{1}{2}$	1	$\frac{3}{2}$	2	$\frac{5}{2}$	2	$\frac{3}{2}$	1	$\frac{1}{2}$
μ (L and S)	3.00	4.47	5.20	5.48	5.92	5.48	5.20	4.47	3.00
μ (spin only)	1.73	2.83	3.87	4.90	5.92	4.90	3.87	2.83	1.73
μ (experimental)	1.7–1.8	2.8–2.9	3.7–3.9	4.8–5.0	5.8–6.0	5.1–5.7	4.3–5.2	2.9–3.9	1.7–2.2

Table 11.11 Spin-Orbit Coupling Constants ζ_{nd} for Single Electrons in Transition-Metal Ions (cm^{-1})

	Ti	V	Cr	Mn	Fe	Co	Ni	Cu
neutral atom:	70	95	135	190	275	390		
charge 1+:	90	135	185	255	335	455	565	830
charge 2+:	123	170	230	300	400	515	630	890
charge 3+:	155	210	275	355	460	580	705	960
charge 4+:		250	355	415	520	650	790	

	Zr	Nb	Mo	Tc	Ru	Rh	Pd	Ag
charge 1+:	(300)	(420)	(670)	(950)			(1,300)	
charge 2+:	(400)	(610)	800	(1,200)	(1,250)		(1,600)	(1,800)
charge 3+:	(500)	(800)	(850)	(1,300)	(1,400)	(1,700)		
charge 4+:			(900)	(1,500)	(1,500)	(1,850)		
charge 5+:								

	Hf	Ta	W	Re	Os	Ir	Pt	Au
charge 1+:		(1,400)	(1,500)	(2,100)			(3,400)	
charge 2+:			(1,800)	(2,500)	(3,000)			(5,000)
charge 3+:			(2,300)	(3,300)	(4,000)	(5,000)		
charge 4+:			(2,700)	(3,700)	(4,500)	(5,500)		
charge 5+:								

Note: Parenthesized values are estimated.

Value in J/mol = tabulated value $\times$ 11.96.

however, we need to know about spin-orbit coupling for the polyelectronic term that constitutes the ground state of the molecule with its particular ligand-field symmetry. If the ground state is an A or E symmetry term, the situation is fairly simple: We can obtain a term spin-orbit coupling constant λ, which is related to the number of unpaired electrons n:

$$\lambda = \frac{\zeta_{nd}}{n}$$

Since λ is essentially an interaction energy, it becomes negative if n is defined by a number of holes instead of by a number of electrons—that is, if it refers to an ion with more than five d electrons. Regardless of the sign of λ, its ratio to the energy gap up to the next excited state whose orbital angular momentum is being mixed in, $10Dq$, determines the extent to which the spin-only magnetic moment μ_{so} is altered by the orbital contribution:

$$\mu_{eff} = \mu_{so}\left(1 - \alpha\frac{\lambda}{10Dq}\right)$$

where α is 4 for A_2 ground states and 2 for E ground states. With a table of ζ_{nd} values for the free ions (which appears here as Table 11.11), μ_{eff} values can be calculated with fair accuracy, or the size of $10Dq$ can be estimated using the above relationship.

Unfortunately, this simple relation cannot be used with ions having a symmetry ground state of T_1 or T_2, because the spin-orbital coupling splits those states by an amount roughly comparable to kT. The effect is thus much more complicated, and the temperature dependence particularly so. This limits the use of the relation to octahedral complexes of d^3, d^4, d^8, and d^9 ions and to tetrahedral complexes of d^1, d^2, d^6, and d^7 ions (see Table 11.6).

Even with its limited applicability, a simple calculation of μ_{eff} may be worthwhile. What should the magnetic moment of VCl_4 be? This is a tetrahedral molecule containing a d^1 species, V^{4+}. The spin-only magnetic moment should be $\sqrt{1(1+2)}$, or 1.73 Bohr magnetons. A tetrahedral d^1 system has an E ground state, so $\alpha = 2$. From Table 11.11 $\zeta_{3d} = 250$ cm^{-1}, and there is one unpaired electron, so $\lambda = +250$ cm^{-1}. From its visible–near-IR spectrum, we know that $10Dq$ for VCl_4 is 8000 cm^{-1}. This is the last piece of information we need to calculate μ_{eff}:

$$\mu_{eff} = 1.73\left(1 - 2\frac{250}{8000}\right) = 1.62 \text{ Bohr magnetons}$$

This is not a bad match for the experimental result of 1.69 Bohr magnetons at 300 K. Note that spin-orbit coupling reduces the magnetic moment from the spin-only value. This is generally true for systems having fewer than five d electrons. In those with more than five, the spin-only magnetic moments are increased because the sign of the spin-orbital term changes.

The temperature dependence of the magnetic susceptibility is an important experimental result because it can give us much information about the electronic interactions occurring inside the transition-metal atom. When its ground state is an A or E term, we have already seen its temperature dependence:

$$\chi_A = \frac{N\beta^2\mu^2}{3kT} \quad \text{or} \quad \chi_A = \frac{C}{T}$$

This temperature dependence is experimentally observed for many compounds. These are said to follow the *Curie law*. When the ground state is a T term, the temperature dependence is much more complicated. At low temperatures, however, it usually follows the *Curie–Weiss law,*

$$\chi_A = \frac{C}{T + \theta}$$

where θ is an empirical constant that usually corresponds to a *negative* absolute temperature. Because of this, the graphical representations of magnetic temperature dependence are usually drawn as $1/\chi_A$ versus T, as in Fig. 11.29. The difference between Curie and Curie–Weiss temperature dependence is sometimes taken as evidence for the nature of the ground state of the ion—A or E versus T. However, this is a shaky judgment, because many electronic-structure parameters can influence the temperature dependence.

Our discussion of magnetic properties in this section has tacitly assumed that each magnetic transition-metal atom is isolated within a molecule or crystal, so that it is influenced by nearby point charges or dipoles or by covalent bonds, but not by other paramagnetic atoms. That is, no other transition-metal atoms are close enough for their wave functions to be appreciably mixed with the wave

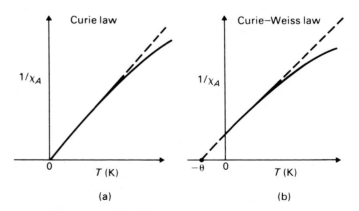

Figure 11.29 Temperature dependence of paramagnetism.

function of the atom being considered. When this is true, the system is said to be "magnetically dilute." Most molecular transition-metal compounds or complexes that contain only one metal atom are magnetically dilute and can be treated by the above methods.

However, in two very important classes of compounds, metal–metal interactions lead to major departures from the simple diamagnetic–paramagnetic behavior discussed so far. In Chapter 10 we have seen that in bulk metals the atoms are, not surprisingly, so close together that there are extensive interactions between their unpaired electrons. Similarly, Chapters 3 and 10 have noted the existence of ferrite magnetic oxides in which the transition metals with unpaired electrons are still close enough to display the offset band structures that allow ferromagnetism.

It is also possible for such spin interactions to compel equal populations of spin-up and spin-down electrons in a crystal even though the electrons are nominally unpaired; such a crystal (or domain) has no net spin and is said to be *antiferromagnetic*. In either case, the behavior of the crystal in a magnetic field is quite different from that of an isolated molecule or magnetically dilute crystal.

The other class of compounds in which transition-metal ions are not magnetically dilute is the large array of metal clusters that are known. Again, Chapters 3 and 10 have considered some of the cluster structures that arise in lattices, and these also have strong metal–metal interactions that alter the lattice's magnetic properties. In future chapters we shall encounter a remarkable variety of molecular species with metal–metal bonds, either between a pair of metal atoms or perhaps in a molecular cluster; often these will have arranged their stoichiometries and structures to allow complete pairing and the resultant diamagnetism. The exceptions are compounds of considerable interest.

PROBLEMS

A. DESCRIPTIVE

A1. Sketch orbital overlaps and construct a MO (LFT) energy-level diagram for the linear $CuCl_2^-$ ion.

A2. In both phosphines, PR_3, and phosphites, $P(OR)_3$, the P atom is a sigma donor and a pi acceptor. What difference should the phosphite O atoms make in the Δ_o for phosphite complexes compared to phosphine complexes?

A3. Should the colors of 5-coordinate (trigonal-bipyramidal or square-pyramidal) transition-metal complexes be more intense or less intense than those of octahedral complexes? Explain your answer.

A4. The H^- ion has no pi-acceptor or pi-donor properties; thus, it is not a particularly good ligand for transition metals. However, mixed complexes can be readily prepared using

H^- and a good pi-acceptor such as *diphos*. By analogy with the appropriate main-group elements, suggest a synthesis for $Fe(diphos)_2H_2$.

Diphos

A5. Mn^{3+} in water is a much stronger oxidizing agent than either of its neighbors Cr^{3+} or Fe^{3+}. Comment on the aspects of electronic structure that combine to make this true.

A6. Should PCl_5 or PCl_3 be a better Lewis-base ligand for a transition metal atom? Why?

A7. Construct an MO energy-level diagram for hypochlorite ion by the methods of Chapter 4. Should OCl^- as a ligand be a pi donor or a pi acceptor? Where, roughly, should it fall in the spectrochemical series?

A8. Could the Tanabe–Sugano diagrams in Appendix C be used to interpret spectra of tetrahedral complexes? If so, briefly describe the procedure necessary; if not, explain the fundamental inconsistency that limits the use of the diagrams to octahedral complexes.

A9. V^{3+} forms octahedral complexes with two ligands L and M: VL_6^{3+} and VM_6^{3+}. The complex with L is yellow, and the complex with M is blue-green. Which ligand yields the higher Dq value in its complex?

A10. Table 11.10 suggests that high-spin octahedral d^6 complexes can show the full orbital contribution to the magnetic moment (i.e., without quenching). What features of the electronic structure make this a particularly favorable electron configuration for an orbital contribution?

B. NUMERICAL

B1. Six-coordinate transition-metal complexes are common, 5-coordinate complexes much less so. Of the metal ions listed in Table 11.2, which would lose CFSE stability on going from octahedral to square pyramidal? Which would gain CFSE stability? Assume the high-spin or low-spin condition does not change on going from 6- to 5-coordinate for each ion, and show calculations for each 5-coordinate species.

B2. Use Jorgensen's f, g, h, and k constants (see Table 11.9) to predict the spin-allowed transitions in the UV–visible absorption spectrum of $Cr(NH_3)_6^{3+}$. What color should solid $\{Cr(NH_3)_6\}Cl_3$ be?

B3. Repeat the calculation of problem B2 for $Cr(NH_3)_5Cl^{2+}$. What color should $\{Cr(NH_3)_5Cl\}Cl_2$ be?

B4. When a crystal of Al_2O_3 is grown in the presence of a low concentration of V^{3+}, the V^{3+} ions occupy octahedral Al^{3+} sites. Such a crystal has UV–visible absorption peaks at 17,400 cm^{-1}, 25,200 cm^{-1}, and 34,500 cm^{-1}. Calculate Dq and B' for V^{3+} in this environment.

B5. The approximately octahedral complex $Ni(dmso)_6^{2+}$ (dmso = dimethylsulfoxide) has UV–visible absorption peaks at 7730 cm^{-1}, 12,970 cm^{-1}, and 24,040 cm^{-1}. Calculate Dq and B' for Ni^{2+} in this environment.

B6. Calculate Dq and B' for Co^{2+} in $CoBr_2$. Co^{2+} has a nearly octahedral environment, and absorbs at 5700 cm^{-1}, 11,800 cm^{-1}, and 16,000 cm^{-1}.

B7. The FeF_6^{3-} complex is nearly colorless, but has four very weak absorption bands at 14,300 cm^{-1}, 19,700 cm^{-1}, 25,350 cm^{-1}, and 28,800 cm^{-1}. Calculate Dq and B'.

B8. $MoCl_6^{3-}$ has weak absorption peaks at 9,600 cm^{-1} and 14,800 cm^{-1}, and much stronger peaks at 19,200 cm^{-1} and 23,900 cm^{-1}. Assign the transitions and calculate Dq and B'.

C. EXTENDED REFERENCE

C1. Iron(II) forms a complex $Fe(phen)_2(NCS)_2$ with o-phenanthroline. Since phenanthroline has two donor atoms, the complex is approximately octahedral. This complex has a magnetic moment of 0.65 Bohr magneton at 80 K, increasing to 5.20 B.M. at 300 K. Most of the increase occurs very sharply near 175 K. The 80 K and 300 K spectra are shown in Fig. 11.30. What's going on here? [See E. König and K. Madeja, *Inorg. Chem.* (1967), *6*, 48.]

C2. Copper(II), in a solution of the amino acid adenine in concentrated hydrochloric acid, crystallizes out $(AdH_2)_2CuCl_6$ [D. B. Brown *et al., Inorg. Chem.* (1977), *16*, 2675]. Estimate $10Dq$ for the $CuCl_6^{4-}$ ion using Jorgensen's f and g constants; then calculate

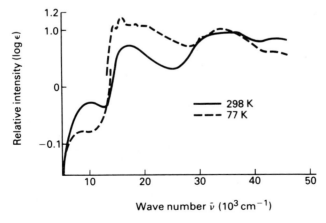

Figure 11.30 *Source:* Reprinted with permission from the source referred to in Problem C1. Copyright © 1967, American Chemical Society.

what μ_{eff} should be for this ion. How does your value compare with the experimental value from Brown's paper? What seems to be happening?

C3. Consider the linear-molecule MX_2 molecular orbitals developed in problem A4. What changes would there be in the orbital energies if the MX_2 molecule were bent? Account for the experimental observation that TiF_2 in the gas phase is bent, but all later difluorides (VF_2 to CuF_2) appear to be linear. [See W. Weltner *et al., Acc. Chem. Res.* (1980), *13,* 242.]

C4. Cooper(II) forms a square-planar complex $(Cu(en)_2^{2+}$, where the four electron-donor atoms are the N atoms from the ethylenediamine (en) molecules. The spectrum of this ion in aqueous solution—where two water molecules probably make it a distorted octahedron—shows a single peak at 17,800 cm^{-1}. Show that this is consistent with a room-temperature magnetic moment of 1.86 Bohr magnetons, as observed by I. Bertini *et al.* [*Inorg. Chem.* (1980), *19,* 1333].

12

Transition-Metal Donor–Acceptor Compounds ━━━━━━━

The previous chapter emphasized that transition metals tend to adopt bonding geometries more or less independent of the charge or formal oxidation state displayed by the metal in its particular molecule or lattice. For example, the ion $Mn(CN)_6^{n-}$ exists in approximately octahedral geometry with $n = 2, 3, 4, 5,$ and 6, corresponding to manganese oxidation states 0, 1+, 2+, 3+, and 4+. Indeed, ordinary manganese cyanide salts may not exist at all because of the overpowering tendency of the manganese ion to add cyanides up to a coordination number of six. This confused the early inorganic chemists who encountered it, because they were used to simple ionic salts in which cations and anions gathered with equal total charges, such as NaCl or $CaSO_4$. The reason for this behavior, of course, is that transition metals are extremely good Lewis acids—electron-pair acceptors— because they can delocalize electron charge back onto the ligands through the pi-donation mechanism shown in Fig. 11.17. The geometry and stoichiometry are thus dictated more by the energy-level relationships deriving from a given symmetry than from net charge. There is a tendency, for instance, to maximize crystal-field stabilization energy. Furthermore, in Chapter 11 we suggested it is specifically the presence of d valence electrons that makes this a dominant effect.

12.1 THE EXPERIMENTAL DEVELOPMENT OF DONOR–ACCEPTOR COMPOUNDS

All this was less than obvious to nineteenth-century inorganic chemists, who regarded $K_3Mn(CN)_6$ as a "double salt," $3KCN \cdot Mn(CN)_3$, between potassium cyanide and manganese(III) cyanide, with no apparent reason for these two "salts" to have an affinity for each other. What's more, simple transition-metal salts such as the chlorides or nitrates, which had presumably satisfied their bonding capability or *valence,* could still add small electrically neutral molecules such as NH_3 or H_2O. The first such observation was probably that of Libavius, who noted in 1597 the formation of the deep blue ion now known to be $Cu(NH_3)_4^{2+}$. In a more modern context (1798), Tassaert reported the formation of what we know as $Co(NH_3)_6^{3+}$ in solution by air oxidation of Co^{2+}, although many others had observed its characteristic red color in solution many years earlier. He did not, however, isolate a salt. Some 50 years later, Frémy showed that Tassaert's compound was indeed an addition compound involving ammonia, but by that time many other salts had also been prepared involving small molecules, particularly ammonia. Since these formed with considerable specificity under given conditions but were otherwise unpredictable from the theory of the day, they were usually simply known by their discoverer's name: *Zeise's salt* ($KPt(C_2H_4)Cl_3 \cdot H_2O$), *Magnus's green salt* ($[Pt(NH_3)_4]^{2+}[PtCl_4]^{2-}$, *Peyrone's salt* (cis-$Pt(NH_3)_2Cl_2$), and many others.

The systematic exploration of coordination chemistry began with the American chemist O. W. Gibbs, who in 1856 published a monograph on the cobalt(III) ammines (ammonia complexes). He prepared 35 salts of four cations, which he distinguished by adding Latin or Greek prefixes describing their colors:

$[Co(NH_3)_6]^{3+}$ luteocobalt (yellow)
$[Co(NH_3)_5H_2O]^{3+}$ roseocobalt (rose-red)
$[Co(NH_3)_5Cl]^{2+}$ purpureocobalt (purple)
$[Co(NH_3)_5NO_2]^{2+}$ xanthocobalt (yellow-orange)

These, of course, were not formulated in this way, but rather as simple additions: $CoCl_3 \cdot 5NH_3 \cdot H_2O$, and so on. Many other similar species were subsequently prepared, such as praseocobalt chloride $[Co(NH_3)_4Cl_2]Cl$ (praseo means "green"). The bonding theory of the time suggested that the geometry around a metal atom was governed by its oxidation state so that chains of ammonia molecules were required in order *not* to increase the number of bonds to the metal.

In 1893, Alfred Werner published the first of 20 papers that, over seven years, established the structural basis for modern coordination chemistry. He suggested the octahedral geometry that we now accept for nearly all six-coordinate complexes, pointing out that either anions or neutral molecules could occupy coordination sites at the corners of the octahedron and that ammonia and water were completely equivalent in their function. Werner carried out a prolonged debate over coordination bonding theory with S. M. Jørgensen (a careful experimentalist with considerable experience in coordination chemistry), in which both men were stimulated to some brilliantly designed experiments. All of these tended ultimately to confirm Werner's theory. The solution conductance, optical activity, and isomer count of series of these compounds all indicate octahedral geometry, as we shall see shortly.

Lewis's theory of bonding in terms of electron pairs, advanced in 1916, was consistent with Werner structures, and in 1937 N. V. Sidgwick suggested that all ligands were electron-pair donors—essentially the present view. He also suggested the Effective Atomic Number, or EAN, rule for complex formation. Under this rule, a metal will acquire ligands until the total number of electrons around it is equal to the number surrounding the next noble gas. This is still a useful rule within certain limits, and we shall develop a quantum-mechanical basis for it later in the chapter. It was also during the 1920s that magnetic studies of the transition metals began to provide direct evidence for electron structure in their ions.

In the 1930s, Linus Pauling proposed the valence-bond theory of bonding in transition-metal complexes. This theory also viewed each ligand as a two-electron donor to a sigma bond with the metal ion. It assumed that the acceptor orbitals on the metal ion were hybrid orbitals equal in number to the coordination number in the complex, formed from the metal's $3d$, $4s$, and $4p$ orbitals (or the equivalent for second- and third-row metals). An octahedral complex required six hybrids, which could be formed in the correct geometry by using $d^2 sp^3$ hybridization. Similarly, square-planar coordination used dsp^2 hybrids, and trigonal bipyramidal coordination used dsp^3 hybrids. These hybrids had to be vacant, since they were to be filled by ligand electrons. The metal's remaining d electrons were to be placed in the remaining unhybridized d orbitals. Unfortunately, many metals had too many electrons for the remaining d orbitals, or the electron spin was so high (from magnetic measurements) that too many d orbitals were needed to allow adequate hybridization within the $3d$ shell. This was handled by postulating hybrids from the $4d$, $4s$, and $4p$ orbitals—that is, by using outer d orbitals. The theory allowed good correlation of magnetic moments, but offered little help with the interpretation of UV spectra, which became widely available in the 1940s and 1950s.

Along with UV spectra, the 1950s brought increased use of infrared spectra to describe the bonding in coordination compounds, as well as the new techniques NMR and ESR for obtaining magnetochemical information. The 1960s saw a number of new spectroscopic techniques: Mössbauer spectra for certain elements (which helped establish coordination symmetry, formal change, and electron delocalization), optical rotary dispersion/circular dichroism, and photoelectron spectroscopy. Many electrochemical techniques, such as cyclic voltammetry, also appeared, along with computerized x-ray diffraction for detailed structural analysis. In fact, more experimental techniques probably exist for the study of transition-metal complexes than for any other class of compounds. This places a heavy burden on the theory that must account for the many results. The existing theory, as outlined in the last chapter, seems to provide a good qualitative correlation of most forms of data, and even reasonable quantitative agreement in many cases.

Historically, coordination compounds have been considered to represent molecules in which a nonmetal atom donates an electron pair to a vacant metal orbital. The metal–nonmetal relationship was assumed long before electrons and orbitals were proposed in explanation. Many metal atoms in low oxidation states also have nonbonding valence electrons, whether they are transition metals with remaining d electrons or main-group metals such as Sn^{2+} with two s electrons. Thus, metal atoms can serve as donors in coordination compounds as well as acceptors. The first coordination compounds in which the donor–acceptor bonding clearly involved a metal–metal bond were prepared about 1940 [Hg–Fe(CO)₄ and compounds of related structures]. In the mid-1960s, extensive exploration of metal–metal bonds involving transition metals began. An enormous number of such complexes have now been prepared, involving both two transition-metal atoms and a transition-metal atom linked to a main-group atom. Many of the most interesting of these are *metal cluster* compounds formed of multiple metal atoms, some of which we have already encountered. We shall defer a detailed consideration of these until Chapter 13. Here we restrict ourselves to fairly simple structures bonded by a series of overlaps of transition-metal acceptor orbitals with electron pairs in donor orbitals, whether from nonmetal atoms or metal atoms.

12.2   COORDINATION NUMBERS AND COORDINATION GEOMETRIES

The most convenient way of classifying transition-metal complexes structurally is simply by their coordination number: the number of electron-donor atoms or donor pairs bonded to a given metal atom. Under varying conditions transition-metal atoms can be isolated with coordination numbers (CN) all the way from 1 to 12. However, CN = 0 and CN = 1 are perhaps suspect as "coordination compounds," and a CN > 8 is rare for *d*-block metals (though not uncommon for the lanthanides and actinides). For CN = 2 and larger, several geometries are usually possible for each coordination number. We shall consider the possibilities in order of increasing coordination number.

Coordination numbers 0 and 1. Coordination number 0 corresponds to an isolated atom. Mass spectrometry shows that such atoms exist at very low pressures in the gas phase for any transition metal. However, when such metal vapors are codeposited with a noble gas such as argon at low temperatures, the resulting solid argon matrix also contains isolated transition-metal atoms. Although the metal atom has a substantial number of nearest-neighbor argon atoms, they are too electronegative to form donor–acceptor bonds with the metal atom. Thus, the solid is held together only by London dispersion forces. Similarly, very simple molecules such as Ni—N≡N, where the coordination number is 1, are stable in solid argon matrices. These simple systems are interesting in that they are amenable to fairly rigorous theoretical treatments. However, the low coordination number makes them extremely reactive toward other Lewis bases, and they cannot be maintained outside the noble-gas crystal matrix. However, such rigorous precautions are not always necessary: In a pattern we shall see for other low coordination numbers, Cu (or Ag) can be stabilized at room temperature in a one-coordinate crystal in which the copper is bonded directly to a carbon atom on the central benzene ring in 1,3,5-triphenylbenzene, and is thus protected from attack by the surrounding phenyl groups. A very small electron-pair donor like CO or NO can add to the Cu or Ag, but a bulkier one such as pyridine does not attack the CN = 1 compound.

Coordination number 2. Complex ions, molecules, or lattices with CN = 2 can be isolated in the more conventional sense for a number of metals. Although two geometries—linear and bent—are possible, nearly all known two-coordinate complexes are linear. The commonest examples of two-coordinate complexes involve d^{10} metal atoms such as Cu(I), Ag(I), and Au(I), most of which are linear. For example, AuI forms crystals containing linear polymers of Au and I atoms. In these polymers, the bond angle is 180° at the gold atom, but is the rather acute angle of 78° at the iodine atom. Isolated two-coordinate ions include $CuCl_2^-$, $Ag(NH_3)_2^+$, and $Au(CN)_2^-$ [but not $Cu(CN)_2^-$—see later discussion]. Neutral ligands can produce linear two-coordinate species if the charge on M^+ is neutralized by a fairly bulky anion, as in the complex $(C_6H_5)_3P—Cu—N(Si(CH_3)_3)_2$. The bis(trimethylsilyl)amide ion, because of its great steric bulk, can effectively shield a central atom or ion from attack by additional Lewis bases. Thus, linear two-coordinate ML_2 neutral species are known where M is Mn, Co, Ni, Zn, Cd, and Hg, and L is the $((CH_3)_3Si)_2N^-$ ion. These compounds extend the possibility of two-coordination to the d^5, d^7, and d^8 configurations. Presumably, others are possible as well. The bent geometry is possible if ligand bulk is not too great; the ligand $P(NMe_2)_3$ forms a two-coordinate Ag complex with a bond angle of 167°, and the $N(SiMePh_2)_2^-$ substituted amide ion forms a two-coordinate Co^{2+} complex with a bond angle of 147°, though two of the phenyl rings are nestling suspiciously close to the Co.

Coordination number 3. This is still a rather rare coordination number, because the metal atom can usually serve as acceptor to more Lewis bases. Such a metal atom is *coordinatively unsaturated*. The copper atom in $Cu(CN)_2^-$ is coordinatively unsaturated, and its crystal structure contains NC—Cu—CN units stacked in such a way that the Cu atom is three-coordinate (planar). The third ligand is the N from another $Cu(CN)_2$ unit. The Cu—N bond is distinctly longer than the Cu—C bond, however (2.05 versus 1.92 Å), so in the $KCu(CN)_2$ crystal, this ion may represent a sort of intermediate stage between CN = 2 and CN = 3. The tricyanomethide ion, $C(CN)_3^-$, forms a silver salt in which each Ag is coordinated by three N atoms (Fig. 12.1a), but the silver atom is about 0.5 Å above the plane of the three nitrogen atoms, so that its geometry is that of a trigonal pyramid. As for two-coordinate systems, three-coordination is promoted by bulky substituted amide ions as anion ligands: Figure 12.1b shows the structure of $Fe(N(SiMe_3)_2)_3$. The iron atom in this compound has trigonal-planar geometry; interestingly, the nitrogen atoms also have planar coordination. The nitrogen geometry is obviously related to that of trisilylamine (Fig. 5.26). However, the diisopropylamide ion, $((CH_3)_2CH)_2N^-$, also has planar geometry in a three-coordinate complex of Cr similar to that of Fe in Fig. 12.1. The planar N geometry thus does not depend on the presence of Si atoms and their presumed $d\pi–p\pi$ overlap. If the N atom has sp^2 hybridization with a sigma-donor pair in one hybrid orbital, then the remaining p orbital is occupied by only one electron and could conceivably serve as a pi acceptor as in Fig. 11.17. Similar complexes exist for d^5, d^7, d^8, and d^9 configurations: $Mn(N(SiMe_3)_2)_2$(thf), where *thf* = tetrahydrofuran; $Co(N(SiMe_3)_2)_2$ (PPh$_3$);

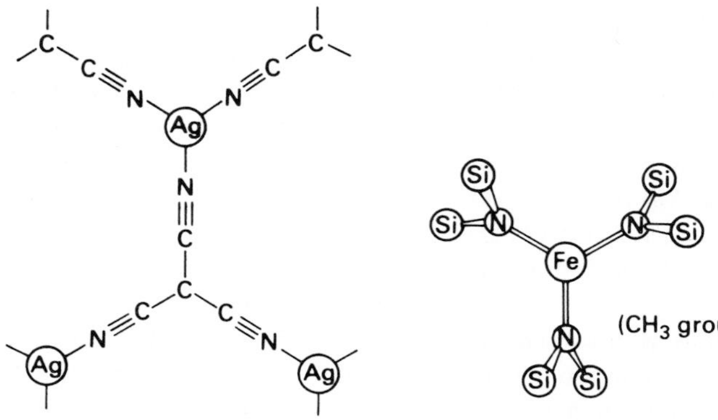

(a) Pyramidal   (b) Planar

Figure 12.1   Three-coordinate complexes.

$Co(N(SiMe_3)_2$ and $[Ni(NPh_2)_3]^-$; and $Ni(N(SiMe_3)_2)(PPh_3)_2$. In the latter cases, the triphenylphosphine presumably stabilizes the complexes in which it appears through its pi-acceptor ability.

Coordination number 4. CN = 4 is the smallest number of ligands commonly found in transition-metal complexes. The tetrahedral and square-planar geometries are both common, tetrahedral for a wide variety of metals and d-electron configurations and square planar primarily for d^8 metals. Let us return to the MO energy-level diagrams of Figs. 11.14b and 11.16b for these two geometries. In each case there are four low-lying sigma bonding orbitals filled by ligand electrons. The discussion associated with Fig. 11.16 gave the MO rationale for the square-planar preference of d^8 systems, but the occurrence of tetrahedral complexes deserves further comment. Four is not really a very high coordination number; in the simplest bonding picture, most metals would still be coordinatively unsaturated in a tetrahedral complex. If steric problems do not arise, the metal will acquire more Lewis-base ligands as long as it has vacant acceptor d orbitals. Since Δ_t is not very large for the energy levels of Fig. 11.14b (that is, the t_2 nonbonding MOs lie at fairly low energy), tetrahedral coordination should be most stable for d^{10} species such as Cu(I), Ag(I), Au(I), or the group IIb metals. Many such complexes are known: $Cu(CN)_4^{3-}$, $(Ph_3P)_2CuONO$ (nitrate), $Ag(SC(NH_2)(CH_3))^{4+}$ (thioacetamide), and others. Gold(I), however, seems to prefer two-coordination; some complexes containing two d^9 Au(II) atoms in square planar complexes will disproportionate to form species with one d^8 square planar Au(III) and one d^{10} linear Au(I). Copper(II), which is d^9, also forms tetrahedral complexes, though Jahn–Teller distortion usually flattens the tetrahedron somewhat toward a square-planar arrangement.

For fewer than nine d electrons, crystal-field stabilization energy usually makes coordination numbers greater than four (particularly octahedral six-coordination) more stable than a tetrahedral complex would be. However, some special circumstances can stabilize four-coordination for fewer d electrons, and the resulting complexes are usually tetrahedral or distorted tetrahedral in geometry. One such circumstance arises when metal ions with low positive charge are coordinated by anions: if the metal is formally 2+, four singly charged anions will give the complex a net 2− charge as MX_4^{2-}. There is a considerable electrostatic penalty if a fifth or sixth anion is brought in to yield a higher coordination number, even if there is no steric problem. Accordingly, halo complexes such as $CoCl_4^{2-}$, $FeCl_4^{2-}$, and MnI_4^{2-} are stable tetrahedral complexes. Even the 3+ ion Fe^{3+} forms $FeCl_4^-$, but also $FeCl_6^{3-}$ under different circumstances. However, if the ligand is high in the spectrochemical series (yielding a large $10Dq$ value), crystal-field stabilization energy can make the octahedral complex preferable in spite of the anionic repulsion, as with ferrocyanide, $Fe(CN)_6^{4-}$, and similar cyanomanganates and cyanochromates. Cobalt and nickel compromise at coordination number five: $Co(CN)_5^{3-}$ and $Ni(CN)_5^{3-}$.

The other fairly obvious circumstance under which the coordination number can be limited to 4 is through the use of bulky ligands. Thus Fe^{2+} forms the tetrahedral complex $Fe(OPPh_3)_4^{2+}$ with triphenylphosphine oxide, Co^{2+} forms tetrahedral $Co(tu)_4^{2+}$ and $Co(HMPA)_4^{2+}$ with thiourea and hexamethylphosphoramide, Co^{3+} forms a square planar complex with the (2,4,6-triisopropyl)benzenethiolate anion, and nickel forms not only the tetrahedral complex $NiCl_2(PPh_3)_2$ with Ni^{2+} but also $NiCl(PPh_3)_3$ with Ni^+ and $Ni(PPh_3)_4$ with neutral Ni—all tetrahedral. Even metal ions as electron-deficient as Cr^{4+} (d^2) can be trapped in four-coordinate complexes (distorted tetrahedral) as, for example, $Cr(OBu^t)_4$ with the very bulky t-butoxide ion. It might be noted that d^2 is an ideal configuration for the MO energy levels shown in Fig. 11.14.

Coordination number 5. In the nineteenth and early twentieth centuries, transition-metal complexes were assumed to have coordination numbers 2 (rarely, as in $Ag(NH_3)_2^+$), 4, or 6 (most common of all). Coordination number 5 was believed highly unusual or nonexistent. However, more recent studies have revealed that there are many stable five-coordinate complexes, particularly for d^7, d^8, and d^9 configurations but for some other cases as well. In a number of complexes, five-coordination is more or less forced on the metal by the geometric requirements of bonding to a *chelate* ligand, in which two or more electron-pair donor atoms are located in such a way that they can donate to the same acceptor metal atom. Figure 12.2 shows the very bulky and relatively rigid ligand tris(o-diphenylarsinophenyl)arsine, or *qas,* in which the four As atoms are donors. Also shown is its five-coordinate complex with Pt^{2+} in which one anion coordinates to the Pt but the *qas* ligand prevents the approach of another donor. In addition to the chelated five-coordinate complexes, however, there are many complexes with five independent donor atoms, such as $CuCl_5^{3-}$, the $Co(CN)_5^{3-}$ and $Ni(CN)_5^{3-}$ mentioned earlier, MnF_5^{3-}, and $CoCl_3(PEt_3)_2$. There are also many in which only two or three coordination sites are occupied by a chelate ligand, but the complex still has a coordination number of only five.

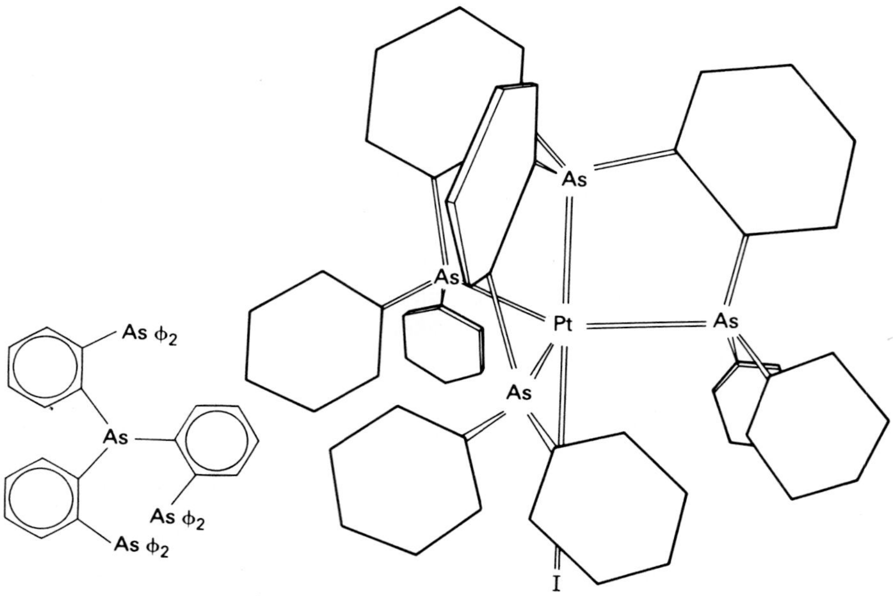

As ϕ_2

As

As ϕ_2

As ϕ_2

qas liqand

Pt(*qas*)I⁺

Figure 12.2. Five-coordination by a bulky ligand.

Two idealized geometries are possible for five-coordinate complexes: square pyramidal and trigonal bipyramidal, as shown in Fig. 12.3. The figure also shows the relatively minor atomic motions necessary to interconvert the two shapes. Since these motions are a natural vibrational mode of a five-coordinate system, intermediate "distorted" geometries are very common, and some complexes can crystallize in either shape. A vivid example of the latter is the complex $Cr(en)_3Ni(CN)_5\cdot3/2H_2O$, in which there are two crystallographically different $Ni(CN)_5^{3-}$ units. One is almost a perfect square pyramid, while the other is a somewhat distorted trigonal bipyramid. The energy balance between the two geometries is obviously a delicate one. In a later section we shall consider the MOs for the two ideal geometries in order to compare their orbital energies. Here, however, we can note that the trigonal bipyramid (whether ideal or distorted) is more common than the square pyramid.

Coordination number 6. The overwhelming majority of all transition-metal complexes are six-coordinate and show octahedral geometry, though (as Fig. 12.4 suggests) a trigonal prism is an alternative geometric possibility shown by a few complexes. Coordination number 6 seems the best compromise between the increased donor–acceptor bond energy shown by high coordination numbers and the decreased ligand–ligand steric repulsion shown by low coordination numbers. In addition, the high symmetry of the octahedron gives low energy to three orbitals (d_{xy}, d_{yz}, and d_{xz} or the MOs arising from them) in either the crystal-field or

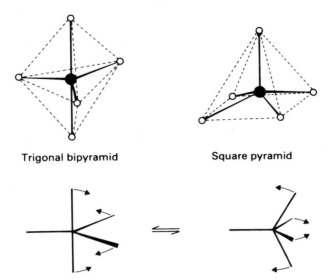

Trigonal bipyramid   Square pyramid

Figure 12.3   Geometries for five-coordination.

molecular-orbital model. This substantially increases the stability of that geometry for six-coordinate complexes with six or fewer d electrons. As this might suggest, four- and five-coordinate complexes are much less common for d^0-d^6 configurations than for d^7-d^{10}. As suggested earlier, the historic foundations of coordination chemistry involved primarily studies of Cr(III) (d^3) and Co(III) (d^6) complexes, both of which are particularly stable as octahedra (see Table 11.2). For all these reasons, inorganic chemists have been more or less conditioned to expect octahedral geometry for most coordination compounds.

Trigonal prismatic geometry, though rare, is an interesting structural variation on the octahedron. An octahedron is actually a regular trigonal antiprism. (A prismatic structure is a polyhedron in which all vertices are defined by two identical

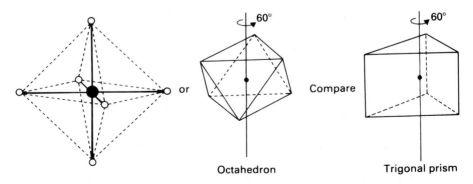

Octahedron   Trigonal prism

Figure 12.4   Geometries for six-coordination.

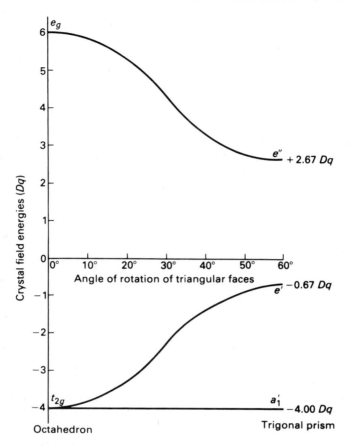

Figure 12.5 Crystal-field energies for *d*-orbitals in an octahedral complex distorting to a trigonal prism.

polygons superimposed in parallel planes. If the polygons are regular and if their vertices are superimposed, the polyhedron is a regular prism, whereas if one regular polygon is rotated in its plane by half the repeat angle with respect to the other regular polygon, the polyhedron is a regular antiprism.) One could transform an octahedral complex into a trigonal prismatic complex by simply rotating three adjacent ligand atoms on a face of the octahedron by 60° (half the 120° repeat angle for an equilateral triangle) relative to the other three ligand atoms. Some complexes also have intermediate geometries—that is, geometries in which the angle of rotation is less than 60°. However, the small class of nearly ideal trigonal prismatic complexes is interesting. One reason so few are known is that the crystal-field stabilization energy is less favorable than that of an octahedron for nearly all d^n configurations. Figure 12.5 shows the energy-level pattern as a function of the rotation angle transforming the antiprism (octahedron) into the prism. From this

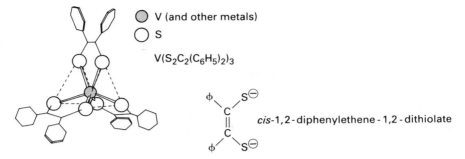

V (and other metals)

S

$V(S_2C_2(C_6H_5)_2)_3$

cis-1,2-diphenylethene-1,2-dithiolate

Figure 12.6 Trigonal-prism coordination in dithiolate complexes.

figure we can calculate that no d^n configuration is more stable in trigonal prismatic geometry than in octahedral geometry. Accordingly, special conditions are necessary if trigonal prismatic geometry is to be observed. However, for d^0 systems with sterically compact ligands it is possible for small distortions from octahedral symmetry to stabilize the p-based t^b_{1u} sigma bonding MOs from the octahedral diagram (see Fig. 11.12a), so that such a system might be more stable as a trigonal prism even though the d-based MOs are destabilized. In fact, after this prediction was made gaseous $W(CH_3)_6$ was found to be a trigonal prism.

The oldest known cases of trigonal-prismatic coordination are the MS_2 and MSe_2 crystals, where M = Nb, Mo, W. These metal ions are d^1 or d^2, which should be the most favorable configurations in terms of CFSE for the prism relative to the octahedron—but the reasons for the crystal packing are still unclear. However, S–S or Se–Se bonding between layers of S or Se atoms might be responsible, especially since a series of dithiolate complexes like the one shown in Fig. 12.6 can be prepared. In these, the sulfur atoms from one dithiolate ligand molecule are considerably closer to each other than the sum of two van der Waals radii for S would suggest (about 3.0 versus 3.6 Å), suggesting that some sort of bond has formed between the sulfurs. (The distance is essentially the same to S atoms on a neighbor ligand molecule, which complicates the argument.)

The other way of forcing trigonal-prismatic coordination is to construct a rigid six-coordinate (*sexadentate*) ligand molecule with that intrinsic geometry. Figure 12.7 shows such a ligand, this one containing six N donor atoms. Several metal complexes have been made from this ligand; they show spectral and magnetic properties related to but significantly different from those of comparable octahedral complexes.

Coordination number 7. Beyond CN = 6, ligand–ligand steric repulsion becomes increasingly prominent in the total electronic energy of the complex, and as a result such complexes are relatively rare. For first-row transition metals with relatively compact $3d$ orbitals, special geometry from polydentate ligands is usually required to force seven-coordination. However, for second- and third-row transition metals with larger valence-orbital radii, simple seven-coordinate complexes

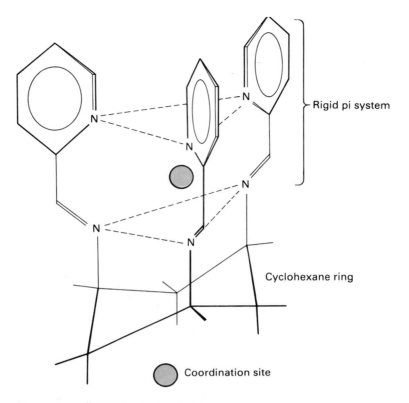

Rigid pi system

Cyclohexane ring

Coordination site

Figure 12.7 Rigid trigonal-prism ligand.

such as MF_7^{3-} (M = Zr, Hf) or MF_7^{2-} (M = Nb, Ta) are found. Apparently, only small ligand atoms such as F or O have low enough repulsion to form stable seven-coordinate complexes. In addition to the problem of increased repulsion, seven-coordinate geometries usually have lower CFSE than the octahedral six-coordinate complexes from which they presumably form. This also reduces the stability of the higher coordination number.

Three idealized geometries are possible for seven-coordination (see Fig. 12.8). Although they appear geometrically distinct, the three can actually be interconverted by only modest distortions, particularly the capped trigonal prism and the capped octahedron. The difference between the latter two is so slight that crystal-packing considerations almost certainly dictate which form is adopted by a given complex. The MF_7^{3-} complexes are usually pentagonal bipyramids (I), whereas the MF_7^{2-} complexes are usually capped trigonal prisms (III). The difference presumably arises from the different numbers of cations that must be accommodated in the lattice. Chelating ligands may have their own spatial requirements because of the bonding within the ligand molecule, which forces a particular geometry on the metal atom. For example, 2,4-pentanedione (acetylacetone or *acac*) is a good

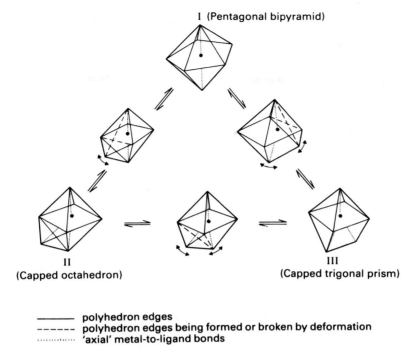

I (Pentagonal bipyramid)

II
(Capped octahedron)

III
(Capped trigonal prism)

—————— polyhedron edges
- - - - - polyhedron edges being formed or broken by deformation
············· 'axial' metal-to-ligand bonds

Figure 12.8   Geometries for seven-coordination.

bidentate chelating ligand through its oxygen atoms as a deprotonated ion $acac^-$. A number of complexes of 3+ metals (both heavier transition metals and rare earths) are known in which the formula is $M(L_2)_3 \cdot S$ and the geometry is seven-coordinate. Here L_2 is *acac* of a related ligand and S is a molecule of the solvent, often H_2O. The capped octahedron (II) is often found in these complexes because the "octahedron" corners provide a good fit for the chelates.

Coordination number 8. Again, the number of examples is quite limited because of the high ligand–ligand repulsion. Apart from a number of more-or-less ionic crystals with CN = 8 in the CaF_2 lattice, most examples involve small ligand atoms such as F (TaF_8^{3-}) or C [$Mo(CN)_8^{4-}$ or $W(CN)_8^{4-}$] or chelate ligands like the *acac* mentioned above [$Zr(acac)_4$]. The same arguments apply to eight-coordinate complexes of the *f*-block elements, the lanthanides and actinides. However, many more *f*-block complexes are known. Eight is apparently a relatively favorable coordination number for these ions, which are both large in size and have a larger number of valence orbitals.

Two geometries are commonly observed for eight-coordinate complexes: the dodecahedron and the square antiprism, as shown in Fig. 12.9. It is interesting that the cube, a rather obvious possibility, is not observed. If a cube is considered as two tetrahedra with a common center (the tetrahedron shown in Fig. 11.4 and

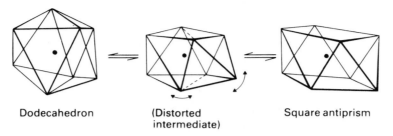

Dodecahedron   (Distorted   Square antiprism
                      intermediate)

Figure 12.9   Geometries for eight-coordination.

another occupying the four vacant corners), the repulsion between the ligands is reduced if one tetrahedron is distorted by stretching it along one coordinate axis of the cube and the other tetrahedron is flattened along the same axis. The resulting figure is the dodecahedron of Fig. 12.9. As in the seven-coordinate geometries, there is a relatively subtle deformation that interconverts the two ideal geometries, essentially by slightly rotating one triangular face of the dodecahedron relative to the rest of the ligand atoms. The two are almost equally stable in the crystal-field sense for most d^n configurations, particularly since the metal ion is usually in a high formal oxidation state and thus has few d electrons.

Higher coordination numbers CN = 9 and CN = 10 are known for a number of complexes, particularly among the f-block elements. In general, the principles already advanced govern the coordination geometry: heavy metals (preferably f-block) and compact ligand atoms. An interesting example is the set of europium(II) halides, in which the Eu atom of $EuCl_2$ is nine-coordinate (the tricapped trigonal prism of the $PbCl_2$ structure, Fig. 3.4), $EuBr_2$ is eight-coordinate (square antiprism), and EuI_2 is seven-coordinate (capped trigonal prism). This is perfectly compatible with the anionic radius trend for the halides, of course; it is a bit disappointing to find that EuF_2 has the eight-coordinate (presumably ionic) CaF_2 structure. However, 10-coordination is perfectly possible. Ethylenediamine (*en*), $H_2NCH_2CH_2NH_2$, is bidentate through its two N atoms, and the complex $[La(en)_4(CF_3SO_3)_2]^+$ is 10-coordinate (eight N and two O atoms) as a somewhat distorted bicapped square antiprism (Fig. 4.28). Other comparable polyamine complexes of La yield nine-coordinate La in either a tricapped trigonal prism or in a capped square antiprism geometry. Here as for nearly all high coordination numbers there is only a modest difference in CFSE (or LFSE) between alternate geometries.

12.3 COORDINATION NUMBERS AND THE 18-ELECTRON RULE

The simplest prediction of coordination numbers says that ligands as electron-pair donors will continue to add to a metal atom, stabilizing the system by forming new bonds, up to CN = 6 in an octahedral complex. Beyond that point, ligand–ligand repulsion costs more in energy than the new bond yields, and higher coordination numbers rarely form. As we have seen, however, coordination numbers other than 6 are frequently observed, even for simple nonchelating ligands. Energy factors other than the simple formation of sigma bonds must be at work. One of the simplest rules for predicting coordination numbers other than 6 is Sidgwick's rule of effective atomic number, mentioned earlier.

The rule of effective atomic number is now usually rephrased as the *18-electron rule:* Many stable transition-metal compounds have stoichiometries or coordination numbers such that the metal atom has 18 electrons in its nine valence orbitals. The 18-electron rule is analogous to the octet rule for first-row *p*-block elements, and has the same quantum-mechanical origin. The maximum possible number of localized bonding MOs that an atom can form in any bond geometry is equal to the number of valence orbitals the atom contributes to the MO basis set. A covalent molecule will in general be most stable when its bonding MOs are all filled and its antibonding MOs are empty. Accordingly, *p*-block elements with four valence AOs ($2s$ and three $2p$) will form their most stable compounds when they are engaged in four bonding (or, at worst, nonbonding) MOs that contain eight electrons. In exactly the same way, *d*-block elements with nine valence orbitals [five $n\,d$, one $(n + 1)s$, and three $(n + 1)p$] will form their most stable compounds when they are engaged in nine bonding (or, at worst, nonbonding) MOs containing 18 electrons.

The 18-electron rule is helpful in predicting the coordination numbers of such complexes as $ZnCl_4^{2-}$ ($10\,d\,e^-$ plus $2\,e^-$ from each of four ligands), $Fe(CO)_5$ ($8\,de^-$ from Fe^0 plus $2\,e^-$ from each of five ligands), and $Co(NH_3)_6^{3+}$ ($6\,d\,e^-$ plus $2\,e^-$ from each of six ligands). However, it is violated more often than it is followed, or at least is violated in a very wide variety of complexes. We shall therefore consider the circumstances under which it is invariably followed and those under which it does not apply.

There are actually three groups of complexes to consider in the context of the bonding theory of Chapter 11: those for which Δ_o ($10Dq$) is not large because the ligands have only a modest sigma energy effect and little or no pi interaction with the metal, those for which Δ_o is large because the ligands have a strong sigma-bonding effect but little pi interaction, and those for which Δ_o is large because the ligands have a strong sigma-bonding effect *and* are strong pi acceptors. MOs for these three cases are shown in Fig. 12.10 for octahedral CN = 6. If the six ligands each donate a pair of electrons, the complex will have a minimum of twelve valence electrons. Since the metal atom/ion can have from 0 to 10 *d*-electrons, the complex could in principle have from 12 to 22 valence electrons. Figure 12.10a shows an energy-level diagram for which this entire range is possible. Because Δ_o is not large, there is only a small energy penalty for filling even the e_g antibonding MOs, and the additional bond energy for six ligands (over five or four) can provide it. In effect, there are six low-energy MOs (sigma bonding), which will be filled by ligand electrons, and five medium-energy MOs (pi nonbonding and sigma antibonding), which may or may not be filled by electrons. So there can be anywhere from 12 to 22 valence electrons in the complex, depending on the *d*

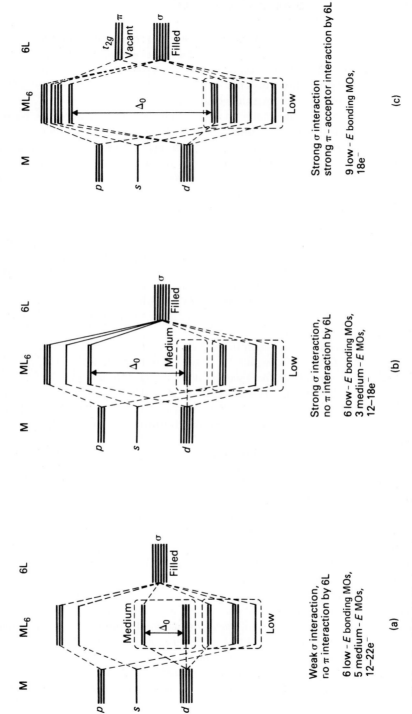

Figure 12.10 Octahedral MOs for varying degrees of sigma and pi interaction.

Table 12.1 Transition-Metal Complexes with Weak Sigma-Bonding Ligands

Complex	d electrons	Total valence electrons
$TiCl_4 \cdot 2(thf)$	0	12
$Ti(H_2O)_6^{3+}$	1	13
$V(urea)_6^{3+}$	2	14
$CrCl_6^{3-}$	3	15
$CrI_2 \cdot 4(dmso)$	4	16
$Mn(H_2O)_6^{2+}$	5	17
CoF_6^{3-}	6	18
$CuCl_5^{3-}$	9	19
$Ni(H_2O)_6^{2+}$	8	20
$Cu(H_2O)_6^{2+}$	9	21
$ZnCl_2 \cdot 2(biuret)$	10	22

Key: thf = tetrahydrofuran, C_4H_4O; urea = $CO(NH_2)_2$; dmso = dimethylsulfoxide, $OS(CH_3)_2$; biuret = $NH_2CONHCONH_2$.

configuration of the metal atom, and the specific number 18 has very little advantage. Table 12.1 indicates that all of these valence-electron configurations are observed, particularly for ligands lying low in the spectrochemical series (which, as Chapter 11 indicated, is linked directly to Δ_o and the pi-acceptor capabilities of the ligands).

Figure 12.10b describes the second case above, in which Δ_o is large because of strong sigma interactions alone. There are still six low-energy MOs (sigma bonding), but now there are only three medium-energy MOs (nonbonding d AOs with pi symmetry) because the sigma antibonding MOs are out of reach. Therefore, such complexes must have 12 electrons and can have up to 6 more—there can be fewer than 18 electrons, but never more. Table 12.2 gives examples of

Table 12.2 Transition-Metal Complexes with Strong Sigma-Bonding Ligands

Complex	d electrons	Total valence electrons
$NbCl_2(NHCH_3)_3 \cdot NH_2CH_3$	0	12
$Ti(en)_3^{3+}$	1	13
$Re(NCS)_6^-$	2	14
$Mo(NCS)_6^{3-}$	3	15
$Os(SO_3)_6^{8-}$	4	16
$Ir(NH_3)_4Cl_2^{2+}$	5	17
ReH_9^{2-}	0	18

Key: en = ethylenediamine, $H_2NCH_2CH_2NH_2$.

Table 12.3   Transition-Metal Complexes with Strong Pi-Acceptor Ligands

Complex	d electrons	Total valence electrons
$Ti(cp)_2(CO)_2$	4	18
$V(CO)_5NO$	5	18
$Cr(C_6H_6)_2$	6	18
$MnH(CO)_5$	7	18
$Fe(NO)_2(CO)_2$	8	18
$Co(NO)(CO)_3$	9	18
$Ni(CO)_4$	10	18

complexes lying in this category, where Δ_o is large either because the metal ion has a high formal oxidation state or uses $4d$ or $5d$ electrons. In fact, if we consider the d^{10} ions Zn^{2+}, Cd^{2+}, and Hg^{2+}, which could only have four electron-pair-donor ligands if limited to 18 electrons, we find that soft Lewis bases (lying high in the spectrochemical series) form five- and six-coordinate complexes much more frequently with Zn^{2+} than with Cd^{2+} or Hg^{2+}. For example, zinc forms $Zn(py)_2Cl_2$, $Zn(py)_3Cl_2$, and $Zn(py)_4Cl_2$ with pyridine, but mercury forms only $Hg(py)_2Cl_2$.

Finally, Fig. 12.10c describes the situation in which pi-acceptor ligands that are also strongly sigma bonding make Δ_o large both by raising the energy of the e_g^* sigma antibonding orbitals and by lowering the energy of the t_{2g} pi-bonding orbitals. In this case, there are nine bonding MOs of low energy (six sigma, three pi), but no medium-energy MOs. In such a molecule, an 18-electron configuration is overwhelmingly more stable than any other number of valence electrons. It is to this sort of transition-metal complex or compound that the 18-electron rule obviously applies; Table 12.3 gives a few examples. As Chapter 11 emphasized, pi-acceptor ligands lie at the very top of the spectochemical series precisely because Δ_o becomes so large in their complexes. Figure 11.19 indicated two ways in which pi-acceptor overlap can occur for some ligands; in Chapter 13 we shall examine another important type of ligand with pi-acceptor qualities, olefins and polyolefins.

Of course, the argument for the 18-electron rule based on Fig. 12.10 applies strictly only to six-coordinate complexes. However, symmetry arguments can be used to show that comparable numbers of bonding/nonbonding MOs are formed for CN = 4, 5, 7, and 8. It is thus reasonable to expect that the coordination number will be governed by the 18-electron rule to the extent that Fig. 12.10 has indicated its validity. Among mononuclear metal carbonyls (which should be governed by the 18-electron rule), for example, the stable first-row examples are as follows: $V(CO)_6$ [17 e^-, very readily reduced to the 18 e^- anion $V(CO)_6^-$]; $Cr(CO)_6$[18 e^-, reduced to the 18-e^- anion $Cr(CO)_5^{2-}$]; $Mn(CO)_6^+$ (18 e^-) and $Mn(CO)_5^-$ (18 e^-); $Fe(CO)_5$ (18 e^-) and $Fe(CO)_4^{2-}$ (18 e^-); $Co(CO)_4^-$ (18 e^-); and $Ni(CO)_4$ (18 e^-).

There is no neutral mononuclear manganese carbonyl, which would necessarily have either 17 or 19 e^-, and similarly no neutral mononuclear cobalt carbonyl. It should be clear that the coordination number in these systems is being determined by the stability of an 18-electron molecule. There are many other examples, some of which we shall consider in Chapter 13.

12.4 THE STABILITY OF METAL COMPLEXES

The inorganic chemist is usually interested in two specific aspects of the structure of a molecular system: the geometry of the bonding and the bond energies. We have considered the coordination numbers and geometries of common complexes, but we still need to examine the energies of ligand-to-metal bond formation. ΔH^0 for the reaction

$$M^{n+}(g) + mL(g) \rightarrow ML_m^{n+}(g) \qquad \Delta H_L$$

is usually called the *ligation energy* by analogy with hydration energy (although hydration energies usually refer to the formation of the hydrated ion dissolved in liquid water). If the ligands can be considered strictly as sigma donors, the ligation energy is equal to the total sigma-bond energy BE $\times$ m, plus the CFSE induced by the formation of the complex. For pi-acceptor ligands, the CFSE should appear as a net pi contribution to the M—L bond energy.

These ligation energies can be quite large. Table 12.4 gives sigma-bonded energies and overall ΔH_L values for first-row M^{2+} ions and a few ligands. These values were calculated from hydration energies and thermochemical data for ligand-substitution reactions in water solution using the appropriate Born–Haber cycles. The sigma-bond energies are on the order of 200 kJ/mol for these M^{2+} complexes, which is comparable to the strength of ordinary covalent bonds. For M^{3+} ions, because the metal atom is a much stronger Lewis acid, the polarizing effect on the ligand electrons is greater, and the sigma-bond energies are on the order of 350 kJ/mol for the hydrates.

Ligation energies and CFSEs have an effect on both the thermodynamic and the kinetic properties of transition-metal complexes. Considering first the thermodynamic properties, it should be obvious that the large donor–acceptor bond energies make it possible to isolate stable complexes. The most familiar of these are the

Table 12.4   Bond Energies and Ligation Energies for Metal Complexes (kJ/mol)

Metal ion	V^{2+}	Cr^{2+}	Mn^{2+}	Fe^{2+}	Co^{2+}	Ni^{2+}	Zn^{2+}
$\Delta H_L(H_2O)$	1146	1155	1096	1163	1230	1301	1272
$BE_{M-L}(H_2O)$	163	176	184	184	197	197	213
$BE_{M-L}(CN^-)$	188	218	222	163	205	—	—
$BE_{M-L}(NH_3)$	—	—	—	—	197	197	222

hydrates, since many common transition-metal salts are crystallized from aqueous solution. However, many other complexes can be formed, even in water solution. The thermodynamic stability constants for such complexes have been actively studied. Many K_n and β_n values are available:

$$ML_{n-1}(H_2O)_m + L \rightleftharpoons ML_n(H_2O)_{m-1} + H_2O \qquad K_n = \frac{[ML_n(H_2O)_{m-1}]}{[ML_{n-1}(H_2O)_m][L]}$$

$$M(H_2O)_m + nL \rightleftharpoons ML_n + mH_2O \qquad \beta_n = \frac{[ML_n]}{[M(H_2O)_m][L]^n}$$

In such reactions (shown here without ion charges for simplicity), the transition-metal ion is coordinated both as a reactant and as a product. Consequently, the *difference* in ligation energies for the two ligands H_2O and L becomes important, as does the difference in CFSE for the two complexes. Figure 11.9 indicates that absolute CFSE values are rather small compared to the total ligation energies we have been discussing (roughly 75 kJ/mol versus some 1200 kJ/mol). However, it is the fact that we are dealing with a difference in CFSEs in the above reactions that makes CFSE considerations the controlling energy factor in the formation of some complexes.

In general, there is a uniform sequence of stability (the *Irving–Williams series*) for the replacement of water by other ligands on M^{2+} ions:

$$M(H_2O)_6^{2+} + L = M(H_2O)_5 L^{2+} \qquad K_1(M)$$

$$K_1(Mn) < K_1(Fe) < K_1(Co) < K_1(Ni) < K_1(Cu) > K_1(Zn)$$

This sequence can be rationalized by considering the effects on CFSE of both trends in ionic radius and the changing number of d electrons. A smaller metal ion polarizes the ligand electrons more strongly into a stronger bond; therefore, the progressively smaller ions form more stable complexes from Mn^{2+} through Ni^{2+} (see Table 2.3). In addition, CFSE increases in the same order for high-spin octahedral complexes (see Table 11.2). Therefore, if the ligand is higher than water in the spectrochemical series (and most ligands are), the increased CFSE also favors replacement of water in the complex. The favored position of Cu^{2+} apparently comes from the increased stability of its six-coordinate complexes due to Jahn–Teller distortion. The Irving–Williams series does not hold for ligands that lead to a Dq value high enough to cause spin pairing, since that can cause major changes in CFSE for the ion. Likewise, it is not valid for chelating ligands that force a certain geometry on the complex (see Problem A1), since that can change CFSE values and even (for sterically strained systems) change sigma-bond energies.

The thermodynamic stability of complexes is, of course, determined by both the enthalpy and entropy changes involved in their formation:

$$\Delta G^0 = -RT \ln \beta_n \qquad \text{and} \qquad \Delta G^0 = \Delta H^0 - T \Delta S$$

Donor–acceptor bond energies and CFSEs determine the enthalpy contribution to the stability of the complex, but there can also be large (sometimes dominant) entropy effects. An important category of entropy-driven reactions is the formation of complexes with chelating ligands. If one compares the complexes formed between a given metal ion and a series of ligand molecules that have the same donor atoms but different chelating or ring-forming abilities, the complex with the greatest number of rings is generally the most stable. This thermodynamic observation is called the *chelate effect*.

Consider the following reaction, in which an octahedral nickel complex with six nitrogen donor atoms is being converted into another octahedral nickel complex with six nitrogen donor atoms:

$$Ni(NH_3)_6^{2+} + 3NH_2CH_2CH_2NH_2 \rightleftharpoons Ni(en)_3^{2+} + 6NH_3$$

$$\Delta G^\circ = -54 \text{ kJ/mol rn}$$

$$\Delta H^\circ = -29 \text{ kJ/mol rn}$$

$$\Delta S^\circ = +88 \text{ J/K mol rn}$$

About half of the free-energy change of the reaction is contributed by the enthalpy change, presumably because *en* (ethylenediamine) has a slightly larger CFSE than ammonia in nickel complexes. However, about half of the free-energy change is provided by the large positive entropy change. The reason for this is not hard to see. The reactants consist of a nickel complex plus three free ligand molecules, whereas the products consist of a nickel complex plus six free ligand molecules. The products have a great deal more translational randomness than the reactants, and therefore have higher entropy.

This effect can be very large indeed for ligands allowing a great deal of chelation. Chapter 2 has already noted the extensive use of $P_3O_{10}^{5-}$, tripolyphosphate, as a detergent builder (complexing agent for M^{2+}). For the reaction

$$Co(H_2O)_6^{2+} + P_3O_{10}^{5-} \rightarrow Co(P_3O_{10})^{3-} + 6H_2O$$

the enthalpy change is actually unfavorable, yet the reaction is strongly entropy-driven to the right: $\Delta G^0 = -45$ kJ/mol rn, $\Delta H^0 = +19$ kJ/mol rn, $\Delta S^0 = +218$ J/K mol rn. In a case like this, which involves highly charged ions, the structure-making or structure-breaking effects of changing the net charge on a complex must be included in the description of the entropy change for the reaction, but the chelate effect is also at work.

In chelated complexes involving organic ligands, the size and preferred bond angles for carbon atoms usually make five-membered rings the most stable. As Fig. 12.11 suggests, chelating ligands such as acetylacetone (*acac*) can also form highly stable six-membered rings if pi delocalization roughly analogous to that in benzene is possible. However, increasing the size of the metal-containing ring usually decreases the extra stability of the chelated complex—the "chelate effect." Table 12.5 shows the thermodynamic functions for the replacement of two three-coordinating ligands in a Co^{2+} complex by an equivalent six-coordinating ligand in which another ring has been formed as the two original ligands are linked (figuratively). For $n = 2$, 3, and 4 in the new ligand—where $n = 2$ corresponds to the familiar ethylenediaminetetraacetate ion, or EDTA—the complex formed contains,

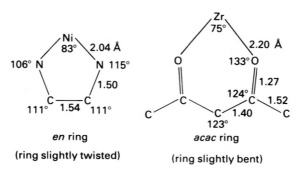

Figure 12.11   Typical chelate-ring shapes.

respectively, a five-, six-, and seven-membered ring. The ΔH values are mildly unfavorable in each case (possibly because some deformation from ideal bond angles is necessary to accommodate the new ring), but the $T\,\Delta S$ values are uniformly favorable, and of almost the same magnitude. However, for $n = 5$ (corresponding to an eight-membered ring) the entropy effect drops rather sharply, and the formation of the complex is no longer thermodynamically favored. Apparently a ligand with a five-membered chain loses about as much rotational entropy on complexing as the system gains in translational entropy.

Beyond the thermodynamic stability of transition-metal complexes, a striking range of kinetic stabilities is observed for complexes with outwardly similar properties and thermodynamic stabilities. Perhaps the single best-characterized reaction that illustrates kinetic stability is the water exchange reaction

$$M(H_2O)_n^{q+} + H_2O^* \rightleftharpoons M(H_2O)_{n-1}(H_2O^*)^{q+} + H_2O$$

Such a reaction obviously has zero thermodynamic driving force and places metal ions in identical environments. Yet, as Fig. 12.12 indicates, rate constants for the exchange of water by different metal ions vary over some 18 orders of magnitude!

Table 12.5   Ring Size and the Chelate Effect

$$Co(mim)_2^{2-} + Y^{4-} \rightarrow CoY^{2-} + 2\ mim^{2-}$$

mim = methyliminodiacetate, $CH_3N(CH_2COO^-)_2$

$$Y = (^-OOCCH_2)_2N\text{---}(CH_2)_n\text{---}N(CH_2COO^-)_2$$

	kJ/mol rn		
	ΔG^0	ΔH^0	$T\Delta S^0$
$n = 2$	−13.4	+5.4	+18.8
$n = 3$	−9.2	+12.1	+21.3
$n = 4$	−9.6	+16.3	+25.9
$n = 5$	+2.9	+9.6	+7.5

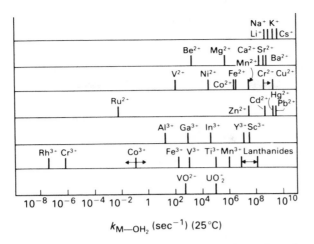

$$k_{M-OH_2} \text{ (sec}^{-1}\text{) (25°C)}$$

Figure 12.12   Rate constants for exchange of coordinated water molecules in aqueous solution. [Reprinted with permission from A. E. Martell, ed., ACS Monograph 174, *Coordination Chemistry*, vol. 2, 1978, Figure 1-1 (corrected). Copyright © 1978, American Chemical Society.]

Note that transition-metal ions show a much greater range of k_{M-OH_2} values than do main-group metal ions. This is true for reactions in general within the coordination sphere of metal ions. In particular, Cr^{3+}, Co^{3+}, and Pt^{2+} are exceptionally slow in reacting, even when the reaction is overwhelmingly thermodynamically favorable. This explains why Cr^{3+}, Co^{3+}, and Pt^{2+} complexes are so prominent in the early history of coordination chemistry—the product complex wouldn't fall apart in water even if that might be the thermodynamic result. Alfred Werner and his coworkers are thought to have prepared over 700 Co^{3+} complexes in their pioneering studies of inorganic stereochemistry.

It is important to distinguish between thermodynamic reactivity and kinetic reactivity. The terms "stable" and "unstable" are commonly reserved for thermodynamic reactivity descriptions, whereas kinetically reactive systems are said to be *labile* and kinetically unreactive systems *inert* (even though either may be thermodynamically unstable). A rule of thumb is that if a complex reacts in less than a minute, it is labile; if it takes longer, it is inert. In terms of Fig. 12.12, the division can be made at about $k_{M-OH_2} = 1$ sec^{-1}.

A complex that is thermodynamically unstable with respect to a particular reaction has a negative ΔG for that reaction. For a complex that is kinetically labile the corresponding statement is that it has a small (though positive) free energy of activation, $\Delta G^{\ddagger}$. By contrast, a kinetically inert complex has a large positive $\Delta G^{\ddagger}$. Without fairly detailed information on the reaction mechanism, we can say little about the electronic-structure reasons for the large differences between a labile and an inert complex, but a few generalizations are possible. Since the electrostatic binding forces between the metal ion and the ligand are larger for a higher cation charge, 3+ ions should generally react slower than 2+ ions. If we view the binding of water molecules as a classical ion–dipole attraction, k_{M-OH_2} correlates reasonably well with $q\mu_0/r^3$ (where the symbols are those used in Fig. 6.1):

$$k_{M-OH_2} = \left(\frac{kT}{h}\right) \cdot \exp\left[\frac{-2qe\mu(\Delta r)}{kTr^3 D}\right]$$

(Here Δr is the size difference between the metal ion and water, and D is the dielectric constant of the liquid.) Figure 12.13 shows this relationship in graphic form. A reasonably good straight line results for ions that have no CFSE or geometric distortion due to the Jahn–Teller effect.

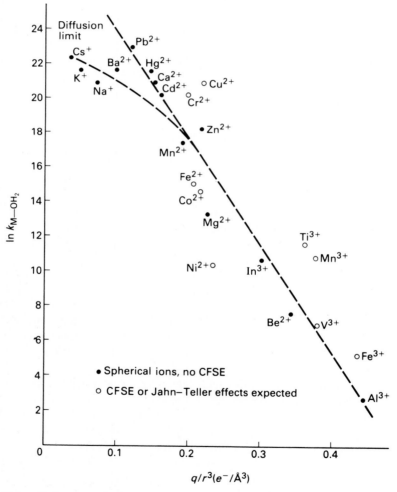

Figure 12.13 Water-exchange rates as a function of the polarizing power of the metal ion.

Table 12.6 CFSE Changes for Octahedral Complexes Changing Geometry during Reaction

d configuration	Dissociative mech. (to 5-coord. sq. pyr.) Dq		Associative mech. (to 7-coord. capped tr. prism) Dq	
d^1	+0.57		+2.08	
d^2	+1.14		+0.68	
d^3	−2.00		−1.81	
d^4	+3.14		+2.80	
d^5	−0.86 LS	0.00 HS	−1.13 LS	0.00 HS
d^6	−4.00 LS	+0.57 HS	−3.68 LS	+2.08 HS
d^7	+1.14		+1.00 LS	+0.68 HS
d^8	−2.00		−1.81	
d^9	+3.14		+2.80	
d^{10}	0.00		0.00	

Note: LS = low spin, HS = high spin. In transition-state orbitals, only electron assignments that produce no change in total spin are considered.

Most transition-metal ions, however, have either a nonzero CFSE or some Jahn–Teller distortion, or both. Frequently, geometric distortion can speed up the exchange of ligands because the rapid interconversion of axial and equatorial ligands through molecular vibration stretches the metal–ligand bonds. On the other hand, the extreme slowness of Cr^{3+}, Co^{3+}, Ru^{2+}, and Rh^{3+} in octahedral complexes and Pt^{2+} in square-planar complexes can be attributed to crystal-field effects. If the transition state in the water-exchange reaction has a significantly lower CFSE than the original complex, that loss of CFSE can form a major part of the activation energy. Even if we do not know the ligand-exchange mechanism of these ions—for instance, whether the transition state is five-coordinate as a result of a dissociative mechanism or seven-coordinate as a result of an associative mechanism—the d-electron configurations in Table 12.6 show that d^3 and d^6 octahedral complexes are considerably destabilized by any reaction that temporarily changes their geometry. This is precisely why Cr^{3+}, Co^{3+}, and the other extremely inert ions are so slow to react: Their $\Delta G^{\ddagger}$ includes a large Δ(CFSE) contribution. (Note that almost all Co^{3+} complexes are low spin, because of the high formal oxidation state of the metal.)

Kinetic and mechanistic studies of transition-metal complexes are a major field of experimental study. We shall return to it in Chapter 14. Here we conclude by reemphasizing the distinction between the thermodynamic stability and the kinetic stability of these compounds. Almost any conceivable complex (within reasonable coordination-number limits) has great thermodynamic stability toward dissociation into a free metal ion and free ligands. On the other hand, it may be quite reactive toward substitution or toward the addition of new ligands. Even where a complex is thermodynamically reactive, however, it may be so inert for mechanistic reasons that it yields no new product even after hours of reaction.

12.5   COMMON LIGANDS AND COMPLEXES

With some background in the possible geometries and stabilities of transition-metal complexes, we can move on to consider the range of such complexes actually observed. We shall group them according to the identity of the electron-donor atom in the ligand, and consider monodentate ligands and chelating ligands separately. We shall limit ourselves here to nonmetal donor atoms, deferring metal clusters until Chapter 13; in addition, we shall also delay carbon-donor species, such as CO and organometallic systems, until Chapter 13.

H donor atoms. The simplest possible electron-pair-donor ligand is the hydride ion, H^-. The hydride ion is an extremely soft base. Therefore, it is usually found in mixed complexes with other soft-base ligands (donor atoms having low electronegativity such as P or C in organometallics). Most transition metals form nonstoichiometric interstitial hydrides; the only well-characterized stoichiometric hydride is CuH. Complex ions containing *only* H^- ligands are also unusual. Technetium and rhenium both form ions with the formula MH_9^{2-}, and Mg_2FeH_6 has been prepared, along with MH_5^{4-} for M = Co,Rh,Ir and NiH_4^{4-}. However, a very large number of mixed hydride complexes has been prepared. Table 12.7 indicates the formulas for the compounds that have the largest number of H atoms per metal atom, but the variety is considerable. Many hydride complexes have been investigated as possible hydrogenation catalysts, a subject we shall take up in Chapter 15. The table suggests that the tendency to take up multiple hydrogen atoms is greater for second- or third-row metals. Stereochemically, H^- occupies a nearly normal site on the coordination polyhedron, but as it is so small other ligands tend to sag toward it, distorting its complexes from ideal geometry.

In rare instances, metallohydride anions serve as donors through an H atom. One of the best characterized of these is $Cu(PPh_3)_2BH_4$, in which the Cu is approximately tetrahedral, coordinated by the two P atoms and two H atoms from the BH_4^- anion. A slightly more complicated case is the bridging of two Th atoms in a binuclear complex by the methyltrihydroborate ion $CH_3BH_3^-$; the ion centers its BH_3 end between the Th atoms so as to form the connection sequence Th–H–B–H–Th. A binuclear Hf complex has been characterized in which two BH_4^- ions donate to a Hf through two H atoms each while another BH_4^- chelates the

Table 12.7   Some Hydride Complexes of Transition Metals

$H_2Ti(cp)_2$		$H_2Cr(CO)_5$	$HMn(PF_3)_5$	FeH_6^{4-}	CoH_5^{4-}	NiH_4^{4-}
$(H_2Zr(cp)_2)_n$		$H_2Mo(cp)_2$	$HTc(cp)_2$	$H_4Ru(PPh_3)_3$	$H_2RhCl(PPh_3)_3$	$HPdBr(PEt_3)_2$
		$H_6Mo(PR_3)_3$	TcH_9^{2-}	$H_6Ru(PPh_3)_2$		
	$H_3Ta(cp)_2$	$H_6W(PMe_2Ph)_3$	$H_7Re(AsEt_2Ph)_2$	$H_4Os(PMe_2Ph)_3$	$H_5Ir(PEt_3)_2$	$H_2PtCl_2(PEt_3)_2$
			ReH_9^{2-}	$H_6Os(PR_3)_2$	IrH_5^{4-}	

Note: cp = C_5H_5, Me = CH_3, Et = C_2H_5, Ph = C_6H_5.

other Hf through three H atoms, and in addition three separate H atoms bridge the two Hf atoms.

This last compound has some rather remarkable bonding in place. We would not be too surprised if a Cl^- ion bridged two metal atoms; it has four pairs of non-bonding electrons in its valence shell, and could presumably donate one pair to one metal atom and a different pair to the other metal atom. But in a Hf–H–Hf bridge three nuclei are necessarily being bound by only two electrons. Of course, we have seen this bonding model before in the boranes and other electron-deficient compounds, but it is interesting that the same pattern is seen in compounds in which the atoms are as different as H and Hf. An even cleaner example is provided by the anion $[(CO)_5W–H–W(CO)_5]^-$, in which the two tungsten halves of the molecule have been shown by a careful neutron diffraction study to be bound *only* by a single bridging H^- ion.

We might pause here to note that standard nomenclature distinguishes between terminal and bridging ligands by using the Greek "mu" (μ) to denote a bridging ligand, with a subscript to indicate the number of atoms it is bridging if the number is more than two. Thus the In_2I_6 dimer shown in Fig. 3.2 would be named di-μ-iodo-bis(diiodoindium) to indicate the presence of the two bridging I atoms. The bridged tungsten carbonyl anion above would be μ-hydridobis(pentacarbonyltungsten). Appendix A includes this in its general treatment of inorganic nomenclature.

Until quite recently hydride complexes to transition metals were considered unusual, and it was thought very unlikely that the sigma-bonded pair of electrons in H_2 could be donated to a metal atom, though some such structure might momentarily exist as a reaction intermediate. However, in 1984 Kubas reported the structure of a stable "dihydrogen" complex, $W(CO)_3(P^iPr_3)_2(H_2)$, with the two H atoms still bonded to each other at about the normal $r_{H–H}$ and equidistant from the W atom. Many dihydrogen complexes have since been reported, and they constitute a research area of great interest because of the importance of transition metal complexes and organometallic systems in catalysis involving hydrogen atom transfer. In many cases the M<H_2 complex also exists as an H–M–H dihydride isomer, in which the H–H bond has been readily cleaved homolytically. This is hardly surprising, but it is also possible to cleave the H–H bond heterolytically so that a dihydrogen complex might, for example, be deprotonated.

Experiments using electrochemical and NMR techniques have shown that in THF solvent a series of $Ru(H_2)$ complexes displays pK_a values from 9.2 to 4.6 (roughly equivalent to acetic acid); structural correlations suggest that with control of other ligands such pK_a values could range from 12 to -6, which would be comparable to sulfuric acid. Obviously detailed control of pK_a of such complexes could have far-reaching catalytic implications for hydrogenations; we will examine a variety of organometallic catalysts and processes in Chapter 15.

X-ray structural studies on hydrogen-containing metal compounds are difficult to obtain in sufficient detail to accurately reveal the position of the hydrogen atoms, because H has no inner core of electrons; an x-ray photon sees a proton

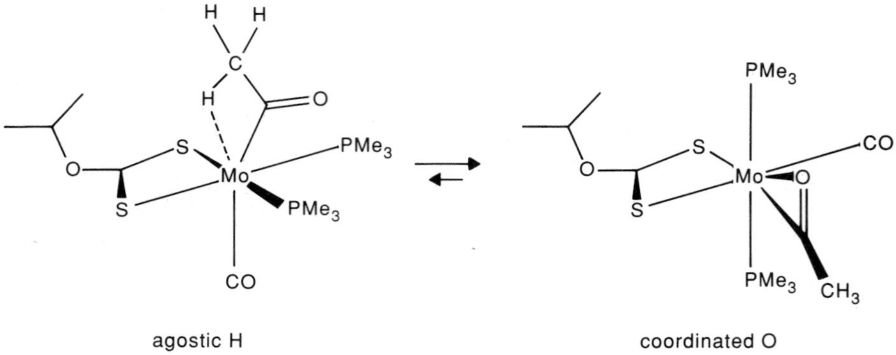

Figure 12.14   The structural equilibrium observed between an agostic hydrogen atom and a coordinated oxygen atom.

as being just an uncommonly slow electron. As a result, it is only in recent years that x-ray studies have reliably characterized hydrogen-nucleus positions. For organometallic systems, one surprising result has been the discovery that ordinary C–H bonds near the metal atom are frequently distorted from "natural" bond angles around the C in such a way that a hydrogen atom approaches the metal atom and is perhaps partially coordinated to it (presumably serving to some degree as a Lewis base through the C–H sigma-bonding electrons). Such hydrogen atoms are said to be *agostic;* studies of agostic hydrogen atoms, like those of dihydrogen complexes, are interesting because of the possibility that they represent a stage in a catalytic process. The agostic interaction is not a casual artifact of crystal packing, as can be seen from the equilibrium in Fig. 12.14. The coordinated acetyl group chelates the Mo in one of two ways: either by a sigma bond to the carbonyl C with the O donating a pair of electrons, or through the same bond from Mo to the carbonyl C and an agostic H on the acetyl methyl group.

N donor atoms. Many, if not most, of the tens of thousands of transition-metal donor–acceptor complexes that have been prepared involve group V donor atoms— N, P, As, and (to some extent) Sb. In low formal oxidation states, these atoms are normally three-coordinate in neutral compounds, with a sigma-symmetry nonbonding electron pair in an orbital distribution approximating an sp^3 hybrid. Nitrogen differs from the other group V donor atoms in that it has no valence d orbitals, so that in saturated compounds it has no pi-acceptor capability. It is also considerably more electronegative than the other donor atoms in its group and thus is a harder base. The other group V donors are very similar to each other as soft-base pi acceptors; they differ mostly in stereochemical requirements because of their different inner-core radii and valence-orbital radii.

Nitrogen as a donor atom is, of course, most familiar in the ammonia molecule, whose complexes are called *ammines.* As we have already noted, the first recorded observation of what we now call a metal complex (1597, Libavius) involved the $Cu(NH_3)_4^{2+}$ ion, whose dark blue color is strikingly different from that of the hydrated Cu^{2+} ion. Because the N atom is a moderately hard base, ammines are particularly stable for metal oxidation states corresponding to moderately hard acids—2+ and 3+. Octahedral $M(NH_3)_6$ hexammines in one or both of these oxidation states are known for every first-row transition metal, though at both ends of the row (Ti, V, Cu, Zn) they very readily lose ammonia. In many cases, complexes with lower $NH_3:M$ ratios are also known. Saturated organic ammines RNH_2, R_2NH, and R_3N are similar in their donor properties, although their bulkier molecules can introduce steric problems that limit the coordination number. We have already seen this behavior in the diisopropylamides (CN = 2,3) earlier in the chapter.

Other N-donor ligands with generally similar properties include pyridine (variously substituted), pyrrole, pyrazole, and imidazole (see Fig. 12.15). As these are also neutral ligands, their complexes are usually cationic. However, pyrazole can be stripped of its N—H proton in basic solution, allowing such neutral complexes as $Co(C_3H_3N_2)_2$ to form. Most complexes are octahedral, with anions such as halide ions occupying several coordination sites for cationic complexes: $Fe(py)_4Cl_2$, $Co(py)_3Cl_3$. However, some interesting structural equilibria can occur for d^7 and d^8 configurations. $CoBr_2$ and CoI_2 engage in the following equilibrium, which can be controlled by the pyridine concentration:

$$Co(py)_2X_2 + 2\,py = Co(py)_4X_2$$

Tetrahedral   Octahedral

The ligand quinoline (benzopyridine) forms two complexes with the formula $Ni(quin)_2Cl_2$, one yellow and the other blue. The first is octahedral with bridging Cl atoms, the second is tetrahedral.

Acetonitrile is a good solvent for many transition-metal compounds, at least partly because it readily forms N-donor complexes. The steric requirements of the linear CH_3CN ligand are very low, and many octahedral $M(MeCN)_6^{n+}$ complexes are known. The complex with empirical formula $Co(MeCN)_3Cl_2$ is one: $[Co(MeCN)_6]^{2+}$ $[CoCl_4]^{2-}$.

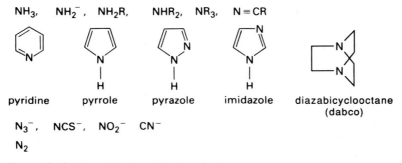

NH$_3$,   NH$_2^-$,   NH$_2$R,   NHR$_2$,   NR$_3$,   N≡CR

pyridine   pyrrole   pyrazole   imidazole   diazabicyclooctane
(dabco)

N$_3^-$,   NCS$^-$,   NO$_2^-$ CN$^-$

N$_2$

Figure 12.15 Some N-donor ligand molecules.

Several pseudohalides containing N form a variety of transition-metal complexes: NO_2^-, CN^-, N_3^-, and NCS^-. The cyanide complexes are usually coordinated through the C atom (though not always, particularly in crystals). Both the NO_2^- and NCS^- complexes can coordinate either through the N or the O or S. We shall consider this form of isomerism in the next section. Note that NO_2^- and CN^-, unlike other N-donor ligands, appear quite high in the spectrochemical series, presumably because of their good pi-acceptor ability.

One final N-donor ligand of great importance is the N_2 molecule. Bacteroids in legume roots fix atmospheric N_2 to ammonium ion by coordinating it in a metallo-enzyme *nitrogenase,* which contains iron and molybdenum. The obvious inference is that N_2 is serving as a sigma-donor, pi-acceptor ligand like the isoelectronic CN^- or CO. However, it has proved quite difficult to prepare model coordination compounds with the N_2 ligand. We shall explore this area further in Chapter 16, but we might note here that many complexes are now known with the N_2 ligand, though a catalytic process for nitrogen fixation is still elusive.

P, As, Sb donor atoms. P, As, and Sb as donor atoms in ligands are softer bases than the corresponding N-donor molecules, and are also generally better pi acceptors. Bi is much less commonly seen as a donor, though a few BiR_3 and BiR_2^- complexes are known. As pi acceptors, their position in the spectrochemical series is quite high, and their complexes are usually more stable if the metal is in a low oxidation state (soft acid). Many PR_3 (substituted phosphine) complexes are known, particularly with aromatic substituents (aromatic phosphines are more resistant to oxidation). In addition, although the nitrogen halides are very poor donors, PF_3 has been used as a donor in many complexes. Like CO, it has the remarkable ability to form nickel(0) complexes from the metal:

$$Ni + 4CO \rightarrow Ni(CO)_4$$
$$Ni + 4PF_3 \rightarrow Ni(PF_3)_4$$

Although phosphorous acid is only dibasic because one H is bonded directly to the P as $HP(O)(OH)_2$, phosphite esters $P(OR)_3$ are usually prepared from PCl_3 and contain three-coordinate phosphorus, with a nonbonding pair available on the P atom. These phosphites have also been extensively studied as ligands, since $P(OR)_3$ is roughly comparable to CN^- in the spectrochemical series.

In N-donor complexes it is frequently advantageous to have a negative charge on the N ligand in order to produce a solid product in which a metal cation has only the N ligand around it and no counterions, as in the diisopropylamides we encountered for low coordination numbers. The net negative charge also strengthens the sigma-donor capabilities of the ligand and makes it a softer Lewis base—qualities that sometimes allow tailoring a synthetic reaction. In exactly the same sense, PR_2^- *phosphides* are sometimes used as ligands; the two-coordinate P atom makes an excellent bridging atom for binuclear complexes. Even a PR^{2-} *phosphinidine* can serve as a donor if the R group is bulky enough to stabilize the M–P bond sterically; usually phosphinidines are bridging ligands, though mononuclear complexes of them are known.

O donor atoms. Group VI donor-atom ligands are also numerous and well studied. By far, the largest number of ligands involve the lightest member of the group (oxygen), as was also true for group V. The most familiar O-donor ligand is, of course, water and its deprotonated forms OH^- and O^{-2}. We have already considered the relationships between these in Chapter 6 under the general heading of hydroxides and oxocations. The O atom in water is a hard base with no pi-acceptor ability. It thus stabilizes intermediate to high formal oxidation states for the metal, though very high oxidation states usually lead to deprotonation (see Chapter 6) and the formation of hydroxides or oxocations. All first-row transition metals form octahedral $M(OH_2)_6^{n+}$ cations in both the 2+ and 3+ oxidation states except for Ti^{2+}, which reduces water to H_2, and Ni^{3+} and Cu^{3+}, which oxidize water to O_2. Some of the octahedra are distorted by the Jahn–Teller effect. In particular, Cu^{2+} normally loses the axial waters as it crystallizes, forming a square-planar 4-hydrate. For the electronic reasons mentioned above, water is quite low in the spectrochemical series even though it is a good Lewis base and hydration energies are, as we have seen, quite high.

Alcohols and ethers ("alkylwater") are relatively weak bases and form relatively fragile complexes. This behavior is consistent with that of N-donor molecules, where alkylamine complexes are less numerous and less stable than ammonia complexes, although presumably comparable in electronic properties. The best known complexes are those of methanol and tetrahydrofuran (*thf*). The ring structure of *thf* reduces its steric requirements, and complexes such as $CrCl_3(thf)_3$ (which has a more or less octahedral coordination geometry) are known. Carbonyl compounds make better O-donor ligands than alcohols or ethers; many complexes are known with aldehydes and ketones, and with some esters and amides as well. Urea forms many stable complexes, and in almost all cases the donor atom is the O from the carbonyl group rather than one of the amide N atoms. It is even possible to use acetone as a bridging ligand, in which the carbonyl O donates a pair to one metal atom and the $C{=}O$ double bond pi electron pair is donated to the other metal atom.

Another large group of complexes involves ligands with a group V or VI donor atom, X, that have been oxidized to contain a $X{=}O$ or $X{\rightarrow}O$ bond so that the O atom is the donor. For example, pyridine forms pyridine 1-oxide or pyridine N-oxide under the influence of 30% H_2O_2, and the new compound is approximately as good a ligand as pyridine itself. Other compounds in this category include trialkyl or triarylphosphine oxides R_3PO, similar arsine oxides, sulfoxides R_2SO (particularly dimethylsulfoxide, *dmso*), and to a lesser extent sulfones R_2SO_2.

A related kind of coordination involves oxyanions such as NO_2^-, NO_3^-, SO_4^{2-}, and even ClO_4^-. All of these species are quite electronegative overall because of their high concentration of O atoms, and they are hard bases and (usually) only weak donors. Furthermore, because of the multiple O atoms, the ion can serve as a bidentate chelating ligand, a complicating factor. Nevertheless, many complexes are known with oxyanions as unidentate O-donor ligands. They are usually more stable the less electronegative the center oxyanion atom is. The NO_2^- ion can bond either through the N or through the O, and it often bridges two metal atoms by bonding to one through the N and the other through the O, as in the polymeric compound $[Ni(en)_2(NO_2)]^+BF_4^-$. Examples of other unidentate oxyanion O-donor ligands include $[Co(NH_3)_5SO_4]^+Br^-$, $Au(NO_3)_4^-$, and $Co(OAsPh_2Me)_4(ClO_4)_2$ (coordinating both through the arsine oxide O and the perchlorate O).

Like dinitrogen, the dioxygen molecule O_2 is a ligand of extreme bioinorganic interest. It coordinates to the Fe in hemoglobin for transport through the blood stream, and to iron and other metals in other metalloproteins. A number of more traditional O_2 complexes have also been prepared. One of these is "Vaska's complex," $IrCl(CO)(PPh_3)_2$, which reversibly absorbs O_2 to form $IrCl(O_2)(PPh_3)_2$. We shall return to the several important kinds of O_2 coordination over the next four chapters, but it is appropriate here to call attention to it as an O-donor ligand.

S donor atoms. Unidentate ligands containing sulfur are relatively rare, though chelating and bridging ligands with multiple sulfurs are widely used. The best known examples of unidentate S donors are the SCN^- and $S_2O_3^{2-}$ ions (note that both of these can coordinate through atoms other than S) and thiourea (*tu*), $SC(NH_2)_2$. Because sulfur has such a low electronegativity, it is a very soft base. Most of the S-donor complexes involve the softest acids among the transition-metal ions, Cu^+ and Ag^+: $Cu(tu)_3Cl$, $Ag(S_2O_3)_2^{3-}$, $[Ag\{SC(NH_2)(CH_3)\}_4]^+Cl^-$. However, ethylenethiourea, a thioketone of imidazole, forms the tetrahedral $Co(etu)_2Cl_2$, and most of the later transition metals form complexes with thiourea or other thioketones. Occasionally the sulfite ion will donate through the S atom: $Pd(SO_3)_2(NH_3)_2$ and $Co(en)_2(NCS)(SO_3)$. In the latter, the Co^{3+} has chosen the harder end of the thiocyanate ligand but, curiously, the softer end of the sulfite. Even dimethylsulfoxide seems to donate through the S atom in Pd and Pt complexes such as $PdCl_2(dmso)_2$, where the metal is one of the softest acids. However, *dmso* as an S donor is fairly bulky, and if other ligands on Pd or Pt are also bulky the *dmso* will donate even to these soft acids through the O atom.

Halogen donor atoms. Group VII elements, the halogens, are more limited in their coordination possibilities as ligands. The halide ions all readily form anionic fluoro-, chloro-, bromo-, and iodometallates, but stable complexes with organic halides are quite rare (though alkyl halides can obviously serve as bases, as in the Friedel–Crafts intermediate $RCl\cdots AlCl_3$). However, methyl iodide forms the complex $[(H)_2Ir(PPh_3)_2(ICH_3)_2]^+$, and alkyl iodides in general can be coordinated to Re in organometallic systems without destroying the RI molecule.

It is important to distinguish F^- as a ligand from the other halides: Fluoride is much more electronegative and therefore a much harder base, and it also has neither pi-acceptor nor pi-donor ability. The fluoride ion is therefore a good ligand with which to stabilize transition metals in their highest oxidation states. As Table 12.8 suggests, many fluorometallates have been prepared, predominantly in the higher oxidation states of metals and particularly so toward the right of the *d*-block. Up to eight fluorides can coordinate a single metal ion, though no first-row metal goes beyond six. Note in the table that as a weak sigma-only ligand, the fluoride ion gives com-

Table 12.8 *d*-Block Fluorometallates

TiF_6^{3-}	VF_6^{2-}	CrF_3^-(?)	MnF_4^{2-}	FeF_4^{2-}	CoF_4^{2-}	NiF_6^{4-}	CuF_4^{2-}	ZnF_4^{2-}
TiF_6^{2-}	VF_6^-	CrF_4^{2-}	MnF_4^-	FeF_4^-	CoF_6^{3-}	NiF_6^{3-}	CuF_6^{3-}	
		CrF_6^{3-}	MnF_5^{2-}	FeF_6^{2-}	CoF_6^{2-}(?)	NiF_6^{2-}		
		CrF_6^{2-}	MnF_6^{3-}					
		CrF_6^-(?)	MnF_5^-					
			MnF_6^{2-}					
ZrF_5^-	NbF_6^-	MoF_4^-	TcF_6^{2-}	RuF_6^{3-}	RhF_6^{3-}	PdF_6^{2-}	AgF_4^-	
ZrF_6^{2-}	NbF_7^{2-}	MoF_6^{3-}	TcF_6^-	RuF_6^{2-}	RhF_6^{2-}			
ZrF_7^{3-}		MoF_6^{2-}		RuF_6^-	RhF_6^-			
ZrF_8^{4-}		MoF_6^-						
		MoF_8^{3-}						
		MoF_7^-						
		MoF_8^{2-}						
HfF_5^-	TaF_6^-	WF_6^-	ReF_6^{2-}	OsF_6^{2-}	IrF_5^{2-}	PtF_6^{2-}	AuF_4^-	HgF_4^{2-}
HfF_6^{2-}	TaF_7^{2-}	WF_8^{3-}	ReF_6^-	OsF_6^-	IrF_6^{2-}	PtF_6^-		
HfF_7^{3-}	TaF_8^{3-}	WF_7^-	ReF_8^{3-}		IrF_6^-			
		WF_8^{2-}						

plexes with valence-electron totals ranging from 12 to 20, corresponding to Fig. 12.10a. This is consistent with the fact that most fluorometallate complexes are high spin and that F^- is near the bottom of the spectrochemical series.

The halogens other than fluoride all have much lower electronegativities, so that the X^- ion (X = Cl, Br, I) has at least modest pi-donor ability. This makes the other halides softer bases than F^- and places them at the very bottom of the spectrochemical series. The X^- ion is larger than F^- and coordination numbers beyond 6 are rare, though WCl_7^{2-} has been reported. The softer X^- ions stabilize lower oxidation states of the metals (often 2+ and 3+) and the buildup of negative charge on the halometallate anion often limits the coordination number to 4. MX_4^{2-} and MX_4^- ions are common, and VCl_3 even forms the dichlorovanadium cation VCl_2^+ when allowed to react with $ZnCl_2$. Presumably because of their lower electronegativities, the X^- ions show a much greater tendency than F^- to donate to two metal ions and form a bridging group. Species such as $Mo_2Cl_9^{3-}$ and $Pd_2Br_6^{2-}$ are frequently observed. Bridging is not inevitable, however; $Os_2Cl_8^{2-}$ is bound only by an Os≡Os bond. For low oxidation states of the metals, the X^- ions participate in some remarkable metal clusters such as $Nb_6Cl_{12}^{2+}$ and $Mo_6Br_8^{4+}$. We shall explore these further in the next chapter (Chapter 13).

Chelating ligands. We have already encountered the chelate effect and noted that it stabilizes complexes with chelating ligands. Many chelating ligands have been synthesized, often in order to force certain stereochemical features such

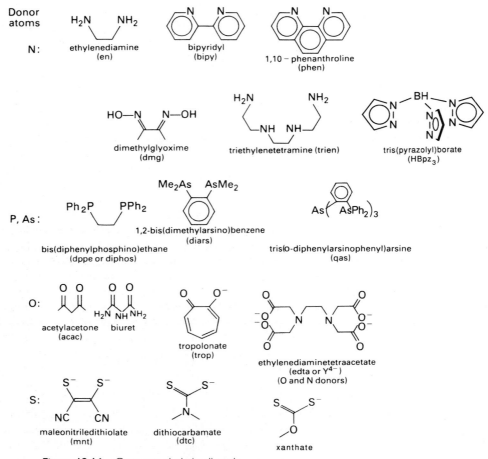

Figure 12.16 Common chelating ligands.

as isomerism or unusual coordination geometries on the resulting complexes. An enormous number of chelated complexes is known. Figure 12.16 shows the structures of some common chelating ligands; some comment on the geometry of their complexes is appropriate.

Among the N-donor ligands, *en* and *trien* and the related molecules with varying numbers of C_2H_4NH groups are relatively flexible because there is no pi bonding in the ligand molecules. They are thus more likely to accommodate themselves to the ideal coordination geometry around the metal than are the rigid molecules *bipy* and *phen* (and the tridentate ligand terpyridyl). Complexes of these ligands have been studied extensively because of the opportunities for isomerism within a given coordination polyhedron, as the next section will show. The P and As-donor ligands have been studied in much the same context, but they are soft enough to form stable complexes in organometallic systems, which we shall encounter in the next chapter.

Figure 12.2 has already indicated that *qas* and other "tripod" ligands can be used to stabilize trigonal-bipyramidal five-coordination. In the same sense, the ligand in Fig. 12.7 is used to stabilize trigonal-prismatic six-coordination, and other rigid chelating ligands have been designed for similar purposes. The tris(pyrazolyl)borate ligands are particularly convenient in this application; there are three N^- donor atoms in a nearly fixed triangular arrangement, and not only can the steric bulk be varied by substituting bulky groups such as *t*-butyl on the adjacent ring carbons, but the overall donor/acceptor properties of the ligand can be varied by substituting electronegative atoms across the ring from the donor N. Unsubstituted $HB(pz)_3$ ligands tend to form octahedral $M(HB(pz)_3)_2$ compounds, but $HB(3\text{-}^tBu\text{-}pz)$ forces tetrahedral coordination on the metal, allowing only one additional coordination site.

Among the O- and S-donors, all of the ligands shown (except *edta*) are rigid when coordinated to a metal ion. Acetylacetone usually loses a proton from the central CH_2 to form an anion with a continuous pi system, equivalent to the tropolonate. However, the *acac* six-membered ring (including the metal atom) is a better fit for most transition-metal ions than the *trop* five-membered ring, and the latter often produces considerably distorted bond angles in the coordination polyhedron. As an earlier section indicated, *mnt* and related dithiolate molecules have been extensively studied because they can provide trigonal-prismatic six-coordination. Note, by the way, that it is difficult or impossible to assign an oxidation state to the metal in these complexes, because it is very difficult to determine the appropriate net charge on the free ligand: Is it ^-S—C=C—S^- or S=C—C=S? In a planar $M(S_2C_2R_2)_2$ complex, there are three possibilities for the metal oxidation state, depending on how much delocalization there is in the five-membered rings:

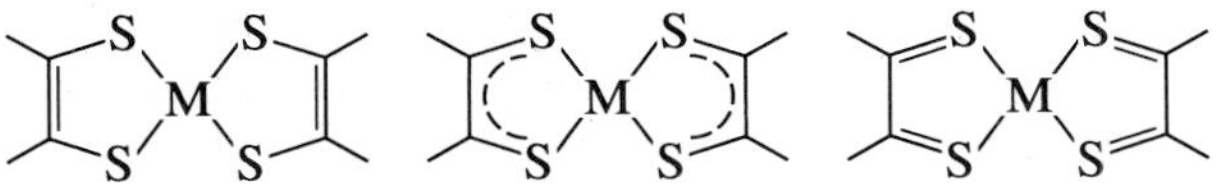

The dithiocarbamate ligand and related species have the unusual property of chelating a metal ion in a four-membered ring. This is more common in inorganic systems than organic, however. We have already noted that the BH_4^- ligand can form four-membered

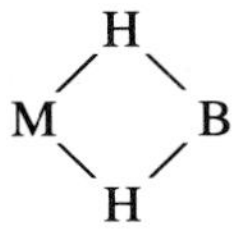

rings, and a number of oxyanions can also form chelate complexes with four-membered

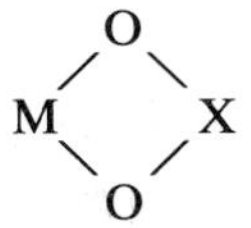

rings. Anhydrous nitrates and carbonates, for example, form chelates more often than they serve as unidentate O-donors.

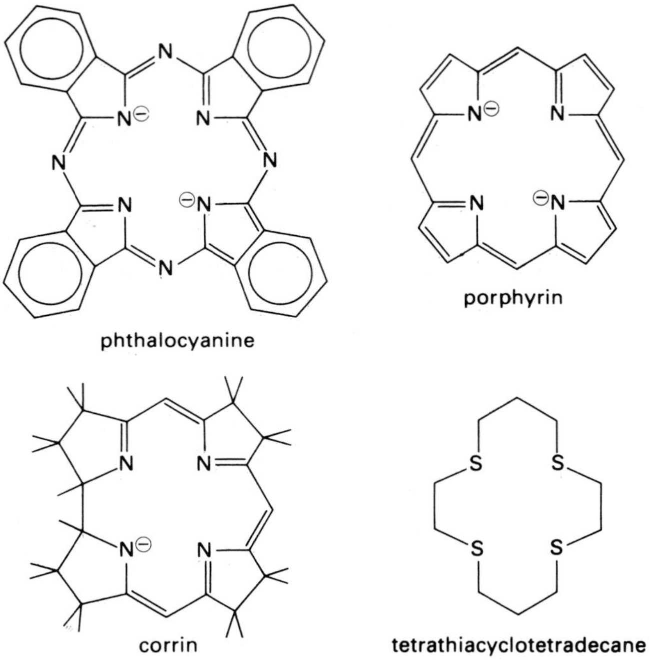

Figure 12.17   Some macrocyclic ligands.

Macrocyclic ligands. A final category of ligand is that of macrocyclic organic molecules. Many metalloproteins (for instance hemoglobin) contain their metal ions in a planar macrocycle, and a great deal of study has been given to both naturally occurring macrocycles and to model compounds. Figure 12.17 shows the structure of a few of the basic macrocycle ring systems that have been studied as ligands. Interest has centered on macrocycles with $\tilde{N}$ donors because of the biological prominence of the porphyrin and corrin systems. Heme is a substituted porphyrin, chlorophyll *a* is a porphyrin with an added saturated ring, and vitamin B_{12} is a corrin ring, containing iron, magnesium, and cobalt respectively. The phthalocyanine ring is a synthetic macrocycle that is actually formed by using a metal ion as a template to link four phthalonitrile molecules:

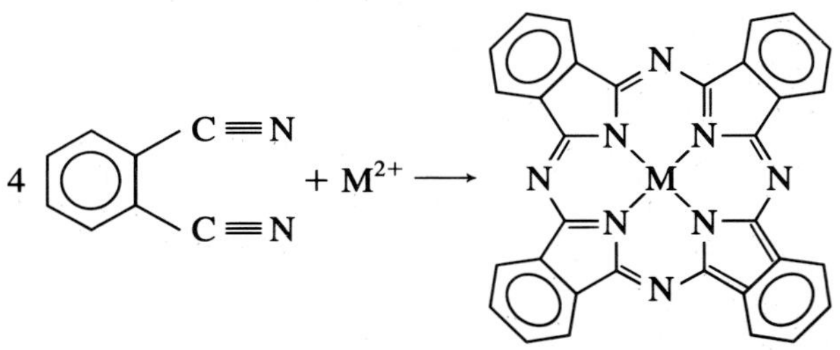

All of these macrocycles are rigidly planar because of the continuous delocalized pi system, although the corrin ring has a small saturated region that allows minor flexibility. Some saturated—and therefore flexible—macrocycles are also good ligands, for instance the crown ethers (see Fig. 7.9 and the associated discussion). Besides the O-donor crown ethers, a series of S-donor macrocycles like the tetrathia compound in Fig. 12.17 have been prepared. The S-donor macrocycles are good ligands; their spectroscopic behavior mimics that of the copper blue proteins. In succeeding chapters we shall return to macrocycle complexes and their reactions, which are among the most interesting in current inorganic chemistry.

12.6 ISOMERISM IN METAL COMPLEXES

The concept of isomerism is familiar from organic chemistry, where both geometric isomers and optical isomers are important to organic reactivity. (Geometric isomers are designated by the prefixes *Z-/E-* or *cis-/trans-;* the latter terms are used by inorganic chemists.) In inorganic chemistry, both types of isomerism are found in transition-metal complexes, along with several others that have no direct analogs in organic compounds.

Considering geometric isomers first, we have already seen that for higher coordination numbers, it is usually fairly easy to interchange two coordination geometries. In particular, five-coordinate complexes can frequently be either trigonal bipyramidal or square pyramidal, depending on the packing forces in their crystal lattice. The chelating ligand *dpe* (see Fig. 12.16) forms the five-coordinate complex $Co(dpe)_2Cl^+$. The salt of this cation with the $SnCl_3^-$ anion crystallizes from hot butanol as a red crystal with the Co atom in square pyramidal coordination, but crystallizes from cold chlorobenzene-ethanol solvent as a green crystal with the Co atom in trigonal bipyramidal coordination. One must conclude that these two forms constitute a pair of geometric isomers. Similarly, the eight-coordinate ZrF_8^{4-} ion is a square antiprism in the crystal $\{Cu(H_2O)_6\}_2ZrF_8$, but a dodecahedron in the crystal $Li_6(BeF_4)(ZrF_8)$. Such complexes are said to be *polytopal*. The mechanisms for the polytopal rearrangements are of some interest.

In a more conventional sense, the kinetically inert octahedral complexes of Cr^{3+} and Co^{3+} and square-planar complexes of Pt^{2+} have been extensively studied for the chemical distinctions between their geometric isomers ever since Alfred Werner. For octahedral complexes with two different ligand atoms A and B, geometric isomers are possible for the formulas MA_2B_4 and MA_3B_3. In the first case, the isomers are called *cis-* and *trans-*, as in Fig. 12.18. In the second case they are called facial (*fac-*) and meridional (*mer-*). Square-planar MA_2B_2 complexes have only the *cis-* and *trans-* possibilities. Clearly, if more than two types of ligand are present, the number of isomers increases. Some effort has gone into devising topological systems for predicting the possible number of isomers.

Chelating ligands generally do not have a large enough "bite" to span the

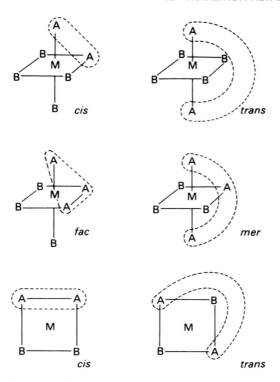

Figure 12.18   Geometric isomers for octahedral and square-planar complexes.

trans- positions on a complex if there are only two donor atoms. However, for a complex such as $Co(en)_2Br_2^+$, *cis-* and *trans-* isomers are still possible because the unidentate ligands can occupy the *trans-* positions:

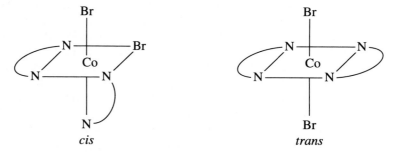

If we make an octahedral complex with three two-coordinate chelating ligands, $M(A_2)_3$, it may seem that no isomerism is possible for the same reason that none is possible in MA_6. However, the chelate complex can be assembled with a left- or right-handed twist:

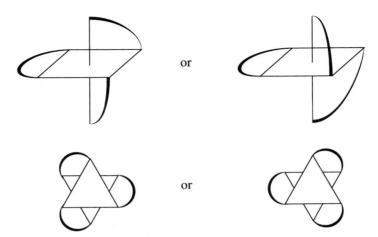

The second set of drawings perhaps makes it more obvious than the first that these are a pair of optical isomers. The traditional statement of the symmetry condition for optical activity or chirality is that there be no mirror plane or center of symmetry in the molecule. The second pair of drawings have the mirror-image relationship.

The ligands need not be chelate ligands to meet the symmetry requirements for optical activity. For example, the *fac-* isomer of the octahedral complex MA_3BCD should be optically active, since the B, C, and D ligands on their triangular face of the octahedron define a chirality. Similarly, certain geometric isomers of $MA_2B_2C_2$, MA_2B_2CD, MA_2BCDE, and MABCDEF should be optically active. However, the resolution of active complexes into their enantiomers is, practically speaking, limited to chelates because they are more stable and their geometric isomers are more specific. To resolve enantiomers, they must be chemically combined with another optically active substance to yield diastereomers—isomers in which a bonding sequence is different, so that the physical properties of the two diastereomers will be different. For example, the enantiomers of the $Co(edta)^-$ complex (Fig. 12.19) were resolved by fractional crystallization of the strychnine-cation salt of the complex. Optically active cations such as those of strychnine, brucine, and morphine are often used to resolve anionic transition-metal complexes, while anions such as tartrate or α-bromocamphor sulfonate are used with cationic complexes. Neutral complexes are more difficult. Selective adsorption on optically active surfaces such as quartz is sometimes effective.

The absolute configuration of an optical isomer can be determined from the manner in which the chiral (helical) molecule interacts with the two helical components of plane-polarized light. In a simple polarimeter, where no optical absorption of the incoming polarized light occurs, this takes the form of a net rotation of the plane of polarization because the refractive indices of the molecule are different for left-handed and right-handed components of the radiation. However, if the plane-polarized light is at a frequency absorbed by the chiral atom—such as the frequency

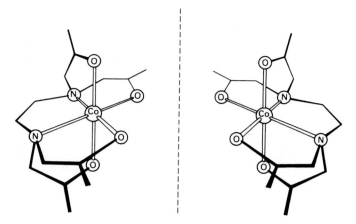

Figure 12.19   Enantiomers of Co(edta)⁻ complex.

of a $d–d$ transition for a complex with an optically active metal atom—the molar absorptivities ϵ_l and ϵ_r for left- and right-circular-polarized light will be different. This is the *circular dichroism* phenomenon (CD), an example of which is shown in Fig. 12.20. The figure also shows the accompanying difference in refractive indices, known as *optical rotatory dispersion* (ORD), which is essentially the first derivative of the CD curve with respect to frequency. Because a given complex can often have a number of $d–d$ transitions, as we have seen, and because the symmetries of excited states can differ, CD–ORD spectra can become quite complex. The sign of the ORD effect (the *Cotton effect*) does not yield the absolute configuration from first principles, but it is the same for peaks of the same electronic origin in complexes of the same configuration. Therefore, it is only necessary to have the absolute configuration of one complex to be able to derive those of others with similar ligands from CD–ORD data. Ordinary x-ray diffraction

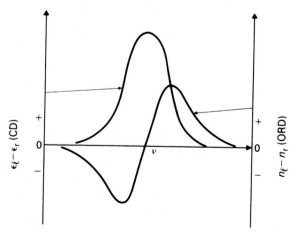

Figure 12.20   The CD and ORD effects.

procedures cannot establish absolute configurations. However, Bijvoet has determined the absolute configuration of $(+)-Co(en)_3^{3+}$ using the anomalous dispersion of x-rays near an x-ray absorption edge (a process physically similar to ORD). Thus, that complex ion can be used as a reference in CD–ORD determinations of other complexes with similar ligands and overall geometry.

Thus far, we have limited ourselves to forms of isomerisms in which all atom–atom bonds are the same. There are several interesting forms of coordination isomerism, however, in which bond sequences differ. The most widely studied form is called *linkage isomerism:* Two linkage isomers differ in that the same ligand molecule is coordinated to the same metal atom through different donor atoms. The oldest recognized example of linkage isomerism involves the nitrite ion, which as we have already noted can donate either through the N atom or through the O atom. Werner and Jørgensen prepared the two compounds $[Co(NH_3)_5$ $NO_2]Cl_2$ and $[Co(NH_3)_5ONO]Cl_2$, which differ in exactly this respect. Ligands for which this is possible are said to be *ambidentate*. Other ambidentate ligands include SO_3^{2-} and SO_2, dmso, thiourea, CN^-, and cyanates OCN^-, SCN^-, and $SeCN^-$. For example, the complexes $Rh(PPh_3)_3NCO$ and $Rh(PPh_3)_3OCN$ are both known; so are $Pd(bipy)(NCS)_2$ and $Pd(bipy)(SCN)_2$ [as well as $Pd(Me_2bipy)(NCS)(SCN)$]; $Ru(NH_3)_5NCS^{2+}$ and $Ru(NH_3)_5SCN^{2+}$; *cis*-$Co(dmg)_2(H_2O)(NCSe)$ and *trans*-$Co(dmg)_2(H_2O)(SeCN)$; and $Ni(NCS)_4^{2-}$ and $Pd(SCN)_4^{2-}$.

We will concentrate briefly on the thiocyanates, which are the most thoroughly studied cases. Several factors seem to influence the ligand's choice of donor atom. For one thing, the steric requirements of the two ends seem to be different. The best shorthand description of the ligand bonding seems to be $N{\equiv}C—S^-$, which implies sp hybridization on the N and requires linear M—N—C geometry, but which also implies sp^3 hybridization on the S and requires bent M—S—C geometry. Experimental evidence shows that most N-donor thiocyanate complexes have M—N—C angles near 180° and always above 160°, whereas most S-donor complexes have M—S—C angles around 100°. Since the linear geometry of M—NCS is sterically less demanding than that of the bent M—SCN, the steric bulk of other ligands can force the complex to coordinate the N end of the NCS⁻ ligand. The other ligands, obviously, also contribute electronic effects. However, it is not obvious exactly what those effects are. The N atom is essentially a sigma donor only, whereas the S atom can also be a pi donor (placing —SCN, as opposed to —NCS, near the bottom of the spectrochemical series). The hardness or softness of the metal as an acid and the hardness or softness of other ligand bases already present on the metal can both influence the pi-acceptor capability of the metal atom. Unfortunately, there are cases like $[Co(CN)_5SCN]^{3-}$ and $[Co(NH_3)_5$ $NCS]^{2+}$ that suggest that soft ligands elsewhere in the complex favor the soft (S) end of NCS⁻ and hard ligands favor the hard (N) end, but also cases like $[Pd(NH_3)_2(SCN)_2]$ and $[Pd(PEt_3)_2(NCS)_2]$ that suggest just the opposite. There are even solvent effects: Other things being equal (a condition notoriously difficult to enforce), the hard–hard and soft–soft condition (symbiosis) will be favored by solvents with high dipole moments and dielectric constants, but will be opposed by less polar solvents. Sorting out these effects is an interesting ongoing process.

Table 12.9 Effects on Donor-Atom Choice for the Ambidentate NCS⁻ Ligand

	Bulky ligands favor	Hard-base ligands (σ-only) favor	Soft-base ligands (π acceptors) favor	Polar solvents favor	Nonpolar solvents favor
Soft-acid metal	—NCS	—SCN	—NCS	—SCN	—NCS
Hard-acid metal	—NCS	—NCS	—SCN	—NCS	—SCN

A rather tentative summary appears as Table 12.9, where the effects are presented in their approximate order of importance.

Sulfur dioxide as a ligand offers a remarkable variety of linkage isomers, as Table 12.10 suggests. The S atom is normally the coordinating atom (only the extremely hard acid SbF_5 coordinates the O), but the molecule should be a good pi acceptor because of its electronegative O atoms, and this can occur in different ways. The bonding is presumably different as the geometry changes. In the planar M—SO_2 configuration, pi back-donation from the metal can occur conveniently into the vacant π^* MO of SO_2, as in Fig. 12.21. In pyramidal M—SO_2 complexes, the bond angles at sulfur are usually somewhat closer to the tetrahedral 109°, and it appears that the sulfur has something close to sp^3 hybridization. But since the sulfur atom in free SO_2 has only one pair of nonbonding electrons, it could only be sp^3-hybridized if the metal had contributed enough electron density to build up an additional nonbonding pair on the sulfur. In that case, then, we can imagine that the SO_2 ligand is serving as a Lewis acid overall, rather than a Lewis base, because of its pi-acceptor qualities. In a third mode of bonding, both the S and O atoms appear to be serving as donors even though only a single coordination site on the metal seems to be involved. Figure 12.21 indicates a possible way of thinking about the bonding in this sort of coordination. One lobe of the pi bond between S and O serves as donor, and the SO_2 molecule then serves as a pi acceptor into the antibonding orbital between S and O. It is perhaps significant that

Table 12.10 Metal Complexes with Sulfur Dioxide

Bond geometry	Complex	Mode of bonding
Planar	$Ru(NH_3)_4Cl(SO_2)$ $Ni(PPh_3)_2(SO_2)_2$	S as sigma donor, S–O pi system as pi acceptor
Pyramidal	$Pt(PPh_3)_3(SO_2)$ I—SO_2^-	S predominantly as sigma acceptor, also pi acceptor
Bidentate	$Mo(phen)(CO)_3(SO_2)$ $Mo(py)(PPh_3)_2(CO)_2(SO_2)$	S—O pi system as sigma donor and pi acceptor
O donor	F_5Sb—OSO (only example)	O as sigma donor
Bridging	$Pt_3(PPh_3)_3(SO_2)_3$ $Fe_2(CO)_8(SO_2)$	S as sigma donor and acceptor, also pi acceptor

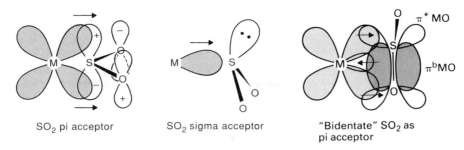

SO$_2$ pi acceptor   SO$_2$ sigma acceptor   "Bidentate" SO$_2$ as pi acceptor

Figure 12.21 Bonding of SO$_2$ ligands.

the S—O distance is significantly greater in the complex than in free SO$_2$, suggesting some reduction in bond strength that can be interpreted as population of the antibonding MO. This pattern of sigma-donation/pi-acceptance should be compared with that shown in Figs. 11.17 and 11.19. The ligand NO also shows a related pattern of ambidentate bonding—it is sometimes linear, and sometimes it is bent relative to the position of the metal atom. We shall take up its bonding in the next chapter.

A final variant of isomerism, *ionization isomerism,* should be considered here. Ionization isomers arise because ligand species can be present in a crystal either in the cation or in the anion, or even sometimes (for neutral ligands) held uncoordinated in the lattice. There are several variations:

1. Two different anions are present, one coordinated and the other present as a counterion. For example, $[Co(NH_3)_5Br]^{2+}SO_4^{2-}$ and $[Co(NH_3)_5SO_4]^+Br^-$ are both known; a slightly more involved case is the pair $[Co(NH_3)_5SO_3]^+NO_3^-$ and $[Co(NH_3)_5NO_2]^{2+}SO_4^{2-}$.

2. An anion and a neutral ligand exchange crystal sites. The best known examples are the hydrated chromium(III) halides

$$[Cr(H_2O)_4Cl_2]Cl\cdot 2H_2O$$

$$[Cr(H_2O)_5Cl]Cl_2\cdot H_2O$$

$$[Cr(H_2O)_6]Cl_3$$

If the cation and the anion are both complexes, one with a neutral ligand and the other with an anion, the ligands can be exchanged: $[Cr(NH_3)_6]^{3+}[Co(CN)_6]^{3-}$ and $[Co(NH_3)_6]^{3+}[Cr(CN)_6]^{3-}$ are both known.

3. A crystal with a cationic complex and an anionic complex in its lattice has the same empirical formula as a neutral compound. Thus

$$[Co(NH_3)_6]^{3+}[Co(NO_2)_6]^{3-}$$

has the same empirical formula as $[Co(NH_3)_3(NO_2)_3]$, and is in that sense an isomer of the latter. (Obviously, other variations with different net ionic charges are also possible.) There is an analogy to polymerization here, and such species are sometimes called *polymerization isomers.*

12.7 STEREOCHEMICALLY NONRIGID SYSTEMS

In discussing isomerism in coordination compounds, we noted that for some coordination numbers two possible geometries are almost equal in stability for at least a few compounds. These polytopal systems can be crystallized in both geometries under the right circumstances, which suggests that it might be possible to tailor the electronic and steric features of such compounds so as to observe a dynamic equilibrium between the two forms. Molecules that interconvert geometries are said to be *stereochemically nonrigid* or *fluxional*. Some very interesting experimental studies have been done on nonrigid systems. Many organometallic molecules are fluxional, and we shall return to that topic in the next chapter.

The ammonia molecule, CN = 3, undergoes the umbrella inversion and may thus be considered nonrigid. However, this is chiefly of interest in explaining why asymmetric amines $R^1R^2R^3N$ are not optically active. From the perspective of coordination chemistry, the simplest nonrigid systems have coordination number 4, for which the two ideal geometries are tetrahedral and square planar. The d-electron configuration most favorable to square-planar coordination is d^8 because of the CFSE arising from the orbital energies shown in Fig. 11.16 and Table 11.1. For the d^8 ion Ni^{2+}, both tetrahedral and square-planar complexes are known. Several complexes have been crystallized in both forms:

$$[Ni(PEtPh_2)_2Br_2] \rightleftharpoons [Ni(PEtPh_2)_2Br_2]$$

Brown square planar   Green tetrahedral

The magnetic properties of these two forms are quite different. Figure 12.22 indicates that the square-planar form should be diamagnetic, whereas the tetrahedral form should have two unpaired electrons. The static magnetic moment, then, pro-

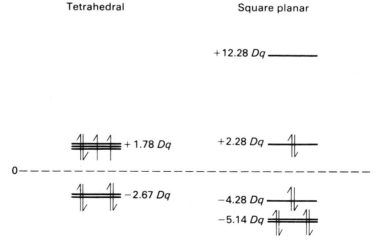

Figure 12.22 Energy levels for four-coordinate Ni^{2+} polytopes.

vides a quick check on which coordination geometry is present in a given crystal.

In solution, the NMR spectrum of the C—H protons in the phosphines provides an equally sensitive test, because the chemical shifts will be enormously exaggerated by contact with the unpaired electrons in the paramagnetic form. Consider the closely related complex [Ni{PMe(p-MeOC$_6$H$_4$)$_2$}$_2$Cl$_2$]. Its diamagnetic (square-planar) form should have three NMR peaks caused by the methyl, methoxy, and phenyl protons; although the phenyl protons are not equivalent, their splitting is not resolved. Its tetrahedral form should show marked differences between phenyl protons *ortho*- and *meta*- to the P atom because of their different degree of interaction with the unpaired Ni d electrons, and the methyl protons (on the P—CH$_3$) will be shifted enormously, perhaps out of the observable range because they are so close to the unpaired electrons. Figure 12.23 shows the NMR spectrum of this complex at three temperatures: At −68°C, both forms are visible in the spectrum, suggesting that the two geometries are frozen into an equilibrium and are not interconverting; at −25°C, kT is high enough relative to the activation energy for interconversion that a given molecule has an average lifetime in each geometry

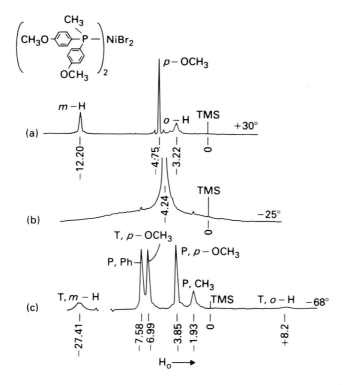

Figure 12.23 Proton NMR spectra (100 MHz) of Ni[(p-MeOC$_6$H$_4$)$_2$MeP]$_2$Cl$_2$ in CDCl$_3$ solution illustrating (a) completely averaged, (b) intermediate, and (c) frozen-out cases. P, planar; T, tetrahedral. [Reprinted with permission from L. H. Pignolet *et al., J. Am. Chem. Soc.* (1970), *92*, 1855. Copyright © 1970, American Chemical Society.]

about equal to the characteristic nmr time scale of $\sim 10^{-4}$ seconds; and at 30°, interconversion is so rapid that nmr reveals only an averaged environment for each proton. This is one of the clearest examples of a four-coordinate complex that is nonrigid or fluxional at room temperature.

Five-coordinate complexes are a good deal less rigid than most four-coordinate complexes are. Except for the d^8 configuration just described, the tetrahedral geometry is so much more stable than the square-planar geometry that usually only the tetrahedral form can be observed. The two ideal geometries for five-coordinate complexes, trigonal bipyramidal and square pyramidal, are so close to each other energetically that the choice between them is fairly subtle regardless of the d-electron configuration of the metal. Furthermore, as Fig. 12.3 suggests, only a very modest deformation with little change in ligand–ligand repulsion is required to interconvert the two shapes. Thus, all five-coordinate complexes might show fluxional behavior, even though the trigonal bipyramidal geometry usually seems to be slightly more stable.

With a little encouragement, this polytopal rearrangement can in fact be observed even in the solid state. We have already noted the five-coordinate $Ni(CN)_5^{3-}$ complex in the $[Co(en)_3][Ni(CN)_5]\cdot 3/2H_2O$ crystal, which for different molecules in the unit cell is both square pyramidal and trigonal bipyramidal. If this crystal is subjected to high pressure (7 kbar) at low temperatures, the $Ni(CN)_5^{3-}$ ions all convert to the square pyramidal geometry, reverting to the original crystal symmetry when the pressure is released.

In solution, NMR provides evidence for fluxional behavior of five-coordinate complexes. Because a trigonal bipyramid is less symmetrical than a tetrahedron or octahedron, the five ligands in such a complex do not all have the same environment. The two axial ligands have three ligand neighbors at 90°, while the three equatorial ligands have two neighbors at 90° and two more at 120°. However, solution NMR spectra of trigonal bipyramidal molecules like PF_5 (^{19}F resonance), $Sb(CH_3)_5$ (^1H resonance), and $Fe(CO)_5$ (^{13}C resonance) reveal that the five ligands always appear identical. This can only be true if there is an efficient mechanism for interchanging ligands from axial to equatorial positions on the trigonal bipyramid. Figure 12.24 (which should be compared to Fig. 12.3) shows the accepted intercon-

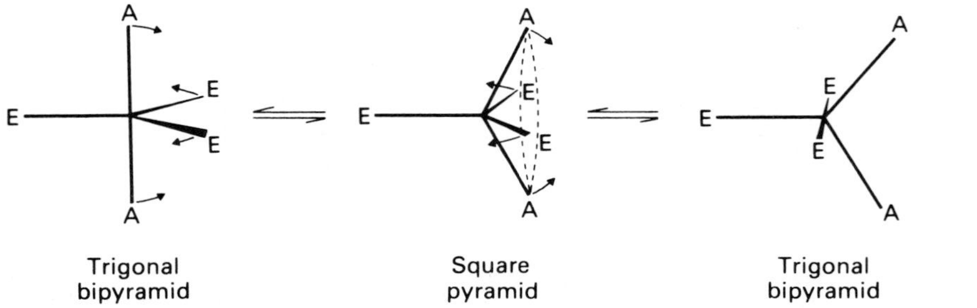

Trigonal   Square   Trigonal
bipyramid   pyramid   bipyramid

Figure 12.24   The Berry mechanism for scrambling axial and equatorial ligands in five-coordinate complexes.

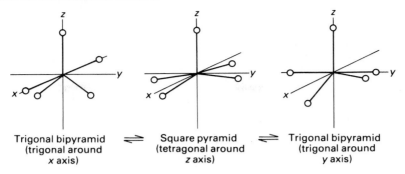

Trigonal bipyramid ⇌ Square pyramid ⇌ Trigonal bipyramid
(trigonal around   (tetragonal around   (trigonal around
x axis)   z axis)   y axis)

Figure 12.25 Coordinate axes for AO overlaps in Berry-mechanism coordination geometries.

version process, the *Berry mechanism:* The two axial ligands bend down toward an open space on the equatorial plane, while the two nearest equatorial ligands spread apart along the equator until they are diametrically opposite and the original axial ligands have a bond angle of only 120° with the central atom. This is now a new trigonal bipyramid, but with the identities of the axial ligands changed. A series of such conversions can completely scramble the five ligands with respect to the axial positions on the complex. Note that the transition state for the Berry mechanism (halfway through the process) is the square-pyramidal geometry. Therefore, even though NMR does not give us direct observation of both geometries in the five-coordinate case, it shows that the polytopal transformation must have occurred.

We can use qualitative MOs for the idealized five-coordinate geometries to explain the very low energy barriers of the Berry mechanism (sometimes called *pseudorotation,* since the effect of the transformation in Fig. 12.24 is to rotate the trigonal bipyramid). Consider the coordinates indicated in Fig. 12.25 for the three geometries in a Berry mechanism. If we consider only sigma overlap as in the MO treatments of Chapter 11, the MO energy levels for these three geometries fall in the order indicated in Fig. 12.26a. Figure 12.26b unites these to suggest that the individual orbital energy changes through a Berry transformation are modest. That is, d-electron energies make only a small contribution to the energy barrier for any Berry mechanism.

By contrast, six-coordinate complexes rarely undergo polytopal rearrangements. The mechanism proposed for scrambling ligands in an octahedral complex, the *trigonal twist,* rotates one triangular face of the octahedron by 120° relative to the parallel opposite face. As Fig. 12.27 indicates, this requires a trigonal prismatic geometry as the intermediate. The energy levels in Fig. 12.5 suggested in the earlier discussion that the change in d-electron energies associated with this geometric change is substantially unfavorable. This accounts for the high barrier to the trigonal twist mechanism and the fact that it is rarely observed. The octahedron is just too stable.

However, ligand–ligand steric repulsion in the trigonal prism also contributes to the energy barrier, and complexes involving hydrides as ligands can minimize this contribution. NMR studies on the complex $[FeH_2\{P(OEt)_2Ph\}_4]$, which can

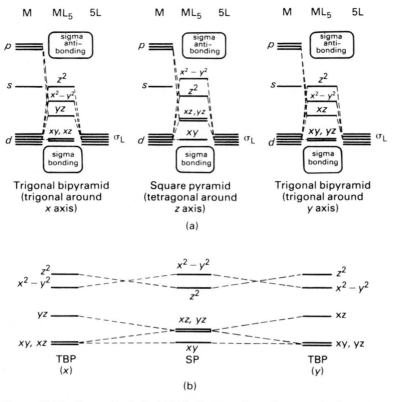

Figure 12.26   Energy levels for MOs in five-coordinate Berry-mechanism geometries.

exist as both *cis*- and *trans*- octahedral isomers, show that at −50°C both isomers exist in solution, with the rearrangement mechanism frozen out. At about 0° the *cis*- and *trans*- peaks have coalesced, indicating that the polytopal rearrangement occurs and that the characteristic lifetime for each isomer is roughly equal to the NMR time scale. At room temperature and above, only a single signal with aver-

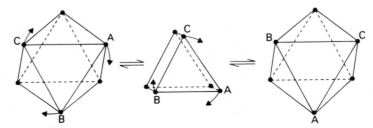

Figure 12.27   The trigonal-twist mechanism for scrambling ligands in six-coordinate geometries.

aged chemical shift and coupling constants is seen. It is not clear that the trigonal-twist mechanism is at work, but some mechanism is permitting *cis-* $\rightleftharpoons$ *trans-* isomerization, presumably through a polytopal rearrangement.

Complexes with coordination numbers higher than 6 are usually very flexible. Figures 12.8 and 12.9 have shown that, as for the five-coordinate systems, the polytopal forms for seven- and eight-coordinate complexes are related by only very subtle rearrangements that presumably require no great changes in *d*-electron energies and only modest changes in ligand–ligand repulsion energies. The axial and equatorial ligand positions are distinct by symmetry for a pentagonal bipyramid just as they are for a trigonal bipyramid, but ReF_7 (which is a pentagonal bipyramid in crystals) shows only one kind of ^{19}F NMR resonance in solution. We therefore assume that the molecule is fluxional and that the ligands are being scrambled by some mechanism analogous to the Berry mechanisms for five-coordinate complexes.

PROBLEMS _____

A. DESCRIPTIVE

A1. The stability constants for the first-row M^{2+} complexes of the strong pi-acceptor ligands *o*-phenanthroline and bipyridyl (see Fig. 12.16) follow the sequence Mn < Fe > Co < Ni > Cu > Zn, which is significantly different from the Irving–Williams series. Why is $Fe(L_2)_3^{2+}$ more stable than $Co(L_2)_3^{2+}$? Why is $Ni(L_2)_3^{2+}$ more stable than $Cu(L_2)_3^{2+}$?

A2. Gustav Magnus prepared a red "double salt" between KCl and $PtCl_2$. Reducing a one-gram sample of the double salt with H_2 yielded 0.825 g of a black solid mixture. When this mixture was extracted with water, the remaining black solid (platinum metal) weighed 0.467 g. What metal complex had he prepared? How did his analytical procedure work?

A3. E. A. Hadow [*J. Chem. Soc.* (1866), *19*, 345] prepared a platinum(IV)–ammonia complex later characterized as $Cl_2OPt_2(H_5N_2)_4 \cdot 4HCl \cdot H_2O$. Reformulate this compound using modern structural ideas about platinum complexes.

A4. Werner proposed octahedral geometry for the complexes MX_6, MX_5Y, MX_4Y_2, and so on because MX_4Y_2 complexes had two isomers, *cis-* and *trans-*. How many isomers would exist if the MX_4Y_2 complex were a planar hexagon? If it were a trigonal prism? Sketch each set.

A5. A 1791 translation of J.-A. Chaptal's *Elements of Chemistry* includes the following discussion. He has previously dissolved copper metal in "the sulphurick acid." "Ammoniack likewise precipitates the copper in a whitish blue; but the precipitate is dissolved nearly the same instant that it is formed; and the result is a solution of a beautiful blue color, known by the name of Aquae Celestis." What's going on here? Why should the final solution have a stronger blue color than the initial one?

A6. A century later (see Problem A5) the situation had not improved much. The 1892 edition of Remsen's *Inorganic Chemistry* notes the behavior Chaptal described, and adds:

When heated the salt loses water and ammonia, until it has the composition $CuSO_4 \cdot 2NH_3$. This is probably analogous to the ammonia compound of cupric chloride, $CuCl_2 \cdot 2NH_3$, and may be regarded as having a similar constitution, that is, to be ammonium sulphate, in which two hydrogen atoms have been replaced by an atom of bivalent copper, as expressed in the formula

$$
\begin{array}{ccc}
 & NH_3 & \\
 & \diagup \quad \diagdown & \\
SO_4 & & Cu \\
 & \diagdown \quad \diagup & \\
 & NH_3 &
\end{array}
$$

It is a curious and interesting, though at present inexplicable, fact, that anhydrous copper sulphate combines with five molecules of ammonia just as it does with five molecules of water, and that by lying in moist air the molecules of ammonia in the compound are successively replaced by water, so that the following series of compounds is formed: $CuSO_4 \cdot 5NH_3$; $CuSO_4 \cdot 4NH_3 \cdot H_2O$; $CuSO_4 \cdot 3NH_3 \cdot 2H_2O$; $CuSO_4 \cdot 2NH_3 \cdot 3H_2O$; $CuSO_4 \cdot NH_3 \cdot 4H_2O$; $CuSO_4 \cdot 5H_2O$.

Is there any modern sense in which Remsen's formulation of the diammine could be correct? What's going on in the final sequence of compounds he mentions?

A7. Set up AO overlaps and from them derive a qualitative sigma-only MO energy-level diagram for planar ML_3 complexes. Assume that the three ligands are in the xy plane and that d_{xy} and $d_{x^2-y^2}$ net overlaps are equivalent. Include $3d$, $4s$, and $4p$ metal orbitals. Compare your energy levels with the crystal-field energies given in Table 11.2.

A8. What coordination geometry would you expect for the ion $NbOF_6^{3-}$?

A9. Metal carbonyl compounds (which involve the ligand CO) almost invariably obey the 18-electron rule. On the other hand, cyanide complexes can have almost any number of valence-shell electrons even though CN^- is isoelectronic with CO. (Consider, for instance, the five cyanomanganates listed at the beginning of the chapter.) Why should these two ligands differ in their electronic requirements?

A10. Why should the $M—OH_2$ bonds in Table 12.4 be generally weaker than the corresponding $M—CN$ bonds?

A11. Dimethylsulfoxide complexes cobalt(II) chloride to give the empirical formula $CoCl_2 \cdot 3dmso$. Suggest a structure with more conventional coordination numbers for this complex.

A12. Assuming that the metal-phosphorus IR stretch frequency is roughly proportional to the bond strength, should PF_3 complexes or PEt_3 complexes yield higher $\nu_{M–P}$?

A13. Suggest probable chlorovanadate complex formulas VCl_n^{q-} by comparison with the fluorovanadates in Table 12.8. Keep in mind the differences between F and Cl.

A14. Sketch all the possible isomers for $\{Co(en)_2(NO_2)Cl\}^+$. Which should be optically active?

B. NUMERICAL

B1. Show that for $d^3 - d^9$ low-spin complexes, CFSE is less favorable for trigonal prismatic geometry than for octahedral geometry.

B2. In trigonal prismatic geometry, complexes of dithiolate ligands seem to have some S—S bonding. Is this because the sulfur atoms are closer together in a trigonal prism than in an octahedron? Consider an octahedron made from seven spheres all of radius R, the outer six tangent to the central sphere. Calculate the separation between nearest-neighbor ligand-sphere surfaces in units of R. Do the same for a trigonal prism with square side faces. In which geometry are the ligand spheres closer? Explain.

B3. Calculate ΔH^0 for dissolving one mole of anhydrous $CoCl_2$ in aqueous ammonia to form $Co(NH_3)_6^{2+}$. Comment on the steps involved, particularly on any approximations. Data are in Table 12.4 and Chapters 6 and 7. Note also that $\Delta H_{soln}(NH_3) = -30.5$ kJ/mol and the diameter of an ammonia molecule is 3.35 Å.

B4. Calculate K_{eq} for the replacement of all six NH_3 molecules in $Ni(NH_3)_6^{2+}$ by three *en* assuming ΔH is zero and the entropy effect is the only driving force.

B5. Calculate the activation energy in kJ/mol for a dissociative substitution mechanism involving the $Co(NH_3)_6^{3+}$ ion. Assume that the ion is low-spin and use data from Table 12.6 and Chapter 11.

B6. Although bonding in high-oxidation-state compounds is certainly not ionic, some interesting and useful stability estimates can be made from ionic Born–Haber cycles. Calculate ΔH on this basis for the reaction $VX_3(s) + X_2(g) \rightarrow VX_5(s)$, where $X = F$ and I (taking the solids to be ionic). What do your results imply for the stabilization of high oxidation states in halide complexes?

C. EXTENDED REFERENCE

C1. What evidence does Werner offer for the planar *cis*- and *trans*- isomeric structures of $Pt(NH_3)_2Cl_2$? [*Z. Anorg. Chem.* (1893), *3*, 267. Translated in G. B. Kauffman, *Classics in Coordination Chemistry,* part I, New York: Dover Publications, 1968, pp. 9–88.]

C2. An interesting case of isomerism is reported by P. J. Peerce *et al.* [*Inorg. Chem.* (1979), *18*, 2593], who synthesized both violet and pink isomers of Cr^{3+} complexed by the ligand TEDTA [thiobis(ethyleneitrilo)-tetraacetic acid, $S(CH_2CH_2N(CH_2COOH)_2)_2)$] in water solution. Keeping the M(edta) structure in mind (Fig. 12.19), suggest isomeric structures that should have different colors.

C3. Suppose that the intermediate-case NMR spectrum of Fig. 12.23b represents a temperature at which the half-life for the first-order planar-tetrahedral interconversion of NiL_2Cl_2 is equal to the NMR time constant of 10^{-4} sec. Calculate the rate constant for the interconversion. Assume an Arrhenius preexponential factor (collision frequency) of 7×10^{10} sec^{-1} (a fairly typical value for solution reactions), and calculate the value of the activation energy for the interconversion. Compare your value to that suggested in L. H. Pignolet *et al.*, *J. Am. Chem. Soc.* (1970), *92*, 1855.

C4. Some of the low-coordination-number complexes mentioned in the chapter undergo substitution reactions that increase the metal coordination number in a predictable way. $Cr(N^iPr_2)_3NO$, CN = 4, reacts with *t*-butyl alcohol to give the four-coordinate $Cr(O^tBu)_3NO$, but reacts with ethyl alcohol to give the six-coordinate polymer $[Cr(OEt)_3]_n$. With isopropyl alcohol, the product analyzes as $Cr(O^iPr)_3NO$, but shows a cryoscopic molecular weight of slightly over 500 g/mol. IR shows a N≡O stretch frequency typical of a terminal—as opposed to bridging—NO group. The proton NMR spectrum shows *two* septets from the —CHR_2 proton, with area ratios 2:1. Suggest a structure for the isopropoxide product. Above 80°C, the NMR spectrum coalesces to a single septet and doublet. What is happening? [D. C. Bradley *et al.*, *Inorg. Chem.* (1980), *19*, 3010.]

C5. S. J. Young *et al.* [*Organometallics* 1985, *4*, 1432] report the interesting compound $(cp)_2(Cl)Zr-CH_2-PMe_2$, in which the Zr is coordinatively unsaturated and the P atom has a nonbonding pair of electrons, yet they do not interact. What is the valence electron count on the Zr atom? What coordination geometry should be expected for the molecule? Why doesn't the P atom donate to the Zr?

13

Transition-Metal Covalent Compounds: Organometallic and Cluster Molecules

Thus far, the discussion of transition-metal compounds has emphasized the chemistry of "ionic" compounds and molecules in which the metal atom or ion is clearly a Lewis acid, accepting electrons from a ligand donor. This is classic coordination chemistry. However, there is a large and increasingly important area of transition-metal chemistry in which the bonding must be considered nearly or completely covalent, even if the molecule is still viewed as a donor−acceptor compound; for instance, there are molecules in which benzene is a ligand for a metal atom. These compounds tend to occur with only soft-base ligands, but there are some intermediate cases such as $(C_6H_6)Ru(H_2O)_3^{2+}$. In this chapter we shall survey the various kinds of molecules that fall into the covalent soft-base category, emphasizing their structural properties and bonding. Some of them are quite important because of their reaction mechanisms and catalytic properties, matters we shall discuss in Chapter 14.

13.1 CLASSES OF COVALENT COMPOUNDS

The simplest view of covalent bonding is that it occurs when neighboring bonded atoms have nearly the same electronegativity. Figure 1.17, which gives the electronegativities of the elements in periodic-table format, indicates that most transition metals have very nearly the same electronegativities as boron, carbon, the group V elements below nitrogen, and the group VI elements below oxygen. An enormous number of compounds have been prepared involving bonds between transition metals and one or several of these elements, particularly carbon (metal carbonyls and organometallics) and to a lesser extent boron (metallaboranes) and phosphorus (phosphine complexes). We shall consider all of these in this chapter, including the phosphine ligands along with the organometallics since both are frequently found in organometallic molecules.

A somewhat newer, extensively studied category of covalent transition-metal compounds involves metal–metal bonds in binuclear complexes and in metal-atom clusters. These compounds are interesting in part because some cluster compounds may be effective homogeneous catalysts for reactions now catalyzed—perhaps only poorly—by heterogeneous metal surfaces. We shall take up reaction mechanisms and the catalytic activity of these compounds in Chapter 15 and some important bioinorganic metal clusters in Chapter 16. In preparation we shall consider here the bonding and structure of these compounds.

Finally, we shall consider heteroboranes in the light of our understanding of metal-cluster molecules. We shall also discuss an interesting and useful (though not infallible) rule for the electronic structure of metal clusters and boranes alike analogous to the 18-electron rule for carbonyls. This rule provides a foundation for understanding the structure of all the types of molecule to be examined in this chapter. We can build on that rule, on molecular orbital extensions of it, and on packing symmetry considerations to begin to understand bonding in very large clusters. The possibility of a unified theoretical treatment of the bonding in these compounds is rapidly emerging.

The organometallic molecules can be subdivided into three categories, based on the type of bonding present between the carbon and metal atoms. The simplest group involves metal alkyls and aryls in which the metal–carbon bond is only a conventional electron-pair sigma bond. These are quite stable thermodynamically in the sense that the metal–carbon bond energy is substantial [Ti—C = 264 kJ/mol in $Ti(CH_2Ph)_4$]. However, there are several kinetically facile mechanisms for their decomposition to even more stable products unless the metal atoms are protected by the right kind of ligands. In consequence, these species were not prepared until the 1950s. The most important decomposition mechanism for alkyls involves the elimination of a hydrogen atom from the carbon β to the metal and the formation of a double bond within the alkyl group. Therefore, the most stable ligands are those that either have no β hydrogens [such as —CH_3, —CH_2C $(CH_3)_3$, and —$CH_2Si(CH_3)_3$] or resist the formation of a double bond at a bridgehead carbon (such as adamantyl $-\langle\rangle$). Most of these are bulky ligands, so coordination numbers tend to be only 3 or 4 for MR_n alkyls with identical (*homoleptic*) ligands. However, CN = 6 is known for methyls in the neutral compound WMe_6, and even CN = 8 for the anion WMe_8^{2-}.

As we shall see, a much larger group of compounds involves M—C bonds in which the ligand molecule serves as both a sigma donor and pi acceptor as in Figs. 11.17 and 11.19. Such electron flow yields little net charge transfer; therefore, the bonds are nearly covalent. Both the CO molecule and the PR_3 molecules have this capability, as Fig. 11.19 suggests. The next section considers some of the many carbonyls and carbonyl–phosphine complexes that have been prepared. Since their stoichiometry is dominated by the 18-electron rule rather than by steric considerations, some unusual coordination geometries are observed—particularly for polynuclear systems.

In the third type of organometallic compound, the ligand serves as a pi acceptor in much the same manner as CO does. In this case, however, the electrons donated by the ligand are not sigma nonbonding electrons, but are pi-bonding electrons as in Fig. 6.2. Again, because the ligand serves as a pi acceptor as well as a sigma donor, there is very little net electron transfer and the bonding must be considered essentially covalent. Bonds of this type are formed not only by isolated double bonds in olefins, but also by extended pi systems. Therefore, a single organic ligand can donate 2, 3, 4, 5, or 6 electrons from its pi system to a single metal atom, depending on the size of that system. (Geometric constraints usually prevent a single organic ligand from donating more than 6 pi electrons.) We shall survey these in a later section of the chapter. Some striking geometries are seen, and the facile rearrangements of the bonding electrons make some very important reactions possible.

13.2 ORGANOMETALLIC SYSTEMS: METAL CARBONYLS

The first metal carbonyl was synthesized in 1890 by Ludwig Mond, who produced the volatile liquid $Ni(CO)_4$ by heating nickel-metal powder under a carbon monoxide atmosphere:

$$Ni(s) + 4CO(g) \xrightarrow{100°C} Ni(CO)_4 (l)$$

(For those who wonder how he could ever have thought to try such a reaction: He had noticed that hot gas mixtures containing CO corroded nickel valves.) $Fe(CO)_5$ was synthesized shortly thereafter, and inorganic chemists have been preparing novel carbonyl compounds ever since. The compounds are often remarkably stable; $Ni(CO)_4$ can be prepared in aqueous solution by bubbling CO through an alkaline Ni^{2+} solution containing ethyl mercaptan. The known metal carbonyls are summarized in Table 13.1. Most of them can be made from the finely divided metal plus CO, but it is often more convenient to reduce the metal from a positive oxidation state:

$$6RuCl_3 + 9Zn + 24CO \rightarrow 2Ru_3(CO)_{12} + 9ZnCl_2$$
$$2CoI_2 + 4Cu + 8CO \rightarrow Co_2(CO)_8 + 4CuI$$
$$OsO_4 + 9CO \rightarrow Os(CO)_5 + 4CO_2$$

All carbonyls are extremely toxic, and many are quite sensitive to air oxidation, so they are normally handled in sealed inert-atmosphere systems.

The sigma-donor/pi-acceptor capability of carbon monoxide produces the large Δ_o and low-energy t_{2g} orbitals for octahedral carbonyls shown in Fig. 12.10c. Therefore, the presence of 18 electrons is strongly favored. Similar arguments apply to other bonding geometries, and the 18-electron rule strongly influences the formation

Table I3.I   Neutral Transition-Metal Carbonyl Molecules

V	Cr	Mn	Fe	Co	Ni
$V(CO)_6$	$Cr(CO)_6$	$Mn_2(CO)_{10}$	$Fe(CO)_5$	$Co_2(CO)_8$	$Ni(CO)_4$
			$Fe_2(CO)_9$	$Co_4(CO)_{12}$	
			$Fe_3(CO)_{12}$	$Co_6(CO)_{16}$	
	Mo	**Tc**	**Ru**	**Rh**	
	$Mo(CO)_6$	$Te_2(CO)_{10}$	$Ru(CO)_5$	$Rh_4(CO)_{12}$	
			$Ru_3(CO)_{12}$	$Rh_6(CO)_{16}$	
	W	**Re**	**Os**	**Ir**	
	$W(CO)_6$	$Re_2(CO)_{10}$	$Os(CO)_5$	$Ir_4(CO)_{12}$	
			$Os_2(CO)_9$	$Ir_6(CO)_{16}$	
			$Os_3(CO)_{12}$		
			$Os_5(CO)_{16}$		
			$Os_6(CO)_{18}$		
			$Os_7(CO)_{21}$		
			$Os_8(CO)_{23}$		

of all the species shown in the table. It is not absolutely sacred, however: Note the presence of $V(CO)_6$, a 17-electron molecule.

The strong influence of the 18-electron rule and the general stability of CO as a neutral molecule mean that the stoichiometry of carbonyls is determined by neither formal oxidation states nor lattice-geometry considerations. Instead, it is determined by the possible number of bonding molecular orbitals that can be formed relative to the number of electrons donated by each ligand. That is, the existence of $Fe(CO)_5$ implies nothing about any particular stability of coordination number 5 for iron—and certainly nothing about any hypothetical 5+ oxidation state. This is just the way the 18-electron rule works out for iron with its eight electrons and the two-electron donor CO. An immediate consequence of the bonding properties of carbonyls and other organometallics is that octahedral geometries are no longer extremely common, as they are for other transition-metal donor–acceptor complexes. The MX_6 octahedral arrangement appears primarily for the group VIb metals and two-electron-donor ligands, as in $Cr(CO)_6$, because that arrangement meets the 18-electron condition. However, the octahedral bond geometry also appears in many metal-cluster compounds that we shall examine in a later section.

Infrared spectroscopy is a particularly powerful tool for investigating carbonyl bonding. A free CO molecule approximates a triple bond and, by virtue of its asymmetric polar nature, has an IR-active stretching vibration at 2143.2 cm^{-1} (for $^{12}C^{16}O$). If transition-metal carbonyls are stable because of the pi-donor quality of the metal d electrons (as has been suggested), then the CO molecule must accept the pi electrons into its π^* MO. That, in turn, should reduce the total bond order of the C—O bond from 3 to some smaller number, and thereby reduce the vibrational frequency of the C—O stretch. This is exactly the experimental result: For $Fe(CO)_5$ there are three IR absorption peaks that can be attributed to C=O stretch vibrations (since there are three possible symmetries for the vibrations of the set of five CO ligands within the molecule). These three peaks occur at 2116 cm^{-1}, 2030 cm^{-1}, and 1989 cm^{-1}. The IR absorption frequencies are similar for the other simple neutral carbonyls containing only one metal atom, regardless of the stoichiometry or symmetry of the molecule. This supports our general view of the bonding in these compounds.

In Chapter 12 it was noted that carbonyl cations and anions can be prepared. Net charges on carbonyl molecules have a strong effect on vibrational frequencies, as is suggested by the two isoelectronic series below (values in cm^{-1}):

$Ti(CO)_6^{2-}$		$V(CO)_6^-$		$Cr(CO)_6$		$Mn(CO)_6^+$
1745–50		2020		2119		2192 cm^{-1}
	$\rightarrow$	1894	$\rightarrow$	2027	$\rightarrow$	2125
		1858		2000		2095
		$Fe(CO)_4^{2-}$		$Co(CO)_4^-$		$Ni(CO)_4$
		1788	$\rightarrow$	2002	$\rightarrow$	2131 cm^{-1}
		1788		1890		2058

In both series, it is clear that the C=O stretch frequencies increase substantially as electrons are drained out of the CO bond toward the metal by the increasing nuclear charge on the metal atom. If the electrons being removed from the COs are in π^* orbitals, as our bonding model suggests, this trend is exactly what we would expect, since removing these electrons actually strengthens the CO bond.

In polynuclear carbonyls (that is, carbonyls having more than one metal atom) two kinds of bonding are possible for the CO ligands. In addition to the terminal COs (like those in mononuclear carbonyls), it is possible to have bridging CO units, which presumably contribute one electron to each metal atom they bridge. Figure 13.1 shows the structures of some of the polynuclear carbonyls from Table

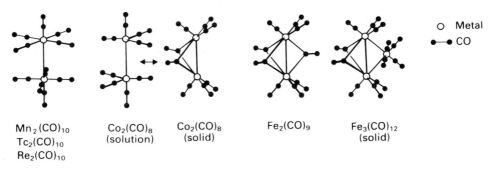

| | o | Metal |
| | ●—● | CO |

$Mn_2(CO)_{10}$ $Co_2(CO)_8$ $Co_2(CO)_8$ $Fe_2(CO)_9$ $Fe_3(CO)_{12}$
$Tc_2(CO)_{10}$ (solution) (solid) (solid)
$Re_2(CO)_{10}$

Figure 13.1 Some polynuclear carbonyl structures.

13.1. Although bridging structures are not necessary, they are quite common. There is some tendency toward roughly octahedral geometries, but it is necessary to postulate metal–metal bonding to account for the fact that all of the compounds shown are diamagnetic. Bridging CO groups have much lower vibrational frequencies than terminal COs, presumably because two metal atoms are donating pi electrons to the CO rather than only one. Thus, for the solid $Co_2(CO)_8$ structure there are five terminal CO stretch peaks at 2075, 2064, 2047, 2035, and 2028 cm^{-1}, but the two C—O stretch peaks for the bridging carbonyls are at 1867 and 1859 cm^{-1}.

The structure of $Co_2(CO)_8$ is markedly different for the solution and solid phases (see Fig. 13.1). These two structures jointly represent an interesting property of metal carbonyls and of organometallic compounds in general: Although the CO groups are fairly strongly bound to the metal atom (125–150 kJ/mol M—C bonds), they are often mobile within the molecule, moving back and forth between terminal and bridging positions. Mononuclear carbonyls are sometimes polytopal or fluxional in the manner described in Chapter 12. $Fe(CO)_5$, for example, has trigonal bipyramidal geometry that should yield magnetically nonequivalent axial and equatorial COs, but its ^{13}C NMR spectrum shows only one line. Presumably this represents a rapid scrambling of the axial and equatorial positions through the Berry mechanism. Polynuclear carbonyls often have different structures in a crystal (by x-ray data) and in solution (by IR data). In addition to the $Co_2(CO)_8$ case, Fig. 13.2 shows two other cases in which CO groups transform their bonding with phase changes. The reasons for these changes are not clear, but it is an important observation that the structures are so flexible and the bonding so mobile.

From the distribution of carbonyls among the transition metals in Table 13.1, we can develop a rationale for the stability of carbonyls beyond the 18-electron rule. First, the metal must have a significant number of d electrons if it is to serve as an adequate pi donor. Accordingly, group IV or V metals (which have only 2 or 3 d electrons) form unstable carbonyls when they can be isolated at all. (This, however, might also be related to the fact that it would require a high coordination number to satisfy the 18-electron rule for these elements.) On the other hand, the d electrons must not be so tightly held that they are not available for pi donation. We remember from Figs. 1.8, 10.2, and the associated discussions that a transition metal's d electrons become much more tightly bound as the atomic number increases across any given row of the periodic table. There are thus few isolable carbonyls beyond Ni/Pd/Pt. The species $Ag(CO)\{B(OTeF_5)_4\}$ shows the metal's very weak pi donor character in its C≡O stretch frequency of 2204 cm^{-1}, an extremely high value indicating little CO π^* population.

The pi-donor requirement also means that the metal cannot be in a high oxidation state, which would also tend to bind the d electrons more tightly. Most carbonyls and other compounds in which CO is a ligand have the metal atom in oxidation states -1, 0, or $+1$ (assuming that CO is electrically neutral). A few such as $Fe(CO)_4^{2-}$ and $Pt(CO)_2Cl_2$, are at -2 and $+2$. Of course, these are unusually low

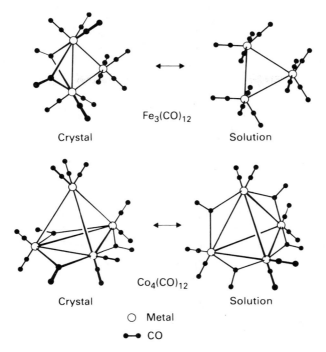

$Fe_3(CO)_{12}$

Crystal

Solution

$Co_4(CO)_{12}$

Crystal

Solution

○ Metal

●—● CO

Figure 13.2 Polytopal forms of metal carbonyls.

oxidation states for metals in general. As Table 10.3 has suggested, carbonyls and other organometallics are the normal host compounds for such low oxidation states. In the right environment, however, the low oxidation states can be quite stable. For instance, gaseous nickel atoms react with CO_2 (not CO!) to yield both NiO and $Ni(CO)_4$—the nickel atom has actually extracted CO out of the very stable CO_2 molecule.

In Chapter 12 we have already seen that for a transition metal with odd atomic number, the 18-electron rule can be satisfied by forming either carbonyl cations such as $Mn(CO)_6^+$ or carbonylate anions such as $Co(CO)_4^-$. The rule can also be satisfied by using ligands other than CO that donate only one electron to the metal–ligand sigma bond. The simplest approach is to allow a metal carbonyl $M(CO)_n$ with 17 electrons to dimerize, forming a metal–metal bond. Each metal atom in the $M_2(CO)_{2n}$ molecule then donates one electron to a sigma bond between the metal atoms, increasing the valence electron total for each metal atom to 18. Thus a $Mn(CO)_5$ unit [with 17 electrons: $7(Mn) + 5 \times 2(CO)$] forms a Mn–Mn bond in $Mn_2(CO)_{10}$. Here, each Mn has 18 electrons [$7(Mn^a) + 5 \times 2(CO) + 1(Mn^b)$]. As Figs. 13.1 and 13.2 indicated, when a metal–metal bond forms, bridging CO groups often form as well. These also may be thought of as donating one electron to each metal. In the solid $(Co_2(CO)_8$ structure (Fig. 13.1),

Table 13.2 Mononuclear and Dinuclear Carbonyl Hydrides and Carbonylate Anions

IVb	Vb	VIb	VIIb	VIII		
$[Ti(CO)_6]^{2-}$	$[V(CO)_6]^-$ $HV(CO)_6$	$[Cr(CO)_5]^{2-}$ $[Cr_2H(CO)_{10}]^-$ $[Cr_2(CO)_{10}]^{2-}$ $[HCr(CO)_5]^-$	$[Mn(CO)_5]^-$ $[Mn_2(CO)_9]^{2-}$ $HMn(CO)_5$ $H_2Mn_2(CO)_9$	$[Fe(CO)_4]^{2-}$ $[HFe(CO)_4]^-$ $[Fe_2(CO)_8]^{2-}$ $[HFe_2(CO)_8]^-$ $H_2Fe(CO)_4$ $H_2Fe_2(CO)_8$	$[Co(CO)_4]^-$ $HCo(CO)_4$	$[Ni_2(CO)_6]^{2-}$ $H_2Ni_2(CO)_6$
$[Zr(CO)_6]^{2-}$	$[Nb(CO)_6]^-$	$[Mo(CO)_5]^{2-}$ $[Mo_2H(CO)_{10}]^-$ $[Mo_2(CO)_{10}]^{2-}$	$[Tc(CO)_5]^-$ $HTc(CO)_5$	$[Ru(CO)_4]^{2-}$	$[Rh(CO)_4]^-$ $HRh(CO)_4$?	(Pd clusters)
	$[Ta(CO)_6]^-$	$[W(CO)_5]^{2-}$ $[W_2(CO)_9]^{4-}$ $[W_2H(CO)_{10}]^-$ $[W_2(CO)_{10}]^{2-}$	$[Re(CO)_5]^-$ $[Re_2(CO)_6H_3]^-$ $HRe(CO)_5$	$[Os(CO)_4]^{2-}$	(Ir clusters)	(Pt clusters)

each Co atom has nine valence electrons of its own, six from the three terminal CO groups, and one more from each bridging CO, for a total of 17. The formation of a Co–Co bond then gives each CO 18 valence-shell electrons.

Metal carbonyl hydrides. Several widely studied one-electron-donor ligands also form classes of mixed carbonyl compounds. Perhaps the most obvious is a hydrogen atom, which forms many *carbonyl hydrides*. In many cases carbonyl hydrides are prepared from carbonylate anions, so one can consider the two species together. Table 13.2 lists the established ions and hydrides. Typically, a carbonylate anion is prepared by reducing a neutral carbonyl molecule with sodium metal in diglyme, tetrahydrofuran, or liquid ammonia. In turn, carbonyl hydrides are often prepared by acidifying carbonylate solutions with H_3PO_4, which has no oxidizing properties. Other preparations involve reduction by $NaBH_4$ and disproportionation in basic aqueous solution to metal cations and carbonylate anions:

$$VCl_3 + 6CO + Na \xrightarrow{\text{Diglyme}} [Na(\text{digly})_2]^+[V(CO)_6]^-$$

$$[Mn(CO)_5]^- + H_3PO_4 \longrightarrow HMn(CO)_5 + H_2PO_4^-$$

$$Cr(CO)_6 + NaBH_4 \longrightarrow Na_2Cr_2(CO)_{10}$$

$$Fe(CO)_5 + OH^- \longrightarrow [HFe_2(CO)_4]^- + Fe^{2+}$$

$$ReH_9^{2+} + CO \longrightarrow [Re_2(CO)_6H_3]^-$$

In the carbonyl hydrides, the hydrogen atom usually occupies an ordinary position in the coordination sphere of the metal atom, though its small size usually allows CO ligands to bend toward it slightly. As might be expected from the fact that they are often prepared from carbonylate anions, they are usually acidic. For $H_2Fe(CO)_4$, $K_1 = 4 \times 10^{-5}$, slightly stronger than acetic acid; for $HCo(CO)_4$, $K_a \simeq 1$—that is, it is essentially leveled by water as a strong acid. The M—H stretch vibration occurs at about $2200-1800$ cm^{-1}. It can be neatly distinguished from the CO stretch in the same region by deuterating the molecule; the doubled mass of the $^2H(D)$ atom lowers the wavenumber of the M—D stretch by 30–50%.

Metal carbonyl halides. Another large category of one-electron donors found in mixed carbonyl complexes is that of the *carbonyl halides,* for example $Mn(CO)_5Br$. If the halogen atoms in these compounds are to be considered one-electron donors, they must be thought of as free radicals X, not as anions X^-. This seems reasonable in view of the preparation reaction

$$Mn_2(CO)_{10} + I_2 \rightarrow 2Mn(CO)_5I$$

However, the carbonyl halides are more commonly prepared by the direct reaction of CO with metal halides:

$$FeI_2 + CO \rightarrow Fe(CO)_4I_2$$

Nearly all elements forming simple carbonyls will form carbonyl halides, usually in the stability order I > Br > Cl. In addition, the electron-rich transition metals

Pd, Pt, and Cu will form the carbonyl halides $[Pd(CO)Cl_2]_2$, $Pt(CO)_2Cl_2$, and $Cu(CO)Br$, even though simple carbonyls are unstable for those elements, which seems to reflect the pi-donor character of the softer halides as ligands. Usually a given compound $M(CO)_nX_m$ can be prepared for X = Cl, Br, and I but not for X = F, presumably because $F \cdot$ is so electronegative and such a hard base. The halogen atoms occupy a normal coordination position on the metal atom, but in polynuclear carbonyl halides the bridging ligand is always the halogen, not the CO.

Metal carbonyl alkyls. The last class of one-electron donors commonly found in mixed carbonyl complexes involves alkyl groups $R \cdot$, which form a conventional sigma bond with the metal but have no pi-acceptor ability. We have already mentioned the transition-metal alkyls such as $W(CH_3)_6$, which tend to be quite unstable not only with respect to oxidation but also with respect to rearrangement within alkyl groups larger than methyl. However, the metal–alkyl bond is stabilized by strong sigma-donor groups or by sigma-donor/pi-acceptor groups such as CO. It is assumed that the facile decomposition mechanisms for alkyls rely on an initial excitation of a sigma-bonding electron to an antibonding orbital. Therefore, if ligands that increase Δ_o, as in Fig. 12.10c, are present they raise the activation energy for the alkyl decomposition and stabilize the metal–alkyl bond. Like the carbonyl hydrides, the alkyl carbonyls are frequently prepared from the carbonylate anion—in this case, by reaction with either an alkyl or acyl halide:

$$RX + Na^+[Mn(CO)_5]^- \longrightarrow NaX + RMn(CO)_5$$

$$RCOCl + Na^+[Mn(CO)_5]^- \longrightarrow NaCl + RCOMn(CO)_5$$

$$RCOMn(CO)_5 \xrightarrow{\text{heat}} RMn(CO)_5 + CO$$

Another useful synthesis involves a carbonyl halide and a Grignard reagent or organolithium compound:

$$C_6F_5Li + Fe(CO)_4I_2 \rightarrow (C_6F_5)Fe(CO)_4I + LiI$$

An interesting reaction occurs between $Co(CO)_4H$ and olefins to produce alkyl–cobalt species:

$$RCH{=}CH_2 + Co(CO)_4H \rightarrow \underset{\underset{\displaystyle Co(CO)_4}{|}}{RCHCH_3}$$

This reaction has important applications in industrial homogeneous catalysis. We shall return to this topic in Chapter 15.

Isonitriles. Besides substituting one-electron donors in transition-metal carbonyls, one can also readily substitute several similar electron-pair donors—that is, sigma-donor ligands like those described in the last chapter—for CO in carbonyls. These yield compounds with very similar properties and, presumably, similar bonding. Perhaps the most closely analogous ligands to CO are the isonitriles, $—C{\equiv}NR$. Isonitrile complexes are prepared in two ways: by direct substitution of the RNC ligand onto the metal (usually in the carbonyl) and by alkylation of cyano complexes:

$$Ni(CO)_4 + 4CNR \rightarrow Ni(CNR)_4 + 4CO$$
$$3Cr(OAc)_2 + 18CNR \rightarrow Cr(CNR)_6 + 2[Cr(CNR)_6](OAc)_3$$
$$Fe(CN)_6^{4-} + 6CH_3I \rightarrow Fe(CNCH_3)_6^{2+} + 6I^-$$

An interesting feature of isonitrile complexes is that the $C\equiv N$ stretch frequency is often raised, not lowered, by coordination to a metal. Since the sigma electrons being donated have some antibonding character within the RNC molecule, we might expect the $C\equiv N$ frequency to increase if pi acceptance by the isonitrile were weak. For alkyl isonitriles, where the organic group has no pi-acceptor capability, the frequency almost always increases significantly. For aryl isocyanides, on the other hand, the organic group increases the pi-acceptor ability of the $C\equiv N$ bond, and the frequency decreases about as much as that of the corresponding carbonyls if the metal is zerovalent. For more highly charged metal ions (poorer donors) the frequency increases, usually less than for the corresponding alkyl isocyanide. The lower electronegativity of the isonitrile N compared to the O in CO means that the isonitrile ligand can serve as a pi donor as well as a pi acceptor, which complicates the task of interpreting sigma and pi interactions by the ligand. Photoelectron spectroscopy suggests that CNR is a weaker sigma donor than CO, perhaps because the C sigma donor orbital is less strongly oriented toward the metal than it is in CO, as well as being a weaker pi acceptor and, to some degree, a pi donor to the metal as well. As a final comparison of M—C bonding in systems analogous to carbonyls, we might note the existence of a $Mo-CF_2$ species, $[(C_5H_5) Mo(CO)_3(CF_2)]^+$. CF_2 has no $C=X$ double bond whose stretch frequency can be monitored to interpret pi acceptor capability, but in the cation the $C\equiv O$ frequencies are much higher than in comparable species with PR_3 in place of CF_2; clearly the CF_2 is draining metal pi electrons from the COs as a very strong pi acceptor.

Phosphines and phosphites. A very large and important class of "carbonyl-equivalent" ligands is that of the group Va donors: phosphines ($:PR_3$), phosphites ($:P(OR)_3$), $:PF_3$, and (to a lesser extent) the equivalent arsenic compounds. In these compounds—though *not* in tertiary amines or NF_3—the group Va atom can both serve as a low-electronegativity sigma donor and accept pi electrons from the metal into vacant *d* orbitals (see Fig. 11.19). Chapter 12 briefly mentioned Group Va ligands, noting that PF_3 is so similar to CO that it will also react directly with nickel metal to form the compound $Ni(PF_3)_4$. In general, increasing the electronegativity of the PX_3 substituents makes the ligand containing phosphorus a weaker sigma donor, but a stronger pi acceptor. It is therefore possible to tailor the sigma/pi bonding qualities of the ligand to achieve specific electronic effects in its transition-metal complexes. PF_3, the strongest pi acceptor, is fully equivalent to CO in a wide variety of compounds, including anions and hydrides. It even forms a few compounds that obviously follow the carbonyl model but have no carbonyl equivalent, such as $Pd(PF_3)_4$ and $Pt(PF_3)_4$. Phosphines, which are softer bases than PF_3, form very strong complexes with low oxidation states of group VIII and later transition metals and substitute freely into zerovalent carbonyls of the earlier groups, since all transition metals are soft acids in the zero oxidation state.

As was true for isonitriles, the aromatic phosphines are better pi acceptors (because of their benzene-ring π^* orbitals) than aliphatic phosphines. Since each PR_3 molecule carries three organic substituents, more flexibility is possible in tailoring pi-acceptor qualities for the ligand. Furthermore, because M—P—C bond angles for M—PR_3 complexes are usually about 120° compared to the 175–180° for the M—C—N angle in isonitriles, it is stereochemically more convenient to make chelating ligands with P sigma-donor atoms. Figure 12.16 has already shown the structures of three commonly used P- and As-donor chelating ligands. In terms, however, of estimating the steric hindrance of adjacent ligands, which is considerable for triphenylphosphine and related molecules, a more useful angle than the M—P—C bond angle is the Tolman *cone angle* illustrated in Fig. 13.3. For PX_3-type ligands, the cone angle ranges from about 180° for triphenyl-

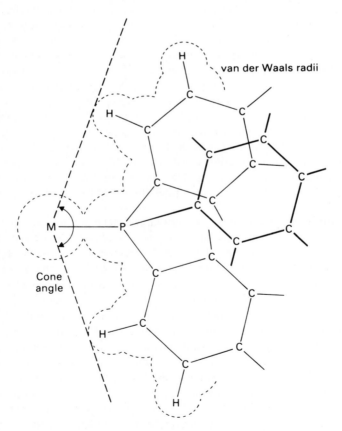

Figure 13.3 Tolman cone angle for PPh_3 ligand.

phosphine to only about 110° for triethylphosphite, a much less sterically demanding molecule.

Carbon dioxide coordination. After an extended discussion of the coordination of carbon monoxide and its analogs, we may digress briefly to consider the possibility of CO_2 coordination. Carbon dioxide complexes with transition metals are unusual and only fairly recently characterized, a contrast to the plentiful and long-known metal carbonyls. Research interest in CO_2 complexes is quite active, however, because of the possibility that an electrocatalytic process using coordinated CO_2 could yield C_1 potential fuel molecules,

$$CO_2 + 5H_2O + 6e^- \rightarrow CH_3OH + 6\,OH^-$$

for example. No such process has yet been demonstrated, but electrocatalytic reduction of CO_2 complexes to formate ion, $HCOO^-$, has; we can expect to see further advances in coming years. Like SO_2, described in Chapter 12, CO_2 can coordinate metals with varied geometry, even though relatively few cases are known. In $RhCl(CO_2)(diars)$ the CO_2 is coordinated only through the C atom; the Rh atom lies in the CO_2 plane and the O–C–O angle is 126°. On the other hand, *trans*-$Mo(CO_2)_2(PMe_3)_4$ has the metal and CO_2 coplanar, but the CO_2 coordinates the Mo through both the C and one O atom. The CO_2 is again bent, with an O–C–O angle of about 130°, and the C and O atoms are nearly the same distance from the Mo (2.02 versus 2.14 Å). A number of research groups are studying the chemistry and electrochemistry of coordinated CO_2, and Chapter 15 will consider some of this chemistry in the context of other known catalytic mechanisms.

13.3 ORGANOMETALLIC SYSTEMS: METAL NITROSYLS

We are used to thinking of electron-pair bonding, so that an atom could bring either one or two electrons to a potential bond. However, we are about to encounter a variety of ligands for transition metals that can contribute more than two electrons to the bonding. The first and in some ways the simplest such bonding is found in compounds with NO, a molecule very similar to carbon monoxide. Among other similarities, it too can coordinate as a pi-acceptor ligand to transition metals. Compounds involving the NO ligand are called *metal nitrosyls*. Only a few binary nitrosyls are known, including $Cr(NO)_4$, $Fe(NO)_4$, and $Co(NO)_3$, but there are many mixed carbonyl–nitrosyls and other complexes involving NO and phosphines or other sigma-donor ligands.

As for carbonyls, the stoichiometry of metal nitrosyls tends to be governed by the 18-electron rule, since the ligand molecule is both a good sigma donor and pi acceptor. However, NO has one more valence electron than CO and, for quick electron bookkeeping, should be regarded as a three-electron donor. The MO energy levels in Fig. 4.6 (which can still be used for the heteronuclear NO molecule, since the VOIPs do not differ greatly) show that NO has the same pair of σ_s^* electrons that CO presumably donates to the metal, but has in addition one π^*

electron to contribute. CO could, in principle, accept four pi electrons from the metal; NO could accept only three. Considering NO as a three-electron donor, we see that $Cr(NO)_4$ satisfies the 18-electron rule: $6(Cr) + 4 \times 3(NO) = 18$. Similarly, mixed carbonyl nitrosyls usually meet the rule as well, as seen in the isoelectronic series $Cr(NO)_4$, $Mn(CO)(NO)_3$, $Fe(CO)_2(NO)_2$, $Co(CO)_3(NO)$, and $Ni(CO)_4$, all of which are known. Permutations that add 3CO and remove 2NO, or vice versa, are also possible: compare $Mn(CO)_4(NO)$, $Fe(CO)_5$, and $Co(NO)_3$ to the previous list.

Most nitrosyl compounds seem to have linear M—N—O bonds (as do carbonyls' M—C—O bonds). By this is meant bond angles at the N atom of 170–180°. However, many well-established examples are known of bent M—N—O bonds with angles of 120–135°. Some effort has therefore gone into understanding the electronic reasons for such a striking geometric difference. For example, in the complex $Mn(NO)(CN)_5^{3-}$ the Mn—N—O angle is 180°, but in $Co(NO)(NH_3)_5^{2+}$ the Co—N—O angle is 119°. Obviously the bonding is different, but how? One proposed rule of thumb is that the NO species in a linear M—N—O arrangement is essentially NO^+, the nitrosyl ion. NO^+ is isoelectronic with CO, and presumably the bonding would be essentially equivalent. On the other hand, bent M—N—O systems would in this model be said to contain NO^-, in which sigma donation occurs either from a filled nonbonding sp^2 hybrid orbital on N or (in the MO equivalent) from a filled π^* MO for the NO^- system. The overlaps for the appropriate orbitals are shown in Fig. 13.4, whereas Table 13.3 lists some examples of both linear and bent nitrosyls. Valence-electron counting shows that the 18-electron

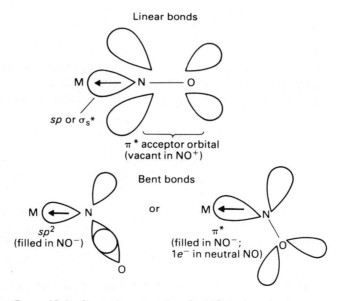

Figure 13.4 Sigma-donor overlaps for NO in nitrosyls.

Table 13.3 The Geometry of Metal Nitrosyls

Linear ($\angle$ M—N—O = 170–180°)		Bent ($\angle$ M—N—O = 120–135°)	
Compound	Valence e^- (NO = $3e^-$ donor)	Compound	Valence e^- (NO = $1e^-$ donor)
CN = 6		**CN = 6**	
$Fe(NO)(CN)_5^{2-}$	18	$Co(NO)(en)_2Cl^+$	18
$Mn(NO)(CN)_5^{3-}$	18	$Co(NO)(NH_3)_5^{2+}$	18
$Cr(NO)(CN)_5^{3-}$	17	$CoCl(NO)(diars)_2^+$	18
$Mo(NO)(CN)_6^{4-}$	18	$Co(NO)(diars)_2(NCS)^+$	18
$Ru(NO)(OH)(NO_2)_4^{2-}$	18	**CN = 5**	
$RuCl_3(NO)(PMePh_2)_2$	18	$Ir(NO)Cl(CO)(PPh_3)_2^+$	16
CN = 5		$Ir(NO)Cl_2(PPh_3)_2$	16
$Fe(NO)(S_2CNEt_2)_2$	19	$IrI(CH_3)(NO)(PPh_3)_2$	16
$Ir(NO)H(PPh_3)_3^+$	18	$IrCl_2(NO)(PPh_3)_2$	16
$Ru(NO)H(PPh_3)_3$	18	$Co(NO)(S_2CNEt_2)_2$	18
$Ru(NO)(diphos)_2^+$	18	$Ru(NO)_2Cl(PPh_3)_2^+$	16
$Co(NO)(diars)_2^{2+}$	18	(1 linear, 1 bent)	($1e + 3e$)
CN = 4			
$Ru(NO)_2(PPh_3)_2$	18		
$Ir(NO)(PPh_3)_3$	18		

rule is satisfied for the six-coordinate "bent" complexes only if NO is considered to be the two-electron donor NO⁻. For the five-coordinate "bent" complexes, similar counting yields 16 electrons for each metal atom, not 18. This should not be surprising; like the square-planar case in Fig. 11.16, both ideal five-coordinate geometries have one very high-energy d-based antibonding MO that can be left vacant in a stable compound. Table 11.1 shows the energy-level relationships for the d orbitals in the crystal-field approximation, which resemble those yielded by MO theory. We can expect, then, that the square-planar, square-pyramidal, and trigonal-bipyramidal complexes will be stable with only 16 valence-shell electrons, although 18-electron systems are also possible.

Assumptions about oxidation states within complexes are usually difficult to establish unequivocally for such diamagnetic molecules as most carbonyls and nitrosyls. This makes the NO⁺–NO⁻ formalism seem a bit artificial. Furthermore, the metal oxidation states that must be assumed are sometimes unusual. A simpler and equally workable approach (see the heading of Table 13.3) is to take NO as an electrically neutral ligand in all nitrosyls, but to assume that it contributes three electrons to the metal valence shell in linear geometry ($2\sigma_s^*$ or sp plus 1 π^*), or only one electron (π^*) in bent geometry. The bent NO thus forms a more or less conventional sigma bond with the metal in which each atom contributes one sigma electron.

An interesting consequence of this relationship between the bond geometry and the number of electrons donated is that the metal atom's coordination needs (particularly the need to avoid filling the high-energy antibonding orbitals beyond 18 electrons) can force geometric changes on coordinated nitrosyls. Thus the linear Co—N—O in the five-coordinate $Co(NO)(diars)_2^{2+}$, which is an 18-electron system, turns into a bent Co—N—O when the ligand NCS^- is added to make the six-coordinate $Co(NO)(diars)_2(NCS)^+$. The geometric change from that for a three-electron donor to that for a one-electron donor holds the complex to 18-electrons, when it would otherwise be expanded to 20 by the NCS^- electron pair. Enemark and Feltham termed this the "stereochemical control of valence."

Infrared spectra are widely used to provide evidence for modes of bonding in nitrosyls. Free gaseous NO has the N—O stretch vibration at 1876 cm^{-1}, and most linear M—N—O nitrosyls in neutral molecules or +1 cations show a slightly reduced frequency, just as in carbonyls. Typical wavenumber values for such systems lie in the range 1800–1900 cm^{-1}. For anions with a substantial negative charge, however, the N—O stretch peak can lie as low as about 1500 cm^{-1} (the peak for $[Cr(CN)_5(NO)]^{4-}$ is at 1515 cm^{-1}). In terms of our bonding model, we assume that the increased net negative charge on the complex allows increased population of the NO π^* orbitals, weakening the bond and lowering its stretch frequency.

For bent M—N—O nitrosyls, the N—O stretch vibration occurs at a much lower frequency than for comparably charged linear systems. Typical wavenumbers are in the range 1500–1700 cm^{-1} for neutral or +1 ions. If we assume that in a linear nitrosyl, σ_s^* electrons are being donated by the NO and π^* electrons are being accepted, it is not surprising that the N—O stretch frequency does not change markedly from that of the free NO molecule. In the one-electron-donor model of the bent nitrosyls, however, only a single π^* electron is being donated, and others are being accepted by the NO bond. This is entirely compatible with the sharp reduction in the N—O stretch frequency.

Bridging nitrosyl structures can also exist, although they are less common than for carbonyls. They are assumed to be three-electron donors to the two metal atoms together, using sp^2 hybrid orbitals as in Fig. 13.5. This limits the π^b overlap to a single pi bond. As the NO molecule can still serve as a pi acceptor into the remaining π^* MO, the overall N—O bond order is substantially reduced and the

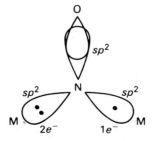

Figure 13.5   Sigma overlap for bridging nitrosyls.

N—O stretch occurs about 1500 cm^{-1} or even lower. Note that both carbonyls and nitrosyls can form triply bridging structures in metal-cluster compounds, with still lower C—O or N—O stretch frequencies: to 1600–1700 cm^{-1} for M_3CO, and as low as 1300 cm^{-1} for M_3NO. We shall look more closely at some of these structures later in the chapter, when we consider metal clusters.

13.4 ORGANOMETALLIC SYSTEMS: PI-ELECTRON DONORS

The first two organometallic compounds prepared were $Pt_2Cl_4(C_2H_4)_2$ and $K^+[PtCl_3(C_2H_4)]^-$, both reported by Zeise in 1831. These are still classic examples of a type of organometallic bonding unique to transition metals: The pi-bonding electrons in an alkene retain their pi-bonding character with respect to the carbon atoms in the alkene, but they are also involved in sigma overlap with a vacant transition-metal acceptor orbital, as in Fig. 6.2b. Figure 13.6 shows the structure of the $PtCl_3(C_2H_4)^-$ ion and also the orbital overlaps considered responsible for the stability of the ethylene–platinum bond. The ligand (ethylene) donates a pair of electrons that have sigma symmetry and some directional character to a vacant orbital on the metal, and the metal avoids the buildup of negative charge from the ligand sigma electrons by donating pi electrons from its d orbitals to a π^* orbital on the ligand. This back-donation of pi electrons can be very important to bonding; some MO calculations indicate that the total energy contribution of bonding ethylene is greater for the pi overlap than for the sigma overlap. The overlap seems exactly

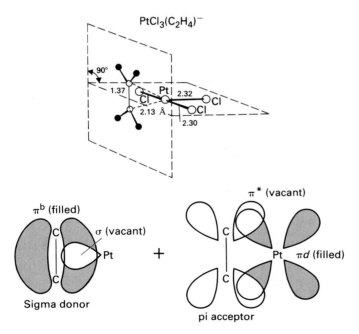

Figure 13.6 Geometry and bonding overlaps for $PtCl_3(C_2H_4)^-$.

the same as that between a transition-metal atom and CO, but here the geometry is different because of the ligand pi electrons that constitute the metal-to-ligand sigma bond. Note that because of the symmetry of the nodes in the ligand π^* orbital (whether CO or $H_2C{=}CH_2$), the pi overlap is essentially identical in both cases.

Both the sigma donation of the $C{=}C$ pi-bonding electrons and the acceptance of metal electrons into the π^* orbital of the alkene reduce the net bonding between the two alkene carbon atoms. This usually shows up as a lengthening of the $C{-}C$ bond. In gaseous ethylene the $C{-}C$ bond length is 1.34 Å, whereas in the Pt complex (Zeise's salt) it is 1.37 Å. The effect is sometimes even greater: In $PtCl_2(NH(CH_3)_2)$ (C_2H_4), a square-planar compound very similar geometrically to $PtCl_3(C_2H_4)^-$, the $C{-}C$ bond length is 1.47 Å.

It should be obvious from our previous discussion that the $C{-}C$ stretch frequency should decrease as the bond weakens. For symmetrically bound C_2H_4, the peak is not IR-active, but it can be seen because it couples with the CH_2 bending mode. The observed result is that the $C{=}C$ stretch at 1623 cm^{-1} for free C_2H_4 is reduced to 1526 cm^{-1} in $PtCl_3(C_2H_4)^-$. For other alkenes, the $C{=}C$ stretch, normally about 1630 cm^{-1}, is sometimes shifted as low as 1410 cm^{-1} on coordination to a metal atom.

Many alkenes give transition-metal complexes with bonding presumably comparable to that discussed above. Ethylene has been coordinated to all the transition metals beyond groups IV and V except for Ru, Os, Co, and Au; propene, cyclohexene, octene, styrene, and many other monoolefins also form complexes. All of these for which the crystal structure have been determined seem to have bonding geometry similar to that in Fig. 13.6. However, tetracyanoethylene (TCNE) complexes platinum with a different bonding geometry (see Fig. 13.7). Here, the two central carbon atoms are in the plane of the Pt and the other two ligands.

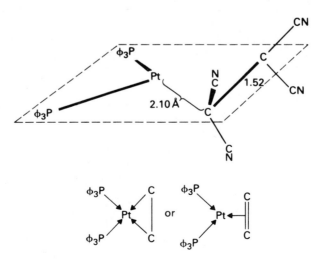

Figure 13.7 TCNE bonding to Pt.

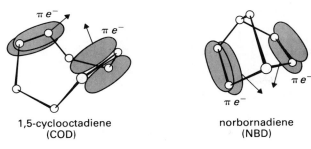

1,5-cyclooctadiene
(COD)

norbornadiene
(NBD)

Figure 13.8 Chelating dienes.

One could view this bonding as equivalent to that in other alkene complexes. However, if one recalls the strong tendency of Pt(II) to form square-planar complexes, it seems more reasonable to assume that the TCNE is a dianion forming conventional sigma bonds from each "ethylene" carbon to the platinum atom, so that a three-membered Pt—C—C ring (a *metallacyclopropane*) is formed. In general, the pi-donor form is more likely if the metal has paired *d* electrons to back-donate (that is, if the metal is toward the right of the *d* block) and if the olefin pi bond is strong, and the metallacyclopropane form is more likely for the reverse conditions. (The student, however, should be aware that most "Tinker-toy" computer structure drawings will automatically show the three-membered ring.)

The failure of ethylene to form a stable, characterizable complex with Ti, V, and their congeners is interesting—perhaps even paradoxical—since a Ti—C_2H_4 complex is almost certainly formed in the low-pressure polyethylene synthesis by the Ziegler–Natta catalyst. We shall return to the Ziegler–Natta process in Chapter 15, but we can suggest here that, as for carbonyls, stable metal–olefin bonding depends on the pi back-donation by the metal. Since Ti has few *d* electrons and a simple hydrocarbon alkene has no strongly electronegative atom to make it a better acceptor, a complex (if formed) would have a very high dissociation pressure because of the relatively weak pi bond between the metal and alkene. In addition, with few *d* electrons the group IV metals will favor formation of a metallacyclopropane structure—but that also requires a weakened C=C double bond, and ethylene (along with other simple straight-chain olefins) has a very strong double bond.

One way to increase the stability of metal–alkene complexes is to take advantage of the chelate effect. In effect, a simple alkene is a two-electron donor analogous to NH_3. If we choose a diene with the double bonds isolated by saturated hydrocarbon chains, as in 1,5-cyclooctadiene (COD) or norbornadiene (NBD) (Fig. 13.8), we can form an organometallic chelate analogous to ethylenediamine complexes. A number of these complexes can be formed simply by allowing the metal carbonyl to react with the diene,

$$[Rh(CO)_2Cl]_2 + C_8H_{12} \text{ (COD)} \rightarrow$$

or with some other labile ligand:

$$PtCl_4^{2-} + C_8H_{12}\ (COD) \rightarrow$$

Other labile ligands are also readily displaced:

$$PdCl_2(NCPh)_2 + C_7H_8\ (NBD) \rightarrow$$

It is often possible to replace a two-electron alkene donor by an alkyne donor—but the alkyne has two pairs of pi electrons, and there is a tendency to form a complex in which the alkyne donates a pair of electrons to each of two metal atoms. The two alkyne C atoms and two metal atoms usually form a rough tetrahedron, and we can view the alkyne as donating its π_x pair to one metal atom and its π_y pair to the other. Figure 13.9 shows such a molecule, [(triphos)Rh(Cl)(C_2H_2) Rh(triphos)]$^+$, where *triphos* is $CH_3C(CH_2PPh_2)_3$. The Rh atoms satisfy the 18-electron rule if C_2H_2 donates a pair to each. However, oxidation of this complex by two electrons changes the geometry: The alkyne changes to a sigma Rh—C bond at each end and a second bridging Cl is added. Instead of donating two pi electrons to each Rh, the acetylene now donates one sigma electron to each Rh, and the bonding is entirely different.

A simple alkene with only one double bond is clearly a two-electron donor in a transition-metal complex. A diene with isolated double bonds should be thought of as a donor of two electron pairs, not as a four-electron donor, because there is little or no interaction between the two pairs of pi electrons in the free ligand. However, a conjugated diene—that is, a delocalized pi system—is a significantly different ligand even though it also contributes four electrons to the metal atom. A wide area of organometallic synthesis and structure/reactivity studies was opened up when chemists realized that planar extended pi systems can contribute varying numbers of electrons to metals in organometallic complexes. We shall consider a

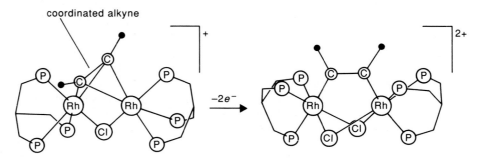

Figure 13.9 Coordinated alkyne geometry and rearrangement from pi donor to sigma donor.

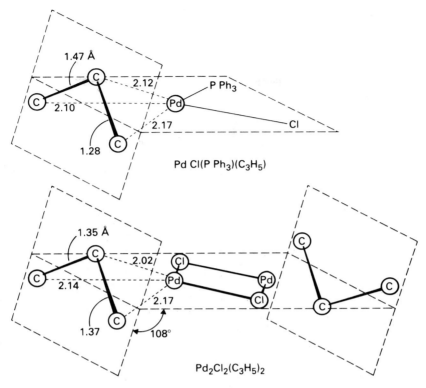

Figure 13.10 Geometry of allyl complexes.

series of these pi-donor ligands in the order of the number of electrons they contribute to the metal.

Three-electron pi donors. The best-known three-electron donor is the allyl radical, $CH_2 \cdots CH \cdots CH_2$. Here the pi overlap extends over all three carbon atoms, and each C contributes one $2p$ electron to the pi system and thus to the metal. Figure 13.10 shows the geometry of the resulting complex $PdCl(PPh_3)(C_3H_5)$. Note that the strikingly different bond lengths for the two C—C bonds make it appear that the allyl group has an isolated single and double bond —CH_2—CH=CH_2, with the isolated bond presumably forming a direct sigma link to the Pd and the pi electrons donating as they would for ethylene. This coordination geometry is unusual for allyls, however. More typical is the dimeric Cl-bridged complex $Pd_2Cl_2(C_3H_5)_2$, also shown in Fig. 13.10. In this and in most other allyl complexes, the C—C bond lengths are essentially equal (and a bit longer than a C—C double bond). The plane of the allyl radical is tilted so that the central C atom is significantly closer to the metal than the end carbons are.

Unlike the pi complexes of ethylene, all transition metals form allyl complexes.

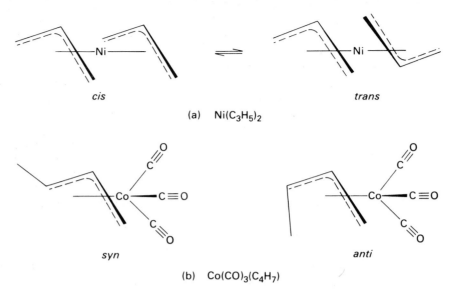

(a) Ni(C$_3$H$_5$)$_2$

(b) Co(CO)$_3$(C$_4$H$_7$)

Figure 13.11 Isomers of allylic compounds.

Zr forms Zr(C$_3$H$_5$)$_4$, and other allyl complexes include V(C$_3$H$_5$)$_3$, CrCl (C$_3$H$_5$)$_2$, Mn(CO)$_4$(C$_3$H$_5$), FeCl(CO)$_3$(C$_3$H$_5$), Co(CO)$_3$(C$_3$H$_5$), and Ni(C$_3$H$_5$)$_2$. It is interesting to note the prevalence of the 18-electron rule for the later but not the earlier transition metals. Some unusual forms of stereoisomerism are seen in allylic complexes: proton NMR spectra show the presence of the *cis*- and *trans*-isomers of diallyl-nickel in solution (Fig. 13.11a), and *syn*- and *anti*- stereoisomers of some complexes can be isolated, particularly for substituted allyls like 1-methylallyl (Fig. 13.11b).

From the coordination chemist's viewpoint, however, a more important kind of isomerism involves the mode of attachment of the allyl group to the metal atom. This can be either sigma or pi, as in the following rearrangement:

$$\underset{CH_2}{\overset{CH}{\diagup}} \underset{CH_2—Mn(CO)_5}{\diagdown} \xrightarrow{h\nu} \diagup\!\!\!\diagdown Mn(CO)_4 + CO$$

Here the mode of bonding is clearly changing from sigma to pi. That distinction is sometimes made in naming the compounds: Mn(σ—C$_3$H$_5$)(CO)$_5$ but Mn(π—C$_3$H$_5$)(CO)$_4$, in the above example. One can also make a distinction between these two compounds in terms of the number of allylic carbon atoms bonded to the metal. The pentacarbonyl has only one allylic C—Mn bond, whereas the tetracarbonyl has three allylic C—Mn bonds. (This, of course, is only a geometric description of touching atoms, not an electronic specification of the bonding.)

This system is commonly followed in naming organometallic compounds with

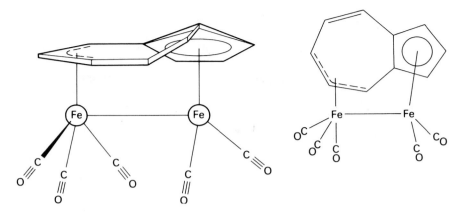

Figure 13.12   Azulenediironpentacarbonyl.

extended pi-system ligands: A Greek numerical prefix and the word *hapto* (from the Greek "haptein," to seize) are used at the beginning of the name to indicate the number of C—M bonds present in the pi system attachment to the metal. The two manganese compounds above would have *monohapto-* and *trihapto-* allyl groups, respectively. In formulas (and sometimes in names) these would be abbreviated (using the Greek "eta") to η^1- and η^3-, respectively: $Mn(\eta^1\text{-}C_3H_5)(CO)_5$ and $Mn(\eta^3\text{-}C_3H_5)(CO)_4$.

The *hapto-* system of nomenclature is particularly useful when one is dealing with organometallic systems in which a larger pi-system ligand donates through only a few carbon atoms. For example, azulene forms a binuclear complex $Fe_2(CO)_5(C_{10}H_8)$ (Fig. 13.12) in which the five-membered ring is clearly bonding all five carbon atoms to the metal, but the seven-membered ring is bonding only three. The formula would be written $(\eta^3,\eta^5\text{-}C_{10}H_8)Fe_2(CO)_5$. It is interesting that the seven-atom ring in azulene is serving as a *trihapto-* or allylic donor even though the three metal-bonded carbon atoms are in no way different from the others. Clearly, the electronic requirements of the iron atoms are controlling the stereochemistry and valence of the azulene ligand.

A number of other pi systems also form *trihapto* complexes with transition metals under circumstances that favor the formation of a three-electron-donor radical. Cyclic alkenes such as cyclohexene and cyclooctene eliminate a hydrogen atom:

$$2PdCl_2 + 2C_6H_{10} \rightarrow Pd_2Cl_2(C_6H_9)_2 + 2HCl$$

Cyclohexene

and dienes can often be protonated:

$$Fe(CO)_3(C_4H_6) + HClO_4 \rightarrow [Fe(CO)_3(C_4H_7)]^+ClO_4^-$$

Butadiene

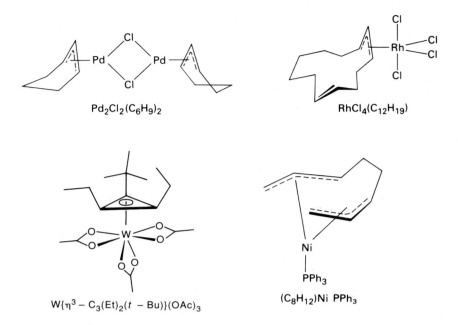

Pd₂Cl₂(C₆H₉)₂

RhCl₄(C₁₂H₁₉)

W{η³ − C₃(Et)₂(t − Bu)}(OAc)₃

(C₈H₁₂)Ni PPh₃

Figure 13.13 Some *trihapto* complexes with large hydrocarbons and with cyclopropenyl.

An interesting symmetrical *trihapto* complex is formed by the cyclopropenyl ring in W{C₃(Et)₂(t-Bu)}(OAc)₃, in which the W and C₃ ring form a tetrahedron similar to that in the alkyne M₂C₂ complexes; the 18-electron rule is satisfied for W^{3+} with 3 d electrons by adding three from the ring and two from each of six acetate O atoms. A few of these structures are shown in Fig. 13.13.

Four-electron pi donors. Many four-electron donors forming *tetrahapto*-complexes have been studied in organometallic chemistry. Conjugated dienes are obvious candidates for such a role; the simplest, butadiene, has been used as a ligand in a number of complexes. The more or less classic butadiene complex is (C₄H₆)Fe(CO)₃, shown in Fig. 13.14. Here, the 18-electron rule is preserved by having butadiene substitute for two carbonyls. The structure has some interesting features that are reasonably easy to correlate with simple MO theory. First, all four Fe—C distances are nearly equal at about 2.1 Å, and the plane of the butadiene molecule is nearly perpendicular to the shortest Fe–ligand axis. As the figure indicates, there are some significant changes in the geometry (and presumably the bonding) of the ligand. In a free butadiene molecule, the terminal C—C bond lengths are 1.36 Å (a bit longer than the 1.34 Å of C=C in ethylene) and the central C—C bond length is 1.45 Å (significantly shorter than the 1.53 Å of C—C in ethane). In terms of simple Hückel MOs (see Fig. 13.15), this is because the four

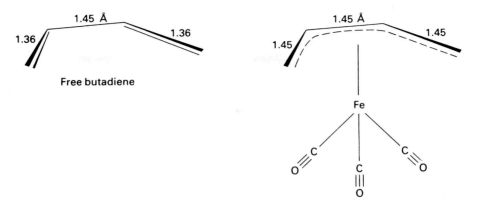

Figure 13.14 The structure of (h^4–butadiene) iron tricarbonyl.

pi electrons are not localized in the two classical double-bond regions. The bond order is about 1.9 for the terminal C—C bond regions, and 1.4 for the central C—C bond. However, when the ligand serves as a pi donor (the ψ_2 electrons are the most readily accessible) it loses bonding electron density in the terminal bond regions and antibonding electron density in the central bond region, judging from the position of the node in ψ_2. On the other hand, if ψ_3 serves as a π^* acceptor for iron $3d$ electrons (pursuing our metal–alkene bonding model) the butadiene ligand gains antibonding density in the terminal bond regions (of ψ_3) and gains

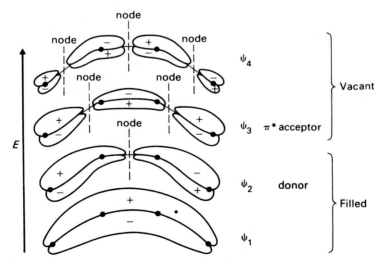

Figure 13.15 Hückel MOs for butadiene.

bonding density in the central bond region. The net change in overall electron density for the coordinated butadiene is that the terminal bond regions lose a substantial degree of pi-bond order while the central bond region gains a small amount of pi-bond order. This correlates nicely with the observed changes in bond lengths, although the central bond region does not seem to change. In other conjugated diene complexes (such as those formed by methylcyclopentadiene), the central bond is actually substantially shorter than the outer bonds, which is qualitatively in accord with this argument.

Figure 13.14 shows coordinated butadiene in the *cis* conformation, which is most often seen. However, the *trans* conformation can donate not only two electrons to each of two metal atoms, seen in Os_2 systems, but to a single metal atom as in $(cp)_2Zr(trans\text{-}C_4H_6)$. In addition, just as coordinated ethylene can bond either as a pi donor or as a metallacyclopropane, butadiene and other dienes can bond either as a *tetrahapto* pi donor or as a metallacyclopentane, where the diene is *dihapto* through the two end carbon atoms, as in $(cp)_2Zr(C_4Ph_4)$. Butadiene does not even have to serve as a four-electron donor, although that is the commonest mode of coordination. In $Fe(CO)_4(C_4H_6)$ it is bound through only one C=C bond in the same manner as ethylene. In this case the conjugated nature of the pi system presumably does not have much (if any) influence over the bonding.

Other four-electron donors are mostly cyclic dienes or polyenes (such as cyclohexa-1,3-diene or cyclooctatetraene), which bond in a manner very similar to butadiene. A particularly interesting case is the molecule cyclobutadiene and its substituted forms. Although the conjugated cyclic hydrocarbon benzene is extremely stable because of its high delocalization energy, the classically equivalent molecules C_4H_4, cyclobutadiene, and C_8H_8, cyclooctatetraene, are much less so. C_8H_8 was synthesized more than 70 years ago by traditional methods, but it proved not to be a planar delocalized molecule. Instead, it adopts a "tub" conformation comparable to saturated cyclic hydrocarbons. Cyclobutadiene was even more difficult; it never proved possible to synthesize even substituted forms C_4R_4 until it was suggested that such cyclobutadienes might be excellent four-electron pi donors. The first cyclobutadiene compound was promptly prepared:

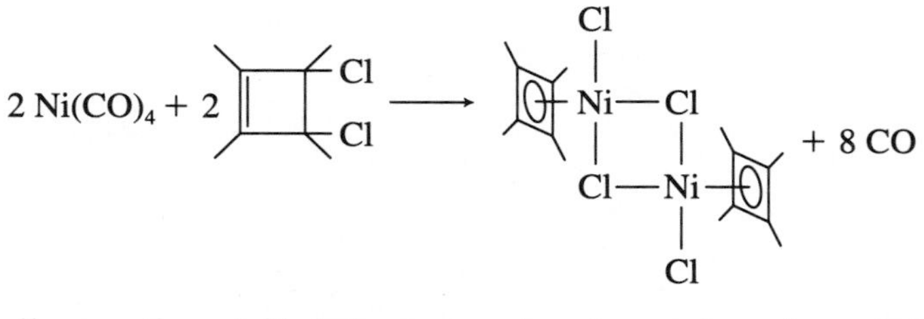

A similar reaction yielded the iron carbonyl complex of unsubstituted cyclobutadiene, which can be oxidized to yield the transient free hydrocarbon C_4H_4.

This in turn condenses with acetylenes to yield substituted Dewar-benzene bicyclic structures:

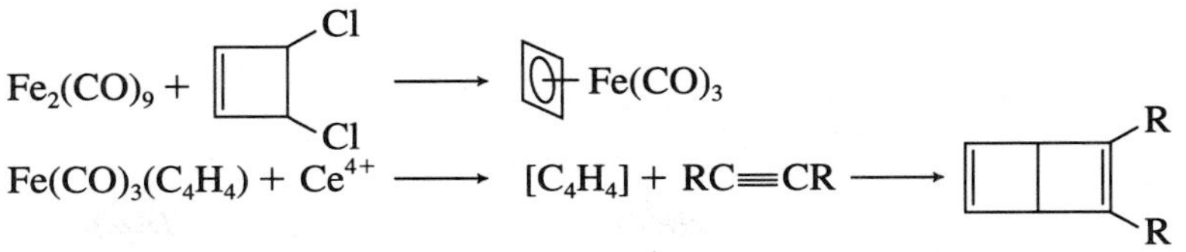

$$Fe_2(CO)_9 + \text{(cyclobutadiene with Cl, Cl)} \longrightarrow \text{(ring)}\!-\!Fe(CO)_3$$

$$Fe(CO)_3(C_4H_4) + Ce^{4+} \longrightarrow [C_4H_4] + RC\!\equiv\!CR \longrightarrow \text{(ring with R, R)}$$

One interesting feature of the structure of cyclobutadiene complexes is that substituted groups on the ring, such as methyls or phenyls, seem to be bent out of the plane of the ring away from the metal atom by roughly 10°. Although the ring is almost exactly square, suggesting true delocalized pi bonding, the out-of-plane substituents suggest that the mirror-image symmetry of the pi electrons that would occur in a free planar aromatic molecule is being substantially modified by the bonding to the metal. (Of course, the instability of the free hydrocarbon compared to the complex shows the same thing.) This deformation can be compared with that of substituted allyls in organometallics such as $RhCl_2$(2-methylallyl)$(AsPh_3)_2$, in which the substituted methyl is bent out of the allyl plane *toward* the metal atom.

A final common four-electron donor is cyclooctatetraene (COT), which bears an interesting relationship to cyclobutadiene in view of the following reaction:

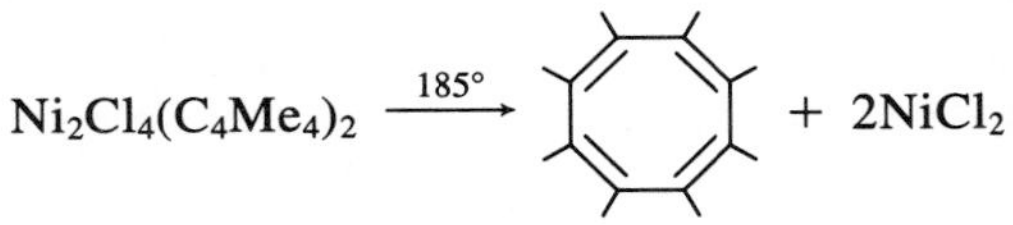

$$Ni_2Cl_4(C_4Me_4)_2 \xrightarrow{185°} \text{(cyclooctatetraene)} + 2NiCl_2$$

Although the free COT molecule is normally found in the tub conformation, it readily serves as a *tetrahapto* ligand, as shown in Fig. 13.16. Metal–COT complexes are usually also in the tub conformation, but they can also adopt the chair conformation when the bonding geometry is more convenient.

Five-electron pi donors. By far the most common five-electron donor is $C_5H_5\cdot$, the cyclopentadienyl radical (cp). In a sense, *cp* started modern transition-metal organometallic chemistry: A tremendous impetus was given to the field by the publication in 1951 of a synthesis of *ferrocene,* bis(cyclopentadienyl)iron or $Fe(cp)_2$. In that initial synthesis, Kealy and Pauson reacted a cyclopentadienyl Grignard reagent with $FeCl_3$ (intending to synthesize a different product, fulvalene). A better synthesis uses cyclopentadiene (C_5H_6), metallic sodium, and the dihalide of a transition metal:

$$2C_5H_6 + 2Na \rightarrow H_2 + 2Na^+C_5H_5^- \xrightarrow{FeCl_2} Fe(C_5H_5)_2 + 2NaCl$$

All but one of the transition metals from Ti across the top row to Ni form $M(cp)_2$ complexes with the "sandwich" structure shown in Fig. 13.17. The one exception is $Ti(cp)_2$, which is polymeric with Ti–Ti bonds. If the Hückel MOs for

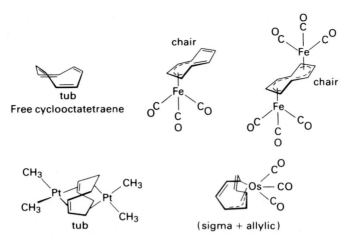

Figure 13.16 Clyclooctatetraene as a four-electron donor.

the cyclopentadienyl radical are taken as the starting point for a set of metallocene MOs, the resulting energy-level diagram (even where details of the calculation differ) has six very stable bonding MOs and three weakly bonding MOs some distance below the antibonding orbitals. There are thus nine stable orbitals, and the 18-electron rule is favored. A quick computation shows that $Fe(cp)_2$ should be the most stable metallocene, since *cp* is a five-electron donor and iron has eight valence electrons as Fe(0). This can be experimentally verified: Ferrocene is stable indefinitely in air at room temperature, and pyrolyzes only above 500°C, but the others are all quite air sensitive.

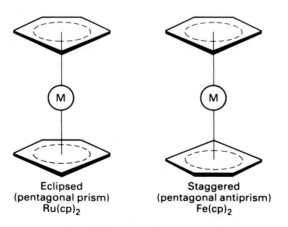

Figure 13.17 Bis(cyclopentadienyl) metal "sandwich" structures.

The bonding of each of the cp groups in metallocenes is thought to follow the same pattern as the other alkene–metal complexes—that is, the ligand serves as a pi donor and a π^* acceptor. Apparently the orbital sizes and orientations are particularly favorable for these purposes, because *pentahapto* metal–cp bonding is quite strong and a single cp ring is often found as a ligand in organometallic complexes for which the bonding of interest involves some other ligand. For example, the only titanium carbonyl known is $Ti(cp)_2(CO)_2$. Other mixed cyclopentadienyl–carbonyl compounds include $V(cp)(CO)_4$, $Cr_2(cp)_2(CO)_6$, $Mn(cp)(CO)_3$, $Fe_2(cp)_2(CO)_4$, $Co(cp)(CO)_2$, and $Ni_2(cp)_2(CO)_2$. The second- and third-row transition metals are somewhat less stable as cyclopentadienyls; the compounds $M(cp)_2$ are known only for the stable iron congeners Ru and Os and for Mo, Rh, and Ir. However, Zr and Hf form $M(cp)_4$ species (though not all the ligands are *pentahapto*) in keeping with the behavior of Ti, which forms $Ti_2(cp)_4$, $Ti(cp)_3$, and $Ti(cp)_4$. The latter compound shows some of the electronic versatility of the cp ligand: Two ligands are η^5 five-electron donors, and two are η^1 one-electron donors. Figure 13.18 shows this and other modes of bonding for $C_5H_5\cdot$, C_5H_6, and $C_5H_7\cdot$.

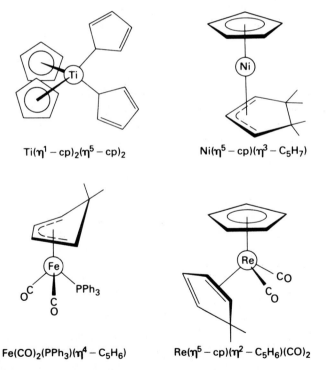

$Ti(\eta^1 - cp)_2(\eta^5 - cp)_2$

$Ni(\eta^5 - cp)(\eta^3 - C_5H_7)$

$Fe(CO)_2(PPh_3)(\eta^4 - C_5H_6)$

$Re(\eta^5 - cp)(\eta^2 - C_5H_6)(CO)_2$

Figure 13.18 Bonding geometries for C_5H_5, C_5H_6, and C_5H_7.

A more recent and very synthetically useful variation on the cyclopentadienyl ligand is pentamethylcyclopentadienyl, $C_5(CH_3)_5$, usually abbreviated *cp**. This species seems at first like a distinction without a difference, but in practice the methyl groups have a significant inductive electron-donating effect, so that *cp** is an even better five-electron base than *cp*. Although the methyl groups make *cp** more sterically demanding than *cp,* its strong donor qualities make some remarkable compounds possible, such as the first η^5 tris(cyclopentadienyl) compound, $Sm(\eta^5-cp^*)_3$. Other poly(alkyl)cyclopentadienyls are sometimes used when great steric bulk is needed to stabilize a metal–cp compound by keeping other bases away from the metal; even without the possibility of *d*-electron back-donation, a calcium sandwich compound can be formed with tetra(isopropyl)cyclopentadienyl: $Ca\{\eta^5-C_5(i-Pr)_4H\}_2$, for which space-filling models indicate almost complete encapsulation of the Ca by the isopropyl groups and the rings.

Other five-electron donors are known, though none offer the stability of cyclopentadienyls. For example, 1,3-pentadiene, C_5H_8, forms a butadiene-like η^4 complex with iron, $Fe(CO)_3(C_5H_8)$, but this can be oxidized to a pentadienyl complex:

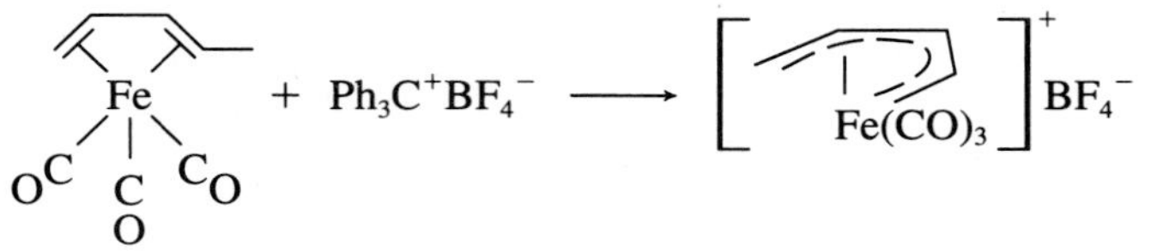

The same reaction forms a planar, "pentadienyl" complex with $Fe(CO)_3$ using the ligands cyclohexadiene or cycloheptadiene.

Because of the Hückel rule for aromaticity $(4n + 2)$ there is a strong tendency for planar-cyclic pi systems to adjust their pi population to six electrons. The cyclopentadienide ion $C_5H_5^-$ is such a system. All metallocenes have some charge separation between $M^{\delta+}$ and $C_5H_5^{\delta-}$, and magnetic evidence suggests that $Mn(cp)_2$ may be predominantly ionic: $Mn^{2+}(C_5H_5^-)_2$. Of course, the valence electron count is not changed if the metal is considered to be M^{2+} and the cp^- ligand is considered a six-electron donor. The geometry tells us nothing about the bond type, since the sandwich arrangement is the most stable geometry for both covalent and ionic systems. The C_5H_5 ligand is often assumed to be cp^- in transition-metal molecules, but it seems more consistent with other olefin ligands to treat it as the five-electron radical *cp*.

Six-electron pi donors. There are many metal-alkene complexes for which there can be no question that the ligand is serving as a six-electron donor. These primarily involve benzene or substituted benzenes, usually 1,3,5-trimethylbenzene (mesitylene) or hexamethylbenzene. Such ligands are known as *arenes,* and the number of arene complexes is quite large. Pursuing the 18-electron rule and the sandwich structure of metallocenes, the most stable dibenzene complex should be dibenzenechromium, $Cr(C_6H_6)_2$ or $Cr(ar)_2$. This was the first arene complex prepared:

$$3CrCl_3 + 2Al + AlCl_3 + 6C_6H_6 \rightarrow 3[Cr(C_6H_6)_2]^+ AlCl_4^-$$

$$[Cr(C_6H_6)_2]^+ + S_2O_4^{2-} + 4OH^- \rightarrow Cr(C_6H_6)_2 + 2SO_3^{2-} + 2H_2O$$

Although we do not normally think of benzene as a good ligand, it is instructive to note that the metal–vapor reaction between $Cr(g)$ and benzene gives dibenzene–chromium in about 60% yield, suggesting considerable stability.

Arene complexes $M(ar)_2$ have been prepared not only for the stable group VI metals Cr, Mo, and W, but also for V, Fe, Co, and Ni. The cationic species $M(ar)_2^+$ have been prepared for every metal from Ti to Co, plus some second- and third-row metals. The dication $M(ar)_2^{2+}$ has been prepared for the 18-electron systems involving Fe, Ru, and Os, plus Co and Rh, and the trication $M(ar)_2^{3+}$ for the 18-electron systems Co, Rh, and Ir. There are a fair number of mixed arene–carbonyl complexes and complexes with arenes and other pi-acceptor ligands.

The structures of metal–arene complexes are consistent with the pattern we have already seen for other pi donors. The metal–ligand axis is perpendicular to the plane of the pi system, and extensive delocalization occurs as indicated by uniform C—C bond lengths in the ligands. Figure 13.19 shows a few arene structures, including η^6 ligands from larger rings such as tropylium, $C_7H_7^+$, and cycloheptatriene, C_7H_8. Although we think of benzene as being rigidly planar, the figure also shows that under the right circumstances hexamethylbenzene can be a *tetrahapto* ligand with a bent conformation, suggesting the power of the 18-electron rule. η^4-Arenes of this sort can in some cases be converted to the more traditional η^6 ligands by electrochemical oxidation of two electrons.

The $C_7H_7^+$ tropylium complex shown in Fig. 13.19 can be thought of as a six- or seven-electron donor, depending on what one assumes about the oxidation state of the metal. The stability of aromatic hydrocarbons with six pi electrons suggests that it is a six-electron donor, but there is no question that all seven carbon atoms are equally bonded. In electrically neutral compounds of this sort, called cycloheptatrienyl compounds, it is more obvious than for the cation in the figure that the ring must be thought of as a seven-electron donor. There are no symmetrical sandwich compounds, but there are several examples of sandwich compounds that contain cycloheptatrienyl and cyclopentadienyl rings—that is, a five-membered ring and a seven-membered ring. They are usually prepared by synthesizing a cationic species, which is then reduced:

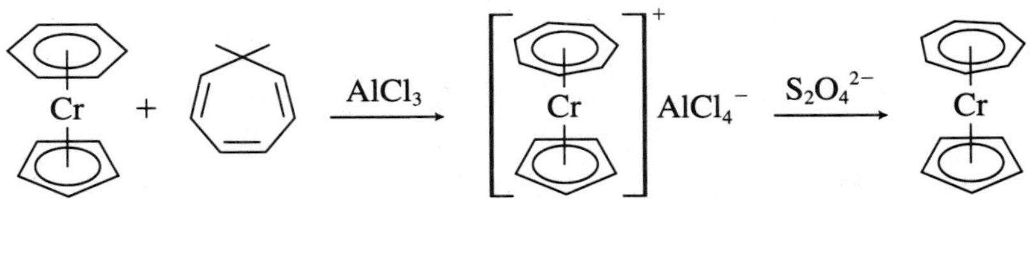

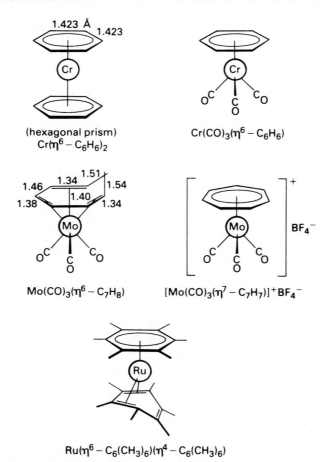

Figure 13.19 Bonding geometries for six-electron arene complexes.

(η^5-Cyclopentadienyl)vanadium tetracarbonyl, on the other hand, will spontaneously displace an H atom:

$$(\eta^5\text{-}C_5H_5)V(CO)_4 + C_7H_8 \rightarrow (\eta^5\text{-}C_5H_5)V(\eta^7\text{-}C_7H_7)$$

This product and the equivalent Cr compound above have very similar structures (see Fig. 13.20). Note that the C—C distances are equivalent in both rings (suggesting extensive delocalization) and that the V—C distances are nearly the same for the η^5- and η^7- rings, even though this means that the distance from the V atom to the ligand planes must be quite different. This is not surprising, since the pi electrons are on the periphery of the ligand rings; but it does remind us not to take the single line drawn to the center of a cyclic ligand as some kind of special 5- or 6- or 7-electron bond.

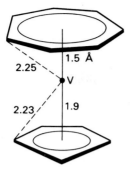

Figure 13.20 Bonding geometry for $(h^5\text{-}C_5H_5)V(h^7\text{-}C_7H_7)$.

One feature of planar donors of large numbers of pi electrons is that they do not seem to form complexes with metals that have large numbers of d electrons. No group VIII cycloheptatrienyls are known, even though the 18-electron rule could be satisfied for Co, for example, by C_7H_7 and one CO. Perhaps the resulting bonding is simply too asymmetric for electrostatic stability, or the increased effective nuclear charge on the d electrons toward the right of the d-block shrinks the d orbitals so much that they cannot overlap effectively with large rings. The latter explanation is nicely consistent with the fact that the only planar cyclooctatetraene (COT) ligand systems are donors to Ti, V, Ce, Th, and U. In the latter three cases, the compounds are $M(COT)_2$ with an almost ideal sandwich structure: parallel rings, equal C—C bond lengths and equal M—C distances. In these three cases, one can presumably assume that the compounds bond by f-orbital overlap with the COT planar ligand, since the f orbitals are relatively diffuse. The rather remarkable structure for one of the titanium compounds is shown in Fig. 13.21. The two outer COT rings are planar, while the inner one is puckered in such a way as to make the ring an η^4 donor to each Ti, with two carbons common to both Ti atoms. Although no simple description for the bonding can be given, it is at least clear that Ti has the largest, most diffuse $3d$ orbitals of all the first-row metals, and can best accept from and donate to the eight-membered ring. COT has, of course, already been indicated

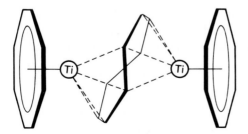

Figure 13.21 Bonding geometry for $Ti_2(COT)_3$.

as a good donor of two η^2- isolated electron pairs, η^4-, η^1/η^3-, and even η^6-. This versatile ligand can apparently alter its geometry extensively in response to different electron-donor opportunities.

Monohapto two- and three-electron donors. At the beginning of the chapter we saw the obvious sigma bonding of an alkyl group to a metal, as in $W(CH_3)_6$. Having expanded our discussion to systems in which multiple pi electrons can be donated to metals, we should now backtrack to note the existence of systems in which M=C double bonds or M≡C triple bonds are formed, so that the organic group is η^1 but nonetheless contributes two or even three electrons to the metal, presumably through $d\pi$–$p\pi$ bonding analogous to that in Fig. 5.26. The M=C (doubly bonded) systems are called *alkylidenes,* and the M≡C (triply bonded) systems *alkylidynes.*

Alkylidenes can be prepared from sigma-alkyls by α-hydrogen abstraction if a replacement ligand is provided:

$$Ta(-CH_2CMe_3)_2Br_3 + 2PMe_3 \rightarrow Ta(=CHCMe_3)(PMe_3)_2Br_3 + CMe_4$$

Note that the starting material is quite electron-deficient, with only 10 electrons in the Ta valence orbitals. It would be perfectly possible to simply add electron-pair ligands like the phosphines to approach 18 electrons, but the coordination number is too high; adding two PMe_3 yields only 14 electrons but is already seven-coordinate with two moderately bulky PMe_3 ligands. That electron count can be maintained, but with more sterically favorable octahedral geometry, by forming the Ta=C double bond and losing one ligand to the solvent by hydrogen abstraction from the remaining sigma-alkyl ligand.

Alkylidynes can be prepared by a somewhat similar α-hydrogen abstraction reaction in which a Grignard reagent replaces an alkoxide:

$$W(OMe)_3Cl_3 + 6Me_3CCH_2MgCl \rightarrow (Me_3CCH_2)_3W(\equiv CCMe_3) + 2CMe_4$$

In an even simpler reaction (stoichiometrically, at least), dinuclear alkoxides can react directly with acetylenes:

$$(MeO)_3W—W(OMe)_3 + RC\equiv CR \rightarrow 2(MeO)_3W(\equiv CR)$$

This reaction would seem to proceed through two intermediates analogous to the two structures seen in Fig. 13.9, with the acetylene first forming a W_2C_2 tetrahedron like the structure at the left of the figure, followed by rearrangement to a planar 1,2-dimetallacyclobutane form like the structure at the right of the figure or to a 1,3-version of the cyclobutane with metal atoms at opposite corners of the square:

(I)   (II)   (III)

Of these three forms, I can separate to the product alkylidyne molecules readily. The separation of the acetylene in this reaction suggests that some interesting catalytic rearrangements of carbon–carbon multiple bonds might be possible. This has been extensively investigated, and Chapter 15 will describe some of the industrially important metathesis reactions that are thought to proceed through an alkylidene or alkylidyne intermediate.

13.5 METAL–METAL BONDING

Perhaps the most obvious form of covalent compound involving transition metals is one involving metal–metal bonds. One such compound, Hg_2Cl_2 (once called calomel), has been known since alchemical times; and it has been known for more than a hundred years that a molecular unit contained two mercury atoms, presumably bonded to each other. However, until recently this was regarded as a highly unusual structural feature, and it seemed remarkable that metal–metal bonds could exist in carbonyls such as $Mn_2(CO)_{10}$. But since about 1960 metal–metal bonding has received a great deal of study. Hundreds of compounds are now known in which metal–metal bonding takes place, both in simple dinuclear systems and in metal clusters where the metal–metal bonding is delocalized over a number of atoms. We have space here for only a brief survey of some of the known systems, bonding patterns, and geometries. In the next section we will take up metal cluster bonding, and in Chapter 15 and 16 we shall also consider the catalytic reactivity of some compounds with metal–metal bonds.

Besides the linear molecule Hg_2Cl_2, a particularly simple form of metal–metal bonding is seen in $Cu_2(CH_3CO_2)_4 \cdot 2H_2O$ (and in the comparable acetates of Cr^{2+} and Rh^{2+}). One might assume that the two copper atoms are simply held together in the dimeric structure (Fig. 13.22) by the difunctional bridging carboxyl groups. However, the Cu–Cu spacing is comparable to that in metallic copper (2.65 Å versus 2.55 Å) and significantly less than the sum of two van der Waals radii for Cu (about 2.86 Å). Internuclear separation is one of the major criteria used to establish metal–metal bonding when it is not obvious from the presence of two metal atoms adjacent to each other with no bridging groups.

The symmetry of coordination about each metal atom is also important: If the two metal atoms are in approximate high-symmetry environments but are nonetheless displaced toward each other, it is usually assumed that a bond exists between the two atoms. A simple example of this is the crystal structure of α-NbI$_4$ (Fig. 13.23). Each Nb has roughly octahedral coordination by 6I, with adjacent octahedra sharing edges. However, in the chain of octahedra there are alternating Nb–Nb distances, so that each niobium is displaced from the center of its I$_6$ octahedron toward an adjacent Nb. Although the closest Nb–Nb distance is significantly longer than the sum of two covalent radii for Nb (2.74 Å) or the metallic bond length (2.86 Å), the distortion of symmetry and the fact that the solid is diamagnetic even though Nb^{4+} would be d^1 (suggesting that a bond must have paired the electrons)

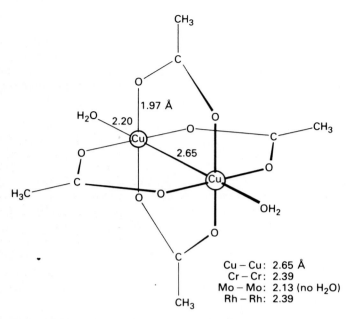

Figure 13.22 Metal—metal bonding in acetates.

lead us to believe that a Nb—Nb bond is present. (Note that all of the acetates indicated in Fig. 13.22 are also diamagnetic, supporting the inference of bonding between the metal atoms.)

There are many other examples of metal–metal bonding in simple binary compounds such as halides, sulfides, selenides, and even phosphides and arsenides (see the NiAs crystal structure in Fig. 2.14). We shall consider a few of the better-established cases in the next few pages. However, a particularly fertile area for encountering metal–metal bonds and metal—atom clusters is the carbonyl and alkene pi-complex group we discussed in the previous section. A number of polynuclear carbonyls are indicated in Table 13.1, and there are even more carbonylate anions. A few of the structures have already been shown, and we shall consider others in surveying clusters.

The appropriate question now is: Under what circumstances can metal–metal bonds be expected to form? A principal criterion is that the metal must be in a low oxidation state, and in general the lower the better. This accounts for the prevalence of metal–metal bonds and clusters in carbonyls and carbonylates, since these are in formal oxidation states near zero or even with negative values. There are several reasons for this requirement. Most obviously, if bonds are to form, there must be d electrons remaining on the metal ion or atom. In addition, however, the d orbitals must not be contracted too much by a large net ionic charge, and (for halides) the stoichiometry MX_n must not require such a large n value that it makes the approach of two M^{n+} ions either sterically or electrostatically unfa-

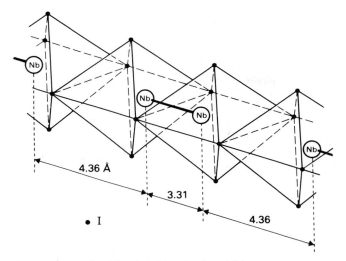

Figure 13.23 Coordination distortion in α-NbI₄.

vorable. The need for "large" *d* orbitals (to allow good metal–metal overlap) tends to favor second- and third-row transition metals, since the 4*d* and 5*d* orbitals are more diffuse relative to the core than the 3*d* orbitals are.

In binary compounds such as halides, there is an even more compelling thermodynamic reason for the formation of metal–metal bonds in low oxidation states. Because of *d–d* overlap, there is extensive covalent bonding in the metal itself, particularly around groups Vb and VIb where the number of *d* electrons is just about equal to the number of valence orbitals. This means that the atomization energy of the metal from the solid is very large, so large that the formation of +1 or +2 halides from the solid element is quite endothermic. Since ΔS is negative for the formation reaction, MX or MX₂ compounds will be impossible to form for metals M with high atomization energies *unless covalent bonds are formed in addition to the ionic attractions.* Consider the following cycle:

$$
\begin{array}{ccc}
M(g) + 2Br(g) & \xrightarrow{\frac{IP_1 + IP_2 +}{2EA}} & M^{2+}(g) + 2Br^-(g) \\[2pt]
\Delta H_{at}\uparrow \quad 2\Delta H_{at}\uparrow & & \downarrow U_2 \\[2pt]
M(s) + Br_2(l) & \xrightarrow{\Delta H} & MBr_2(s)
\end{array}
$$

For this cycle, we have

$$\Delta H = \Delta H_{at}(M) + 2\Delta H_{at}(Br) + [IP_1 + IP_2](M) + 2EA(Br) + U_2(MBr_2)$$

If the metal is niobium,

$$\Delta H = 772 + 2(112) + [2004] + 2(-325) + (-2044)$$
$$= +306 \text{ kJ/mol}$$

where U_2 has been calculated by the Kapustinskii approximation for an ionic lattice. The lattice energy and the sum of the ionization energies almost exactly offset each other, but the large atomization energy for Nb is more than enough to make the overall reaction substantially endothermic. Hypothetical M^+ halides such as NbBr are even worse.

We can expect, then, that if halides such as $NbBr_2$ form, there will be a substantial covalent-bonding contribution from Nb–Br bonds, and perhaps also from Nb–Nb bonds. There are two more or less classic examples of such compounds: $MoCl_2$ and $NbCl_{2.33}$. Structurally, these are metal clusters $\{Mo_6Cl_8\}^{4+}(Cl^-)_4$ and $[Nb_6Cl_{12}]^{2+}(Cl^-)_2$, with the structures shown in Fig. 13.24. Halides with oxidation states near 2+ of niobium and tantalum, molybdenum and tungsten, and to some extent palladium and platinum tend to form one of these two structures. The clusters are quite stable: If $MoCl_2$—that is, Mo_6Cl_{12}—is dissoved in ethanol, only one-third of the chloride ions can be precipitated by Ag^+. In aqueous HCl, six more Cl^- can be coordinated to the Mo atoms to produce $Mo_6Cl_{14}^{2-}$. The metal–metal bonding is not limited to finite clusters. The M_6X_{12} structure, for instance, is the unit cell for NbO, which has a metallic luster and significant electrical conductivity precisely because the Nb–Nb bonding runs throughout the crystal. In the same way, the M_6X_8 structure is the Mo_6S_8 cube seen in the Chevrel phases discussed in Chapter 10, with their valuable electrical and magnetic properties. Sometimes the octahedron of metal atoms can accept an interstitial atom—indeed, may require an interstitial atom to be stable. The Zr_6Cl_{12} clusters seen in Fig. 3.23 and the associated discussion contain interstitial Be, but many other atoms can be accommodated with appropriate adjustments for electron count; as we shall see, there are electron counts ideal for specific cluster geometries.

On first encountering compounds containing metal–metal bonds, one tends to think of the bonds as single pairs of electrons, but in fact a wide range of multiple-bond possibilities exists. Many of these have been observed, if we agree that unusually short bond lengths and a substantial range of inferred bond energies constitute the observation of multiple bonds. However, to interpret these observations it is necessary to have some sort of model for the possible orbital overlaps that can

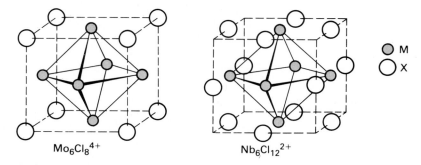

Figure 13.24 Octahedral cluster structures for MX_2.

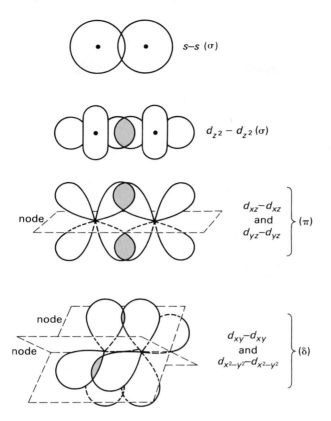

s–s (σ)

$d_{z^2} - d_{z^2}$ (σ)

node

$\left.\begin{array}{c} d_{xz} - d_{xz} \\ \text{and} \\ d_{yz} - d_{yz} \end{array}\right\}$ (π)

node

node

$\left.\begin{array}{c} d_{xy} - d_{xy} \\ \text{and} \\ d_{x^2-y^2} - d_{x^2-y^2} \end{array}\right\}$ (δ)

Figure 13.25 Transition-metal AO overlaps for M—M bond formation.

lead to multiple bonding. If we consider the qualitative MO energy levels for M_2, a diatomic system in which both atoms have low-energy d and s orbitals at least partially populated by electrons and higher-energy vacant p orbitals, we can see that it is theoretically possible to have a sextuple bond. The possible orbital overlaps are shown in Fig. 13.25.

Because d–d overlap is possible, we can have not only σ and π but also δ bonding (two nodes containing the bond axis). Just as π overlap produces a smaller energy effect than σ overlap at the same internuclear distance because of the angular properties of the atomic orbitals, δ overlap produces an even smaller effect than π overlap because of the substantial directionality of the d orbitals that must necessarily be involved. Depending on the d–s atomic-orbital energy spacing, as many as six bonding MOs can arise strictly from metal–metal overlap before any antibonding orbital must be filled, as Fig. 13.26a indicates.

The figure is strictly schematic; it should not be taken literally as to the spacing—or even the precise order—of energy levels. In particular, when the two

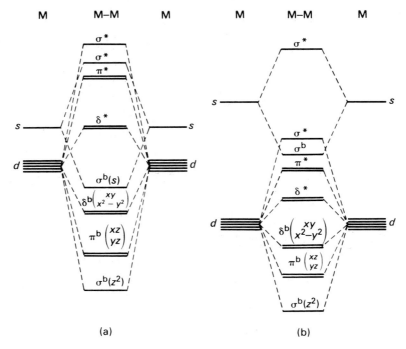

Figure 13.26 MO energy levels for diatomic M—M bonding.

metal atoms have bridging atoms such as Cl or S, the p orbitals of the bridging atoms that lie parallel to the bond axis will turn the $M_2\delta^*$ MO into a σ^b MO, and vice versa for the δ^b; as a result, the "δ^*" for a bridged M_2 system will frequently lie at lower energy than the "δ^b", unusual as that may seem.

The unbridged molecules Cr_2 and Mo_2, which are obviously accessible only at very high temperatures, have 12 valence electrons and should on the basis of these energy levels have a sextuple bond. Mass-spectrometric measurements on these two systems indicate bond energies of 150 kJ/mol for Cr_2 and 406 kJ/mol for Mo_2. The Mo_2 value seems high enough to indicate significant multiple bonding, but the Cr_2 value is surprisingly low.

A molecule for which multiple bonds between two metal atoms have been fairly well demonstrated is the $Re_2Cl_8^{2-}$ ion (and the isoelectronic $Mo_2Cl_8^{4-}$ ion), whose structure is shown in Fig. 13.27. It is particularly interesting that the two sets of four Cl^- ligands are in the eclipsed configuration, which gives an important clue to the bonding. For simplicity, let us assume that the M—Cl bonding at each metal atom is square planar and involves dsp^2 hybrid orbitals formed from the $d_{x^2-y^2}$ orbital (using the coordinate axes shown in Fig. 13.27). The four d orbitals remain for metal–metal bonding using the overlaps shown in Fig. 13.25: one σ pair, two π pairs, but now only one δ pair. Since the formal oxidation state of Re

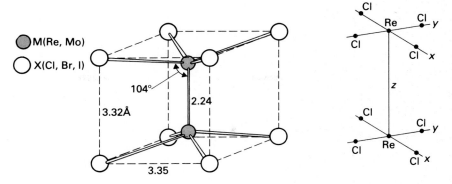

Figure 13.27 Metal–metal bonding in $M_2X_8^{n-}$.

in the complex is 3+, each Re has four valence electrons: The total of eight electrons just occupies all four bonding $d–d$ orbitals and produces a quadruple bond. Experimental support for this assignment lies in the very short Re–Re distance (2.24 Å versus 2.75 Å in the metal and a covalent-radius sum of 3.18 Å) and in the high bond energy. The latter is still only poorly determined from vibrational spectra, but is believed to be 480–540 kJ/mol. Perhaps a more compelling argument, however, lies in the eclipsed geometry of the Cl atoms. Only in the eclipsed configuration can the two Re $5d_{xy}$ orbitals overlap to form a δ bond, whose energy contribution has been estimated at 46 kJ/mol bonds; in the sterically preferable staggered configuration, the d_{xy} orbitals have exactly zero overlap and no δ bond is possible.

Figure 13.28 shows an interesting structure comparison between $Mo_2Cl_8^{4-}$ (with the octachlorodirhenate structure), $Mo_6Cl_8^{4+}$, and the bridging acetate structure shown earlier. The Cl_8 cube is deformed remarkably little on going from

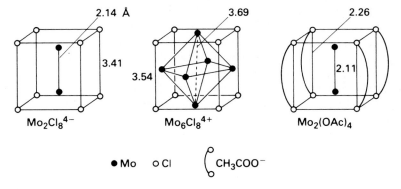

Figure 13.28 Cubic coordination of molybdenum clusters.

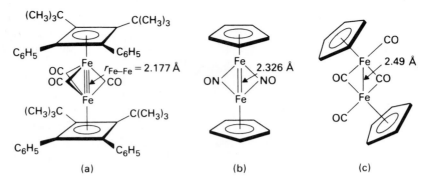

Figure 13.29 Multiple metal–metal bonding in comparable organoiron structures.

containing two Mo atoms to containing six Mo atoms. The symmetry of the two-centered structure is quite like the bridged molybdenum acetate.

Besides the halide and carboxylate species that show multiple bonds between metal atoms, a number of such bonds are thought to exist in carbonyl, nitrosyl, and alkene complexes. In general, such bonding is assumed in order to satisfy the 18-electron rule for both metal atoms. This is a legitimate principle, but not an infallible one. Almost the only experimental evidence to be had is the metal–metal bond length, since bond-dissociation energies are complicated by the almost invariable presence of bridging ligands. However, the limited evidence seems to support multiple bonding in a number of cases. For example, the cyclobutadiene complex in Fig. 13.29a can be compared with $Fe_2(CO)_9$ (Fig. 13.1), since both molecules have three CO groups bridging the two iron atoms. The 18-electron rule would require a triple bond in the cyclobutadiene complex but only a single bond in the carbonyl. Experimentally, the Fe–Fe distance in the cyclobutadiene complex is 2.177 Å, which is substantially shorter than the equivalent distance in the carbonyl, 2.46 Å. This seems to confirm multiple bonding. Similarly, the nitrosyl in Fig. 11.29b requires a double bond to satisfy the 18-electron rule, whereas the carbonyl of similar structure in Fig. 11.29c presumably has only a single bond. The Fe–Fe bond length for the double bond is significantly shorter: 2.326 Å versus 2.49 Å. Unfortunately, the bond length of formally multiply bonded compounds is influenced a good deal by the nature of the ligands present. There can be differences of as much as 0.2 Å between the lengths of bonds that would seem to have the same bond order for the same metal atoms. Some caution is clearly indicated when interpreting multiple-bond lengths in species with significantly different stoichiometry or structure.

13.6 METAL CLUSTERS

So far, the discussion of metal–metal bonding has dealt primarily with complexes or lattices containing only two bonded metal atoms—that is, with an isolated bond region. Many species, however, contain three or more metal atoms, all bonded to each other. The term *metal cluster* is usually reserved for species with three or more metal atoms, though we should not assume that the bonding is formally any different for clusters than for two-atom (binuclear) species.

Figure 13.1 has already illustrated the fact that an $Fe(CO)_4$ unit can replace a single CO in $Fe_2(CO)_9$ to yield the trinuclear triangular cluster $Fe_3(CO)_{12}$. Most trinuclear clusters are triangular, with each metal atom bonded to both the others. However, clusters are also known in which a seemingly open chain of three metal atoms is bridged by other atoms, frequently from groups V or VI (see Fig. 13.30f). Note, however, that even though sulfur is unquestionably nonmetallic, this cluster and others like it are probably better thought of as five-atom heteronuclear clusters—that is, the nonmetal atoms should be included in the cluster. The figure shows three modes of bonding for CO in trinuclear clusters: terminal, bridging, and triply bridging. In the latter, a CO is normal to the triangular surface of the M_3 cluster and presumably contributes two electrons to the whole cluster total. This suggests that the 18-electron rule needs to be extended to cover the electron total for a cluster. In the next section some interesting proposals are discussed for such rules as they relate to cluster geometry.

One final note on the trinuclear clusters in Fig. 13.30 is that the $Re_3Cl_{12}^{3-}$ ion is one of the most important halide clusters (along with $Mo_6Cl_8^{4+}$ and $Nb_6Cl_{12}^{2+}$ clusters). A number of related structures are known that preserve the M_3 triangle for Re and a few other metals, particularly Nb and Ta.

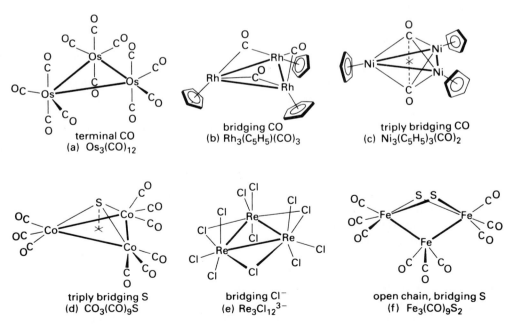

Figure 13.30 Typical three-atom cluster geometries.

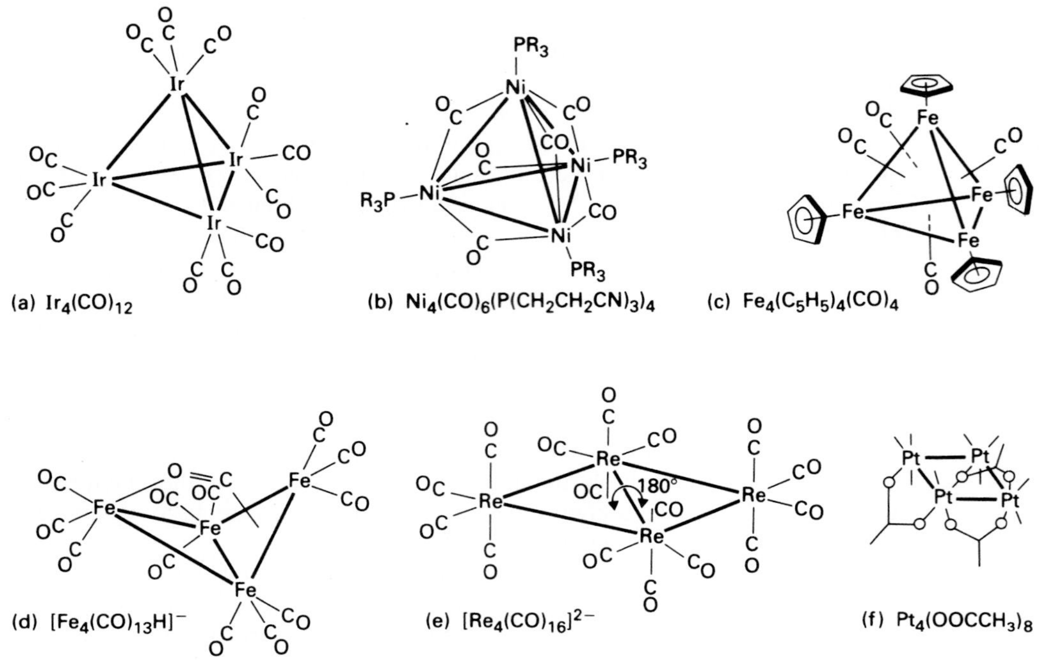

Figure 13.31 Typical four-atom cluster geometries.

Four-atom (tetranuclear) clusters occur in three fundamentally different geometries. Most of these are near-regular tetrahedra, but a significant group take on the "butterfly" structure, in which the four metal atoms form two equilateral triangles sharing an edge (see Fig. 13.31). In a special case of the butterfly geometry all four atoms are in a plane, with a dihedral angle of 180°. Finally, there is at least one case of a square-planar cluster. In the butterfly geometry, there is frequently a bridging ligand over the hinge (the shared edge) of the cluster. This bridge bonds to both outer atoms. A unique case of this bonding is shown in Fig. 13.31d, where a CO molecule bonds through both the C and the O, thus serving as a four-electron donor. In other tetranuclear clusters, however, CO is found in ordinary coordination geometry; structures (a), (b), and (c) show terminal, bridging, and triply bridging CO ligands.

Five-atom (pentanuclear) clusters are normally trigonal bipyramidal in their geometry (see Fig. 13.32). However, there is one rather striking square-pyramidal cluster, $Fe_5(CO)_{15}C$, also shown in Fig. 13.32. The carbide carbon atom is five-coordinate, since the Fe—C distances vary only from 1.89 Å to 1.96 Å. Its principal function seems to be contributing four electrons to the cluster-bonding total. This can have the same cluster-stabilizing role as the interstitial atom in the clusters seen in Chapter 3 (for instance, Fig. 3.21), which we shall return to in the next section.

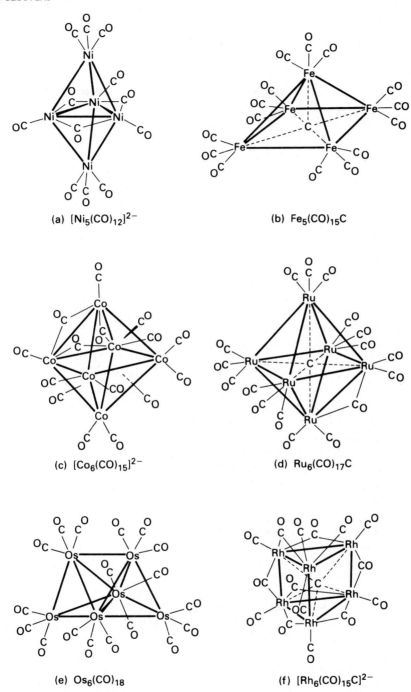

(a) $[Ni_5(CO)_{12}]^{2-}$

(b) $Fe_5(CO)_{15}C$

(c) $[Co_6(CO)_{15}]^{2-}$

(d) $Ru_6(CO)_{17}C$

(e) $Os_6(CO)_{18}$

(f) $[Rh_6(CO)_{15}C]^{2-}$

Figure 13.32 Typical five-atom and six-atom cluster geometries.

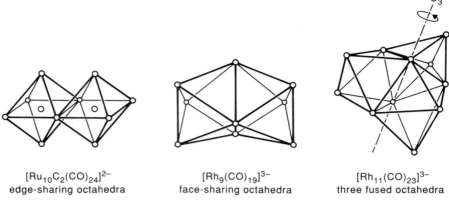

$[Ru_{10}C_2(CO)_{24}]^{2-}$
edge-sharing octahedra

$[Rh_9(CO)_{19}]^{3-}$
face-sharing octahedra

$[Rh_{11}(CO)_{23}]^{3-}$
three fused octahedra

Figure 13.33 Linked M_6 octahedra in larger cluster cores.

Hexanuclear clusters are also shown in Fig. 13.32. They are most commonly octahedral, but four other geometries are known: a capped square pyramid, a bi-capped tetrahedron, a trigonal prism, and a triangular planar network. In the next section, we shall qualitatively outline a molecular-orbital treatment that is reasonably successful in predicting these changes in geometry. The figure shows an octahedral complex in which a carbon atom is held within the metal-atom cage in a six coordinate position. The distinction between this *encapsulated*-bonding geometry and the usual rules for carbon bonding is striking (see Fig. 13.32b, d, and f). A number of similar compounds with other encapsulated atoms have been prepared: C in such clusters as $[Co_8(CO)_{18}C]^{2-}$ (a square antiprism of Co atoms), P in $[Rh_9(CO)_{21}P]^{2-}$, and H in $[Ni_{12}(CO)_{21}H]^{3-}$. We have seen other encapsulating clusters earlier, of course; it is likely that for some clusters such interstitial atoms are necessary in that the cluster geometry would be unstable without them.

These formulas for encapsulated-atom heteronuclear clusters suggest correctly that the number of metal atoms in a cluster can become quite large. Examples are known in which the nuclearity (number of metal atoms) ranges from 7 to 15, and many larger clusters are known in which 20 to about 50 metal atoms adopt certain stable geometries. Figure 13.33 shows three cluster structures that consist of linked octahedra. The Ru_{10} cluster adopts a geometry in which two Ru_6 octahedra share an edge; the Rh_9 cluster shows two Rh_6 octahedra sharing a face; and the Rh_{11} cluster shows three Rh_6 octahedra sharing two faces each in a triangular fused array. The octahedral geometry can continue to large systems, as seen in Fig. 13.34. The Pt_{38} cluster is a truncated octahedron, but the $Ni_{38}Pt_6$ cluster not only has overall octahedral symmetry with four Ni atoms along each edge but a Pt_6 octahedron at the core of the large cluster.

The octahedron is by no means sacred, however. Figure 13.35 shows three clusters with trigonal symmetry with interesting nonoctahedral geometries. Note that in the Rh_{13} cluster the central atom is in a hexagonal close-packed site equivalent to

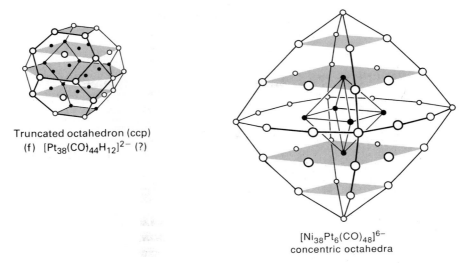

Truncated octahedron (ccp)
(f) $[Pt_{38}(CO)_{44}H_{12}]^{2-}$ (?)

$[Ni_{38}Pt_6(CO)_{48}]^{6-}$
concentric octahedra

Figure 13.34 Large octahedral cluster cores containing close-packed metal atoms.

that for each Pt in the $Ni_{38}Pt_6$ cluster of Fig. 13.34, even though no overall octahedral symmetry exists. There is also an intriguing tendency for icosahedral clusters to form, along with others also having fivefold symmetry; Fig. 13.36 shows two clusters with constraining fivefold symmetries; the Pt_{19} cluster is prismatic, while the $Au_{13}Ag_{12}$ cluster has its five-membered rings arranged antiprismatically, so that the overall geometry is that of three fused icosahedra.

There is a sense, of course, in which a particle of the bulk metal is just a very large cluster—or, alternatively, we may say that a large cluster molecule is a model of the bulk metal. One difference is that the cluster is usually coated with covalently bound CO or phosphine molecules; another is that the curvature of the "metal" surface is much higher for the cluster than for a bulk metal surface unless the cluster is extremely large. The high curvature reflects weaker metal–metal bonding in the cluster surface than in the bulk metal. This in turn influences the

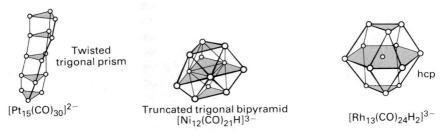

Twisted
trigonal prism

$[Pt_{15}(CO)_{30}]^{2-}$

Truncated trigonal bipyramid
$[Ni_{12}(CO)_{21}H]^{3-}$

hcp

$[Rh_{13}(CO)_{24}H_2]^{3-}$

Figure 13.35 Cluster core geometries with threefold symmetry.

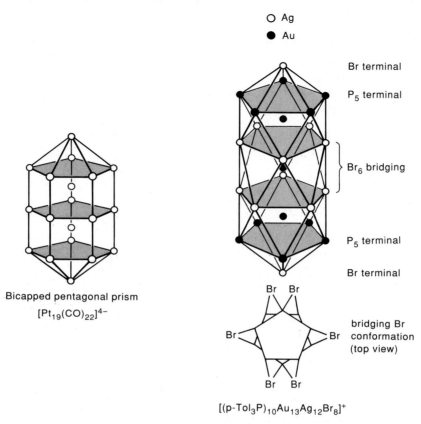

Bicapped pentagonal prism
$[Pt_{19}(CO)_{22}]^{4-}$

$[(p\text{-}Tol_3P)_{10}Au_{13}Ag_{12}Br_8]^+$

Figure 13.36   Cluster core geometries with fivefold symmetry. Ligand conformations are indicated for the $Au_{13}Ag_{12}$ cluster.

bonding of other atoms to the surface. As has already been suggested, a major focus of current research is developing homogeneous catalysts for both organic and inorganic reactions by assembling clusters that simulate the surface properties of known heterogeneous catalysts (particularly the platinum metals). It is convenient that the platinum metals Pd, Pt, Rh, and Ir are both the most versatile heterogeneous catalysts and the easiest elements to prepare in large clusters.

How large must a cluster be to have surface metal atoms with the properties of metal atoms in a surface plane of bulk metal? Unfortunately, the answer is larger than you think. Figure 13.36 has hinted at a tendency for late transition metals to form clusters in which fivefold symmetry is maintained, and there is an interesting relationship between such symmetry and the threefold symmetry characteristic of the close packing seen in bulk metals. True close packing yields coordination number 12 for each metal atom: six neighbors in the plane of the atom, three above it, and three below it. A metal atom in the center of an icosahedron of similar atoms is also 12-coordinate, but the arrangement of neighbors (in planes) is 1-5-5-1 instead

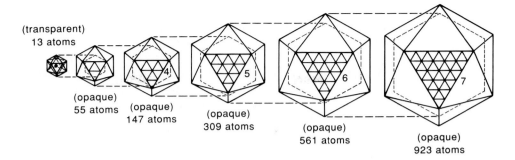

Figure 13.37 Giant cluster growth sequence from M_{13} icosahedral core. Metal atoms are at each fivefold or sixfold intersection. Numbers on clusters indicate metal atoms along each side of icosahedron. Only the largest cluster shown has facial atoms none of whose neighbors are on an edge.

of the 3-6-3 seen for a close-packed atom. As a result, the central atom pushes the 12 neighbors out slightly, and the space filling is not truly close-packed. Rearrangement of such a 13-atom core can yield a close-packed arrangement, as in the Rh_{13} cluster of Fig. 13.35. But if rearrangement does not occur before further metal-atom condensation onto the core does, the icosahedral core can grow by adding layers like an onion; each face of the M_{13} icosahedron is an almost-close-packed M_3 triangle, which can add additional atoms.

Figure 13.37 shows schematically the stages in the growth of such a cluster. In the figure, the smallest (13-atom) icosahedron is drawn as if it were transparent, with the central metal atom and the back-side metal atoms visible through the network of atoms on the front side—but only the network of lines between nuclei is shown, so that an atom must be visualized at each intersection. Putting one more layer of metal atoms on yields a 55-atom cluster core (the first 13 plus 42 surface atoms), which is shown as if it were opaque, with only the front-side atoms visible. Another layer of atoms yields a 147-atom cluster core with four atoms along each edge. This cluster is the smallest that has a surface atom that is in the center of a face, not on an edge. But all the neighbors of that atom are edge atoms, and we must conclude that even this core does not have surface metal atoms that are a good model for a bulk metal. The next stages of growth of the icosahedral cluster have 309 atoms, 561 atoms, and 923 atoms respectively, and it is only in this last giant cluster that we see surface atoms whose neighbors are not on an icosahedral edge. Most of these cluster stoichiometries have been prepared, though because the x-ray diffraction difficulties for large molecules are formidable not all have good structure data. For example, $Co_{55}(PMe_3)_{12}Cl_{20}$ has been prepared, as have $Ni_{147}(PPr_3)_{12}Cl_{24}$, $Pt_{309}(phenSO_3)_{36}O_{30\pm10}$, and $Pd_{561}(phen)_{36}O_{190-200}$ and $Pd_{561}(phen)_{60}(OAc)_{180}$. As some of the formulas indicate, we have only tentative information about some of these superclusters. In these formulas *phen* is *o*-phenanthroline and *phenSO_3* is *p*-sulfonylphenanthroline. The Pd_{561} acetate cluster has been investigated more than some others; it has phenanthrolines bridging

each icosahedral edge and an acetate shell around the rest of the core. It is a catalyst for some organometallic reactions that we shall encounter in Chapter 15, and
that catalytic ability is maintained if the acetates are replaced by other ions such as
oxide or PF_6^-.

Although most of the atoms in these large cluster cores are approximately close-
packed, there remains the initial misfit around the central atom and resulting strain.
As a result, the larger versions of these clusters tend to develop cracks not unlike
tectonic plate boundaries. The larger the core becomes, the more likely it is to
rearrange to a close-packed form, at which point it seems to agglomerate to bulk
metal particles and the solubility advantage of the cluster is lost. However, as a
final note on these clusters we can call attention to mass-spectral evidence for a
gold cluster with molecular weight above 140,000 that is surmised to be $(Au_{13})_{55}$,
in which the 55-sphere cluster of Fig. 13.37 has been formed from "spheres" of
Au_{13} icosahedra.

The synthesis of metal clusters has not been fully systematized, but several
general approaches have been developed. The condensation of smaller carbonyls is
quite endothermic, because strong M—CO bonds are broken and replaced by relatively weak M—M bonds:

$$2Co_2(CO)_8 \rightarrow Co_4(CO)_{12} + 4\ CO \qquad \Delta H = 138\ kJ/mol\ rn$$

$$3Rh_4(CO)_{12} \rightarrow 2Rh_6(CO)_{16} + 4\ CO \qquad \Delta H = 268\ kJ/mol\ rn$$

Generation of gaseous CO makes ΔS positive, however, and condensations such as
these can be entropy-driven at higher temperatures. Pyrolysis is thus an effective
route to some clusters:

$$Os_3(CO)_{12} \xrightarrow{200°C} Os_4(CO)_{13} + Os_5(CO)_{16} + Os_6(CO)_{18}$$

$$+ Os_7(CO)_{21} + Os_8(CO)_{23} + Os_8(CO)_{21}C$$

$$Rh_4(CO)_{12} \xrightarrow{80°C} [Rh_{15}(CO)_{27}]^{3-} \qquad (50\%)$$

Carbonylate anions frequently undergo a kind of redox condensation in which
the formation of a larger cluster spreads the negative charge out over a larger number of atoms, thus reducing electrostatic repulsion and partially overcoming the
bond-energy problem:

$$[Fe_3(CO)_{11}]^{2-} + Fe(CO)_5 \rightarrow [Fe_4(CO)_{13}]^{2-} + 3\ CO$$

Mild oxidation of carbonylate anions also sometimes produces coupling or condensation:

$$[Rh_6(CO)_{15}C]^{2-} + Fe^{3+} \rightarrow Fe^{2+} + [Rh_{15}(CO)_{28}C_2]^-$$

Mixed-metal clusters can be prepared through the redox condensation above, but
also by direct substitution:

$$[Fe_4(CO)_{13}]^{2-} + Co_2(CO)_8 \rightarrow [Fe_3Co(CO)_{13}]^- + Co(CO)_4^- + Fe(CO)_5$$

This approach can also build clusters by one atom in the core:

$$[Fe_3(CO)_{11}]^{2-} + Co_2(CO)_8 \rightarrow [Fe_3Co(CO)_{13}]^- + Fe(CO)_5 + ?$$

The whole field of metal-cluster chemistry is developing and changing so rapidly that it is difficult even to characterize what may be possible. We can be sure, however, that coming years will produce even more remarkable structures and will find them adapted to both organic and inorganic synthesis and to catalysis in industry.

13.7 CLUSTER-BONDING THEORIES

In Chapter 10, we developed the 18-electron rule on a molecular-orbital basis to explain the stoichiometry and (to some extent) the geometry of mononuclear carbonyls and alkene complexes—or, more generally, the complexes of a single transition-metal atom with good sigma-donor/pi-acceptor ligands. If we postulate the formation of single and multiple metal–metal bonds the 18-electron rule is still an excellent guide to the stoichiometry of dinuclear complexes and even small clusters. However, at about hexanuclear-cluster size, the 18-electron rule becomes unwieldy and even misleading. Some effort, therefore, has gone into developing reasonably simple theoretical models for clusters that will allow the prediction of the number of CO ligands and the net charge for a given cluster size, as well as the prediction of the cluster geometry where several shapes are available. There are two general approaches that have shown good ability to correlate these quantities, though neither in its present form constitutes a completely accurate predictive model. Both are extensions of simple molecular-orbital calculations, and they have some features in common although their emphases are different. Both yield electron-counting rules or structure correlations analogous to the 18-electron rule or to the VSEPR geometry prediction rules for p-block compounds.

The simpler approach is to extend Wade's rules, which we developed in Chapter 4 (section 4.9) for main-group clusters. For our purposes in this chapter, the interesting property of Wade's rules is that they can readily be extended to transition-metal metallaboranes—and even to transition-metal clusters with no main-group elements present in the cluster. All that is necessary is to adjust the number of orbitals participating in cluster bonding to take into account the added d orbitals present for each metal atom. For main-group elements, we assumed that one valence orbital would be used to bind external ligands and that three would be available for cluster overlap. For transition metals, Wade assumes that three of the nine valence orbitals will be used as external-ligand acceptor orbitals and that three more will be low-energy and essentially nonbonding (so that they are filled before any bonding begins). This leaves three cluster-bonding orbitals and a variable number of cluster-bonding electrons, depending on how many are donated by the external ligands. Since six orbitals are going to be filled by electrons before any cluster bonding begins, the cluster-bonding electron contribution for any given transition-metal atom becomes the number of its valence electrons v, plus the number x donated by external ligands on that atom, minus six pairs: $v + x - 12$. Table 13.4 indicates the number contributed

Table 13.4 Cluster-Bonding Electrons Contributed by Transition Metals ($v + x - 12$)

v	Transition metal M	Cluster unit				
		$M(CO)$	$M(CO)_2$	$M(cp)$	$M(CO)_3$	$M(CO)_4$
5	V, Nb, Ta	−5	−3	−2	−1	1
6	Cr, Mo, W	−4	−2	−1	0	2
7	Mn, Tc, Re	−3	−1	0	1	3
8	Fe, Ru, Os	−2	0	1	2	4
9	Co, Rh, Ir	−1	1	2	3	5
10	Ni, Pd, Pt	0	2	3	4	6

in this way for various metals and ligands. Just as with bridging H atoms in boranes, bridging ligands such as CO in metal clusters are assumed to contribute their Lewis-base electrons directly to the cluster bonding. If the cluster contains encapsulated atoms, all their valence electrons are considered part of the cluster bonding. With this cluster-electron counting procedure, the structural argument is exactly the same as that for boranes: pair the cluster electrons, assume a deltahedron with one fewer vertices than the number of cluster pairs, and compare the deltahedron size with the number of cluster atoms to decide on a *closo-, nido-,* or *arachno-* geometry.

Consider the metallaborane $MnB_3H_8(CO)_3$. If we assume that the core is made up of three B−H units and a $Mn(CO)_3$ unit, then from Table 4.6 each BH contributes two cluster electrons and from Table 13.4 the $Mn(CO)_3$ contributes one cluster electron; the remaining five bridging H atoms contribute five cluster electrons, for a total of 12 cluster electrons. Such a molecule has six cluster pairs, indicating a trigonal bipyramid deltahedral geometry. There are only four cluster core atoms, so the observed tetrahedral molecular geometry can be interpreted as a *nido* trigonal bypyramid. In the same sense, $Fe_2B_2H_6(CO)_6$ has 12 cluster electrons and is, again, tetrahedral. We can, in the electron counting scheme, replace BH by $M(CO)_3$ even when M has a varying electron count, because the stoichiometry will adjust itself to yield a fixed electron count for the whole cluster. The BH fragment and the $M(CO)_3$ fragment are said to be *isolobal,* meaning that they contribute equivalent orbital lobes to the symmetry of the molecular orbitals formed by the cluster. The concept of isolobal groups is very helpful within the electron counting scheme when considering heteronuclear clusters.

For an example of a pure transition metal cluster, consider $[Co_6(CO)_{15}]^{2-}$. We can initially assume that we do not know how the CO ligands are bonded to the cluster. The simplest assumption is that there are six $Co(CO)_2$ units and three bridging COs. Then from Table 13.4 we have

$$6Co(CO)_2: 6 \times 1 = 6$$

$$3CO: 3 \times 2 = 6$$

$$2- \text{net charge}: 2 \times 1 = \underline{2}$$

$$14 \text{ cluster electrons}$$

This corresponds to seven pairs, which requires a deltahedron with six vertices—an octahedron. Since there are six cluster core atoms (the Co) the cluster has a *closo-* form, octahedral Co_6. In fact, the structure is that of Fig. 13.32c: three $Co(CO)_2$ units, three $Co(CO)$ units, three edge-bridging CO, and three triply bridging CO, but the cluster electron count is unchanged! This is a strength and a weakness of Wade's rules: They cannot predict ligand-bonding patterns, but they frequently arrive at the correct cluster geometry anyway.

The $v + x - 12$ stipulation for d-electron clusters within Wade's rules is effective for metals with numerous d electrons, as in the Co_6 cluster we have just seen, but artificial and misleading for the early transition metals; note that Ti/Zr/Hf do not even appear in Table 13.4. However, it is still useful to consider, as we did in Figure 4.21 and the associated discussion, that an octahedral cluster will have seven bonding MOs and should therefore have 14 cluster electrons, in keeping with Wade's basic electron count. Corbett has prepared a number of $[Zr_6Cl_nZ]^{q-}$ clusters with octahedral Zr_6, bridging Cl atoms, and interstitial main-group Z, as for instance $[Zr_6Cl_{15}C]^-$ and $Zr_6Cl_{12}Be \cdot 6PEt_3$. The cluster electron count for these is similar to the calculation for a main-group cluster: four for each Zr minus one for each Cl (filling the Cl valence orbitals outside the cluster) plus n_Z valence electrons from the encapsulated atom plus q net charge adjustment, or $(4 \times 6) - (1 \times 15) + (4 \times 1) + 1 = 14$ for $[Zr_6Cl_{15}C]^-$. Most of these Zr_6 clusters with encapsulated atoms prove to have the Wade count of 14 cluster electrons, though encapsulated transition metals such as Fe or Cr tend to have 18 cluster electrons because the encapsulated atom has a pair of d AOs that are nonbonding (and therefore low energy and filled) with respect to the Zr_6 orbitals.

From Fig. 4.21 it can be seen that an encapsulated atom's s orbital will interact very strongly with what is described in the figure as the "B–B cluster" MO, producing a strongly bonding MO that is responsible for the seventh low-energy electron pair. In first-row clusters such as boranes, the small inner core of each B atom also makes this centered MO strongly bonding because overlap is good. But in clusters of heavier metals that do not have encapsulated atoms the longer internuclear distances mean weaker overlap, and this centered MO may not be as strongly bonding. If outer bridging ligands make the six "sigma ring" and "pi ring" MOs of Fig. 4.21 more stable, those six MOs can easily drop below the centered MO. In that event, the cluster electron count might well be 12 or 13 rather than 14, and the molecules $Zr_6Cl_{12}(PR_3)_6$ with no encapsulated atom have a 12-electron count. For other geometries and other metals, similar overlap considerations can yield other complications in the simple electron count, which is essentially the reason that Wade's rules are an excellent guide to borane geometries but less useful for transition metal clusters.

The second cluster model, that of Lauher, also relies on electron counting through a molecular-orbital calculation for each possible geometry for a given cluster. However, Lauher does a separate MO calculation for each possible geometry of the metal atoms in the cluster without considering ligands at all, using rhodium as a typical cluster-forming metal. The resulting MOs number nine per metal atom, since each Rh (or other transition metal) has five d orbitals, one s

Table 13.5   Electron-Counting Data for Metal Clusters (Lauher MO Model)

Geometry	Metal AO Total MO	Vacant HLAO	Filled CVMO	Cluster e^-	Example
(3) Triangle	27	3	24	48	$Os_3(CO)_{12}$
(4) Tetrahedron	36	6	30	60	$Rh_4(CO)_{12}$
(4) Butterfly	36	5	31	62	$[Re_4(CO)_{16}]^{2-}$
(4) Square planar	36	4	32	64	$Pt_4(CH_3CO_2)_8$
(5) Trigonal bipyramid	45	9	36	72	$Os_5(CO)_{16}$
(5) Square pyramid	45	8	37	74	$Fe_5(CO)_{15}C$
(6) Bicapped tetrahedron	54	12	42	84	$Os_6(CO)_{18}$
(6) Octahedron	54	11	43	86	$Ru_6(CO)_{17}C$
(6) Trigonal prism	54	9	45	90	$[Rh_6(CO)_{15}C]^{3-}$
(7) Capped octahedron	63	14	49	98	$[Rh_7(CO)_{16}]^{3-}$
(8) Triangular dodecahedron	72	16	56	112	—
(8) Square antiprism	72	15	57	114	$[Co_8(CO)_{18}C]^{2-}$
(8) Bicapped trigonal prism	72	15	57	114	$[Co_8(CO)_{18}C]^{2-}$ intermediate
(8) Cube	72	12	60	120	$Ni_8(PC_6H_5)_6(CO)_8$
(9) Tricapped octahedron	81	18	63	126	—
(9) Tricapped trigonal prism	81	17	64	128	—

orbital, and three p orbitals. For the transition metals toward the right of the d block, the d orbitals have only modest overlap and lie at quite low energies; furthermore, the p orbitals are at rather high energies, even compared to the s orbital. The result is that the antibonding orbitals composed primarily of p AOs lie at such high energies for the bare metal-atom cluster that they cannot serve even as acceptors for ligand electrons. Lauher proposes that the metal-cluster MOs up to about the metal p-orbital energy serve as acceptors for ligand electrons as well as bonding the cluster internally. These MOs are called *cluster valence molecular orbitals* (CVMOs). Those MOs significantly higher than the Rh p AO (and there is normally a substantial gap near that energy) cannot be occupied by electrons and are thus called *high-lying antibonding orbitals* (HLAOs). The total number of MOs for a cluster with n metal atoms must be $9n$. The critical question, however, is how many of these are HLAOs, since the electron capacity of the cluster will be $2 \times$ (no. of CVMO). It should be clear that, inasmuch as all metal valence electrons are being counted even for late transition metals, this electron count will be much higher than the Wade count for a given geometry; in general the Lauher count will be about $12n$ greater than the Wade count for the same deltahedral geometry of n atoms.

Table 13.5 gives the total number of AOs/MOs, the number of HLAOs, the number of CVMOs, and the resulting Lauher cluster-electron count for a number of possible geometries, along with some examples of clusters whose geometry is correctly predicted by Lauher's approach. As with Wade's rules, Lauher's model does not predict the correct electron count (or, alternatively, the geometry) for some systems. It is particularly inaccurate for Pt clusters, which have p orbitals at such high energy that not all the CVMOs are filled. However, the model is obviously powerful and gives us added understanding of cluster stability.

PROBLEMS

A. DESCRIPTIVE

A1. Write a balanced equation for a reaction that could yield $Fe_2(CO)_9$, starting with a halide. Discuss the probable composition of the product mixture and indicate how you would optimize the yield of $Fe_2(CO)_9$.

A2. Remembering that Si—H bonds are much more reactive than C—H bonds, suggest a method of preparing R_3Si—$Co(CO)_4$. Suggest reasons for the shift in IR stretch frequency on changing from R=Cl to R=C_2H_5:

R	Wavenumber of C—O stretch (cm^{-1})
Cl	2120, 2070, 2040, 2030, 2000
C_2H_5	2090, 2030, 2000, 1960

A3. No cyclohexyltitanium compound has ever been isolated, but the norbornyl compound $Ti(C_7H_{11})_4$ seems to be quite stable. Why is one stable and not the other?

A4. Suggest a synthesis of the mixed-metal carbonyl $(OC)_5Mn$—$Re(CO)_5$ that is strongly favored thermodynamically.

A5. Figure 13.1 calls attention to the structural similarities between $Co_2(CO)_8$, $Fe_2(CO)_9$, and $Fe_3(CO)_{12}$. Given the fact that $Fe_2(CO)_9$ can be formed from $Fe(CO)_5$ by irradiation with UV light and that $Fe_3(CO)_{12}$ is a thermal decomposition product of $Fe_2(CO)_9$, predict the reaction product when $[Mn(CO)_5]^-$ is irradiated with $Fe(CO)_5$ in an inert solvent. Sketch the structure of the product.

A6. The P—F IR stretch frequency is higher for $Ni(PF_3)_4$ than for free PF_3, but it is lower for $[Co(PF_3)_4]^-$ than for free PF_3. Why?

A7. Four complexes with the formula $FeCl_2(PR_3)_2$ have been prepared where PR_3 is respectively PEt_3, PEt_2Ph, $PEtPh_2$, and PPh_3. The complexes of the last two R groups are much more stable than the first two, although all four are high-spin tetrahedral complexes. Why is there a difference in stability?

A8. Sketch the π^* acceptor orbital for an NO molecule in a bent nitrosyl complex.

A9. In the Pt complex with TCNE, why are the CN groups bent back from the C–C bond axis and the metal atom (Fig. 13.7)?

A10. Zirconium cyclopentadienyl amides, $(cp)_2Zr(NHR)_2$, undergo an equilibrium in which RNH_2 is released and an imide with a Zr$=$N bond is formed. The imide can react with alkynes such as PhC$\equiv$CPh to form an adduct. Propose a structure for the

1:1 adduct, and discuss its electron count and your reasons for choosing that geometry and bonding pattern.

A11. Ni(CO)$_4$ reacts with allyl bromide to form a complex in which there are no carbonyl groups and only one allyl per Ni atom. Propose a structure for the product, and write a balanced equation for its formation. Should there be Ni–Ni bonds in the complex? Why or why not?

A12. Why is it necessary to *oxidize* the 1,3-pentadiene complex of iron, Fe(C$_5$H$_8$)(CO)$_3$, to get the *pentahapto-* planar pentadienyl complex?

A13. The reaction between Bu$_3$SnCl and Fe(CO)$_5$ yields, among other things, Sn(Fe(CO)$_4$)$_4$. What general bonding geometry do you expect around the Sn atom? If the 18-electron rule is applied to the Fe atoms, will it modify the Sn geometry?

A14. In the complexes *cis*-(R$_3$P)$_3$Mo(CO)$_3$, where the PR$_3$ groups are those listed below, there is a clear trend in the CO stretch IR absorption. What changes in bonding are occurring?

PR$_3$	IR absorption (cm^{-1})
PF$_3$	2074, 2026
PCl$_3$	2041, 1989
PCl$_2$Ph	2016, 1943
PClPh$_2$	1977, 1885
PPh$_3$	1949, 1835

B. NUMERICAL

B1. It has been suggested that WCl$_5$ is a trimeric cluster [W$_3$Cl$_{12}$]$^{3+}$(Cl$^-$)$_3$ on the basis of its solution conductivity and its magnetic moment (which corresponds to one unpaired electron per three W atoms). The sequential ionization energies of tungsten are not known beyond IP$_2$. Use a Born–Haber cycle to calculate a lower limit for the sum of the third through fifth ionization energies for W, assuming that the formation of a cluster halide represents instability of the ionic lattice.

B2. The complex (OC)$_4$Cr(μ–PMe$_2$)$_2$Cr(CO)$_4$ has the butterfly geometry for the two Cr and two P atoms, but the cluster is planar (the dihedral angle is 180°). The related complex (OC)$_3$Fe(μ–PMe$_2$)$_2$Fe(CO)$_3$ has the bent butterfly structure (dihedral angle < 180°). The proton NMR spectra of the two compounds are different, because the coupling constants are different for the two geometries. When both compounds are reduced electrochemically to dianions, the NMR spectrum of [Fe cluster]$^{2-}$ is similar to that of the neutral compound, but the [Cr cluster]$^{2-}$ spectrum now resembles that of the Fe cluster rather than that of the neutral Cr cluster. Show that all of these structural data are consistent with Wade's rules for heteronuclear cluster geometry.

B3. The interpretation of C–C stretch frequencies in the IR spectra of metal–alkene complexes is complicated by the interaction of the C=C stretch mode with a CH$_2$ bending mode near the same frequency. However, comparisons are possible for closely related structures. The square-planar complexes *trans*-PtCl$_2$(C$_2$H$_4$)(X-pyridine N-oxide) show changes in the C=C frequency as substituents in the 4-position on the pyridine ring are varied. These changes can be correlated with the pK_a of the substituted pyridine oxide. Plot the data below to show the extent of correlation, and relate the results to the bonding in the complexes.

4-Substituent on pyO	pK_a	C=C stretch (cm^{-1})
—OCH$_3$	2.05	1490
—CH$_3$	1.29	1500
—H	0.79	1510
—Cl	0.36	1515
—COOCH$_3$	−0.41	1528
—NO$_2$	−1.7	1545

B4. Why is no Ni–Ni bond proposed for the structure of the cyclobutadiene complex ((C$_4$Me$_4$)NiCl$_2$)$_2$ shown on p. 668?

B5. The cycloheptadienyl complex Mo(CO)$_2$(cp)(C$_7$H$_7$) has a proton NMR peak at τ = 5.21 due to the C$_7$H$_7$ protons. At room temperature the peak is a sharp singlet, but at −40°C the peak is much broader. Why?

B6. Apply Wade's rules to Fe$_3$(CO)$_9$S$_2$ (Fig. 13.30f). Do the rules predict the correct structure? What assumption must be made about the number of electrons contributed by each S? Is this a reasonable assumption?

B7. Compare the structural predictions of Wade's model and Lauher's model for Fe$_4$(cp)$_4$(CO)$_4$(Fig. 13.31c) and [Fe$_4$(CO)$_{13}$H]$^-$ (Fig. 13.31d). In the latter case, do both theories support the idea that CO is a four-electron donor?

B8. Triphenylphosphine reduces Cu(II) to Cu(I). When it reacts with a solution of CuBr$_2$, the product is a tetramer [(PPh$_3$)CuBr]$_4$, in which Cu and Br atoms are located at alternate corners of a cube, each Cu having a terminal PPh$_3$ ligand. Assume that Br atoms have no d orbitals and show that this structure satisfies Lauher's cluster model but not Wade's rules. When the tetramer reacts with Tl$^+$C$_5$H$_5^-$, the product is (cp)Cu(PPh$_3$). Show that Lauher's model satisfies no cluster structure, so that the complex should be a monomer.

C. EXTENDED REFERENCE

C1. D. J. Darensbourg [*Inorg. Chem.* (1980), *20*, 1911] has studied tetrahedral Co$_4$ clusters, in particular the substitution of other ligands for CO in Co$_4$(CO)$_{12}$. This cluster, unlike the isoelectronic Ir$_4$(CO)$_{12}$ in Fig. 13.31a, has three bridging CO units and less than complete tetrahedral symmetry. Sketch the possible isomers of Co$_4$(CO)$_{11}$(PPh$_3$) (where one CO has been removed for a triphenylphosphine), and the possible isomers of Co$_4$(CO)$_{10}$(PPh$_3$)$_2$. Assume that the substitution is made in a terminal-ligand position and that there is no change in the bridging symmetry. How many isomers of each sort would there be for the Ir$_4$(CO)$_{12}$ symmetry?

C2. M. C. Jennings *et al.* [*Organometallics* (1991), *10,* 580] report a series of Pt_3Sn clusters that they propose as a model for bonding of Sn to metallic Pt surfaces in heterogeneous catalysts. Their complexes **3** and **4** are Pt_3 triangles, each edge bridged by $Ph_2PCH_2PPh_2$ (dppm), with a SnX_3 forming a Pt_3Sn tetrahedron and the opposite side of the Pt_3 triangle either bare (**3**) or having a μ_3 CO (**4**); each complex is a singly charged cation. They describe the complexes as having 42 and 44 electrons respectively. Derive these numbers and comment on their relevance to the Lauher model. What geometry would be predicted by the Wade model? The authors note that "these clusters are very crude models for a platinum surface." Comment on the basis for this cautionary statement.

C3. R. Weiss and R. N. Grimes [*J. Am. Chem. Soc.* (1977), *99,* 8087] have prepared two isomers of the ferraborane $(cp)FeB_5H_{10}$. Do both isomers conform to the geometry predicted by Wade's rules for the formula above? Explain.

C4. The nitrosyl complex $Co(NO)(PEt_3)_2Cl_2$ has only one NO group, but it shows two N—O stretch peaks in its IR spectrum at 1720 and 1650 cm^{-1} (in CH_2Cl_2 solution). In the solid state, x-ray data show only one coordination geometry, a trigonal bipyramid with the NO in an equatorial position and a Co—N—O angle at 165°. Where do the two IR peaks come from? [J. P. Collman *et al., J. Am. Chem. Soc.* (1971), *93,* 1788.]

C5. Triangular Ru_3 complexes usually have 48 cluster electrons, in keeping with the Lauher model. Lavigne and others [*Inorg. Chem.* (1991), *30,* 4112] report a 46-electron species with the alkyne diphenylacetylene coordinated in an unusual way (their compound **2**). Under CO **2** rapidly changes to **3** with very different alkyne coordination. Describe the bonding role of the alkyne for each compound, including electron counts and overlap symmetries.

Part **V**

Transition-Metal Reactions ───────

14

Reaction Mechanisms for Donor–Acceptor Compounds ─────

The previous four chapters have provided an introductory survey of compounds formed by the *d*-block transition metals, emphasizing bonding and structural aspects of these compounds. In this last part of the book, we shall inquire into what is known about the way transition-metal compounds react. Inevitably, this means focusing on reaction mechanisms; unfortunately, this area of inorganic chemistry was rather late to develop. Organic chemists understand a great deal about the patterns of chemical reactivity shown by organic compounds—they can systematically think in mechanistic terms about almost any reaction observed in the laboratory. Inorganic chemists are only beginning to approach that habit of thought with respect to the stunning array of diverse reactions of inorganic compounds. In this chapter we shall consider a few of the better-established mechanisms for transition-metal compounds, emphasizing primarily changes in the direct coordination of the metal atom. As previous chapters have noted, these mechanisms are interesting because metal coordination can substantially change the reactivity of a ligand, so that important kinds of catalysis are possible under favorable circumstances. We shall reserve ligand reactivity and catalysis for Chapters 15 and 16, for industrial and biochemical catalysis respectively, though a complete separation is not possible—some ligand reactivity changes and catalytic potential will be obvious at several points in the discussion of this chapter.

Three principal reaction types have become part of the inorganic chemist's mechanistic thinking. Perhaps the best-established are those for ligand-substitution reactions:

$$L_nM—D + E \rightarrow L_nM—E + D$$

where D is the departing ligand and E the entering ligand. Ligand-substitution reactions of octahedral and square planar complexes have received particular attention.

703

Less attention has been given to other coordination geometries, but some of these are unquestionably important in, for example, biochemical reactions in which a ligand is coordinated to a metal as a substrate for a metalloenzyme.

Another major reaction type for which some mechanistic information is available is that of solution redox reactions, particularly those in which both the oxidant and the reductant are transition-metal complexes. As Chapter 8 has already indicated, the most important distinction to be made in these mechanisms is between inner-sphere and outer-sphere attack. We shall look at both of these, along with some of the ingenious experiments that have been devised to demonstrate mechanisms.

Just as organic reactive species—carbonium ions or carbocations, carbanions, free radicals—dictate certain reaction mechanisms when they are formed, certain combinations of valence-electron count and coordination number yield comparable reaction types for transition-metal complexes. We shall look briefly at several of these. In current inorganic mechanistic thinking, the most significant is the class of reactions called *oxidative additions,* which resemble carbene reactions:

$$ML_4 + X_2 \rightarrow ML_4X_2$$

The reaction is obviously an addition, but it is also oxidative if X is a reasonably electronegative species that would normally be assigned a negative oxidation state in the product complex (for example, Cl^- if X = Cl in the above reaction). Thus, if M were in oxidation state 1+ initially, it would be assigned oxidation state 3+ in the product, obviously being oxidized. Many processes involving transition-metal complexes as catalysts operate through an oxidative addition step, and the study of these reactions (and their reverse, *reductive eliminations*) is a particularly exciting area of coordination chemistry.

14.1 LIGAND-SUBSTITUTION REACTIONS IN GENERAL

The net stoichiometric result of a ligand-substitution reaction is that the coordination number of the metal does not change:

$$L_5M—D + E \rightarrow L_5M—E + D$$

In this rather general case, the metal remains six-coordinate. It should be obvious, however, that within the substitution mechanism there are three possible sequences: D leaves first (giving a five-coordinate intermediate); E arrives first (giving a seven-coordinate intermediate); or a cooperative interchange occurs in which D leaves as E arrives (in which case the coordination number may not be well defined). It is helpful to categorize these possibilities in terms of the metal-to-ligand internuclear distance when the entering ligand and the departing ligand are at the *same* distance R from the metal: $r_{M—D} = r_{M—E} \equiv R$. Figure 14.1 shows a transition-metal atom or ion with its inner and outer coordination spheres. The inner sphere involves the metal-to-ligand bonding we have described in previous chapters; the outer sphere represents potential ligands held loosely in place by ion-pairing forces

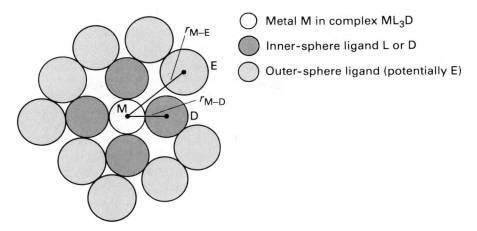

Figure 14.1 Inner and outer coordination spheres.

or ion–dipole attractions. The inner-sphere distance r_{M-D} is indicated; if E is in the outer coordination sphere, $r_{M-E} \simeq 2 \times r_{M-D}$.

Now consider the four possibilities shown in Fig. 14.2. In (a), E has arrived before D leaves, and the condition of equal metal–ligand distance R is reached when that distance is 1.0 times the equilibrium distance for a M—L bond. This is clearly a higher coordination number than existed before in the ML_3D reactant. In (b), ligand D is loosened a bit as E arrives, and the critical distance R is perhaps 1.2–1.5 times r_{M-L}. The coordination number is probably larger than in the reactant, but it

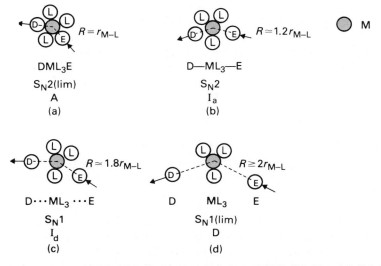

Figure 14.2. Possible ligand-substitution mechanisms: $ML_3D + E \rightarrow ML_3E + D$.

is less clear-cut. In (c), the bond between M and D has stretched so far before E approaches to the same distance (R) that D is nearly in the outer sphere—perhaps $R = 1.8–2.0 \times r_{M—L}$. Here the coordination number is probably smaller than in the reactant, but again it is difficult to specify. Finally, in (d), the equal-distance R is not reached until D has separated from M by a greater distance than even the outer sphere would correspond to: $R > 2r_{M—L}$, which unequivocally has a smaller coordination number than the reactant. There is, of course, a complete range of intermediate cases.

Two sets of mechanistic nomenclature are used to describe these cases. In the more common terminology, the reaction is said to be a *nucleophilic substitution*. The metal atom is assumed to be at least partially positively charged, and the incoming ligand E is therefore a nucleophile. In Fig. 14.2a there is a characterizable intermediate of increased coordination number that could only be formed in a bimolecular, second-order process. Such a mechanism is $S_N2(lim)$, since it represents a limiting case. Figure 14.2b represents a substantially similar case, and is usually known simply as S_N2. In a symmetrical fashion, we assume that E is so far away that it can have no influence on the reactant at all (see Fig. 14.2d). The reaction must therefore be unimolecular and first-order, or $S_N1(lim)$. In Fig. 14.2c, E has only a modest influence, so that the reaction mechanism is said to be S_N1. These terms are appropriated directly from organic mechanisms, though the definitions are altered slightly.

In the other nomenclature, case (a) is said to represent an *associative* process A, case (d) represents *dissociative* process D, and cases (b) and (c) represent interchanges (I) between inner-sphere ligand D and outer-sphere ligand E. Case (b) is predominantly an associative process, termed an *associative interchange* I_a. Case (c) is a *dissociative interchange* I_d. The boundaries between these cases, obviously, are somewhat arbitrary, although the limiting cases D or A, however, can be firmly established if unequivocal evidence for a well-characterized intermediate with a smaller (case d) or larger (case a) coordination number, can be derived from experiment.

Establishing a substitution mechanism is a remarkably difficult task—many mechanisms are suggested, but few are established. To do so, we must undertake kinetic studies and derive a rate law for the reaction. Regardless of the stoichiometry of the reactants, we very often find a two-term rate law:

$$\text{Rate} = k_1[ML_nD] + k_2[ML_nD][E]$$

Here k_1 and k_2 are the rate constants for the first-order and second-order paths for the reaction. Clearly, at least two mechanisms are at work, but usually one is dominant under most experimental conditions, particularly at a given pH in aqueous studies. The relationship between the kinetic order of the reaction and the reaction mechanism is complicated by the fact that the solvent is often a reactant even if it does not appear in the overall stoichiometric equation. Most simple donor–acceptor complexes are charged and are therefore soluble primarily in polar Lewis-base solvents. Such a solvent molecule is a nucleophile in its own right, and enjoys an enormous concentration advantage over other nucleophile ligands in the solution. Water, for example, is 55.5 M in the Lewis-base ligand H_2O. Because this concentration is so high, it does not change appreciably as the reaction proceeds. If the solvent concentration appears in the rate law, it will appear to be part of the rate constant in a pseudo-first-order rate law. This often means that instead of measuring the rate of replacement of D by E in the complex L_nMD, we are really measuring the rate of replacement of H_2O (or other solvent ligand) by E in the complex $L_nM(H_2O)$:

$$L_nMD + H_2O \rightarrow L_nM(H_2O) + D \qquad (14.1)$$

$$L_nM(H_2O) + E \rightarrow L_nME + H_2O \qquad (14.2)$$

Process (14.2), particularly where E is an anion, is often called *anation,* and many of the rate studies we shall see are anation rates.

14.2 LIGAND SUBSTITUTIONS IN OCTAHEDRAL COMPLEXES

In previous chapters, we pointed out that the most common coordination geometry by far is the octahedron of six ligands. It follows that the ligand-substitution reactions of octahedral complexes should be a major mechanistic concern. Since this requires kinetic studies, the earliest systems chosen for experiment were, of course, those that react at convenient rates. The range of ligand-exchange rates is very broad (see Fig. 12.12), but the only systems accessible to traditional techniques are the inert complexes, particularly Co^{3+}, Cr^{3+}, and Pt^{2+}. Of these three, Co^{3+} and Cr^{3+} normally form octahedral complexes, and studies of Co^{3+} complexes in particular dominated this area of kinetic studies until relatively recently. A few octahedral complexes of other metals are known to react by significantly different mechanisms, but in general the substitution mechanisms proposed for Co^{3+} complexes are thought to apply fairly broadly to other octahedral systems.

Cobalt(III) complexes (and by inference most other octahedral complexes) react by an essentially dissociative mechanism. Kinetic studies of a number of complexes in acidic aqueous solution show first-order or pseudo-first-order kinetics and no influence by the entering ligand. The sequence mentioned above of acid hydrolysis followed by anation is the only pattern seen for these complexes:

$$[Co(NH_3)_5(NO_3)]^{2+} + H_2O \rightarrow [Co(NH_3)_5(OH_2)]^{3+} + NO_3^-$$

$$[Co(NH_3)_5(OH_2)]^{3+} + SCN^- \rightarrow [Co(NH_3)_5(NCS)]^{2+} + H_2O$$

We shall return shortly to the evidence for a dissociative process. However, it is worth noting that there are obvious steric reasons for preferring a dissociative process (with CN < 6) to an associative process. The rarity of seven-coordinate complexes suggests a kind of steric saturation at CN = 6. This might well cause a substantially increased activation energy and therefore a noncompetitively slow rate for a seven-coordinate intermediate. Furthermore, for Co^{3+} a complex with six electron-pair donors yields 18 valence-orbital electrons for the Co atom. Therefore, an added ligand in an intermediate (or even a transition state with increased coordination number) would have a substantially unfavorable electronic energy for its donated electrons

Table 14.1   Kinetic and Thermodynamic Data for the Acid Hydrolysis of $[Co(NH_3)_5X]^{2+}$

$$[Co(NH_3)_5X]^{2+} + H_2O \underset{k_{-1}}{\overset{k_1}{\rightleftharpoons}} [Co(NH_3)_5(OH_2)]^{3+} + X^-$$

$$K_{eq} = k_1/k_{-1}$$

X^-	k_1	K_{eq}
NCS^-	4.1×10^{-10}	3.7×10^{-4}
N_3^-	2.1×10^{-9}	1.2×10^{-3}
$HC_2O_4^-$	2.2×10^{-8}	2.9×10^{-2}
F^-	8.6×10^{-8}	4×10^{-2}
$H_2PO_4^-$	2.6×10^{-7}	1.3×10^{-1}
Cl^-	1.7×10^{-6}	9.0×10^{-1}
Br^-	6.5×10^{-6}	2.9×10^{0}
I^-	8.3×10^{-6}	8.3×10^{0}
NO_3^-	2.7×10^{-5}	1.3×10^{1}
$OSO_2CF_3^-$	2.7×10^{-2}	Large
SO_3F^-	2.2×10^{-2}	Large

(see Fig. 12.10b). Of course, the electronic energy preference for a dissociative mechanism does not necessarily apply to octahedral complexes other than those of Co^{3+}, but the steric energy preference usually does.

The reaction equations above for the substitution of SCN^- for NO_3^- in $[Co(NH_3)_5(NO_3)]^{2+}$ are stoichiometric equations that imply nothing about the mechanism. However, there are several pieces of evidence for a dissociative mechanism. First, entering-group effects on the rate constant are very small, as might be expected if the entering group approaches only after the rate-determining step of the mechanism (dissociation). For the anation of $[Co(NH_3)_5(OH_2)]^{3+}$ by Cl^-, Br^-, NCS^-, N_3^-, NO_3^-, $H_2PO_4^-$, and NH_3, the rate constants at 25°C all lie in the range 1.3×10^{-6} to 2.5×10^{-6} M^{-1} sec^{-1}, a very small range for rate constants. If the mechanism were associative, the bonding properties of the incoming ligand would profoundly influence the rate constant.

On the other hand, leaving-group effects are large, as they should be if M—D bond breaking is the rate-determining step. Table 14.1 gives some rate constants for the acid hydrolysis of $[Co(NH_3)_5X]^{2+}$ ions, where X is an anion with a 1– charge; it can be seen that in these cases the rate constants vary over eight orders of magnitude. However, an even more specific correlation with a dissociative mechanism involves the thermodynamic equilibrium constants for the hydrolysis reactions, also given in the table. There is a very clean linear free-energy relationship between these rate constants and the equilibrium constants, as Fig. 14.3 shows. For the straight line shown, the slope is 1.03, or almost exactly 1. It should be clear from the reaction-coordinate diagram in Fig. 14.4 that if $\Delta G^{\ddagger}$ is very much like

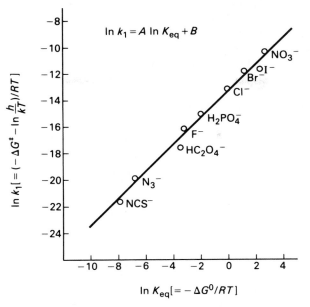

Figure 14.3   Linear free-energy relationship between rate constant and equilibrium constant for hydrolysis of $[Co(NH_3)_5X]^{2+}$: $\ln K_1 = A \ln K_{eq} + B$.

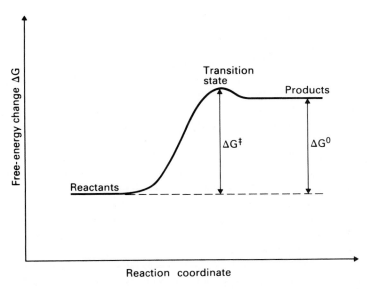

Figure 14.4   Reaction-coordinate diagram for a reaction with a similar free energy of activation and overall free-energy change.

Table 14.2 Kinetic Data for Acid Hydrolysis of
Co^{3+} Complexes with Different Net Charges

Complex	Ligand replaced	Rate constant (sec^{-1})
Net charge 2+		
$[Co(NH_3)_5Cl]^{2+}$	Cl^-	6.7×10^{-6}
$[Co(NH_3)_4(OH_2)Cl]^{2+}$	Cl^-	2.2×10^{-6}
$[Co(en)_2(NH_3)Cl]^{2+}$	Cl^-	4.0×10^{-7}
$[Co(NH_3)_5Br]^{2+}$	Br^-	6.3×10^{-6}
$[Co(en)_2(NH_3)Br]^{2+}$	Br^-	1.2×10^{-6}
Net charge 1+		
$[Co(NH_3)_4Cl_2]^+$	Cl^-	1.8×10^{-3}
$[Co(en)_2Cl_2]^+$	Cl^-	3.2×10^{-5}
$[Co(en)_2(N_3)Cl]^+$	Cl^-	2.5×10^{-4}
$[Co(en)_2Br_2]^+$	Br^-	1.4×10^{-4}

Note: Values are given for *trans*-configuration complexes, except
for $[Co(NH_3)_5Cl]^{2+}$ and $[Co(NH_3)_5Br]^{2+}$.

ΔG^0 for the reaction, the transition state must be very much like the reaction products in both energy and structure. That is, in this case the transition state must be a configuration of the system in which the X^- ion has already separated from the Co^{3+}—just what a dissociative mechanism would require.

There is even more evidence, however. If an anion is to separate from a positively charged complex, the reaction rate should decrease for ions of comparable structure as the net positive charge increases *if* that bond breaking and separation is the rate-determining step (as a dissociative mechanism requires). Table 14.2 shows that there is a substantial effect of just this sort for acid hydrolysis reactions—but if there were an associative mechanism, the higher charge should attract the polar H_2O molecule in more rapidly and actually speed up the reaction.

The steric effects that we suggested must favor dissociative mechanisms for six-coordinate complexes can be used to demonstrate the dissociative mechanism in a more quantitative way. Consider the rate data in Table 14.3 for the acid hydrolysis of the first Cl^- from *trans*-$[Co(diamine)_2Cl_2]^+$, where the diamines are the indicated methyl-substituted ethylenediamines. Since methyl groups have very little inductive effect, there can be little electronic difference between the ligands. Their bulk, however, increases as methylation increases. Increased ligand bulk would make it easier to drive away a leaving group in a dissociative mechanism and therefore speed up the reaction. In an associative mechanism, on the other hand, increased ligand bulk should actually slow down the reaction. Therefore, the data shown are consistent only with a dissociative mechanism. The difference between *dl*- and *meso*-butylenediamine is particularly interesting, because those two differ only in the steric prominence of their two CH_3 groups.

The situation should seem pretty clear by this time, but transition-state theory gives us two more indicators of a dissociative mechanism: the entropy of activation, $\Delta S^{\ddagger}$, and the volume of activation, $\Delta V^{\ddagger}$. A positive entropy of activation indicates increasing randomness as the complex reaches its transition state. This would presumably be true of a dissociative mechanism, but not of an associative mechanism. However, in a polar solvent such as water, changes in ion charge can orient or disorient surrounding solvent molecules enough to mask this effect. Still, for a wide range of $[CoN_4X_2]^+$ complexes (where N_4 is a variety of amines and X_2 is a variety of halides and pseudohalides) acid hydrolysis yields $\Delta S^{\ddagger}$ values from -20 to $+80$ J/mol K. Bringing a neutral water molecule into the Co complex in an associative mechanism should not increase the entropy of the transition state, so this evidence also suggests a dissociative mechanism.

Negative entropies of activation can indicate associative mechanisms, of course. An interesting example is the reaction of the organometallic system Fe(Cp)(Tol)

Table 14.3 Kinetic Data for Acid Hydrolysis of $[Co(Diamine)_2Cl_2]^+$ Complexes with Different Degrees of Steric Hindrance

Diamine	Structure	Rate constant (sec^{-1})
Ethylenediamine	$H_2N-CH_2-CH_2-NH_2$	3.2×10^{-5}
Propylenediamine	$H_2N-CH_2-CH-NH_2$ $\quad\quad\quad\quad\quad \mid$ $\quad\quad\quad\quad\quad CH_3$	6.2×10^{-5}
dl-Butylenediamine	CH_3 $\mid$ $H_2N-CH-CH-NH_2$ $\quad\quad\quad\quad\quad \mid$ $\quad\quad\quad\quad\quad CH_3$	1.5×10^{-4}
meso-Butylenediamine	$H_2N-CH-CH-NH_2$ $\quad\quad\quad \mid \quad\quad \mid$ $\quad\quad\quad H_3C \quad CH_3$	4.2×10^{-3}
Tetramethylethylenediamine	$H_3C \quad\quad CH_3$ $\quad \mid \quad\quad\quad \mid$ $H_2N-CH-CH-NH_2$ $\quad\quad\quad \mid \quad\quad \mid$ $\quad\quad\quad H_3C \quad CH_3$	3.3×10^{-2}

with trimethylphosphite in THF to yield a "piano-stool" complex in which the toluene ligand has been replaced by two phosphines and a phosphide:

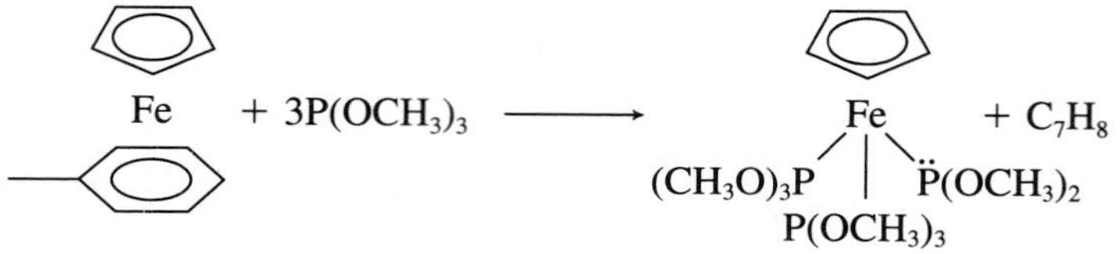

The organoiron sandwich compound starting material has an electron count of 19, which would seem to preclude the approach of another electron-pair donor. However, rate data yield $\Delta H^{\ddagger} = +51.5$ kJ/mol rn and $\Delta S^{\ddagger} = -92$ J/mol rn K; here the substantial negative entropy of activation clearly indicates an associative reaction even with the 19-electron starting material. Figure 14.5 shows the suggested mechanism, in which the iron, having only the unsatisfactory possibilities of $17e^{-}$ or $19e^{-}$ coordination, systematically strips itself of toluene pi electrons as phosphites come in; note the flexure of the toluene molecule, similar to that seen in Fig. 13.19. The first step having a stoichiometry change (step 1) adds the 2-electron donor P(OMe)$_3$ to the 17-electron iron structure in an associative process, and $17e^{-}/19e^{-}$ alternation occurs after that until the final approach to an 18-electron count.

Experimental data on the volume of activation can be obtained by running the reaction at varying pressures, since

$$\Delta V^{\ddagger}(P_1 - P_2) = RT \ln k_2/k_1$$

Solvent molecules are compressed by increasing charge (*electrostriction*), so charge effects can also dominate volume-of-activation data. However, this can be offset by comparing the volume of activation to the overall volume change for the stoichiometric reaction. For a variety of $[Co(NH_3)_5X]^{n+}$ hydrolysis reactions (corrected for electrostriction), $\Delta V^{\ddagger}$ is more positive than ($\Delta V/2$) by about 1.2 mL/mol rn. That is, the (partial molal) volume of the transition state is about a milliliter larger than the average of the volumes of the reactants and products. The overall volume change ranges from $+3$ to -30 mL/mol rn, with the shrinkage being an increase in electrostriction when an anion leaves and the neutral H$_2$O enters. If the mechanism were a limiting dissociative case [S_N1(lim) or D] we might expect the volume of activation to be positive and roughly equal to the volume of a mole of the leaving group—usually about 10–20 mL. If the mechanism were an exact interchange, in which the departing D and entering E are equally important, we would expect $\Delta V^{\ddagger} = \frac{1}{2}\Delta V$. Since $\Delta V^{\ddagger}$ is slightly larger than the mean volume change, we interpret the mechanism as a dissociative interchange I$_d$, in which the leaving group has partially transferred from the inner sphere to the outer sphere before the entering water molecule has formed equivalent bonding.

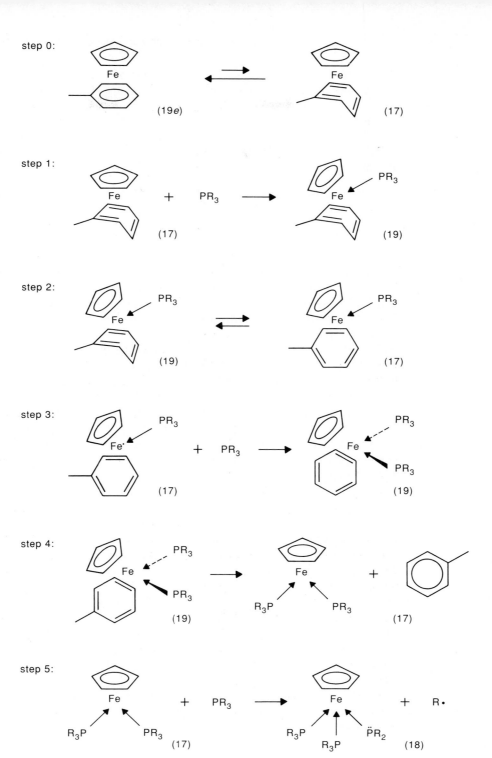

Figure 14.5 Associative mechanism for substitution on an organoiron species initially having 19 electrons.

All of the arguments presented for a dissociative mechanism have dealt specifically with Co^{3+} complexes. It does seem to be true that the dissociative-interchange I_d mechanism applies to essentially all of these, and to many others. However, associative mechanisms are thought to govern some octahedral substitutions. For example, the anation of $[Rh(NH_3)_5(OH_2)]^{3+}$ occurs faster for some ions than for simple water exchange, which implies a specific bond-making role for the anion in the rate determining step:

$$[Rh(NH_3)_5(OH_2)]^{3+} + X^{n-} \rightarrow [Rh(NH_3)_5X]^{(3-n)+} + H_2O$$

$$k_{SO_4^{2-}}/k_{H_2O} = 1.0$$

$$k_{Cl^-}/k_{H_2O} = 2.6$$

$$k_{Br^-}/k_{H_2O} = 4.9$$

A few other complexes show comparable behavior. Some [notably $Cr(OH_2)_6^{3+}$] have negative values for the volume of activation $\Delta V^{\ddagger}$, which implies an associative-interchange I_a model for the water-exchange reaction. In particular, one study uses ^{18}O-labeled water to study H_2O interchange for $Cr(OH_2)_6^{3+}$, where no electrostriction change can possibly occur, and yields $\Delta V^{\ddagger} = -9.6$ mL/mol rn, implying a fairly substantial degree of association in the process. The same measurements for $Fe(OH_2)_6^{3+}$ also yield a negative volume of activation (-5.4 mL/mol rn), but the smaller value suggests a less strongly associative process.

In the preceding discussion of *aquation* or hydrolysis reactions of octahedral complexes, we have consistently limited ourselves to acid hydrolysis. For hydrolysis by base—and in particular by the OH^- ion in water solution—a completely different mechanism applies. Above about pH 4, most Co^{3+} complexes with ammonia or amine ligands hydrolyze more rapidly than in acid. The rate law is always second-order:

$$\text{Rate} = k[Co(amine)][OH^-]$$

At 1 M OH^-, in fact, the hydrolysis of these "inert" complexes is so rapid that flow methods must be used to follow the kinetics. However, the suggested mechanism is not associative, in spite of the second-order rate law. One of the major pieces of evidence supporting a dissociative mechanism is the steric effect. As the discussion associated with Table 14.3 indicated, a reaction proceeding by a dissociative mechanism should be accelerated by increased bulk of the nonreacting ligands in the transition-state complex, whereas an associative mechanism should be retarded. In base hydrolysis, the rate constant for the hydrolysis of $[Co^i(BuNH_2)_5Cl]^{2+}$ is 175,000 times as great as the rate constant for the comparable reaction with $[Co(NH_3)_5Cl]^{2+}$. This increase is so great that it requires bond breaking as the rate-controlling event. On the other hand, this dissociative mechanism must be compatible with the second-order rate law. The accepted mechanism involves a dissociative rate-determining step, preceded by a rapidly established equilibrium.

This base-hydrolysis mechanism uses the OH^- to react with one of the protic hydrogen atoms on the amine ligand, forming the conjugate base of the amine as an amide ion. For example,

1. $[Co(NH_3)_5Cl]^{2+} + OH^- \overset{K_{eq}}{\rightleftharpoons} [Co(NH_3)_4(NH_2)Cl]^+ + H_2O$

2. $[Co(NH_3)_4(NH_2)Cl]^+ \overset{k_2}{\longrightarrow} [Co(NH_3)_4(NH_2)]^{2+} + Cl^-$

3. $[Co(NH_3)_4(NH_2)]^{2+} + H_2O \overset{fast}{\longrightarrow} [Co(NH_3)_4(NH_2)(OH_2)]^{2+}$

4. $[Co(NH_3)_4(NH_2)(OH_2)]^{2+} \overset{fast}{\longrightarrow} [Co(NH_3)_5(OH)]^{2+}$

This mechanism is compatible with a second-order rate law. Since (2) is the rate-determining step,

$$\text{Rate} = k_2[Co(NH_3)_4(NH_2)Cl^+]$$

This species, however, appears in the equilibrium-constant expression for K_{eq}:

$$K_{eq} = \frac{[Co(NH_3)_4(NH_2)Cl^+]}{[Co(NH_3)_5Cl^{2+}][OH^-]}$$

or

$$[Co(NH_3)_4(NH_2)Cl^+] = K_{eq}[Co(NH_3)_5Cl^{2+}][OH^-]$$

Substituting this into the rate law, we have

$$\text{Rate} = k_2K_{eq}[Co(NH_3)_5Cl^{2+}][OH^-]$$

which is properly second-order, even though the rate-determining step itself is first-order or unimolecular. The mechanism is termed S_N1CB, where "CB" indicates the intermediate formation of the *Conjugate Base* of the initial ligand.

Such a mechanism obviously requires a protic hydrogen on an amine ligand in order to form the conjugate base. If, for example, we attempt the basic hydrolysis of *trans*-$Co(py)_4Cl_2^+$, where the pyridine ligands have no protic hydrogens, we find that the hydrolysis rate is *not* affected by $[OH^-]$ up to about pH 9 (where the complexes decompose completely). This is convincing evidence for the S_N1CB mechanism.

It is worth noting that the acid–base reactivity of the NH_3 ligand has been substantially modified by coordinating it to Co^{3+}. Chapter 7 (in particular the discussion accompanying Fig. 7.8) showed that the OH^- ion in water solution should not be able to deprotonate ammonia to the NH_2^- ion; hydroxide simply isn't a strong enough base. Yet when the intrinsic basicity of the NH_3 molecule has been modified by coordinating it to a Co^{3+} ion, the reaction to produce a coordinated NH_2^- is quite rapid. The reason is simple enough—the presence of the positively charged (Lewis-acid) metal ion drains electron density away from the NH_3, making it more acidic. This, of course, is exactly the sort of modification of ligand reactivity that a transition-metal catalyst or metalloenzyme could be expected to cause.

14.3 LIGAND SUBSTITUTIONS IN SQUARE-PLANAR COMPLEXES

Just as the inert (and therefore kinetically convenient) $Co^{3+}L_6$ complexes have been widely studied as a model for octahedral substitution, the four-coordinate square-planar complexes of d^8 Pt^{2+} have been used as model compounds for ligand substitution in that geometry. These are also inert, with substitution rate constants close to those of Co^{3+} complexes. For the substitution of a single ligand, a particularly convenient model compound for study is $[Pt(dien)X]^+$, where X^- is a halide ion and *dien* (diethylenetriamine) locks up the other coordination sites:

Kinetic studies of square-planar Pt^{2+} complexes usually show a two-term rate law, suggesting two simultaneous mechanisms:

$$\text{Rate} = k[L_3PtD] = k_1[L_3PtD] + k_2[L_3PtD][E]$$
$$k = k_1 + k_2[E]$$

The k_1 (first-order) mechanism presumably resembles the mechanism for octahedral substitution in being solvent-controlled, though it may not be dissociative. We can get separate values for k_1 and k_2 by plotting k against $[E]$ as in Fig. 14.6. The intercept (k_1) is the same for both entering groups ($E = Br^-$ and $E = Cl^-$), but the slope (k_2) is quite different. This suggests that there is a strong bond-making or associative quality for the second-order mechanism. Furthermore, k_2 is usually 10–100 times as large as k_1. Therefore, the experimental situation is that square-planar complexes undergo substitution predominantly by an associative mechanism. Table 14.4 substantiates this in two ways: There is a large entering-group effect (but only a modest leaving-group effect for the two leaving groups shown), and all the reactions have substantial negative entropies of activation. Since an associative mechanism brings the two reactants together in the transition state without significant bond breaking, the negative $\Delta S^{\ddagger}$ is a good diagnostic of the increased order in such a mechanism.

An associative mechanism seems much more reasonable for square-planar complexes than it does for octahedral complexes. The coordination number is lower, so that there is at least potential coordinative unsaturation. Furthermore, the planar arrangement leaves access open even for an entering ligand with a fairly large cone angle unless the planar ligands themselves are quite bulky. The obvious geometry for a five-coordinate transition state is a square pyramid with the entering ligand E at the apex. This may well be the first structure formed. But we always observe that substitution in Pt^{2+} square-planar complexes occurs with retention of configuration (*cis-* / *trans-*, etc.), so that some geometric shift such as that in

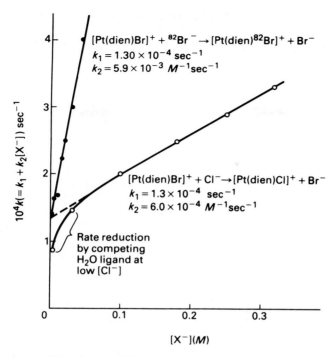

Figure 14.6 Rate data for first- and second-order substitutions on $[Pt(dien)Br]^+$.

Table 14.4 Rate Constants and Entropies of Activation[a] for $L_3PtD + E \rightarrow L_3PtE + D$

L_3PtD	E	k_2 (M^{-1} sec^{-1})	$\Delta S\ddagger$ (J/mol K)
$[Pt(dien)Br]^+$	H_2O	3.6×10^{-6}	-71
	Cl^-	1.0×10^{-3}	-46
	Br^-	9.4×10^{-3}	-63
	I^-	0.32	-105
	NCS^-	0.68	-113
	$SC(NH_2)_2$	1.3	-121
$[Pt(dien)Cl]^+$	H_2O	2×10^{-7}	-75
	Cl^-	1.4×10^{-3}	-17
	N_3^-	5×10^{-3}	-71
	Br^-	7×10^{-3}	-105
	I^-	0.170	-105
	NCS^-	0.270	-117
	$SC(NH_2)_2$	0.580	-130

[a](at 30°C)

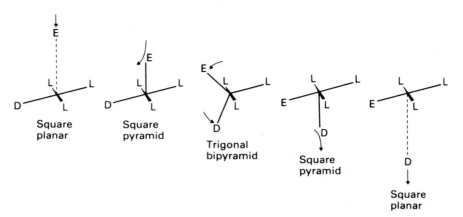

Figure 14.7 Substitution mechanism for square-planar complexes with retention of configuration.

Fig. 14.7 must be occurring. This in turn suggests that if there is a stable five-coordinate intermediate in an A mechanism (as opposed to I_a), it should probably be a trigonal bipyramid, and so should the transition state in an I_a mechanism. The difference between these two essentially associative mechanisms is shown in the reaction-coordinate diagrams of Fig. 14.8.

Although there seems to be no question that square-planar Pt^{2+} complexes substitute ligands by an associative mechanism, there is still a significant leaving-group effect (bond breaking) when a wide variety of leaving groups is considered. Table 14.5 compares some leaving groups, where the initial L_3Pt group and the entering E group are held constant. It can be seen that there is an effect of some five orders of magnitude or more. This means that the transition state must be a geometry that has modified the bonding of the leaving ligand significantly, even though it has not fully departed. This is not inconsistent with an associative mechanism;

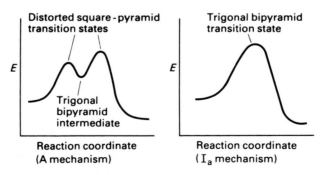

Figure 14.8 Reaction-coordinate diagrams for A and I_a mechanisms (with and without a characterizable five-coordinate intermediate).

Table 14.5 Rate Constants for
$[(dien)PtD]^{+} + py \rightarrow [(dien)Pt(py)]^{2+} + D^{-}$

Leaving group D	Rate constant (25°C) (sec^{-1})
H_2O	1.9×10^{-3}
Cl^-	3.5×10^{-5}
Br^-	2.3×10^{-5}
I^-	1.0×10^{-5}
N_3^-	8.3×10^{-7}
SCN^-	3.0×10^{-7}
NO_2^-	5.0×10^{-8}
CN^-	1.7×10^{-8}

a ligand in an equatorial position in a trigonal bipyramid certainly has different orbital-overlap possibilities from a ligand in a square-planar complex, particularly with respect to pi overlap. So actual bond breaking need not occur in the transition state for a mechanism to show a leaving-group effect.

Besides the observable effects of entering and leaving groups on the kinetics of substitution in square-planar complexes, there is a strong and theoretically interesting effect due to the nonreacting ligands present. Specifically, groups that serve as ligands in square-planar complexes can be placed in order of their increasing ability to labilize the group *trans* to themselves in the complex. The effect can be quite large. For instance, pyridine replaces Cl^- in the following complexes (with the indicated *trans-* ligands) with the indicated rate constants. There is obviously an effect of at least six orders of magnitude, considering the temperatures:

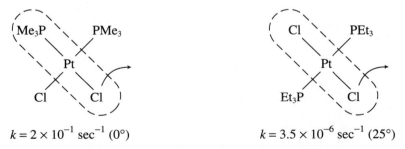

$$k = 2 \times 10^{-1}\ sec^{-1}\ (0°) \qquad k = 3.5 \times 10^{-6}\ sec^{-1}\ (25°)$$

Although the entering and leaving groups modify it somewhat, the general order of *trans-* labilizing ability is

$$CN^- \simeq C_2H_4 \simeq CO \simeq NO > PR_3 \simeq H^- \simeq SC(NH_2)_2 > CH_3^- > C_6H_5^-$$
$$> SCN^- > NO_2^- > I^- > Br^- > Cl^- > py \simeq NH_3 > OH^- > H_2O$$

This effect of nonreacting ligands in square-planar complexes is called the *trans-effect*. Ligands with high *trans-* labilizing ability are said to be *trans-directing*. The reason for this name is apparent: By choosing the right sequence of ligand substitutions, the experimenter can use kinetic control of the reaction products to get a desired isomer of a square-planar complex. For example, suppose we want to obtain separately the *cis-* and *trans-* isomers of $[PtCl_2(NO_2)(NH_3)]^-$. We can start with the readily available $PtCl_4^{2-}$ ion, then use the fact that NO_2^- is more strongly *trans-*directing than Cl^-, but NH_3 is less *trans-*directing than Cl^-. The NH_3 will be forced to enter *trans-* to the strongly *trans-*directing NO_2^- if the NO_2^- is added first:

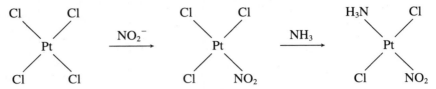

On the other hand, when we want the *cis-* isomer it is only necessary to reverse the order of addition, since now the Cl^- opposite another Cl^- will leave in preference to the Cl^- opposite the weakly *trans-*directing NH_3 ligand:

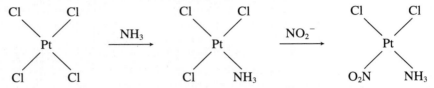

Because of the great differences in rate constants (in favorable cases), substitution reactions can often be run with great stereospecificity. This is obviously a synthetically useful tool. We shall see in Chapter 16 that there are profound physiological differences between *cis-* and *trans-*$PtCl_2(NH_3)_2$; the former is the anticancer drug cisplatin, while the latter is medicinally useless. Stereospecific synthesis is obviously extremely important for this drug.

The ligands in the *trans-*directing series are all serving as Lewis bases within the complex. To a good approximation, the softer bases they are, the more strongly *trans-*directing they are. Both CO and H^- are quite soft bases and are strong *trans-*directors, whereas H_2O is a hard base with very little *trans-*directing effect. There are two points to consider if we want to explain the series in terms of the bonding present within the complex. The first is that soft bases are good sigma donors and are quite polarizable. Since the d^8 metal ions that predominantly form square-planar complexes are fairly soft acids, they are also polarizable, and the bond to the *trans-* ligand will be weakened somewhat by the reverse polarization of the metal ion, as in Fig. 14.9.

The second consideration in explaining the *trans-*directing series is the changes that take place in the pi bonding between the initial square-planar complex and the trigonal-bipyramidal intermediate (or transition state), which has nonequivalent axial and equatorial positions. Figure 14.7 shows that the ligand in L_3PtD that is *trans-* to D lies in the equatorial plane of the trigonal bipyramid L_3PtDE, but that

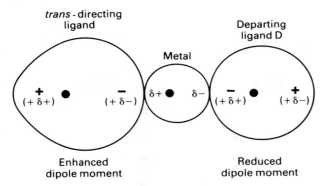

Figure 14.9 Sigma polarization theory of the *trans* effect: δ+ and δ− arise from polarization by the *trans*-directing ligand.

the other two nonreacting ligands are axial. Now the geometric opportunities for pi overlap are excellent in the square-planar geometry: If we apportion the metal orbitals capable of pi overlap among the ligands, each ligand has access to 0.75 metal pi orbital (see Problem B1). But in the trigonal-bipyramidal intermediate, by contrast, the axial ligands have access to 0.57 metal pi orbital while the equatorial ligands have access to 0.95 metal pi orbital. Thus, there is a clear preference for a pi-bonding ligand to remain in the equatorial plane as the intermediate forms, which means that the ligand *trans-* to it must be the one to depart. Good pi-acceptor ligands such as CO and C_2H_4 tend to force the ligand *trans-* to them off the complex both by being good polarizable sigma donors and by forming pi bonds to the metal that are maintained in the five-coordinate intermediate only if the ligand *trans-* to them is the departing species. Note also that the leaving-group effect seen in Table 14.5 shows that good pi acceptors are most strongly bound and least reactive as departing ligands. This is consistent with the kinetic-intermediate argument for the *trans-* effect.

14.4 REDOX-REACTION MECHANISMS

In Chapter 8 we have already called attention to the fundamental mechanistic distinction between inner-sphere and outer-sphere redox reactions. Transition-metal complexes react by both routes. Usually, the first task in establishing a mechanism is to determine which is applicable to the given stoichiometric reaction. An outer-sphere mechanism corresponds fairly well to the naive assumption that in a redox reaction the two reactants simply float up to each other, exchange electrons, and float away. Such a reaction can be very rapid if no substantial geometric changes are called for in either of the reactant molecules as they become product molecules. However, that is a relatively rare condition.

An inner-sphere mechanism, on the other hand, requires a certain degree of ligand substitution. The bridging ligand (even a single-atom reactant, as in $Fe^{3+} + I^- \rightarrow Fe^{2+} + \frac{1}{2} I_2$) must form a new bond to the transition-metal atom, by either a dissociative or an associative mechanism. As we shall see, the study of some inner-sphere redox mechanisms has been aided greatly by the use of reactants or products that are inert toward substitution.

The simplest kind of redox reaction is an electron exchange reaction. In this case, no net chemical change occurs, but the process can be followed by using isotopic labeling. Essentially all of the exchange reactions that have been studied involve one-electron transfer. The range of possible rates is enormous:

$$*Co(NH_3)_6^{3+} + Co(NH_3)_6^{2+} \rightleftharpoons *Co(NH_3)_6^{2+} + Co(NH_3)_6^{3+}$$
$$k < 10^{-8} M^{-1} \, sec^{-1}$$
$$*Ru(OH_2)_6^{3+} + Ru(OH_2)_6^{2+} \rightleftharpoons *Ru(OH_2)_6^{2+} + Ru(OH_2)_6^{3+}$$
$$k = 20 \; M^{-1} \, sec^{-1}$$
$$*Fe(phen)_3^{2+} + Fe(phen)_3^{3+} \rightleftharpoons *Fe(phen)_3^{3+} + Fe(phen)_3^{2+}$$
$$k > 10^7 M^{-1} \, sec^{-1}$$

These reactions can occur by either an inner- or outer-sphere mechanism. When electron transfer occurs much more rapidly than substitution, an outer-sphere mechanism is obviously involved. This, however, is not too common, though it is seen for the $Ru^{2+/3+}$ exchange above. Similarly, if a bridged intermediate is detected that has oxidation states different from those of the reactants, an inner-sphere mechanism is clearly at work. This too is only occasionally possible. Unfortunately, the rate law does not help—both mechanisms are usually second-order:

$$Rate = [oxidant][reductant]$$

As a rough generalization, if the ligands in an outer-sphere exchange reaction are pi systems, as in *phen* or *bipy* (Fig. 12.16) or CN^-, or if they are soft (polarizable) bases such as Br^-, the exchange will occur relatively rapidly. On the other hand, if they are hard bases with no pi-overlap possibilities, such as H_2O or NH_3, the exchange will be relatively slow. This is qualitatively reasonable: In an outer-sphere mechanism, the electron must be transferred through two layers of ligands. Therefore, the reaction will be faster the more mobile the electrons in the ligand are. Geometric changes between reactant and product also make a substantial difference in the rate: The greater the change in bond lengths, the slower the exchange. For a symmetrical reaction such as the electron-exchange reactions being considered here,

$$*ML_6^{3+} + ML_6^{2+} \rightleftharpoons *ML_6^{2+} + ML_6^{3+}$$

the transition state must have equal M—L bond lengths for both reactants. If the bond lengths are initially quite different, there will be a substantial activation energy corresponding to the energy required to stretch one bond and compress the other. If the initial bond lengths are similar, correspondingly less energy is required to make the transition state symmetrical, and the low activation energy leads to a faster reaction. For unsymmetrical outer-sphere reactions such as

$$Fe(CN)_6^{4-} + IrCl_6^{2-} = Fe(CN)_6^{3-} + IrCl_6^{3-}$$

the rate is usually greater than for either symmetrical exchange. In this case, k for $Fe(CN)_6^{4-/3-}$ is 7.4×10^2 M^{-1} sec^{-1}. For $IrCl_6^{3-/2-}$, $k = 1 \times 10^3$ M^{-1} sec^{-1}; and for the unsymmetrical reaction $k = 1.2 \times 10^5$ M^{-1} sec^{-1} (all at 25°C). Where the reduction potentials of the two half-reactions are not too different, these rate constants can be related through a simplified form of the *Marcus equation:*

$$k_{12} = (k_{11} \cdot k_{22} \cdot K_{12})^{1/2}$$

where subscripts 11 and 22 refer to the two symmetrical exchange reactions, and K_{12} is the equilibrium constant for the unsymmetrical reaction. In this case, the Marcus equation predicts a k value of 2.3×10^5 from the E^0 value of 0.66 V—a fairly good match.

Inner-sphere exchange reactions have also been studied extensively, particularly those between Cr^{2+} and Cr^{3+} in aqueous solution. Figure 12.12 shows that Cr^{2+} complexes in water are quite labile, so the reacting species must be $Cr(OH_2)_6^{2+}$. On the other hand, Cr^{3+} complexes are inert and that reactant can be varied. Ignoring coordinated water molecules to emphasize the inner-sphere process, all the reactions in Table 14.6 proceed by the mechanism

$$Cr—X^{2+} + Cr^{2+} \rightarrow C—X—Cr^{4+} \rightarrow Cr^{2+} + X—Cr^{2+}$$

Since the rate constants vary by six orders of magnitude, the nature of the bridging ligand is obviously extremely important. Since the oxidant is inert, it supplies the bridging ligand, which is transferred to the reductant. This atom transfer or ligand transfer usually occurs in an inner-sphere mechanism, but not always. For example, the redox reaction between VO_2^+ and $Fe(CN)_6^{4-}$ to yield VO^{2+} and $Fe(CN)_6^{3-}$ occurs by an inner-sphere mechanism (as shown by the UV spectrum of a bridged intermediate containing V and Fe in their final oxidation states). However, the bridging CN^- ligands remain on the iron atom in the products. It is important to note that there are exceptions to the pattern of bridging-ligand transfer in inner-sphere mechanisms, because the bridging ligand can serve several roles other than simply carrying an electron along in its own orbitals as it moves from one metal atom to another. Other possibilities, all of which seem likely to occur under varying circumstances, include (1) *direct exchange,* in which the ligand holds the metal atoms close enough that overlap can occur between the orbitals of the metal atoms M_1 and M_2; (2) *superexchange,* in which the ligand provides vacant orbitals to delocalize the electron from M_1 and bring it into overlap with M_2; (3) *double exchange,* in which the ligand simultaneously accepts an electron from the reductant and donates one of its own to the oxidant; and (4) *chemical exchange,* in which the ligand is temporarily oxidized or reduced as one step in the mechanism.

Table 14.6   Kinetic Data for
$[*Cr(OH_2)_6]^{2+} + [Cr(OH_2)_5X]^{2+} \rightarrow [*Cr(OH_2)_5X]^{2} + [Cr(OH_2)_6]^{2+}$

X	k (25°C) $(M^{-1}\ sec^{-1})$	$\Delta H\ddagger$ (kJ/mol rn)	$\Delta S\ddagger$ (J/mol K)
H_2O	$<2 \times 10^{-5}$	—	—
OH^-	7×10^{-1}	54	−67
NCS^-	1.4×10^{-4}	—	—
SCN^-	4×10^1	—	—
N_3	6.1×10^0	40	−96
F^-	2.4×10^{-3} (0°C)	57	−84
Cl^-	9×10^0 (0°C)	—	—
Br^-	$>6 \times 10^1$ (0°C)	—	—
CN^-	7.7×10^{-2}	39	−134

A now-classic series of redox-mechanism experiments carried out primarily by Taube involves the reaction between Cr^{2+} (a good reductant) and Co^{3+} (a good oxidant). For example, consider the following reaction:

$$[Cr(OH_2)_6]^{2+} + [Co(NH_3)_5Cl]^{2-} + 5H_3O^+ \rightarrow [CrCl(OH_2)_5]^{2+} + [Co(OH_2)_6]^{2+} + 5NH_4^+$$

This reaction is rapid ($k = 6 \times 10^5\ M^{-1}\ sec^{-1}$) and involves the bridged intermediate $[(H_2O)_5Cr\text{—}Cl\text{—}Co(NH_3)_5]^{4+}$. Even remembering that Cr^{3+} and Co^{3+} complexes are inert while Cr^{2+} and Co^{2+} complexes are labile, it is not obvious that an inner-sphere mechanism is necessary. Ultimately the $[CrCl(OH_2)_5]^{2+}$ product partially hydrolyzes, so that the thermodynamic products are $[CrCl(OH_2)_5]^{2+}$, $[Cr(OH_2)_6]^{3+}$, $[Co(OH_2)_6]^{2+}$, Cl^-, and NH_4^+. These could form through either an outer-sphere or an inner-sphere mechanism:

Outer-sphere

$$[Cr(OH_2)_6]^{2+} + [Co(NH_3)_5Cl]^{2+} \xrightarrow[\text{transfer}]{\text{direct } e^-} [Cr(OH_2)_6]^{3+} + [Co(NH_3)_5Cl]^+$$

$$[Co(NH_3)_5Cl]^+ + 5H_3O^+ + H_2O \xrightarrow{\text{Rapid}} [Co(OH_2)_6]^{2+} + 5NH_4^+ + Cl^-$$

$$[Cr(OH_2)_6]^{3+} + Cl^- \xrightarrow{\text{Slow}} [CrCl(OH_2)_5]^{2+} + H_2O$$

Inner-sphere

$$[Cr(OH_2)_6]^{2+} + [Co(NH_3)_5Cl]^{2+} \xrightarrow[\text{mech.}]{\text{Bridged}} [CrCl(OH_2)_5]^{2+} + [Co(NH_3)_5(OH_2)]^{2+}$$

$$[Co(NH_3)_5(OH_2)]^{2+} + 5H_3O^+ \xrightarrow{\text{Rapid}} [Co(OH_2)_6]^{2+} + 5NH_4^+$$

$$[CrCl(OH_2)_5]^{2+} + H_2O \xrightarrow{\text{Slow}} [Cr(OH_2)_6]^{3+} + Cl^-$$

The key difference is whether the $CrCl^{2+}$ species is formed rapidly (by the redox mechanism) or slowly (by the ligand-substitution step). Since the $CrCl^{2+}$ species can be separated from the reaction mixture by ion-exchange and identified by UV methods much more rapidly than it forms by simple ligand substitution, the inner-sphere mechanism is shown to apply. Note that this cleverly chosen model reaction requires a reductant (hydrated Cr^{2+}) whose oxidized species (product) is inert. Other such reductants include $Co(CN)_5^{3-}$ and $[Ru(NH_3)_5(OH_2)]^{2+}$, and these also react with ligand transfer.

Some of the constraints on inner-sphere mechanisms can lead to interesting changes of mechanism—and even of stoichiometric products—as reactant ligands are changed. The most obvious constraint is that the bridging ligand must be able to serve as a Lewis base (electron-pair donor) to two metal atoms at once. Where all the ligands on the inert reactant are amines, the redox reactions must proceed by an outer-sphere reaction because the N has only one pair of nonbonding electrons and no bridging structure is possible. For the Cr^{2+} reaction above in which the oxidant was $[Co(NH_3)_5Cl]^{2+}$, the rate constant is $6 \times 10^5 \ M^{-1} \ sec^{-1}$; but when the oxidant is $[Co(NH_3)_6]^{3+}$, an outer-sphere mechanism must operate and k is only $8 \times 10^{-5} \ M^{-1} \ sec^{-1}$—10 orders of magnitude slower.

As soon as we consider polyatomic ligands with a demonstrated bridging capacity the question arises: Does the bridge consist of a single atom within the XY ligand

$$M_1 \diagdown \atop M_2 \diagup XY$$

or do two different atoms within the ligand bridge the two metal atoms M_1—XY—M_2? The first mechanism is called *adjacent attack;* the second, *remote attack.* One interesting comparison has already been presented in Table 14.6, where the rates for Cr—NCS^{2+} and Cr—SCN^{2+} are given. The fact that the rate for Cr—NCS^{2+} is much lower is taken to mean that the moderately hard acid $Cr(OH_2)_5^{2+}$ bonds more favorably to the hard base atom N (remote attack on Cr—SCN^{2+}) than to the soft base atom S (remote attack on Cr—NCS^{2+}). Since —NCS and —SCN lie at quite different points in the spectrochemical series (Chapter 11), UV spectra can readily distinguish the two resulting linkage isomers. For $[Co(NH_3)_5NCS]^{2+}$, attack by Cr^{2+} produces exclusively the product Cr—SCN^{2+}—that is, the reaction occurs only through remote attack. But for $[Co(NH_3)_5SCN]^{2+}$, the reaction with Cr^{2+} produces roughly 30% Cr—SCN^{2+} (adjacent attack) and 70% Cr—NCS^{2+} (remote attack). The Cr^{2+} seems to have a modest preference for attack on the S atom, even though the Cr—SCN^{2+} isomer is thermodynamically less stable than the Cr—NCS^{2+} isomer. One possible reason is that double exchange or chemical exchange is occurring in the bridged intermediate, and S is more easily reduced than N by a neighbor Cr^{2+}.

Another comparison involving rates for remote attack involves the isoelectronic, isostructural bridging ligands —NCS and —NNN (azide). If the reduction mechanism is inner-sphere by remote attack, a hard-acid reductant should prefer azide with its harder-base atom N. That is, the rate constant should be significantly

Table 14.7   Rate Constants for —NCS and —NNN
as Potential Bridging Ligands

Oxidant	$Cr(OH_2)_6^{2+}\ k$ $(M^{-1}\ sec^{-1})$	$Cr(bipy)_3^{2+}\ k$ $(M^{-1}\ sec^{-1})$
$[Co(NH_3)_5NCS]^{2+}$	1.9×10^1	1.0×10^4
$[Co(NH_3)_5N_3]^{2+}$	3×10^5	4.1×10^4
	Inner sphere	Outer sphere

higher for M_1—NNN than for M_1—NCS. But since the ligands are isoelectronic and similarly bonded, they should conduct electrons about equally if no new bond needs to be formed. Therefore, in an outer-sphere mechanism Co—NCS^{2+} and Co—NNN^{2+} should react at very nearly the same rate. Table 14.7 gives some rate constants that support this interpretation.

Some inner-sphere redox reactions involving large organic ligands show remote attack that is very remote indeed. When the oxidant is $[Co(NH_3)_5L]^{3+}$ and L is pyridine, no possible bridging ligand exists and the reaction with Cr^{2+} proceeds slowly by an outer-sphere mechanism. However, when L is nicotinamide (pyridine carboxamide) the reaction with Cr^{2+} produces the inner-sphere product below much more rapidly ($k = 1.7 \times 10^1\ M^{-1}\ sec^{-1}$ versus $4.1 \times 10^{-3}\ M^{-1}\ sec^{-1}$):

$$\left[(H_2O)_5Cr-\underset{\underset{NH_2}{|}}{O}C-\hspace{-4pt}\bigcirc\hspace{-4pt}NH \right]^{4+}$$

Since the Co^{3+} was originally coordinated to the pyridine-ring nitrogen, the remote attack has occurred so far away that direct exchange is surely impossible and even superexchange seems unlikely. Here we expect to see chemical intervention by the ligand, either as double exchange or chemical exchange. The more reducible the pi-system ligand is—that is, the more readily it can accept an electron from the incoming reductant M_2—the more rapidly this reaction is likely to occur.

14.5 STRUCTURE-REACTIVITY CORRELATIONS

The mechanisms we have considered so far have applied to rather traditional reaction types—substitution and oxidation–reduction. If we draw an organic analogy, these are the kinds of reactions we might expect for a saturated hydrocarbon. Can we further adapt the mechanistic theory of organic chemistry to transition-metal reactions? This would be particularly useful with respect to transition-state structures as they influence stoichiometric reaction products. For example, we expect carbonium ions to add nucleophiles rapidly. This is analogous to the behavior of a five-coordinate intermediate in a dissociative octahedral substitution reaction, which rapidly adds another Lewis-base ligand once formed. There are other typical organic reactive species that have direct counterparts in transition-metal complex chemistry. By considering the more familiar organic-reaction patterns, we can make some specific predictions for reactive d-electron configurations and coordination geometries that will point us toward the catalytic reactivity dealt with in the next two chapters.

Table 14.8, adapted from a comparison made by Halpern, yields just such predictions; it deserves some explanation. First, "ligand substitution" is the reaction typical of a saturated hydrocarbon. A carbon atom undergoing substitution will have a coordination number of four (its own version of coordination saturation, since it has only four valence orbitals) and eight electrons in its valence-shell orbitals or in the MOs that arise from them. All its bonding orbitals are filled, and all its antibonding orbitals are vacant. This is comparable to CoL_6^{3+}, the system we examined most closely when we considered octahedral ligand substitution. Such complexes have reached an effective coordination saturation at CN = 6, and since Co^{3+} has six d electrons and the six ligands each donate a sigma pair of electrons, there are 18 valence-shell electrons that fill all the bonding MOs but no antibonding MOs.

Obviously, two key considerations are at work: How many additional ligands can the carbon (metal) atom take on before reaching coordination saturation, and how many electrons can it accept into bonding orbitals? In organic systems, the key numbers for carbon are eight electrons (the octet rule) and four ligands, since both numbers represent a particular form of saturation. For a d-block transition metal, the key numbers are 18 electrons and six ligands, though neither number is absolutely inviolable or even as reliable as the corresponding numbers for carbon. Reactive systems, then, must have a lower coordination number than the value representing coordination saturation if any new bond is to be formed. In the ligand-substitution case we just noted, a new bond is formed, but only by breaking an old one. If the coordination number is less than the normal maximum, one or more ligand atoms will be added to reach coordination saturation. If the number of valence electrons is less than the maximum, the incoming atoms will donate electrons to reach electronic saturation.

The nucleophilic addition comparison in the table is the first example of a reactive species. A classical carbonium ion or carbocation is one ligand short of coordination saturation and two electrons short of electronic saturation. Therefore, it is a good target for an incoming Lewis base, from which a single atom can donate two electrons. Organic chemists call such a base a nucleophile; the characteristic reaction of a carbocation is nucleophilic addition. In exactly the same sense, a transition-metal complex with coordination number five and 16 electrons is a prime candidate for attack by a Lewis base. $[Co(CN)_5]^{2-}$ in the table is a well-characterized intermediate in the dissociative mechanism for the substitution of $[Co(CN)_5(OH_2)]^{2-}$. It reacts as shown in the second step of the overall substitution reaction.

A carbanion has the maximum number of electrons, but it is one atom short of the maximum coordination number. Therefore, it can only form a bond to an atom

Table 14.8 Characteristic Reactive Structures for Organic and Transition-Metal Systems

Reaction type	Carbon or metal atom		Organic species	Examples	Reaction examples
	Valence shell e^-	Coordination number			
Ligand substitution	$\begin{cases} 8 \\ 18 \end{cases}$	$\begin{matrix} 4 \\ 6 \end{matrix}$	Saturated C	CH_3I $[Co(NH_3)_5Cl]^{2+}$	$CH_3I + 2Li \rightarrow CH_3Li + LiI$ $Co(NH_3)_5Cl^{2+} + H_2O \rightarrow Co(NH_3)_5(OH_2)^{3+} + Cl^-$
Nucleophilic addition	$\begin{cases} 6 \\ 16 \end{cases}$	$\begin{matrix} 3 \\ 5 \end{matrix}$	Carbonium ion	Ph_3C^+ $[Co(CN)_5]^{2-}$	$Ph_3C^+ + H_2O \rightarrow Ph_3COH + H_3O^+$ $Co(CN)_5^{2-} + I^- \rightarrow Co(CN)_5I^{3-}$
Electrophilic addition	$\begin{cases} 8 \\ 18 \end{cases}$	$\begin{matrix} 3 \\ 5 \end{matrix}$	Carbanion	$CH_3CH_2^-$ $[Mn(CO)_5]^-$	$3EtLi + GaCl_3 \rightarrow GaEt_3 + 3LiCl$ $Mn(CO)_5^- + H_3O^+ \rightarrow Mn(CO)_5H + H_2O$
Dimerization, abstraction	$\begin{cases} 7 \\ 17 \end{cases}$	$\begin{matrix} 3 \\ 5 \end{matrix}$	Free radical	$Ph_3C\cdot$ $[Co(CN)_5]^{3-}$	$2Ph_3C\cdot + I_2 \rightarrow 2Ph_3CI$ $Co(CN)_5^{3-} + CH_3I \rightarrow Co(CN)_5I^{3-} + CH_3\cdot$
Addition, insertion	$\begin{cases} 6 \\ 16 \end{cases}$	$\begin{matrix} 2 \\ 4 \end{matrix}$	Carbene (diradical)	$\cdot CH_2\cdot$ $IrCl(CO)(PPh_3)_2$	$CH_2 + HCl \rightarrow CH_3Cl$ $IrCl(CO)(PPh_3)_2 + HCl \rightarrow IrHCl_2(CO)(PPh_3)_2$

having a vacant orbital—that is, an electrophile. The comparable electrophilic addition to a transition-metal system occurs with five-coordinate metals that have 18 electrons. For instance, $Mn(CO)_5^-$ reacts with protic acids to form $Mn(CO)_5H$, a weak acid ($pK_a = 7$). The metal complex is undergoing electrophilic addition, but another view of its role is that the metal is serving as a base. We have not previously emphasized this aspect of transition-metal chemistry, but several subsequent topics will view metals at least partly in this role.

We are used to metal ions that behave as Lewis acids, however, by accepting pairs of electrons. Indeed, we have concentrated so much on electron pairs that it is difficult to imagine a transition-metal complex in the role of a free radical, which has one free site in its coordination sphere and is one electron short of electronic saturation. A comparable complex, however, is $Co(CN)_5^{3-}$. In this complex, the ligands yield a large ligand-field splitting and thus stabilize the 18-electron configuration, but the complex has only 17 electrons. Characteristic reactions of free radicals include dimerization and the abstraction of one-electron ligands from other molecules. An abstraction reaction for this ion is shown in the table, but dimerization often occurs as well:

$$2[Co(CN)_5]^{3-} \xrightarrow{\text{water}} [(CN)_5Co\!-\!Co(CN)_5]^{6-}$$

The abstraction reaction occurs so readily that $Co(CN)_5^{3-}$ rapidly absorbs H_2 to produce what is almost certainly $[Co(CN)_5H]^{3-}$ in aqueous solution. This solution is a powerful hydrogenating agent (reducing agent) for a number of organic systems, so that the $Co(CN)_5^{3-}$ ion is a catalyst for these hydrogenations. For example, nitrobenzene can be reduced to aniline and styrene to ethylbenzene by H_2 in the presence of $Co(CN)_5^{3-}$. These reactions do not occur to any measurable extent if a catalyst is not present.

Another interesting example of a d-electron "free radical" is the compound $[Ir_2(tmb)_4]^{2+}$, whose structure is shown schematically in Fig. 14.10. In this binuclear complex, it is reasonable to assume the presence of an Ir—Ir single bond, given the internuclear separation of 3.199 Å. Under that assumption the electron count per Ir is $9(Ir) + 4 \times 2(\text{isonitrile C}) + 1(Ir\text{–}Ir \text{ bond}) - 1(\text{net } 1 \oplus \text{ per Ir}) = 17$ electrons. The figure makes it clear that the coordination number of each Ir is 5, with an open site at each end of the molecule. When this complex is irradiated in the presence of cyclohexadiene, each Ir abstracts an H from the diene to yield $[Ir_2(tmb)_4H_2]^{2+}$; note that the net charge has not changed, so that transfer of a hydrogen atom and not a proton has occurred. The dihydride will slowly reduce (hydrogenate) styrene to ethylbenzene, returning to the original Ir_2 complex, so that a hydrogenation catalysis is occurring not unlike that seen for $Co(CN)_5^{3-}$.

The final category listed in Table 14.8 is that of compounds comparable to carbenes. Carbenes are carbon diradicals that have two vacant coordination sites and are two electrons short of electronic saturation. We expect this behavior from transition-metal complexes that are square-planar four-coordinate (allowing room for attack by two incoming atoms) and that have 16 electrons, although it is also true as

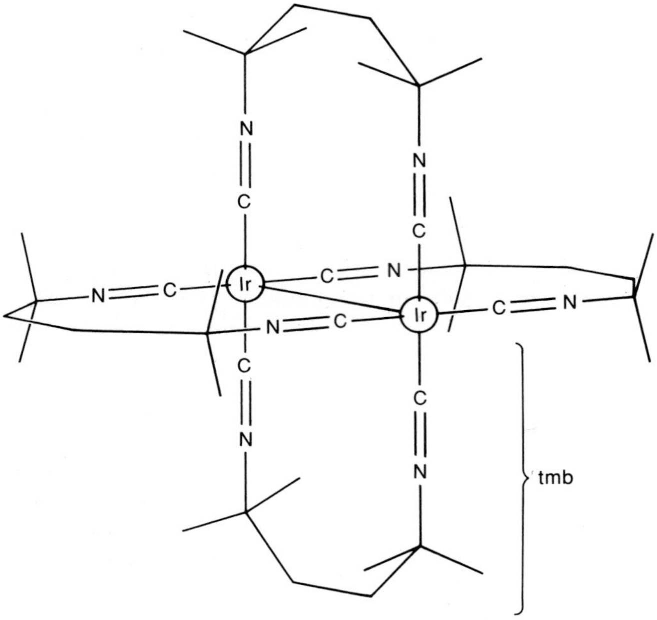

tmb = 2,5-diisocyano-2,5-dimethylhexane

Figure 14.10   A binuclear "free radical" complex in which each metal atom has C.N. = 5 and 17 valence electrons.

we have seen that square planar d^8 systems have a considerable stability of their own. A number of d^8 complexes meet this electron count condition. However, not all undergo the characteristic carbene reactions of two-atom addition or insertion equally well. For transition-metal complexes, the addition reactions often take place with relatively electronegative atoms, so that the addition produces an increase of two in the formal oxidation state of the metal atom. Unless the initial formal oxidation state is quite low, the final oxidation state can be so high that stabilization of the complex by pi back-bonding is difficult—which means, of course, that the addition reaction is less likely to occur. For this reason, reactions with d^8 complexes are usually called *oxidative additions*. In the next section we shall look at oxidative additions more carefully. Note, however, that the example in the table involves Ir(I), a metal in a suitably low oxidation state. It is also true that the tendency toward an 18-electron count is enhanced by strong pi-acceptor ligands (see Fig. 12.10), so that the nature of existing ligands influences a 16-electron metal's capability for oxidative addition.

Another example of carbene-like behavior from a 4-coordinate 16-electron system is the chemistry of the ruthenium(0) porphyrin, $[Ru(por)]^{2-}$, whose geometry has the Ru in the porphyrin plane (see Fig. 12.16). Figure 14.11 indicates some of the observed reactions of $[Ru(por)]^{2-}$; it is straightforward enough to think of the reaction with CH_3CH_2I as an oxidative addition if the coordinated ethyl group is taken to be an ethide ion, but the diethyl product is a neutral molecule with only 16 electrons. The reaction with CH_3CHCl_2 to yield an alkylidene is even less obviously an oxidative addition in that it is more difficult to think of the product as having a negative ligand—but the neutral molecule now clearly contains Ru^{II}, because the porphyrin is a dianion. The reaction with $BrCH_2CH_2Br$ yields a conventional ethylene pi complex, but it also now contains Ru^{II}. Finally, the slow conversion of coordinated $-CH_2CH_3$ to $=CHCH_3$ to coordinated $CH_2=CH_2$ involves hydrogen atom transfers that much more closely resemble reactions of a radical or diradical than oxidation-reduction. However, the production of cyclopropane from 1,3-dichloropropane surely involves its elimination from a metallacyclobutane intermediate, in which the Ru is six-coordinate and has 16 electrons. The opposite of an oxidative addition is a reductive elimination, but no reduction has occurred here because the $Ru(por)(thf)_2$ product is still Ru^{II}.

As Table 14.8 suggests, it is possible to regard an oxidative addition as inserting the diradical analog ML_4 into the $X—Y$ bond to produce XML_4Y. However, inorganic and organometallic chemists usually reserve the term "insertion" for a reaction in which a nonmetal molecule such as CO or SO_2 with a pair of nonbonding

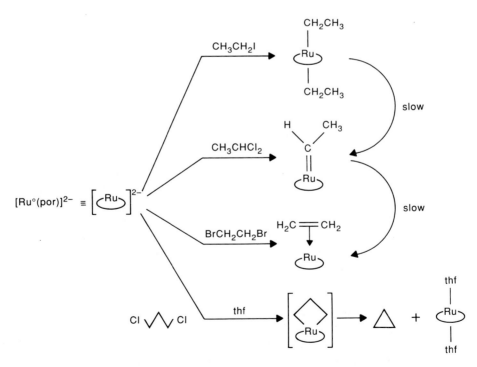

Figure 14.11 Oxidative additions and carbene-like reactions seen for a Ru(0) porphyrin.

electrons and at least two vacant coordination sites on the potential donor atom inserts itself into a metal-ligand bond:

$$CH_3 \text{—} Mn(CO)_5 + CO \rightarrow CH_3 \text{—} \overset{\displaystyle }{\underset{\displaystyle O}{\underset{\|}{C}}} \text{—} Mn(CO)_5$$

The metal-complex reactant for such insertion reactions is almost always a carbonyl or alkene complex obeying the 18-electron rule. Thus, the reactant is electronically saturated and it is also frequently six-coordinate (if a *dihapto-* alkene is thought to occupy one coordination site). From the metal's point of view, therefore, the reaction is only a ligand substitution. When CO is being inserted in a metal–alkyl bond to form a metal–acyl bond (as in the above reaction), however, the inserted CO is one that was already coordinated to the metal in the reactant complex. The incoming CO only replaces it in the metal coordination sphere, so it can be replaced by other electron-pair donors and still yield CO insertion:

$$R \text{—} M \text{—} CO + L \rightarrow R\overset{\displaystyle O}{\overset{\|}{C}}M \text{—} L$$

Depending on the reactive nature of L, further reactions can sometimes occur:

$$CH_3Mn(CO)_5 + HC \equiv CH \rightarrow [CH_3CO \text{—} Mn(CO)_4(\eta^2 \text{—} C_2H_2)]$$

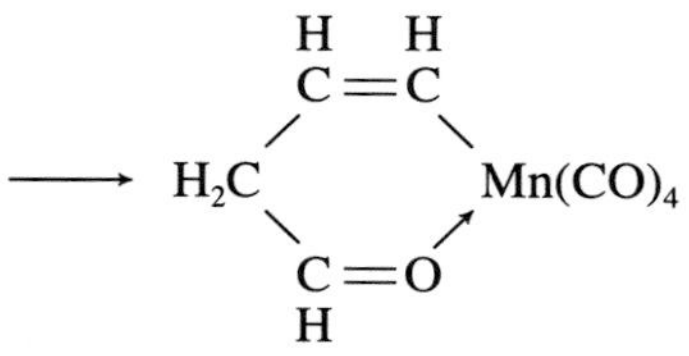

$$\longrightarrow \quad \overset{\displaystyle \begin{array}{c} H \quad H \\ C = C \end{array}}{\underset{\displaystyle \begin{array}{c} C = O \\ H \end{array}}{\overset{/}{H_2C}\underset{\backslash}{} \quad \overset{\backslash}{} \underset{/}{Mn(CO)_4}}}$$

There are obvious catalytic uses for reactions of this sort, and we shall return to them in the next chapter. A final note is that these "carbonylation" reactions can usually be driven in reverse—that is, in the high-entropy direction—to release CO from an acyl complex simply by heating:

$$EtCO \text{—} Mn(CO)_5 \xrightarrow{\text{30°C}} Et \text{—} Mn(CO)_5 + CO$$

Such *decarbonylation* reactions proceed by a dissociative mechanism, with a rate constant very near that for ordinary ligand substitution on the same reduced-coordination-number intermediate.

14.6 OXIDATIVE-ADDITION REACTIONS

Carbenes are notoriously difficult species to keep handy for synthetic purposes. However, the comparable transition-metal system, the square-planar d^8 complex, is in some cases so stable that the "insertion" reaction, or oxidative addition, will not

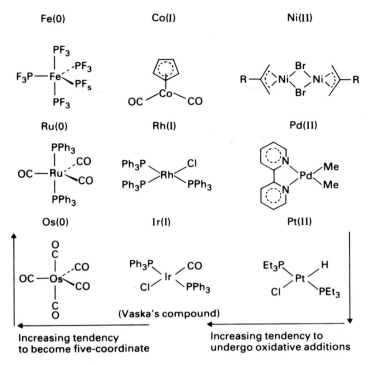

Figure 14.12 d^8 Complexes of group VII metals. [From J. P. Collman and W. R. Roper, *Advances in Organometallic Chemistry*, vol. 7, p. 56: Academic Press, New York, 1968. By permission from Academic Press, Inc.]

even occur. By the proper choice of metal and ligands, we can tailor the electronic energies of the square-planar complex to achieve almost any degree of reactivity. For example, the M^{2+} d^8 ions Ni^{2+}, Pd^{2+}, and Pt^{2+} have a fairly high effective nuclear charge, which makes the d electrons so stable that oxidative addition to give coordination number six is not very energetically favorable, particularly for Pd and Pt. This is similar to the reason there are no simple Pd and Pt carbonyls. On the other hand, the neutral d^8 atoms Fe, Ru, and Os are good enough pi donors to make a five-coordinate system such as $Fe(CO)_5$ vastly preferable in energy to a square-planar complex. So against an increasing tendency to undergo oxidative addition must be balanced a tendency to become five-coordinate (see Fig. 14.12). Five-coordinate complexes can also undergo oxidative addition, as we shall see; but the reaction is more complicated, since a ligand must dissociate while the addition is occurring.

The term "oxidative addition" was introduced by Vaska, who studied some addition reactions of the complex *trans*-IrCl(CO)(PPh$_3$)$_2$, a compound that attracted so much attention it became generally known as "Vaska's compound." A number of gaseous ligands or di-ligands form oxidative-addition products with

Vaska's compound. Some of the products are readily reversible, but others are permanent:

$$IrCl(CO)(PPh_3)_2 + O_2 \rightleftharpoons IrCl(O_2)(CO)(PPh_3)_2$$

$$IrCl(CO)(PPh_3)_2 + Cl_2 \rightarrow IrCl_3(CO)(PPh_3)_2$$

If Cl in Vaska's compound is replaced by CH_3, the O_2 oxidative addition becomes irreversible, again indicating the influence of other ligands. The difference in reversibility between O_2 and Cl_2 suggests that the degree of oxidation occurring in the reaction may not be equivalent for the two cases. One way of monitoring the degree of oxidation is to follow the C—O stretch frequency in the IR spectrum of each compound. Remember from Chapter 13 that increasing the positive charge on a metal atom to which a CO is bonded raises the frequency of the vibration. Table 14.9 shows clearly that the more stable and less reversible the product of the oxidative-addition reaction, the higher its C—O stretch wavenumber and, presumably, the greater the positive charge on Ir and the more complete the oxidation.

As Fig. 14.12 suggested, not all addition reactions involving Vaska's compound are oxidative. Some simply add one electron-pair-donor ligand to satisfy the 18-electron rule:

$$IrX(CO)(PPh_3)_2 + CO \rightleftharpoons IrX(CO)_2(PPh_3)_2$$

These compounds (X = Cl, Br, I) are trigonal bipyramidal with axial PPh_3 units. Both COs are bonded in the conventional fashion. The softness of the halogen base affects the stability of the five-coordinate dicarbonyl. The iodide is stable as the five-coordinate species at room temperature and must be heated to drive off CO, whereas the chloride decomposes to Vaska's compound at room temperature even under 1 atm CO pressure.

When an oxidative addition of the molecule XY forms new bonds to both X and Y to make the product approximately octahedral, there are two geometric possibilities. If X and Y remain bonded to each other (as for O_2), they must be *cis-* to each other in the six-coordinate product. But if X and Y are separate (as for HBr) they can be either *cis-* or *trans-*; HBr produces an equilibrium mixture of both isomers, but most additions are stereospecific. Methyl halides add to $IrX(CO)(PPh_3)_2$ in a kinetically controlled *trans-* geometry. We know it is kinetically controlled because if X is Br and CH_3Cl is added, the product (with CH_3 and Cl *trans-* to each other) spontaneously isomerizes to an arrangement with CH_3 *trans-* to Br. This is the same as the product obtained when X = Cl and the methyl halide is CH_3Br. The addition reaction is second order:

$$Rate = [IrX(CO)(PPh_3)_2] \cdot [CH_3Y]$$

Beyond the second-order kinetics, most oxidative addition reactions are thought to be predominantly associative because $\Delta S^{\ddagger}$ is usually quite negative—roughly -150 J/K mol rn. It seems likely that the mechanism is a fairly traditional S_N2 attack on CH_3Y by the Ir complex, with the metal acting as nucleophile. This means

Table 14.9 IR Stretch Frequencies of CO in IrCl(CO)(PPh$_3$)$_2$ Adducts

Incoming ligand	Adduct (IrCl(CO)(PPh$_3$)$_2$ = IrL$_4$)	CO stretch wavenumber (cm^{-1})	Reversibility
None	IrL$_4$	1967	
O$_2$	L$_4$Ir(O)(O)	2015	
SO$_2$	L$_4$Ir—SO$_2$	2021	Easy
H$_2$	L$_4$Ir(H)(H)	2034 (deuterium)	
HCl	H—IrL$_4$—Cl	2046	
CH$_3$I	CH$_3$—IrL$_4$—I	2047	
C$_2$F$_4$	L$_4$Ir(CF$_2$)(CF$_2$)	2052	
C$_2$(CN)$_4$	L$_4$Ir(C(CN)$_2$)(C(CN)$_2$)	2057	Stable reversible
BF$_3$	L$_4$Ir—BF$_3$	2067	
I$_2$	I—IrL$_4$—I	2067	
Br$_2$	Br—IrL$_4$—I	2072	Irreversible
Cl$_2$	Cl—IrL$_4$—Cl	2075	

that if the mechanism is correctly understood, oxidative-addition reactions involve a metal atom acting as a base, not an acid.

Much of the initial interest in Vaska's compound came from its reversible addition of O$_2$, which implied that it might be a reasonable model compound for the function of iron in hemoglobin. The complex is shown in Fig. 14.13. The O—O bond length is comparable to that in superoxide ion, O_2^-. This corresponds to an Ir oxidation state of 2+, which is reasonably consistent with the Ir stretch frequency of CO in the complex (Table 14.9). The O_2 complex is diamagnetic, which corresponds nicely to coupling of the one unpaired electron in octahedral Ir^{2+} with the one unpaired electron in O_2^-. This mode of bonding is analogous to that of ethylene to transition metals, as in Fig. 13.6. Unfortunately for the model, in hemoglobin the O_2 is bonded to the Fe in a bent end-on manner, much like NO when it is serving as a one-electron donor. A bond geometry so different necessarily reflects a quite different electronic arrangement. An interesting similarity, however, is that both bond-length and vibrational-frequency measurements indicate that in oxyhemoglobin, the coordinated O_2 is probably in superoxide form. Chapter 16 will consider biological coordination of O_2 in some detail.

If O_2 coordinated to Vaska's compound has a bond length appropriate for O_2^-, it is clear that the Ir atom has, on a net basis, served as a base, since the O_2 electron density has actually increased. Olefin or alkene complexes are also formed by Vaska's compound and several other comparable square-planar d^8 complexes. They are more stable the more electronegative the substituents on the alkene; thus, C_2F_4 and $C_2(CN)_4$ form quite stable complexes. This trend also suggests how important the electron donation from the metal to the ligand is—that is, how important it is that the metal function as a Lewis base. Perhaps even more convincing, however, is the formation of a complex between Vaska's compound and BF_3, a classical Lewis acid. The rhodium equivalent, $RhCl(CO)(PPh_3)_2$, seems to be a weaker base, since it forms a 1:1 adduct with BCl_3 (a stronger Lewis acid than BF_3) but not with BF_3 itself. In these compounds, it is difficult to imagine any electronic interaction other than the metal serving as a simple sigma donor and the

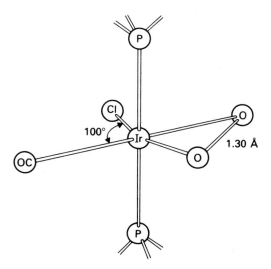

Figure 14.13 $IrCl(CO)(PPh_3)_2$—O_2 complex.

boron serving as a traditional sigma acceptor. Thus, we see metal basicity in its simplest form. The fact that BF_3 appears more or less in the middle of the ligands in Table 14.9, however, indicates that the Ir—B bonding is not unusual relative to the other compounds present and that the metal-base character must be important to the stability of all the compounds. It is not even necessary to have an 18-electron complex to see metal-base character, surprisingly: $W(CO)_3(PCy_3)_2$ (Cy = cyclohexyl) is a 16-electron complex, though a cyclohexyl C-H hydrogen is agostic. It reacts with fluoboric acid HBF_4 etherate in toluene to yield a protonated tungsten, with coordinated BF_4: $WH(BF_4)(CO)_3(PCy_3)_2$. Here the metal atom is sufficiently electron-rich to serve as a base even from a formal count of 16 electrons.

Most of this discussion has dealt with $IrCl(CO)(PPh_3)_2$, since it has served as a sort of model compound for oxidative-addition reactions to square-planar d^8 complexes. One other such compound should be mentioned because of its catalytic usefulness: $RhCl(PPh_3)_3$, known as *Wilkinson's catalyst*. It readily adds H_2 to form $RhClH_2(PPh_3)_2$ in solution; one PPh_3 dissociates, but is replaced by the solvent (benzene or $CHCl_3$). The H atoms then add in *cis*- geometry. This solution directly hydrogenates alkenes to alkanes and even alkynes to alkanes. Both the two H atoms and the pi-donor hydrocarbon are probably coordinated to the Rh in the transition state, since the hydrocarbon can replace the coordinated solvent molecule. Figure 14.14 shows the stereochemical course of the catalytic hydrogenation reaction at the Rh atom, established predominantly by proton NMR studies. The alkene–dihydrido complex is clearly the product of an oxidative-addition step, whereas the separation of the alkane product requires a reductive elimination at the Rh atom to yield a cyclic process. We shall take up other important related processes in the next chapter.

Five-coordinate complexes can also undergo oxidative addition. However, if two new ligand atoms are to be added to the system, one ligand in the reactant complex normally must leave for steric reasons. The reaction thus takes place in two steps: First, the metal (acting as a base) donates an electron pair to X^+ from the incoming X—Y ligand, leaving Y^- uncoordinated in the outer sphere. Then one ligand L from the initial ML_5 complex leaves, and the Y^- enters the inner coordination sphere to bond with the metal, which is now serving as an acid. Of course, if X and Y are the same, as in Br_2, the M—X bonding is ultimately the same as the M—Y bonding. The intermediate can sometimes be isolated. It behaves like a 1:1 electrolyte in solution, showing the nonequivalence of the inner-sphere and outer-sphere new ligands. An example is the trigonal bipyramidal compound $Os(CO)_3(PPh_3)_2$, which undergoes an oxidative-addition reaction with Br_2:

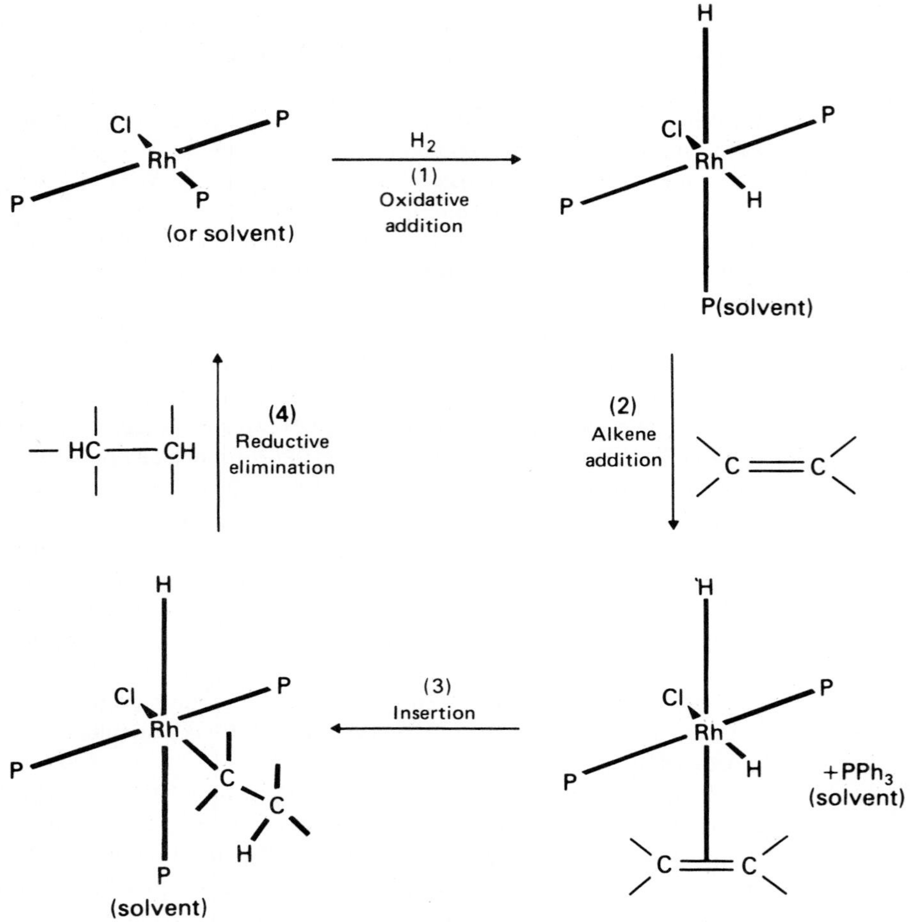

Figure 14.14   Catalytic hydrogenation by RhCl(PPh$_3$)$_3$.

The overall reaction is

$$Os(CO)_3(PPh_3)_2 + Br_2 \rightarrow OsBr_2(CO)_2(PPh_3)_2 + CO$$

in which the oxidation state of Os has increased from 0 to 2+.

It is less common for first-row transition metals to undergo oxidative addition than for heavier metals, even with the same d^8 configuration. However, an interesting organic coupling reaction involves the oxidative addition of an alkyl iodide to an allylnickel(II) bromide dimer [(h^3–C$_3$H$_5$)$_2$Ni$_2$Br$_2$], which has the same structure as the palladium dimer shown in Fig. 13.10. The Lewis-base solvent dimethylformamide (DMF) displaces a bridging Br and separates the dimer into (C$_3$H$_5$) NiBr(DMF) monomers. This square-planar Ni^{II} compound undergoes oxidative addition of RI to yield an octahedral complex of Ni^{IV} (though that oxidation state should not be taken too seriously). The alkyl and allyl groups then couple to form the organic product, leaving solvated Ni^{2+} in solution:

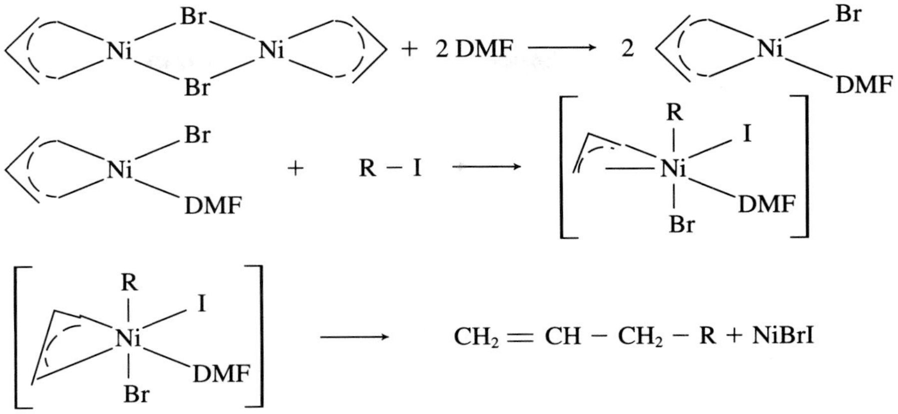

This sort of coupling reaction, like many of those recently discussed, has obvious possible extensions to catalytic applications. In the next two chapters, we shall consider the demonstrated catalytic properties of transition-metal complexes both in industrial organometallic catalysis and in metalloenzyme biocatalysis, and discuss the altered reactivities of ligand species in the context of the structural and mechanistic arguments of this and earlier chapters.

PROBLEMS

A. DESCRIPTIVE

A1. What mechanism for reaction (a) below do you infer from the rate constants given for the two reactions? Explain your answer.

(a) $Ru(NH_3)_6^{3+} + NO + H_3O^+ \rightarrow Ru(NH_3)_5(NO)^{3+} + NH_4^+ + H_2O$
$k = 2 \times 10^{-1}\ M^{-1}\ sec^{-1}$

(b) $Ru(NH_3)_6^{3+} + H_2O \rightarrow Ru(NH_3)_5(OH_2)^{3+} + NH_3$
$k < 1 \times 10^{-10}\ M^{-1}\ sec^{-1}$

A2. The rate constant for the acid hydrolysis of $Co(tn)_2Cl_2^+$ is $2 \times 10^{-2}\ sec^{-1}$, where $tn = H_2NCH_2CH_2CH_2NH_2$ and the other conditions are the same as those in Table 14.3. Why is this rate so high?

A3. The thermodynamic acid-dissociation constants (K_a) for increasingly chelated Pt^{4+} amine complexes are

$$Pt(NH_3)_6^{4+} \qquad K_a = 1.2 \times 10^{-8}$$
$$Pt(en)(NH_3)_4^{4+} \qquad K_a = 7.1 \times 10^{-7}$$
$$Pt(en)_3^{4+} \qquad K_a = 3.5 \times 10^{-6}$$

What is the relationship between these constants and the following base-hydrolysis rate constants ($M^{-1}\, sec^{-1}$)?

$$Co(NH_3)_5Cl^{2+} \qquad k_b = 0.86$$
$$cis\text{-}Co(en)_2(NH_3)Cl^{2+} \qquad k_b = 54$$
$$Co(tetren)Cl^{2+} \qquad k_b = 35{,}000$$

Note: Tetren =

A4. One isomer of $Pt(NH_3)_2Cl_2$ reacts with pyridine to give a single isomer of

$$[Pt(NH_3)_2(py)_2]Cl_2$$

which in turn decomposes when heated to yield a single isomer of $Pt(py)(NH_3)Cl_2$. If the other isomer of $Pt(NH_3)_2Cl_2$ is given the same treatment, it yields the other isomer of $[Pt(NH_3)_2(py)_2]Cl_2$, and finally a mixture of $Pt(NH_3)_2Cl_2$ and $Pt(py)_2Cl_2$. Identify the starting isomers and the product isomers as *cis-* or *trans-*.

A5. In the complex $[Pd_2(CNCH_3)_6]^{2+}$ each Pd is square planar with three isonitriles and another Pd around it. Reflux of this species with CCl_4 in acetonitrile for six hours shows no reaction. However, photolysis of the complex breaks the Pd–Pd bond, yielding $[Pd(CNCH_3)_3]^+$. This species does react rapidly with CCl_4. Consider the preferred geometry and electron count for palladium complexes and suggest what the reaction with CCl_4 should be (after photolysis).

A6. When $(RCOO)Co(NH_3)_5^{3+}$ is allowed to react with Cr^{2+} solutions, the net reactions and rate constants ($M^{-1}\, sec^{-1}$ at 25°C) are

$$Cr^{2+} + (CH_3COO)Co(NH_3)_5^{2+} \rightarrow CrOOCCH_3^{2+} + Co^{2+}$$
$$k = 0.18\ M^{-1}\, sec^{-1}$$
$$Cr^{2+} + (HOOCCH_2CH_2COO)Co(NH_3)_5^{2+} \rightarrow CrOOCCH_2CH_2COOH^{2+} + Co^{2+}$$
$$k = 0.27\ M^{-1}\, sec^{-1}$$
$$Cr^{2+} + (HOOCCH=CHCOO)Co(NH_3)_5^{2+} \rightarrow CrOOCCH_2CH_2COOH^{2+} + Co^{2+}$$
$$k = 4.87\ M^{-1}\, sec^{-1}$$

Suggest a mechanistic reason for the higher rate of the fumarate complex.

A7. How could you show experimentally that an insertion reaction requires a CO already coordinated to the metal if the incoming ligand is also CO?

A8. The $[Ir(CO)_2I_3]^-$ ion can be prepared, but only as a dimer $K_2[Ir_2(CO)_4I_6]$. Why is the monomer unstable?

A9. In the reaction

$$IrX(CO)(PPh_3)_2 + CH_3I \rightarrow CH_3IrXI(CO)(PPh_3)_2$$

should the rate be faster for $X = I$ or for $X = Cl$? Why?

A10. Redox reactions proceed by steps involving one-electron transfer or two-electron transfer. Sometimes, however, no convenient combination of oxidation states exists for such steps in a bimolecular elementary process if one element wants to transfer one electron and the other wants to transfer two electrons (a *noncomplementary* reaction). In the initial step of the stoichiometric reaction

$$2Fe^{2+} + Tl^{3+} \rightarrow Tl^{+} + 2Fe^{3+}$$

either the Fe^{4+} or the Tl^{2+} ion has to be made. Neither of these is a stable oxidation state. Write a two-step mechanism for each of these possibilities. Given that the rate law is

$$\text{Rate} = \frac{k[Fe^{2+}]^2\,[Tl^{3+}]}{[Fe^{2+}]+k\,'[Fe^{3+}]}$$

decide which mechanism is correct.

A11. What reaction, if any, should occur between Vaska's compound and $SnCl_4$?

B. NUMERICAL

B1. Consider the square-planar AO overlaps shown in Fig. 11.15 for sigma bonding. Which *d* orbitals can engage in pi overlap with pi-symmetry orbitals? How many ligand pi orbitals overlap each individual d_π orbital? Apportion each d_π orbital equally to its overlapping individual ligand pi orbitals, sum the fractional d_π–L_π overlaps for each ligand atom L, and show that each ligand in ML_4 has access to $0.75d_\pi$ orbital. Now consider a trigonal-bipyramidal ML_5 complex with axial ligands on the *z* axis and one equatorial ligand on the *x* axis. Sketch possible pi overlaps for this geometry (analogous to Fig. 11.17b). Apportion metal d_π orbitals as in the square-planar case, and show that the axial ligands have access to 0.57 metal d_π orbital, but that the equatorial ligands have access to $0.95d_\pi$ orbital.

B2. Calculate $\Delta V^{\ddagger}$ for the aquation of $Cr(H_2O)_5(NO_3)^{2+}$ at 25°C from the data below (some fiddling with units will be required):

P (kbar)	0.020	0.456	0.963	1.520	2.026
$10^5 k\ (sec^{-1})$	6.82	8.05	10.5	14.9	18.5

Comment on the probable significance of your $\Delta V^{\ddagger}$ value for the mechanism.

B3. The macrocycle *tet* has the formula

Its complex *trans*-Co(tet)X_2^{+} hydrolyzes the first X^- [to Co(tet)X(OH$_2$)$^{2+}$] at 25°C with rate constants 2.6×10^{-4} sec^{-1} (X = Cl), 3.8×10^{-2} sec^{-1} (X = Br), and 6 sec^{-1} (X = NCS):

$$Co(tet)X_2^{+} + H_2O \xrightarrow{\ 25°C\ } Co(tet)X(OH_2)^{2+} + X^-$$

The thermodynamic equilibrium constant for the reaction is 3.0×10^{-3} (X = Cl), 3.0×10^{-2} (X = Br), and 4.0 (X = NCS). Construct a linear free-energy relationship graph for these data and obtain a slope using a linear regression treatment. What is the significance of the results for the mechanism of the substitution?

B4. The $Fe(CN)_6^{3-/4-}$ potential is 0.68 V and the corresponding electron-exchange rate is 7.4×10^2 M^{-1} sec^{-1}. The $W(CN)_8^{3-/4-}$ potential is 0.54 V and the corresponding electron-exchange rate is 7×10^4 M^{-1} sec^{-1}. Calculate a value of the rate constant for the oxidation of $W(CN)_8^{4-}$ by $Fe(CN)_6^{3-}$. Compare your results with the observed value of 4.3×10^4 M^{-1} sec^{-1}. Is an outer-sphere mechanism reasonable?

B5. For the redox reaction

$$V^{2+} + Co(NH_3)_5(SO_4)^+ \rightarrow V^{3+} + Co^{2+} + 5NH_3 + SO_4^{2-}$$

the rate constant as a function of temperature is as follows:

$T(°C)$	k (M^{-1} sec^{-1})
20	17.7
30	36.4
40	69.1
50	125.

Calculate $\Delta H^{\ddagger}$ and $\Delta S^{\ddagger}$ from these data at 25°C. Do the values suggest an inner-sphere or outer-sphere mechanism?

B6. The table below gives log k values for the substitution of various ligands into high-spin octahedral $M(OH_2)_6^{2+}$ ions at 25°C. Plot these values against the Δ(CFSE) values for a dissociative mechanism in Table 12.6. Comment on the relationship and on approximations in the treatment.

Entering ligand	V^{2+}	Cr^{2+}	Mn^{2+}	Fe^{2+}	Co^{2+}	Ni^{2+}	Cu^{2+}	Zn^{2+}
H_2O	2.0	8.5	7.5	6.5	6.3	4.3	9.3	7.5
NH_3					5.1	3.7	8.3	6.6
F^-	1.5		6.5	6.0	5.3	3.9	8.7	

C. EXTENDED REFERENCE

C1. Plot the CO stretch frequencies in Table 14.9 against the electron affinity of the incoming adduct species. (Atomic, molecular, and radical electron affinities may be found in the *CRC Handbook of Chemistry and Physics*). Decide on a consistent way to treat XY adducts that retain a X—Y bond compared to those that separate (X—Ir—Y). Is the relationship shown on the graph physically reasonable? Are any points completely unrelated? Compare to R. N. Scott *et al.* [*J. Amer. Chem. Soc.* (1968), *90,* 1079.]

C2. We have seen that there are two paths for the aquation of octahedral complexes, acid hydrolysis and base hydrolysis:

$$k = k_1[ML_6] + k_2[ML_6][OH^-]$$

For several complexes, there is a third term in the rate law:

$$k = k_1[ML_6] + k_2[ML_6][OH^-] + k_3[ML_6][H_3O^+]$$

For $Cr(H_2O)_5CN^{2+}$, $k_1 = 9.7 \times 10^{-3}$ sec^{-1}, $k_2 = 6 \times 10^{-7}$ M^{-1} sec^{-1}, and $k_3 = 8.0 \times 10^{-3}$ M^{-1} sec^{-1} at 55°C. What sort of reaction pathway does this third term represent? How does H$^+$ help hydrolyze CN$^-$? [J. P. Birk and J. H. Espenson, *Inorg. Chem.* (1968), 7, 991; D. K. Wakefield and W. B. Schaap, *Inorg. Chem.* (1969), 8, 512.]

C3. P. F. Meier *et al.* [*Inorg. Chem.* (1979), 18, 610] report on two substitution reactions of five-coordinate nickel complexes. For halide exchange

$$[Ni(PMe_3)_4X]^+ + {}^*X^- \rightarrow [Ni(PMe_3)_4^*X]^+ + X^-$$

they suggest an I_a mechanism because the reaction is first-order in both the complex and X$^-$, and there is an entering-group effect for X$^-$. For phosphine exchange

$$[Ni(PMe_3)_4X]^+ + {}^*PMe_3 \rightarrow [Ni(PMe_3)_3(^*PMe_3)X]^+ + PMe_3$$

they suggest a D mechanism because the reaction is zero-order in $^*PMe_3$ and $\Delta S^{\ddagger}$ is positive. Which of these mechanisms involves the least stable intermediate in electronic terms? Suggest additional experimental evidence for the unstable-intermediate mechanism.

C4. If the PPh$_3$ ligands in Vaska's compound are replaced by related phosphines PR$_3$, the second-order rate constants for CH$_3$I addition

$$IrCl(CO)(PR_3)_2 + CH_3I \rightarrow CH_3IrClI(CO)(PR_3)_2$$

are as follows:

PR$_3$	k (M^{-1} sec^{-1})	$\Delta H^{\ddagger}$ (kJ mol^{-1})
PMe$_2$Ph	4.7	46.4
PMe$_2$(O-MeOC$_6$H$_4$)	530	27.6
PMe$_2$(P-MeOC$_6$H$_4$)	6.8	38.1

Why does the *ortho* methoxy group on the coordinated phosphine enhance the rate of substitution so greatly, even though a similar *para-* group does not? [E. M. Miller and B. L. Shaw, *J. Chem. Soc.* (*Dalton*) (1974), 480.]

C5. Swiss workers [Y. Ducommun, G. Laurenczy, and A. E. Merbach, *Inorg. Chem.* (1988), 27, 1148] have performed comparable kinetic studies for the first-row ion Zn^{2+} and the second-row ion Cd^{2+} in the following reaction:

$$M(H_2O)_6^{2+} + bpy \rightarrow M(H_2O)_4(bpy)^{2+} + 2H_2O$$

For M = Zn, they find $\Delta S^{\ddagger} = -4$ J/K mol rn and $\Delta V^{\ddagger} = +7.1$ mL/mol rn; for M = Cd, they find $\Delta S^{\ddagger} = -40$ J/K mol rn and $\Delta V^{\ddagger} = +5.5$ mL/mol rn. What mechanism(s) do these values suggest for the two reactions? The two metal ions are both d^{10}; what difference between them might be responsible for these differing values for comparable stoichiometric reactions?

15

Ligand Reactions and Catalytic Mechanisms in Industry———

In the previous chapter, we looked at some of the reactions of coordinated transition-metal atoms or ions essentially from the viewpoint of the metal atom itself. Some of the most important and interesting studies of transition-metal compounds, however, are being done in order to understand and use changes in ligand reactivity that occur when a free ligand molecule or radical is coordinated to the metal. In this chapter, we shall focus our attention on reactions of coordinated ligand molecules that are governed or aided by the metal atom they are coordinated to. There are numerous synthetic organic applications in which the transition-metal compound is a stoichiometric reactant, and others in which it serves as a catalyst. A number of the catalytic processes are very important industrially.

——— *15.1 CHANGES IN LIGAND REACTIVITY ON COORDINATION*

In conventional coordination compounds, the metal atom is in a positive oxidation state and has at least a partial positive net charge. As the last several chapters have emphasized, the metal serves as a Lewis acid in these compounds and the ligand as a Lewis base. If we limit ourselves initially to compounds for which the above is a complete or nearly complete description of the bonding—that is, to compounds in which pi overlap or metal–base behavior is not important—the electronic effect on the ligand's reactivity is easy to describe. For species like $Co(NH_3)_6^{3+}$ or the various metal hydrates, the effect is that the coordinated ligand is now a weaker electron donor to other atoms. The atom within the ligand molecule that actually donates an electron pair to the metal experiences an increase in its net positive charge, which draws electrons from surrounding atoms in the ligand molecule. If the ligand is protic, the acidity of the hydrogens is significantly increased. We saw this behavior in the discussion accompanying Fig. 6.9 for metal-ion hydrates, and again in Chapter 14 in discussing the S_N1CB mechanism for the base hydrolysis of coordinated NH_3 molecules. So for a coordinated ligand where only sigma bonding is important—for simplicity, a hard-acid/hard-base complex—the ligand's acidity always increases. To give a different example of this behavior, consider the experimental K_a for the copper(II) hydrate:

$$Cu(OH_2)_4^{2+} + H_2O = Cu(OH_2)_3OH^+ + H_3O^+ \qquad K_a = 1.0 \times 10^{-8}$$

Each water molecule coordinated to the Cu^{2+} is roughly a million times as acidic as the same molecule would be if uncoordinated in solution, where $K_w = 10^{-14}$. Of course, the higher the formal charge on the metal ion, the greater this effect becomes, so that high oxidation states often yield oxocations or oxoanions.

The increased acidity of the ligand upon coordination need not be proton acidity, of course. In the above reaction, the incoming water molecule serves in a general sense as a nucleophile. Its nucleophilic attack on the coordinated water molecule is enhanced by the draining of electrons from the coordinated ligand. This is generally true for the attack of any nucleophile on any coordinated ligand. Consider the hydrolysis of an amino-acid ester such as ethyl glycinate:

$$H_2N\text{—}CH_2\text{—}COOEt + OH^- \rightarrow H_2N\text{—}CH_2\text{—}COO^- + EtOH$$

The OH^- nucleophile attacks the carboxyl C, but the reaction is slow if no transition metals are present to coordinate the ester. When the ester is coordinated to Co^{3+} through the amine N atom, the hydrolysis is more rapid, but not much. The intervening CH_2 group tends to isolate the carboxyl C from the electron-withdrawing effect of the Co. However, when two coordination sites on the Co^{3+} are opened up to allow the ester to chelate the Co through the carbonyl O, nucleophilic attack on the carboxyl C by the OH^- is speeded up by at least a factor of 100:

$$cis\text{-}(en)_2CoBr(NH_2CH_2COOEt)^{2+} + Hg^{2+} \longrightarrow$$

$$(en)_2Co(NH_2CH_2COOEt)^{3+} + HgBr^+$$

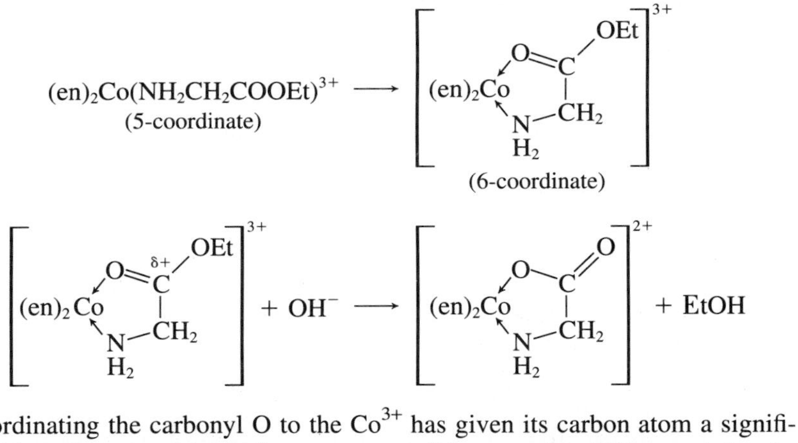

Coordinating the carbonyl O to the Co^{3+} has given its carbon atom a significant positive charge and made it more susceptible to nucleophilic attack by the

OH^-. With respect to acid–base reactions, then, a coordinated ligand that serves as an electron donor to the metal atom will be (1) more acidic if it contains protic hydrogens; (2) less basic with respect to reactions with an external acid; and (3) more susceptible to nucleophilic attack by an external electron donor.

For redox reactions the situation is similar, but more complicated. The coordinated ligand can be reduced or oxidized either by an external reagent or, perhaps, by the metal itself. Ligands are generally harder to oxidize by an external oxidant if they are coordinated to a metal atom, because the electrons that the ligand would otherwise lose to the oxidant are tied up in the metal–ligand bond. On the other hand, if the metal itself is the oxidant, having the ligand coordinated makes the redox reaction kinetically much more facile. For example, benzene is quite resistant to oxidation under ordinary circumstances. Note, for instance, the formation of "purple benzene" when $KMnO_4$ is dissolved in benzene with the aid of a crown ether; the very strong oxidant MnO_4^- does not attack the solvent benzene. However, benzene is oxidized readily by lead(IV) trifluoroacetate precisely because an aryllead intermediate can be formed:

$$C_6H_6 + Pb(OOCCF_3)_4 \rightarrow C_6H_5Pb^{IV}(OOCCF_3)_3 + CF_3COOH$$

$$C_6H_5Pb^{IV}(OOCCF_3)_3 \rightarrow C_6H_5OOCCF_3 + Pb^{II}(OOCCF_3)_2$$

Here the highly electronegative Pb^{IV} atom has withdrawn enough electrons through the C—Pb bond from the coordinated phenyl group that a coordinated trifluoroacetate can attack the positively charged phenyl carbon. Since the natural flow of electrons in simple donor–acceptor compounds is away from the ligand, oxidation is a natural consequence if the metal's thermodynamic oxidizing ability is great enough.

Another controlled organic oxidation that is industrially important is the oxidation of ethylene to acetaldehyde by $PdCl_4^{2-}$. Written as a single stoichiometric redox equation, the reaction is

$$C_2H_4 + PdCl_4^{2-} + 3H_2O \rightarrow CH_3CHO + Pd^0 + 2H_3O^+ + 4Cl^-$$

However, it can be made catalytic in Pd (the *Wacker process*) by adding the cocatalyst $CuCl_2$:

$$Pd^0 + 2CuCl_2 + 2Cl^- \rightarrow PdCl_4^{2-} + 2CuCl$$

$$4CuCl + 4H_3O^+ + 4Cl^- + O_2 \rightarrow 4CuCl_2 + 6H_2O$$

This yields the overall catalyzed reaction equation

$$C_2H_4 + \tfrac{1}{2}O_2 \xrightarrow{\ PdCl_4^{2-},\ CuCl_2\ } CH_3CHO$$

Primarily on the basis of kinetic studies, the following mechanism has been proposed:

1. $PdCl_4^{2-} + C_2H_4 \rightleftharpoons [PdCl_3(C_2H_4)]^- + Cl^-$

2. $[PdCl_3(C_2H_4)]^- + H_2O \rightleftharpoons cis\text{-}[PdCl_2(C_2H_4)(OH_2)]+ Cl^-$

3. $PdCl_2(C_2H_4)(OH_2) + H_2O \rightleftharpoons [PdCl_2(C_2H_4)(OH)]^- + H_3O^+$

4. $[PdCl_2(C_2H_4)(OH)]^- + H_2O \xrightarrow{slow} \underset{\text{(Sigma alkyl)}}{[HOCH_2CH_2-PdCl_2(OH_2)]^-}$

5. $[HOCH_2CH_2-PdCl_2(OH_2)]^- \rightleftharpoons [HOCH=CH_2)PdHCl_2]^-$

$\rightleftharpoons [(HOCH(CH_3)PdCl_2(OH_2)]^-$

6. $[(HOCH(CH_3)PdCl_2(OH_2)]^- \rightleftharpoons [(CH_3CH=O)PdHCl_2]^-$

$\rightarrow CH_3CHO + [PdHCl_2(OH_2)]^-$

The complex $[PdCl_3(C_2H_4)]^-$ formed in step 1 is presumably a conventional alkene complex like those of Chapter 13 or, indeed, like the very similar Zeise's salt. Step 2 is a further square-planar ligand substitution reaction that gives a *cis* geometry in which the ligand O atom is next to the ethylene. Step 3 deprotonates the coordinated H_2O to a hydroxyl group. All of these first three steps are rapidly established equilibria, and the products Cl⁻ and H_3O^+ show up as inverse concentration dependence in the rate law for the overall reaction. Step 4 is the rate-determining step, a rearrangement in which the OH shifts to one of the ethylene carbons and the other carbon adopts a sigma-bonded arrangement with the Pd. Step 5 is a sequence of hydrogen-transfer equilibria in which, by shifting the H atom onto the Pd and back, the 2-hydroxyethyl ligand is converted into a 1-hydroxyethyl ligand. Finally, in step 6, another hydrogen transfer leaves acetaldehyde as the organic ligand. This dissociates, and the remaining Pd(0) complex is destroyed by hydrolysis unless the $CuCl_2$ co-catalyst regenerates $PdCl_4^{2-}$. The extensive involvement of the Pd atom in the necessary OH shift and H atom transfers should make it clear that this oxidation depends on the metal not only as an electron sink but also for the detailed geometry of its coordination.

In the ethylene oxidation above, metal coordination was used to activate a small molecule with respect to a certain kind of reaction. We shall consider a number of other reactions in which coordination activates the ligand molecule, but there are also important cases in which coordination is used to deactivate or mask the reactivity of a molecule. There are several possible ways the metal can make a particular reaction difficult or impossible for a ligand molecule. Three of these we can call electronic: (1) Coordination can tie up an electron pair necessary for the other reaction; (2) coordination can induce a partial electric charge of the wrong polarity at the reactive site; and (3) coordination can change the redox potential of the molecule enough to make the other reaction no longer thermodynamically possible. Two other possible influences are essentially conformational: (4) Coordination can force the ligand molecule into a geometry favorable for coordination but unfavorable for the other reaction, and (5) coordination can shift the molecule into a tautomeric form that undergoes the other reaction less readily because either the conformation or the partial charges in the new tautomer are wrong.

A particularly useful form of masking is the protection of terminal carboxyl groups on amino acids when a reaction is desired elsewhere on the amino acid or polypeptide chain. Cu^{2+} is ideal for this purpose, since it chelates both the amino group and carboxyl O:

$$R-\underset{\underset{\underset{O}{\parallel}}{\overset{\overset{H_2}{N}}{\underset{C-O}{\mid}}}}{CH} \ Cu \underset{\underset{H_2}{N}}{\overset{O-C\overset{O}{\diagup}}{\diagup}} CH-R$$

For example, other amine groups on the amino acid can be made to react:

$$\underset{H_2}{\overset{Cu}{\underset{N}{\diagup}}} \overset{O-C\overset{\diagup O}{\diagdown}}{\underset{}{\mid}} -CHCH_2CH_2CH_2NH_2 \quad \xrightarrow{\ NH_2CONH_2\ }$$

$$\underset{H_2}{\overset{Cu}{\underset{N}{\diagup}}} \overset{O-C\overset{\diagup O}{\diagdown}}{\underset{}{\mid}} -CHCH_2CH_2CH_2NHC\overset{\diagup O}{\underset{NH_2}{\diagdown}}$$

Other carboxyl groups can react as well:

$$\underset{H_2}{\overset{Cu}{\underset{N}{\diagup}}} \overset{O-C\overset{\diagup O}{\diagdown}}{\underset{}{\mid}} -CHCH_2CH_2C\overset{\diagup O}{\underset{OH}{\diagdown}} \quad \xrightarrow{\ C_6H_5CH_2Cl\ } \quad \underset{H_2}{\overset{Cu}{\underset{N}{\diagup}}} \overset{O-C\overset{\diagup O}{\diagdown}}{\underset{}{\mid}} -CHCH_2CH_2C\overset{\diagup O}{\underset{OCH_2C_6H_5}{\diagdown}}$$

After these or other possible reactions, the amino acid or peptide is readily regenerated by simply adding H_2S to precipitate CuS, freeing the carboxylic acid.

We are more often interested in activation than in masking, however. The ethylene oxidation to acetaldehyde is a good example of organometallic systems used as reactants or intermediates for useful organic syntheses. Very commonly these reactions involve an edge-on alkene-to-metal bond, whether of an isolated double bond or a larger delocalized pi system. This alkene complex frequently shifts its bonding from pi to sigma by acquiring a hydrogen atom or some other one-electron radical coordinated next to it on the metal, just as the C_2H_4 picked up the OH coordinated to Pd in the acetaldehyde synthesis. In particular, H atoms are usually quite mobile, particularly if the metal is a soft enough acid to form stable hydride complexes. Another useful shift is the insertion of a metal-bonded CO, retaining the M—CO bond but forming a C—CO bond as described in the last chapter. The acetaldehyde synthesis showed many of these features, and the catalytic reactions we shall examine in the next four sections use them all, in various ways, to very good effect.

15.2 METAL CARBONYL ACTIVATION: THE WATER-GAS SHIFT REACTION

The industrial production of elemental hydrogen H_2 is extremely important to our economy. The Haber process for nitrogen fixation (Chapter 8) requires enormous amounts of H_2 for ammonia production, but in addition the production of synthetic chemical fuels will require H_2 as a reactant in any thermodynamically feasible process that does not begin with a hydrocarbon. On the earth's surface, the great reservoir of hydrogen is in liquid water, but O–H bonds are so strong that electrolysis, for example, requires a prohibitively large energy input. The earliest Haber plants used coke to make *water gas* through the reducing power of elemental carbon:

$$C(s) + H_2O(g) \rightarrow CO(g) + H_2(g)$$

This is still the only economic process producing H_2 without using a hydrocarbon reactant. Unfortunately, it is enormously endothermic (see Table 8.6) and can be entropy-driven to the right only at very high temperatures—well over 1000°C. At that temperature the following equilibrium is shifted well toward the left because of its negative $\Delta S°$, using up the hydrogen just produced:

$$CO(g) + H_2O(g) \rightleftharpoons CO_2(g) + H_2(g) \quad \Delta S^0 = -42.4 \text{ J/K mol rn}$$

This reaction (to the right) is the *water-gas shift reaction (WGSR)*. At low temperatures it would be enthalpy-driven to the right, yielding more hydrogen ($\Delta H^0 = -41.0$ kJ/mol rn), but the mixture is kinetically inert below about 600°C. However, it can be catalyzed by solid Fe_3O_4/Cr_2O_3 at about 350°C, and hydrogen production plants all use WGSR reactors following the main reaction to enhance their yield, whether they start with coke or natural gas (*steam reforming*).

The WGSR could be driven even more efficiently to the right if there were a catalyst effective at lower temperatures. In the 1970s several groups began experiments based on the idea that if the CO were present as a metal carbonyl, nucleophilic attack on the C by a water O atom to yield CO_2 or carbonate might be favored. This is in fact the case, and a number of metal carbonyl species have been shown to catalyze the WGSR; indeed, in basic solution so many metal carbonyls show at least modest catalysis in base that it has been suggested the catalysis is a general property.

Two widely differing WGSR catalyst systems and their proposed mechanisms are outlined in Fig. 15.1 and 15.2. Figure 15.1 shows an example (perhaps the best studied) of catalysis in basic solution, using $Ru_3(CO)_{12}$. It has been proposed that the first step is nucleophilic attack by solution OH^- on the carbon atom from one of the carbonyls, yielding a negatively charged cluster with a COOH carboxyl ligand. This species can eliminate CO_2 (which in a basic solution would become HCO_3^-) by transferring a hydride ion to Ru: $[HRu_3(CO)_{11}]^-$. This is a well-known species that can serve as a hydride donor in the presence of CO; reaction of H^- with water yields H_2 and regenerates the OH^- used in the first step, and CO is absorbed to maintain the proper cluster electron count. Both $Ru_3(CO)_{12}$ in solution

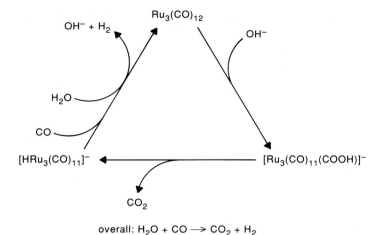

overall: $H_2O + CO \longrightarrow CO_2 + H_2$

Figure 15.1   A catalytic cycle in basic solution for the water-gas shift reaction over Ru_3 $(CO)_{12}$.

and $HRu(CO)_{11}^-$ anchored to a silica support have been shown to be catalytic under mild reaction conditions.

Figure 15.2 shows a proposed mechanism for another catalyst, this one effective in acidic solution. The catalyst consists of $PtCl_4^{2-}$ and $SnCl_4$ in a mixture of acetic and hydrochloric acids. CO can reduce Sn^{IV} to Sn^{II}, which in HCl solution

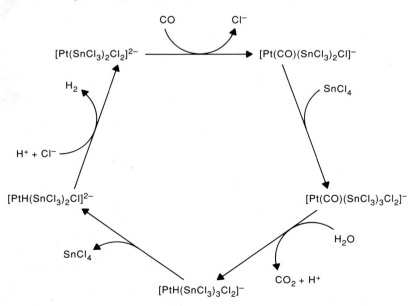

Figure 15.2   A catalytic cycle in acidic solution for the water-gas shift reaction over [Pt $(SnCl_3)_2Cl_2]^{2-}$.

will be in the form $SnCl_3^-$. This species is a good sigma-donor ligand for Pt (and pi acceptor into Sn p orbitals), and substitution yields the $[Pt(SnCl_3)_2Cl_2]^{2-}$ species shown at the top of the figure. This is a square planar, 16-electron Pt^{II} complex, which can undergo ligand substitution again, replacing another Cl^- by CO. Oxidative addition of $SnCl_4$ yields a six-coordinate 18-electron complex of Pt^{IV}, $[Pt(CO)(SnCl_3)_3Cl_2]^-$; the strong electron withdrawal by Pt^{4+} from the coordinated CO makes it susceptible to nucleophilic attack by water, releasing CO_2 and a proton. Reductive elimination of $SnCl_4$ gives a four-coordinate 16-electron species with a less positive metal atom, $[PtH(SnCl_3)_2Cl]^{2-}$. This species, finally, has a hydridic hydrogen that can release H_2 when attacked by Cl^-, yielding the starting $PtSn_2$ species.

Both of these catalytic cycles contain speculative structures, and they involve very different structures and solution conditions, but they share the key feature of attack by a solution nucleophile—OH^- or H_2O—on the carbon atom in a coordinated CO molecule. This is consistent with the general features of ligand–molecule reactivity outlined in the previous section. Other metal-catalyzed reactions of CO are known, such as methanation, but for such a small molecule the variety is necessarily limited. But we have seen that olefins and unsaturated organic molecules in general are capable of a striking variety of modes of coordination, and catalytic systems have been developed to take advantage of many of these.

15.3 ORGANOMETALLIC CATALYSIS: POLYMERIZATION

The tremendous surge of interest in organometallic chemistry since the 1940s comes predominantly from two discoveries: the synthesis of ferrocene in 1952, and the low-pressure catalytic polymerization of ethylene, discovered by Ziegler in the same year. The latter, together with its extensions into other polyolefins and copolymer rubbers, has been applied so extensively in industry that it has influenced all our lives. Although the formation of ferrocene was something of a surprise, Ziegler's catalytic process was a logical extension of many years of work on alkylaluminum compounds AlR_3 and their reactions with alkenes. AlR_3 will undergo insertion reactions with ethylene:

$$R_2Al\!-\!R + H_2C\!=\!CH_2 \rightarrow R_2Al\!-\!CH_2\!-\!CH_2\!-\!R$$

The product of this insertion is itself an alkylaluminum molecule and can undergo the reaction again, lengthening the alkyl chain. However, the product is either the dimer 1-butene or a small polymer, not the thermoplastic solid polyethylene.

Ziegler, Natta, and later many others studied the effect of altering the triethylaluminum catalyst by adding transition-metal compounds. The first successful catalyst that yielded the white solid high polymer polyethylene was a mixture of triethylaluminum and zirconium(III) acetylacetonate. Subsequent research has found that catalytic properties are shown by mixtures of RLi, R_2Be, or R_3Al and transition-metal compounds (usually halides) from groups IV, V, or VI in lower-than-maximum oxidation states—that is, with one or two d electrons remaining. Both

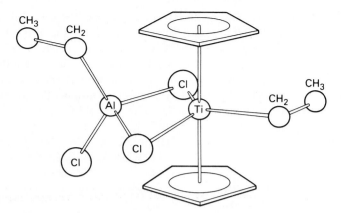

Figure 15.3   Homogeneous Ziegler–Natta catalyst EtClAlCl$_2$Ti(cp)$_2$Et.

heterogeneous and homogeneous solution catalysts are known. The most widely used Ziegler–Natta catalyst is solid TiCl$_3$ with the surface covered by a compound between the surface Cl atoms (coordinated to Ti) and AlEt$_3$. The active sites appear to be surface titanium atoms exposed so as to have a coordination number less than 6. It is difficult to write a specific formula for the heterogeneous catalytic site, but the importance of the alkylaluminum seems to be that it alkylates the TiCl$_3$ surface, replacing the Cl atoms and forming very labile Ti—C bonds. Rupturing one of these leaves an exposed Ti with the necessary lowered coordination number, so that catalysis can begin.

A solution species that shows Ziegler–Natta catalytic ability is the molecule EtClAlCl$_2$Ti(cp)$_2$Et (see Fig. 15.3). It is formed in toluene solution by reaction between Ti(cp)$_2$Cl$_2$, Ti(Cp)$_2$ClEt, and EtAlCl$_2$. Except for the cyclopentadienyl groups, it may be a reasonable model of the active sites on the heterogeneous catalyst; note that the Ti coordination number is only 5 in the proposed structure of the solution species.

Several somewhat similar mechanisms have been proposed for the catalytic process. The most widely accepted of these is outlined in Fig. 15.4. The alkene coordinates to the Ti atom as a pi donor in its previously vacant coordination site. The electron influx and the electropositive nature of the Ti atom allow electrons in the Ti—C$_2$H$_5$ bond to flow toward the ethyl group. This weakens that bond, which is labile in any case and is *cis*- to the newly arrived alkene pi bond. In a concerted four-center rearrangement that is thought to be the rate-determining step, the ethyl (or alkyl) group shifts to one carbon of the alkene double bond while the other carbon develops a sigma bond to the titanium atom. When the rearrangement of the alkyl onto the alkene carbon is complete, a new coordination site is opened up on the Ti atom *cis*- to the new, longer alkyl group. This allows the whole cycle to begin again. It is worth noting that detailed quantum mechanical calculations support a four-membered titanacyclobutane ring as the transition state; in that state,

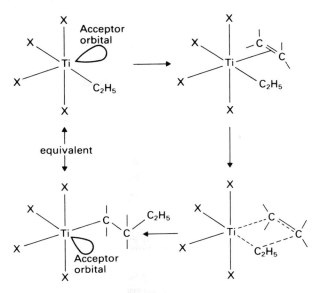

Figure 15.4 Ziegler–Natta catalytic mechanism.

the ethylene C=C bond has stretched from 1.34 Å to 1.41 Å, but the adjoining alkyl (CH$_3$ in the calculation) is still 2.22 Å away from the carbon that it will form a bond to. This implies that the closure of that bond to a typical alkane value (1.54 Å) occurs after the primary energy-consuming geometry change, which is presumably the stretching of the C=C bond coupled with the shift of the olefin away from ideal pi-donor overlap with the Ti acceptor orbital.

Two years after Ziegler's discovery, Natta produced polypropylene using a Ziegler-type catalyst. Such a polymer, unlike the strictly linear polyethylene, should be methylated on alternate carbon atoms in the polymer chain,

$$\underset{\overset{|}{\text{CH}_3}}{} \qquad \underset{\overset{|}{\text{CH}_3}}{} \qquad \underset{\overset{|}{\text{CH}_3}}{}$$

$$-\text{CH}_2-\text{CH}-\text{CH}_2-\text{CH}-\text{CH}_2-\text{CH}-\text{CH}_2-$$

although one might expect random conformations. Part of Natta's product was rubbery, presumably such a random-conformation (*atactic*) chain. Part of it, however, was a crystalline high-melting polymer with a repeat distance along the fiber axis less than the expected zigzag C—C—C distance. In the crystalline polymer, all of the methyl groups had the same conformation relative to the chain; steric repulsion between methyl groups thus forced a helical conformation. This polymer geometry was termed *isotactic* (see Fig. 15.5). Experimentation produced the TiCl$_3$-based catalyst now used to produce high yields of isotactic polypropylene. Presumably the regular conformation arises because the slight polarity and steric irregularity of the now-unsymmetrical alkene monomer, coordinated *cis-* to the alkyl group on the

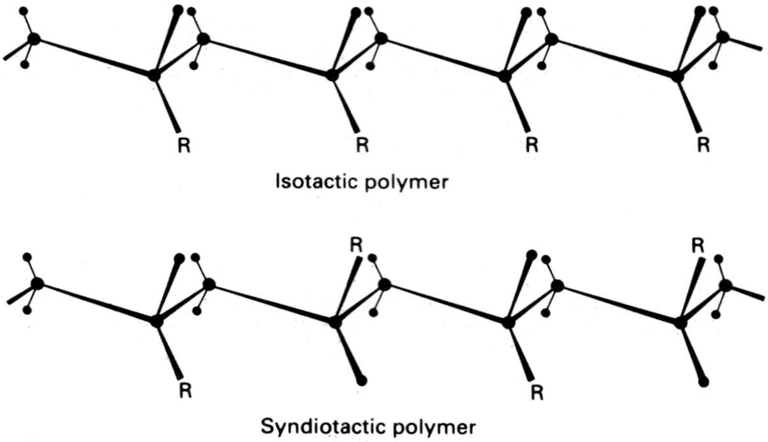

Figure 15.5 Crystalline polymer stereoregularity.

Ti at the surface of the heterogeneous catalyst, forces the methyl into the same conformation at the rate-determining step of each successive insertion reaction. Substituting VCl$_4$ for TiCl$_3$ in the catalyst has the remarkable effect of yielding *syndiotactic* polypropylene, also shown in Fig. 15.5. In syndiotactic polypropylene, the branching methyl groups occupy opposite configurations along the chain. A very wide variety of substituted alkenes has been polymerized using catalysts in this series, including essentially all of the geometrically possible stereoisomers. Because dienes undergo Ziegler–Natta polymerization to yield polycycloalkanes (see Fig. 15.6), it is even possible to synthesize optically active polymers in which the ring orientations make the polymer chiral. Besides polyethylene and polypropylene, many other commercial polymers are in production, including a

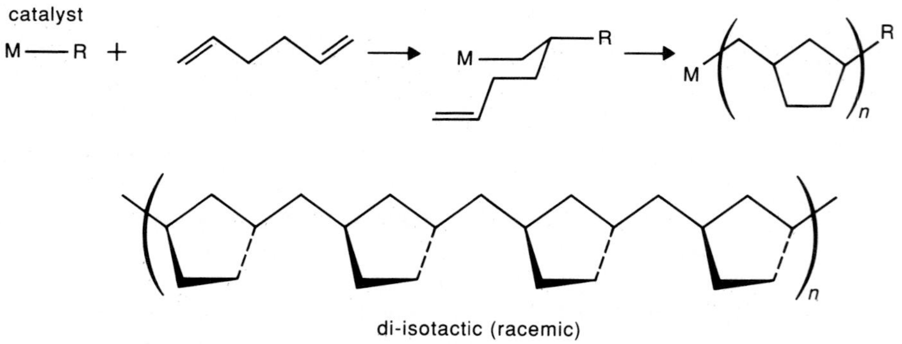

Figure 15.6 Ziegler–Natta polymerization of 1,5-hexadiene to poly(methylcyclopentane). Stereoregular polymerization yields several symmetries; only the di-isotactic conformation shown is optically active, due to the twist of the cyclopentane rings relative to the chain axis.

polyisoprene identical to natural rubber and a number of copolymers in which the physical properties of the polymer are regulated by controlling the proportions of two different alkenes.

Some of the common features of most organometallic catalysts are visible in the Ziegler–Natta catalyst mechanism. The alkene substrate coordinates the Ti edge-on through its pi electrons, shifts to a metal–carbon sigma bond by acquiring the one-electron substituent (the alkyl) that is *cis*- to it in the complex through a temporary four-membered ring, and repeats. In this case, it seems to be necessary to use an early transition metal (d^1 or d^2) to prevent the η^2-metal–alkene bond from being too stable to repeatedly undergo the insertion reaction. However, d^8 metals can also be used to catalyze alkene polymerization (in conjunction with a Lewis acid), and most other transition-metal organometallic catalysts use later transition metals, particularly those of group VIII.

15.4 ORGANOMETALLIC CATALYSIS: ALKENE METATHESIS AND ISOMERIZATION

The metathesis of alkenes, indicated in the general reaction below, is catalyzed by a variety of transition-metal systems, some bearing a strong resemblance to Ziegler–Natta catalysts. The net reaction

$$X_2C{=}CX_2 + Y_2C{=}CY_2 \rightleftharpoons \underset{\text{(Equilibrium mixture)}}{2X_2C{=}CY_2}$$

does not destroy pi unsaturation, however, although with cyclic alkenes such as cyclobutene or norbornene it can lead to *ring-opening metathesis polymerization* (ROMP):

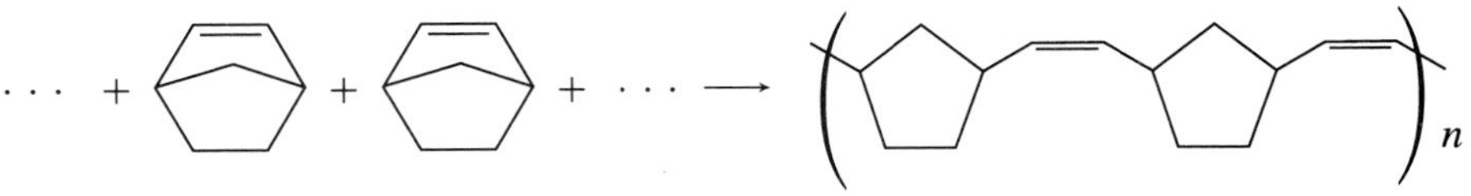

Poly(norbornene) is used commercially in rubber formulations that must be extremely soft, but is used in relatively small quantities. Perhaps the most important industrial application of alkene metathesis is also one of the simplest to visualize:

$$
\begin{array}{c}
CH_2{=}CH{-}CH_3 \\
+ \\
CH_2{=}CH{-}CH_3
\end{array}
\rightarrow
\begin{array}{c}
CH_2 \\
\| \\
CH_2
\end{array}
+
\begin{array}{c}
CH{-}CH_3 \\
\| \\
CH{-}CH_3
\end{array}
$$

When the reaction is written in this way, it suggests a mechanism involving a square intermediate with a cyclobutane structure that arises from a pairwise interaction of the molecules. This mechanism is forbidden by the Woodward–Hoffmann orbital symmetry rules, but it could be mediated by a transition-metal catalyst that coordinated both alkenes and the cyclobutane intermediate.

However, it seems clear that this is not the mechanism. For one thing, true cyclobutanes are not cleaved into alkenes by metathesis catalysts, either by themselves or when added to catalyzed metathesis mixtures. Rather, the reaction mechanism appears to require the formation of an intermediate alkylidene—the metal–carbene complex M=CH$_2$, sometimes called a *Fischer carbene*. The first catalysts discovered for alkene metathesis (and catalysts are definitely required: The thermal reaction occurs only above 700°C) were heterogeneous, consisting of MoO$_3$ deposited on alumina and partially reduced by triisobutylaluminum—a process obviously resembling the preparation of Ziegler–Natta catalysts. Subsequently, homogeneous catalysts were developed. These involve a number of transition-metal systems, but perhaps the most widely used are WCl$_6$–EtOH–EtAlCl$_2$ and [MoCl$_2$(PPh$_3$)$_2$(NO)$_2$]–Me$_3$Al$_2$Cl$_3$. It appears that two coordination sites must be vacant on the transition-metal catalyst atom so that the atom can coordinate first a carbene intermediate and then an alkene pi donor:

$$L_4M \xrightarrow{[CH_2]} L_4M{=}CH_2 \xrightarrow{R_2C=CR_2}$$

$$\begin{array}{ccccccc}
L_4M{=}CH_2 & \rightarrow & L_4M{-}CH_2 & \rightarrow & L_4M & & CH_2 \\
\uparrow & & | \quad | & & \| & + & \| \\
R_2C{=}CR_2 & & R_2C{-}CR_2 & & CR_2 & & CR_2
\end{array}$$

Since the L$_4$M=CR$_2$ produced is itself an alkylidene, a catalytic cycle is obviously possible. The function of the alkylaluminum cocatalysts seems to be to produce the initial carbene complex. It is known, for example, that WCl$_6$ reacts with Me$_2$Zn to produce methane, for which a speculative mechanism is

$$WCl_6 + (CH_3)_2Zn \rightarrow ZnCl_2 + Cl_4W \overset{\displaystyle CH_3}{\underset{\displaystyle CH_3}{\rightleftharpoons}} Cl_4W{=}CH_2 \rightarrow Cl_4W{=}CH_2 + CH_4$$

It is not necessary to have a =CH$_2$ carbene (that is, a methylene complex) to initiate catalysis. The alkylidene (OC)$_5$W=C(Ph)$_2$ is itself a catalyst for alkene metathesis, with no cocatalyst required. This alkylidene is prepared by an interesting series of reactions:

$$W(CO)_6 \xrightarrow[\text{2. } (CH_3)_3O^+BF_4^-]{\text{1. PhLi}} (OC)_5W{=}C\begin{smallmatrix} \diagup Ph \\ \diagdown OCH_3 \end{smallmatrix}$$

$$\xrightarrow[\text{2. HCl}(-78°C)]{\text{1. PhLi}(-78°C)} (OC)_5W{=}C\begin{smallmatrix} \diagup Ph \\ \diagdown Ph \end{smallmatrix}$$

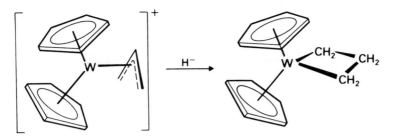

Figure 15.7   Metallocyclobutane synthesis by allyl reduction.

The intermediate in the mechanism outlined above is an unusual structure, a four-membered ring involving tungsten (a tungstacyclobutane). This structure was shown to be chemically feasible by the synthesis of the stable tungstacyclobutane shown in Fig. 15.7. When heated, this compound decomposes to yield cyclopropane and propene; when irradiated, it yields ethylene and methane. The parallels to the alkene metathesis process are obvious.

As in the Ziegler–Natta case, it is very difficult to describe the structure of the active sites of heterogeneous catalysts for alkene metathesis. However, there is some structural evidence for the pentacarbonyltungsten alkylidene shown above. The compound exchanges CO ligands at room temperature, so there is a possible vacant coordination site on the W next to the carbene, as the proposed mechanism would require. Furthermore, the comparable compound $(OC)_5W=CHPh$, which is a much more active catalyst for metathesis than the diphenyl compound, does not exchange CO ligands at $-78°C$—but it does continue to react with alkenes, yielding substituted cyclopropanes rather than metathesized alkenes. This suggests that when no coordination site on the W atom is available for bonding by the alkene pi donor (so that the tungstacyclobutane intermediate cannot form), the metathesis reaction cannot occur even if the coordinated carbene is still reactive enough to attack a neighbor alkene.

The group VI transition metals are ideal candidates for stabilizing a carbene complex. If the $·CR_2·$ is assumed to be a two-electron donor, molecules like $(OC)_5W=CR_2$ not only have 18 electrons, but have reached steric saturation at $CN = 6$. However, catalytic activity for metathesis has been shown by a suitably cocatalyzed element from each transition-metal group, all the way from Ti to Au. Table 15.1 shows some systems that catalyze metathesis, even including some f-block elements. The d-electron count is obviously not too critical, but the Mo and W systems are in general much more active than those of other metals.

Industrial interest in alkene metathesis and isomerization is intense, because the products are used in a wide variety of petrochemical processes. The first plant to use the metathesis reactions was making polymerization-grade ethylene and high-purity butene from propylene less than three years after the first publication describing the catalysis. A number of the olefins are used in Ziegler–Natta polyolefin processes; others are alkylated to the branched-chain molecules used in high-octane alkylate additives to unleaded gasoline. Another very large use is in

Table 15.1 Alkene Metathesis Catalytic Systems

Group IVb	Vb	VIb	VIIb
$TiCl_4(py)_2\text{-}EtAlCl_2$			
$TiCl(cp)_2\text{-}Me_3Al_2Cl_3$			
$Zr(acac)_4\text{-}Me_3Al_2Cl_3$	$NbCl_5\text{-}NO\text{-}Me_3Al_2Cl_3$	$MoCl_4\text{-}NO\text{-}Me_3Al_2Cl_3$	
		$MoO_3\text{-}Al_2O_3$	
		$WCl_6\text{-}EtAlCl_2$	$ReCl_4(PPh_3)_2\text{-}EtAlCl_2$
		$WO_3\text{-}SiO_2\text{-}AlEt_3$	$ReCl_5\text{-}SnBu_4$
			$[Re(CO)_3OH]_4\text{-}SiO_2$

VIII			Ib
$[Fe(NO)_2Cl]_2\text{-}Me_3Al_2Cl_3$	$CoCl_2(4\text{-vinylpy})_2\text{-}Me_3Al_2Cl_3$		$CuCl(PPh_3)_3\text{-}Me_3Al_2Cl_3$
			$Cu_2Cl_2(PPh_3)_2\text{-}EtAlCl_2$
	$[RhCl(NO)_2]_2\text{-}Me_3Al_2Cl_3$	$PdBr_2\text{-}PPh_3\text{-}EtAlCl_2$	$AgBr(PPh_3)\text{-}EtAlCl_2$
	$[RhCl(C_3H_5)_2]_2\text{-}Me_3Al_2Cl_3$		
$OsOCl_3(PPh_3)_2\text{-}Me_3Al_2Cl_3$	$IrCl_2(PPh_3)(NO)\text{-}Me_3Al_2Cl_3$		$AuCl(PPh_3)\text{-}Me_3Al_2Cl_3$

f-block: $SmCl_3\text{-}Me_2Al_2Cl_3$; $ThCl_4\text{-}Me_3Al_2Cl_3$; $UCl_4\text{-}Me_3Al_2Cl_3$

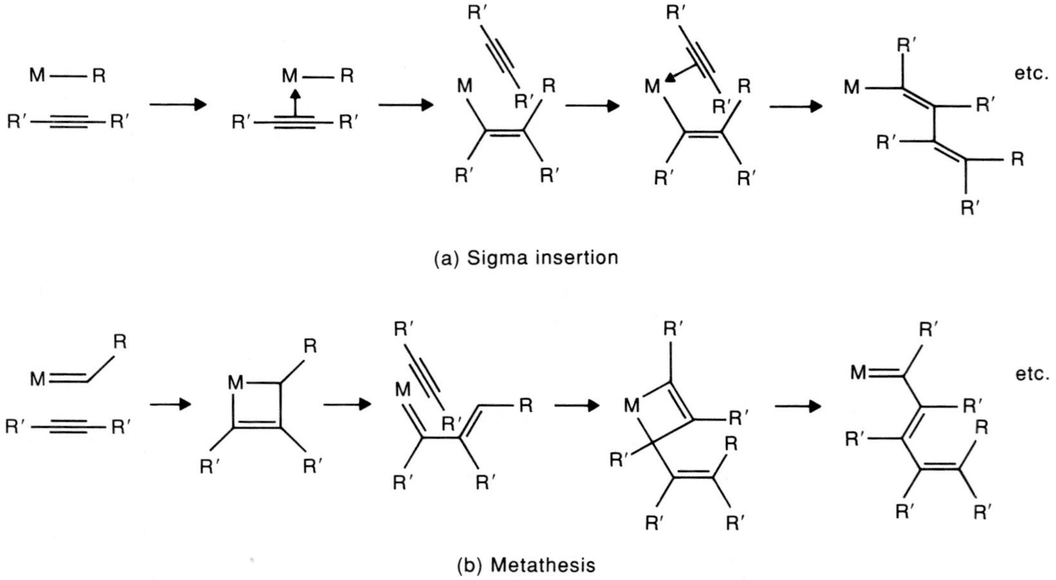

(a) Sigma insertion

(b) Metathesis

Figure 15.8 Two mechanisms for alkyne polymerization (M is d-block metal). (a) Sigma insertion mechanism begins with M—C single bond in metal alkyl. (b) Metathesis mechanism begins with M=C double bond in metal alkylidene.

the hydroformylation reaction to be discussed in the next section, which makes aldehydes from olefins by CO insertion.

The alkylidene catalysts for olefin metathesis have another interesting property: They catalyze the polymerization of acetylenes. Just as polymerization of ethylene breaks one pi bond to form a new sigma bond at each end of the molecule, polymerization of acetylene does the same thing, but leaves one of the original two pi bonds intact. Polyacetylene thus contains a series of conjugated pi bonds, and the resulting band of pi energy levels is readily partially oxidized or reduced to yield an electrically conducting polymer. This is a novel and important kind of material, and the mechanism of its formation is of considerable interest.

Good evidence exists for two different mechanisms of alkyne polymerization. One mechanism is very similar to the Ziegler–Natta alkene polymerization mechanism in that an alkyl–metal single bond is broken by a pi-coordinated alkyne so that the alkyne inserts into a M–C single bond, as seen in Fig. 15.8a. However, it is also observed that some metal alkylidenes catalyze alkene metathesis only if traces of an alkyne are present, and these systems will also catalyze polymerization of the alkyne, as in Fig. 15.8b. Although a metal-containing four-membered ring is common to both mechanisms, in the "sigma-insertion" mechanism the original M–C single bond must break as the ring forms (so that the ring is only a transition state, not a characterizable species). On the other hand, in the "metathesis" mechanism the existence of a M=C double bond in the catalyst means that a stable, perhaps even isolable metallacyclobutane species is formed even though it subsequently reacts further. In the context of this mechanism it will not be surprising that W≡C alkylidynes also catalyze alkyne polymerization, at least for terminal and methyl alkynes, yielding 20–50% of polymeric material with an average molecular weight on the order of 100,000.

There are two other classes of metal-catalyzed reactions that convert alkenes into other alkenes: oligomerization and isomerization. These are industrially important for essentially the same reasons that metathesis is. Oligomerization, or limited polymerization, is primarily used with ethylene from natural gas. It is possible to control the catalyzed reaction to form either almost pure dimers (mostly 2-butene), or higher oligomers [$n > 2$ in $(H_2C=CH_2)_n$]. It is not presently possible, however, to selectively produce a single higher oligomer, such as the trimer or tetramer.

The catalysts for alkene oligomerization predominantly involve group VIII transition metals. Among these, nickel compounds seem to be the most catalytically active. A variety of Ni compounds are catalytic, both Ni^{2+} salts that have been alkylated by alkylaluminum compounds (like Ziegler–Natta catalysts), and alkyl- or allylnickel(0) compounds that have been treated with $AlCl_3$ or some other Lewis acid. Figure 15.9 shows the structure of one effective allylnickel catalyst. It is likely, however, that regardless of the initial formulation of the catalyst, the true catalytic species is a more or less square-planar complex in which Ni is surrounded by a hydrogen atom, a PR_3 pi-acceptor ligand, a halide (or other Lewis-acid) ligand, and the reactant alkene as a pi donor: $HNiX(PR_3)(C=C)$. The H

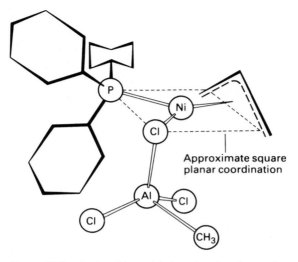

Figure 15.9 A phosphine–nickel catalyst for alkene oligomerization:

$$(P(C_6H_{11})_3)(C_3H_5)NiClAlCl_2CH_3.$$

cyclohexyl   allyl

ligand is *cis*- to the alkene, which is probably *trans*- to the phosphine. During catalysis, the H shifts to one carbon of the alkene while the other forms a sigma bond to the Ni, much like step 3 of the hydrogenation shown in Fig. 14.14. The hydridonickel complex thus becomes an alkyl-nickel complex, as shown in the proposed catalytic mechanism of Fig. 15.10.

In this mechanism, as in Ziegler–Natta polymerization, the key process is inserting $H_2C=CH_2$ into a metal–alkyl bond. The preliminary step of inserting $H_2C=CH_2$ into the Ni—H bond is reversible, allowing the H to migrate back to the Ni; careful kinetic studies suggest that the transition-state ring is nearly non-polar but that the H migrates more nearly as a hydride than as a proton: $H^{\delta-}$. On the other hand, the insertion into Ni—R is irreversible. The reversibility of the Ni—H insertion is important because separation of the oligomer product requires it (see step 3 of the mechanism suggested in the figure, in which the starting alkyl presumably has a strongly agostic H on its β carbon). Steps 4 and 5 constitute an isomerization of 1-butene to 2-butene through another H-atom shift and an allylic intermediate. The sequence shown forms only the dimer 2-butene, but in fact the composition of the products will depend on the relative rates of the insertion reaction (steps 1 and 2, with overall rate constant k_i) and the displacement reaction (steps 3–6, with overall rate constant k_d). If $k_i \gg k_d$, the product of step 2 (a butyl-nickel complex, which is an alkylnickel complex itself comparable to catalytic reactant A) will insert another ethylene before an ethylene could displace it, leading to the trimer and higher oligomers. On the other hand, if $k_d \gg k_i$, the butene dimer

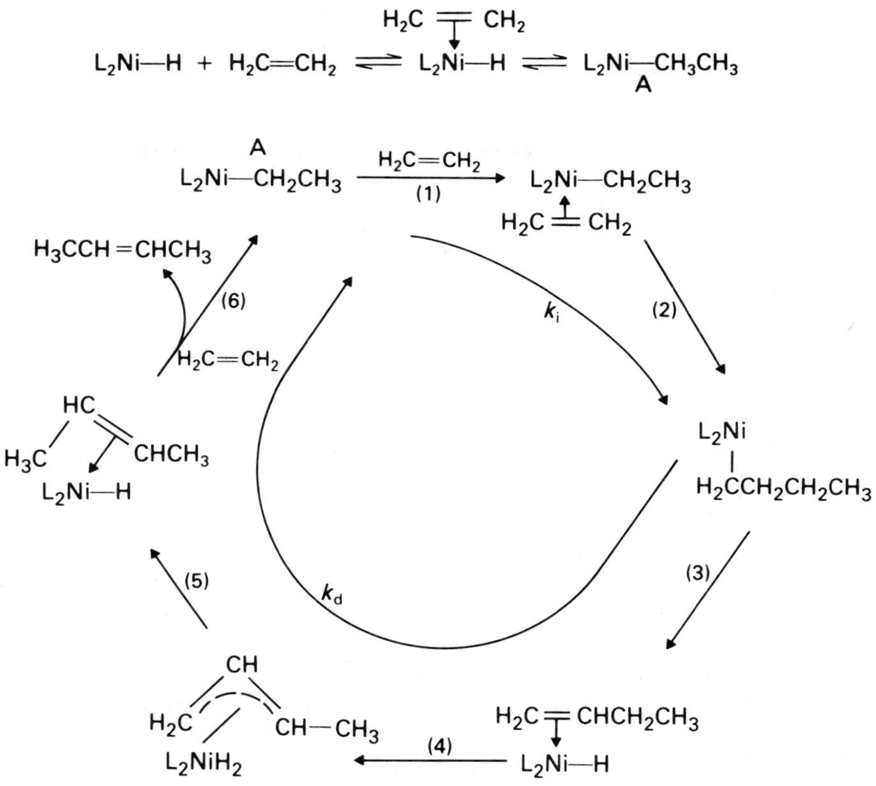

Figure 15.10 A catalytic mechanism for the dimerization of ethylene to 2-butene.

will dissociate before a third or fourth ethylene can insert, and the product will contain very little trimer or other oligomers.

It should not be too surprising that we can control the relative rates k_d/k_i by tailoring the stereochemistry of the nickel catalyst. This allows the chemist to control the reaction products by the proper choice of catalytic molecule. Table 15.2 shows the effect of changing the identity of the phosphine ligand. It is clear that one can, to a considerable extent, choose the product one prefers. It is less clear why the phosphine has this marked effect, but one speculation is that the very bulky phosphines near the bottom of the table suppress the increasing coordination number for Ni required in steps 3 and 4 of the mechanism, thereby lowering k_d. Of course, when k_i becomes more important than k_d, higher oligomers or even high polymers are preferred in the overall catalysis.

When the alkene reactant is not ethylene, the possibility of isomerization arises. The choice of k_d/k_i ratio offered by the different phosphines again allows

Table 15.2 Products from Ethylene Oligomerization over $(R_3P)(C_3H_5)NiBrAl(Et)Cl_2AlCl(Et)_2$, as R Is Varied

PR$_3$	Dimer (C$_4$) (%)	Trimer (C$_6$) (%)	Higher oligomers ($\geq$C$_8$) (%)
P(CH$_3$)$_3$ (methyl)	>98	1	—
P(C$_6$H$_5$)$_3$ (phenyl)	90	10	—
P(C$_6$H$_{11}$)$_3$ (cyclohexyl)	70	25	5
P(C$_2$H$_5$)(C$_4$H$_9$)$_2$ (ethyl di-*t*-butyl)	65	25	10
P(C$_3$H$_7$)(C$_4$H$_9$)$_2$ (isopropyl di-*t*-butyl)	25	25	50
P(C$_4$H$_9$)$_3$ (tri-*t*-butyl)	—	—	polyethylene

a useful tailoring of reaction products. Compact phosphines such as P(CH$_3$)$_3$ favor the displacement portion of the mechanism and can be used to isomerize alkenes with very little coupling (oligomerization); bulky phosphines favor the insertion part of the mechanism and can be used to produce dimers and even higher oligomers with little isomerization. Table 15.3 indicates the composition of the product mixture when 4-methyl-1-pentene is allowed to isomerize in the presence of the catalyst (PMe$_3$)(C$_3$H$_5$)NiClAlCl$_2$Et, whose structure is much like that shown in Fig. 15.9 but with a much more compact phosphine. Very little of the starting compound remains; the predominant product has undergone a 1,3 shift of the

Table 15.3 Products from Isomerization of 4-Methyl-1-Penetene over $((CH_3)_3P)(C_3H_5)NiClAlCl_2Et$

Alkene	% in Products
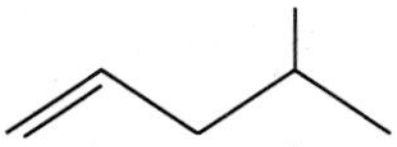 Starting material	0.2
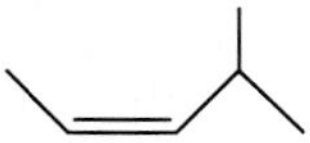	1
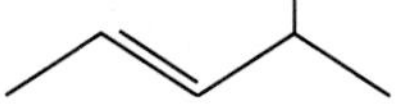	7
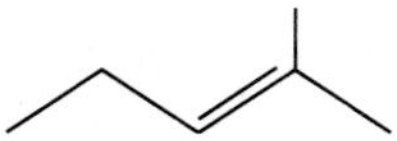	84
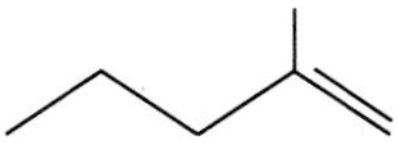	8

double bond.

A variety of transition-metal systems other than Ni complexes can catalyze alkene isomerization. A limited and quite incomplete list would include the molecules $Fe(CO)_5$, $HCo(CO)_4$, and $IrHCl_2(PEt_2Ph)_3$ and the cocatalyzed combinations $RuCl_3$(allyl alcohol), $Rh_2Cl_2(C_2H_4)_4$(HCl or H_2), or $Li_2PdCl_4(CF_3COOH$ or $H_2)$. Several mechanisms have been proposed, all of which feature the formation of a metal–alkene complex (usually with a mobile M—H bond). One mechanism, that shown in steps 4 and 5 of the mechanism of Fig. 15.10, involves an allylic intermediate. Another possible isomerization mechanism with experimental support (primarily from deuteration studies) involves an oxidative addition followed by reductive elimination, with isomerization occurring in the oxidized form, again presumably through an agostic H. This mechanism is outlined in Fig. 15.11 for the catalyst $Rh_2Cl_2(C_2H_4)_4$, cocatalyzed by HCl. Only one Rh(I) is shown, and its approximate square-planar coordination in the starting catalyst is indicated as RhL_4.

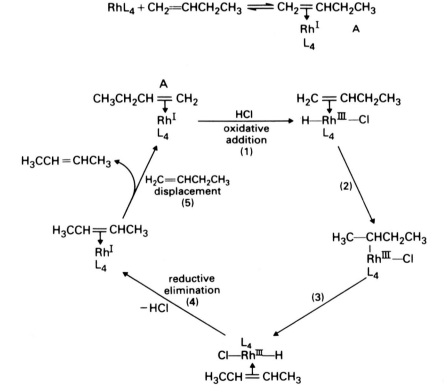

Figure 15.11 A catalytic mechanism for the isomerization of 1-butene.

It should be emphasized that these isomerization mechanisms are still somewhat speculative. Steps 2 and 3 of this mechanism could equally well be substituted for steps 4 and 5 of the allylic process in Fig. 15.10.

It is worth calling attention to the substantial similarities between the mechanisms for Ziegler–Natta polymerization, alkene metathesis, alkene oligomerization, and alkene isomerization. All involve coordination of an alkene to a transition metal through the alkene's pi electrons, followed by insertion of the two pi-bonded carbons into a metal–ligand bond with sigma-bond formation. If the metal-ligand bond is M—H, isomerization can occur; if it is M—C, the carbon chain is extended by a C_2 unit; and if it is M=C, the intermediate is cyclic because it still contains a M—C, and it can fragment to allow metathesis. In Chapter 14 we saw very similar behavior in reductions with Wilkinson's catalyst RhCl$(PPh_3)_3$. There are still more comparable reactions, as we shall see.

15.5 ORGANOMETALLIC CATALYSIS: ALKENE HYDROCARBONYLATION AND REDUCTION

The hydrocarbonylation or hydroformylation reaction, sometimes industrially called the "oxo reaction," is quite remarkable viewed strictly as an organic reaction:

$$RCH=CH_2 + CO + H_2 \rightarrow RCH_2CH_2CHO$$

At this point in our discussion, however, it bears a certain resemblance to several preceding reactions. We can therefore expect to see a related mechanism, particularly when we learn that the catalyst when the reaction was first discovered (by Roelen in 1938) was $Co_2(CO)_8$. This catalyst is still widely used industrially. Some variants, however, seem to be superior, particularly rhodium carbonyls, phosphines, and phosphites. Because an aldehyde is a particularly flexible synthetic tool, the reaction is very important industrially. Some 10 billion pounds of aldehydes are produced each year. Accordingly, close study has been given to the mechanism. Whether the metal is cobalt or rhodium, the mechanism outlined in Fig. 15.12 is believed to be essentially correct. Some mechanistic information can be derived from the rate law, which is first-order in alkene and H_2 but inverse first-order in CO—which is unusual for a reactant. This requires that the key catalytic species be produced in a dissociative equilibrium in which CO is lost, as shown for $HCo(CO)_3$ at the top of Fig. 15.12. $HCo(CO)_3$ is a 16-electron system, so it is a good electron-pair acceptor for the alkene pi donor in step 1 of the mechanism. Step 2 is the same H-atom shift or insertion of the C=C into the M—H bond that we have seen in other mechanisms, but we may note that here it produces a 16-electron product. In step 3 the 16-e metal again accepts an electron pair, this time from CO, to yield an 18-e product. Step 4 is a carbonyl insertion, like the reaction we saw for $CH_3Mn(CO)_5$ in Chapter 14. It yields, again, a 16-electron product. This can react with H_2 as shown to yield the aldehyde product

$$Co_2(CO)_8 + H_2 \rightleftharpoons 2\ HCo(CO)_4$$

$$HCo(CO)_4 \rightleftharpoons HCo(CO)_3 + CO$$

$$A$$

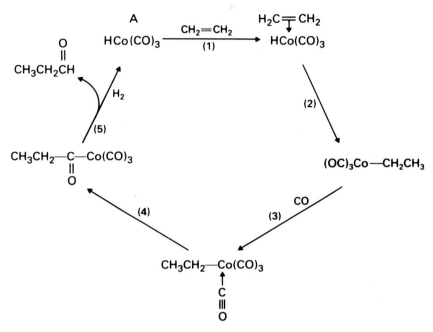

Figure 15.12 A catalytic mechanism for the hydroformylation of ethylene over $Co_2(CO)_8$.

and catalyst A, or in a side reaction with more CO to yield $RCOCo(CO)_4$. The latter can then be hydrogenated to yield the aldehyde and the catalyst precursor $HCo(CO)_4$.

As the reaction is shown in the figure, with ethylene as the model aldehyde, the product propionaldehyde is more or less inevitable. However, with longer-chain alkenes, isomerism is possible in the H atom shift (step 2), resulting in branched-chain aldehydes as the final product. Since linear aldehydes are much more useful for industrial purposes, considerable effort has gone into tailoring the catalyst and reaction conditions to produce increased yields of linear aldehyde product. Isomerization of the alkene itself can also occur while it is coordinated to the cobalt, by a mechanism like those proposed in the last section. Operating at high CO pressures minimizes both isomerization problems, though it slows the reaction. Another problem is that the coordinated alkyl group sometimes is reduced to an alkane, in lieu of step 3 of the mechanism. To some extent this and the isomerism problem can be ameliorated by tailoring the catalyst. If $HRh(CO)_4$ is the starting catalyst and phosphines are added to the reaction mixture, a complicated series of equilibria tie together the various phosphine-substituted carbonyls:

$$HRh(CO)_4 + PR_3 \rightleftharpoons HRh(CO)_3(PR_3) + CO$$

$$HRh(CO)_3(PR_3) + PR_3 \rightleftharpoons HRh(CO)_2(PR_3)_2 + CO$$

(and so on). Changing the donor properties of the phosphine affects the catalyzed rate: A triphenylphosphine-modified Rh catalyst takes the hydroformylation reaction to completion six or eight times as fast as a tributylphosphine-modified catalyst. Further, raising the phosphine concentration to drive the substitution equilibria like those above to the right increases the linearity of the product aldehydes.

In the original hydroformylation reaction the $Co_2(CO)_8$ catalyst was effective but suffered from three problems: (1) The rate was only marginally acceptable and still required fairly high temperatures; (2) under the reaction conditions, the catalyst was not completely stable; and (3) the selectivity of the catalyst for specific products was only fairly good. The introduction of rhodium phosphine catalysts, even though the rhodium cost roughly $6000/oz, improved all three problems. A further catalyst modification by Union Carbide, to rhodium phosphites $Rh[P(OR)_3]_2(CO)_2$, improved the reaction rate by roughly another factor of 100 and, if the nature of the phosphite R groups is optimized, can lead to linear/branched ratios as high as 80:1. Metal phosphites are susceptible to hydrolytic decomposition, but using bulky R groups increases both the rate of the reaction and the stability of the catalyst molecule. Figure 15.13 shows two phosphites used by Union Carbide, with their rates relative to triphenylphosphine and the linearity ratios of their products.

An interesting variation on homogeneous catalysts is the use of a solid-supported heterogeneous catalyst with a similar ligand environment for each metal atom. If polystyrene supported on silica gel has its phenyl groups brominated, then treated with $K^+PPh_2^-$, the resulting "polytriphenylphosphine" shown in Fig. 15.14 can coordinate a Rh atom at each phosphorus. (Usually, however, only a few phosphine groups, perhaps 5%, do coordinate a metal.) Under the best experimental conditions, this solid catalyst is capable of improved yields and much better linear-aldehyde selectivity (up to about 92% linear) than comparable homogeneous catalysts.

Two related hydrocarbonylations deserve brief mention. The first is the *homologation* of methanol,

$$CH_3OH + 2H_2 + CO \rightarrow CH_3CH_2OH + H_2O$$

catalyzed in acid by $[Co(CO)_4]^-$. CH_3OH is protonated by acid to $CH_3OH_2^+$, which is attacked by the Co anion to split out water and yield $CH_3Co(CO)_4$. One of the coordinated carbonyls inserts in the CH_3—Co bond (like step 4 in Fig. 15.12); the resulting acylcobalt carbonyl, a 16-e species, undergoes oxidative addition of H_2 to the Co, then further reduction by another H_2 of the acyl $C\!=\!O$ to $CH-OH$. This final product is $CH_3CH(OH)-Co(H)_2(COO)_3$; adding CO transfers a metal hydride to

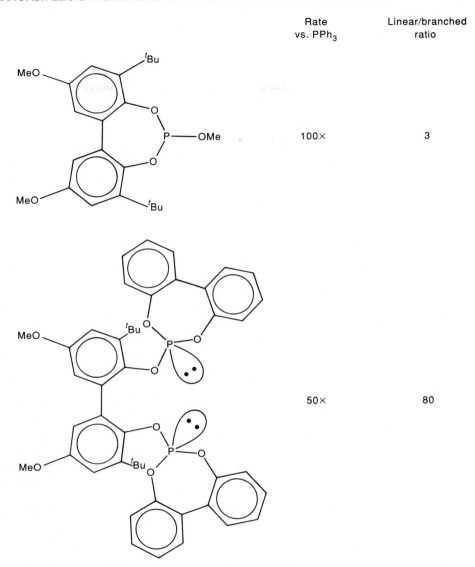

	Rate vs. PPh$_3$	Linear/branched ratio
	100×	3
	50×	80

Figure 15.13 Phosphite ligands for rhodium hydroformylation catalysts in Union Carbide process. *t*-Butyl groups increase steric bulk, slowing hydrolysis of the catalysis and tending to force specific orientation on the coordinated substrate.

the alcohol, splits off CH_3CH_2OH, and leaves $HCo(CO)_4$ to begin the process again.

The second hydrocarbonylation is only a carbonylation inasmuch as H_2 is not a reactant. Methanol can be carbonylated to acetic acid:

$$CH_3OH + CO \rightarrow CH_3COOH$$

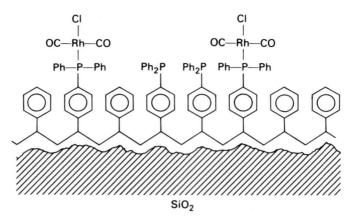

Figure 15.14 Heterogeneous Rh catalyst for hydroformylation.

This reaction, not fermentation, is the source of most industrial acetic acid (there are competing processes). Acetic acid is widely used in acetate esters as solvents and as vinyl acetate monomer for PVA polymer. Monsanto's low-pressure process for the carbonylation (below 500 psi) catalyzes the conversion using the square-planar $16\text{-}e^-$ complex $[RhI_2(CO)_2]^-$; Fig. 15.15 shows the structures in the catalytic cycle. The catalysis is iodide-dependent, and the rate-controlling step is the oxidative addition of CH_3I (formed from $CH_3OH + HI$) to yield the octahedral $18\text{-}e^-$ complex $[RhI_3(CO)_2(CH_3)]^-$. Sigma insertion of an adjacent carbonyl gives a $16\text{-}e^-$ acetyl complex (possibly present as a dimer), which adds CO to give $[RhI_3(Co)_2(COOCH_3)]^-$. Water hydrolyzes the acetyl group to free acetic acid, leaving a rhodium hydride. Reductive elimination of HI completes the cycle.

Hydrogenations using transition-metal catalysts without CO involvement are extremely important, both in research-scale organic synthesis and in industry. We saw catalytic hydrogenation using Wilkinson's catalyst $RhCl(PPh_3)_3$ in Fig. 14.14, and have drawn parallels between that mechanism and those of this chapter. There are other catalysts used in the same contexts. Perhaps the most widely used alternative to Wilkinson's catalyst is the electronically equivalent $HRuCl(PPh_3)_3$, which seems to react by a different mechanism in that the alkene pi complex forms before the H_2 oxidative addition:

$$HRuCl(PPh_3)_3 + R_2C{=}CR_2 \rightleftharpoons HRuCl(PPh_3)_2(R_2C{=}CR_2) + PPh_3$$

$$HRuCl(PPh_3)_2(R_2C{=}CR_2) \rightleftharpoons RuCl(PPh_3)_2({-}CR_2CHR_2)$$

$$\overset{PPh_3}{\rightleftharpoons} RuCl(PPh_3)_3(CR_2CHR_2)$$

$$RuCl(PPh_3)_3(CR_2CHR_2) + H_2 \rightleftharpoons HRuCl(PPh_3)_3 + HCR_2CHR_2$$

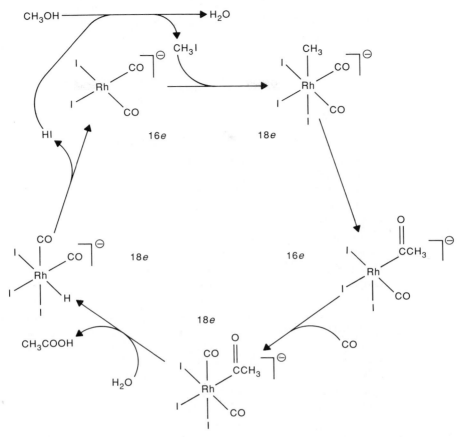

Figure 15.15 A catalytic mechanism for the carbonylation of methanol to acetic acid by the Monsanto process. Catalyst is iodide-assisted $[RhI_2(CO)_2]^-$.

This is possible, of course, because of the coordinated hydride in the original catalyst molecule. If the last step of this mechanism involves an oxidative addition of the H_2 as with Wilkinson's catalyst, the five-coordinate nature of the starting alkyl complex makes it likely that there is a dissociative step somewhere within the addition, but this has not yet been shown.

Wilkinson's catalyst, the ruthenium analog above, and several other similar species such as $[Rh(diene)(PR_3)_2]^+$ are powerful and often very selective catalysts for hydrogenating unsaturated organic systems, particularly C=C bonds but also epoxides to alcohols, aldehydes to alcohols, nitro compounds to amines, some ketones to alcohols, and cyclic anhydrides to lactones, among other reactions. In recent years, great interest has developed in using asymmetric (optically active) catalysts to produce optically active products from symmetric reactants. Species that could be rendered optically active or chiral by addition of a symmetrical reactant are said to be *prochiral*. For example, in the reaction

$$\underset{\displaystyle \text{Me}}{\overset{\displaystyle \overset{\text{O}}{\underset{\displaystyle \|}{}}}{\text{C}}}\text{—Et} + \text{H}_2 \ \rightarrow \ \text{Me—}\overset{\text{OH}}{\underset{\text{H}}{\overset{|}{\underset{|}{\text{C*}}}}}\text{—Et}$$

the 2-butanol product is chiral at the starred carbon, and the butanone reactant is prochiral. Normally such a reaction would produce a racemic mixture, but an asymmetric catalyst can produce an asymmetric product. With the right catalyst, the reaction above can produce 71% *R* enantiomer and 29% *S* enantiomer. The difference between these two percentages is termed the *optical purity* or *enantiomeric excess* (ee) of the reaction, so that the product above is *R, 42% ee*.

To achieve asymmetric synthesis using a system comparable to Wilkinson's catalyst, the most obvious approach is to use asymmetric phosphines PRR'R" in the RhCl(PR$_3$)$_3$ catalyst. Since the pi-bond insertion into the Rh—H bond must take place between a hydride and a pi ligand *cis*- to each other, asymmetric phosphines might force a specific orientation on the incoming pi ligand and therefore a specific optical configuration on the product. Some complexes with phosphines such as PPh-MePr were prepared, and as catalysts they did indeed lead to asymmetric reduction with 20–30% ee; 90% ee in amino acid synthesis was achieved at Monsanto using the ligand P(Me)(cyclohexyl)(*o*-anisyl), or *CAMP*. A simpler approach synthetically is to prepare a chiral diphosphine in which the asymmetry is in the organic link between the phosphines rather than in the P atoms themselves. A number of such chelating ligands have been prepared. Perhaps the most widely used is the DIOP ligand, which is shown in Fig. 15.16 both as a free ligand and in its catalytically active Rh complex as the (+) form. It is conveniently prepared by reacting (−)-tartaric acid with 2,2-dimethoxypropane to give the five-membered ring, reducing the carboxyl groups to alcohols with LiAlH$_4$, converting these to tosylates, and finally adding Na$^+$PPh$_2^-$. The DIOP-Rh catalysts yield optical purities of 60–80% and occasionally higher in a wide variety of reductions. A particularly important asymmetric reduction is used in synthesizing the Parkinson's disease drug L-DOPA, where 90% ee is achieved both with DIOP-Rh and with CAMP-Rh. In general, a Rh-DIOP catalyst will hydrogenate a *Z*- isomer of an alkene faster than an *E*- isomer, and the *Z*- isomer will have greater optical purity. This is obviously related to the stereospecific coordination of the alkene to the Rh atom. In general, optical purity for an asymmetric hydrogenation will be greater if the prochiral alkene has polar groups nearby; the starting material for the L-DOPA synthesis is an unsaturated amide. Unsubstituted alkenes usually yield only modest asymmetry, perhaps 20% ee, though isolated examples of more successful asymmetric hydrogenation have been demonstrated.

Other asymmetric reactions have been demonstrated in good optical yield. Sharpless has optimized a catalytic asymmetric epoxidation of allylic alcohols using

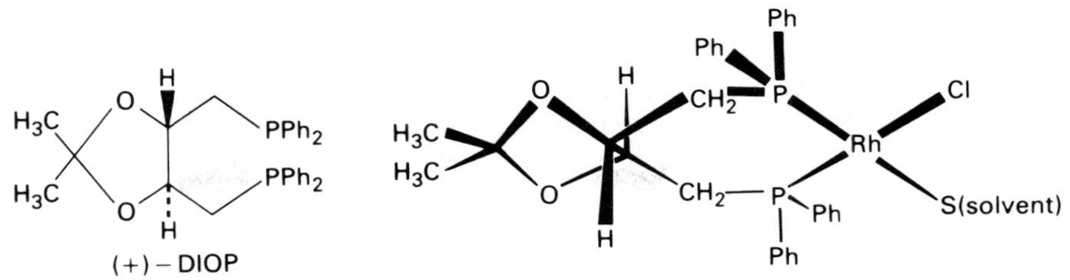

Figure 15.16 The DIOP asymmetric ligand and catalyst.

a titanium alkoxide, asymmetric dialkyl tartrates coordinating the Ti, and ROOH hydroperoxides. Asymmetric C–C bond formation can also be achieved: ZnR_2 alkyls can add R to an aldehyde asymmetrically to yield a secondary alcohol R–CHOH–R' with up to 100% ee in favorable cases, using chiral pyrrolidinyl-methanols as coligands on the Zn atom.

We can close this very brief survey of organometallic catalysis by reviewing the common features of the mechanisms. Where the stoichiometric reaction involves an alkene, we often see the formation of a complex with the transition metal in which the alkene pi bond serves as a donor to the metal, with the metal in turn presumably back-donating d-electron density to the alkene π^* orbitals. This pi complex can often insert the C—C into a M—H or M—C bond with *cis*-geometry, forming M—C—C—H or M—C—C—C. The fate of this complex and the stoichiometric products of the overall reaction then depend on the reaction conditions. The insertion can continue indefinitely to a polyolefin; the inserted C—C in a four-membered ring formed from a metal alkylidene can itself fragment to a new alkene and a new metal carbene; the inserted C—C can shift a hydrogen atom back to a new M—H bond and a longer alkene chain; a coordinated CO can insert into the new M—C alkyl complex to form an acyl group; or the inserted C—C can pick up a second H atom from another coordinated M—H to form a reduced alkane. There are other possibilities and other specific catalyzed reactions, and more will undoubtedly be discovered. However, there are enough well-characterized cases before us to reveal a consistent outline.

PROBLEMS

A. DESCRIPTIVE

A1. Benzoic acid has a pK_a of 5.68, but in the compound $(PhCOOH)Cr(CO)_3$ (where it is an η^6-donor), it has a pK_a of 4.77. Is this consistent with the usual change of acidity for sigma-coordinated ligands? Are there differences between the electronic changes in sigma-bonded ligands and this ligand coordinated through its pi electrons?

A2. An approximate expression for the entropy of activation for two spherical reactants in water is $\Delta S^{\ddagger} = -10Z_N Z_L$, where Z_N and Z_L are the charges on N (an attacking nucleophile) and L (a molecule undergoing nucleophilic attack that can serve as a ligand for a transition-metal ion M): N + L = Products. If the molecule L is coordinated as M—L, the entropy expression changes to $\Delta S^{\ddagger} = -10Z_N(Z_M + Z_L)$, where $Z_M + Z_L$ is now the net charge on the complex. Why, physically, does the first expression predict negative entropy of activation when N and L have the same charge, and predict positive entropy of activation when they have opposite charges? In terms of charges, what kind of nucleophilic attack will be enhanced if the L molecule is coordinated to a metal ion?

A3. Industrial water-gas shift reactors used on feed gas with relatively high CO concentration are usually split into two reactors: the first operates at a higher temperature than the second, although the catalysts are the same in both. Gas from the first reactor is cooled before entering the second reactor. Why is it advantageous to split the WGSR reactors in this way? What's going on thermodynamically and kinetically?

A4. Suggest electronic reasons why early transition metals (Ti, Zr, V, and so on) are uniformly used as Ziegler–Natta catalysts rather than such metals as Rh Pd, or Pt.

A5. In the Wacker process, the proposed rate-determining step involves the formation of a four-membered ring

$$
\begin{array}{ccc}
\text{O} & \!\!-\!\! & \text{Pd} \\
| & & | \\
\text{C} & \!\!-\!\! & \text{C}
\end{array}
$$

from a Pd—C_2H_4 complex in which the C=C axis is perpendicular to the initial Pd—OH bond. This seems to require that the C_2H_4 rotate about the bond to the Pd. Compare this to the ^{1}H NMR spectrum of Pt(acac)Cl(cis-CH$_3$CH=CHCH$_3$), which shows two CH$_3$ peaks below −25°C, but only one CH$_3$ peak at room temperature. Is this NMR change consistent with rotation of the cis-2-butene group about the

$$
\begin{array}{c}
\text{C}=\text{C} \\
| \\
\text{Pt}
\end{array}
$$

axis? Does the low-temperature spectrum correspond to a structure with the C=C axis in the Pt plane, or to one with the C=C axis perpendicular?

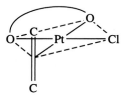   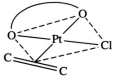

A6. Suggest a mechanism for the carboxylation of propylene to a mixture of *n*-butyric and *iso*-butyric acids over a NiI_2 catalyst at 250°C and 200 atm CO in the presence of aqueous HI.

A7. Suggest a mechanism for the following carbonylation:

A8. Suggest a mechanism for the dimerization of norbornadiene:

$$2 \quad \text{(norbornadiene)} \quad \xrightarrow{\text{Ni(CH}_2\text{=CHCN)}_2} \quad \text{(dimer)}$$

A9. What products would be formed during the metathesis of a mixture of propylene and cyclohexene over a group VI catalyst?

A10. What products can result from the isomerization of 1,2-dimethylcyclopentene, if the oxidative–addition mechanism controls the reaction?

A11. Suggest a mechanism for the reduction of propionaldehyde to propanol by H_2 over the catalyst $HRuCl(PPh_3)_3$.

A12. CO can be catalytically converted to methanol by hydrogenation over solid CuO/ZnO:

$$CO + 2H_2 \rightleftharpoons CH_3OH \quad \Delta H^0 = -91 \text{ kJ/mol rn}$$

Assume that the catalyst surface initially contains M−OH groups and that the reaction proceeds through a surface formate complex; suggest a mechanism for the overall production of methanol.

A13. Sketch three other conformational isomers of the cyclopentane polymer in Fig. 15.6 that are not chiral.

A14. Suggest a mechanism for the isomerization of 3-heptyne to 3-hexyne and 4-octyne by the alkylidyne $(^tBuO)_3W(\equiv C^tBu)$.

B. NUMERICAL

B1. Assume a structure for the $Ru_3(CO)_{12}$ WGSR catalyst in Fig. 15.1 like that for the best-fit isoelectronic molecule in Fig. 13.30. Show that the catalyst molecule satisfies the 18-electron rule for each Ru, and that it satisfies both the Wade and the Lauher theoretical electron counts for a triangular cluster.

B2. Assuming that none of the ΔH or ΔS values change with temperature, calculate the value of the gas equilibrium constant for the steam reforming reaction shown as reaction 8 in Table 8.6 at 298 K and at 1300 K. Calculate the value for the equilibrium constant of the water-gas shift reaction (reaction 7) at the same two temperatures.

B3. From the rate data in the table below for the carbonylation of methanol to acetic acid, calculate the Arrhenius activation parameters E_a, $\Delta H^{\ddagger}$, and $\Delta S^{\ddagger}$. Comment on the relationship between the entropy of activation and the mechanism suggested in the chapter.

T (K)	Rate const k (L/mol sec)
383	0.015
400	0.043
424	0.12
442	0.30
460	0.55

C. EXTENDED REFERENCE

C1. In *J. Am. Chem. Soc.* (1988), *110,* 310, H. Yamamoto and others report high stereospecificity for a series of hetero-Diels–Alder reactions using an asymmetric alkylaluminum catalyst; they achieve up to 97% ee for 15 different reactions. They offer a suggested structure for the transition state complex in which a diene approaches an aldehyde coordinated to the aluminum atom. If their speculation is correct, what coordination geometry is required for the aluminum in the transition state? What electronic factors affect the choice of a metal for the catalyst, assuming no change in ligands? Is there a *d*-block metal that might prove superior to Al in this catalyst?

C2. We have consistently represented alkene–metal complexes as η^2 pi complexes, and this is consistent with most structural studies. However, Stoutland and Bergman [*J. Am. Chem. Soc.* (1985), *107,* 4581] report the formation of a stable sigma-bound H–Ir–CH=CH₂ from ethylene and an iridium reactant; the sigma complex is believed to be the initial product and an intermediate on the way to an η^2 pi complex. Both species are formed in the same ratio (sigma 66:pi 34) over a range of temperatures. How does Bergman know the sigma complex is formed first in the mechanism? If the sigma complex is thought of as the product of an oxidative addition, is Ir more likely or less likely than other metals to form a sigma complex?

C3. R. R. Schrock [*J. Am. Chem. Soc.* (1982), *104,* 4291] reports the metathesis of a combination of a W≡W triple bond and an alkyne C≡C triple bond to yield a W≡C alkylidyne and a different alkyne. Suggest a mechanism for the reaction (which is stoichiometric, not catalytic).

C4. The binuclear complex Rh₂Cl₂(μ-CO)(dppm)₂ serves as a hydrogenation catalyst for acetylenes; phenylacetylene yields styrene [M. Cowie and T. G. Southern, *Inorg. Chem.* (1982), *21,* 246]. Acetylenes RC≡CR with electronegative R substituents are not reduced, however. Cowie pursues model compounds for the catalysis in a subsequent publication [*Organometallics* (1990), *9,* 1594]. Suggest structures and a mechanism for the hydrogenation.

16

Bioinorganic Molecules and Mechanisms ─────────────

In considering the reaction mechanisms of transition metals, we have most recently looked in some detail at the industrial reactions of coordinated organic ligands, usually with coordinated pi systems. In a remarkable variety of processes, a transition metal atom is able to catalyze a desired reaction for some organic ligand; the focus is on ligand reactions. We now turn our attention to bioinorganic systems, with a particular interest in the roles of metals in biological systems. Again we shall see that many of the most important roles involve ligand (or substrate) catalysis by metalloenzymes—but there are other roles, and other important metals beyond the *d*-block.

We can place metals in five different roles in bioinorganic systems, though of course sometimes a given system will have a metal atom with more than one role:

1. Acid/base or redox catalysis for ligands
2. Ligand carriers, metal carriers, or metal storage
3. Structure formers for an organic environment
4. Charge carriers
5. Photoreceptors

In this chapter we will consider the first four categories, and the fifth will be developed in the final chapter on photochemistry (Chapter 17).

──────── ## 16.1 THE OCCURRENCE AND STUDY OF METALS IN BIOLOGICAL SYSTEMS

Figure 16.1 indicates the metals that are important in biological systems. Some, like Na and Ca, are present in high concentrations, while others such as Co are

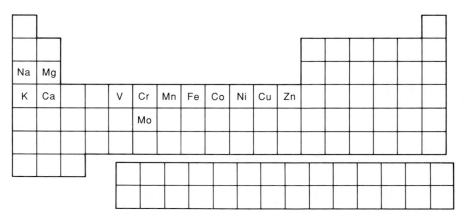

Figure 16.1   Metals occurring with known function in biological systems.

present in microscopic concentrations in any living system but are essential for their purpose nonetheless. A few general observations may help put the pattern in perspective. Perhaps the simplest observation is that, except for molybdenum, all important metals are no heavier than zinc. This is a result of the evolutionary processes of biochemistry seeking out metals that are abundant in the environment. Remember from Chapter 1 that elements past iron are in general much less abundant than the lighter ones because they cannot be formed by exothermic fusion, which is reflected in the abundances shown in Table 10.6.

We may wonder why evolution did not take advantage of the large amount of titanium or zirconium available, but that leads us directly to another generalization. In general, transition metals occur in biological systems because they have multiple oxidation states separated by only modest potentials, so that they can conveniently transfer electrons within the system. On the other hand, the electropositive s-block metals occur in ionic environments with a fixed, relatively low charge. In biochemical systems these metal ions form structures around them through ion–dipole attractions that are stronger than kT thermal agitation but not strong enough to break bonds in the organic environment (by O atom abstraction, for example). Ti and Zr have only one stable oxidation state under normal biochemical potentials (though some highly reducing biochemical environments could reduce Ti^{4+} to Ti^{3+}), and at +4 charge that state is so high that it would disrupt some more delicate biochemical bonds.

The introduction to Chapter 12 noted the historical sequence of development of both theory and experimental tools for investigation of metal complexes. Because transition metals are almost always a very small atom or weight fraction of bioinorganic species in which they appear, the study of these species had to await development of sophisticated techniques. With the advent of a number of these since about 1960, many studies have been published and interest is rapidly expanding. As usual, inorganic chemists are interested in geometric structures, energetics, and reaction mechanisms; both of the latter two are strongly influenced by the first, and for most bioinorganic species the most immediate task has been to establish the stoichiometry, then the coordination geometry of the metal or metals in the species.

The most direct structural tool is x-ray diffraction (XRD). Unfortunately, large biomolecules have thousands of atoms per unit cell and array repeat distances that may be hundreds of bonds long. This means that exposures must be quite long in order to develop enough diffracted intensity to measure accurately, and the long exposure to x-rays sometimes damages the crystal. It is also, in many cases, extraordinarily difficult to obtain crystals of crystallographic quality—and it is often not certain that the molecular configuration in the crystal is the same as that in the living system. As a result, bioinorganic chemists use other techniques more often than XRD; but some stunningly beautiful XRD studies have been done, and when available they are always the strongest evidence for structure.

In many cases in which the immediate coordination environment of the metal is sought, EXAFS (Extended X-ray Absorption Fine Structure) studies are used. Each element has a threshold energy for core electron ejection by an incoming x-ray, and just above this threshold energy the ejected electron, with low kinetic energy and relatively long wavelength, will be back-scattered by the neighbor atoms. The degree of back-scattering and resulting interference in the outgoing electron wave will depend on the atomic number of the neighbor atom and on its distance from the metal atom—but not on any bond angles. Fourier transformation of the ejected electron intensity as a function of its wavenumber gives a radial distribution curve for the metal atom's neighbors from which the number of neighbors, the elemental identity of each, and the internuclear distance can be determined. For example, in nitrogenase from *Azotobacter vinelandii* the radial distribution function for Fe reveals that each Fe has 2.3 ± 0.9 Fe neighbors at 2.66 Å; 0.4 ± 0.1 Mo neighbor at 2.84 Å (that is, about half the Fe atoms have one Mo neighbor); 3.4 ± 1.6 S neighbors at 2.25 Å; and 1.2 ± 1.0 neighbors that are either O or N at 1.81 Å. Obviously, EXAFS studies are strongly complementary to XRD studies; this interpretation should be compared to the XRD structure described in section 16.4.

NMR studies can also reveal structural information about bioinorganic molecules, but in general the molecules are so complex that an ordinary spectrum is uninterpretable. However, the advent of two-dimensional and three-dimensional NMR techniques, in which (at least at the highest attainable frequencies) specific atom–atom couplings can be extracted, has permitted solution structure determinations for proteins in which the primary amino acid sequence was known.

Mössbauer spectroscopy is a nuclear phenomenon in which gamma rays emitted by a standard isotope source of a given element are absorbed by an atom of the same isotope in a sample to be studied; because the incoming gamma ray is extremely monochromatic and the absorbing atom's energy levels are extremely narrow, the gamma ray can only be absorbed if the emitting and absorbing atoms have exactly the same electronic (coordination) environment, *or* if relative motion of the two atoms provides enough Doppler shift in the gamma-ray energy to match the necessary absorption energy difference. Doppler shifts for different coordination environments can be tabulated in much the same sense that NMR chemical shifts are, and as a result Mössbauer spectra (intensity versus Doppler velocity) can yield direct information on coordination geometry. In bioinorganic systems, the most useful nucleus by far is ^{57}Fe; but because iron is so important in a wide variety of bioinorganic studies, that can be very helpful.

In general, vibrational (IR or Raman) spectra are nearly meaningless for biomolecules because of the enormous variety of bonds present. Here, Raman spectra are obtained by irradiating the sample with visible light, which is not absorbed—it carries energy suitable for electronic, not vibrational, spectra—but is instead scattered with vibrational frequencies added to or subtracted from the irradiation frequency. A separate technique, resonance Raman spectroscopy, irradiates the sample with a frequency that will be absorbed by a $d-d$ transition of the metal. For such irradiation, the scattering frequencies for vibrations involving the metal atom will be greatly enhanced in intensity so that those frequencies alone can be selected. The resulting bonding information is similar to that from ordinary vibrational spectroscopy: bond identity and force constants. For large biomolecules, this is the only practical way to obtain vibrational data for the metal atom.

Other techniques useful on large bioinorganic molecules directly include electron paramagnetic resonance (EPR or ESR) and a variety of electrochemical techniques, but particularly cyclic voltammetry (CV). EPR is analogous to NMR for unpaired electrons; it yields information on the distribution within the biomolecule of unpaired d electrons from the metal, normally in frontier orbitals. CV yields characteristic potentials for reduction of the biomolecule, which may or may not be linked to the metal atom but in any event yields information on general availability of the electron pool in the bioinorganic molecule.

Because of the size of bioinorganic molecules and the low density of metal atoms in them, inorganic chemists have taken two major approaches to their study. One approach that does not help with the low-density problem but that does provide additional experimental tools is to substitute a paramagnetic transition-metal atom for a diamagnetic atom, particularly for Zn. This substitution not only allows magnetic measurements such as EPR, it allows observation of $d-d$ transitions in the visible–UV spectrum and the structural information from that source.

Another very important approach is the synthesis of model compounds: compounds in which the metal atom is coordinated with an immediate environment similar to that of the biomolecule but with the surrounding organic biopolymer missing, usually simulated by small bulky groups such as t-butyl. Depending on what is known about the real bioinorganic system being studied, model compounds tend to fall in one of three classes, though the best models combine properties of two or even all three of the classes. The most important class is that of the *functional models,* in which the model compound engages in the same chemical processes as the biomolecule: catalyzes the same reactions as a metalloenzyme, for example. In a later section we shall see that several model heme compounds have been prepared that mimic hemoglobin or myoglobin in reversibly taking on O_2, with an equilibrium partial pressure very similar to that of myoglobin.

When the biological function of the molecule is uncertain or cannot be directly duplicated, but the metal atom's coordination geometry is at least partially known, *structural models* are sometimes prepared. These have the metal atom in a coordination environment more or less equivalent to that in the biomolecule, as determined by one or more experimental techniques.

Often structural models attempt to match an unusual UV–visible or EPR spectrum seen in the biomolecule. An example is that of the copper blue proteins, which are known to contain a copper atom or atoms in an environment not known with good accuracy, but which yield an unusual UV–visible spectrum for the copper atom. Many model compounds have been prepared, some of which yield good fits for the unusual spectrum; this presumably provides structural information about the copper environment, but offers no guidance as to the function of the copper atom.

In some cases, particularly when metal clusters are involved, inorganic chemists are still at the stage of preparing *stoichiometric models*. In the nitrogenase case already mentioned, the iron–molybdenum cofactor (FeMoco), it was clear from many experiments that both metals are contained in an iron–molybdenum–sulfur cluster, but of poorly determined stoichiometry until the recent XRD determination. Various determinations yielded Fe:Mo = 5, 7, or 8 and S:Mo all the way from 4 to 9. Fe:Mo:S is now known to be 7:1:6, but it is easy to see that cluster geometry will be sensitive to these values. Cluster synthesis is currently proceeding in several laboratories in efforts to prepare Fe–Mo–S clusters with approximately 7 Fe:1 Mo. It should be clear that such stoichiometric models are precursors to structural and functional models, though there is a reasonable likelihood that achievement of the right stoichiometry for the cluster will also yield the right structure, and possibly even the function.

16.2 LIGAND CATALYSIS: METALLOENZYMES AS ACID/BASE CATALYSTS

We have already seen in Chapter 15 that coordination of a Lewis base ligand (or, for metalloenzymes, substrate) to a transition metal will in general reduce the basicity of the coordinating ligand atom or even increase the Lewis acidity of a nearby atom; we saw for ethyl glycinate that coordinating the carbonyl O made the carbonyl C more receptive to nucleophilic attack. We can expect comparable Lewis acid/base behavior in biocoordination. If the metal atom has only one stable oxidation state, such as Zn, its metalloenzymes cannot catalyze redox reactions but can catalyze acid–base reactions comparable to the hydrolysis of ethyl glycinate. A number of zinc enzymes are known, all of which appear to act through the zinc atom to coordinate substrates that are Lewis bases toward the Zn acid. Coordination is not the only reaction-determining factor, because major conformational changes are often seen for the substrate when coordinated as a result of hydrogen-bonding opportunities between the substrate and the protein part of the enzyme; but the Zn atom is always the key to the active site, and we shall focus on its behavior here.

Carboxypeptidase A. In the digestive process, carboxypeptidase A (CPA) is a zinc enzyme that catalyzes the hydrolysis of the amide bond at the C terminal of the protein being digested; the C-terminal amino acid should have an aromatic or L-configuration alkyl side-chain. The overall reaction is

$$H_2N \cdots \underset{NH}{\overset{\overset{\displaystyle O}{\|}}{C}} \underset{COO}{\overset{\overset{\displaystyle Ar}{|}}{CH}} + H_2O \longrightarrow H_2N \cdots \underset{}{\overset{\overset{\displaystyle O}{\|}}{C}}O^- + H_3N^+ \underset{COO^-}{\overset{\overset{\displaystyle Ar}{|}}{CH}}$$

CPA contains 307 amino-acid units and one Zn, with a molecular weight of 34,600 daltons (1 dalton = 1 g/mole). The zinc atom, probably best thought of as Zn^{2+}, is bound to the protein by three donor groups: two imidazoles from histidines and a carboxyl from glutamic acid. In the absence of a substrate, the Zn^{2+} is approximately tetrahedral (yielding an 18-electron count), with the fourth ligand a water molecule. The arrival of a substrate peptide coordinates the carbonyl O from the C-terminal amide group onto the Zn^{2+}; it is not clear whether the Zn loses the water or becomes temporarily five-coordinate (but see the discussion of the cobalt-substituted enzyme below). The zinc is at the bottom of a crevice in the enzyme, and the substrate deforms the crevice on its arrival (and therefore the enzyme), a process known as *induced fit*. The electronic effect of coordination on the peptide is similar to the effect on ethyl glycinate (section 15.1). The additional positive charge on the C from the coordinated carbonyl induces attack by a water molecule. The negative O is stabilized by a nearby cationic guanidinium group from arginine 127 in the enzyme, and the positive C by nearby glutamate 270; the glutamate may even form a temporary acid anhydride with the CO that is hydrolyzed. Figure 16.2 schematically indicates this sequence. It should be clear that the network of stabilizing interactions is so extensive that the organic portion of the enzyme must be flexible enough to allow the substrate to enter and the products to leave; a rigid enzyme structure would bind the substrate permanently.

Carboxypeptidase A has been subject to some elegant metal-replacement experiments. The Zn^{2+} can be removed by a chelating ligand, leaving behind the unmetalated *apoenzyme*. A number of M^{2+} ions have been substituted into the apoenzyme, of which the Co^{2+} enzyme is most important. The Co enzyme is actually a better catalyst for the peptidase reaction than the native Zn enzyme by 30–500%, so that the Co^{2+} must be geometrically compatible with the catalytic Zn conformation. The Co enzyme has electronic transitions at 500, 555, and 572 nm in the visible spectrum [molar absorptivities about 150] and 940 and 1570 nm in the near infrared [molar absorptivity about 20]. This can be compared with tetrahedral $Co(OH)_4^{2-}$ (600 nm [150], 1400 nm [50]) and distorted trigonal bipyramidal

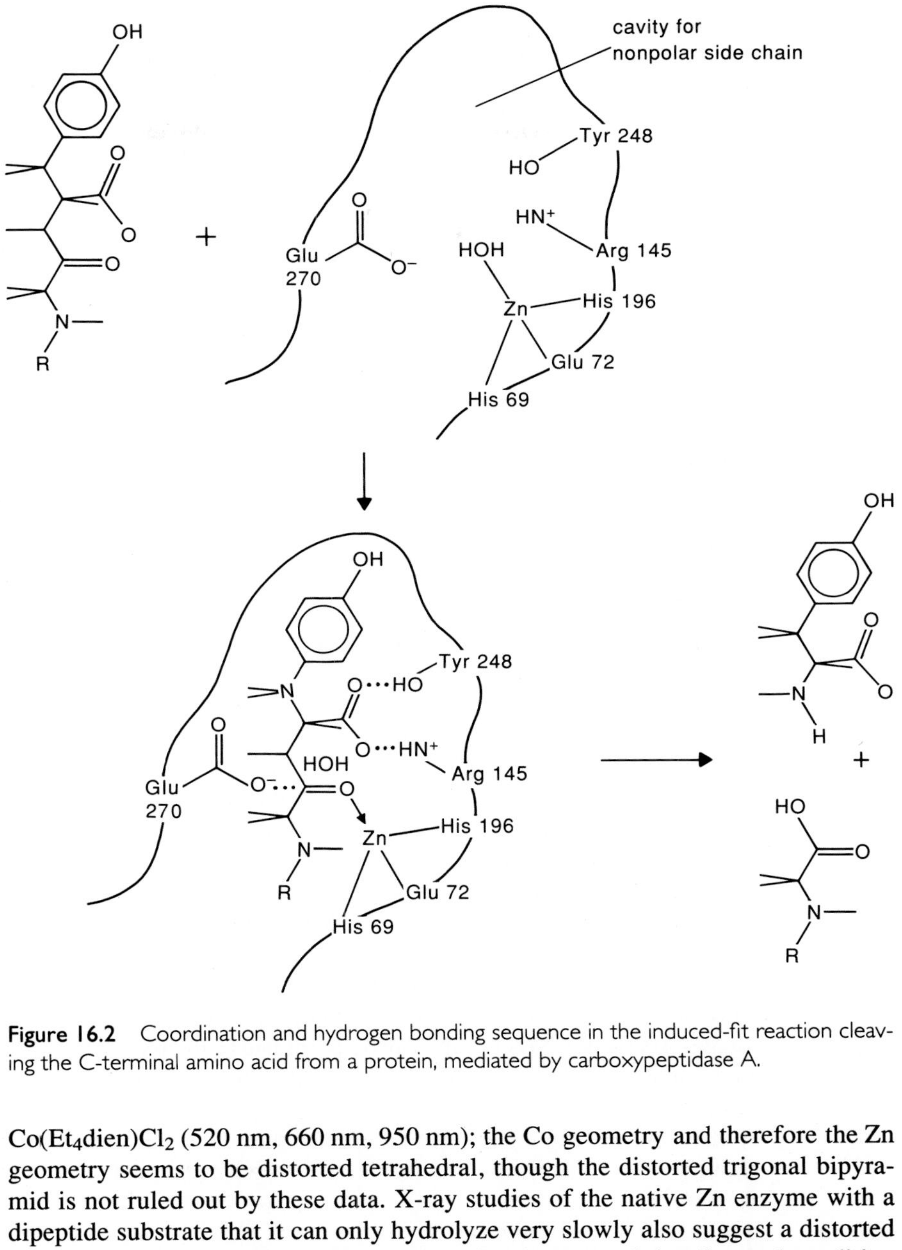

Figure 16.2 Coordination and hydrogen bonding sequence in the induced-fit reaction cleaving the C-terminal amino acid from a protein, mediated by carboxypeptidase A.

$Co(Et_4dien)Cl_2$ (520 nm, 660 nm, 950 nm); the Co geometry and therefore the Zn geometry seems to be distorted tetrahedral, though the distorted trigonal bipyramid is not ruled out by these data. X-ray studies of the native Zn enzyme with a dipeptide substrate that it can only hydrolyze very slowly also suggest a distorted tetrahedral geometry around the metal. Vallee has termed the distorted condition the "entatic state" and suggested that the distortion is electronically important to the metal's role in catalysis.

Carbonic anhydrase. Carbonic anhydrase is also a zinc enzyme, with a molecular weight of approximately 29,000 daltons and one zinc atom per molecule. Again the Zn^{2+} is the active site: the apoenzyme is completely noncatalytic. Both x-ray diffraction and Co^{2+}-substitution UV spectra indicate a distorted tetrahedral geometry for the metal, bound by three histidine groups to the protein and having a water molecule on the fourth ligand. The enzyme catalyzes the hydration of dissolved CO_2 to HCO_3^-:

$$OH^- + CO_2 \rightleftharpoons HCO_3^-$$

As written, the reaction requires a hydroxide ion, which is uncommon at biological pH; the reaction can equally well be written to use H_2O instead of OH^-, but it seems likely that attack involves OH^- coordinated to Zn, produced by proton transfer from the coordinated water molecule. UV spectra of the CO^{2+} enzyme show a marked dependence of the spectrum on pH, suggesting that in more basic solutions a metal ligand is being deprotonated, as might be true for $Zn-OH_2^{2+} \rightarrow Zn-OH^+$. The pK_a value for the reacting group is about 7.1, which is lower than is seen for most $L_3Zn-H_2O^{2+}$ complexes (about 8.5), but the pK_a is affected by both geometry and the nature of the L groups, and the observed value does not seem outside the range of possibility.

The catalyzed reaction is important in both directions. If dissolved HCO_3^- could not be rapidly converted into gaseous CO_2 in the alveoli of the lungs, CO_2 removal by respiration would be inadequate. Although the dehydration proceeds at a reasonable uncatalyzed rate, the enzyme has a turnover rate (the fundamental reversible molecular process rate constant) of 10^6 sec^{-1}, which is one of the highest known biological rates.

Some mechanistic information can be obtained from the study of inhibitors of the enzyme catalysis. The catalysis is inhibited by the anions I^-, HS^-, N_3^-, CN^-, CNO^-, and by the neutral species $R-SO_2NH_2$ (sulfonamides) and imidazoles. The anions are all soft bases and might be expected to displace the harder base OH^- from the moderately soft acid Zn^{2+}; EXAFS studies have shown that the I^- is indeed coordinated directly to Zn^{2+} in the inhibited enzyme, and UV spectra of inhibited Co^{2+} enzyme indicate distorted tetrahedral geometry so that the OH^- must have been displaced. X-ray studies of the sulfonamide-inhibited enzyme also suggest that a SO_2NH_2 group is also coordinated directly to Zn^{2+} through either O or N, and the Co^{2+} enzyme has a pH-insensitive spectrum suggesting that the water has been displaced. Only imidazole is a competitive inhibitor, coordinating to Zn^{2+} but leaving the Zn five-coordinate with its water ligand still in place. Although the mechanism is not absolutely established, the best current representation seems to be

1.

2.

3.

Other hydrolytic zinc enzymes (alkaline phosphatase, DNA polymerase, thermolysin). Not all acid/base metalloenzymes contain zinc, but these three are interesting examples that do. Alkaline phosphatase has a molecular weight of approximately 94,000 daltons, with four zinc atoms per molecule. If the zinc ions are progressive removed, catalytic activity is lost after the first two Zn^{2+} are gone, suggesting that the last two are present to maintain the structure of the enzyme but are not directly involved in catalysis. In addition, approximately two Mg^{2+} per molecule are necessary for catalysis, but these also seem to have only a structure-forming role.

The enzyme catalyzes the reaction

$$R\text{---}OPO_3^{2-} + H_2O \rightarrow R\text{---}OH + HOPO_3^{2-}$$

The reaction proceeds by transferring the phosphate to a serine OH from the enzyme protein, then hydrolyzing that bond to yield free phosphate. X-ray studies reveal that the phosphate ester is coordinated as a bridging ligand between two Zn^{2+}, which are held at the critical spacing to accept the phosphate by the overall enzyme structure; deprotonated serine $-O^-$ attacks the coordinated phosphate, releasing the ROH, and the serine phosphate is finally hydrolyzed to free HPO_4^{2-} by a water molecule coordinated to one of the Zn^{2+} ions. The distinction of interest is that we see zinc ions in a structure-forming role as well as a Lewis acid role, and s-block metals in the same role.

DNA polymerases are the mechanism through which DNA and RNA replication and error repair occurs. Their biochemical function is beyond the scope of this discussion, but they have the net effect of removing the pyrophosphate from the next nucleotide unit to be placed in the growing nucleic acid strand, and binding it to the sugar OH from the growing end of the strand. This involves hydrolyzing a phosphate ester bond as in the phosphatases and, as in those enzymes, both Zn^{2+} and Mg^{2+} are required. Mn^{2+} can be substituted for Mg^{2+} (retaining enzyme activity), which allows magnetic investigations of Mg^{2+} coordination sites. Zn^{2+} appears to be bound to the enzyme at the bottom of a cleft into which the nucleic acid strand fits, with a phosphate group binding next to the zinc. Little is yet known about the specific role of the metals in DNA polymerases, but their fundamental importance guarantees that these metalloenzymes will be comprehensively investigated in coming years.

Thermolysin is an enzyme from *Bacillus thermoproteolyticus* that has essentially the same function as carboxypeptidase. It has a molecular weight of 34,600 daltons and one Zn^{2+} ion, bound as in carboxypeptidase by two histidines and one glutamate, with water as a fourth ligand. Its unusual feature is the ability to function at temperatures up to 80°C, an extremely high temperature for most biochemical processes. The thermal stability of the conformation is due to four Ca^{2+} ions, which have a strong structure-forming effect on the protein; removing only the zinc has no thermal effect on structure (though catalysis stops), so Zn^{2+} bears no structure-forming role. We shall return to the structure-forming role of *s*-block metals in a later section.

16.3 LIGAND CATALYSIS: METALLOENZYMES AS REDOX CATALYSTS

Transition metals with multiple more-or-less equally stable oxidation states are often used as electron pools in redox-catalyzing enzymes, particularly Fe, Cu, Mn, Mo, and Co. Iron, perhaps because of its high abundance, is particularly widespread in this role, and we can begin by considering some iron proteins.

Ferredoxins. The name *ferredoxin* (Fd) is given generically to a series of iron proteins in which each iron atom is more or less tetrahedrally coordinated by four sulfur atoms. A sulfur atom is a relatively soft base, which tends to stabilize the lower oxidation state Fe^{2+} in the iron redox couple. However, slight shifts in coordination geometry can have large effects on the iron reduction potential, and it is observed that $\mathscr{E}^0$ for ferredoxins varies over almost a volt. As a result, nature has been able, in effect, to design a ferredoxin to suit any desired chemical potential for electrons. Ferredoxins are thus found as electron pools in a wide variety of biochemical processes.

The simplest ferredoxin is *rubredoxin,* in which a single Fe is coordinated by four cysteine sulfur atoms in a slightly distorted tetrahedron, with S−Fe−S bond angles ranging from 104° to 114°. The $Fe(SR)_4^{1-/2-}$ potential is −0.06 V, and its self-exchange rate is high—about 10^9 electron transfers per second. The high exchange rate implies that there is not a great geometric difference between the Fe^{2+} and Fe^{3+} forms of rubredoxin, which is consistent with medium-resolution x-ray data on the two forms. Several model systems have been prepared that mimic these structures; detailed spectroscopic studies (polarized absorption spectra and magnetic circular dichroism) of one model compound in which the cysteine R−S⁻ is modeled by 2-phenylthiophenoxide suggest that the relative stability of the two Fe oxidation states will be affected significantly by the dihedral angle between the S−Fe−S plane and the Fe−S−C plane involving the same Fe and S. This indicates that the reduction potential can be adjusted by tailoring this angle, which is consistent with the general flexibility of ferredoxin $\mathscr{E}^0$ values.

Beyond rubredoxin, there are also iron–sulfur clusters in other ferredoxins, including 2-Fe, 3-Fe, and 4-Fe clusters. In addition to cysteines, these contain "labile sulfurs" that are released as H_2S in acid. Figure 16.3 shows the Fe–S cluster geometry of some ferredoxins. The 2-Fe clusters contain two labile sulfurs and are often designated Fe_2S_2 proteins, although four other cysteine sulfurs also coordinate the Fe atoms. In the oxidized state, both irons seem to be Fe^{3+}, and reduction involves only one electron, yielding Fe^{3+}—S_2—Fe^{2+}. The oxidized form is diamagnetic, indicating that the Fe^{3+} ions (each of which has an odd number of electrons) are thoroughly antiferromagnetically coupled to yield no net spin; the reduced form is paramagnetic with $S = \frac{1}{2}$. The 3Fe–4S cluster is less common, but has been characterized in aconitase and other systems; apparently it also only undergoes a one-electron transfer at about -0.3 V, with a paramagnetic oxidized state and diamagnetic reduced state. Its structure is easier to visualize as a variant of the 4Fe–4S cluster, which is described as a "cubane-type" structure but actually consists of a tetrahedron of sulfur atoms concentric with a tetrahedron of iron atoms. These eight atoms form an approximate cube, but their different radii distort the cube's symmetry. The 3Fe–4S clusters maintain a tetrahedron of sulfur atoms, but one iron is missing from the cube leaving what is sometimes called a "voided-cubane" or "metal-depleted cuboidal" geometry. Some 3Fe–4S clusters readily add another Fe^{2+} (or other M^{2+}) to fill out the cube, but others are inert in this sense. It is possible that iron storage within the cell is regulated by such equilibria, but this remains to be shown.

Many examples of the 4Fe–4S cluster are known; the first ferredoxin described was thought to have a 8Fe–8S cluster, but crystallography showed it to contain two 4Fe–4S structures separated by 12 Å. The 4Fe–4S clusters usually undergo a one-electron transfer, but CV studies show that the cluster structure is stable in three oxidation states, separated by two electron transfers. To the extent that isolated-atom oxidation states can be assigned, most 4Fe–4S clusters seem to be $(Fe^{3+})_2(Fe^{2+})_2$ in the oxidized form (diamagnetic) and $(Fe^{3+})(Fe^{2+})_3$ in the reduced form (paramagnetic, $S = \frac{1}{2}$), with a reduction potential of approximately -0.40 V. However, a category of 4Fe–4S clusters known as HiPIPs (High Potential Iron Proteins) because of their potentials approximating $+0.35$ V seems to have oxidation states $(Fe^{3+})_3(Fe^{2+})$ in the oxidized form (paramagnetic) and $(Fe^{3+})_2(Fe^{2+})_2$ in the reduced form (diamagnetic). This is only a convenient shorthand notation, however, because a variety of experimental techniques indicate that the four iron atoms are equivalent in the cluster to a considerable extent. On the other hand, trigonal distortion can yield a single Fe in the cubane-type structure that is chemically distinct from the other three, which may have some bearing on the origin of the 3Fe–4S proteins.

Successful model compounds involving alkyl or aryl thiolate ions RS^- as replacements for cysteine sulfurs have been prepared for rubredoxin (1Fe) and for the 2Fe–2S and 4Fe–4S ferredoxins, primarily by Holm. The model clusters are reducible in much the same pattern as is seen for ferredoxins, but their potentials are more negative and usually less widely separated for successive reductions. Full

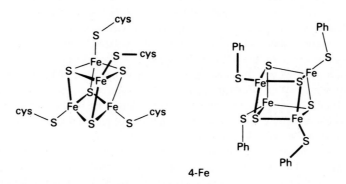

Figure 16.3   Coordination geometry in Fe−S proteins and model compounds.

control of model-compound potentials may require tailoring of the Fe—S—C dihedral angle relative to the cluster as indicated above for rubredoxin. The chemically distinct iron atom in the distorted $Fe'Fe_3S_4$ mentioned above has been modeled by Holm using the tridentate ligand shown in Figure 16.4; the exposed Fe shows a number of reactions that distinguish it from the other three, but it has not yet been stripped out of the cluster to yield a 3Fe–4S model compound.

Cytochromes. The ferredoxins are classed together because they contain iron in a sulfur coordination environment. Similarly, there is a class of iron-based redox systems that all contain iron in a heme environment, called *cytochromes;* heme is

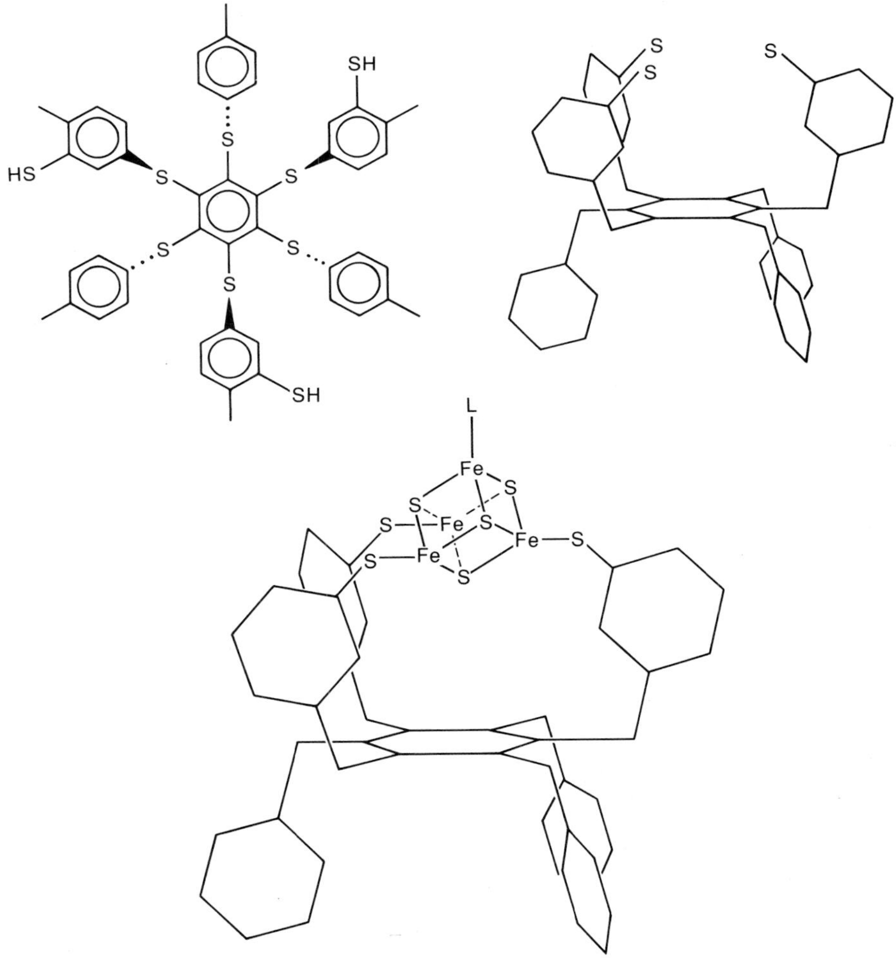

Figure 16.4 A three-sulfur ligand yielding axial $Fe'Fe_3$ symmetry in 4Fe–4S clusters.

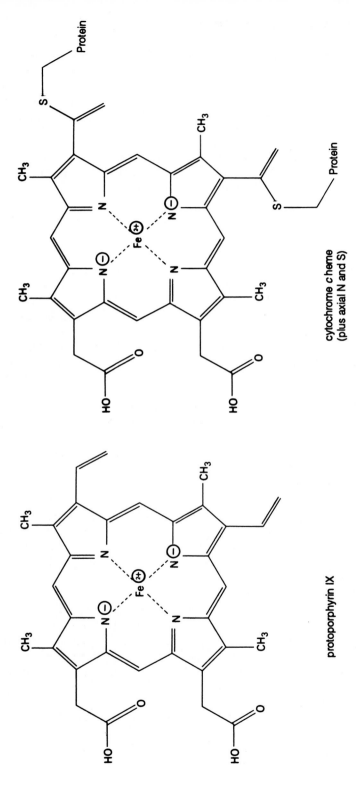

protoporphyrin IX

cytochrome *c* heme
(plus axial N and S)

Figure 16.5 Fe–protoporphyrin IX and its modification in the heme group of cytochrome *c*.

the porphyrin macrocycle shown in Fig. 12.17. The planar porphyrin provides both rigid square planar coordination for Fe and an extensive pi system that can take part in electron transfer itself. This tidy situation is complicated by three additional factors. First, the heme is substituted around its periphery by organic groups, varying somewhat from one heme system to another; the cytochromes have the substituents in Fe-protoporphyrin IX, seen in Fig. 16.5. Second, the iron is not square planar but octahedrally coordinated, with ligands above and below the porphyrin ring, often histidines but sometimes methionine sulfurs or other protein constituents; some cytochromes have one site open for coordination of O_2. Third, the ferriheme unit is bound by one or both of these additional two ligands to a protein environment. The cytochromes thus differ in details of the protein sequence from one organism to another—but the heme-binding groups (both the histidines binding Fe and other groups covalently bonded to the porphyrin) have been conserved in evolution through over a billion years for a variety of organisms.

The cytochromes are electron-pool molecules that take part in the O_2-burning process that provides free energy for ATP synthesis and other cell processes. O_2 burning is a four-electron process: $O_2 + 4H^+ + 4e^- \rightarrow 2H_2O$. The electron pool must therefore provide four electrons in a single O_2 encounter, because it is highly undesirable to release intermediate O_2^{q-} products such as superoxide or hydrogen peroxide. We have seen in Chapters 8 and 14 that elementary redox processes involve one or at most two electrons, so a sequence of electron transfers will be necessary for cytochromes in their function. To understand that sequence, we need the structure of the relevant cytochromes.

The cytochromes occur primarily attached to the inner wall of the cell, though to some extent in the cell solution (the cytosol). They were discovered by the spectroscopic study of intact cell walls and divided into groups described as cytochrome a, cytochrome b, and cytochrome c respectively, based on the frequency of their UV-visible absorption peaks. However, later studies have shown a variety of structures in each of the three categories. Cytochrome c is easiest to detach from the cell wall intact, and has been characterized in greatest detail. It has histidine N and methionine S ligands in addition to the porphyrin, and is also shown in Fig. 16.5 as attached to its protein. With six ligands it cannot coordinate O_2, but one edge of the heme is accessible from outside the protein, and it is thought that the Fe electron pool can be transferred through the pi system edgewise, in an outer-sphere mechanism. However, the cytochrome c data do not get us to the location of molecular O_2 reduction.

The dioxygen molecule is complexed directly by cytochrome a within the enzyme *cytochrome oxidase,* which contains two stoichiometrically identical cytochrome a units known as a and a_3 that differ slightly in reduction potential because their enzyme environments differ. Both have five-coordinate Fe and a nearby Cu that can also serve as an electron pool and a coordination site. In practice, cytochrome a_3 coordinates O_2 between Fe^{2+} and Cu^+, reducing it to O_2^{2-} and oxidizing each metal. Cytochrome a (nearby within the oxidase) donates an electron to a_3, and the dioxygen is protonated, allowing the O–O bond to break yielding a

Cu^{2+}–OH^- and a Fe^{4+}=O^{2-}. Another electron, acquired by cytochrome a (through its Cu) from a cytochrome c, and two more protons allow the Cu-coordinated OH^- to escape as H_2O, protonate the O^{2-} to OH^-, and reduce Fe^{4+} to Fe^{3+}; at this point the remaining O from the original O_2 is coordinated between the Fe^{3+} and Cu^{2+}. Another cytochrome c electron reduces Cu^{2+} to Cu^+, which is a sufficiently soft acid to release the OH^- so that it is now coordinated only to Fe^{3+}, where it is very tightly held. Finally, another electron from cytochrome c reduces Fe^{3+} to Fe^{2+}, allowing the OH^- to be protonated and lost as H_2O. The a_3 metal oxidation states are now correct to begin the cycle again, as indicated in Fig. 16.6. This cyclic process is the central energy source for animals; we should consider that the cytochrome a_3 strips electrons from a and c and protons from its environment, yielding high-energy systems that drive other redox processes and ultimately yield ATP. Its importance is attested to by the fact that the poisons CO and CN^- act by coordinating to the cytochrome a Fe, blocking it from the electron transfer process.

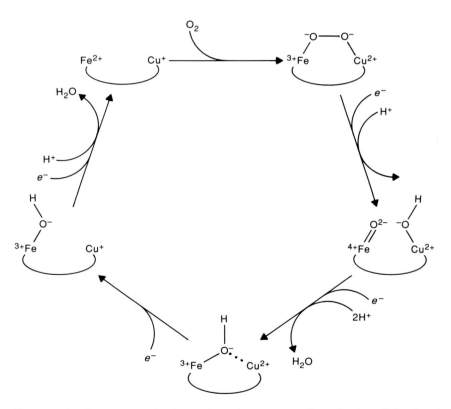

Figure 16.6 Catalytic coordination and reaction sequence for reduction of O_2 directly to H_2O by cytochrome oxidase.

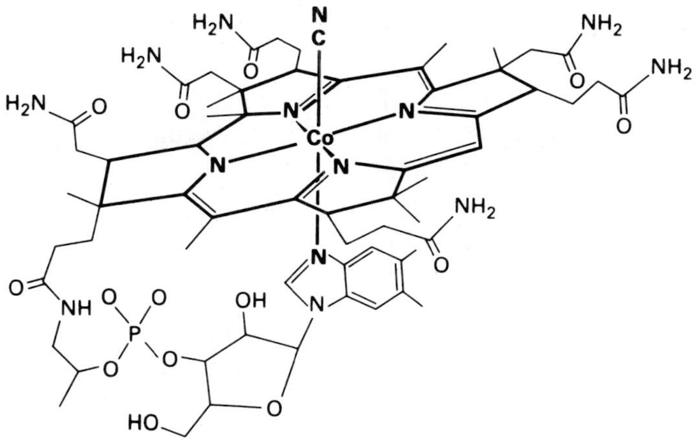

Figure 16.7   Cobalt coordination in cyanocobalamin, shown in schematic form.

Cobalamin. The two previous redox systems have primarily involved iron as the electron pool, though cytochromes also have copper present. Cobalt appears (at least for higher animals) only in vitamin B_{12} coenzyme, which has cobalt bound in a corrin macrocycle ligand (see Fig. 12.17) usually generically known as *cobalamin*. The Co coordination is approximately octahedral; in addition to the four corrin N donors one site is filled by a benzimidazole N donor pendant from the corrin ring. The sixth site has a variable ligand. In vitamin B_{12} the sixth ligand is CN^- (*cyanocobalamin*), while in coenzyme B_{12} the sixth ligand is deoxyadenosine bound to the Co through the 5' carbon atom; this Co–C bonding is the only metal–alkyl organometallic bonding known in biological systems. Figure 16.7 shows the general structure of cyanocobalamin; unlike heme systems there is no surrounding protein, although the coenzyme B_{12} functions by binding to a protein. If the fifth-ligand benzimidazole is removed by hydrolysis at the phosphate group shown in the figure, the remaining system is known as a *cobinamide*. Cobinamides permit studies of coordination on Co–C reactivity.

So far, we have avoided assigning an oxidation state to the cobalt, which can vary from 3+ to 2+ to 1+ in different systems as redox reactions occur. In cyanocobalamin the cobalt is Co^{3+}, constituting a six-coordinate 18-electron complex. As Chapter 14 has indicated (see Table 14.8), such a system will react primarily by ligand substitution in a dissociative mechanism. The five-coordinate intermediate will, in the terms of Table 14.8, resemble a carbocation and react as an electrophile. When a cobalamin is reduced to Co^{2+} (B_{12r}), it would be a 19-electron species if all six ligands remained; in electrochemical studies the Co^{2+} species that results is five-coordinate with 17 electrons, reacting as a free radical. Further reduction to Co^+ (B_{12s}) yields a four-coordinate 16-electron species, undergoing oxidative addition reactions (like a carbene) or serving as a nucleophile. The pattern of linkage of coordination number with oxidation state means that

Co–C bonds (to the sixth ligand) can be broken in three fundamentally different ways:

1. $Co^{III} - CR_3 + H_2O \rightarrow Co^{III} - OH_2 + R_3C^-$ (heterolytic, carbanion)
2. $Co^{II} - CR_3 \rightarrow Co^{II}\cdot + R_3C\cdot$ (homolytic, free radical)
3. $Co^{I} - CR_3 \rightarrow Co^{I} + R_3C^+$ (heterolytic, carbocation)

Because homolysis to a free radical occurs with a substantial positive volume of activation (about +15–20 mL/mol rn), while heterolysis usually shows small or negative $\Delta V^{\ddagger}$, the pressure dependence of reaction rate can be used to choose between these three reaction modes.

Biochemical reactions of cobalamins fall in three categories. Best known are isomerases, in which a hydrogen atom and another small group on adjacent carbons in an organic substrate exchange positions:

$$
\begin{array}{ccc}
\text{H} \quad \text{Z} & & \text{Z} \quad \text{H} \\
| \quad | & & | \quad | \\
R_2C\!-\!CR_2 & \rightleftharpoons & R_2C\!-\!CR_2
\end{array}
$$

Diol dehydrase uses coenzyme B_{12} to interchange H and OH on a vicinal diol $RCH(OH)-CH(OH)H$ to yield an intermediate geminal diol $RCH_2-C(OH)_2H$, which eliminates water to form an aldehyde RCH_2-CHO. The coenzyme binds a protein whose purpose appears to be the steric destabilization of the Co–C bond (to the adenosyl group) toward homolysis. Without the protein this bond is surprisingly stable for an organometallic bond in an aqueous environment—at room temperature $t_{1/2}$ for its homolysis is about 10 years, but in the presence of the enzyme protein and substrate this can be speeded up by a factor of 10^{13}. In this case the purpose of the enzyme protein is not to change the reactivity of the substrate but to change the reactivity of the metal active site.

A second category of B_{12}-mediated reactions is methyl transfers involving methylcobalamin. In higher animals this leads to synthesis of the amino acid methionine (with an $S-CH_3$ group) from a thiol precursor, while in bacteria methyl transfer reactions lead to methane formation by methanogens and to methylmercury formation in aquifers polluted by mercury metal. The mobile character of the Co–C bond is emphasized by these mechanisms, which appear to require the release of CH_3^+ to form methionine, but (in a different environment) the release of CH_3^- to form methane or methylmercury.

The third category of B_{12}-mediated reactions is the reduction of ribonucleotides (with —OH on the ribose 2′ C) to deoxyribonucleotides (—OH replaced by —H). Less is known about this mechanism, though there are obvious resemblances to the diol dehydrase reactions in general stoichiometric terms.

Cobalamins are complex molecules, even in the absence of a protein. Several different model systems have been prepared that yield good structural models for B_{12} species and which have shown some success as functional models in that steric features of the model compounds can influence the Co–C bond energy and, presumably, its lability. The best known model compounds are square planar

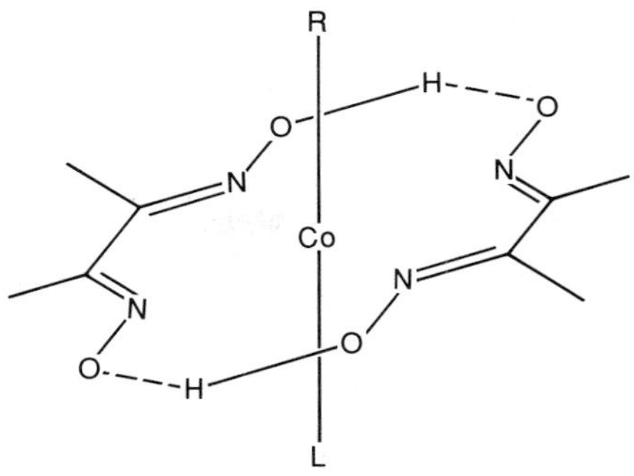

R = alkyl
L = N- or P-donor, pi acceptor ligand

Figure 16.8 Octahedral cobalt coordination in cobaloxime model compounds (dimethyl-glyoxime shown) mimicking cobalamins.

complexes of Co^{3+} with two deprotonated dimethylglyoxime molecules, as in Fig. 16.8, known as *cobaloximes*. If a base L such as *py* occupies the fifth coordination site and an alkyl group R the sixth site, the Co–C bond length and energy can be measured in the context of these structural parameters. These bond properties should relate to B_{12} reactions, because the cobaloximes yield some of the same reactions; for instance, methylcobaloxime will transfer a methyl group to yield methionine from its precursor just as the coenzyme does (though less efficiently). By varying the base L and alkyl group R, the Co–C bond length can be varied from 1.92 Å [L=CN⁻, R=CN⁻] to 2.21 Å [L=P(OMe)₃, R=adamantyl], which is an enormous range for what is formally a single bond. The two hydrogen-bonded halves of the equatorial plane can also be bent up to 10° with respect to each other by bulky R groups such as adamantyl. Both of these effects are thought to be present in the protein-bound coenzyme, weakening the Co–C bond to labilize it.

Blue copper proteins. We have seen that copper appears in cytochrome *a* electron transfer processes, but there are also a number of electron-transfer metalloproteins containing only Cu. A number of these show an intense blue color (absorbing at 600 nm) and have been grouped as the blue copper proteins. Most are small proteins containing only a single Cu atom, but some contain multiple Cu atoms in a variety of environments. *Type 1* copper atoms, present in most single-Cu proteins, show molar absorptivity roughly 100 times greater than would be expected for Cu^{2+} in complexes governed by the Laporte selection rule, and they show relatively high reduction potentials. *Type 2* copper atoms may be present along with Type 1 in multicopper proteins; they have normal Cu^{2+} spectroscopic properties. *Type 3* copper atoms occur in pairs in some proteins; they are quite close, with the d^9 Cu^{2+} spins interacting so strongly as to be diamagnetic. Strong electronic absorption by Type 3 coppers occurs in the UV at about 330 nm.

One important blue copper protein is *plastocyanin,* with a molecular weight of 10,500 daltons and one Type 1 copper. It serves an electron-transfer role in the photosynthesis process in higher plants, which as we shall see in the next chapter is essentially a photochemically induced redox sequence. X-ray diffraction studies of plastocyanins from a number of plants yield a common coordination geometry for their Cu atoms: trigonally distorted tetrahedral with two histidine N donor atoms (Cu–N: 1.94 Å, 2.02 Å) and two S donors from cysteine and methionine (Cu–S: 2.07 Å, 2.82 Å). The tetrahedron of donor atoms is substantially compressed toward a trigonal arrangement; this is a useful structural compromise for ready electron transfer, and is quite stable. The intense blue color is maintained even if the distant methionine S ligand atom is removed.

Although the plastocyanin protein is not large, the copper atom is buried inside it, which raises the question of how electron transfer is to occur. Ever since the origin of the orbital theory of electronic structure it has been tacitly assumed that electron transfer from A to B could only occur through direct orbital overlap of A and B (inner-sphere) or by transfer through adjacent orbitals (outer-sphere). In recent years inorganic chemists have become aware, as a result of studies primarily by Williams and by Gray, that long-range electron transfer, over distances of 10 to 20 Å between formally reacting atoms, can occur if the electron is passing through stabilizing channels inside a large folded molecule such as a protein that is otherwise redox-inert. Such channels can consist of hydrogen bonds, sigma bonds, pi bonds, or even electrostatically stabilized regions of free space analogous to the potential well that stabilizes "free" electrons in liquid ammonia. With this recognition, studies of metalloprotein electron transfer mechanisms take on a requirement that the pathways be specified. In the absence of special circumstances such as a pi system lying directly on the closest path from outside the protein to the metal center, electron transfer rates can be calculated directly from the transfer distance, as the next chapter will show.

Superoxide dismutase. As we have noted above, cytochrome oxidase and other enzymes carry out reduction of O_2 to water without releasing any intermediate oxidation states for the O atoms. This is an important precaution for preserving living cells in an aerobic world, because the intermediate oxidation states represented by superoxide, O_2^-, and hydrogen peroxide are strong enough oxidizing agents to damage cell molecules. But there are biomolecules that undergo oxidation by O_2 directly and release either O_2^- or H_2O_2, and the cell needs a chemical mechanism to eliminate these species. Fortunately, the reduction potentials are such that both species can spontaneously disproportionate, and enzymes have evolved to catalyze this disproportionation in order to avoid oxidative cell damage.

The enzymes eliminating O_2^- are known as superoxide dismutases (SODs). Three evolutionary families exist inside cells. The best characterized is a copper–zinc protein occurring in cells with a nuclear membrane (eucaryotes). Cu,ZnSOD is a dimeric protein with a molecular weight of 32,000 daltons, containing one Cu and one Zn in each identical subunit. The Zn is buried in the protein with respect to any approaching species, but the Cu—as Cu^{2+} in the absence of O_2^-—is exposed at the bottom of a narrow cleft in the protein. Presumably the Zn^{2+} is structural, because it can be replaced by other M^{2+} cations without loss of catalytic activity although the Cu^{2+} is essential. The Cu^{2+} is bound to the protein by four histidines, one of whose imidazole rings also coordinates the Zn^{2+} along with two other histidines and an aspartate.

The overall reaction for superoxide dismutation (disproportionation) is

$$O_2^- + O_2^- + 2H^+ \rightarrow H_2O_2 + O_2$$

It occurs in the enzyme in two steps: In the first, Cu^{2+} is reduced to Cu^+ by an O_2^-, yielding O_2 and releasing the bridging imidazole as a very basic imidazolate anion still coordinated to Zn^{2+}. Water protonates the imidazolate, and when a second O_2^- approaches, the Cu^+ is oxidized to Cu^{2+} and the reduced superoxide is protonated by the imidazole to yield HO_2^-, which is released into solution to be protonated to H_2O_2. At biological pH both of these steps have rate constants on the order of $10^9 \ M^{-1} \ s^{-1}$, which is near the rate at which diffusion of reactants can occur.

Less is known about the other two SODs, which are related in an evolutionary sense (through their amino acid sequence) though they are quite different from Cu,ZnSOD. Both are usually found in single-celled organisms. One form contains iron, the other manganese; both are thought to function by cycling between M^{II} and M^{III} just as Cu,ZnSOD cycles between Cu^{II} and Cu^I.

Catalases and peroxidases. Superoxide is the product of the first one-electron reduction of O_2, and SODs eliminate it efficiently. However, SODs disproportionate superoxide to O_2 and H_2O_2, and hydrogen peroxide is also a sufficiently vigorous oxidizing agent to damage biological molecules; witness the use of 3% H_2O_2 as an antiseptic. Catalases disproportionate H_2O_2:

$$H_2O_2 + H_2O_2 \rightarrow O_2 + 2H_2O$$

while peroxidases use H_2O_2 to oxidize another substrate:

$$H_2O_2 + H_2A \rightarrow A + 2H_2O$$

Catalases are found in all plant species, and the enzymes are all similar in having four heme irons as Fe^{3+} in separate protein environments, with a total molecular weight of about 250,000 daltons. The mechanism of its disproportionation is not known in detail, but is extremely effective: The activation energy for the reaction above is about 75 kJ/mol rn if uncatalyzed, but the reaction proceeds at approximately the diffusion-limited rate at catalase, suggesting an activation energy on the order of kT, about 3 kJ/mol rn. It is known that one molecule of H_2O_2 binds to one catalase subunit forming a species known helpfully as Compound I, which has the ability to oxidize other species by two electrons. A second H_2O_2 approaches Compound I and reacts directly with it to produce the products O_2 and two H_2O. Compound I may or may not involve a direct $Fe-O_2H_2$ bond, inasmuch as it can be produced from catalase by oxidants other than H_2O_2. However, like the cytochrome oxidase intermediate in Fig. 16.6, Compound I is thought to have iron present as Fe^{4+}.

As the general peroxidase reaction above suggests, peroxidases reduce H_2O_2 to water but do not disproportionate it to O_2; instead, another chemical substance is oxidized, usually not involving an oxygen from the peroxide. Of the large class of peroxidases, perhaps the best studied is horseradish peroxidase (HRP). Like the others, HRP has one Fe^{3+} heme iron; its molecular weight is about 40,000 daltons. The heme iron is bound to the protein by a histidine and is thought to be displaced out of the porphyrin plane toward the histidine so as to be five-coordinate in the absence of peroxide. An approaching H_2O_2 is initially coordinated to the sixth Fe site yielding an $Fe^{4+}{=}O$ radical cation grouping and splitting off a water molecule. This "ferryl" radical is again termed Compound I. It is thought to abstract a hydrogen atom H· from H_2A reactant, forming HA· and a neutral $Fe^{4+}{=}O$ species, Compound II. Abstraction of another H· from another H_2A allows the ferryl oxygen to be released as water, returning the HRP to its initial condition as in Fig. 16.9. The two HA· species produced undergo hydrogen-atom transfer to yield product A and the starting material H_2A. Both catalase and the peroxidases provide efficient removal of H_2O_2; the peroxidases have the additional advantage of coupling peroxide removal with oxidation of another organic substrate.

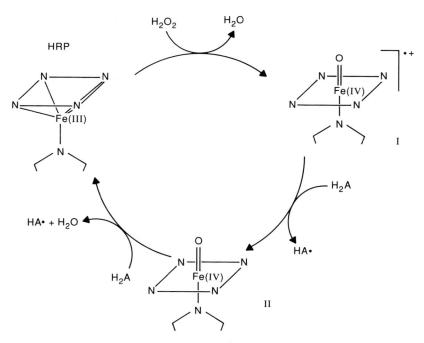

Figure 16.9 The horseradish peroxidase catalytic sequence for the reaction of H_2O_2 with species H_2A, yielding H_2O and A.

16.4 REDOX SYNTHESIS: NITROGENASE

To transition metal chemists the coordination chemistry of N_2 has been somewhat frustrating. The diatomic element, of course, is quite inert both thermodynamically and kinetically because the small inner core of the nitrogen atom allows a very strong triple bond to form. But N_2 is isoelectronic with CO, which also has a strong triple bond—stronger, in fact, than the bond in N_2 (1071 versus 941 kJ/mol). Yet as we have seen, CO has a very extensive coordination chemistry with transition metals, whereas until the mid-1960s no transition metal complex of the N_2 ligand molecule was known.

On the other hand, even before the isolation of an N_2 complex it seemed clear that such coordination could be accomplished under the right conditions, because metals (Fe and Mo) were known to exist in the *nitrogenase* enzyme that certain bacteria use in the root nodules of legumes and some other plants, or free in the soil, to fix atmospheric N_2 (which we shall call dinitrogen as a ligand). This fixation produces NH_4^+ from N_2 and H_2O in the presence of O_2 at soil temperatures and 1 atm pressure—rather better than the Haber–Bosch process. Since N_2 and H_2O will not react under any ordinary circumstances, it seemed likely that one or the other of the metals in the enzyme (now known to include V as an alternative to Mo for some species) was coordinating the dinitrogen to induce chemical reactivity. Therefore, studies of N_2 complexation proceeded in two areas. One area sought to produce synthetic N_2 complexes, reduce them to NH_3 if possible, and extend the reduction to a catalytic cyclic process as a final goal. The other area sought to establish the structure of nitrogenase (particularly the structure of the metal centers) and perhaps to mimic it with model compounds that would have N_2-complexing ability. We will look separately at these areas.

The first synthetic N_2 complex was made by Allen and Senoff in 1965: $[Ru(NH_3)_5(N_2)]^{2+}$. This was prepared not from atmospheric or pure N_2 but from hydrazine and Ru^{3+} in water. Other N_2 complexes have subsequently been prepared by more or less analogous routes involving the oxidation of coordinated hydrazine (for instance, $Mn(cp)(CO)_2(N_2)$) or azide (for instance, $[Ru(diars)_2Cl(N_2)]^+$.

Although such compounds are structurally interesting, they do not carry us very far toward the direct coordination of dinitrogen or nitrogen fixation. However, following Allen and Senoff's discovery, many complexes have been prepared using N_2 directly. Usually each N_2 ligand is monohapto- and replaces a single-atom ligand such as Cl^- or H^-. If the resulting complex is to be a neutral compound or at least have the metal in a low oxidation state, some reduction of the metal is necessary. (Remember that in carbonyls a low oxidation state is necessary if the metal is to serve as a pi donor to the CO.) Active-metal reduction of halides is usually necessary:

$$MoCl_3(thf)_3 \xrightarrow[\text{diphos, } N_2]{\text{Na–Hg or Mg}} Mo(diphos)_2(N_2)_2$$

but hydrides usually undergo reduction by releasing H_2:

$$MoH_4(diphos)_2 + 2N_2 \rightarrow Mo(N_2)_2(diphos)_2 + 2H_2$$

Several hundred dinitrogen complexes have been isolated since 1965, including every transition metal, except perhaps Pd and Pt, and at least one lanthanide. As for carbonyls, the complexes seem to be more stable with d^4–d^9 metals than with either d^2, d^3, or d^{10} metals, presumably for the same reasons involving the metal as a pi donor (see Chapter 13). With only a few exceptions, dinitrogen complexes are linear in their M—N≡N geometry, whether they are mononuclear like the ones already mentioned or binuclear with bridging dinitrogen (M—N≡N—M) such as $[\{Ru(NH_3)_5\}_2N_2]^{4+}$ or $(cp)_2Ti$—NN—$Ti(cp)_2$. An interesting dinitrogen complex with both terminal and bridging N_2 groups is the zirconium complex shown in Fig. 16.10. No bent M—N≡N complex has been reported, but several side-on $N\overset{M}{\underset{M}{\equiv}}N$ complexes such as $[(cp^*)_2Sm]_2N_2$ and $[\{^iPr_2$ $PCH_2SiMe_2)_2N\}ZrCl]_2N_2$ have been characterized, with the four M_2N_2 atoms forming either an approximate tetrahedron or a parallelogram.

As with carbonyls, there are substantial shifts of the N≡N stretch frequency in the IR spectrum of N_2 complexes. For free N_2 the vibration occurs at 2331 cm^{-1} (Raman); for terminal complexes it seems to lie in the range 2200–1850 cm^{-1}. This lowering could occur either because bonding electrons are donated or because antibonding electrons are accepted. Some effort has gone into establishing the separate σ-donor and π-acceptor capacities of the N_2 ligand. Mössbauer spectra of ^{57}Fe–N_2 complexes have been combined with IR wavenumber and intensity data to suggest that N_2 is an extremely weak sigma donor but a fairly good pi acceptor—perhaps better than a nitrile, though not as good as CO. This means that the stability of dinitrogen complexes will be strongly influenced by the pi-donor ability of the metal and by the pi donor/acceptor characteristics of the other li-

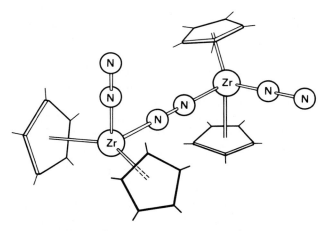

Figure 16.10   Terminal and bridging dinitrogen coordination in $[(N_2)(cp^*)_2Zr]_2N_2$.

gands on the metal atom. For example, a stable complex $IrCl(PR_3)_2(N_2)$ is formed when PR_3 is triphenylphosphine, but not when PR_3 is methyldiphenylphosphine. This is a striking dependence on coligands; it probably caused much of the long delay in discovering stable dinitrogen complexes.

The pi-acceptor capability of a dinitrogen ligand means that when it is coordinated as a terminal complex $M—N{\equiv}N$, the ligand gains electron density and becomes a somewhat better base than it originally was. Allen and Senoff's original complex, for example, will displace a water ligand from another Ru^{II}:

$$[Ru(NH_3)_5(N_2)]^{2+} + [Ru(NH_3)_5(OH_2)]^{2+} \rightleftharpoons [(NH_3)_5Ru—NN—Ru(NH_3)_5]^{4+} + H_2O$$

This increased basicity makes it possible to protonate coordinated N_2, at least under some circumstances, which is an important goal of nitrogen-fixation research. There are various possible reduction (hydrogenation) products—N_2H_2 and N_2H_4 as well as NH_3—but all are much more reactive than N_2 and would be useful industrial products.

There have been two major approaches to the reduction of coordinated N_2. The earliest involved complexes of the powerful reducing agent Ti^{II}. Either potassium metal or sodium naphthalenide will reduce $TiCl_4$ in the presence of alkoxides to $Ti(OR)_2$, which will reduce gaseous N_2 at 1 atm to NH_3 using hydrogens from an ethereal solvent such as *thf*. Similarly, titanocene [the dimer or polymer of $Ti(cp)_2$] forms N_2 complexes analogous to the one shown in Fig. 16.10. When treated with HCl at low temperatures, these yield either N_2 and N_2H_4 or N_2 and NH_3. Several systems of this sort are catalytic for NH_3 in that they produce more than 100% molar yield of ammonia, given an adequate supply of strong reducing agent to maintain the low oxidation state for the titanium or zirconium. However, the process does not seem industrially promising, and obviously does not resemble Fe-Mo nitrogenase chemistry.

The other dinitrogen reduction process seems a little closer to the nitrogenase system in that it relies on group VI metals (usually Mo, W, or both) to complex N_2. The complex $Mo(N_2)_2(diphos)_2$ has already been mentioned; treating this or its tungsten counterpart with HBr or HI at room temperature gives the protonated system $(diphos)_2X_2Mo{=}NNH_2$ plus N_2.

The N_2H_2 ligand is difficult to reduce further, however, though a small yield of NH_3 results if either complex is treated with excess HBr in a Lewis-base coordinating solvent such as propylene carbonate (Table 6.16). In keeping with the sensitivity of the N_2 ligand to the pi character of other ligands, the very similar complex *cis*-$[M(N_2)_2(PMe_2Ph)_4]$ converts half its bound N_2 to ammonia on acidification with H_2SO_4 in methanol:

$$M(N_2)_2(PMe_2Ph)_4 \xrightarrow[CH_3OH]{H_2SO_4} N_2 + 2NH_4^+ + M^{VI} \text{ products} \qquad (M{=}Mo,W)$$

Unfortunately, the M^{VI} products are difficult to characterize or to reconstitute as a dinitrogen complex, so the system does not have direct catalytic potential. The problem is that six electrons are being stripped from one metal atom, which drastically changes its solubility, ligation, and redox properties. We can split the six-electron change between two metal atoms:

$$2Mo(N_2)_2(triphos)(PPh_3) + 8HBr \rightarrow 2NH_4Br + 3N_2 + 2PPh_3 + 2MoBr_3(triphos)$$

where *triphos* is $PhP(CH_2CH_2PPh_2)_2$. Here the molybdenum product is the precursor to the starting dinitrogen complex, which obviously raises the possibility of a catalytic cycle. One problem with this pattern of direct reduction is that only a quarter of the initial coordinated N_2 is actually fixed, which may be an inefficiency arising from the separate three-electron metal centers. If we bind two Mo or W atoms together, the fixed-nitrogen yield can be improved, and the mixed-metal compound $Cp*Me_3Mo-NN-WMe_3Cp'$, where Cp' is C_5Me_4Et, can be protonated by the lutidinium ion (with zinc amalgam as a reducing agent) to yield over 90% of the coordinated N_2 as ammonium ion. Although no genuinely catalytic system has yet been demonstrated, it is clear that much of the fundamental knowledge necessary is at hand.

The synthetic N_2 complexes and the reducing Ti and Mo/W systems above represent a sort of industrial-catalysis approach to the function of nitrogenase. What do we understand of the biological system itself? First, biofixation of dinitrogen occurs only through a nitrogenase enzyme, of which there are several although the structures seem to be closely related. Nitrogenases occur only in prokaryotic cells (those having no nuclear membrane or mitochondria) such as bacteria and blue-green algae, also called cyanobacteria. A number of species, both aerobic and anaerobic, contain a nitrogenase and can fix nitrogen. The most prominent groups are *Azotobacter, Clostridium,* and *Rhizobium. Rhizobium* invades legume roots and is the most important agricultural nitrogen fixer, but the others can fix nitrogen free in the soil.

Nitrogenase can be extracted from many species of nitrogen-fixing bacteria. All the extracts seem to be much the same chemically except for the striking fact that while most (and the best characterized) contain molybdenum, a few seem to contain vanadium instead. However, in the present state of our chemical knowledge, this discussion will follow the molybdenum species. Each of these nitrogenases is an association or complex between three proteins: a molybdenum–iron protein, an iron protein, and a ferredoxin. The ferredoxin is present only as a reducing agent; its role can be filled by nonbiological reducing agents such as dithionite, $S_2O_4^{2-}$. Neither of the other two proteins can fix nitrogen alone, though they can be separated and still fix nitrogen when recombined. The iron–molybdenum protein is thought to be the actual site of N_2 coordination, because its mutants with a defective Fe–Mo cluster cofactor can still reduce acetylene (see Fig. 16.11) but N_2 only poorly. All nitrogenases are extremely sensitive to oxygen, which means that the bacteria must have a mechanism for protecting the nitrogenase from atmospheric O_2.

The nitrogenase reduction of N_2 to NH_4^+ is curiously inefficient thermodynamically. Although the hypothetical reaction

$$N_2(g) + 8H_3O^+(aq) \rightarrow 2NH_4^+ + 8H_2O(l)$$

is thermodynamically both exothermic and spontaneous ($\Delta H^0 = -132.8$ kJ/mol rn and $\Delta G^0 = -79.5$ kJ/mol rn), the process that has evolved for its catalysis requires a large energy *input* in the form of ATP consumed:

$$N_2 + 4S_2O_4^{2-} + 16MgATP + 46H_2O \xrightarrow{\text{Nitrogenase}}$$
$$2NH_4^+ + H_2 + 8SO_3^{2-} + 16MgADP + 16HPO_4^{2-} + 22H_3O^+$$

One of the sources of the inefficiency, of course, is the reduction of water to H_2. However, this is an intrinsic part of the overall process, because the system will produce H_2 even if no N_2 is present to be reduced. Figure 16.11 indicates schematically the requirements for the reaction and the species that have been shown to be reduced by the enzyme. One interesting reduction is that of methyl isocyanide: Organic reductions of this species normally produce dimethylamine, but reduction of isonitrile transition-metal complexes produces the methane-methylamine mixture.

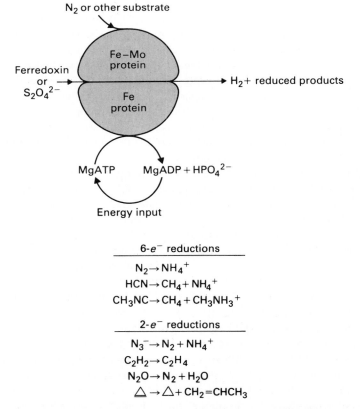

Figure 16.11   Schematic representation of nitrogenase reaction requirements and reaction products.

This strongly suggests that the enzyme directly coordinates N_2 and the other substrates to one or more metal atoms.

The molybdenum–iron protein has a molecular weight of roughly 230,000. It appears to be an $\alpha_2\beta_2$ dimer of two $\alpha\beta$ units, where α and β refer to separately folded sections of the protein. Each unit contains a paramagnetic Fe–Mo cluster cofactor (FeMoco) that can be extracted from the rest of the protein by dithionite, and two diamagnetic P clusters that are 4Fe–4S clusters with one iron different from the other three (as simulated in the model compound of Fig. 16.4).

In the only x-ray structure available for the Mo–Fe protein, by Rees, the level of refinement does not allow detailed judgments of cluster geometry. However, it indicates that the P clusters are probably aggregates of two 4Fe–4S clusters bridged by two more sulfurs. The P cluster aggregate is about 14 Å from the iron-molybdenum cluster, which as we shall see in the next chapter is not too far to allow electron transfer. The P clusters may thus serve as an electron pool for the reduction of N_2. The FeMoco cluster seems from Rees' x-ray determination to have two voided-cubane M_4S_4 units, each missing a sulfur atom rather than a metal to leave an open Fe_3 face, facing each other so that the sequence of atoms from top to bottom of the cluster is $FeS_3Fe_4\cdots Fe_3S_3Mo$. The six central iron atoms are probably arranged as a trigonal prism, bridged from one M_4S_3 to the other by two more cysteine sulfurs and a lighter atom Y (such as oxygen in H_2O). It is tempting to assume that Y represents the site of N_2 coordination—from one Fe to another—but this is as yet only an assumption. It does seem likely that the Mo cannot coordinate N_2, because it is already six-coordinate (distorted octahedral), coordinated by three cluster S plus three light atoms, probably a side-chain N and two O atoms from a homocitrate –CH(OH)–COOH group. In the oxidized form, the Mo oxidation state is probably 4+.

The iron protein seems to be only the energy-transfer mechanism for the reduction. It has a molecular weight of about 64,000 with one 4Fe–4S cluster bridging two similar protein units. Its spectroscopic and redox behavior are strongly influenced by the presence of ATP, and the cluster is thought to bind two ATP units in order to deliver one electron to the Mo–Fe protein. Both a reducing agent and an ATP-binding agent are required to drive the reaction.

As we have suggested earlier, modeling of the FeMoco cluster is still at the level of stoichiometric model compounds—and the native stoichiometry is not absolutely certain. Figure 16.12 shows idealized geometries of the cluster cores of several possible model compounds, some prepared in aqueous solution and some through organometallic syntheses. Several "double-cubane" structures have been prepared even before Rees' recent x-ray structure, but other recent efforts have tended to yield Mo environments of lower symmetry. Of these structures, the one best fitting the $FeS_3Fe_3\cdots Fe_3S_3Mo$ double-voided-cubane sequence suggested by the x-ray data is $Mo_2Fe_6S_6$, where both the top and bottom half of the cluster are MoS_3Fe_3 units. However, this model cluster binds the two halves too closely, with no bridging atoms but rather direct S–Fe coordination between the halves, and it also maintains the iron atoms in higher symmetry than is indicated for the iron

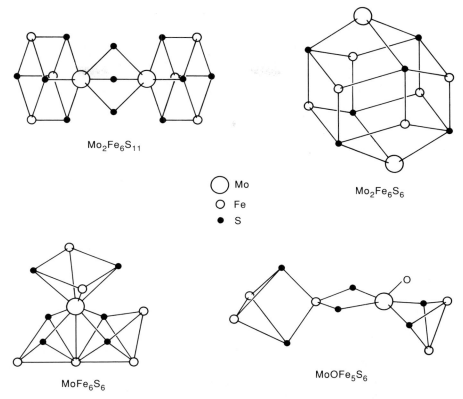

$Mo_2Fe_6S_{11}$

○ Mo
○ Fe
● S

$Mo_2Fe_6S_6$

$MoFe_6S_6$

$MoOFe_5S_6$

Figure 16.12   Idealized core geometries for some Fe–Mo–S clusters that have been prepared as possible nitrogenase cluster models. Outer Fe and Mo atoms have other ligands.

atoms by EPR results. As the x-ray structures are refined and more model compounds are prepared, the various interpretations of experiments are likely to converge on a well-defined structure, and beyond that on a mechanism.

16.5 METALLOPROTEINS: LIGAND CARRIERS, METAL CARRIERS, AND METAL STORAGE PROTEINS

Not all metals in proteins serve as catalysts. Transport and storage proteins are also important in regulating concentrations of bioligands and of the metals themselves in an organism. We shall consider ligand carriers in the particular context of O_2, dioxygen, as a ligand. However, it is also important to recognize that there are specialized proteins for the transport of iron and calcium ions, and for the storage of iron and calcium. Let us begin with O_2 carriers.

Metals as bioligand carriers must be able to coordinate the ligand, but reversibly so that it can be removed unchanged at its destination. Because it must be bound but not react (at least not in any irreversible fashion) with the metal, the metal–ligand bonding is of critical interest, both with respect to coordination geometry and with respect to free-ligand/coordinated-ligand equilibrium constants. Dioxygen is a particularly simple but particularly important bioligand, which we have delayed in previous discussions in order to take it up here. Like dinitrogen, the primary interest in O_2 as a ligand lies in its character and reactions as a coordinated ligand rather than in its effect on the metal atom to which it is coordinated. We have seen several diatomic molecules as ligands: CO, NO, CN^-, N_2, and now O_2. What geometric structures do they show as ligands?

There are four geometric possibilities for a diatomic ligand X_2 or XY bonding to a single transition metal atom:

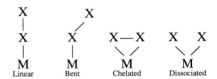

We have seen examples of all of these: Terminal η^1carbonyls are linear; nitrosyls are either linear or bent ($\approx 130°$); O_2 is chelated when coordinated to Vaska's compound (Chapter 14); and H_2 and halogens are dissociated upon oxidative addition to Vaska's compound. We can correlate the observed bonding geometries with the identity of each ligand by considering its diatomic MOs in more detail than was presented in Chapter 4.

Assume (somewhat arbitrarily) that each of the two nonhydrogen atoms in the ligand is bonded to the other using sp hybrid orbitals for sigma overlap and the other two unchanged p orbitals for pi overlap. The resulting MO energy level diagrams, following our previous qualitative MO guidelines, are shown in Fig. 16.13 for X_2 and XY, where Y is the more electronegative of the two atoms. The relative positions of the energy levels depend, of course, on the s—p energy separation for each atom and on the difference between the valence-orbital ionization potentials (VOIPs) of the two atoms. The positions shown are fairly realistic for CO, NO, O_2, and N_2, the ligands we shall be primarily concerned with.

We next assume that the diatomic ligands serve as sigma-donor Lewis bases, and that the bonding geometries for X_2 or XY as a ligand will be those that give the best possible overlap with metal acceptor orbitals *for the ligand HOMO*—the most available electrons on the ligand. To predict the geometry from these assumptions, we only have to fill the energy levels of Fig. 16.13 with the correct number of electrons, see which is the HOMO, and orient the ligand toward the metal atom to optimize that orbital's overlap with the metal acceptor orbital. CO, for example, has 10 valence electrons. It will fill the XY MOs in Fig. 16.13 up through the σ_x nonbonding level, which is the sp hybrid on the C atom that points away from the

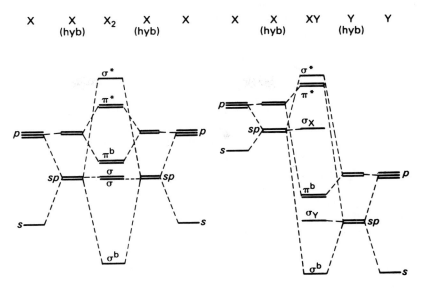

Figure 16.13 Qualitative molecular orbital energy levels for X_2 and XY molecules as possible ligands. X and Y have only s and p valence orbitals and are assumed to be sp hybridized.

O. The characteristic 180° angle of sp hybrids assures that a terminal M—C≡O system will always have linear geometry. NO, by contrast, has 11 electrons. One electron is in the π^* MO and the next most available electrons are in the σ_x. If NO is to be a one-electron donor, it will be bent as shown in Fig. 13.4; if it is to be a three-electron donor it must use the sp electrons on the nitrogen atom, which dictate a linear geometry.

Now let us consider O_2 as a ligand. In the X_2 diagram of Fig. 16.13, O_2 fills the levels through the two π^* levels with its 12 electrons. The π^* MOs contain one electron each, making O_2 paramagnetic (as of course it is). The O_2 molecule can donate one electron by adopting the bent geometry like NO, or it can donate two electrons by adopting the chelated geometry so that a lobe of each π^* orbital can overlap a metal d orbital. Note that the σ nonbonding electrons (sp on either O) are now buried relatively deep in the O_2 energy-level diagram, so that there is very little likelihood of a linear M—O≡O complex. We expect, then, to see either bent or chelated metal–dioxygen complexes. These are indeed the observed geometries in a wide variety of actual complexes. Vaska has described the two possibilities as Type I (bent) and Type II (chelated), and has further subdivided them by considering the O—O bond order as revealed by the vibrational stretch frequency of the O–O bond. Remember from Fig. 4.22 and the associated discussion that the O–O bond strength decreases with the addition of electrons to form O_2^- and O_2^{2-}. Weaker bonds have a lower stretch frequency, and ν_{O-O} is 1580 cm^{-1} for O_2, 1097 cm^{-1} for O_2^-, and 802 cm^{-1} for O_2^{2-}. Dioxygen can coordinate a metal atom at each end of the molecule, or only at one end. The bent Type I complexes that have only one metal atom formally resemble superoxide; these are Type Ia. A number of these are known, predominantly with Fe and Co as we shall see below, and they do in fact have ν_{O-O} comparable to that of O_2^-: about $1105-1195$ cm^{-1} for nine examples, with an average near 1135 cm^{-1}.

The Type I complexes with a metal atom at each end of the O_2 are Type Ib. From IR evidence they still have a "superoxo" quality, with ν_{O-O} between 1075 and 1125 cm^{-1}. Type II complexes with only one metal atom formally resemble a cyclic peroxide, and these (Type IIa) do have ν_{O-O} in the peroxide range. Well over a hundred of these are known, for every transition metal except Hf, Fe, and the group VIIb metals. All have ν_{O-O} between 800 and 930 cm^{-1}. Type IIb complexes, those with a metal at each end, are generally similar to Type Ib in their geometry (that is, they are bent at each O atom), but they have longer O$-$O bond lengths (1.44 versus 1.31 Å) characteristic of a weaker bond and "peroxo" stretch frequencies ($790-840$ cm^{-1}). Some examples of each type will be discussed below when we consider hemoglobin model compounds.

Note that although dioxygen serves formally as an electron donor to the metal atom regardless of what bond geometry it adopts, the IR wavenumbers are characteristic of O_2 with a significant net negative charge (presumably through pi acceptance). On balance, then, the metal must be serving as a net base—perhaps even as a reducing agent. This is consistent with the experimental fact that metal–dioxygen complexes form only with the metal in a reduced oxidation state, allowing it to be oxidized by the O_2; we have seen a well-established Fe^{4+} as an intermediate in the cytochrome iron–dioxygen system.

Myoglobin and hemoglobin. The zero oxidation state for O in O_2 is a quite high and quite unstable oxidation state for an element as electronegative as oxygen. This is the key to its biological function. At biological pH (about 7.4) the half-reaction

$$O_2 + 4H_3O^+ + 4e^- \rightarrow 6H_2O$$

has an associated ΔG of -305 kJ/mol rn, which is a lot of free energy. Stepwise oxidations mediated by cytochromes provide the fundamental biological energy source. However, any animal larger than the distance for rapid diffusion of O_2 through water must develop a biochemical mechanism for picking up O_2 at its surface and delivering it, still in a high oxidation state, to its cells. That is, each organism needs an oxygen-carrier molecule in its circulatory system. The vast majority of animals rely on hemoglobin and myoglobin, both of which (like cytochromes) have a heme unit containing Fe^{II} that actually coordinates the O_2, bound into a largely helical globin protein with a molecular weight of 16,000–18,000. The heme unit in each is protoporphyrin IX (see Fig. 16.5). A histidine imidazole N provides a fifth ligand for the Fe, binding the heme to the globin protein, and the sixth position can coordinate O_2, water, or other ligands. The Fe is out of the porphyrin plane by some distance, perhaps 0.5 Å, as seen in Fig. 16.14. The figure also shows the structure of a β unit of hemoglobin, which is a tetramer

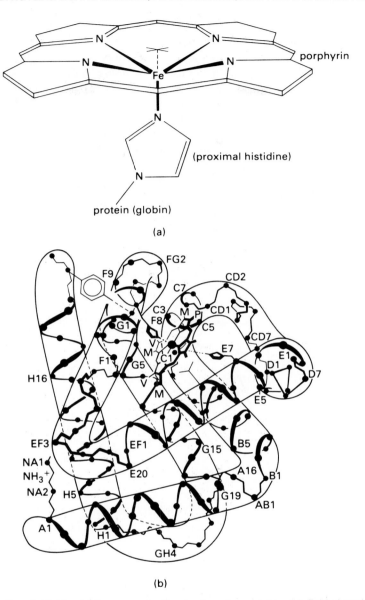

Figure 16.14 (a) Iron coordination in deoxyhemoglobin. (b) Folded tertiary structure of a hemoglobin monomer unit. [From M. F. Perutz and L. F. Ten Eyck, Cold Spring Harbor Symposium on Quantitative Biology, vol. 36, p. 296, Cold Spring Harbor Laboratory, Cold Spring Harbor, New York, 1971. By permission from Cold Spring Harbor Laboratory.]

of four globin units, two α and two β. Each of the four globin units can coordinate an O_2 ligand to its heme iron in the sixth position.

It should be apparent from the folded tertiary structure in Fig. 16.14 that the ease with which an O_2 coordinates to a hemoglobin unit (oxygenation) will depend on how tightly the globin is folded around the heme. In the hemoglobin tetramer, a more or less tetrahedral assembly of four units, the O_2 affinity is much lower than it is for an isolated monomer, perhaps only 1% as great. However, as soon as partial oxygenation has occurred, the O_2 affinity of the remaining deoxyhemoglobin units increases substantially through a "heme–heme interaction" that helps open a neighbor globin structure when the first unit is oxygenated. When the tetramer is almost saturated with O_2, its affinity is almost the same as for an isolated monomer. Figure 16.15 compares the behavior of the monomer and the tetramer on exposure to O_2. The S-shaped (sigmoid) character of the tetramer curve indicates the heme–heme interaction.

The bonding geometry of O_2 within hemoglobin has been a matter of long theoretical debate and ingenious experimental design. Numerous arguments have been put forward based on the magnetic properties of oxyhemoglobin (diamag-

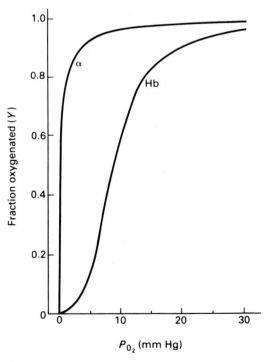

Figure 16.15 O_2 absorption by a hemoglobin tetramer (Hb) and by an α monomer unit as a function of P_{O_2} (20°C). Y is fraction of heme sites oxygenated.

netic), its vibrational spectrum ($v_{O-O} = 1106$ cm^{-1}), its electronic spectrum, and its chemical properties (Cl$^-$ attack on oxyhemoglobin releases O$_2^-$). These are consistent with medium-resolution x-ray studies; they indicate that the Fe−O−O complex in oxyhemoglobin is a bent superoxo complex Fe^{3+}−O$_2^-$ with a Fe−O−O angle somewhere near 120° (Vaska's Type Ia).

It thus seems to be true that, although hemoglobin is a carrier of a molecule that is to be delivered unchanged, a redox reaction has occurred upon coordination. Obviously, to be such a carrier, oxyhemoglobin must be able to reverse this reaction at the cell to which the O$_2$ is to be delivered. In effect, this reversibility is provided by the globin environment of the heme unit, which is generally nonpolar and can flex some of its heme-neighbor groups in response to changing coordination of the Fe. In the absence of the protein, a Fe^{2+}−heme complex in water will react irreversibly with O$_2$ to give an oxygen-bridged Fe−O−Fe product. If the out-of-plane ligands are L and the heme porphyrin is *Por,* the reaction is

$$2Fe^{2+}(Por)L_2 + \tfrac{1}{2}O_2 \rightarrow (Por)Fe^{3+}-O-Fe^{3+}(Por) + 4L$$

The mechanism of this decomposition has been studied extensively. It is thought to involve attack by one deoxy heme unit on a dioxygen−heme complex to yield a Type IIb peroxo complex, which breaks apart symmetrically to yield a ferryl Fe$^{4+} = O^{2-}$ complex like that in the cytochrome a_3 of Fig. 16.6:

$$Fe^{2+}(Por)L_2 \rightleftharpoons Fe^{2+}(Por)L + L$$

$$Fe^{2+}(Por)L + O_2 \rightleftharpoons L(Por)Fe^{2+}-O_2$$

$$L(Por)Fe^{2+}-O_2 + Fe^{2+}(Por)L \rightleftharpoons L(Por)Fe^{3+}-O_2^{2-}-Fe^{3+}(Por)L$$

$$L(Por)Fe^{3+}-O_2^{2-}-Fe^{3+}(Por)L \xrightarrow{fast} 2L(Por)Fe^{4+}=O^{2-}$$

$$L(Por)Fe^{4+}=O^{2-} + Fe^{2+}(Por)L \xrightarrow{fast} (Por)Fe^{3+}-O-Fe^{3+}(Por) + 2L$$

The globin protein in the biological complex, then, prevents irreversible oxidation of the Fe^{2+}−heme by keeping each pair of heme units so far apart that they cannot form the bridged oxidation intermediate. In recent years, many efforts have been made to synthesize iron-based model compounds for hemoglobin that can engage in reversible oxygenation—that is, functional models as well as stoichiometric and structural models. These efforts have focused on this feature of the globin's function by surrounding one side of the macrocyclic complex with bulky nonpolar groups.

The adequacy of a functional model of hemoglobin is its ability to reversibly complex O$_2$, preferably with Type Ia geometry as in a true structural model. However, the model should also be five-coordinate in the absence of O$_2$ and should have a nonpolar environment for coordinated O$_2$. One clever iron−porphyrin complex that has all these qualities is the "picket-fence" porphyrin Fe(TpivPP)(L), where *TpivPP* refers to tetraphenylporphyrin with four pivalamide substituents and

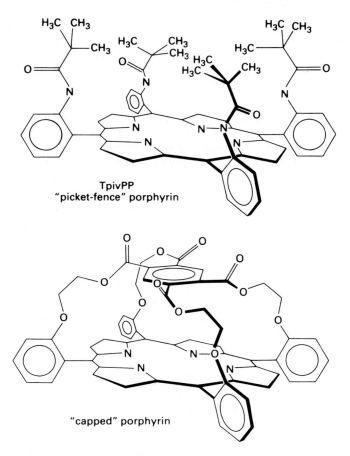

TpivPP
"picket-fence" porphyrin

"capped" porphyrin

Figure 16.16   Hemoglobin model dioxygen carrier ligands.

L refers to a base such as pyridine or imidazole. Another successful model is the "capped" porphyrin. Both are shown in Figure 16.16.

    Collman and others have studied the picket-fence porphyrin in some detail. Its Fe^{II} complex is an excellent functional equivalent for myoglobin, the monomeric biological oxygen carrier. The free ligand TpivPP has four isomers with respect to the location of the *t*-butyl groups relative to the porphyrin plane. Fortunately, they do not interconvert rapidly in solution and the $\alpha\alpha\alpha\alpha$ "picket-fence" isomer can be separated. To mimic myoglobin, the Fe^{2+} complex needs an imidazole base ligand on the open side of the porphyrin plane, and the steric bulk of 1-methylimidazole causes it to add preferentially to the open side. Steric bulk is an important consideration, as we shall see.

    The resulting five-coordinate Fe^{2+} adds O_2 reversibly both as a crystalline solid and in benzene solution, in an O_2:Fe ratio of 1.0. The complex $Fe(TpivPP)(1\text{-MeIm})(O_2)$

gradually and irreversibly oxidizes to a dimeric Fe—O—Fe product analogous to that of an unprotected Fe—heme in solution. However, multiple cycles of reversible oxygenation can be carried out before the irreversible oxidation is too extensive to continue. The thermodynamics of O_2 coordination to solid Fe(TpivPP)(1-MeIm) are very nearly the same as those measured for various myoglobins, as Table 16.1 indicates. In solution, the picket-fence system is a model for myoglobin, but remarkably in the solid state the O_2 affinity shows a cooperative effect analogous to that for the tetramer hemoglobin. X-ray and EXAFS studies show that there are two alternate binding modes for O_2 on the model, in which r_{Fe-O} is either 1.77 Å or 1.90 Å depending on the position of the Fe relative to the porphyrin plane. If the steric bulk of the imidazole base is increased (2-methylimidazole), the Fe is about 0.08 Å toward the imidazole, as in Fig. 16.14, while the less-hindered 1-methylimidazole allows the Fe to be about 0.03 Å out of the plane toward the O_2 instead, yielding the shorter Fe—O bond. The cooperative effect on O_2 affinity is seen only for the hindered 2-MeIm model with the longer Fe—O bond, suggesting that hemoglobin may show its cooperative effect by altering the histidine environment through conformation changes.

X-ray data also indicate a bent Fe—O—O linkage with an O—O distance nearly the same as that of free O_2 and shorter than that of most other dioxygen complexes. The O—O stretch vibration occurs at 1159 cm^{-1}, similar to the 1103 cm^{-1} for oxymyoglobin (MbO$_2$) and suggesting that the complexed O_2 is in Type Ia superoxo form. The higher frequency suggests slightly less reduction for the model than for MbO$_2$, but the difference is probably not too significant. The Fe—O—O angle varies for the four disordered orientations of the O—O axis within the complex, but it is near 130° and apparently is not constrained by contact with the *t*-butyl groups. This angle is also a good fit for the best reported value for sperm-whale myoglobin, 121°.

Many other oxygen-carrier model compounds have been synthesized that complex O_2 reversibly, even though they do not mimic the molecular properties of

Table 16.1 Thermodynamic Data for O_2 Binding by Myoglobins and the "Picket-Fence" Model Complex

	Fe(Por) + O_2 $\rightleftharpoons$ Fe(Por)(O_2)		
O_2 binding species	$P_{1/2}$ (20°C) (torr)	$\Delta H°$ (kJ/mol rn)	$\Delta S°$ (J/K mol rn)
Fe(TpivPP)(1-MeIm)	0.31	−65	−159
Human Mb (reconst.)	0.72	−56	−134
Adult ox Mb	0.55	−63	−155
Tuna Mb	0.90	−55	−134
Horse Mb	0.70	−57	−138

Note: $P_{1/2}$ is partial pressure of O_2 at which coordination sites are half filled in the equilibrium shown.

hemoglobin or myoglobin nearly as well as the two just described. Cobalt(II) is particularly good at forming dioxygen complexes; Werner characterized the complex $(H_3N)_5Co—O_2—Co(NH_3)_5$ before the turn of the century, the first example of what we would now call a Type IIb dioxygen complex. This complex does not reversibly release O_2, but a number of Co^{2+} complexes with Schiff-base ligands do. Schiff-base ligands are planar chelating or even macrocyclic ligands, usually with four donor atoms, that are formed by condensing a primary amine with a carbonyl group:

salicylaldehyde   en   Co (salen)

As in this example, the symbols for the Schiff-base ligands are usually developed from those used for the carbonyl compound and the amine. Square-planar complexes such as Co(salen) reversibly absorb O_2 by forming a dimeric Co–O–O–Co system in the solid phase, usually with $P_{1/2}(O_2)$ about 5 torr and ΔH roughly -80 kJ/mol rn. This is a somewhat weaker interaction than is seen with myoglobin or the protected porphyrin models. Five-coordinate Schiff bases can be formed, for example, by using diethylenetriamine instead of ethylenediamine in the example above. These also coordinate O_2, but only as a monomeric Co–O–O unit. The same behavior is seen if a ligand such as pyridine is first coordinated to a four-coordinate Schiff-base complex. The 2:1 complexes seem to be Type IIb, with a peroxo O_2, but the 1:1 complexes, on the basis of their magnetic properties, seem to be Type Ia superoxo.

An interesting special category of Schiff-base complexes is the group of *lacunar* complexes formed by macrocyclic ligands in which the four N donor atoms are part of two saturated rings and two aromatic rings when complexed, as in Fig. 16.17. The saturated six-membered rings shown favor the chair conformation, which leads the planar aromatic rings to bend away from the CoN_4 plane in the same direction. The result is a saddle-shaped complex that begins to suggest steric protection for one side of the Co as in the sterically protected porphyrins. If each end of the two aromatic rings contains a reactive functional group, they can be bridged by a hydrocarbon chain, forming a *cyclidene* ligand complex. As the bridging chain lengthens, the O_2 affinity of the complex increases—from 10^{-5} for a C_4 bridge to 0.65 for a C_8 bridge—even though the saddle shape of the Co-macrocycle assembly does not change significantly. The center of the C_8 chain folds back away from the metal when the O_2 enters, which may relate to the stronger Co–O_2 bonding that is seen for the longer bridging chain.

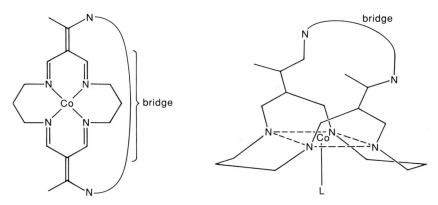

Figure 16.17 General structure for lacunar Co dioxygen carrier complexes. The bridge chain need not be present.

Hemerythrin. Marine worms and some similar invertebrates use hemerythrin instead of hemoglobin as an oxygen carrier. Hemerythrin is a nonheme protein with a molecular weight of about 108,000 daltons consisting of eight identical subunits, each of which contains two Fe^{2+} near each other, binding one O_2. Within a subunit, each Fe is roughly octahedrally coordinated and the two octahedra share a face; the three bridging ligands are two —COO⁻ groups and a water molecule. The other three ligands are histidine imidazoles for one iron and two histidines and a tyrosine —OH group for the other iron. In spite of this difference, Mössbauer spectra show no distinction between the two Fe atoms for the deoxy form. In oxyhemerythrin, however, the iron atoms are different, suggesting that the O_2 is not symmetrically bound to both irons. Resonance Raman spectra suggest the dioxygen is in the peroxo form ($\nu_{O-O} = 844$ cm^{-1}), and electronic spectra suggest the iron atoms in oxyhemerythrin are both Fe^{III}.

Hemocyanin. Molluscs and anthropods transport O_2 using a copper protein, hemocyanin. All hemocyanins are enormous molecules with molecular weights above 10^6 daltons, consisting of many subunits that are not necessarily identical. The degree of aggregation depends on the Ca^{2+} concentration, a topic we shall return to in the next section. A basic subunit usually has a molecular weight in the 50–100,000 range containing two copper atoms to bind one O_2; the subunits show a strong cooperative effect in their oxygen affinity, like hemoglobin. The oxy form is blue, suggesting Cu^{II}, with the Cu atoms bridged both by O_2 (in the peroxo form) and by another unidentified group, perhaps a tyrosine phenoxide. The Cu coordination geometry is tetragonal (probably square planar), bound to the protein by two histidines. The Cu···Cu distance is 3.6 Å and does not change significantly from oxy to deoxy forms of the protein. The deoxy form is colorless, presumably Cu^I, with three binding histidines for each Cu and probably a small bridging molecule such as water yielding tetrahedral coordination. It is interesting that the coordination geometry seems to change quite significantly on addition of O_2; this is probably related to the cooperative effect seen for O_2 affinity.

We do not have definitive information on the mode of peroxo binding to the two copper atoms, except that the two O atoms must be symmetrically bound to the two Cu atoms. Two families of model compounds have been prepared meeting this criterion, one by Solomon with a Cu–O–O–Cu linkage and another by Kitajima and by Karlin with the O atoms at the hinge of a Cu_2O_2 butterfly cluster, either bent or planar. The butterfly arrangement seems to yield better agreement with resonance Raman O–O stretch frequencies, but this is still an active area of research; in addition to hemocyanin there are several other copper oxygenases to which structural results may be extended.

Iron carriers (transferrins). In the body, new hemoglobin is synthesized in the bone marrow while the spleen and liver destroy old red blood cells. An iron carrier is needed to transfer these iron atoms from the liver to the bone marrow; the protein involved is serum transferrin, one of a family of comparable molecules from different species. Serum transferrin is a protein with a molecular weight of about 77,000 daltons, binding two iron atoms at similar but not identical sites, one near the C terminal of the protein and the other near the N terminal. Each iron is bound to the protein by a histidine N, two tyrosine phenoxide —O^-, and an aspartate —COO^-. These are relatively hard bases, suggesting a preference for Fe in a high oxidation state, and Fe^{3+} is bound much more strongly than Fe^{2+} ($K_{stab} = 10^{20}$ versus 10^7). The four protein-binding ligands are arranged octahedrally around the Fe^{3+} with two *cis* sites remaining open, at which the iron coordinates a CO_3^{2-} or HCO_3^- ion; in the absence of carbonate (or an equivalent anion), transferrin will not bind iron.

A series of model compounds has been prepared that successfully mimics the electronic spectra, electrochemistry, and ligand-substitution chemistry of the transferrins. The components and structure of one of these are shown in Fig. 16.18. In this structural model, the histidine donor in the protein is replaced by the imidazole N in *Bymp* [2-(benzimidazol-2-ylmethylphenoxide], one tyrosinate by the *Bymp* phenoxide, the second tyrosinate by the phenoxide in *Msal* [3-methylsalicylate], and the aspartate by the salicylate carboxyl. To maintain an electrically neutral complex to allow nonaqueous studies, the carbonate was replaced by two methyl-imidazole neutral ligands. The reduction potentials of both native transferrins and of the model compounds appear to be so low ($\mathscr{E}^0 \approx -500$ mV) that ordinary biochemical reducing agents could not yield Fe^{2+}. This means that release of iron from the transferrin in the bone marrow cannot rely on stripping out less-tightly bound iron(II) by reducing it. It appears that manipulation of iron binding by carbonate removal or protonation is probably necessary to remove the iron from transferrin, but this remains to be demonstrated.

Iron storage (ferritin). The evolution of Earth's atmosphere from reducing to oxidizing with the release of photosynthetic O_2 provided a powerful biochemical

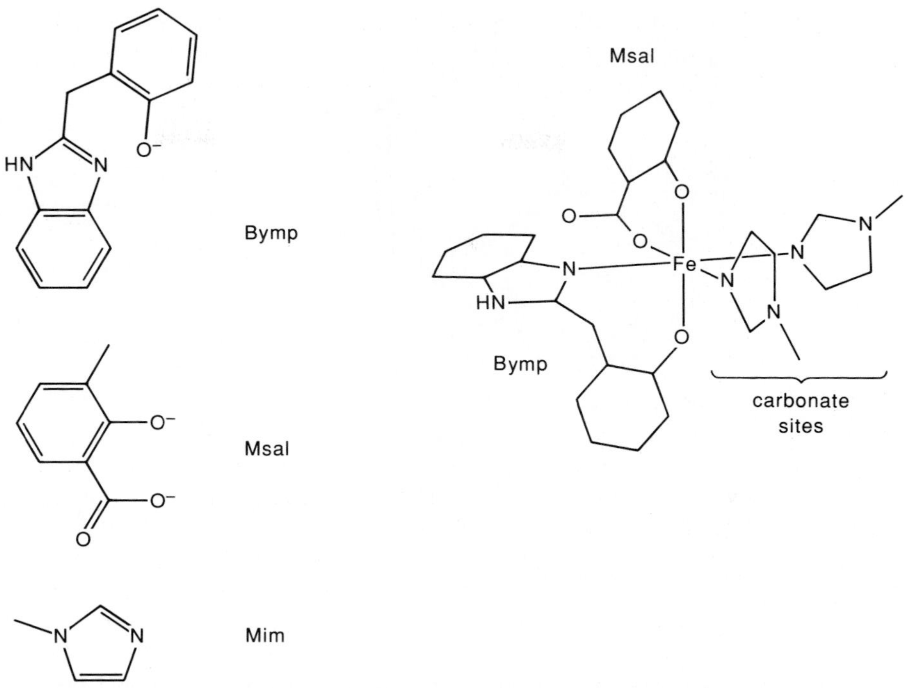

Figure 16.18   Transferrin model compound.

energy source through carbon oxidation, but it presented iron-dependent systems with a problem. O_2 will oxidize Fe^{2+} to Fe^{3+}, and as Chapter 6 has suggested $Fe(OH)_3$ is so insoluble that at biological pH far too little $Fe(H_2O)_6^{3+}$ can be dissolved to be usable in biochemical processes. Accordingly, a system had to evolve to store iron until apotransferrin could collect it for delivery to the bone marrow. This is done by ferritin, which occurs in most types of cells for higher animals. Ferritin is essentially a protein overcoat for a sphere of $Fe(OH)_3$; the protein is an assembly of 24 subunits, each of which is approximately cylindrical, into a hollow sphere with internal diameter about 70 Å. This assembly is shown schematically in Fig. 16.19. On each cylinder N refers to the N-terminal end of the protein and E is the designation for a short helical segment. Note that at eight sites on the sphere three N ends meet to form a polar channel through the protein that can accommodate Fe^{3+} ions being transferred in or out. There are also six nonpolar channels formed by the junction of four E helixes; their function is unknown. Each cylindrical subunit has a molecular weight of about 20,000 daltons, though for most species the subunits are not identical, falling into two or three categories of mobility during electrophoresis.

Within the apoferritin capsule, EXAFS indicates that each iron is surrounded

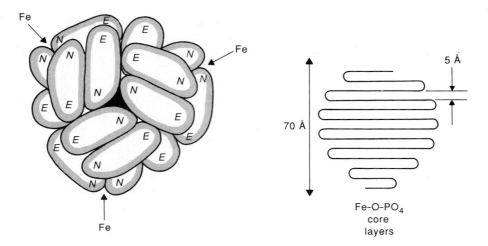

Figure 16.19   View of ferritin 24-subunit protein sheath down three-fold axis of channel used for Fe transfer into and out of the core; three other Fe channels are shown. Symmetry of layered FeO(OH)–PO$_4$ core is shown to same scale. N refers to N-terminal end of protein in sheath subunit consisting of four parallel helical segments A, B, C, D; E refers to fifth short helical segment at opposite end of sheath subunit from N-terminal.

by 6.4 ($\pm$0.6) oxygen atoms at a distance of 1.95 Å, which is exactly equal to the shorter of two Fe–O distances in Fe$_2$O$_3$. The stoichiometry (Fe$_9$O$_{21}$P) and density of the core suggest a layer-of-octahedra structure like that of CdI$_2$ (Fig. 2.14), with each sandwich layer approximately nine FeO$_6$ octahedra wide and terminated on each side by a phosphate; this phosphate-edged ribbon, slightly narrower than the cavity diameter of 70 Å, is thought to Z-fold like computer paper into a compact, roughly spherical core for the ferritin protein. This arrangement is also seen in Fig. 16.19 in a schematic fashion. The cavity can be only partially filled by the iron oxide, but if it is filled about 4500 iron atoms can be stored.

Calcium storage (calsequestrins). As we shall see in the next section, large and abrupt changes in calcium concentration within muscle tissue are necessary for muscle contraction to occur. The reserves of Ca^{2+} are held within the resting muscle within a protein known as *calsequestrin,* whose structure is not known in detail even though it is quite abundant (about 10% by weight of the intramuscular fluid). For most species and muscle sources, calsequestrin has a molecular weight of about 40,000 daltons and binds about 50 Ca^{2+} per molecule. In the absence of Ca^{2+}, calsequestrin has an extended structure, which becomes much more compact when it binds ions (Tb^{3+} can be used as a fluorescent substitute for Ca^{2+}). The protein is quite acidic, suggesting that many side groups have carboxyls, like glutamate or aspartate. Such side groups would coordinate Ca^{2+} through O donor atoms, presumably as CaO$_6$ or CaO$_7$ units; most biologically bound calcium is seven-coordinate. The hard-base character of carboxylate oxygens is consistent with the hard-acid nature of Ca^{2+}. It is not hard to imagine in a general way what the pattern of calcium binding is like, but it would be extremely interesting to have detailed structural data on the Ca^{2+}-binding (compact) form and the Ca^{2+}-free (extended) form of this protein.

16.6 METAL IONS AS STRUCTURE FORMERS AND STRUCTURE PROBES

Up to this point, we have focused primarily on transition metals in bioinorganic systems. We have seen them as Lewis acids (or even as bases) and in multiple oxidation states as redox catalysts. In all of these applications, one consistent role has gone more or less unnoticed: the structure-forming ability of the multiply charged metal cation, with the strong directional forces of its coordinating bonds. We have noted for alkaline phosphatases, for example, that zinc ions appear in an acid/base catalytic role but also in a structure-forming role—but the catalytic zinc atoms have no less structure-forming ability than the zinc atoms that have only that role. There are many metal applications in biological systems for which only a structure-forming ability is needed, and for these evolution has primarily chosen the abundant group IIa ions Ca^{2+} and Mg^{2+}. In this section we shall look at some of these species, but also at some transition metal species that bind to nucleic acids with strongly defined geometry, either to establish the nature of that geometry for the nucleic acid or to alter it for therapeutic purposes.

Ca^{2+} in muscle contraction. Skeletal muscle consists of elongated cells called muscle fibers, which in turn consist of clusters of parallel *myofibrils*. These clusters show a repeating pattern of interruption by discs across the cluster; the region between two discs is a *sarcomere*. In muscle contraction, each sarcomere shortens itself by up to 30%. Within the sarcomere, there is again a cluster of parallel linear units: two sets of thin filaments end to end, each anchored at one end to the disc at the end of the sarcomere, interspersed with a set of thick filaments that are not anchored. This arrangement is shown schematically in Fig. 16.20.

Each thick filament in the sarcomere is primarily made up of molecules of *myosin,* which is a protein whose conformation is a rod (made up of a long twisted-pair α helix) with two small globular heads at one end, each made up of two small polypeptide chains. In Fig. 16.20, the projections shown on the thick filaments are these myosin heads. The thin filaments in the sarcomere are made up primarily of *actin,* which is a small globular protein that polymerizes into chains of actin "beads"; each thin filament resembles a twisted pair of strings of these beads. As the figure indicates, the thick filaments penetrate both sets of thin filaments in the sarcomere; the myosin assembly in each thick filament is directional, so that the myosin heads at the left end of the thick filament have a reversed orientation from those at the right end.

The myosin heads have the ability to bind to the actin units in the thin filaments, but in the relaxed muscle this binding is suppressed so that the filaments

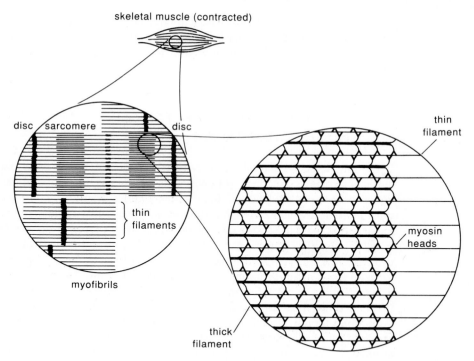

Figure 16.20 Schematic representation of skeletal muscle structure. Parallel anchored thin filaments are interpenetrated by parallel thick filaments. Myosin heads protruding from thick filaments bind to actin beads along thin filaments.

simply lie next to each other, bathed in a network of fluid called the *sarcoplasmic reticulum,* which contains a high concentration of calsequestrin. The myosin does not bind to the actin in this context because the binding sites on the actin are protected by the other component of the thin filament, a polypeptide *tropomyosin* that is entwined with the polyactin chains and excludes the myosin by steric hindrance. If the tropomyosin were not present, it has been shown that the pairs of myosin heads would "walk" along the actin chain forming increasing numbers of myosin-actin bonds as the thick filament moves along the thin filament. This relative motion of the thick and thin filaments within each sarcomere is the basic process of muscle contraction.

The dedicated inorganic student will by this time be desperate for some metal ions. Here they come. How is the tropomyosin protection of the actin-binding sites released so that myosin-to-actin binding can occur and the muscle can contract? Every seven actin units along the actin–tropomyosin thin filament, the tropomyosin strand is interrupted by a troponin complex, which is a set of three small polypeptides: troponin I (which binds to actin beads), troponin T (which binds to tropomyosin), and troponin C (which binds to Ca^{2+}). When calcium ions are supplied in significant numbers, they bind to troponin C. Because of their strong structure-forming effect, they change the geometry of troponin C, which in turn changes the geometry of the associated troponin I, pulling it away from the actin and uncovering the myosin-binding sites. To have quick, coordinated muscle movement, it must thus be possible to supply relatively high concentrations of Ca^{2+} suddenly. The total calcium concentration in the sarcoplasmic reticulum is high, perhaps 10^{-2} M, but it is virtually all bound to calsequestrin. However, the calsequestrin can release hydrated Ca^{2+} so rapidly that pCa can change from 8 to 5 in milliseconds—a factor of 1000 in free Ca^{2+} concentration. These ions can flood the troponin sites along the thin filaments simultaneously and allow very rapid muscle contraction.

Ca^{2+} in blood clotting. Calcium ions are equally important to blood clotting, which is one of the most dramatic examples of biological structure forming. Blood clots when exposed to substances foreign to the body because a protein *fibrinogen* in the blood polymerizes and crosslinks into *fibrin,* the basic clot material, when protective groups on the fibrinogen reactive sites are removed by another peptide *thrombin.* To prevent instant clotting, thrombin does not exist as such in the blood normally. It is a fragment of a larger protein *prothrombin,* which does not have the catalytic effect on fibrinogen. To produce thrombin, prothrombin must be bound to the membranes of injured blood platelets, which contain two protein enzymes that catalyze the fragmentation of prothrombin.

Binding one protein to another is a job for a structure former, and prothrombin contains modified glutamates that have two carboxyls at the end of the side chain, thereby strongly chelating about 10 Ca^{2+} per prothrombin molecule, and these Ca^{2+} also bind to the platelet membranes. This binding is critical, because without it the inactive prothrombin will not stay in contact with the catalysts that produce thrombin from it. Any substance that prevents this calcium binding will in effect be an anticoagulant to prevent clotting. The extra carboxylate group that yields strong Ca^{2+} chelation is introduced to the prothrombin by vitamin K, and species that interfere with binding vitamin K to prothrombin, like dicoumarol or warfarin, yield abnormal prothrombin that cannot bind calcium ions strongly enough to yield thrombin for normal clotting. Dicoumarol in limited doses is administered to heart attack and stroke victims to prevent the formation of clots within the circulatory system, and warfarin in large doses kills rodents by internal hemorrhage that cannot be stopped by normal clotting.

Mg^{2+} in phosphate transfer. The phosphate ion and its polymers and esters are critical life components in two contexts. The free energy source for nonspontaneous processes in all higher life forms is the hydrolysis of one PO_4 group from ATP to yield ADP, and the fundamental genetic materials, polynucleic acids, are phosphate esters. The mechanisms of phosphate transfer are thus extremely widespread in living systems. Unfortunately for simplicity, they are also mediated by a wide variety of enzymes, which do not all function in the same way even for the relatively small number that have been studied. An important feature that we can briefly address here is the fact that most phosphate-transfer enzymes either are metalloproteins or require a metal cofactor. Most prominently, but not uniformly, the metal is magnesium. We have already mentioned alkaline phosphatase, which uses both Zn^{2+} and Mg^{2+} in structure-forming roles.

The polyphosphates and phosphate esters that are biologically important are not very reactive under ordinary biological solution conditions. Apart from enzyme systems, hydrated divalent metal ions are known to assist hydrolysis of these systems. Presumably the effects of the metal ion are those suggested at the beginning of the previous chapter: increasing the acidity of, in this case, the P atom in coordinated phosphate; and organizing the geometry of the coordinated phosphate(s) for intramolecular reaction. We can add, for this case, several more: neutralization of the phosphate negative charge to allow anion approach; delivery of a coordinated OH^- nucleophile to the phosphate by the metal; and orientation of the biophosphate species to fit the appropriate enzyme. For biological Mg^{2+}-dependent systems it is difficult to study the relative importance of these possible functions for two reasons. One is that the s-block Mg^{2+} ion does not respond to most experimental techniques; the other is that Mg^{2+} complexes are so labile in solution that a particular structure is very difficult to isolate. As a result, most of the investigations have used a d-block metal ion to model the Mg^{2+}.

One simple model-compound approach is to consider the effect of metal ions on the hydrolysis of simple organic phosphates. Figure 16.21 shows a phosphate ester of the strong ligand o-phenanthroline, which hydrolyzes in solution only very slowly, and the proposed structure of a Cu^{2+} complex, which hydrolyzes about 300 times faster. This effect presumably represents the increased acidity of the P atom, because the coordinated H_2O in the square planar complex is sterically inaccessible. By comparison, p-nitrophenylphosphate (NPP) hydrolyzes at least 10^5 times faster as cis-$[(en)_2Co(OH_2)(NPP)]^+$ than alone in solution; a speculative mechanism for this process is also shown in Fig. 16.21. The greater rate enhancement (than is seen for the Cu-phen ester) seems to be due to the convenient attack of the coordinated nucleophile OH^-, which is sterically convenient on a cis octahedral site. An interesting variation that may be biologically significant is the transfer of the phosphoryl group from O to N, which is seen in a number of biological mechanisms. $(NH_3)_5Co(NPP)^+$ reacts by what should be recognizable as a classic S_N1CB mechanism to hydrolyze the ester and yield a phosphoramide product. The rate of this process appears to be enhanced over that of the metal-free system by 10^8.

The electron withdrawal from the phosphate P by a metal ion is enhanced, of course, if more than one metal ion coordinates the phosphate. The dimeric complex $[(tn)_2Co(NPP)]_2$, where each NPP phosphate group bridges two Co atoms (tn = trimethylenediamine), hydrolyzes 500 times as fast as $(tn)_2Co(OH)(NPP)$, which is comparable to the species in Fig. 16.21. This suggests that in general it will be advantageous to have multiple metal atoms coordinate a phosphate in a phosphatase enzyme. This is the pattern seen for alkaline phosphatase in section 16.2.

The multiple-metal advantage is seen for polyphosphate hydrolysis as well as for ester hydrolysis. Pyrophosphate, $P_2O_7^{4-}$, is very slow to hydrolyze at pH 7 and room temperature, but the hydrolysis rate can be enhanced by up to 10^7 by cis-

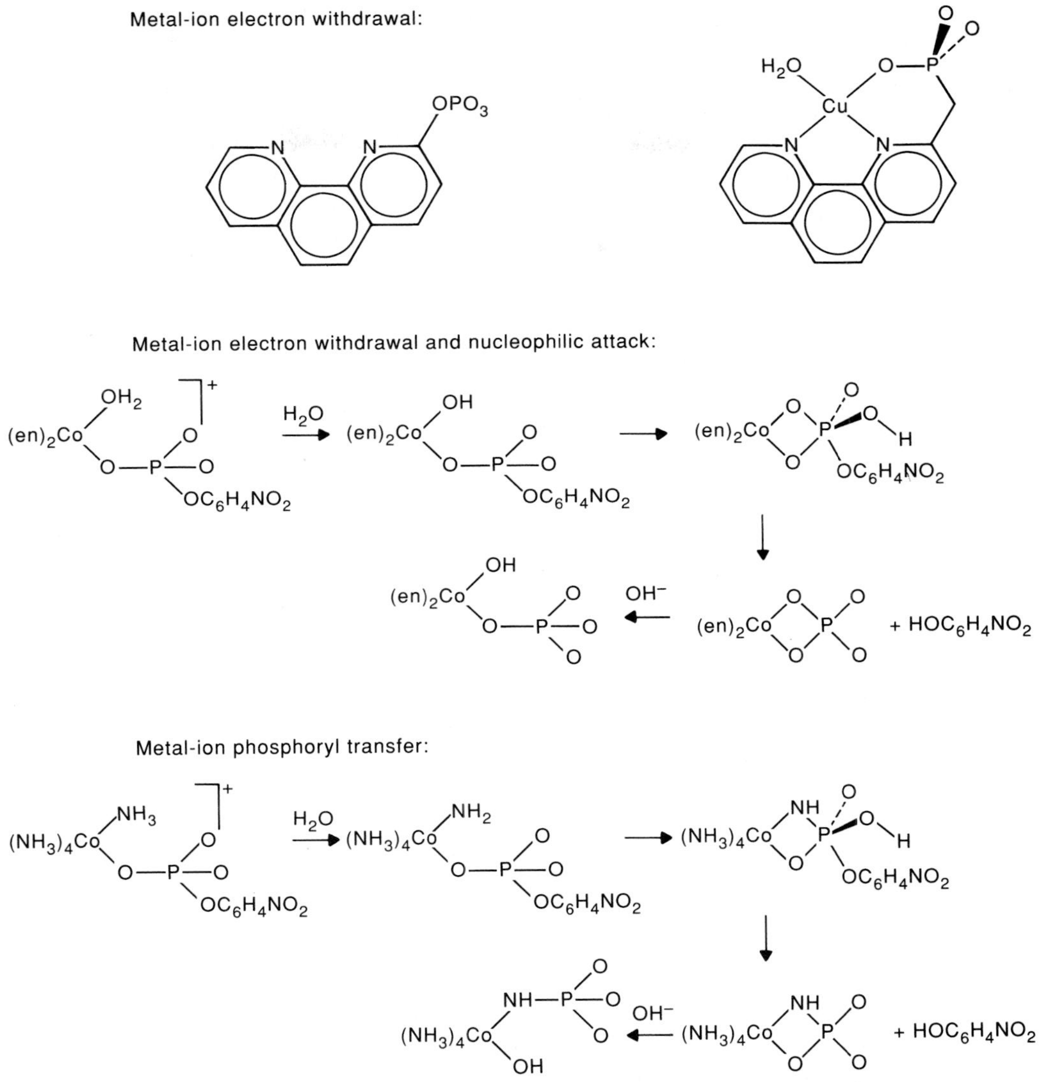
Metal-ion electron withdrawal:
OPO3
H2O
Cu
Metal-ion electron withdrawal and nucleophilic attack:
(en)2Co
H2O
OC6H4NO2
OH−
+ HOC6H4NO2
Metal-ion phosphoryl transfer:
(NH3)4Co
NH3
NH2
NH
H2O
OC6H4NO2
OH−
+ HOC6H4NO2

Figure 16.21 Model compounds and reactions illustrating the role of Mg^{2+} in phosphate hydrolysis and transfer.

$[(tn)_2Co(OH)(OH_2)]^{2+}$. It is not simply a matter of coordinating the $P_2O_7^{4-}$, however. At a stoichiometry of one Co to one $P_2O_7^{4-}$, little enhancement occurs. Two Co per $P_2O_7^{4-}$ is better, but greatest enhancement occurs either by adding three $(tn)_2Co(OH)(OH_2)^{2+}$ per $P_2O_7^{4-}$ *or* by adding two $(tn)_2Co(OH)(OH_2)^{2+}$ to one chelated $(en)_2CoP_2O_7^-$. The structure implied by these results is shown in Fig. 16.22. Note that chelated phosphates exist even after the hydrolysis; without them, the

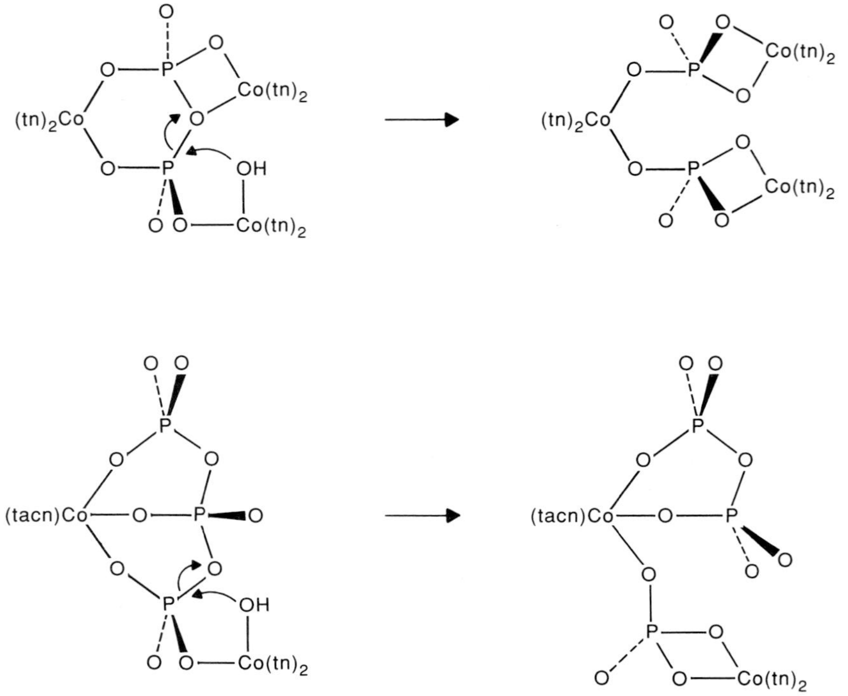

Figure 16.22 Model compounds and reactions illustrating the role of Mg^{2+} in polyphosphate hydrolysis analogous to the ATP $\rightarrow$ ADP + HPO_4^{2-} reaction.

initial chelated pyrophosphate would be too thermodynamically stable to react relative to coordinated but unchelated products. Triphosphates behave similarly: The figure also shows the reaction of a tridentate $P_3O_{10}^{5-}$ coordinated to Co^{III}-tacn (*tacn* = 1,4,7-triazacyclononane). With added $(tn)_2Co(OH)OH_2)^{2+}$, the hydrolysis rate (using the coordinated OH^-) is 10^6 times that for the Co-tacn-P_3O_{10} complex itself in water.

Many *kinase* enzymes mediate the transfer of a terminal PO_3 from nucleoside triphosphates to a phosphoryl acceptor molecule, yielding a free energy release. Nearly all these require Mg^{2+}, and it is assumed that the Mg is chelated by the last two (β, γ) phosphates in the triphosphate, with a mechanism like that indicated in Fig. 16.22. This is supported by the fact that some kinases can function in the absence of Mg^{2+} and nucleoside triphosphates if, instead, the β, γ chelate $[(NH_3)_4 Co(ATP)]^-$ is supplied. Such chelates are chiral at the β P, and the kinases are uniformly able to hydrolyze only one of the two diastereoisomers, indicating strong stereospecificity.

Pt in cisplatin. Metals do not only bind to proteins in the body, and their structure-forming effects are not limited to amino acid side groups and small-molecule

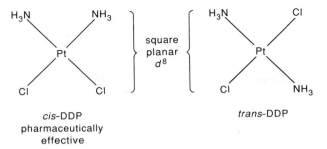

Figure 16.23   The two isomers of diamminedichloroplatinum(II). The *cis* isomer is the anticancer drug cisplatin, while the *trans* isomer has virtually no anticancer effect.

ligands. In 1969 Rosenberg and collaborators described their somewhat serendipitous but dramatic discovery that the century-old complex *cis*-diamminedichloroplatinum(II) (DDP), as seen in Fig. 16.23, was an effective anticancer drug; it was soon approved for clinical use as *cisplatin* and is now a very effective treatment for testicular and ovarian cancers, and for several other types. The effectiveness of *cis*-DDP is due to the fact that it blocks DNA synthesis in cancer cells that would otherwise rapidly divide in a growing tumor. The fact that the *trans* isomer of the same complex is almost completely ineffective in slowing tumor growth indicates a strong stereospecificity and therefore a strong structure forming or recognition on the part of the *cis* isomer.

Lippard and other workers have been able to investigate the process of *cis*-DDP interaction with DNA at an almost unbelievable level of molecular detail, using elegant biochemical techniques that are beyond our scope here. The basic double-helix structure of B-DNA should be generally familiar but is shown in Fig. 16.24 for reference; the double helix has a major groove and a minor groove, both lined with the core of paired bases whose hydrogen bonding holds the double helix together, while the raised "threads" on the helix are the phosphate-sugar ester polymer chain that forms each individual molecular strand. When an intact double helix interacts with a coordinating molecule, that molecule can bind to the phosphates in the outer ridges of the helix, to the bases in the major groove (predominantly to the imidazole N), to the bases in the minor groove (primarily to the pyrimidine N), or for molecules with a large planar segment by intercalating within the axial stack of base pairs (implying a pi acid–base interaction like those described in Chapter 7). Although *cis*-DDP is square planar, it does not intercalate within the base stack. Rather, it coordinates primarily to guanine imidazole N donors. To maintain the square planar coordination that is strongly favored for d^8 metals, DDP must lose at least one ligand. In dilute aqueous solution in the cytosol it hydrolyzes Cl^-, and the resulting weak $Pt-OH_2$ bond can be substituted by a guanine or adenine imidazole N:

$$(NH_3)_2PtCl_2 + H_2O \rightarrow (NH_3)_2PtCl(H_2O)^+ + Cl^-$$
$$(NH_3)_2PtCl(H_2O)^+ + NdnaN \rightarrow (NH_3)_2PtCl(NdnaN)^+ + H_2O$$

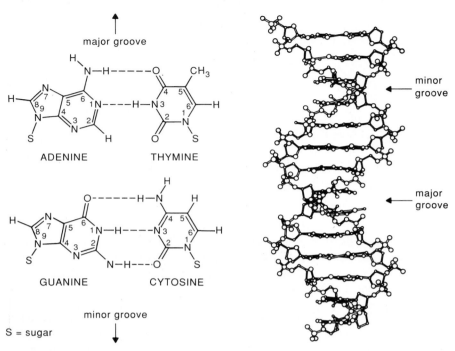

Figure 16.24 The base pairs in a DNA double helix and the helix itself (B conformer). The top of each base pair is open to the major groove and the bottom to the minor groove. [From Bruhn, Toney, and Lippard, "Biological Processing of DNA Modified by Platinum Compounds," *Progress in Inorganic Chemistry* vol. 38, (S. J. Lippard, ed.), Wiley Interscience, 1990. By permission of Wiley Interscience.]

where NdnaN refers to the multiple N donors present within a groove of the DNA double helix. Here only one N donor is bound to Pt, but the second Cl can also be replaced by another N donor so that the Pt cross-links two bases in the double helix:

$$(NH_3)_2PtCl(NdnaN)^+ + H_2O \rightarrow (NH_3)_2Pt(H_2O)(NdnaN)^{2+} + Cl^-$$

$$(NH_3)_2Pt(H_2O)(NdnaN)^{2+} \rightarrow (NH_3)_2Pt(NN'dna) + H_2O$$

Here the NN'dna indicates that two NdnaN bases are chelating the Pt. Although for *cis*-DDP both intrastrand cross-linking and interstrand cross-linking are possible, very little interstrand interaction appears to occur (perhaps 1–7% of total Pt by different estimates) and the very low concentrations of Pt present at normal dosages suggest that DNA synthesis may be blocked almost entirely by intrastrand chelated Pt(NH_3)_2 units. The preferred chelating ligands for the Pt are two adjacent guanine bases on the same strand, which introduces a low level of specificity to the location of Pt along the strand. In general, the Pt centers will bind to the most negative sites along the groove, which suggests a fairly straightforward Lewis acid–base interaction.

The effect of this bonding by (often) adjacent guanine bases into two *cis* sites in the strongly preferred square-planar coordination around Pt is to sharply change the angle between the planes of the base molecules, formerly parallel planes. The axis of the helix kinks by an angle that is variously estimated at 40–70° toward the major groove—a strong structure-forming effect. By comparison, although *trans*-DDP can form intrastrand crosslinks, its geometry will not allow bonding with adjacent bases, and the predicted bend angle from molecular modeling studies is only about 18°. The sharp kink angle for the clinically effective *cis* isomer essentially sterically prevents the DNA polymerase enzymes, which normally move along a strand assembling its complement, from passing the Pt coordination site on the strand. Adducts in which the Pt is coordinated to only one base show little or no bending and do not in general block DNA synthesis, although at least one singly coordinated Pt complex has recently been reported to have this effect. We thus see again a situation in which a coordinated metal has altered its ligands' geometry to change their ability to serve as a substrate for an enzyme.

Because DNA sequencing is such a vital part of biological function, the body has evolved sophisticated DNA repair mechanisms. These fall in several categories; if any of them can "program around" the Pt kink, the therapeutic effect of the cisplatin dosage will fail. The interaction of the repair mechanisms with Pt–DNA complexes is less well understood than the coordination itself, but it appears that *cis*-DDP blocks repair perhaps five or six times as well as *trans*-DDP. Part of the overall repair mechanism is a *damage recognition protein,* which binds to abnormal DNA at the site of damage. Unfortunately for clinical purposes, the *cis*-DDP kink in DNA binds the damage recognition protein effectively while the *trans*-DDP adduct does not.

The major defect of cisplatin as an anticancer drug is that, after treatment and remission of the cancer's growth, resistance to the drug can be acquired. Even less is understood about the origins of this acquired resistance, but at least three changes are known to occur. There is some reduction in uptake of DDP by the cell DNA, though this is much too small to account for the necessary increased doses. After coordination, cross-linking can be prevented by intervention of an S-donor ligand, which as a soft base binds Pt^{II} strongly; it is known that cisplatin-treated cells develop increased concentrations of –SH groups, particularly in glutathione, which is a tripeptide containing cysteine. Finally, increased efficiency of the repair mechanisms that eliminate Pt from the DNA could block its action. It is known that cell sensitivity to *cis*-DDP is proportional to the number of Pt-DNA adducts unrepaired after a period of hours, and that resistant cancer cell lines show increased concentrations of DNA polymerases known to be involved in DNA repair. In designing improved analogs to cisplatin, all of these resistance mechanisms will have to be considered.

Transition metal complexes as nucleic acid structural probes. The discussion of cisplatin has completely, if regretfully, omitted any mention of experimental techniques by which nucleic acid structures and sequences may be established. But in view of the extreme importance of such structural information for genetic as well as medicinal purposes, we cannot leave the topic of metals as structure formers without a brief account of the use of various transition metal complexes to indicate the presence of certain DNA/RNA bases, or the location of protein/nucleic-acid complexes, or the conformation of the double helix itself.

The cisplatin discussion has indicated the various modes by which metals can bind to DNA. Of these, the most useful are major-groove binding to bases, minor-groove binding to bases, and intercalation. As the discussion of Chapters 12 and 13 has indicated, transition metals can adopt a wide variety of coordination geometries and in at least some cases maintain those geometries quite rigidly. It follows that if we choose a test complex properly, we can arrange for it to bind to a nucleic acid in almost any fashion we prefer, making the metal–DNA complex susceptible to the specialized spectroscopic or magnetic techniques that apply to transition metals. Given the geometric preferences of the metals, we can for instance design a metal species that will bridge the two strands of a double helix across the minor groove but not the major groove, then test the complex by spectroscopic or chemical means for the immediate molecular environment of the metal. If the metal species is chiral, we normally find that only one enantiomer will bind to the nucleic acid, giving us a check on the chirality of the nucleic acid itself.

A wide variety of such ingenious probes has been developed and described by Barton and others since about 1975. For example, $Ru(phen)_3^{2+}$ will intercalate a phenanthroline ligand between stacked base pairs, and a charge-transfer electronic transition from metal to *phen* will have its energy and excited-state lifetime altered. $Ni(phen)_3^{2+}$, which binds within the minor groove and is paramagnetic, will broaden the ^1H NMR peak of a nearby base hydrogen atom, indicating its conformation with accuracy. If the double strand has partially separated enough to allow access to the individual unpaired bases, OsO_4 will add across the 5,6 double bond (see Fig. 16.24) of pyrimidine rings and can be seen through electron microscopy. Within the major groove, $Ru(phen)_2Cl_2$ will bind to guanines much as cisplatin does.

A number of structure probes bind to the double helix and cleave it, leaving fragments whose molecular weight can identify the binding sites. Some are photoactivated in that their normal ground state does not cause reaction but the excited state of the metal species does; these are convenient in that the cleavage reaction can be started at any convenient time, and even in some cases inside the cell. $Co(phen)_3^{3+}$ and $Rh(phen)_3^{3+}$ have this ability, and cause sugar cleavage quite generally along the strands. On the other hand, $Ru(phen)_3^{2+}$ binds generally to all bases but cleaves the DNA by photochemical formation of $^1O_2$, which reacts far more with guanine than with any of the other three bases, so that cleavage occurs at guanine sites. A technique called "footprinting" binds chemical cleaving groups along the strand, except that they cannot bind where a protein is already bound; when general cleavage occurs, it will occur everywhere except at the protein site, and this yields structural information on that binding.

DNA double strands have a natural twist caused by the geometric requirements of the base pairing and the sugar-phosphate ester chains. However, there is more than one way to construct such a double helix. Figure 16.25 shows the three known conformations of the DNA double helix, known as the *A, B,* and *Z* confor-

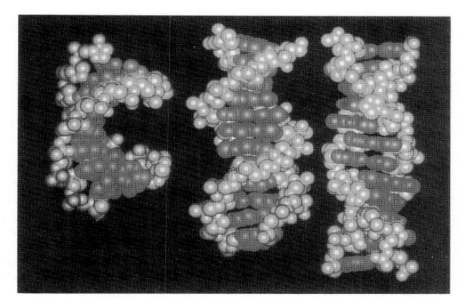

Figure 16.25 The A, B, and Z conformers of DNA double-helix strands. Central dark atoms are the base pairs, light atoms the sugar-phosphate chains. [From Pyle and Barton, "Probing Nucleic Acids with Transition Metal Complexes," *Progress in Inorganic Chemistry* vol. 38 (S. J. Lippard, ed.), Wiley Interscience, 1990. By permission from Wiley Interscience.]

mations. The A and B forms are right-handed, while Z is left-handed. A has the base pairs bent out toward the minor groove side and tilted with respect to the polymer axis, yielding a wide, shallow minor groove, a very deep major groove that is relatively inaccessible from a surrounding solution, and a generally twisted conformation; it is often seen for RNA double strands. The B form is common for DNA double strands. It has centered base pairs perpendicular to the helix axis, with the familiar major and minor groove conformations. The Z form spaces the stacked base pairs farther apart than the others and yields a very broad, flat major groove and narrow minor groove.

A number of metal species can cause redox cleavage of nucleic acid strands, but only if they can bind to the strands. $Fe(EDTA)^{2+}$, $Cu(phen)_2^+$, and $Co(NH_3)_6^{3+}$ have this effect; they can be tethered to groups that bind to specific sites for fragmentation as described above, mostly for the B conformation. Some other groups, however, have been shown to bind only to specific alternate conformations of the double strand and cleave them, such as $Ru(tmp)_3^{2+}$ (where *tmp* is 3,4,7,8-tetramethylphenanthroline). The four methyl groups prevent intercalation and the whole assembly is too large to fit the minor groove of the B conformation. It does bind, however, to the broader minor groove of the A conformation, so that cleavage by this reagent is a check for the presence of the A conformation. Similarly, $Co(dip)_3^{3+}$ binds to and cleaves only the Z conformation. These and other species promise to yield patterns of structural information about nucleic acids that have never before been available; they represent a prospectively invaluable application of the structure-forming property of metal ions for their complexes.

16.7 METAL IONS AS CHARGE CARRIERS

The last remaining category to be discussed here of biochemical roles for metal ions from the list at the beginning of the chapter is that of charge carrier. Although the net charge on any metal ion is important in determining structures and stabilities, biochemical systems do not usually use transition metal ions for this purpose. Charge transport is usually carried out by H^+, Na^+, K^+, Mg^{2+}, and Ca^{2+}, with the divalent ions showing the structure-forming property as well. The whole charge transport role for a species the size of a human is substantial. All of the free energy for life processes comes from an electron-transfer process: $O_2 + 4e^- + 4H^+ \rightarrow 2H_2O$. If this electron transfer is viewed for simplicity as an electrical current, we can calculate the total amount of charge transport that must occur for the human organism. A resting human will use O_2 equivalent to about 0.2 L per minute (STP). This is about 9 millimoles of O_2, equivalent to 36 millimoles of electrons per minute, to 0.6 millimole of electrons per second, or to 60 amperes of current. Many houses have a main circuit panel that has a maximum capacity of 60 amperes, so this is a substantial current. In the body there is no electronic conduction, only ionic conduction, so all of this electrical transport is being carried out by the ions listed above as charge carriers.

Much of the charge transport in the body is carried out by proton transfer, but that is a topic so extensive and diverse as to be beyond our scope here; we shall limit ourselves to an examination of metal ion transport. The central fact of group Ia and group IIa ion distribution in the body is that cell wall membranes use much of the overall ATP energy expenditure of the body to maintain a steep concentration gradient between Na^+ and K^+, and between Mg^{2+} and Ca^{2+}. Although the actual concentrations differ substantially among different types of cells, it is quite generally true that $[Na^+_{inside}]/[Na^+_{outside}]$ for a typical cell membrane is roughly 0.1, but conversely $[K^+_{inside}]/[K^+_{outside}]$ is about 20. Similarly, $[Mg^{2+}_{inside}]/[Mg^{2+}_{outside}]$ is very roughly 100, while $[Ca^{2+}_{inside}]/[Ca^{2+}_{outside}]$ is about 0.001.

Sodium and potassium are thought to be exchanged in opposite directions through the cell membrane by a single structure, the Na–K pump, driven by an integral enzyme *Na–K ATPase* which uses the energy yield from ATP hydrolysis to move Na^+ and K^+ in opposite directions. Figure 16.26 suggests a schematic structure for the pump; it is known to involve two subunits α with molecular weight about 112,000 and two more (β) with molecular weight about 40,000. The larger unit contains the ATPase. In its function, three Na^+ from inside the cell bind to sites on the α units consisting probably of six hard-base O atoms. This coordination changes local polarities so that it is favorable to coordinate an ATP molecule into the ATPase, which hydrolyzes it to ADP and a covalently bound phosphate ester. Phosphorylation changes the conformation of the α unit, presumably through

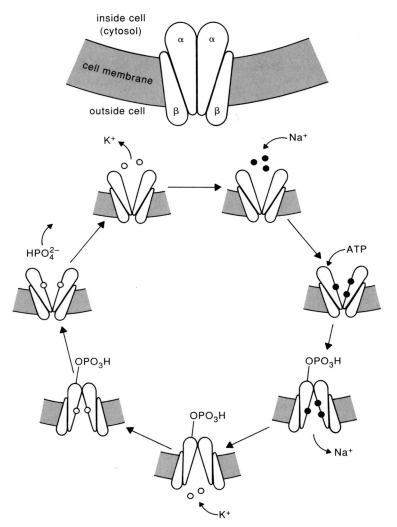

Figure 16.26 Schematic representation of the sodium pump structure in cell walls and of the process through which the pump maintains opposite Na^+ and K^+ concentration gradients across the cell wall.

the strong structure-forming effect of the internal Mg^{2+} associated with the ATPase. The new conformation is open to the outside of the cell and does not bind Na^+ as strongly as before, so the Na^+ ions are lost to the extracellular fluid. The open channel binds two K^+ from solution, probably to seven O atoms or to seven O and one N. Potassium binding causes dephosphorylation of the α unit, losing monophosphate to the intracellular fluid and changing the bistable conformation of the α_2 system back to the form in which it is open at the inner end. In this conformation K^+ is less strongly bound, and it is lost to the cytosol inside the cell leaving the $\alpha_2\beta_2$ channel in the conformation to accept Na^+ again. The free energy change necessary to maintain the two overall concentration gradients can be calculated ($\Delta G = -RT \ln Na_{out}K_{in}/Na_{in}K_{out}$), and it is slightly smaller than the free energy release from the hydrolysis of one ATP—so the process is thermodynamically consistent.

The best-known application of Na^+ and K^+ ions in charge transport processes is the transmission of nerve impulses through the network of nerve cells, *neurons*. Nerve impulse transmission obviously requires transfer of information by chemical means from one cell to another, so neurons have properties adapted to this purpose. A neuron is an irregularly star-shaped cell with a number of extensions (*dendrites*) that can receive chemical information and one transmitting extension known as an *axon*. The junction between an axon from one cell and a dendrite from another (only about 500 Å wide) is a *synapse,* which is the location of actual chemical transfer. The nerve signal is an electrical potential, carried by the movement of the group Ia ions.

At rest a neuron, like other cells, has a large K^+ concentration gradient from inside to outside the cell, which can be expressed as the electrical potential equivalent to the free energy change for the gradient, about −70 mV. Its cell wall contains numerous ion channels, which unlike the Na−K pump are specific for a single ion each: Na^+, K^+, Ca^{2+}. The cell wall of the resting neuron is strongly polarized, which effectively binds these ion channels shut. The Na^+ concentration outside the cell is high, and if the membrane becomes partially depolarized sodium ions will begin to flow into the channels, depolarizing it still further and creating a rapid cascade of depolarization along the neuron wall. This can last only until the sodium channels are completely open and the inner and outer concentrations have equalized, bringing the cell potential up to +30 mV. At this potential the sodium channels close and potassium ion channels open, releasing K^+ from the neuron and restoring the negative potential. Normal ion pump processes rapidly rebuild the normal concentration gradients. But in the process, electrical depolarization has spread rapidly from one end of the long nerve cell to the other, in particular to the axon. Different junctions have different transmitter molecules, but perhaps the best characterized are the *cholinergic* junctions, which use acetylcholine, a quaternary ammonium ion: $H_3CCOOCH_2CH_2N(CH_3)_3^+$. Depolarization releases the ammonium ion, which diffuses across the synapse and lodges in receptors, partially depolarizing them. This in turn can trigger depolarization of the second nerve cell by Na^+, continuing the impulse. The acetylcholine is removed from the receptors by reaction with *acetylcholinesterase,* which leaves the receptors free to accept another signal. Although the literal chemical transfer of charge from one neuron to another has involved a quaternary ammonium ion, most of the spatial transfer of electrical potential has been due to Na^+ and K^+ transfer through the neuron wall.

Cation carrier species (valinomycin and nonactin). The alkali−metal cation transfer processes we have just described have involved movement of bare or par-

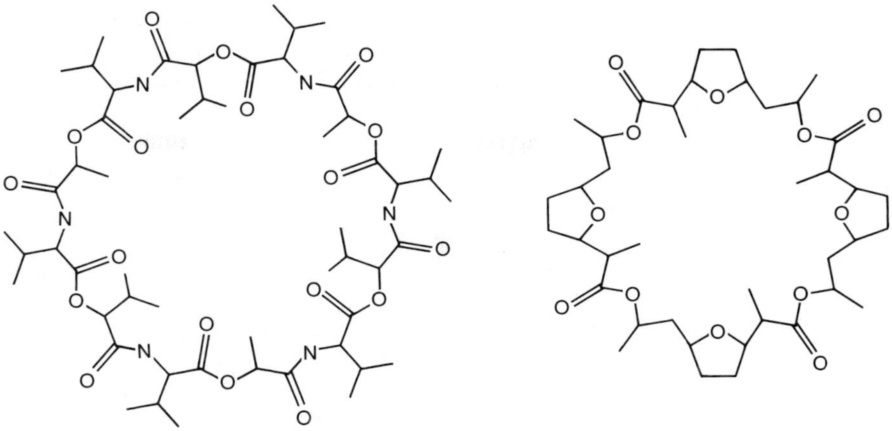

Figure 16.27 Structures of ionophore antibiotics valinomycin and nonactin.

tially hydrated M^+ ions through ion channels in the cell wall. However, some naturally occurring molecules are known that have the ability to serve as phase-transfer catalysts like the crown ethers, complexing M^+ in an organic sheath that allows it to dissolve in nonpolar regions of the cell. Two of these that have been characterized are *valinomycin* and *nonactin,* with the structures shown in Fig. 16.27.

Valinomycin is a cyclic protein with 12 peptides, consisting of three repeated sequences L-valine/D-α-hydroxyisovaleric acid/D-valine/L-lactic acid: [L-Val-D-HyV-D-Val-L-Lac]$_3$. This yields a 36-membered flexible macrocycle that can adopt quite varied geometries depending on the polarity of its surroundings and the presence or absence of a metal ion coordinated within it. Just as a helical protein is held in its conformation by hydrogen bonds between nonadjacent amide groups, valinomycin can form a variety of intramolecular hydrogen bonds across the macrocycle giving it different conformations. In nonpolar solvents analogous to lipid phases it has a ring of six amide-to-amide N–H···OC hydrogen bonds with the rest of the carbonyls pointing outward and the macrocycle looping up and down around the hydrogen bonds. In moderate-polarity solvents three such hydrogen bonds squeeze the molecule into a triangular, approximately planar form with nonpolar isopropyl groups forming a hydrophobic core to the molecule; in water or strongly polar solvents the molecule loses intramolecular hydrogen bonds entirely, forming them instead to solvent molecules and leaving the flexible chain with no fixed conformation; and in the crystal the unmetallated molecule forms a square of four internal hydrogen bonds, with two more hydrogen bonds bridging loops of the molecule on opposite sides of the square, keeping polar atoms inside the molecule and nonpolar groups such as isopropyl around the outside. Finally, in the 1:1 K^+ complex valinomycin loops around the metal ion so as to form an approximate octahedron of carbonyl O atoms as hard-base donors, with the macrocycle folding into approximate S_6 symmetry similar to the cryptand geometry shown in Fig. 7.10 (but missing one side chain). This set of donors and ligand-molecule geometry resembles that of the crown ethers also discussed in Chapter 7.

Nonactin, also seen in Fig. 16.27, is also a naturally occurring cyclic polyoxygen donor ligand, but is a polyester rather than a polyamide. The unmetallated form, in the crystal, has the four oxygens from the tetrahydrofuran rings in a square near the center of the molecule, with the four *thf* ring planes forming a sort of square box open at the top and bottom; the macrocycle chain loops around this box with approximate S_4 symmetry, the carbonyl oxygens pointing outward from the periphery. When nonactin coordinates an alkali metal ion (even NH_4^+), its conformation changes dramatically so that the four carbonyl O and four ether O atoms form a nearly perfect O_8 cube around the metal ion, the 32-member macrocycle adopting an S_4 baseball-seam symmetry. In this conformation the outside of the molecule is extremely nonpolar.

These are not the only naturally occurring molecules that coordinate alkali—metal ions much as the crown ethers do; quite a number, known generally as *ionophores,* have been described, though for most the structures have not been as well characterized as for valinomycin and nonactin. Because they are able to serve as phase-transfer catalysts, all of these can carry both Na^+ and K^+ through the nonpolar cell membrane in either direction. As a result, the concentration gradients described above are destroyed and for a one-celled animal such as a bacterium the result is fatal. These species are thus antibiotics. However, they have virtually no pharmaceutical applications because they have as bad an effect on the host animal's cells as on the bacteria, for exactly the same reason.

Calcium ion carrier (calmodulin). Calcium ions are believed to trigger many biochemical processes by arriving at an enzyme (for example) that is in an inactive conformation but is changed to an active conformation by the structure-forming effect of the Ca^{2+} ion. Calcium transport is thus very important in almost every kind of cell. However, it will not do to have hydrated calcium ions floating about within the cell; not only would this lead to a permanent structure-forming effect by the free Ca^{2+}, but the very low solubility of calcium phosphates would destroy many of the phosphate esters present in many roles within the cell. The cell needs a calcium carrier species, something like calsequestrin but operating on a smaller scale so as to have the capability to deliver one or two calcium ions to a Ca^{2+}-dependent system. In most eukaryotic cells this is met by the protein *calmodulin,* which is able to maintain the hydrated Ca^{2+} concentration below 10^{-7} M even though the total Ca concentration in the cell is usually about 10^{-3} M.

Calmodulin is a small protein with a molecular weight of about 17,000 daltons. It has a very large number of carboxylates (aspartate and glutamate) present and is quite acidic as a result. Its amino acid sequence is known, and there are sites along the chain for four Ca^{2+} ions. When four Ca^{2+} are bound to calmodulin, it adopts a dumbbell-shaped conformation in which two Ca^{2+} are bound near each other at each end and the two Ca_2 segments are connected by a helical strand of protein. However, the calmodulin geometry changes substantially on binding (or removing) each successive Ca^{2+}, and as a result the molecule is able to bind with Ca-dependent enzymes that receive several different geometries. As a result of this binding the enzyme is activated:

$$nCa^{2+} + CaM \rightarrow CaM \cdot (Ca^{2+})_n \rightarrow \underset{\text{Activated form}}{CaM^* \cdot (Ca^{2+})_n}$$

$$CaM^* \cdot (Ca^{2+})_n + Enz \rightarrow CaM \cdot (Ca^{2+})_n \cdot Enz^*$$

For example, the enzyme *myosin kinase,* which mediates the myosin-actin interaction in muscle described earlier, is calmodulin-dependent in just this fashion. A very wide variety of enzymes are calcium/calmodulin-dependent in this way. Phenothiazine tranquilizers and a variety of insect venom proteins appear to work by inhibiting calmodulin's action.

PROBLEMS

A. DESCRIPTIVE

A1. Plants contain significant amounts of Si in addition to the metals we have described. Based on the elemental properties of silicon, what coordination would you expect for Si within the plant? What role should Si play in the biochemical system? In view of silicon's abundance in the earth's crust, why is it not seen in more applications or in animals?

A2. A previously unknown protein extracted from cells in the spleen is found to have a molecular weight of about 66,000 daltons and contains 2.2 ± 0.6 Fe, 3.5 ± 0.7 Zn, and very roughly 16 Mg by elemental analysis. What experimental techniques could be applied to the determination of the structure and function of the protein? Comment on the information available from each technique. What experimental problems can be expected to arise?

A3. What kinds of experimental data could lead to the construction of a structural model for a nickel enzyme?

A4. What kinds of structural or electronic effects could cause pK_a for the Zn^{2+} reactive group in carbonic anhydrase to be lower than the values commonly seen for $Zn^{2+}—OH_2$?

A5. If Fe^{3+} in oxidized rubredoxin is bound ionically to its S ligand atoms (to keep the electronic bookkeeping simple), should it be high-spin or low-spin? What about Fe^{2+} in the reduced form? What geometric differences between the two forms might be expected from this analysis? Is this consistent with observation? How does this analysis bear on our understanding of the system?

A6. For the octahedral Fe^{4+} ferryl complex postulated in the cytochrome a_3 O_2 reduction cycle, what magnetic properties should be expected? Comment on your reasoning.

A7. Suggest a mechanism at the coenzyme B_{12} Co atom for the diol dehydrase interchange of H and OH in $RCH(OH)–CH(OH)H$, by analogy with organometallic catalytic rearrangements in previous chapters.

A8. If a single-Cu blue-copper protein undergoes reduction, what should happen to its UV–visible absorption peaks?

A9. Suggest a mechanism for the Cu,ZnSOD disproportionation of O_2^- in which the first step is oxidation of Cu^{2+} to Cu^{3+}. How could this mechanism be experimentally distinguished from the accepted one?

A10. Show that the coordination geometries seen in $M—N_2$ complexes are consistent with predictions from simple MO theory.

A11. Sketch a trigonal prism corresponding to the probable geometry of the six iron atoms at the center of the FeMoco iron-molybdenum cluster in *A. vinelandii* nitrogenase. Indicate bridging groups as described in the chapter. Assuming that the Y bridging atom described can be replaced by N_2 for the reduction process that fixes nitrogen, describe the probable binding geometry for N_2 with respect to the trigonal prism. Use data in this and earlier chapters to estimate the dimensions of the prism with N_2 bound.

A12. What sort of AO overlap would account for the 75° bond angle at S in the 2-Fe ferredoxin model compound in Figure 16.3?

A13. In lacunar O_2-carrier Co complexes, what effects can a bridging hydrocarbon chain have on the strength of O_2 binding?

A14. Several hemocyanins show an O–O stretch absorption about 745 cm^{-1}. What type of bonding does this suggest? Are the hemocyanin Cu atoms stronger or weaker reductants than the Fe atoms in hemerythrin?

A15. What amino acids are likely to be present at the Ca^{2+} binding sites of troponin C in skeletal muscle?

A16. Cisplatin is a classic d^8 square-planar complex, *cis*-$PtCl_2(NH_3)_2$. Cis geometries of $NiCl_2(NH_3)_2$ and $PdCl_2(NH_3)_2$ are both known, but the Ni complex is thought to be bridged to a distorted octahedral form in the solid state, and the Pd complex is more labile than Pt with respect to ligand substitution. What differences should the changes bring to DNA binding by the Ni and Pd complexes?

B. NUMERICAL

B1. Analyze the spectrum of Co^{2+}-carboxypeptidase A by assuming it to be tetrahedral, making reasonable simplifying assumptions, and solving for Dq and B'. Comment on your assumptions. For the groups known to coordinate Co^{2+}, how does Dq compare with the value that could be predicted by crystal field theory (Chapter 11)? Does the B' value indicate predominantly ionic or covalent bonding of Co^{2+} to the ligand groups? How do we know from the spectrum that the coordination is distorted?

B2. If in the ferritin core each Fe^{3+} occupies an octahedral hole in a close-packed array of oxides and the stoichiometry (ignoring phosphates) is FeO(OH) [i.e., 1 Fe:2O], calculate the density of the core. Use Shannon crystal radii.

B3. A cell membrane can be thought of as a single-ion electrode for K^+. Calculate the concentration gradient equivalent to a 70-mV potential from inside to outside the cell.

B4. The irreversible oxidation of unprotected heme in solution to Fe^{III}—O—Fe^{III} is kinetically first-order in O_2 and inverse-second-order in the axial ligands L [in $Fe^{II}(Por)L_2$]. Show that the mechanism proposed in the chapter is consistent with this finding.

B5. Lactate that accumulates in muscle tissue after exercise must be oxidized to pyruvate for elimination. Ultimately this is the redox responsibility of O_2, but in an intermediate sense it is carried out by cytochrome c. From the $\mathscr{E}^0$ data below (pH 7) calculate ΔG^0 and the equilibrium constant for

$$2\text{cyt } c^+ + \text{lact} + 2\text{H}_2\text{O} \rightleftharpoons 2\text{cyt } c + \text{pyruv} + 2\text{H}_3\text{O}^+$$

$$\text{cyt } c^+ + e^- \rightarrow \text{cyt } c \qquad\qquad \mathscr{E}^0 = -0.25 \text{ V}$$

$$\text{pyruv} + 2\text{H}_3\text{O}^+ + 2e^- \rightarrow \text{lact} + 2\text{H}_2\text{O} \qquad \mathscr{E}^0 = -0.19 \text{ V}$$

C. EXTENDED REFERENCE

C1. A. Desbois *et al.* [*Inorg. Chem.* (1989), *28,* 825] report resonance Raman studies of O_2 binding to Fe in hemoglobin/myoglobin model compounds similar to picket-fence porphyrin but with a bridging hydrocarbon chain across the top of the porphyrin. Shortening the bridging chain raises the Fe—CO stretch frequency for the CO complex, but has no effect at all on the Fe—O_2 frequency for the O_2 complex. Why should these two ligands behave differently? As $\nu_{\text{Fe-CO}}$ rises, $\nu_{\text{C}\equiv\text{O}}$ falls; account for this behavior in electronic terms.

C2. In *Acc. Chem. Res.* (1990), *23,* 419, Malcolm Chisholm reviews a series of d^3–d^3 dimer complexes with a Mo_2 or W_2 core. These have a suggestive relationship to structures that might be required for a six-electron N_2-fixing catalyst, though Chisholm does not discuss this possibility. Of the structures reviewed in his article, which appear to be best suited to N_2 coordination and reduction? Comment on structural and electronic reasons for your answer.

C3. Oxygenases have chemical properties similar to peroxidases except that they start with molecular O_2 rather than H_2O_2. The best characterized oxygenase is cytochrome P-450, which has no Cu atom to aid the heme Fe as in cytochrome a. Nonetheless, it can convert alkanes to alcohols:

$$\text{RH} + O_2 + 2\text{H}^+ + 2e^- \rightarrow \text{ROH} + \text{H}_2\text{O}$$

In *Science* (1988), *240,* 433, Dawson outlines a reaction sequence for this process; compare his cycle and discussion to that in Figure 16.6. What structure does he propose as the functional equivalent of Cu in splitting the O_2O–O bond?

C4. W. F. Nijenhuis *et al.* [*J. Am. Chem. Soc.* (1991), *113,* 7963] report on a series of possible model compounds for valinomycin, comparing their selectivity for K^+ over Na^+. Their studies indicate that, from an aqueous phase with $[K^+] = 0.01$ M, $[Na^+] = 0.10$ M, valinomycin has a K^+ flux into a lipid membrane at least 50 times the comparable Na^+ flux. Dibenzo-18-crown-6 had a K^+ selectivity of about 7×, while calix[4]arenes (shown in the article) would not transport K^+ but a calix[4]crown-5 carrier had a K^+ selectivity of over 20×. From the structures shown in the article, what reasons can you propose for this pattern of selectivities?

17

Photochemical Reactions of Transition Metals

All chemical reactions must be carried out under conditions that make them thermodynamically possible, but they must also (of course) be kinetically feasible. The right atoms must be able to meet each other (the stereochemical aspect of the reaction's mechanism) and the necessary activation energy must be supplied. Up to this point we have concentrated on reactions and mechanisms for which the activation energy was supplied thermally—that is, by raising the temperature to pump kT into the reaction. We now turn to a rapidly developing experimental technique, photochemistry, in which the activation energy for a reaction is supplied as radiation energy $h\nu$. We shall briefly sketch some of the basic theoretical background for the photochemistry of transition-metal atoms, and then consider a few important cases.

Thermal excitation of reactions is a statistically broad, almost "shotgun" method of providing energy to a molecule. There is no way to bring a collection of molecules to a temperature of 643 K without first passing through 529 K and all the other intermediate temperatures, and for any statistically significant group of molecules there will be a wide spread of individual molecular "temperatures." Photochemical excitation can be much more precise, however. Because $E = h\nu$, a monochromator can provide a molecule with any desired amount of energy, very specifically and with very little spread. It is thus possible to have a "hot" molecule without ever having had a "warm" molecule. There is an obvious limitation on photochemical excitation: The radiation does no good unless the molecule absorbs it, which means that the molecule must have an accessible electronic excited state at an appropriate energy above the ground state in order to absorb the incoming $h\nu$. This in turn means that the selection rules for spectroscopic transitions described in Chapter 11 limit the kinds of photochemical excitation that can occur.

In favorable cases, the selectivity of photochemical excitation can be used to control the course of a reaction for a transition-metal complex. This is surprising at first glance, but is only an example of the kinetic control of reaction products. Consider the complex $[Co(NH_3)_5(N_3)]^{2+}$. As a d^6 Co^{3+} complex, it has several d–d transitions, though all but the lowest-energy transition are hidden under an intense charge-transfer transition in which the excited state has in effect transferred an electron from the azide ligand to the Co atom. If the d–d peak is irradiated, aquation of a NH_3 ligand occurs:

$$[Co(NH_3)_5(N_3)]^{2+} + H_2O \xrightarrow{LF} [Co(NH_3)_4(OH_2)(N_3)]^{2+} + NH_3$$

where LF indicates irradiation at a frequency corresponding to the isolated d–d or ligand-field absorption peak. However, if the charge-transfer band is irradiated instead, the products are quite different:

$$[Co(NH_3)_5(N_3)]^{2+} \xrightarrow{CT} Co^{II} (\text{probably } [Co(NH_3)_4]^{2+} \text{ initially}) + \cdot N_3$$

We have thus changed a substitution reaction to a redox reaction by simply changing the exciting frequency from green to ultraviolet. This degree of choice is unusual, but we shall briefly survey the kinds of photoreactions that can be observed.

Perhaps an even more remarkable feature of inorganic photochemical reactions is that they often produce entirely different products from those obtained with thermal excitation from the same reaction system. For example, the bromopentamminechromium(III) complex is kinetically inert, but it aquates the bromide ion if heated without photochemical excitation (a so-called *dark reaction*):

$$[Cr(NH_3)_5Br]^{2} + H_2O \xrightarrow{kT} [Cr(NH_3)_4(H_2O)]^{3+} + Br^-$$

However, the photochemical reaction of the same starting aqueous solution aquates predominantly the coordinated ammonia ligand:

$$[Cr(NH_3)_5Br]^{2+} + H_2O \xrightarrow{h\nu} [Cr(NH_3)_5(H_2O)Br]^{2+} + NH_3$$

In other cases the products are stoichiometrically the same, but the photochemical reaction yields different geometric isomers from the thermally excited reaction (*cis*- instead of *trans*-, for instance). Such photochemical reactions are said to be *antithermal* (even though light is not the opposite of heat!).

There are also some important applications of photochemistry, both actual and potential, in which the absorption of light rather than the chemical products are of interest. The capture of solar energy, for instance, is a photochemical process. Biochemical systems achieve it through chlorophyll, but some interesting attempts have been made to decompose water into H_2 and O_2 using solar energy and transition-metal catalysts. We shall examine both of these, and the photochemistry of the silver photographic process as well.

17.1 BASIC PHOTOCHEMICAL PROCESSES

We shall briefly survey photophysical and photochemical processes before we consider specific photochemical reactions. As we suggested, the first step in any

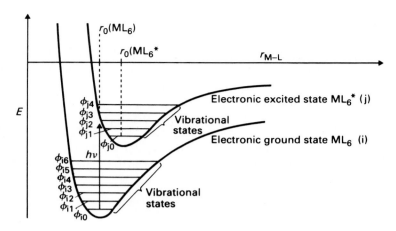

Figure 17.1 Potential-energy functions and energy levels for an electronic transition of an ML_6 complex.

photophysical process must be the absorption of the incident light. This means that the atom or molecule undergoes a transition from an electronic ground state to an electronic excited state (some of the possible states were described in Chapter 11). It is likely, however, that with the changed electron distribution, the excited state will not have the same equilibrium geometry as the ground state. The electronic transition is very rapid compared to the speed of vibrational nuclear motion (the Franck–Condon principle). Therefore, the excited electronic state will usually be in a vibrationally excited geometry, corresponding to a vibrationally excited state of the electronic excited state. Figure 17.1 shows a fairly typical case. Each potential-energy well corresponds to an electronic state, but the complex ML_6 has a different equilibrium M—L distance in the two states. Although the electronic ground state is also in its vibrational ground state, the electronic transition is so rapid that it is drawn as a vertical line (and is thus called a *vertical transition*). The vertical transition leads to a vibrational excited state of the electronic excited state. Of course, rotational excitation is also possible, but rotational energy levels are normally so close together that they are not considered on the scale of the energies involved here. In the usual notation, both the electronic and vibrational energy levels are designated as in Fig. 17.1, where ϕ_{i0} indicates the vibrational ground state of the ith electronic state. The transition shown is primarily to the state ϕ_{j2} from the state ϕ_{i0}, and is indicated as $2 \leftarrow 0$.

Figure 17.2 compares the potential-energy diagrams of two complexes: Part (a) is the diagram for a complex MX_6 like that in Fig. 17.1, and part (b) is the diagram for a complex MY_6 in which the excited-state geometry is essentially the same as the ground-state geometry for the complex. It can be seen that the $0 \leftarrow 0$ transition is most probable if the geometries are the same, and much less likely if the geometries differ. If the frequency $\nu_{0\leftarrow 0}$ is considered the true energy differ-

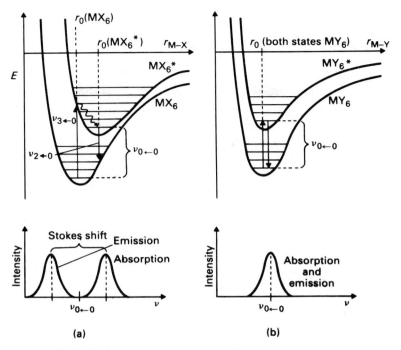

Figure 17.2 Radiation absorption and emission for complexes with different ground-state/excited-state geometric relationships.

ence between the two electronic states, the observed absorption peak will in general be at higher frequencies unless the electronic states have the same geometry.

Once the excited state has been achieved, several things can happen. Immediate relaxation back to the ground state by radiation is unusual, particularly for molecules in a solid or liquid. In such dense media, collisions with neighbors redistribute vibrational energy so effectively that emission almost always occurs from the vibrational ground state of the electronic excited state (Fig. 17.2a). Just as absorption occurs at a higher frequency than $\nu_{0\leftarrow 0}$, emission will usually occur at a lower frequency because it returns the molecule to a vibrationally excited state of the electronic ground state. The energy or frequency difference between an absorption peak and the corresponding emission peak is known as the *Stokes shift*. It should be apparent that the magnitude of the Stokes shift is a fairly sensitive guide to the geometric differences between the ground state and the excited state.

The vibrational ground state of the electronic excited state from which emission occurs in condensed phases is an important intermediate, because it often lives long enough to be a characterizable chemical species. It is sometimes referred to as the *thexi state* (from *thermally equilibrated exci*ted state). It has obvious spectroscopic importance, but it can also be the key precursor to a photochemical reaction (though these can also occur directly from the initial Franck–Condon excited state).

When there are several possible excited electronic states (as is usually the case) other processes can occur that involve transitions between excited states. These are governed by the same selection rules that limit absorption by the ground state. For example, interconversions between excited states having the same spin are facile. These are often radiationless transitions that lead to the lowest-energy excited state of a given spin multiplicity. A generalization to cover this behavior is *Kasha's rule:* The principal emission for a molecule will occur from the lowest-energy excited state of a given spin multiplicity. These radiationless transitions between states of the same spin are called *internal conversions;* precisely because they do not involve radiation, they are not limited in rate by the quantum-mechanical factors that limit radiative processes.

Radiationless transitions between excited states of different spin are called *intersystem crossings.* They are distinguished from internal conversions because they have a much lower probability and are thus much slower. Because the original excited state can usually disappear rapidly by other transitions, possible intersystem crossings are often not observed, though their rates are of some interest. The possibility of an intersystem crossing seems to depend on the extent of spin-orbit interaction, which can allow the spin pairing or unpairing to occur. Figure 17.3 indicates (somewhat schematically) a sequence of molecular events that can lead to an intersystem crossing: At *A,* the molecule is in its singlet electronic

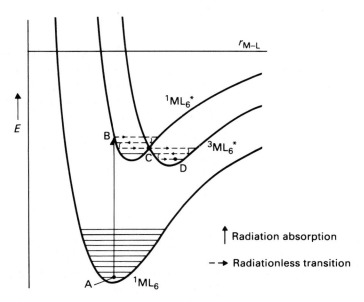

Figure 17.3 Potential-energy functions for an ML_6 complex undergoing an intersystem crossing from $^1ML_6^*$ to $^3ML_6^*$.

ground state. It undergoes a vertical transition to B, an excited vibrational state of the singlet electronic excited state. It then vibrates and relaxes to a lower vibrational state until, at C, it reaches an energy and molecular conformation for which the singlet excited state and a triplet excited state are equivalent. At C, spin-orbit coupling allows the spin conversion, and the molecule continues to relax to D, the vibrational ground state of the triplet electronic excited state.

Emission or luminescence from excited states can occur either from an excited state with the same spin as the ground state (*fluorescence*) or from an excited state with a different spin from the ground state, produced by an intersystem crossing. Because this emission is forbidden by the spin selection rule, it tends to take a much longer time and is called *phosphorescence*.

Another possibility for the disappearance of excited states (and the one we shall be most interested in) is a photochemical reaction from the reactant's excited electronic state into an electronic state characteristic of a new molecule, the product. We would, of course, like to see a photochemical reaction occur for each reactant molecule that arrives in the excited state, but in the perverse nature of things this rarely occurs. Photochemical reactions are usually characterized by their *quantum yield* Φ, defined as

$$\Phi = \frac{\text{Molecules reacting in the desired manner}}{\text{Photons absorbed by the reactant molecules}}$$

"Photons absorbed" refers to the production of a specific excited state of the reactant molecule. Therefore, the quantum yield should be described in terms of the specific wavelength or frequency of the exciting radiation and also in terms of the intensity of the incident exciting radiation (since the efficiency of absorption is also of interest in characterizing the overall reaction).

For transition-metal molecules in general, several categories of photochemical reaction are commonly observed. We shall briefly summarize them here with a few examples to suggest the overall range of possibilities, then consider each in more detail in later sections of the chapter. Perhaps the simplest reaction is the *photodissociation* of a ligand, which is widely observed for carbonyls:

$$Cr(CO)_6 \xrightarrow{h\nu} Cr(CO)_5 + CO$$

This reaction can be continued to yield $Cr(CO)_4$ and $Cr(CO)_3$. From our earlier discussion of carbonyls it should be obvious that the influence of the 18-electron rule on the bonding makes all of these product molecules spectacularly coordinatively unsaturated, so much so that only the greatest experimental care will prevent the reaction from becoming a *photosubstitution:*

$$Cr(CO)_6 \xrightarrow{h\nu} CO + [Cr(CO)_5] \xrightarrow{thf} Cr(CO)_5(thf)$$

$$\underset{\text{1-pentene}}{W(CO)_5(NH_3) + C_5H_{10}} \xrightarrow{h\nu} W(CO)_5(C_5H_{10}) + NH_3$$

Photosubstitution is not limited to carbonyls or other organometallics, and we shall consider other well-established examples in the next section. Closely related

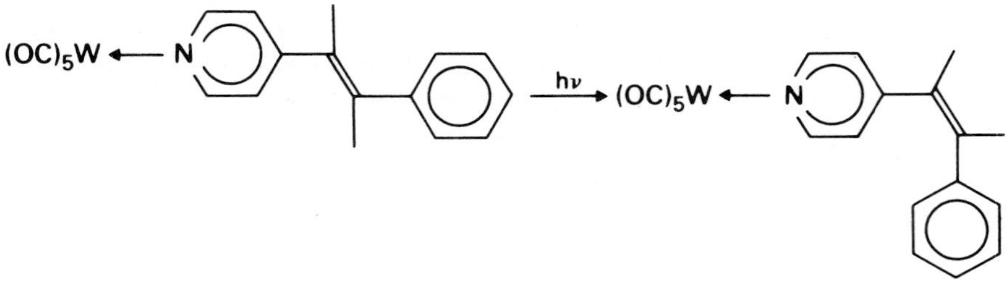

Figure 17.4 An intraligand reaction: the isomerization of 4-styrylpyridine.

to photosubstitution is *photoisomerization,* in which a linkage isomer of the reactant forms that is thermodynamically unstable with respect to the reactant [though not with respect to the hypothetical $Mo(cp)(CO)_3$ intermediate]:

$$Mo(\eta^5\text{-}C_5H_5)(CO)_3(NCS) \underset{h\nu \text{ or heat}}{\overset{h\nu}{\rightleftharpoons}} Mo(\eta^5\text{-}C_5H_5)(CO)_3(SCN)$$

Other isomerizations are known, such as *cis-trans* rearrangements of octahedral molecules. Some of these require preliminary photodissociation while others proceed through twist mechanisms. An interesting variation that seems to involve neither dissociation nor substitution is photoisomerization within a ligand that does not affect the metal–ligand bond. An example of this sort of *intraligand* photoreaction is shown in Fig. 17.4 for the ligand *trans*-4-styrylpyridine, which undergoes *trans-cis* isomerization when it is coordinated to $W(CO)_5$ and irradiated.

As an alternative to photosubstitution and related mechanisms, we have already mentioned *photoredox* reactions. These are almost always one-electron reductions (or oxidations), because absorption to the excited state is almost always a one-electron process. Often a free-radical ligand species is released:

$$[Rh(NH_3)_5(NCS)]^{2+} \xrightarrow{h\nu} [Rh(NH_3)_4]^{2+} + NH_3 + \cdot NCS$$

$$IrCl_6^{2-} + H_2O \xrightarrow{h\nu} IrCl_5(OH_2)^{2-} + \cdot Cl$$

However, reductive eliminations with a net two-electron change in the metal-atom oxidation state can also occur:

$$[IrH_2(diphos)_2]^+ \xrightarrow{h\nu} [Ir(diphos)_2]^+ + H_2$$

Although the net change is from Ir(III) to Ir(I), the mechanism of this reaction is not well enough known that one can say whether it proceeds by a two-electron transfer or by a sequence of two one-electron transfers. (Free $\cdot H$ atoms, however, have not been observed.)

The various processes we have mentioned for creating, interchanging, and eliminating excited states can be summarized in what is called a *Jablonski diagram.* A Jablonski diagram is an energy-level diagram analogous to those shown so far, but with the horizontal axis representing not the geometry of the complex, but the various possible spin multiplicities for the electronic ground and excited states. Jablonski diagrams are sometimes drawn to show only electronic-state energy levels. In this case, all singlet states are listed in a vertical column, triplet states in a neighboring column, and so on. Alternatively, the diagram can be drawn to show vibrational levels for each electronic state, in which case the singlet "column" (or other multiplicity) must usually be staggered to avoid overlapping the vibrational levels. Figure 17.5 gives an example of both forms of Jablonski diagrams. The radiative transitions marked *a* are absorptions from the ground state to the geometrically appropriate vibrational level of the excited state. The radiative transitions marked *b* are fluorescence emissions from the nearest excited state with the same spin as the ground state. The radiative transitions marked *c* are phosphorescence emissions from the nearest excited state with different spin. The wavy lines are nonradiative or radiationless transitions involving either simple vibrational relaxation by collision, internal conversion to another state with the same spin,

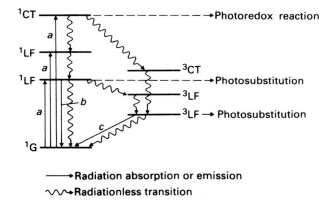

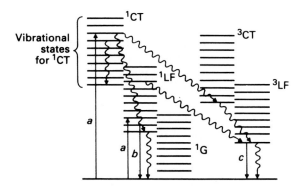

Figure 17.5   Jablonski diagrams.

or intersystem crossing to a state with a different spin. Finally, horizontal dashed lines represent photochemical reaction exits for the molecule from reactant excited states into product states.

17.2 PHOTOSUBSTITUTION REACTIONS

In our brief sketch of photochemical reactions, it should have become apparent that there are three basic categories:

1. Substitution or isomerization reactions
2. Redox reactions
3. Ligand-based reactions

Each of these reaction types must be electronically excited in the appropriate way. A useful generalization is that substitution photoreactions will be activated by irradiating the complex at a ligand-field-transition wavelength, redox photoreactions will be activated by irradiating the complex at a charge-transfer-transition wavelength, and ligand-based photoreactions will be activated by irradiating the complex at a wavelength corresponding to a transition between ligand-based orbitals. This, of course, is an oversimplification—as we have seen, excited states can interconvert, and absorption bands often overlap. However, it is a very useful guide to general photochemical reactivity, and we shall explore its consequences.

The idea that photosubstitutions are promoted by light absorption at ligand-field or $d-d$ transition frequencies is consistent with the picture of bonding in complexes that we developed in Chapter 11. If we consider Cr(III) complexes under the simplest crystal-field model, an octahedral d^3 complex will have the three valence electrons in the low-energy d_{xy}, d_{xz}, and d_{yz} AOs in its electronic ground state; parallel spins make the ground state a quartet. These orbitals (the t_{2g} set) maintain electron density between donor-ligand electron pairs and thus stabilize the complex. The lowest-energy doublet excited state (which can be reached only by a spin-forbidden transition) pairs two of these electrons. This leaves one t_{2g} orbital vacant, thereby making the metal atom more accessible to an incoming potential ligand. The lower-energy quartet excited states move one t_{2g} electron to an e_g orbital, strongly repelling an existing ligand and promoting dissociation. By either form of $d-d$ excitation, the *angular* distribution of the excited electron is changed, but not the *radial* distribution. This seems to be characteristic of photosubstitution reactions. On the other hand, a charge-transfer excitation that moves an electron from, for instance, a metal-based orbital to a ligand-based orbital, is essentially changing the radial distribution of the electron. To the extent the transfer is complete, a redox reaction has occurred. Accordingly, if we wish to promote photosubstitution, it is usually necessary to avoid irradiating the complex at the frequency of its CT bands.

For photosubstitution or any other kind of photoreaction, the observed chemical reaction in a given system will be the sum of the photoreaction and any "dark reactions"—thermally initiated reactions—that can occur. We already saw in Chapter 12 that many classical complexes are extremely labile toward thermal substitution in solution. This extensive dark reaction makes it difficult to study any photosubstitution reactions but those of the more inert complexes. Accordingly (as for the thermal mechanistic studies of Chapter 14) experimental attention has been focused on the inert Cr(III), Co(III), and low-spin Fe(II) complexes. Furthermore, ligand-field bands usually have a relatively low molar absorptivity because of the Laporte selection rule for $d–d$ transitions. Therefore, much of the incident light is not absorbed, and the fraction absorbed changes as the reaction proceeds, making the quantum yield difficult to calculate. Thus, the observable range of photosubstitution reactions is somewhat limited by experimental complications. Still, some useful results and generalizations exist that deserve our attention.

Cr(III) is a particularly favorable system for the study of photosubstitution because the other oxidation states Cr^{2+} and Cr^{4+} are so much less stable than Cr^{3+} that complications from redox reactions are unlikely, particularly in water solution. Solvation (aquation in particular), anation, and isomerization reactions have all been studied. The results are sometimes quite different from thermal reactions. Perhaps the simplest case is photoaquation of symmetrical CrL_6 complexes:

$$CrL_6 + H_2O \xrightarrow{h\nu} CrL_5(OH_2) + L$$

Two LF transitions are usually visible, corresponding to the excited states $^4T_{1g}$ and $^4T_{2g}$ in Fig. 11.26. For a given ligand, the quantum yield (usually 0.1 to 0.4) is independent of which band is irradiated. This suggests that the photoreaction occurs from the lowest excited state, a rule analogous to Kasha's rule for radiation emission. The quantum yield tends to rise with Dq for the ligand L. For less symmetrical complexes such as CrX_5Y or CrX_4Y_2, the complex can displace either ligand:

$$CrX_5Y + H_2O \nearrow CrX_4Y(OH_2) + X$$
$$\searrow CrX_5(OH_2) + Y$$

When $X = NH_3$ and $Y =$ halide or pseudohalide, both photoreactions occur, but NH_3 is preferentially released by a factor of 20–50 in relative quantum yields. This is the opposite of the thermal reaction, which displaces only the Y^- ion. Where it is possible to define *cis-* and *trans-* isomers for the products of Cr(III) photosubstitution reactions, the products always seem to have the *cis-* configuration, even though isotope-labeling studies have shown that there is a clear *trans-* labilization effect. This antithermal behavior suggests that the excited states of Cr(III) complexes are relatively stereomobile, in contrast to the ground states, which are stereorigid in thermal reactions. Substitution lability can be predicted fairly well by *Adamson's rules:*

1. Consider the six ligands to lie in pairs at the ends of three mutually perpendicular axes. The axis with the *weakest* average crystal field will be the one labilized, and the total quantum yield will be about that for a CrL_6 complex of the same average crystal field.
2. If the labilized axis contains two different ligands, then the ligand with the *greater* field strength will preferentially aquate.

Note that the rules specify which of the six ligands is displaced, but not the final stereochemistry of the product.

Co(III) complexes also appear to undergo photosubstitution according to Adamson's rules, but there are major experimental difficulties. In aqueous solution, Co^{2+} is much more stable than Co^{3+}. Because the CT bands overlap the LF bands to some extent, it is very difficult to prevent the photoredox reaction from producing the labile Co^{2+} complex as a product. Crystal-field arguments have been used to account for the reactivity pattern summarized by Adamson's rules for both Cr^{3+} and Co^{3+}, and to account for the stereochemistry of the photosubstitution products for the two metals.

An interesting variety of photosubstitution reactions exists for organometallic compounds. We already noted that metal carbonyls tend to lose a CO on irradiation and to undergo immediate substitution. This tendency is so pronounced in $Ru(CO)_5$ that the compound decomposes in daylight to form $Ru_3(CO)_{12}$, which presumably represents trimerization of a $Ru(CO)_4$ intermediate. If the carbonyl is unsymmetrical, as in $M(CO)_nL_m$, substitution can involve either CO or L. Usually CO loss is favored:

$$Mn(CO)_5I \xrightarrow{h\nu} [Mn(CO)_4I]_2 + 2\ CO$$

$$Mn(CO)_5H \xrightarrow[PF_3]{h\nu} Mn(CO)_{5-n}(PF_3)_nH + n\ CO \qquad (n = 1-4)$$

$$Mn(cp)(CO)_3 \xrightarrow[py]{h\nu} Mn(cp)(CO)_2(py) + CO$$

$$Fe(CO)_2(PF_3)_3 \xrightarrow[PF_3]{h\nu} Fe(PF_3)_5 + 2\ CO$$

$$Co(CO)_3(NO) \xrightarrow[PPh_3]{h\nu} Co(CO)_2(NO)(PPh_3) + CO$$

However, metal–metal bonds are preferentially labilized [the $Ru_3(CO)_{12}$ noted above exists in a photochemical equilibrium with $Ru(CO)_5$, so that monomer decomposition can be suppressed by maintaining a pressure of CO over the monomer]. Some interesting heteronuclear systems result from M–M bond labilization. There are even a few cases in which a metal–alkyl bond is activated:

$$Mn_2(CO)_{10} \xrightarrow[HBr]{h\nu} Mn(CO)_5Br$$

$$Mn_2(CO)_{10} \xrightarrow[Fe(CO)_5]{h\nu} (OC)_5MnFe(CO)_4Mn(CO)_5$$

$$Mn_2(CO)_{10} \xrightarrow[Re_2(CO)_{10}]{h\nu} (OC)_5MnRe(CO)_5$$

$$[Mo(cp)(CO)_3)]_2 + Co_2(CO)_8 \xrightarrow{h\nu} (OC)_4CoMo(cp)(CO)_3$$

$$CH_3Mn(CO)_5 + F_2C\!=\!\!CF_2 \xrightarrow{h\nu} CH_3CF_2CF_2Mn(CO)_5$$

A homonuclear M–M bond will normally break to yield two M· 17-electron radicals, which corresponds to excitation of an electron from the singlet sigma-bonded ground state to a triplet excited state in which one electron from the M–M bond pair occupies the σ MO and the other the σ^* MO with parallel spin. However, excitation at higher frequencies can yield a singlet excited state in which the antibonding electron has opposed spin to the bonding electron; this arrangement separates at long $M\cdots M$ to M^+ and M^-, a heterolytic bond breaking. Heterolysis is not usually seen, but has been observed for the reaction of $Cp_2Mo_2(CO)_6$ with bulky phosphines:

$$(cp)_2Mo_2(CO)_6 + 2PR_3 \xrightarrow{\lambda<290\ nm} [(cp)Mo(CO)_2(PR_3)_2]^+ + [(cp)Mo(CO)_3]^- + CO$$

In general, photosubstitution reactions of carbonyls seem to proceed by a dissociative mechanism. Because in solution the separated ligand is initially held in place by a solvent-molecule cage, one would expect recombination to occur frequently. Indeed, many complexes that appear to be photoinert probably do dissociate, but then recombine with high efficiency. In a stoichiometric sense, such a process obviously has no net chemical effect. However, if the intermediate with reduced coordination number is not rigid, the recombined product complex can be isomeric with the original complex. We have already seen that Cr(III) complexes usually adopt *cis-* configurations for CrX_4Y_2 substitution products, regardless of the original geometry, so *trans* → *cis* isomerization can occur along with photosubstitution. However, photoisomerization can occur without any net reaction: *trans*-$Cr(NH_3)_4(OH_2)Cl^{2+}$ efficiently produces the *cis-* isomer (though water exchange may occur), and $Cr(C_2O_4)_3^{3-}$ racemizes from either optical isomer. The *cis-* configuration is not a necessary result, since *cis*-$Rh(en)_2Cl_2^-$ is photoisomerized to the *trans-* isomer with some aquation. All of these isomerizations may involve at least temporary solvent coordination, but electrically neutral acetylacetonates photoisomerize in hexane solution (Fig. 17.6), where no solvent coordination seems possible.

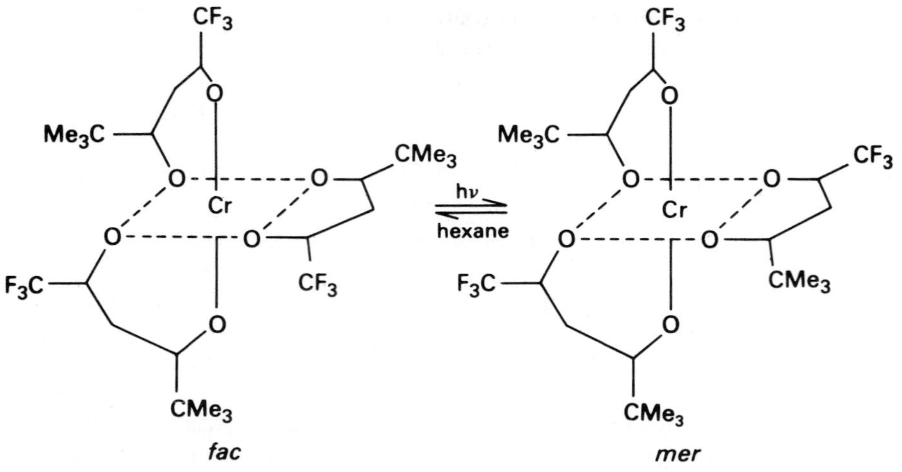

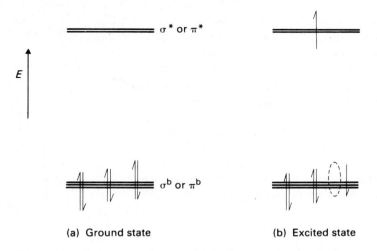

Figure 17.6 Photoisomerization of an unsymmetrical acetylacetonate complex.

17.3 PHOTOREDOX REACTIONS

Photoredox reactions represent a particularly powerful synthetic tool for the inorganic chemist. Consider a metal complex whose sigma-bonding orbitals are filled, but whose antibonding orbitals are all empty (see Fig. 17.7a). One-electron excitation as in Fig. 17.7b produces a species that is both a better oxidant than the original species (because it now has a vacant low-energy orbital) and a better reductant (because it has a high-energy antibonding electron). If the original species

Figure 17.7 Electron-transfer potential for the ground and excited states of ML_6 complexes.

already was a good oxidant, as for example Co^{3+}, the photoexcited state can be an extremely powerful oxidant. In general, the electronic transitions that actually promote photoredox reactions are charge-transfer transitions, as previously suggested. These transitions primarily change the radial distribution of the electron involved relative to the metal nucleus, in contrast to the angular rearrangement that occurs in $d-d$ transitions. Of course, the electron distribution can move either closer to the metal nucleus or farther away. If the electron originates in a MO that is essentially ligand-based and moves to a metal-based MO, the metal has been reduced and the ligand has been oxidized. Such a transition is called either ligand-to-metal charge transfer (*LMCT*) or charge-transfer to metal (*CTTM*). On the other hand, if the electron moves from a metal-based MO to a ligand-based MO, the transition is a metal-to-ligand charge transfer (*MLCT*) or charge-transfer to ligand (*CCTL*) transition. A variation of the latter is charge-transfer to solvent (*CTTS*). Either of these last two oxidizes the metal atom and reduces the ligand or solvent molecule.

Of the two possible CT transitions, the observed photoreactive excited states tend to be those in which the metal oxidation state conforms to the usual (thermal) redox stability of that element. That is, a readily reducible metal ion such as Co^{3+} will usually be photoreduced in a LMCT transition, whereas a readily oxidizable metal ion such as Ru^{2+} will usually be oxidized in a MLCT transition. Table 10.3 may be helpful in correlating the direction of charge transfer. Two inferences can be drawn from the table: (1) Second- and third-row transition metals are generally stable in somewhat higher oxidation states than the first-row metal with the same d^n configuration. Thus, second- and third-row transition metals are more likely to show photoreactive MLCT (oxidizing) transitions for their coordination compounds in intermediate oxidation states than first-row metals are. Conversely, LMCT (reducing) transitions are more probable for the same oxidation state of a first-row metal. (2) Organometallic compounds, which are almost always in very low oxidation states, are unlikely to undergo LMCT transitions in which the metal is reduced, but they can be expected to show photoreactive MLCT transitions in which the metal is oxidized.

Much effort has been given to the study of the redox photochemistry of Co^{3+} complexes and other d^6 systems. Co^{3+} complexes characteristically undergo LMCT transitions to yield labile Co^{2+} complexes, so that substitution accompanies the photoredox reaction. Even when the coordination of the Co^{2+} photoreaction product can be established, it is often found that one ligand has aquated (been replaced by water)—usually the oxidized radical redox product. The oxidized ligand, now a free radical or radical ion, can often be identified by flash photolysis. Some representative reactions following LMCT excitation are

$$Co(NH_3)_5NCS^{2+} \xrightarrow{h\nu} Co(NH_3)_5^{2+} + \cdot NCS$$

$$Co(CN)_5N_3^{3-} \xrightarrow{h\nu} Co(CN)_5^{3-} + \cdot N_3$$

$$Co(NH_3)_5(NO_2)^{2+} \xrightarrow{h\nu} Co(NH_3)_5(ONO)^{2+}$$

(In the third reaction, the linkage isomer is reformed from the radical $\cdot NO_2$.) Other classical complexes also undergo photoredox reactions following LMCT excitation:

$$PtCl_6^{2-} \xrightarrow{h\nu} PtCl_4^- + Cl^- + \cdot Cl$$

$$Fe(OH_2)_5Br^{2+} + H_2O \xrightarrow{h\nu} Fe(OH_2)_6^{2+} + \cdot Br$$

Organometallic systems are much more likely to undergo MLCT transitions with oxidation of the metal, because of the low oxidation state of the metal in the initial molecule. The resulting 17-electron species, however, are extremely reactive, so a single simple oxidized product does not usually result. Similar complications occur if an MLCT transition for a carbonyl causes a CO to photodissociate, leaving a 16-electron species. In many cases, the result is an oxidative addition reaction:

$$Mn(cp)(CO)_3 + HSiCl_3 \xrightarrow{h\nu} MnH(cp)(CO)_2(SiCl_3) + CO$$

$$PPh_4^+ + Mn(CO)_5^- \xrightarrow[THF]{h\nu} PhMn(CO)_4(PPh_3) + CO$$

$$Fe(CO)_5 + \underset{\text{Allyl bromide}}{C_3H_5Br} \xrightarrow{h\nu} (\eta^3\text{-}C_3H_5)Fe(CO)_3Br + 2CO$$

Sometimes the reactivity of the organometallic fragment results in a disproportionation reaction:

$$2[Mn(CO)_4Br]_2 \xrightarrow{h\nu} Mn_2(CO)_{10} + 2MnBr_2 + 6CO$$

$$Rh_4b_8Cl_2^{4+} \xrightarrow{h\nu} Rh_2b_4Cl^+ + Rh_2b_4Cl^{3+} \qquad (b = CNCH_2CH_2CH_2NC)$$

Interestingly, it seems likely that MLCT excitation is responsible for the following reductive elimination—the *opposite* of oxidative addition:

$$IrH_2(diphos)_2^+ \xrightarrow{h\nu} Ir(diphos)_2^+ + H_2$$

This mechanism is uncertain (though free $\cdot H$ atoms are not formed). However, if the initial Ir—H bond is covalent, the MLCT transition would yield Ir^+—H^- and the very strong base H^- could abstract a proton from the other Ir—H bond:

$$L_4Ir \underset{H}{\overset{H}{\diagup}} \rightarrow L_4Ir^+ \underset{H}{\overset{H^-}{\diagup}} \rightarrow L_4Ir + H^+H^-$$

One particularly interesting complex that undergoes CT photoactivation is the bipyridyl complex of Ru(II), $Ru(bpy)_3^{2+}$. This d^6 complex can also exist with the same stoichiometry as a 1−, 0, 1+, and 3+ species. It thus has a rich redox chemistry and many opportunities for CT transitions. Figure 17.8 shows the structure of the complex and gives a modified Latimer diagram for the three principal Ru oxi-

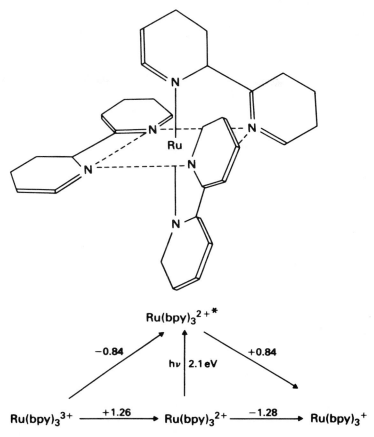

Figure 17.8 Structure and reduction potentials for $Ru(bpy)_3^{n+}$.

dation states; the modification is worth considering. The bottom line is the sequence of conventional reduction potentials, comparable to those in Table 8.2. The species $Ru(bpy)_3^{2+*}$ at the top, however, is the excited state resulting from an MLCT transition that absorbs a frequency corresponding to an excitation energy of 2.1 eV. This lowers the 3+/2+* potential by 2.1 V and raises the 2+*/1+ potential by 2.1 V. That is, the excited state is both a better oxidant and a better reductant, as suggested in Fig. 17.7. So although the 2+ complex is stable with respect to disproportionation in water, the 2+* excited state spontaneously disproportionates to the strong oxidant 3+ species and the strong reductant 1+ species (though this usually happens through a reaction between the 2+* species and the other acceptor or donor molecules in solution). An interesting sidelight is that the 3+ species in aqueous solution can be reduced by N_2H_4 or a hydrated electron to give the 2+* excited state chemically. This is a triplet state that phosphoresces orange light on returning to the 2+ ground state. Since the excited state was produced

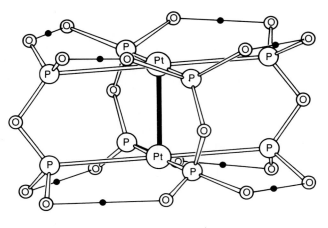

$$[Pt_2(\mu - P_2O_5H_2)_4]^{4-}$$

Figure 17.9 Structure of Pt–pop photochemical reactant. Solid dots represent H atoms; single lines represent hydrogen bonds.

chemically rather than by radiation absorption, the phosphorescence is called *chemiluminescence.*

An interesting variation on $Ru(bpy)_3^{2+}$ as a photochemical redox reactant is the $H_2P_2O_5^{2-}$ (pyrophosphite) complex of platinum(II): $[Pt_2(\mu\text{-}P_2O_5H_2)_4]^{4-}$, whose structure is shown in Fig. 17.9 to be a face-to-face combination of square-planar PtP_4 units bridged at each P by the P–O–P unit (hence sometimes known as Pt–pop). MOs for a single square-planar unit are shown in Fig. 11.16; in Pt–pop the 16 electrons on each Pt fill the MOs through the a_{1g} (HOMO), but interaction of the nonbonding $a_{2u} p_z$ yields a LUMO that is a p_z bonding MO. Photoexcitation thus is neither MLCT nor LMCT, but yields a triplet state Pt_2^* in which one electron is localized on each Pt atom. Pt_2^* reacts as a free radical Pt· yielding atom transfer with many reactants (Pt_2 represents the initial pyrophosphite 4– ion):

$$2Pt_2^* + 2Bu_3SnH \rightarrow Pt_2H_2 + Pt_2(SnBu_3)_2$$

where the new groups have added to the Pt_2 in the open axial positions. The hydrophilic character of the phosphite oxygens gives the Pt_2^* free radical different electronic properties from organic free radicals, which in future work could yield photochemical activation of primary C–H bonds in hydrocarbons even in preference to tertiary C–H bonds.

17.4 PHOTOREDOX REACTIONS AND LONG-RANGE ELECTRON TRANSFER IN PROTEINS

The $Ru(bpy)_3^{2+}$ complex is an excellent photoreceptor, and its excited state $Ru(bpy)_3^{2+*}$ is a strong selective oxidant and reductant. As a result, it can be used to study electron transfer in systems for which conventional thermal oxidants or reductants would react with the system of interest in undesirable ways or in an unspecific manner. Chapter 16 referred at several points to the question of the mechanism of electron transfer for metalloproteins in which the metal atom undergoing oxidation or reduction appears to be buried in the protein to such an extent that contact between the metal atom and another reactant in the ordinary way seems unlikely or impossible. Is long-range electron transfer possible in such a molecule, over perhaps 10 Å internuclear distance or more? If it is possible, how does the rate of electron transfer depend on the environment through which the electron must move? Gray and others have performed an extensive and elegant series of experiments using $Ru(bpy)_3^{2+}$ as a selective photoreductant or photooxidant to trace electron flow through a protein from a specific site on its periphery to a metal atom at its center.

In Chapter 14 we saw that the Marcus equation allowed us to correlate electron-transfer reaction rates with the thermodynamic equilibrium constant for the reaction. A fuller version of the Marcus treatment predicts that the rate constant for electron transfer will be proportional both to $e^{-(\Delta G^{0}+\lambda)}$ (λ is the energy to reorganize the reactants into the transition state geometry) and to $(H_{ab})^2$, the square of the quantum mechanical interaction energy between reactants a and products b. With respect to long-range electron transfer, we would expect that the exponential decay of radial wavefunctions with distance (and therefore of overlap) would lead to an exponential decay of H_{ab}. As the distance r between electron donor and electron acceptor increases we expect the electron transfer rate constant k_{ET} to decrease:

$$k_{ET} = k_0 \cdot e^{-\beta(r-r_0)}$$

where k_0 and r_0 refer to the rate and distance for intimate contact of the donor and acceptor atoms. In this expression β describes the electron-carrying efficiency of the medium between donor and acceptor; in a vacuum $\beta=1$, while a very good electronic conductor between the two atoms would yield a very small value of β, perhaps 0.01 or smaller. It seems to be generally true that extensive intervening pi-electron systems facilitate electron transfer, often yielding β values of 0.2–0.5, while sigma-bond networks help less ($\beta=0.5–0.9$). However, because β is appearing in an exponent it should be clear that electron transmission through even a sigma bond network is strongly preferable to transmission across free space, or even across a van der Waals contact.

As an example of Gray's studies we can consider his metal-modified sperm-whale myoglobin, in which an electron is transferred from the metal atom (normally Fe^{II}) at the center of the heme group to a specific site on the outside of the globin. Control of the outer site is achieved by coordinating a Ru^{II} to a histidine imidazole ring on the outside of the globin:

$$Ru(H_2O)_5(NH_3)^{2+} + His\text{-}Mb \rightarrow (H_2O)_5Ru^{II}\text{-}His\text{-}Mb + NH_3$$

Figure 17.10 shows the four specific locations on the outside of a Mb molecule at which a histidine can coordinate a Ru^{II}, along with the minimum distances from the

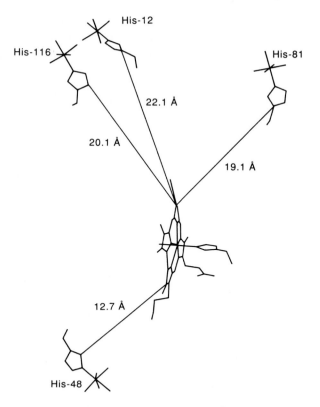

Figure 17.10 Computer-generated view of the central heme group and four ruthenated surface histidines for sperm whale myoglobin. Closest heme-edge/histidine-edge distances are
shown. [From B. E. Bowler, A. L. Raphael, and H. B. Gray, "Long-Range Electron Transfer in
Donor (Spacer) Acceptor Molecules and Proteins," *Progress in Inorganic Chemistry: Bioinorganic
Chemistry, vol. 38*, S. J. Lippard, ed., 1990. By permission from John Wiley & Sons, Inc.]

imidazole edge to the porphyrin edge; if electron transfer can occur over any of
these distances it will be long-range indeed.

In fact, electron transfer does occur, but for the more distant Ru locations it is
too slow to observe with the low driving force (0.02 V) between Fe^{II} and Ru^{III}. Accordingly, the Fe–heme electron donor is replaced by another metalloporphyrin
MP that can be photoexcited to yield a strong reductant $^3MP^*$. Magnesium mesoporphyrin in this photoexcited state yields a potential of about 0.8 V for transfer
to the outer Ru^{III}, which is produced from the coordinated Ru^{II} by reaction with
photogenerated $Ru(bpy)_3^{2+*}$, as in Fig. 17.11. The resulting electron-transfer rates
are seen as a function of edge–edge distance for the four histidine locations in
Fig. 17.12. It is clear that very-long-range electron transfer is occurring at reasonable rates, with β values corresponding to transmission along a sigma-bond network. The slightly lower value of β for Ru-His-12 may represent participation by

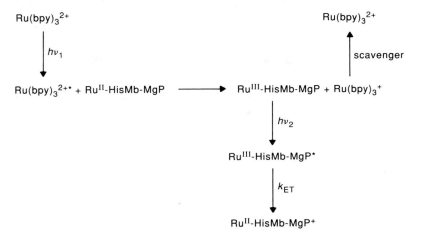

Figure 17.11 Reaction sequence for photoinitiated study of long-range electron transfer rates in ruthenated myoglobin. Photon ν_1 prepares the oxidized Ru sites, photon ν_2 triggers the redox reaction by exciting an electron on the central Mg porphyrin, and k_{ET} measures the rate of transfer of that electron to a surface Ru.

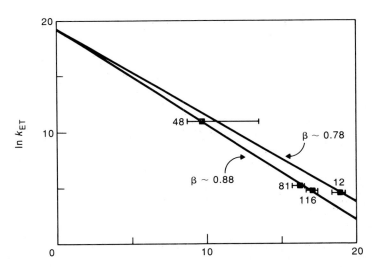

Figure 17.12 Plot of electron-transfer rate constant versus edge-to-edge distance for four different surface Ru–His locations on sperm whale myoglobin. Numbers indicate histidine sequence as indicated in Fig. 17.10. Electron transfer is from excited triplet of central Mg–mesoporphyrin to Ru–His. [From B. E. Bowler, A. L. Raphael, and H. B. Gray, "Long-Range Electron Transfer in Donor (Spacer) Acceptor Molecules and Proteins," *Progress in Inorganic Chemistry: Bioinorganic Chemistry, vol. 38,* S. J. Lippard, ed., 1990. By permission from John Wiley & Sons, Inc.]

the pi electrons in a tryptophan benzene ring that lies along the direct route indicated in Fig. 17.10, but this is still unclear.

This procedure may seem an elaborate, complex way in which to study electron-transfer rates through redox proteins—but each of the unusual features is necessary to the study. Although an iron- or copper-containing metalloprotein has a localized electron source or sink at the center of the molecule, without the specific metallated sites on the periphery of the protein it is not possible to say precisely where the electron is coming from or going to. The metallated sites must be kinetically inert with respect to substitution so that they stay put during the redox experiment (hence low-spin d^6) and inert with respect to electron transfer until the beginning of the experiment (hence Ru^{2+}, which Fig. 17.8 shows to have over a volt of redox stability in either direction in the *bpy* environment). But it must be possible to turn the Ru–His sites into good electron acceptors or donors at a precise time, which can be done by oxidizing the Ru^{II}–His to Ru^{III}–His by photogenerated $Ru(bpy)_3^{2+*}$ in solution. Finally, given the great distance over which electron transfer is suspected to occur, a measurable rate can only be achieved by having a high potential difference between the central metal and the Ru^{III}–His; this requires photoexcitation of an electron from a closed-shell porphyrin at the center in place of the Fe–heme porphyrin. The control of all these conditions yields an elegant series of experiments that shed light on an unusual feature of redox behavior in metalloproteins.

As Chapter 16 indicated, natural myoglobin is not an electron-transfer or redox-catalyzing protein, and it is probably not surprising that the electron-transfer rates for various routes through modified myoglobin in the experiments just described are pretty closely related to straight-line distance rather than to any particular bond pattern. However, there are redox-catalyzing proteins (such as the cytochromes) with an iron-porphyrin core, and it might well be that these proteins contain convenient preferred routes for efficient electron transmission. Gray has studied Ru-modified cytochromes in which the outer Ru(II) atoms have coordination $Ru(bpy)_2$(imidazole)(histidine), with the histidine being a group on the cytochrome protein. These Ru atoms are robustly bound and can be conveniently photoactivated like $Ru(bpy)_3^{2+}$, yielding a potential of about 1 volt for electron transfer to the central iron atom.

Because the presence of a bond network, whether sigma or pi, helps transmit electrons (lowers β), the various possible electron entry sites on the outside of a cytochrome have been compared for shortness (and thus efficiency) of sigma-network communication with the central Fe. It seems generally to be true that short, continuous sigma chains provide most efficient electron transmission. As a rule of thumb, a route through a hydrogen bond interrupting the sigma chain adds the equivalent of about 5 more sigma bonds to the chain length, and a route through a nonbonded space of 3.7 Å adds the equivalent of about 10 more sigma bonds to the chain length. When electron-transfer rate constants from the outside of Ru-modified cytochromes are compared for different Ru locations, they correlate poorly with absolute distance through space but quite well with what might be called "most favorable sigma-equivalent chain length." There are only a few highly efficient routes through the cytochrome; it is almost as if there were molecular wires through the insulating protein from certain locations on the protein surface. These studies offer the possibility that synthetic proteins may be able to control multi-electron redox reactions as, for instance, nitrogenase does—or even by changing protein conformation to provide molecular on/off switches for electron transfer reactions; the design and study of these possibilities is an exciting area in the future of bioinorganic chemistry.

17.5 LIGAND PHOTOREACTIONS

We have already touched on an important category of ligand photoreaction in that a photoredox reaction must oxidize something as well as reduce something. As we have seen, ligand free radicals are often released, but in general some sort of ligand redox must occur unless a disproportionation is occurring. A variation on this is the phenomenon of *quenching* in which a free (uncoordinated) donor or acceptor molecule is oxidized or reduced by the photoexcited state of the complex:

$$ML_6^{2+} \xrightarrow{h\nu} ML_6^{2+*} + A \rightarrow ML_6^{3+} + A^-$$

$$ML_6^{2+*} + D \rightarrow ML_6^{+} + D^+$$

Quenching is possible without actual electron transfer, and the resulting excited state of the quencher molecule, Q*, may be quite chemically reactive:

$$ML_6^{2+} \xrightarrow{h\nu} ML_6^{2+*} + Q \rightarrow ML_6^{2+} + Q^*$$

Quenchers can be other complexes (such as $Cr(CN)_6^{3-}$), simple ions (such as I^-), or organic molecules (particularly extended pi systems, such as phenanthrene or paraquat):

Quenching studies can yield valuable information about the electronic states of the photosensitive species.

In addition, reactions of the quenching molecules themselves can be interesting; these reactions constitute a sort of "outer-shell" ligand-based reaction. Strained polycyclic organic molecules can be photochemically oxidized as quenchers, following which they can rearrange. Figure 17.13 shows the structure of a square-planar Pd^{II} complex with norbornadiene that catalyzes the rearrangement of quadricyclene *Q* to norbornadiene *NBD* by the mechanism shown, which obviously involves the oxidation of *Q*. When electron-withdrawing substituents such as –COOR are attached to *Q* to increase its reduction potential (making it harder to

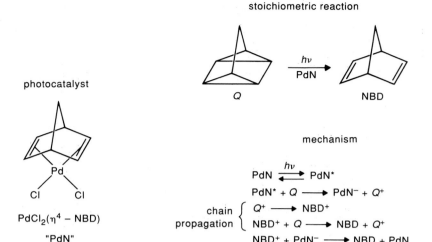

Figure 17.13 Reaction pattern for strained-ring quencher reacting with excited state of PdCl$_2$–norbornadiene complex.

oxidize) the quantum yield of the photoreaction drops dramatically from over 200 for unsubstituted Q to effectively zero for QR_2.

We have generalized earlier that ligand photoreactions tend to arise from transitions between ligand-centered MOs. This implies that the metal-to-ligand bonding is changed little or not at all by the ligand reaction. This is certainly true for reactions such as the isomerization in Fig. 17.4. Other such isomerizations are known:

$$W(CO)_6 + PhCH{=}CHPh \xrightarrow{h\nu} (OC)_5W(stil) \qquad [40\% \; E(trans), \; 60\% \; Z(cis)]$$
$$\text{Stilbene}$$

Here the W(CO)$_6$ is photoassisting the stilbene, since it will photoisomerize without metal coordination. However, the 40% E/60% Z mixture is produced regardless of which isomer of stilbene is used as the reactant. Many other alkenes undergo similar isomerization as organometallic ligands, usually more readily for terminal double bonds than for internal ones. Some interesting mechanisms have been proposed, but not convincingly demonstrated. One suggestion is that light simply dissociates a CO from a metal carbonyl, allowing alkene coordination followed by a rearrangement through an allylic intermediate related to that in steps 4 and 5 of Fig. 15.10. Another suggestion is that radiation lowers the rotation barrier for the coordinated double bond by removing a π electron ($\pi \to \pi^*$, $\pi \to M$, or $\pi_{alk} \to \pi^*_{CO}$) or adding a π^* electron ($\pi \to \pi^*$, $M \to \pi^*$, or $\pi_{CO} \to \pi^*_{alk}$).

On the other hand, some ligand photoreactions clearly do change the nature of the ligand-to-metal bonding. An unusual pi-to-sigma shift occurs for the ligands 1,2-dicyanoacetylene and tetracyanoethylene (TCNE):

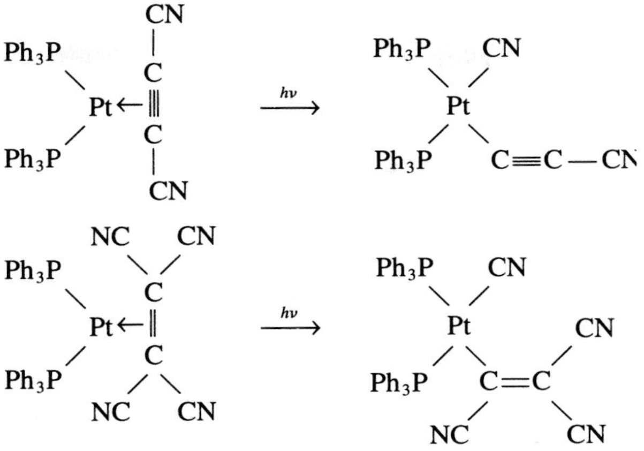

These are formally similar to insertion reactions. According to Table 14.8, these are likely for species with four ligands and 16 valence-shell electrons—that is, a species missing two electrons and two ligands form the stable configuration. As we have seen, however, d^8 systems are particularly stable in square-planar complexes where they have a total of 16 valence-shell electrons. So for the Pt(II) d^8 product of the above reactions a comparable insertion-reactive species might be a two-coordinate 14-electron complex. This corresponds to Pt(PPh$_3$)$_2$, an intermediate that might form if the photoreaction simply dissociated the pi ligand (as CO frequently does from carbonyls). An unusual but somewhat similar reaction yields an ortho-metallated triphenylphospine:

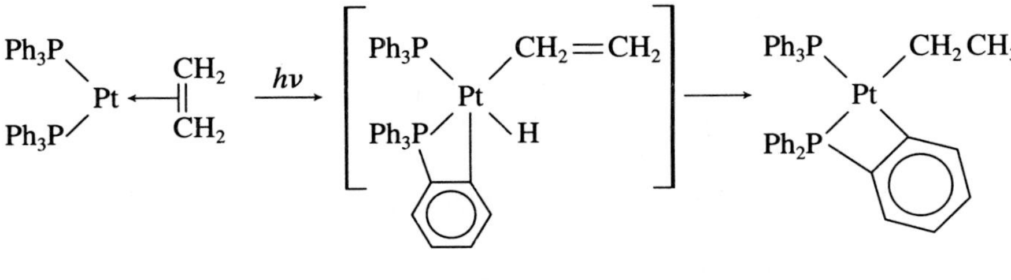

Other ligand photoreactions involve free potential-ligand molecules that appear to combine with transition-metal photocatalysts, react, and dissociate either as an excited state that forms the product or as the product itself. Butadiene is hydrogenated by H$_2$ when irradiated in the presence of Cr(CO)$_6$:

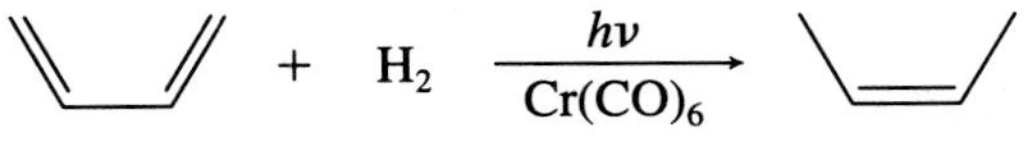

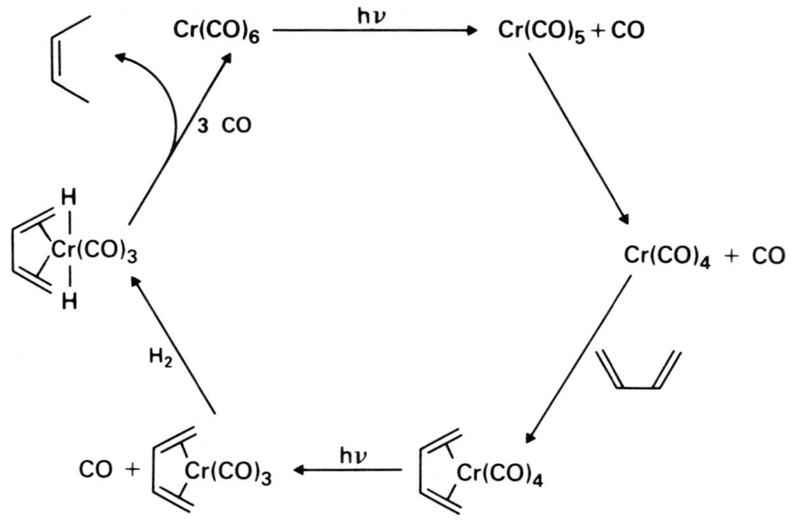

Figure 17.14   Possible mechanism for the photochemical hydrogenation of butadiene by $Cr(CO)_6$.

The mechanism for this reaction probably involves repeated photodissociation of CO, followed by diene coordination and H_2 coordination (see Fig. 17.14). The photodissociation of $Cr(CO)_6$ has been extensively studied, and the original excitation and dissociation to $Cr(CO)_5$ yields a molecule that is sufficiently vibrationally excited to dissociate a second CO rapidly. The $Cr(CO)_4$ product of this dissociation is stable toward further dissociation, but can accept two more pairs of electrons as in the figure.

In somewhat similar fashion, CuCl photocatalyzes the rearrangement of cyclooctadiene (COD) to tricyclooctane:

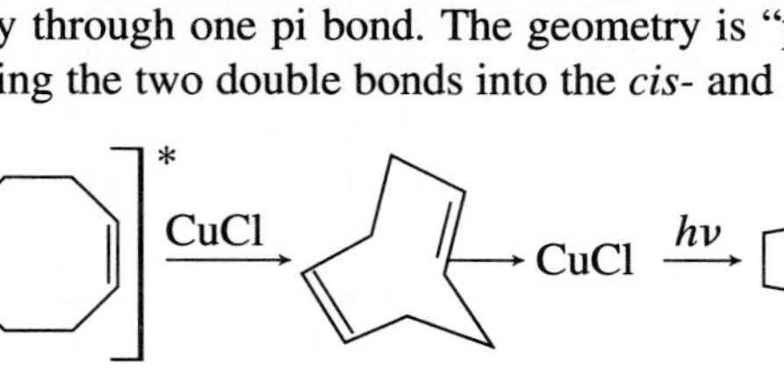

By curling up, COD can use both its pi bonds to donate four electrons to a single metal atom. The dimer $[(COD)CuCl]_2$ is known to have this structure, with bridging Cl atoms. However, the intermediate in this reaction appears to have the COD molecule coordinated only through one pi bond. The geometry is "prepared" for the rearrangement by moving the two double bonds into the *cis-* and *trans-* arrangements:

Note that this pi-to-polycyclic-sigma rearrangement is the opposite of the quadricyclene/norbornadiene rearrangement seen earlier in Fig. 17.13.

The relatively high oxidation state Pt^{IV} in $PtCl_6^{2-}$ can be photoreduced to Pt^{II} and ultimately to Pt metal, but only in the presence of alcohols; the alcohol is oxidized to the corresponding aldehyde (primary) or ketone (secondary). Intercepting the Pt reduction with $CuCl_2$ before the Pt oxidation state reaches 0 yields a catalytic cycle in which $RCH(OH)R'$ becomes $RR'C=O$:

$$RCH(OH)R' + \tfrac{1}{2}O_2 \xrightarrow[H_2PtCl_6,\,CuCl_2]{h\nu} RC(=O)R'$$

The reaction is thought to proceed through a stage in which photosubstitution of RO^- for Cl^- on $PtCl_6^{2-}$ yields HCl and ultimately $PtCl_5^{2-}$ and $RO\cdot$. The role of $CuCl_2$ is reminiscent of its function in the Wacker process described in Chapter 15, but the overall mechanism must differ because no metal–alkene bonding is possible here. Good yields are possible: 94% acetaldehyde from ethanol, 98% cyclopentanone from cyclopentanol.

In chlorinated solvents, ferrocene undergoes photosubstitution reactions on the cyclopentadienide rings through a CTTS transition:

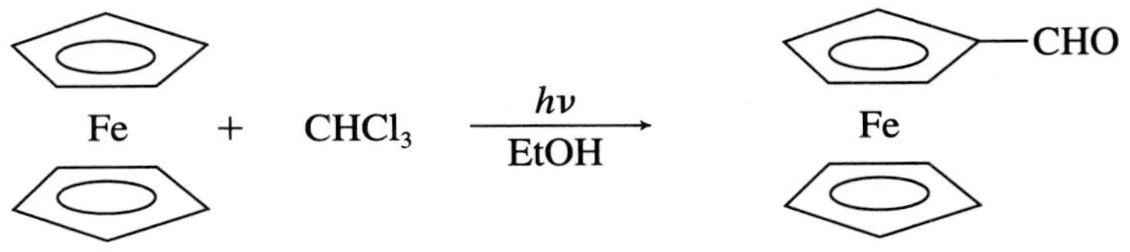

The CTTS transition produces $Fe(cp)_2^+$, Cl^-, and the $\cdot CHCl_2$ radical. The radical attacks the C_5H_5 ring, a base removes H^+ from the temporarily saturated ring C atom, and the two remaining Cl atoms solvolyze by reacting with the ethanol. An interesting combination of the previously discussed reductive elimination of H_2 and this aromatic sigma-substitution is seen in the photolysis of $Mo(cp)_2H_2$, which produces the transient, very reactive $Mo(cp)_2$. This in turn dimerizes by sigma substitution:

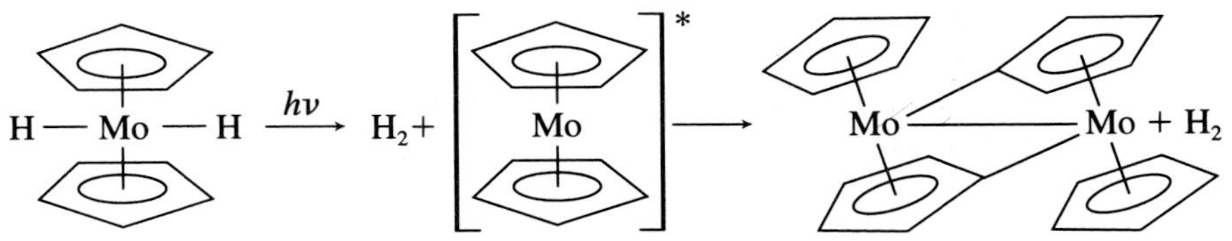

This remarkable product contains *cp* as both an η^5- and an η^1- ligand, together with an Mo—Mo bond.

17.6 PHOTOREACTIONS AND METAL REDOX CONTROL IN PHOTOSYNTHESIS

Solar-energy conversion, currently one of the most exciting areas of inorganic photochemical research, has been going on biochemically for hundreds of millions of years. The photosynthetic process is responsible for all of the fossil fuels we use (with the possible exception of some of the methane in natural gas). Before we look, in the next section, at some of the current efforts to use the photoredox reactions of transition-metal systems for solar-energy conversion, we shall sketch the broad features of photosynthesis for purposes of comparison.

   In addition to the green plants, some bacteria also carry out photosynthesis, but the process does not release O_2. Both kinds of photosynthesis, however, have a net stoichiometric effect that can be represented by the general *Van Niel equation:*

$$CO_2 + \underset{\text{Reductant}}{2H_2A} \xrightarrow{h\nu} \underset{\text{Carbohydrate}}{(CH_2O)} + 2A + H_2O$$

For halobacteria, the H_2A reductant is usually H_2S, but for higher plants, H_2A is water. The net effect is to use solar energy photochemically to split water into its elements; the O_2 is evolved, and the free H is used to reduce CO_2 to a carbohydrate. (Note that O_2 evolution normally requires a powerful oxidant.)

   Photosynthesis is an enormously complex process that is only beginning to be understood; we cannot begin to describe the full pattern of reactions, but will concentrate on the areas that are currently receiving greatest attention from bioinorganic chemists. Even there it will become apparent that existing data do not allow an unequivocal description of the system and its chemical mechanisms. Still, we know a number of things about the structures involved and even about the individual reaction steps.

   In higher plants, photosynthesis occurs in all the leaf cells except the epidermis covering both sides of the leaf; it can also occur in any green part such as flower bud coverings and even in some forms of bark that do not appear green because of other chromophores present. Specifically, photosynthesis occurs inside tiny organelles (called *chloroplasts*) within the leaf cells, which are surrounded by their own membrane and contain flat envelope-shaped sacs, each surrounded by a *thylakoid membrane;* these sacs are stacked like envelopes in a box, into compact bunches called *grana,* which are often connected to each other by the thylakoid membrane. The fluid within the chloroplast but outside the thylakoid membrane sacs is called the *stroma.* Figure 17.15 schematically indicates this arrangement in views *a* and *b.*

   In the figure, views *a* and *b* more or less resemble what a microscope would show in a leaf cross section. Views *c* and *d,* although they are presented as magnifications, are more schematic and are intended only to represent the sites of certain kinds of photochemical behavior within the cell and chloroplast. The photosynthesis process is carried out within the thylakoid membrane and seems to be energetically driven by light but chemically driven by proton transfer across the membrane.

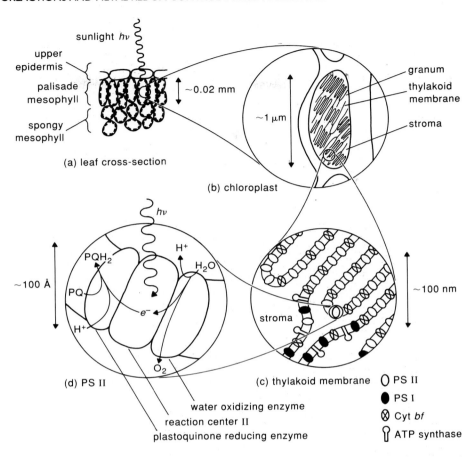

Figure 17.15 Progressive enlargement (in schematic form) of photosynthetic structures in leaf cells.

In view c, we see a representation of the four functional units within the membrane: two photosystem units *PS I* and *PS II*, an electron-transfer unit *cytochrome bf*, and an *ATP synthase* unit that spans the membrane and protrudes into the stroma fluid. We shall return to examine these shortly. View d focuses on a PS II photosynthetic unit, which has attracted most bioinorganic studies. In that unit light is absorbed, providing energy to strip electrons from water molecules within the thylakoid sac or *lumen* to produce O_2; the electrons are absorbed by a *plastoquinone-reducing enzyme*, which provides a reductant ultimately yielding ATP.

It is important for light-gathering efficiency that a photosynthetic unit be able to absorb the photons near the right frequency that hit it. As a result, evolution has yielded on the order of 10^3 light-sensitive chlorophyll molecules (see Fig. 17.16) as *antenna pigments* for each chemical *reaction center*, so that one of them will be

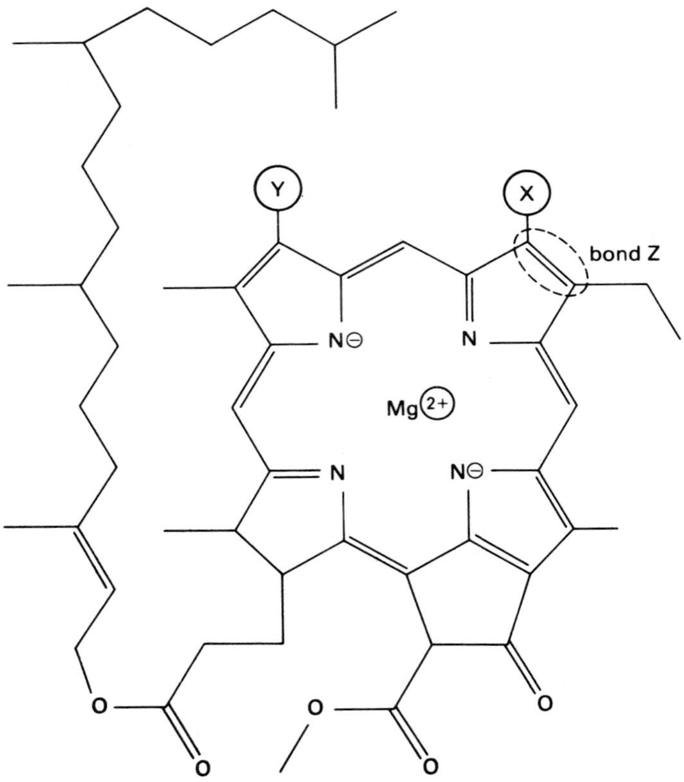

Figure 17.16 Molecular structures of chlorophylls.

Chlorophyll a: X = —CH$_3$, Y = —CH=CH$_2$;

Chlorophyll b: X = —CHO, Y = —CH=CH$_2$;

Bacteriochlorophyll: X = —CH$_3$, Y = —COCH$_3$, single bond Z.

in the right place to absorb almost any incoming photon of the right frequency (roughly 14,000–17,000 and 20,000–25,000 cm^{-1}). In only about a picosecond the photoexcitation energy is transferred to the reaction center, from which point the resulting chemistry is, strictly speaking, no longer photochemistry but redox chemistry with accompanying proton transfer.

There are two basic kinds of antenna pigments, chlorophyll *a* and chlorophyll *b*, as indicated in Fig. 17.16. Their visible-radiation absorption peaks do not overlap much, though both fall within the indicated frequency ranges; this increases the efficiency of light harvesting. Absorption by both is necessary for optimum photosynthesis. In PS II reaction centers a chlorophyll *a* molecule or dimer of molecules absorbs strongly at about 680 nm and is called pigment 680 (P680). The absorption seems to be essentially π→π* from the porphyrin ring. Its photo-

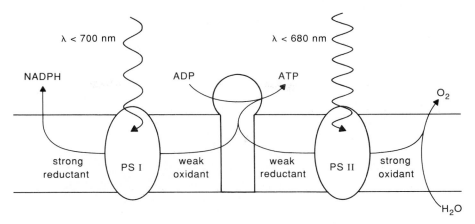

Figure 17.17 Schematic representation of photochemical reaction products from PS I and PS II, yielding O_2 through PS II, NADPH through PS I, and ATP through ATP synthase.

chemical products are a weak reductant, which assists in ATP formation, and a strong oxidant that produces O_2 from water. PS I reaction centers have P700 pigment (absorbing wavelengths below 700 nm) and yield a strong reductant that is responsible for the reduction of CO_2 through coupled redox reactions, plus a weak oxidant that assists in ATP formation. This cooperative effect is indicated in Fig. 17.17.

Figure 17.17 very schematically indicates a function for three of the four thylakoid membrane units shown in Fig. 17.15c. The remaining unit is the cytochrome *bf* complex, which carries electrons from PS II to PS I; it contains a ferredoxin protein that serves as the electron pool, reducing a Cu^{II} in plastocyanin (PC) (bound by two histidine N atoms and two S from cysteine and methionine as CuN_2S_2) to Cu^I. The ferredoxin is reduced by the PQH_2 formed on the reducing side of the PS II unit, regenerating PQ; the reduced ferredoxin reduces PC and in the process transfers two protons to the inside of the thylakoid membrane. Figure 17.18 schematically shows these linked processes within and across the thylakoid membrane.

The discussion to this point and several figures have suggested proton transfer across the thylakoid membrane from the stroma fluid to the inside of the membrane (the thylakoid lumen). In effect, this proton transfer is driven by the absorption of light energy; it yields a pH in the thylakoid lumen of about 4—some 3.5 pH units more acidic than cell fluids generally. The proton concentration gradient drives ATP synthesis outside the thylakoid membrane in the stroma, with the protons being conducted through the ATP synthase unit present in the membrane along with the PS I, PS II, and Cyt *bf* units.

The presence of a ferredoxin and of $Cu^{I/II}$ in the plastocyanin indicate a significant redox-chain role for metal ions in the photosynthetic process, but there is an even more critical role. The water-oxidizing enzyme that forms the oxidative

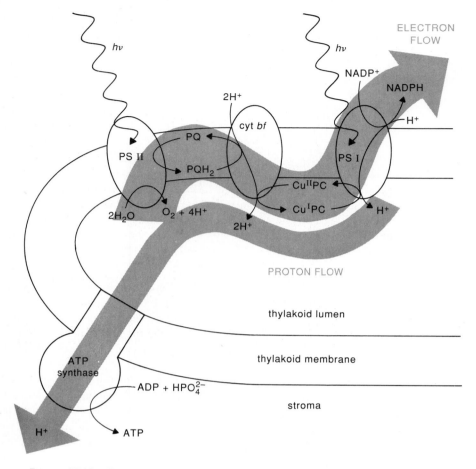

Figure 17.18 Electron flow and proton flow across thylakoid membrane in coupled processes leading to photosynthesis and O_2 production.

side of the PS II unit must contain a powerful oxidant to be able to take water to O_2; further, as was the case with the peroxidases and SODs in Chapter 16, it is important to be able to conduct this four-electron oxidation without releasing any toxic intermediates such as O_2^- or O_2^{2-}. This means that a single molecular system must be able to bind two oxygen atoms and keep them in close proximity while stripping four electrons out of that two-atom system. Few if any monatomic redox systems can absorb four electrons without undergoing a massive change in bonding and/or coordination. The water-oxidizing complex (*WOC*) contains four Mn atoms, associated in some way not yet certain, that accommodate this four-electron change. It is interesting that in other biochemical roles for manganese single Mn atoms frequently catalyze one-electron processes, and Mn_2 clusters frequently catalyze two-electron processes; perhaps there is a direct relationship between the number of electrons to be transferred in a biological redox process and the number of Mn atoms involved.

Joliot and Kok established the redox model currently viewed as the cyclic catalytic process for the WOC, involving five redox states of the WOC, S_0–S_4, separated by four one-electron oxidations and driven by four photon absorptions at the reaction center as suggested in Figure 17.19. The WOC accumulates energy from four photons (evidenced by peaks in O_2 evolution every fourth flash supplied to dark-adapted WOC), progressively moving through higher oxidation states of Mn as the S_n states advance. There seem to be three possibilities for the four Mn atoms within the WOC, not yet sorted out: a Mn_4 cluster of four nearly equivalent atoms, a Mn_3Mn cluster of three atoms with a fourth nearby, or a Mn_2Mn_2 dimer of dimers. The oxidation states of individual Mn atoms are not yet established, but are probably limited to Mn^{II}, Mn^{III}, and Mn^{IV}. EXAFS results indicate Mn$\cdots$Mn separations of 2.7 Å and 3.3 Å, though it is at least possible that the longer distance involves Mn$\cdots$Ca (calcium is known to be present). Such a short distance indicates that the Mn atoms can only be bridged by a small atom, and the high oxidation states in the most oxidized form make a hard base likely. Accordingly, attention has been focused on oxygen bridges and Mn_mO_n clusters as models for the WOC active site, particularly Mn_4O_n.

Clearly, if Mn oxidation states are changing by one electron per photon absorbed varying states of paramagnetism should be observed at different S_n states.

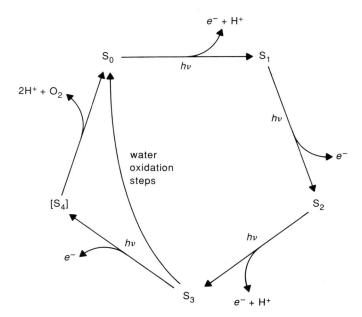

Figure 17.19 Joliot–Kok five-state mechanism for water oxidation within PS II.

In practice, the S_3 and S_4 states are difficult to isolate for study because they react extremely rapidly once formed, and in the same way the S_0 state is difficult to isolate because it slowly oxidizes to S_1 even in the dark (S_2 and S_3 revert to S_1 in the dark, which may be a comproportionation reaction between Mn_m clusters). For the S_1 and S_2 states, however, EPR data are available; S_1 shows a broad absorption at about $g_{eff} = 4.8$, and S_2 shows about 20 sharp lines near $g_{eff} = 2$ *or* a broad absorption at about $g_{eff} = 4.1$, both from Mn and interconvertible. Inasmuch as EXAFS edge structure is related to oxidation state as well as to ligand atoms, it should be possible to interpret these data to yield established oxidation states. However, the uncertainty as to the nuclearity of the Mn clusters and the difficulties associated with the techniques are such that no consensus yet exists. Perhaps the most widely held view is that the S_1 state is a Mn_4 cluster with all atoms in the same oxidation state (III, III, III, III), although some EPR data seem to favor (III, IV, III, IV). Pursuing the first of these, the S_2 state seems to be (III, III, III, IV), and speculation yields (III, IV, III, IV) for S_3 and (III, IV, IV, IV) for S_4.

If the manganese atoms are in a Mn_4 cluster it cannot be fully symmetrical, because two of the atoms are bound by a protein of about 33,000 daltons MW that is not part of the thylakoid membrane while the other two are bound by one (or more) intrinsic proteins. Still, two tentative mechanisms have been proposed using Mn_4 units, seen in Fig. 17.20. Brudvig and Crabtree, in 1986, proposed the "adamantane" mechanism in Fig. 17.20a, in which a cubane-type Mn_4O_4 cluster is progressively oxidized until, at S_3, a threefold rotation of the top three oxygens and attack by two H_2O on the bottom of the cluster produces a Mn_4O_6 adamantane-type cluster. Christou, in 1987, proposed the "double-pivot" mechanism in Fig. 17.20b, in which a Mn_4O_2 butterfly is progressively oxidized (with proton transfer from a base such as histidinium), with two H_2O attacking S_1 and closing with deprotonation to a cubane-type Mn_4O_4 structure at S_3, forming an O–O bond across a cube face at S_4, and eliminating O_2. Both mechanisms match the deprotonation sequence and the magnetic data reasonably well; the active study currently under way in a number of laboratories will, in years to come, yield reasons to prefer one of these or another not yet proposed.

A number of Mn_mO_n model compounds have been prepared. Figure 17.21 shows a few, emphasizing cluster shapes that appear in the mechanisms of Fig. 17.20. No Mn_4O_4 cubane-type cluster has yet been prepared, but with respect to the Mn_4O_3Cl cube it may be noted that the adamantane geometry corresponds to higher Mn oxidation states as required by mechanism *a,* and the butterfly geometry corresponds to lower Mn oxidation states as required by mechanism *b.* The remaining structure bears some resemblance to the butterfly geometry except that the positions of Mn and O have been interchanged; it is interesting in that its dimer-of-dimers character seems appropriate to the protein Mn binding in the PS II unit, where two Mn atoms are bound by an extrinsic protein and two by an intrinsic protein. It should be emphasized that these are first approaches to stoichiometric models, perhaps to structural models; none serve as functional models with respect to O_2 evolution. As we noted above for the proposed mechanisms,

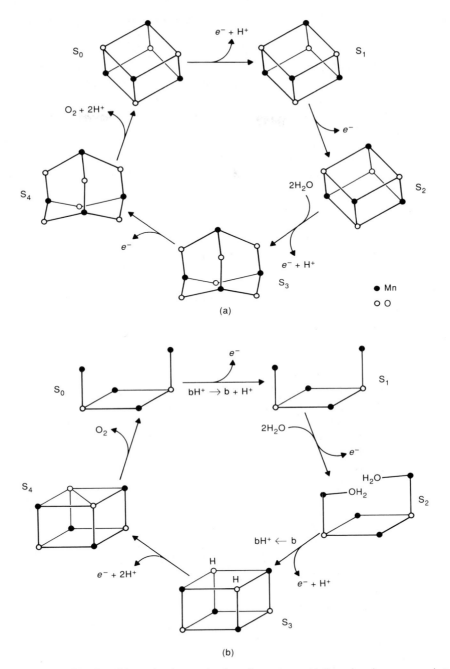

Figure 17.20 Possible molecular mechanisms for water oxidation, showing proposed structures for each of the S_n levels. (a) Brudvig and Crabtree "adamantane" mechanism, named for the proposed S_3–S_4 structure; (b) Vincent and Christou "double-pivot" mechanism, named for the pattern of motion of the coordinated H_2O molecules at S_2; *b* represents a neighboring base such as histidine, influenced by the photoredox process.

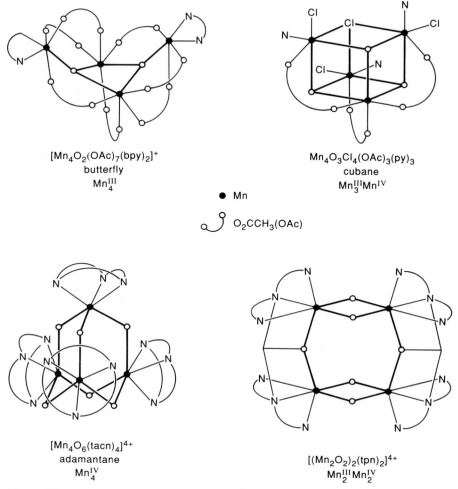

$[Mn_4O_2(OAc)_7(bpy)_2]^+$
butterfly
Mn_4^{III}

$Mn_4O_3Cl_4(OAc)_3(py)_3$
cubane
$Mn_3^{III}Mn^{IV}$

● Mn

$O_2CCH_3(OAc)$

$[Mn_4O_6(tacn)_4]^{4+}$
adamantane
Mn_4^{IV}

$[(Mn_2O_2)_2(tpn)_2]^{4+}$
$Mn_2^{III}Mn_2^{IV}$

Figure 17.21   Model compounds for Mn_mO_n clusters involved in the PS II water-oxidation complex. Symmetry of core structures is emphasized, and net Mn oxidation states are shown. *bpy* = 2,2′-bipyridyl; *tacn* = 1,4,7-triazacyclononane; *tpn* = N,N,N′,N′-tetrakis(2-pyridylmethyl)-2-hydroxypropane-1,3-diamine.

this is an active area of research in which models will be refined and mechanisms explored for years to come.

17.7 PHOTOREACTIONS AND SOLAR ENERGY CONVERSION

In a schematic sense—without implying the real existence of the bracketed species—the green-plant photosynthetic reaction can be represented as

$$H_2O \qquad [2H] + \tfrac{1}{2}O_2$$

$$+ \qquad \xrightarrow{h\nu} \qquad + \qquad \rightarrow [CH_2O] + O_2$$

$$CO_2 \qquad [CO] + [O]$$

If we focus on the top line of this reaction sequence, we can speculate that if the right kind of photocatalyst were available, it might be possible to divert the photosynthetic reducing capability to produce H_2 without ever involving carbon:

$$H_2O \xrightarrow{h\nu} H_2 + \tfrac{1}{2}O_2$$

Drawing an analogy with the two green-plant photosystems, we might also speculate that the photosensitive molecule (the "antenna pigment") might not directly oxidize or reduce H_2O itself. Instead, it may activate donors and acceptors in the reaction mixture, each of which is an effective oxidant or reductant when activated. If the overall process is to be photocatalytic, these donor–acceptor reactants must react to yield a form that can again be activated by the "antenna pigment." Research on such a process, however, will probably require stoichiometric or "sacrificial" regents to establish the best photosensitive system.

An effective photocatalytic system for splitting water into elemental hydrogen and oxygen using unassisted sunlight could be enormously significant in the long-range shift from reliance on fossil fuels. Although the storage of hydrogen as a fuel may remain too complex or too hazardous for use in automobiles or other small portable systems, stationary electric power plants using a hydrogen-air flame would be at least as thermally efficient as fossil-fuel plants, and would yield essentially no pollutants other than NO_x (which is inescapable in an air-oxidized flame). The intensity of solar radiation at ground level in the United States is about 800 W/m^2; most of the United States gets about 3000 hours of sunlight per year. From these facts, one can calculate that with perfect photochemical conversion of the incident sunlight, an 800-megawatt power plant would only require an array about 9000 feet on a side to operate year round—less than two miles by two miles. Perfect photochemical conversion will not be achieved, of course, but the calculated result is promising enough to deserve the close attention of photochemists.

A photocatalytic route to "artificial photosynthesis," the photolytic production of H_2 and O_2 from water, requires in its simplest terms only that an absorbing species be found that, when excited, is a good enough reductant to reduce H_2O to H_2 and also a good enough oxidant to oxidize H_2O to O_2. Since under the right circumstances an excited state is both a better oxidant and a better reductant than the ground state of the same molecule, this seems possible. Consideration of Fig. 8.1 indicates that at pH 7 this requires only that the oxidant have a potential more positive than +0.815 V and that the reductant have a potential more negative than −0.415 V. In the previous section we saw that the complex $Ru(bpy)_3^{2+}$ meets this requirement, as the Latimer diagram in Fig. 17.8 indicates for the excited state $Ru(bpy)_3^{2+*}$. Figure 17.22 shows schematically the reduction-potential relationships. Furthermore, the light absorption occurs in the visible range of light, where

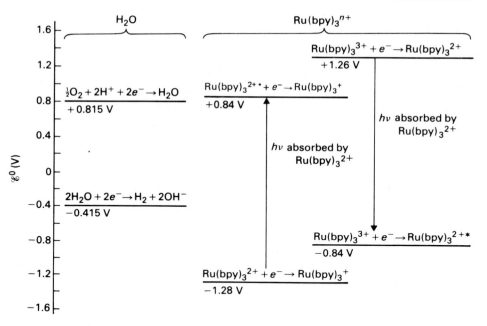

Figure 17.22 Reduction-potential relationships for water and $Ru(bpy)_3^{n+}$ species.

the sun's power output at the earth's surface is high. The following sequence of reactions is thus thermodynamically possible:

$$4Ru(bpy)_3^{2+} \xrightarrow{h\nu} 4Ru(bpy)_3^{2+*} \qquad \text{(4 photons)}$$
$$4Ru(bpy)_3^{2+*} \longrightarrow 2Ru(bpy)_3^{+} + 2Ru(bpy)_3^{3+}$$
$$2Ru(bpy)_3^{+} + 2H^+ \longrightarrow 2Ru(bpy)_3^{2+} + H_2$$
$$\underline{2Ru(bpy)_3^{3+} + H_2O \longrightarrow 2Ru(bpy)_3^{2+} + 2H^+ + \tfrac{1}{2}O_2}$$

Net:
$$H_2O \xrightarrow{4h\nu} H_2 + \tfrac{1}{2}O_2$$

Unfortunately, the process is not that simple. It is necessary to physically separate the production of O_2 and the production of H_2. Otherwise, they simply recombine as they are produced, releasing the energy as heat in the solution. There must be at least one *relay molecule* in solution to quench part of the $Ru(bpy)_3^{2+*}$, carrying away either its oxidizing or its reducing capacity so that one of the two electron-transfer processes can occur at a different location from the other. Usually the relay compound becomes the reductant, as is the case for the compound paraquat (PQ^{2+}, also sometimes known as *methylviologen*), which we have already mentioned as a quencher:

$$Ru(bpy)_3^{2+*} + PQ^{2+} \xrightarrow{h\nu} Ru(bpy)_3^{3+} + PQ^+$$

$$PQ^+ + H_2O \xrightarrow{Pt^0} PQ^{2+} + OH^- + \tfrac{1}{2}H_2$$

Note that a platinum–metal catalyst is now needed in the water reduction, which means that the reaction is no longer homogeneous in solution. A similar kinetic problem exists for the oxidation reaction of $Ru(bpy)_3^{3+}$ with water, which can be catalyzed by solid RuO_2. Unfortunately, the system is now complicated enough that efficiency is lost through cross-reactions such as the oxidation of $Ru(bpy)_3^{2+}$ by RuO_2.

An alternate possibility for separating the oxidation and reduction sites is to use a semiconductor solid as the relay compound. If the photosensitive compound is adsorbed on the surface of a small particle of solid semiconductor suspended in solution, it can release an electron from its photoexcited state into the semiconductor, which can release the electron to H_2O at another surface location. In this case, the solid particle can also serve as the support for the required catalysts. Semiconducting TiO_2 particles have been successfully used as the support for Pt and RuO_2 in the system above, with free H_2 and O_2 generated directly by irradiating the aqueous suspension. Converting $Ru(bpy)_3^{2+}$ to an anion by using as a ligand 2,2′-bipyridyl-4,4′-dicarboxylate (yielding $[Ru(bpd)_3]^{4-}$) greatly improves the electrical interaction of the photosensitive Ru species with the TiO_2 semiconductor; such a system has yielded 44% photon-to-current efficiency for a given frequency of irradiation.

To understand the electronic role of the semiconductor in this application and in the closely related *photoassisted electrolysis* of water, it is necessary to return to the basic band theory of solids, as in the discussion associated with Fig. 4.32. In that figure the Fermi energy or Fermi level was indicated for *p*-type and *n*-type semiconductors. The filled lower-energy band is usually called the *valence band* and the empty (at 0 K) upper band is called the *conduction band*. Figure 17.23 shows these drawn for a vertical energy axis; compare them with Fig. 4.32.

Thermal excitation populates the conduction band with a few electrons from the valence band, leaving holes behind. If there are more electrons than could be

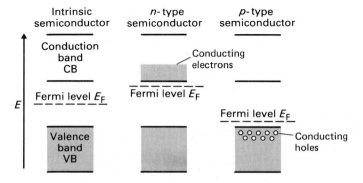

Figure 17.23 Band-energy relationships for semiconducting solids.

accommodated in the valence band (because of doping with an electron-rich element), the Fermi level lies closer to the conduction band and there is a net surplus of negative or n-type charge carriers. On the other hand, if there are fewer electrons than would fill the valence band, the Fermi level lies closer to the valence band and there is a surplus of holes or positive charge carriers.

In a conventional nonphotochemical silicon solar cell, a $p-n$ junction is formed by doping the Si with three-electron and five-electron atoms, respectively. Since these layers are now shorted together electrically, their Fermi levels—which represent the chemical potential of the electrons in the solid—must be the same. This bends the band energies sharply at the junction (see Fig. 17.24). If the junction is irradiated by light for which $h\nu$ is greater than the energy width of the band separation, electrons will be excited from the valence band to the conduction band. If luminescence does not occur, the excited electron will relax back to lower energy in the conduction band, which means that it will move into the n-type region away from the junction. Similarly, the newly created hole (which relaxes to a *higher-energy* orbital in its band) will move into the p-type region. This creates a charge separation that can be used to drive electrons through an external circuit and do electrical work, as the cell cross-section in Fig. 17.24 suggests.

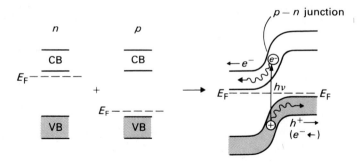

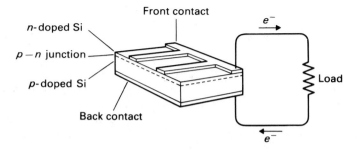

Figure 17.24   Band energies and construction of a silicon solar cell.

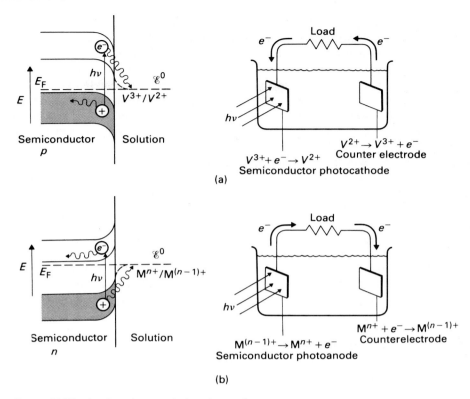

Figure 17.25 Semiconductor-solution photocells.

Now consider the formation of a different kind of junction, one in which a p-type semiconductor is immersed in a solution containing a dissolved redox couple such as V^{3+}/V^{2+}. The reduction potential of the dissolved couple is the chemical potential of electrons in the solution medium—the equivalent of a Fermi level for the solution. Therefore, the solution/semiconductor interface is a junction. Furthermore, the two phases are electrically shorted so that the redox potential must equal the Fermi level, just as the two Fermi levels must be equal across a $p-n$ junction. On the semiconductor side of the junction, the bands must bend in the appropriate direction; on the solution side, the redox potential must "bend" by changing the relative concentrations of the oxidized and reduced ionic species in a layer at the junction (the solid surface). This behavior is diagrammed in Fig. 17.25a. If the semiconductor is p-type, it has a relatively low Fermi level; so if it is immersed in a solution of a fairly high reduction potential there will be a substantial bending downward of the semiconductor bands so that the two chemical potentials can match. Conversely, in Fig. 17.15b, if the semiconductor is n-type and is immersed in a solution with a low reduction potential (that is, a general tendency to serve as an oxidant), the equalization of potentials causes the bands to bend in the opposite sense.

If this junction—the semiconductor/solution interface—is irradiated to excite a valence-band electron to the conduction band, it will relax by moving toward or away from the junction, depending on the direction of bending. In Fig. 17.25a, the excited electron relaxes down into the redox couple in solution. The couple is then reduced, and the junction behaves as a photocathode. In Fig. 17.25b, the electron relaxes by moving away from the junction and the hole moves toward the junction, stripping electrons from the redox couple in solution and oxidizing it. Here the junction serves as a photoanode. In either case, a counterelectrode must be supplied in the solution to allow electron removal or return, serving the same function as the back contact on the silicon solar cell in Fig. 17.24. If a reversible redox couple is chosen, such as $Fe(CN)_6^{3-}/Fe(CN)_6^{4-}$, oxidation at one electrode will be just balanced by reduction at the other, and no net chemical reaction will occur. One simply has a two-phase solar cell.

We can go a step farther, however. If we eliminate the reversible redox couple from the solution, the solvent itself—water, in particular—can serve as the redox couple. H_2O will be reduced to H_2 at the semiconductor electrode if the semiconductor is a p-type photocathode, and will be oxidized to O_2 at the counterelectrode. A good example of this sort of cell uses p-type InP as the photoelectrode, an HCl–KCl solution as the electrolyte, and a Pt counterelectrode. However, the band gap for InP is small enough that the light energy $h\nu$ is not great enough to force the reduction and oxidation of water. Instead, a voltage is maintained across the two electrodes externally. The water is simply electrolyzed, but instead of requiring 1.23 V for decomposition, the cell with the photocathode irradiated by sunlight requires only 0.64 V, so that much less external electrical energy is consumed. This is the photoassisted electrolysis we spoke of earlier.

The final stage in the conceptual simplification of the semiconductor-solution system for the photolysis of water to H_2 and O_2 is to choose a semiconductor for which the band gap is so large that the light absorbed yields enough energy to split water without an added external voltage. For such a system, water would be reduced by the very-high-energy conduction-band electron at a more or less anionic site on the semiconductor surface, while a hole in the low-energy valence band would oxidize water at a somewhat positively charged site. The semiconductors TiO_2 (n-type) and GaP (p-type) combined in a single particle have this ability; so does n-type $SrTiO_3$ partially coated with Pt metal. Oxidation occurs at the n-type surface and reduction at the p-type or metal surface. The major drawback of these systems is that the high-energy radiation required uses only the ultraviolet portion of the sun's radiant energy, which is only a small fraction of the total energy input at the earth's surface.

To summarize briefly, there are four photochemical processes for solar-energy conversion (besides the Si solar cell, which is nonchemical) using the photoexcitation of redox couples in water:

1. A semiconductor/solution junction with a redox couple, yielding no net chemical change: n-CdS/S^{2-}, S_2^{2-}, OH^-/C

2. A semiconductor/solution junction with water decomposition, using photo-assisted electrolysis: p-InP/H_3O^+, Cl^-/Pt

3. A semiconductor/solution junction with water decomposition using high-frequency radiation: In$-$SrTiO$_3$/Pt

4. A solution species yielding a strong oxidant and reductant on irradiation: RuO$_2$/Ru(bpy)$_3^{2+}$/Pt

None of these is yet a practical commercial means of storing solar energy. The last three produce H_2 and O_2, which seems to be an effective means of storing the daytime solar energy rather than only converting it, but each has its difficulties. However, efficiencies as high as 16% have been shown in the best cases, which makes it likely that practical systems may not be too far in the future.

17.8 PHOTOCHEMISTRY AND PHOTOGRAPHIC SYSTEMS

By far the most familiar example of commercial inorganic photochemistry to all of us is the black-and-white photographic process. In its simplest essentials the process is familiar: A photographic emulsion containing finely divided AgBr is exposed to light, which forms a *latent image* on the emulsion that is developed to a visible image formed by aggregates of black silver atoms. The developer is a reducing agent that reacts more readily with AgBr grains that have been exposed to light than with grains that have not been exposed. But a century and a half of study, sometimes by very sophisticated techniques, has yielded a rather sophisticated model of photochemical events within the emulsion, and we shall consider the model for exposure and development.

To begin with, the word "emulsion" is somewhat misleading. The photosensitive layer on photographic film or paper is not really an emulsion (defined as a stable mixture of two immiscible liquids, suspended as particles in the size range 10^{-2}–10^{-3} mm). The term, however, is firmly attached to its photographic use, and we will continue it here. In the emulsion, fine particles of AgX (usually a mixture of halides) are suspended in a protective colloid (usually gelatin) and small amounts of stabilizers and sensitizers are added. Negative film is usually about 40% AgX by weight; the halide composition is anywhere from 99 AgBr:1 AgI to 90 AgBr:10 AgI. Photographic paper has an emulsion with about 30% AgX; the halide composition varies from 95 AgCl:5 AgBr to 30 AgCl:70 AgBr. In general, the emulsion is more photosensitive or "faster" the more heavy halide ion it contains. In negative film, the AgX particles are roughly one micron in diameter, though the grain sizes are different for films with different purposes. In general, film speed varies directly with grain area, so high-speed films have large grains of AgX.

Silver chloride and silver bromide have the NaCl crystal lattice shown in Fig. 2.4, but AgI has the wurtzite structure shown in Fig. 2.13. The mixtures used in emulsions are normally solid solutions of the two AgX halides, and all of these solid solutions have the NaCl structure. For reasons discussed in Chapter 3, none of the silver halides is particularly ionic, and this is reflected in the wide departures of the Ag—X crystal internuclear distances from those predicted from the crystal radii of Table 2.3: AgI has an experimental distance of 2.81 Å compared to a radius sum of 3.20 Å, for instance. However, the movement of silver ions (or atoms) within the lattice is an unusual and important feature of these halides; the discussion of Frenkel defects in Chapter 3 has already noted this ability. AgI carries Ag atom mobility so far that it transforms at 146°C to an arrangement of *bcc* iodine atoms within which the silver atoms/ions move so freely that the high-temperature form has an ionic conductivity of 1.3 ohm^{-1} cm^{-1} at the transition temperature. It is almost as if the crystal had half melted, with the I$^-$ still crystalline and the Ag$^+$ liquid.

Several kinds of crystal defects in the AgX grains are important in forming the latent image. These are common to most more-or-less-ionic crystals as we have seen in Chapter 3, but (given the silver-ion mobility just mentioned) might be expected to form especially easily in these crystals. If the defect shows a residual positive charge, the defect site will trap electrons, whereas if the defect shows a residual negative charge, it will trap mobile Ag$^+$ ions. As a guide to the approximate frequency of defects, there is roughly one interstitial Ag$^+$ for every 10^8 Ag$^+$–Br$^-$ ion pairs in pure AgBr at room temperature. Although interstitial Ag$^+$ ions can move through the crystal, they are much more common near the surface because they tend to form near surface kinks, as Figure 3.8 suggests.

For the AgX crystal, the orbital band structure is relatively simple. At room temperature, the halides are all insulators, which implies an energy gap between the valence band and the conduction band that is large compared to kT at room temperature (see the discussion associated with Figs. 4.32, 10.5, and 17.23). Experimentally, the gap seems to be about 2.6 eV for AgBr, which compares to $kT = 0.025$ eV at 298 K. The 2.6 eV gap corresponds to a wavenumber of 21,000 cm^{-1}, or a wavelength of about 480 nm for radiation that must excite an electron from the valence band to the conduction band. This excitation is the key step in the photochemical formation of the latent image. Since light of lower frequency or longer wavelength cannot supply enough energy to excite a photoelectron, it follows that unmodified AgBr in an emulsion will be sensitive only to blue light.

When an electron is excited by light falling on the emulsion, it becomes a substantially delocalized electron in the conduction band. Its mobility is about 60 cm^2/V sec, which may be compared with the value given in Chapter 8 for the solvated electron in NH$_3$ (1.2×10^{-2} cm^2/V sec). The photoelectron leaves behind a "hole," or positive-charge equivalent in the valence band. This hole can also contribute to charge conduction (though its mobility is only about 2 cm^2/V sec). The photoelectron has available to it essentially all the options discussed at the beginning of this chapter for an excited-state electron. If it were to luminesce, there would presumably be no latent-image formation, since the electron and hole would simply recombine. However, extensive evidence suggests that the photoelectron combines with a silver ion in a photoredox reaction:

$$Ag^+ + e^- \rightarrow Ag^0$$

Since in the valence band the photoelectron was originally associated with the bromide ions, we have the overall photoreaction:

$$Ag^+ + Br^- \xrightarrow{h\nu} Ag^0 + \tfrac{1}{2}Br_2$$

This reaction is strongly favored over luminescence, so the quantum yield is essentially 1.00 for low levels of illumination. However, elemental bromine attacks silver metal thermally, so the quantum yield will fall off because of a back-reaction unless a halogen scavenger such as acetone semicarbazone, $(CH_3)_2C{=}N{-}NH{-}CO{-}NH_2$, is added to the emulsion.

The silver atoms formed by this photoredox reaction constitute the latent image that is subsequently developed. However, it is possible to have such low levels of illumination that no latent image is formed even though some photons reach the emulsion and are absorbed. For ordinary high-speed film such as Kodak Tri-X with a known AgX grain size, it appears to be necessary to have, on the average, about 10 photons absorbed per grain (which is to say 10 Ag atoms formed per grain) for the grain to be developable preferentially to unexposed grains. However, there is a wide spread of grain sensitivity even in a single emulsion; the most sensitive grains seem to be developable after the formation of only four Ag atoms. A crucial point in understanding the role of silver in the formation of a latent image is that these silver atoms are not scattered throughout the crystal structure, but are aggregated into particles. Studies of image bleaching by monochromatic light show clearly that the silver in the latent image is present not as monodisperse atoms, but in colloidal particles of varying sizes. This, of course, requires that there be some mechanism for concentrating the silver atoms into these particles. So the overall photolytic efficiency of latent-image formation will depend on the quantum yield of Ag^0 formation, the efficiency of the concentration mechanism that forms Ag_n particles, and the minimum value for n in Ag_n.

A model for the latent-image formation process was suggested by Gurney and Mott using the properties described thus far:

1. The photoelectron produced when light strikes the emulsion enters the conduction band and diffuses through the AgX crystal until it is bound by a trap, usually a crystal defect with an effective partial positive charge. This yields a net negative charge at the trap site.
2. A mobile silver ion, attracted by the opposite charge, diffuses through the crystal lattice up to the trapped photoelectron and combines with it to form a silver atom Ag^0.
3. With the charge neutralization in step 2, the crystal defect is again positively charged and can again trap a photoelectron.
4. Repeated diffusion of silver ions to the trap and its photoelectron builds up an Ag_n particle.

Note that, besides photochemical reactivity on the part of the silver ion, two kinds of mobility are necessary. The photoelectron must be able to diffuse through the crystal to a trap, and—equally important—silver ions must also be able to diffuse through the crystal to meet the electrons at a single location and form the Ag_n aggregate. It is believed that a single Ag^0 atom at a trap (defect) would be unstable with respect to ionization and the formation of more crystal lattice: $Ag^0 \rightarrow Ag^+(s) + e^-$. However, this process requires a finite time. If another silver ion can diffuse up to the trap and combine with another photoelectron to form Ag_2 before the first Ag^0 atom can decompose, the Ag—Ag binding energy will stabilize the diatomic particle for further photoreaction and aggregation. The two-atom particle is called a *subimage*. It is not developable, but it is indefinitely stable for further aggregation which, at $n = 4$ to 10, can yield a developable latent image.

It must be emphasized that the latent image is invisible—given the average grain size of film emulsions, only about one out of 10^8 silver atoms in a grain is in an Ag_n particle. The image must be developed to be visible. A developer is a reducing agent that reacts with Ag^+ ions in an exposed grain (though *not* with the Ag_n nucleus) to completely reduce the Ag^+ in that grain to Ag^0 *before* reducing any Ag^+ in grains that do not contain a Ag_n latent-image nucleus. This is clearly a kinetically controlled process, since thermodynamically all of the silver ions present in the emulsion should be reduced. The developing reaction

$$Ag^+ + Dev \rightarrow Ag^0 + Dev^+$$

is autocatalytic in Ag^0 for many developers, which of course means that particles containing the Ag_n nucleus have a head start on those that do not. The metallic silver that is formed is present initially as filaments, though recrystallization of the Ag filaments to more stable crystal forms can occur slowly. A very small grain may yield only one filament on developing, but a large grain will usually yield a tangled mass of filaments. The reason for filamentous growth is not well understood, though it may be related to the electrical and surface-energy asymmetry of a metal fiber.

A general electrochemical mechanism has been proposed as follows for the development process: The latent-image nucleus strips an electron from the developer molecule, acquiring a negative charge that attracts nearby silver ions. These absorb on the surface of the nucleus and are reduced:

$$Ag_n + Dev \rightarrow Ag_n^- + Dev^+$$
$$Ag^+ + Ag_n^- \rightarrow Ag_{n+1}$$

Unfortunately, the heterogeneous nature of the reaction means that many physical as well as chemical processes can contribute to governing the rate. Besides the diffusion of *Dev* and Ag^+ to the nucleus, the rate of absorption of each species may be rate-controlling, as well as the rate of desorption of Dev^+. Beyond these physical processes, it is possible that the rate is controlled by the two direct electron-transfer processes, either singly or in a coordinated manner. At present, no model that assumes a single rate-controlling step can account for all the features of the developing kinetics.

There are usually several minor constituents in a commercial film emulsion besides the photosensitive AgX and the protective binder gelatin. Sensitization (an increase in effective film speed) can be achieved by any of several additions. (1) Sulfur sensitization involves adding thiosulfate or thiourea at about 10^{-5} mole per mole of AgX. (2) Gold sensitization uses comparable amounts of AuSCN. (3) Reduction sensitization uses Sn(II) compounds, SO^{2-}, or H_2, at only about 10^{-6} mole per mole AgX. Studies of these sensitization mechanisms suggest that both sulfur and gold add photoelectron traps and stabilize single Ag^0 atoms so that they do not ionize within the lattice before an Ag_2 subimage can form. On the other hand, reduction sensitization produces very small quantities of metallic silver in the grain, which traps holes created by photoelectron excitation so that they cannot recombine with the photoelectrons:

$$Ag^0 + h^+ \rightarrow Ag^+$$

We have already mentioned the use of halogen scavengers to prevent the reoxidation of photolytically formed Ag^0. These can also improve sensitivity to some extent by eliminating surface effects that destroy latent image nuclei at very low exposures.

A critical additive in commercial films is a sensitizing dye (or combination of dyes) adsorbed on the surface of the AgX grains. We noted that unmodified AgX is sensitive only to blue light, since lower frequencies do not provide enough energy

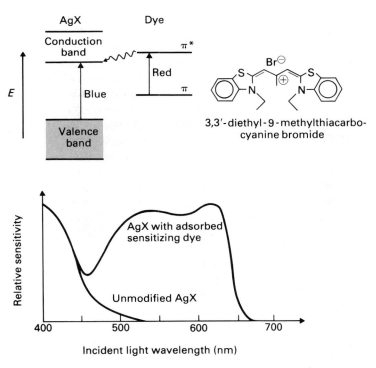

Figure 17.26   Spectral sensitization by dyes in black-and-white emulsions.

to excite an electron to the conduction band. This is why photographs from the Civil War period and before have a rather contrasty, bleached appearance. If the right dye, however, is adsorbed on the AgX surface, a photoelectron resulting from a lower-energy $\pi \rightarrow \pi^*$ transition within the dye can populate the AgX conduction band. Figure 17.26 shows the necessary energy-level relationships. The π^* level must be as high as the bottom of the conduction band, which means the π level must be somewhat higher than the top of the valence band. The figure also shows the structure of a typical cyanine dye used for this type of spectral sensitization, and indicates the increased spectral range obtainable.

PROBLEMS

A. DESCRIPTIVE

A1. The magnitude of the Stokes shift indicates the geometric differences between MX_6 and MX_6^* that lead to different r_0 values, but it does not indicate whether r_0 is larger for the ground state or for the excited state. In general, which state should have the greater r_0? Why? What electron conditions might lead to zero Stokes shift?

A2. Table 11.11 indicates that spin-orbit coupling constants are much larger for second- and third-row transition metals than for first-row metals. What does this suggest about the rate at which phosphorescence emission decays for $[IrCl_2(bpy)_2]^+$, as compared to $[CoCl_2(bpy)_2]^+$?

A3. In the Jablonski diagrams of Fig. 17.5, why are only one fluorescence transition and one phosphorescence transition shown?

A4. What product should result from LF irradiation of $trans$-$[Cr(NH_3)_4Cl_2]^+$ in water?

A5. What product should result from LF irradiation of $trans$-$[Cr(en)_2(OH)(OH_2)]^{2+}$ in water?

A6. What product should result from irradiating $[W(cp)(CO)_3]_2$ and $Mn_2(CO)_{10}$ in hydrocarbon solution?

A7. What product should result from irradiating $[W(cp)(CO)_3]_2$ and PPh_3 in hydrocarbon solution?

A8. Photolysis of $Fe(cp)(CO)_2Cl$ in a polar solvent yields $[Fe(cp)(CO)_2]_2$, free C_5H_6 and CO, and solvated $FeCl^{2+}$. Suggest a mechanism for the photolysis.

A9. It is particularly easy to break Si—Si bonds photolytically. In light of this, what products might you expect from irradiating $R_2Si(SiMe_3)_2$ (where R is the bulky group 2,4,6-trimethylphenyl) in hydrocarbon solution?

A10. Irradiating $CoH(PF_3)_4$ gives a product with the formula $Co_2HP_7F_{20}$. Suggest a structure consistent with the 18-electron rule.

A11. Suggest a detailed mechanism (with electron counts for each species) for the Pt $(C_2H_4)(PPh_3)_2$ reaction on page 859.

A12. Why should d^6 complexes be particularly well suited for the photochemical decomposition of H_2O to H_2 and O_2?

A13. How can a kink at the surface of an AgX crystal serve as a trap for electrons? As a trap for Ag^+ ions? What structural properties are necessary in each case?

B. NUMERICAL

B1. Calculate ΔCFSE for spin-allowed photoexcitation of Cr^{3+} and Co^{3+} (low-spin). What fractional change in CFSE occurs in each case? What inferences can you draw about the stereorigidity of the excited state?

B2. To establish a photochemical quantum yield Φ, it is necessary to know the number of photons absorbed by the reacting system. One technique uses a *chemical actinometer* for this determination, in which a reaction with a known quantum yield is used for comparison. A widely used system involves the photoredox reaction

$$2Fe(C_2O_4)_3^{3-} \xrightarrow{h\nu} 2Fe^{2+} + 2CO_2 + 5C_2O_4^{2-}$$

$\Phi_{Fe^{2+}} = 0.86$ at 510 nm, 0.93 at 468 nm, 1.21 at 366 nm, and 1.25 at 254 nm. Explain how this system could be used to establish the number of photons absorbed by *another* photoreaction.

B3. For the Cr(III) isomerization shown in Fig. 17.6, calculate ΔCFSE for a dissociative mechanism involving a five-coordinate intermediate, making reasonable assumptions about the size of Dq. How does the resulting energy requirement compare with the observed maximum wavelength for photoisomerization of 366 nm and with the excitation ΔCFSE of problem B1?

B4. What wavenumber and wavelength of light must be absorbed by $Ru(bpy)_3^{2+}$ to provide the 2.1 V shift in redox potential indicated in Fig. 17.8?

B5. A MLCT absorption in aqueous $Fe(CN)_5L^{3-}$ (where L is an aromatic heterocycle) forms free L and $Fe(CN)_5^{3-}$. However, the $Fe(CN)_5^{3-}$ hydrates extremely rapidly to $Fe(CN)_5(OH_2)^{3-}$, which in turn reacts with free L to regenerate $Fe(CN)_5L^{3-}$. If this last reaction proceeds by a dissociative mechanism

$$Fe(CN)_5(OH_2)^{3-} \underset{k_{-1}}{\overset{k_1}{\rightleftharpoons}} Fe(CN)_5^{3-} + H_2O$$

$$Fe(CN)_5^{3-} + L \xrightarrow{k_2} Fe(CN)_5L^{3-}$$

show that the rate of regeneration should be

$$\text{Rate} = \frac{k_1 k_2 [Fe(CN)_5(OH_2)^{3-}][L]}{k_{-1}[H_2O] + k_2[L]}$$

B6. How many cm^{-1} wavenumber increments correspond to a temperature increase of $1°C$ if a given activation energy is supplied either by light absorption or by heating? What does the magnitude of this number suggest about the activation energies supplied by light absorbed in the visible range of the spectrum around $20,000\ cm^{-1}$?

C. EXTENDED REFERENCE

C1. Describe the chemical process responsible for the *electrochemiluminescence* of $Ru(bpy)_3^{2+}$. [See A. J. Bard, *J. Amer. Chem. Soc.* (1972), *94*, 2862.]

C2. In electronic terms why should the photolytic product of $FeH_2(N_2)(PEtPh_2)_3$ be a good hydrogenating agent for olefins? [See E. K. von Gustorf *et al., Angew. Chem., Int. Ed. Engl.* (1972), *11*, 1088.]

C3. How can $Cr(OH_2)_5H^{2+}$ be prepared photochemically? Write a detailed mechanism. What do the values of $\Delta H^{\ddagger}$ and $\Delta S^{\ddagger}$ for

$$CrH^{2+} + H_3O^+ \rightarrow Cr^{3+} + H_2 + H_2O$$

indicate about that mechanism? [D. A. Ryan and J. H. Espenson, *Inorg. Chem.* (1981), *20*, 4401.]

C4. Karl Wieghardt [*Angew. Chem. Int. Ed. Engl.* (1989), *28*, 1153] describes some manganese biocomplexes other than the photosynthetic PS II water oxidation complex. One species he describes is a catalase using two Mn^{III} rather than the usual iron redox centers (his section 2.3.1). In his Scheme 1 he shows a Mn–O–Mn unit bridged by two carboxylates. In this chapter's Fig. 17.21 four Mn_4 model compounds for the WOC core are shown; which of these has a Mn_2 subunit most closely resembling Wieghardt's Mn_2 unit? Comment on the structural comparison and on the functionality of the two systems.

Appendix A

Inorganic Nomenclature

As inorganic chemistry has grown and new modes of bonding have been explored, the problem of providing structurally descriptive and unequivocal names for the thousands of new compounds has become increasingly severe and the naming systems increasingly complex. This short appendix cannot begin to give systematic coverage to all of the systems or to all of the possibilities within each system. We shall only attempt to outline common nomenclature usage in major American journals, and even that is not entirely uniform.

There are (approximately) four general systems of inorganic nomenclature. The oldest and least systematic is that used by most chemical manufacturers. It is an outgrowth of Berzelius's assumption that all compounds were a combination of a cation and an anion. Particularly for complex anions, manufacturers' names are often as much as a century out of date (for instance, hydrosulfite for $S_2O_4^{2-}$, now known as dithionite). Trivial names are also particularly common, for instance sulfolane for $(CH_2)_4SO_2$, which in other systems is either tetrahydrothiophene-1, 1-dioxide or (more commonly) tetramethylene sulfone. The use of -ic and -ous for higher and lower oxidation states of metals is common in manufacturers' catalogs [for instance, stannic and stannous for tin(IV) and tin(II)]. Unfortunately, translation from manufacturers' nomenclature to an approximate journal standard will probably be necessary for some years.

A second nomenclature system—or partial system—that is convenient but not very systematic is that of contemporary trivial names for important classes of compounds. These tend to be created when a complex molecular structure or family of related structures has an important function, such as the crown ethers of Chapter 7. The dibenzo-18-crown-6 of Fig. 7.9 should really be named 2,3,11,12-dibenzo-1,4,7,10,13,16-hexaoxacyclooctadeca-2,11-diene, but this is clearly not a convenient term for routine use. Similarly, the vitamin B_{12} structure of Fig. 16.7 can presumably be named systematically, but for convenience it is called a cobalamin. Because the CN group can be replaced by other ligands, it is included in the trivial name as cyanocobalamin. Several of these contemporary trivial names are introduced at appropriate points in the text. They are often used in journal articles and in suppliers' catalogs, but usually not in formal indexing services such as *Chemical Abstracts (CA)*.

The last two inorganic nomenclature systems are closely related, but not identical: the International Union for Pure and Applied Chemistry (IUPAC) system and the Chemical Abstracts referencing system. IUPAC maintains a continuing review of formal inorganic nomenclature that is the official international standard for journal use, although journals often use contemporary trivial names and also sometimes show minor variations on the IUPAC standard that reflect national preferences. The IUPAC system is systematic and complete, but is not in every case unambiguous. Some compounds can be given more than one correct IUPAC name, which a name-indexing service such as *CA* cannot tolerate. Accordingly, *CA* has extended the IUPAC system by removing some alternatives. Therefore, the student should familiarize himself with the details of the *CA* system for the specific area of interest before attempting a literature search through *CA*.

A brief outline of the IUPAC system with minor U.S. variations follows; it is quite incomplete, and for details the student should consult IUPAC reports, which are published every few years in the journal *Pure and Applied Chemistry* and in book form most recently by the publisher Blackwell Scientific Publications under the title *Nomenclature of Inorganic Chemistry—Recommendations 1990*. At their most basic, the rules derive from Berzelius's assumption that all inorganic compounds consist of a positively charged part and a negatively charged part. In general, the names of the elements are given; these names are shown in the periodic table on the front endpapers.

In compounds between a metal and a nonmetal, the name of the metal is given first (aluminum iodide); if two nonmetals are involved, the more electropositive is named first. Officially, the more electropositive is the earlier element in the following list: Rn, Xe, Kr, B, Si, C, Sb, As, P, N, H, Te, Se, S, O, At, I, Br, Cl, F. S_4N_4 is thus properly called tetranitrogen tetrasulfide, rather than a name in the reverse order. The Greek prefixes in Table A.1 are used to indicate the number of atoms or identical groups present in a molecule or formula unit, or, alternatively, to indicate the degree of polymerization of an acid, anion, or hydride: triphosphate, $P_3O_{10}^{5-}$, and diborane (4), B_2H_4. The prefixes ending in *-is* are used for polyatomic groups where ambiguity could result if the simpler prefix were used. For instance, if three methylamine molecules are to serve as ligands to a transition-metal ion, the name should contain "tris(methylamine)" to avoid confusion with the known compound trimethylamine. In this same context, the joining of root name units often creates long combined words, and parentheses should be used to isolate structural units, as in "tris(methylamine)" above. In general, parentheses should follow any *-is* ending such as tetrakis, but they should also be used for clarity in any name: trichloro(ethylene)platinate(II) for $[PtCl_3(C_2H_4)]^-$.

Cations are named by simply giving the element's name with no change in spelling; for instance, barium for Ba^{2+} or iron(II) for Fe^{2+}. The Roman numeral in

Table A.1   Numerical Prefixes in IUPAC Names

Number of structure units	Prefixes for simple names	Prefixes for extended names
1	Mono-	
2	Di-	Bis-
3	Tri-	Tris-
4	Tetra-	Tetrakis-
5	Penta-	Pentakis-
6	Hexa-	Hexakis-
7	Hepta-	Heptakis-
8	Octa-	Octakis-
9	Nona- *or* ennea-	Nonakis-
10	Deca-	Decakis-
11	Undeca- *or* hendeca-	Undecakis-
12	Dodeca-	Dodecakis-

the latter name is part of the Stock system for naming elements with variable oxidation states: The Roman numeral is placed in parentheses immediately following the end of the element's name. If the oxidation state is zero, (0) is used; if the oxidation state is negative, a minus sign is used with the Roman numeral inside the parentheses, as in pentacarbonylmanganese(−I) for $Mn(CO)_5^-$. Cations that consist of protonated mononuclear neutral hydrides are named by dropping the terminal *-ine* and adding *-onium:* phosphonium for PH_4^+ (by analogy with the trivial name ammonium, which is standard). For consistency, H_3O^+ is officially called oxonium, though hydronium is common. Protonated polynuclear bases simply drop a final *-e* and add *-ium:* pyridinium for $C_5H_5NH^+$. The cations NO^+ and NO_2^+ are sometimes called nitrosonium and nitronium, respectively, but they should by IUPAC standards be given the same name as the neutral radical: "nitrosyl cation" and "nitryl cation."

Anions consisting of a single element are named by abbreviating the element name and adding *-ide,* as in oxide for O^{2-} and triiodide(1−) for I_3^-; note that the polyatomic ion should have its net charge appended. A number of protonated anions and a few special cases are also given the *-ide* ending, as in Table A.2. Anions containing more than one element are given the ending *-ate,* although *-ite* is permitted for a few anions that contain a central element in a lower oxidation state, as shown in Table A.3. By tradition, oxygen atoms are not named or counted in anion names: aluminate is AlO_2^-, but AlF_6^{3-} is hexafluoroaluminate. In these names, the central atom is named last, prefixed by the names of nonoxygen outer or ligand atoms. Oxygen is also not named or counted in the names of some neutral radicals or cationic species. These are given a *-yl* suffix, as in the nitryl and nitrosyl cases above and as shown in Table A.4. Other group VIa nonmetals can

Table A.2   Special Cases of Anion Names Ending in "-ide"

Formula	Name
HF_2^-	Hydrogen difluoride
OH^-	Hydroxide
O_2^{2-}	Peroxide
O_2^-	Superoxide
O_3^-	Ozonide
NH_2^-	Amide
NH^{2-}	Imide
$N_2H_3^-$	Hydrazide
N_3^-	Azide
CN^-	Cyanide
C_2^{2-}	Acetylide

sometimes substitute for oxygen in these radicals, in which case the name is prefixed by *thio-, seleno-,* etc. An alternative systematic set of names is constructed in the same way except that the number of oxygens and the charge are designated: SO_4^{2-} is tetraoxosulfate(2−).

The large number of phosphorus, sulfur, and, to a lesser extent, arsenic oxyanions makes their nomenclature difficult. Usage still varies to some extent, but Table A.5 gives official names of the oxyacids of these anions. In general, to name the anion, "-ous acid" is replaced by *-ite* and "-ic acid" is replaced by *-ate*. For these and other oxyacids, the molecular form in which the central atom is hydroxylated up to its normal coordination number is the *ortho-* form. If one H_2O

Table A.3   Special Cases of Anion Names Ending in "-ite"

Formula	Name
NO_2^-	Nitrite
$N_2O_2^{2-}$	Hyponitrite
NOO_2^-	Peroxonitrite
PO_3^{3-}	Phosphite (but see Table A.5)
AsO_3^{3-}	Arsenite
SO_3^{2-}	Sulfite
$S_2O_6^{2-}$	Disulfite
$S_2O_4^{2-}$	Dithionite
SeO_3^{2-}	Selenite
XO_2^-	Halite (hal- = group VIIa minus "-ine")
XO^-	Hypohalite

Table A.4   Oxygen-Containing Radical and Cation Species

Formula	Name
HO	Hydroxyl
CO	Carbonyl
NO	Nitrosyl
NO_2	Nitryl
PO	Phosphoryl
SO	Sulfinyl or thionyl
SO_2	Sulfonyl or sulfuryl
S_2O_5	Disulfuryl
SeO	Seleninyl
SeO_2	Selenonyl
VO	Vanadyl
CrO_2	Chromyl
UO_2	Uranyl (and other transuranium elements)
XO	Halosyl (X = halogen)
XO_2	Halyl (hal = fluor-, chlor-, brom-, iod-)
XO_3	Perhalyl

molecule is stoichiometrically eliminated from the ortho- form, the resulting molecule is the *meta-* form of the acid; if two molecules of the ortho- form condense to a dimer by eliminating one H_2O molecule, the result is the *pyro-* form of the original acid: $(HO)_3PO$ is orthophosphoric acid, $(HO)PO_2$ is metaphosphoric acid, and $(HO)_2(O)POP(O)(OH)_2$ or $H_4P_2O_7$ is pyrophosphoric acid. Table A.6 gives the names of some other oxyacids, the anions of which take the *-ate* ending. If some protic hydrogens remain on the partially deprotonated acid, the number of remaining hydrogens is indicated in the name: $NaHSO_4$ is sodium hydrogen sulfate or sodium hydrogensulfate.

Some of the acids and, by implication, anions in Table A.5 are dimers, and it is necessary to name isopoly and heteropoly anions in general. For isopoly anions in which the electronegative element is oxygen and the electropositive characteristic element is in its normal (group number) oxidation state, the anion is simply named by using the Greek prefix for the number of electropositive atoms, as in triphosphate for $P_3O_{10}^{5-}$. Only in a few of the best-known cases, however, is it possible to assume that the number of oxygen atoms is known to the reader of the name. When the isopoly anion is less familiar, the net charge on the whole ion is added to the name as for the previously mentioned triiodide(1−) or tetraoxosulfate(2−). This is the Ewens–Bassett system for naming ions: $V_{10}O_{28}^{6-}$ is named decavanadate(6−). The presence of 28 oxygen atoms can be inferred from the presence of 10 (deca) vanadium atoms in oxidation state 5+ (the group number) and the net charge of 6−. Heteropoly anions with coordination polyhedra of one

Table A.5   Phosphorus- and Sulfur-Based Oxyacids

Formula	Name
$(HO)_3P$	Phosphorous acid
$(HO)_2PH$	Phosphonous acid
$(HO)PH_2$	Phosphinous acid
HOPO	Phosphenous acid or metaphosphorous acid
$(HO)_3PO$	Phosphoric acid
$(HO)_2PH(O)$	Phosphonic acid
$(HO)PH_2(O)$	Phosphinic acid
$HOPO_2$	Phosphenic acid or metaphosphoric acid
$(HO)_2POP(OH)_2$	Diphosphorous acid or pyrophosphorous acid $H_4P_2O_5$
$(HO)_2POP(O)(OH)_2$	Diphosphoric(III,V) acid $H_4P_2O_6$
$(HO)_2(O)PP(O)(OH)_2$	Hypophosphoric acid $H_4P_2O_6$
$(HO)(O)HPOP(O)(OH)_2$	Isohypophosphoric acid $H_4P_2O_6$
$(HO)_2(O)POP(O)(OH)_2$	Diphosphoric acid or pyrophosphoric acid $H_4P_2O_7$
$(HO)_2SO$	Sulfurous acid
$(HO)_2SO_2$	Sulfuric acid
$(HO)_2SO(O_2)$	Peroxomonosulfuric acid
$(HO)_2SO(S)$	Thiosulfuric acid $H_2S_2O_3$
$(HO)(O)SS(O)(OH)$	Dithionous acid $H_2S_2O_4$
$(HO)(O)SS(O)_2(OH)$	Disulfurous acid $H_2S_2O_5$
$(HO)(O)_2SS(O)_2(OH)$	Dithionic acid $H_2S_2O_6$
$(HO)(O)_2SOS(O)_2(OH)$	Disulfuric acid or pyrosulfuric acid $H_2S_2O_7$
$(HO)(O)_2SOOS(O_2)(OH)$	Peroxodisulfuric acid $H_2S_2O_8$

Note: Arsenic acids are named by analogy to phosphorus acids using the stem ars- (or arsen- to avoid duplication) instead of phosph-.

characteristic atom surrounding another characteristic atom are named with the outer atoms first—since the number of outer characteristic atoms must also be specified, an Arabic numeral prefix is used: $MnMo_9O_{32}^{6-}$ is 9-molybdomanganate(6−).

Hydrides represent a special case: Unless the electronegativity differences between the components are quite large, hydrides are usually not named as a combination of an electropositive and an electronegative element. Metal hydrides are named as such: CaH_2 is calcium hydride. At the other end of the electronegativity scale, the halogen hydrides are called hydrogen halides, and this is continued to some of the pseudohalides: HBr is hydrogen bromide, HCN is hydrogen cyanide, and HN_3 is hydrogen azide. However, nonmetal hydrides are in general given

Table A.6   Oxyacids of "-ate" anions

Formula	Name
H_3BO_3	Boric acid
H_2CO_3	Carbonic acid
$H_2C_2O_4$	Oxalic acid
HOCN	Cyanic acid
HSCN	Thiocyanic acid
HNCO	Isocyanic acid
HONC	Fulminic acid
H_4SiO_4	Orthosilicic acid
$(H_2SiO_3)_n$	Metasilicic acid (polymers)
HNO_3	Nitric acid
HEO_4	Sulfuric or selenic acid (E = S, Se)
HXO_4	Perhalic acid (X = Cl, Br, I)
H_5IO_6	Orthoperiodic acid
HXO_3	Halic acid (X = Cl, Br, I)

Note: Lower oxidation states are named as "-ous acid" (see Table A.3).

names of one word. The most familiar have traditional names: water, ammonia, hydrazine, and of course methane. By analogy with methane, most other nonmetal hydrides are named by using a stem from the name of the element and the suffix *-ane:* SiH_4 is silane and SnH_4 is stannane. For polymers, the Greek prefixes are used. If the stoichiometry is not simple and obvious by analogy with hydrocarbon chains (boranes in particular are not), the number of hydrogens is indicated by an Arabic numeral in parentheses at the end of the name: Si_2H_6 is disilane and H_2S_4 is tetrasulfane, but B_4H_{10} is tetraborane(10). The group Va hydrides are exceptions to the *-ane* suffix rule; since organic usage has made the term "amine" well known, ammonia in a combined form (that is, as a ligand in complexes) is called ammine (note the two *m*'s!). Extending this analogy, the other group Va hydrides take the ending *-ine,* so that SbH_3 is stibine and P_2H_4 is diphosphine.

Coordination compounds are named in conformity with the rules given above for cations and anions. However, since their formulas are often considerably more complicated than those of simple ions, some additional rules are required. Ligands are named first and the central method atom last. Greek prefixes are used to designate the number of each kind of ligand, and if the ligands are different, they are listed in alphabetical order. (In older rules, negatively charged ligands were listed first, neutral ligands next.) Either a Stock-system Roman numeral or a Ewens–Bassett number can be used to specify the oxidation state/charge relationship, but for very well-known complexes neither is necessary. Neutral ligands are simply given their molecular name (but note ammine for NH_3 and aqua for H_2O). However, anionic ligands replace the final *-e* in the ion name by *-o:* A coordinate

Table A.7 Special Cases of Anionic Ligand
Names in Coordination Compounds

Group	Prefix
—F	fluoro-
—Cl	chloro-
—Br	bromo-
—I	iodo-
—O	oxo-
—H	hydrido-
—OH	hydroxo-
—O—O	peroxo-
—CN	cyano-

acetate ion becomes acetato, for example. Table A.7 lists some anionic ligand
names that vary slightly from this rule.

In organometallic compounds, hydrocarbon ligands RH (such as ethylene) are
given the molecule name, but hydrocarbon radicals R· (such as allyl or cyclopentadienyl) are given the radical name even if they are in fact serving as anions: Ferrocene, for instance, can be made using $FeCl_2$ and $Na^+C_5H_5^-$, sodium cyclopentadienide, but the compound is bis(cyclopentadienyl)iron following this naming
rule. Actually, ferrocene and the other bis(cyclopentadienyl)metal compounds
are officially called metallocenes: $V(C_5H_5)_2$ is vanadocene and $U(C_5H_5)_2$ is ura-
nocene, for example. The *hapto-* system, normally using the Greek letter η (eta)
but occasionally *h,* is used to designate the location of the pi-electron-to-metal
bond with respect to the pi ligand. The *hapto-* nomenclature is used in two ways
that are not equivalent. One usage (described in Chapter 13) is to specify only the
total number of carbon atoms whose pi electrons are involved in pi-to-metal bonding. The other usage specifies the atom numbers of the double bonds within
the organic pi system that are serving as donors if not all the pi-system atoms
are equivalent: $TiCl_2(C_5H_5)_2$ is dichlorobis (η-cyclopentadienyl)titanium, but a
methylallyl or 2-butenyl radical serving as an allyl trihapto donor through the
C_1—C_3 atoms would be 1-3-η-2-butenyl, as in bis(1-3-η-2-butenyl)nickel for
$Ni(C_4H_7)_2$.

In polynuclear complexes, bridging ligands are given the prefix μ- (mu) and
are set off from the other ligands in the name even if they are the same molecule
as other ligands. They are listed in alphabetical order, but before terminal ligands
of the same sort. The $Fe_2(CO)_9$ molecule in Fig. 13.1 becomes tri-μ-carbonyl-
bis(tricarbonyliron) though it is often simply called enneacarbonyldiiron, with a
knowledge of the structure assumed. If more than two metal atoms form a cluster
and its geometry is known, an italicized prefix such as *triangulo-, tetrahedro-,* or
octahedro- is inserted before the name of the metal. For instance, the $Nb_6Cl_{12}^{2+}$

cluster in Fig. 13.24 becomes dodeca-μ-chloro-*octahedro*-hexaniobium(2+). Note that for cluster compounds, Stock designation of the oxidation state of each metal atom often becomes impossible, so that Ewens–Bassett numbers are preferable. If bridging ligands are known to bridge more than two metal atoms, the number is indicated by a subscript on the μ symbol. For instance, the $Mo_6Cl_8^{4+}$ cluster of Fig. 13.24 becomes octa-μ_3-chloro-*octahedro*-hexamolybdenum(4+).

Some examples of coordination-compound names may be helpful:

$[Fe(CN)_6]^{4-}$	Hexacyanoferrate(II) *or* hexacyanoferrate(4−)
$[Cr(NH_3)_5Cl]^{2+}$	Pentamminechlorochromium(III) *or* pentamminechlorochromium(2+)
$[Co(NH_3)_5NO]^{2+}$	Pentamminenitrosylcobalt(2+) (oxidation state uncertain)
$[Ni(NCO)_4]^{2-}$	Tetraisocyanatonickelate(II) *or* tetraisocyanatonickelate(2−)
$[Co(CO)_3(PPh_3)I]$	Tricarbonyliodo(triphenylphosphine)cobalt
$[Co(C_6Me_6)_2]^{+}$	Bis(η^6-hexamethylbenzene)cobalt(I) *or* bis(η^6-hexamethylbenzene)cobalt(1+)

$$CH_3 \longrightarrow \!\!\!\! \overset{Cl}{\underset{Cl}{< Pd \quad Pd >}} \!\!\!\! \longrightarrow CH_3$$

Di-μ-chlorobis(1-3-η-2-methylallyl-palladium)

Appendix B

VSEPR Geometry Prediction and Hybridization

Quantitative molecular-orbital calculations have reached a stage of refinement in which the details of a molecule's geometry—the bond lengths and bond angles—can in most cases be predicted in fair detail by a series of MO calculations to yield the lowest-energy conformation. This process, however, is neither convenient nor pictorial. It would be very convenient to be able to produce reasonably accurate structure predictions for covalent molecules on the back of an envelope. Such a scheme must necessarily be simple, and we must expect that occasional exceptions to it will be found. R. J. Gillespie's valence shell electron pair repulsion (VSEPR) scheme is the most successful yet proposed. It is described in most general chemistry texts and in a number of organic chemistry texts as well, so a student in inorganic chemistry may already be familiar with the VSEPR arguments. However, a review of the VSEPR scheme will be helpful, along with its relation to hybridization patterns. Both are used at various points in this text.

The basic premise of VSEPR is that the influence of the Pauli exclusion principle on electron–electron repulsion is dominant in establishing the geometry of main-group covalent molecules. The scheme does not work for transition-metal molecules in general, nor for ionic crystal geometries. It assumes that bond angles are established by (and bond lengths are modified by) the repulsions between the pairs of electrons doing the bonding in the molecule. As such, it represents an extension of Lewis dot structures, which should be familiar. The sequence of arguments involved in predicting the geometry of a molecule by VSEPR can most helpfully be presented as a list. At the beginning, we have the stoichiometry of the molecule and nothing more.

1. Establish the central reference atom about which the bond angles are to be predicted. Sometimes the stoichiometry makes this obvious, as in NH_3 or PCl_5. In

questionable cases, we should avoid chain structures if possible and place the least electronegative atom at the center of the structure. Neither H nor F is ever at the center, other halogens are not at the center unless combined with oxygen or more electronegative halogens, and oxygen is not at the center unless it is combined with two other atoms, as in H_2O or OF_2.

2. Draw the best Lewis dot structure for the molecule, including double or triple bonds if necessary to minimize formal charges and noting possible resonance between equivalent Lewis structures.

3. Count the valence electrons around the central atom, including its own electrons (VE), the electrons contributed to bonds by the ligand atoms (LE), and any adjustment to the number of central-atom valence electrons required by the net charge (if any) on the molecule as a whole (CA). If the molecule is a cation, subtract the net charge from the valence-electron total. If it is an anion, add the net charge since there is an electron surplus. The total is thus VE + LE + CA.

4. Convert the valence-electron total to the number of electron groups present around the central atom in the Lewis structure. In general these will be pairs, either bonding or nonbonding lone pairs. However, multiple bonds can yield groups of four or six electrons, and odd-electron free radicals will have a one-electron nonbonding group or a three-electron bonding group.

5. Predict the *electron geometry* (which is not yet the molecular geometry) by arranging the electron groups around the central atom according to Table B.1, disregarding for the moment any differences between the number of electrons

Table B.1   Preferred Electron Group Geometries

Number of electron groups	Minimum-repulsion ideal electron-group geometry	Number of nonbonding electron pairs	Molecular example	Molecular geometry
2	Linear	0	$HgCl_2$	Linear
3	Trigonal planar	0	BF_3	Trigonal planar
		1	$SnCl_2$	Bent
4	Tetrahedral	0	$SiCl_4$	Tetrahedral
		1	NH_3	Trigonal pyramidal
		2	H_2O	Bent
5	Trigonal bipyramidal	0	PCl_5	Trigonal bipyramidal
		1	SF_4	See-saw
		2	BrF_3	T-shaped
		3	XeF_2	Linear
6	Octahedral	0	SF_6	Octahedral
		1	IF_5	Square pyramidal
		2	IF_4^-	Square planar
7	Pentagonal bipyramidal	0	IF_7	Pentagonal bipyramidal
		1	$SbCl_6^{3-}$	Distorted octahedral

in each group. For example, if a central atom has four groups of electrons around it, they will be directed toward the corners of a tetrahedron, whether they are bonding or nonbonding groups. The electron-group polyhedra in Table B.1 with their implied inter-electron-group angles, minimize repulsion among a given number of electron groups if all groups are equally charged and of the same size. This is the simplest possible approximation, and will be refined later.

6. Decide how many kinds of electron groups are present and the number of each kind. For example, PCl_5 has only one kind of electron group about the P atom, and five of that kind – that is, there are five P—Cl bond pairs. SF_4, on the other hand, has four S—F bond pairs and one nonbonding pair, and BrF_3 has three Br—F bond pairs and two nonbonding pairs. $POCl_3$ has three P—Cl bond pairs and one P=O four-electron bond group, but NOCl has one N—Cl bond pair, one N=O four-electron bond group, and one nonbonding pair on the N atom.

7. Place the specific electron groups around the electron-group polyhedron so as to minimize the overall repulsion between groups in the molecule. If all the groups are equivalent (as in PCl_5 above) it is immaterial how they are arranged, and the electron-group geometry becomes the molecular geometry. If there are two, three, or four electron groups, the nature of each still does not matter because every position in those polyhedra has the same relationship to every other position. However, if there are five or more electron groups, the distribution must be chosen to minimize repulsion in the molecule as a whole. The largest or most highly charged groups must be placed in the most favorable, lowest-repulsion locations. In general, nonbonding pairs are larger than bonding pairs, because in bonding pairs an outer-atom nucleus is pulling the valence electrons away from the immediate neighborhood of the central atom where the repulsion of the neighbor groups is strongest. For the same reason, very electronegative ligand atoms yield smaller bonding pairs than less electronegative ligand atoms. Four-electron or six-electron groups are larger and more highly charged than bonding pairs. It is important to minimize lone-pair/lone-pair repulsions if the central atom has more than one lone pair, and similarly lone-pair/four-electron repulsions, and so on. For five electron groups, larger electron groups such as nonbonding pairs should occupy equatorial positions, since repulsion by two neighbors at $90°$ and two at $120°$ is less than the polar-position repulsion by three neighbors at $90°$. For six groups, the initial positions are all equivalent and the first lone pair may be placed anywhere. However, the second should be placed *trans-* to the first to minimize repulsion by two large neighbor groups. For seven-electron groups, a single larger group should again occupy an equatorial position. However, as Fig. 12.8 and the associated discussion indicate, geometric interconversion is so facile for seven-coordinate systems that prediction becomes somewhat uncertain.

8. After properly locating the specific electron groups within the idealized electron-group polyhedron, modify the bond angles and bond lengths to take into account the following departures from ideal symmetry.

(a) The angle between neighboring electron groups will be larger than the ideal value if the neighbor groups are larger than other adjacent groups. Two lone pairs in a tetrahedral geometry will open up the angle between them and thereby crowd the angle between the other two bonding pairs. Water thus has an H—O—H angle less than the tetrahedral ideal of 109°27′ (experimentally 104°).

(b) Bond angles next to very electronegative ligand atoms will be smaller than the ideal angle, since the ligand atom shrinks its bond pair. Conversely, if the general structure stays the same but the central atom is more electronegative in one of two molecules (such as NH_3 versus PH_3), the more electronegative central atom will draw bond pairs in, increase their repulsion, and open up the bond angles. NH_3 has an H—N—H angle of 107° and PH_3 has an H—P—H angle of 93°, illustrating the working of this rule.

(c) As we noted before, the five-electron-group and seven-electron-group geometries have nonequivalent polar and equatorial positions. Therefore, ligand atoms with greater numbers of neighbors will be farther from the central atom. PCl_5, for instance, has an axial (polar) bond length of 2.14 Å, but an equatorial bond length of 2.02 Å. This suggests that the two neighbors at 120° for the equatorial positions have very little influence on the overall repulsion within the molecule. On the other hand, IF_7 in the gas phase seems to have very nearly equal bond lengths. We can therefore surmise that the axial bonds with five neighbors at 90° are nearly equal in repulsion to the equatorial bonds with two neighbors at 72° and two at 90°.

9. Having made detailed adjustments to the idealized electron-group geometry, describe the *molecular* geometry in terms of the positions of nuclei or bonds, ignoring nonbonding electron pairs. Although for water the electron-group geometry is distorted tetrahedral, the molecular geometry is bent or V-shaped. This is an important distinction to make. Table B.1 indicates the general possibilities.

One or two examples will help to illustrate the procedure. Consider $POCl_3$, phosphoryl chloride. The central atom is P, the least electronegative atom. The Lewis structure with no formal charge is

There are ten valence electrons around the P atom in four groups: a four-electron group to the O atom and three bonding pairs to the Cl atoms. From Table B.1, the electron-group geometry should be tetrahedral. Since there are no nonbonding pairs, and since all positions on a tetrahedron are equivalent, we can say that the molecular geometry is tetrahedral. But the four-electron P=O bond group will compress the surrounding P—Cl bond pairs, so the Cl—P—Cl angle should be somewhat less than 109°. Experimentally, the angle is 103.3°.

For a slightly more complex example, consider $Te(CH_3)_2Cl_2$, dichlorodimethyltellurium. The central atom is Te, the least electronegative atom. [In addition, the methyl groups can only occupy terminal (ligand-group) positions.] The Lewis structure, somewhat abbreviated, is

$$
\begin{array}{c}
:\ddot{\text{Cl}}: \\
\text{H}_3\text{C}:\ddot{\text{Te}}:\text{CH}_3 \\
:\ddot{\text{Cl}}:
\end{array}
$$

There are 10 valence electrons in five groups around the Te atom: two Te—C bonding pairs, two Te—Cl bonding pairs, and a nonbonding pair. From Table B.1, the electron-group geometry should be trigonal bipyramidal. The largest electron group is the nonbonding pair, which should be placed in an equatorial position to give the "see-saw" shape of Table B.1. Carbon is less electronegative than chlorine, so the Te—C bond pairs will be larger than the Te—Cl pairs. Therefore, the methyl groups should also be in equatorial positions. The large lone pair will repel the methyl groups, so the C—Te—C angle should be less than the ideal 120° and the Cl—Te—Cl angle should be less than the ideal 180°. Experimentally, the methyl groups are in the equatorial positions of a "see-saw" shaped molecule. The C—Te—C angle is near 100° and the Cl—Te—Cl angle is near 180°.

If we can establish the electron-group geometry of a molecule, it is often convenient to describe localized sigma bonding in the molecule by creating hybrid orbitals of the correct geometry. Thorough molecular-orbital calculations provide a more flexible (and presumably more accurate) description of how atomic orbitals are involved in the overall bonding within the molecule, but again the computations are difficult and not very pictorial. Consequently, at various points in the text we started an MO treatment by assuming one of the forms of hybrid atomic orbitals familiar from organic chemistry—sp, sp^2, or sp^3. These are the only hybrid orbitals that can be created if the basis set of atomic orbitals is limited to s and p. However, in earlier treatments of the bonding in transition-metal complexes under valence-bond theory a fairly extensive set of hybrid orbitals was developed using s, p, and d orbitals as the basis set. These are somewhat less familiar since they are not normally used in organic-chemistry applications, and they are no longer used in formal descriptions of transition-metal complexes. They do provide an easy and pictorial description of sigma bonding in many main-group compounds in which the coordination number of the central atom is greater than 4, however, and still find use in that context.

The four-lobed character of d orbitals, and the fact that there are five different ones, means that hybrid orbitals with almost any desired geometry can be constructed from appropriate combinations of s, p, and d orbitals. Table B.2 gives a list of hybrid-orbital sets that have been constructed for central-atom coordination numbers up through CN = 8. It has been shown mathematically that these all yield sigma hybrid-orbital lobes in the appropriate directions for the bonding geometries shown in the table. Most, however, are rarely used. The hybrid sets appearing in boldface type are the sets most commonly used for qualitative sigma-bonding descriptions in main-group molecules such as BF_3, PCl_5, and SF_6—and even for d^0 complexes such as ZrF_7^{3-}. The table also lists the remaining orbitals from the s,

Table B.2   s-, p-, and d-Based Hybrid Orbitals Appropriate to Common Bond Geometries

Coordination number	Molecular geometry	Sigma-symmetry hybrid orbitals	Remaining pi-symmetry AOs
2	**Linear**	sp	p^2d^2
		dp	p^2d^2
3	**Trigonal planar**	sp^2	pd^2
		dp^2	pd^2
4	**Tetrahedral**	sp^3	d^2
		d^3s	d^2
	Square planar	dsp^2	d^3p
		d^2p^2	d^3p
5	**Trigonal bipyramidal**	dsp^3	d^2
		d^3sp	d^2
	Square pyramidal	d^2sp^2	d
		d^4s	d
	Pentagonal planar	d^3p^2	pd^2
6	**Octahedral**	d^2sp^3	d^3
	Trigonal prismatic	d^4sp	p^2d (weak)
7	Pentagonal bipyramidal	d^3sp^3	d^2 (weak)
	Capped trigonal prismatic	d^4sp^2	dp (weak)
8	Dodecahedral	d^4sp^3	d
	Square antiprismatic	d^5p^3	—

p, d basis set that have the right symmetry to form pi bonds between the atoms that are sigma-bonded by overlap with the central-atom hybrids. For hybrids formed from s and p AOs only, we are used to having all the remaining p orbitals available for pi bonding, but there are some important differences when d AOs are added to the basis set. Central atoms with tetrahedral sp^3 hybrids can still engage in some pi bonding through two of the d orbitals, though the ligand atoms will have to serve as pi donors. This possibility complicates the simple model that sees, for instance, the Si—Cl bonds in $SiCl_4$ as pure single sigma bonds. Nonetheless, for simple bonding arguments it is helpful to see which sets of hybrid orbitals can engage in localized sigma bonds.

Appendix C

Tanabe–Sugano Diagrams

These diagrams for the analysis of d-block visible–UV–near-IR spectra are adapted from those presented by Yukito Tanabe and Satoru Sugano in *J. Phys. Soc. Japan* (1954), *9*, 766. Of the two repulsion integrals B and C, only one can vary in a two-dimensional diagram; for each diagram a best-fit value of C/B has been chosen and is indicated. Spectral frequency fits will in general be good to about two significant figures, occasionally three.

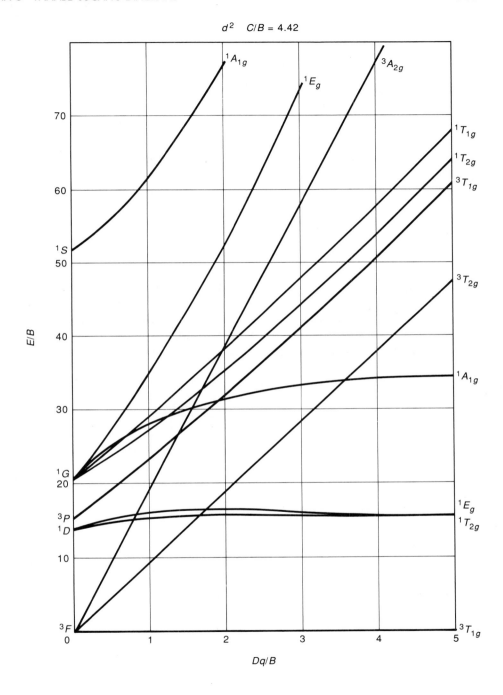

d^2 $C/B = 4.42$

d^3 C/B = 4.50

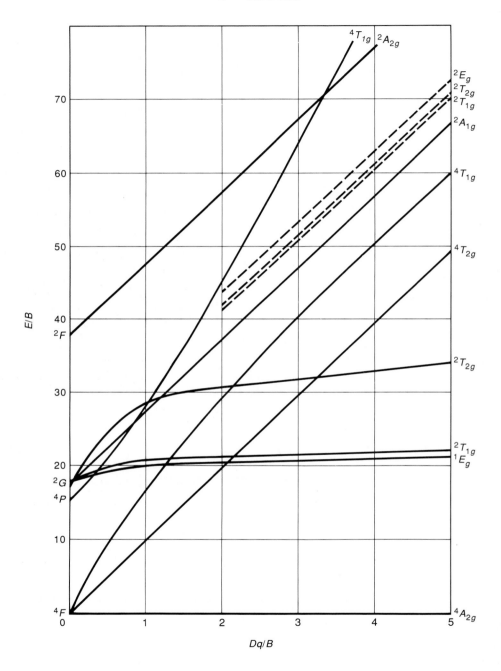

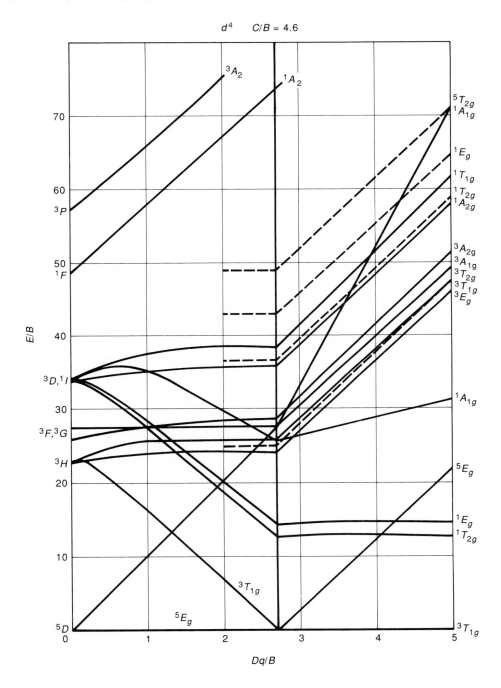

d^4 $C/B = 4.6$

d^5 $C/B = 4.48$

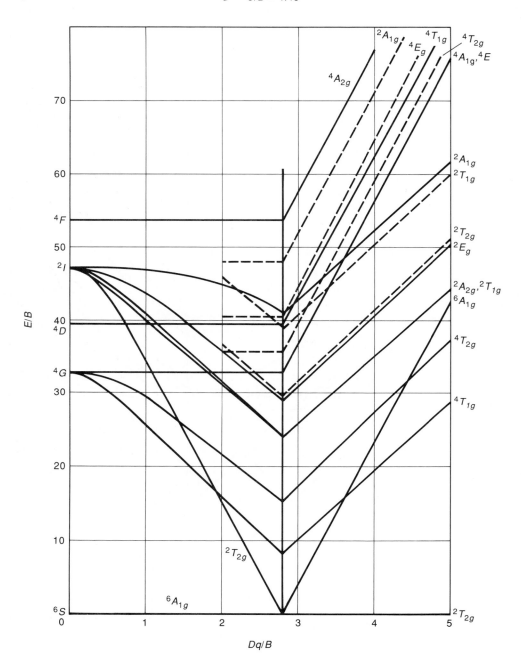

d^6 $D/B = 4.81$

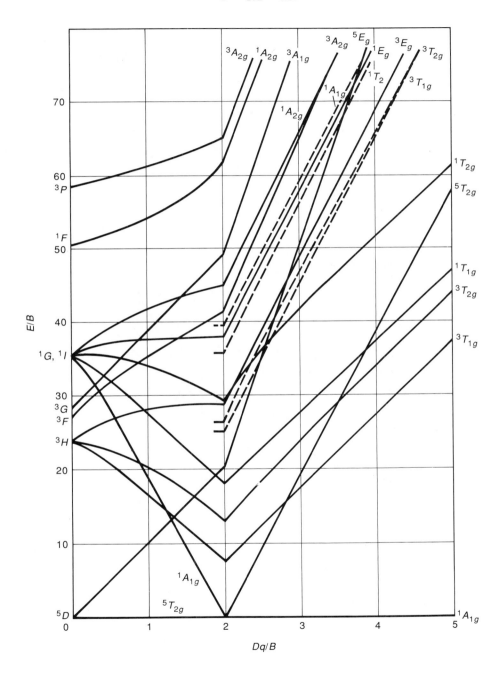

d^7 $C/B = 4.63$

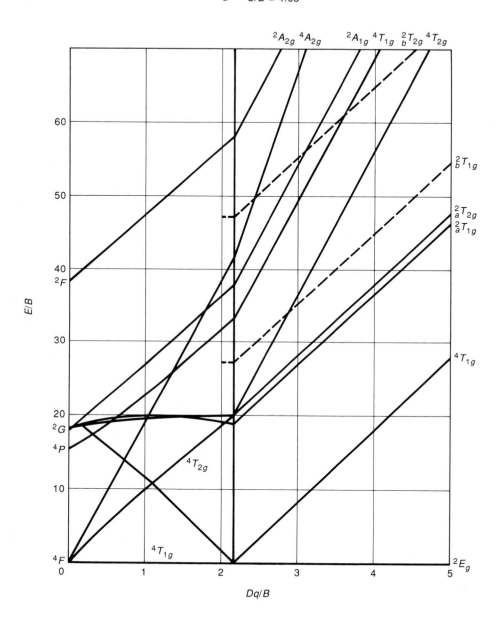

d^8 $D/B = 4.71$

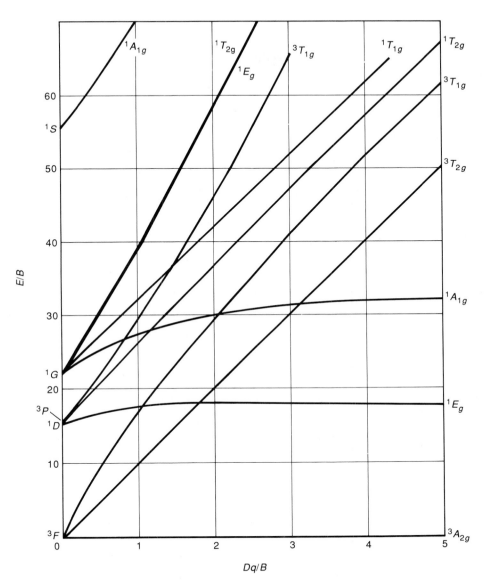

Index

909

Values of Atomic Constants

Physical quantity	Symbol	SI units	Common units
Charge on electron	e	1.60219×10^{-19} C	4.80325×10^{-10} esu
Planck's constant	h	6.62618×10^{-34} J s	6.62618×10^{-27} erg s
Mass of electron	m_e	9.10953×10^{-31} kg	9.10953×10^{-28} g
Avogadro's number	N_0	6.02205×10^{23} mol^{-1}	
Boltzmann's constant	k	1.38066×10^{-23} J K^{-1}	1.38066×10^{-16} erg K^{-1}
Thermal energy at 298 K	kT	2.479 kJ mol^{-1}	0.592 kcal mol^{-1}
			208 cm^{-1} molecule^{-1}
Vacuum permittivity	$4\pi\epsilon_0$	1.11265×10^{-10} C^2 J^{-1} m^{-1}	
Bohr radius	a_0	5.29177×10^{-11} m	0.529177 Å
Mass of proton	m_p	1.67265×10^{-27} kg	1.67265×10^{-24} g
Bohr magneton	μ_B	9.27408×10^{-24} J T^{-1}	9.27408×10^{-21} erg gauss^{-1}

Energy Conversion Factors (to five significant figures)

Base units	J/mol	cal/mol	L atm/mol	cm^{-1}/molecule	eV/molecule
1 J/mol	1	0.23901	9.6892×10^{-3}	0.083594	1.0364×10^{-5}
1 cal/mol	4.1840	1	0.041293	0.34976	4.3363×10^{-5}
1 L atm/mol	101.33	24.217	1	8.4702	1.0501×10^{-3}
1 cm^{-1}/molecule	11.963	2.8591	0.11806	1	1.2398×10^{-4}
1 eV/molecule	96487.	23061.	952.25	8065.7	1